Humanoid Robots

New Developments

Edited by
Armando Carlos de Pina Filho

Humanoid Robots, New Developments
Edited by Armando Carlos de Pina Filho

Republished by InTech
Janeza Trdine 9, 51000 Rijeka, Croatia

Technical Editor Teodora Smiljanic

Cover Designer Dino Smrekar

Additional hard copies can be obtained from orders@intechopen.com

Humanoid Robots, New Developments, Edited by Armando Carlos de Pina Filho
p. cm.
ISBN 978-3-902613-00-4

For EU product safety concerns:
IN TECH d.o.o., Prolaz Marije Krucifikse Kozulić 3, 51000 Rijeka, Croatia,
info@intechopen.com or visit our website at intechopen.com.

Preface

For many years, the human being has been trying, in all ways, to recreate the complex mechanisms that form the human body. Such task is extremely complicated and the results are not totally satisfactory. However, with increasing technological advances based on theoretical and experimental researches, man gets, in a way, to copy or to imitate some systems of the human body.

These researches not only intended to create humanoid robots, great part of them constituting autonomous systems, but also, in some way, to offer a higher knowledge of the systems that form the human body, objectifying possible applications in the technology of rehabilitation of human beings, gathering in a whole studies related not only to Robotics, but also to Biomechanics, Biomimmetics, Cybernetics, among other areas.

This book presents a series of researches inspired by this ideal, carried through by various researchers worldwide, looking for to analyze and to discuss diverse subjects related to humanoid robots. The presented contributions explore aspects about robotic hands, learning, language, vision and locomotion.

From the great number of interesting information presented here, I believe that this book can offer some aid in new research, as well as stimulating the interest of people for this area of study related to the humanoid robots.

Editor
Armando Carlos de Pina Filho

Contents

1

Design of Modules and Components for Humanoid Robots

Albert Albers, Sven Brudniok, Jens Ottnad,
Christian Sauter, Korkiat Sedchaicharn
University of Karlsruhe (TH), Institute of Product Development
Germany

1. Introduction

The development of a humanoid robot in the collaborative research centre 588 has the objective of creating a machine that closely cooperates with humans. The collaborative research centre 588 (SFB588) "Humanoid Robots - learning and cooperating multi-modal robots" was established by the German Research Foundation (DFG) in Karlsruhe in May 2000. The SFB588 is a cooperation of the University of Karlsruhe, the Forschungszentrum Karlsruhe (FZK), the Research Center for Information Technologies (FZI) and the Fraunhofer Institute for Information and Data Processing (IITB) in Karlsruhe.
In this project, scientists from different academic fields develop concepts, methods and concrete mechatronic components and integrate them into a humanoid robot that can share its working space with humans. The long-term target is the interactive cooperation of robots and humans in complex environments and situations. For communication with the robot, humans should be able to use natural communication channels like speech, touch or gestures. The demonstration scenario chosen in this project is a household robot for various tasks in the kitchen.
Humanoid robots are still a young technology with many research challenges. Only few humanoid robots are currently commercially available, often at high costs. Physical prototypes of robots are needed to investigate the complex interactions between robots and humans and to integrate and validate research results from the different research fields involved in humanoid robotics. The development of a humanoid robot platform according to a special target system at the beginning of a research project is often considered a time consuming hindrance. In this article a process for the efficient design of humanoid robot systems is presented. The goal of this process is to minimize the development time for new humanoid robot platforms by including the experience and knowledge gained in the development of humanoid robot components in the collaborative research centre 588.
Weight and stiffness of robot components have a significant influence on energy efficiency, operating time, safety for users and the dynamic behaviour of the system in general. The finite element based method of topology optimization gives designers the possibility to develop structural components efficiently according to specified loads and boundary conditions without having to rely on coarse calculations, experience or

intuition. The design of the central support structure of the upper body of the humanoid robot ARMAR III is an example for how topology optimization can be applied in humanoid robotics. Finally the design of the upper body of the humanoid ARMAR III is presented in detail.

2. Demand for efficient design of humanoid robots

Industrial robots are being used in many manufacturing plants all over the world. This product class has reached a high level of maturity and a broad variety of robots for special applications is available from different manufacturers. Even though both kind of robots, industrial and humanoid, manipulate objects and the same types of components, e.g. harmonic drive gears, can be found in both types, the target systems differ significantly. Industrial robots operate in secluded environments strictly separated from humans. They perform a limited number of clearly defined repetitive tasks. These machines and the tools they use are often designed for a special purpose. High accuracy, high payload, high velocities and stiffness are typical development goals.

Humanoid robots work together in a shared space with humans. They are designed as universal helpers and should be able to learn new skills and to apply them to new, previously unknown tasks. Humanlike kinematics allows the robot to act in an environment originally designed for humans and to use the same tools as humans in a similar way. Human appearance, behaviour and motions which are familiar to the user from interaction with peers make humanoid robots more predictable and increase their acceptance. Safety for the user is a critical requirement. Besides energy efficient drive technology, a lightweight design is important not only for the mobility of the system but also for the safety of the user as a heavy robot arm will probably cause more harm in case of an accident than a light and more compliant one. Due to these significant differences, much of the development knowledge and product knowledge from industrial robots cannot be applied to humanoid robots.

The multi-modal interaction between a humanoid robot and its environment, the human users and eventually other humanoids cannot fully be simulated in its entire complexity. To investigate these coherences, actual humanoid robots and experiments are needed. Currently only toy robots and a few research platforms are commercially available, often at high cost. Most humanoid robots are designed and built according to the special focus or goals of a particular research project and many more will be built before mature and standardized robots will be available in larger numbers at lower prizes. Knowledge gained from the development of industrial robots that have been used in industrial production applications for decades cannot simply be reused in the design of humanoid robots due to significant differences in the target systems for both product classes. A few humanoid robots have been developed by companies, but not much is known about their design process and seldom is there any information available that can be used for increasing the time and cost efficiency in the development of new improved humanoid robots. Designing a humanoid robot is a long and iterative process as there are various interactions between e.g. mechanical parts and the control system. The goal of this article is to help shortening the development time and to reduce the number of iterations by presenting a process for efficient design, a method for optimizing light yet stiff support structures and presenting the design of the upper body of the humanoid robot ARMAR III.

3. Design process for humanoid robot modules

The final goal of the development of humanoid robots is to reproduce the capabilities of a human being in a technical system. Even though several humanoid robots already exist and significant effort is put into this research field, we are still very far from reaching this goal. Humanoid robots are complex systems which are characterized by high functional and spatial integration. The design of such systems is a challenge for designers which cannot yet be satisfactorily solved and which is often a long and iterative process. Mechatronic systems like humanoid robots feature multi-technological interactions, which are displayed by the existing development processes, e.g. in the VDI guideline 2206 "design methodology for mechatronics systems" (VDI 2004), in a rather general and therefore abstract way. More specific development processes help to increase the efficiency of the system development. Humanoid robots are a good example for complex and highly integrated systems with spatial and functional interconnections between components and assembly groups. They are multi-body systems in which mechanical, electronic, and information-technological components are integrated into a small design space and designed to interact with each other.

3.1 Requirements

The demands result from the actions that the humanoid robot is supposed to perform. The robot designed in the SFB 588 will interact with humans in their homes, especially in the kitchen. It will take over tasks from humans, for example loading a dish washer. For this task it is not necessary, that the robot can walk on two legs, but it has to feature kinematics, especially in the arms, that enable it to reach for objects in the human surrounding. In addition, the robot needs the ability to move and to hold objects in its hand (Schulz, 2003).

3.2 Subdivision of the total system

The development of complex systems requires a subdivision of the total system into manageable partial systems and modules (Fig. 1). The segmentation of the total system of the humanoid robot is oriented on the interactions present in a system. The total system can be divided into several subsystems. The relations inside the subsystems are stronger compared to the interactions between these subsystems. One partial system of the humanoid robot is e.g. the upper body with the subsystem arm. The elements in the lowest level in the hierarchy of subsystems are here referred to as modules. In the humanoid robot's arm, these modules are hand-, elbow-, and shoulder joint. Under consideration of the remaining design, these modules can be exchanged with other modules that fulfil the same function. The modules again consist of function units, as e.g. the actuators for one of the module's joints. The function units themselves consist of components, here regarded as the smallest elements. In the entire drive, these components are the actuator providing the drive power and the components in the drive train connected in a serial arrangement, e.g. gears, drive belt, or worm gear transferring the drive power to the joint.

3.3 Selection and data base

Many components used in such highly integrated systems are commonly known, commercially available and do not have to be newly invented. However, a humanoid robot consists of a large number of components, and for each of them there may be a variety of technical solutions. This leads to an overwhelming number of possible combinations, which cannot easily be overseen without help and which complicates an efficient target-oriented development. Therefore it is helpful to file the components of the joints, actuators and sensors as objects in an object-oriented classification. It enables a requirement-specific access to the objects and delivers information about possible combinations of components.

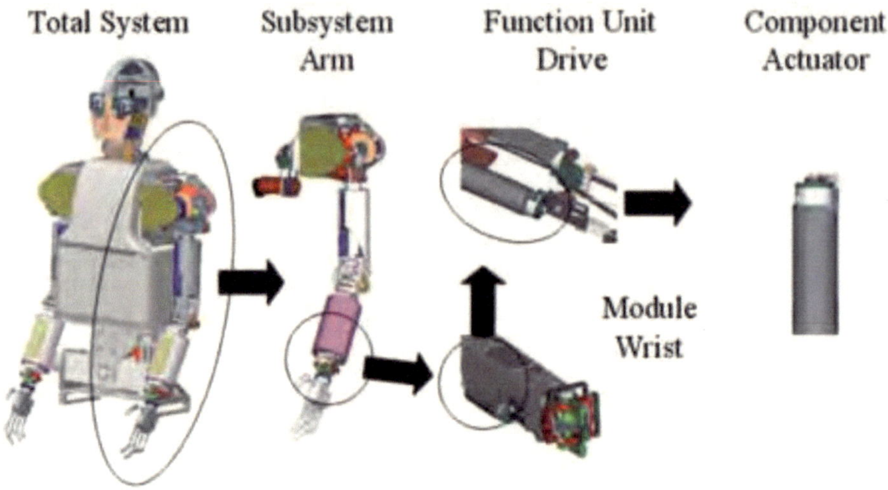

Fig. 1. Subdivision of the total system.

3.4 Development sequence

The development sequence conforms to the order in which a component or information has to be provided for the further procedure. The development process can be roughly divided into two main sections. The first section determines the basic requirements for the total system, which have to be known before the design process. This phase includes primarily two iterations: In the first iteration, the kinematics of the robot is specified according to the motion space of the robot and the kinematics again has to be describable in order to be controllable. In the second iteration, the control concept for the robot and the general possibilities for operating the joints are adjusted to the requirements for the desired dynamics of the robots. The second sector is the actual design process. The sequence in which the modules are developed is determined by their position in the serial kinematics of the robot. This means that e.g. in the arm, first the wrist, the elbow joint and then finally the shoulder joint are designed.

Since generally all modules have a similar design structure, they can be designed according to the same procedure. The sequence in this procedure model is determined by the interactions between the function units and between the components. The relation between the components and the behaviour of their interaction in case of a change of the development order can be displayed graphically in a design structure matrix (Browning, 2001). Iterations, which always occur in the development of complex systems, can be limited by early considering the properties of the components that are integrated at the end of the development process. One example is the torque measurement in the drive train. In the aforementioned data base, specifications of the components are given like the possibility for a component of the drive train to include some kind of torque measurement. It ensures that after the assembly of a drive train, a power measurement can be integrated.

3.5 Development of a shoulder joint

The development of a robot shoulder joint according to this approach is exemplarily described in the following paragraphs. For the tasks that are required from the robot, it is sufficient if the robot is able to move the arm in front of its body. These movements can be performed by means of a ball joint in the shoulder without an additional pectoral girdle. In the available design space, a ball joint can be modelled with the required performance of the actuators and sensors as a serial connection of three single joints. The axes of rotation of these joints intersect at one point. A replacement joint is used which consists of a roll joint, a pitch joint, and then again of another roll joint. The description of the kinematics can only be clarified together with the entire arm, which requires limiting measures, especially if redundant degrees of freedom exist (Asfour, 2003).
Information about the mass of the arm and its distribution are requirements for the design of the shoulder joint module. In addition, information about the connection of elbow and shoulder has to be available. This includes the components that are led from the elbow to or through the shoulder, as e.g. cables or drive trains of lower joints. The entire mechatronic system can be described in an abstract way by the object-oriented means of SysML (System Modelling Language) (SysML, 2005) diagrams, with which it is possible to perform a system test with regard to compatibility and operational reliability. It enables the representation of complex systems at different abstraction levels. Components that are combined in this way can be accessed in the aforementioned classification, which facilitates a quick selection of the components that can be used for the system. In addition, it makes a function design model possible at every point of the development.

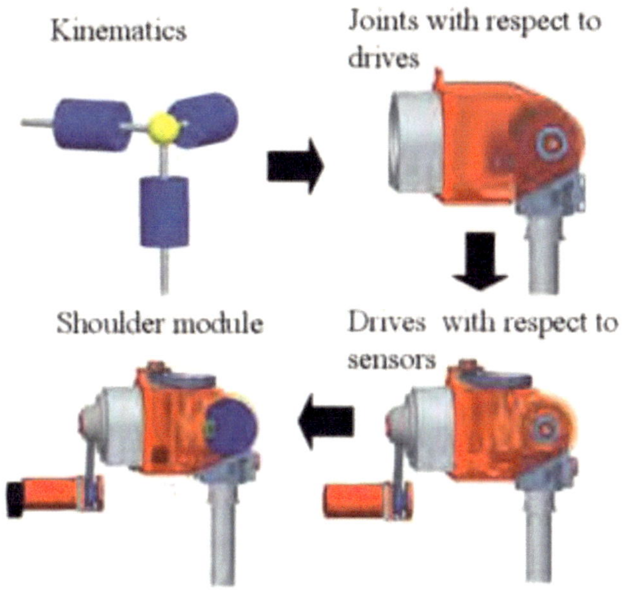

Fig. 2. Design of the shoulder module.

In the development of the shoulder module (Fig. 2), at first the function units of the joints for the three rotating axes are selected according to the kinematics. Then, the function unit drive, including the actuators and the drive trains, are integrated. Hereafter, the sensors are selected and integrated. In order to prevent time consuming iterations in the development, the components of the total system, integrated at a later stage, are already considered from the start with regard to their general requirements for being integrated. Examples for this are the sensors, which can then be assembled without problems since it is made sure that the already designed system offers the possibility to integrate them. During the next step the neighbouring module is designed. Information about the required interface of the shoulder and the mass of the arm and its distribution are given to the torso module.

4. Topology optimization

Topology optimization is used for the determination of the basic layout of a new design. It involves the determination of features such as the number, location and shape of holes and the connectivity of the domain. A new design is determined based upon the design space available, the loads, materials and other geometric constraints, e.g. bearing seats of which the component is to be composed of.
Today topology optimization is very well theoretically studied (Bendsoe & Sigmund, 2003) and applied in industrial design processes (Pedersen & Allinger, 2005). The designs obtained using topology optimization are considered design proposals. These topology optimized designs can often be rather different compared to designs obtained with a trial and error design process or designs obtained upon improvements based on experience or intuition as can be deduced from the motor carrier example in Fig. 3. Especially for complex loads, which are typical for systems like humanoid robots, these methods of structural optimization are helpful within the design process.

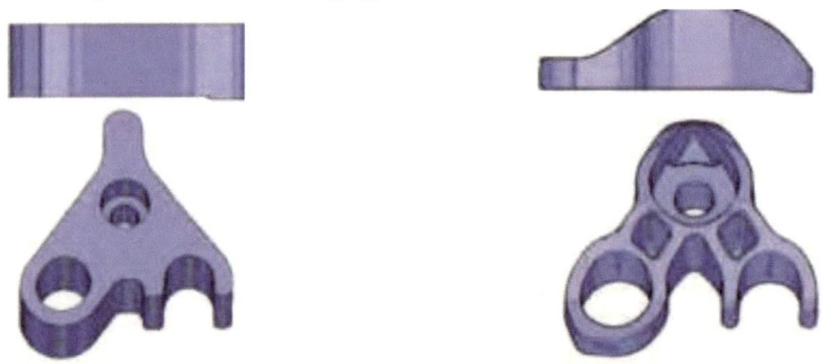

Design space for topology optimization   Constructional implementation

Fig. 3. Topology optimization of a gear oil line bracket provided by BMW Motoren GmbH.

The standard formulation in topology optimization is often to minimize the compliance corresponding to maximizing the stiffness using a mass constraint for a given amount of material. Compliance optimization is based upon static structural analyses, modal analyses or even non-linear problems e.g. models including contacts. A topology optimization scheme as depicted in Fig. 4. is basically an iterative process that integrates a finite element solver and an optimization module. Based on a design response supplied by the FE solver like strain energy for example, the topology optimization module modifies the FE model.
The FE model is typically used together with a set of loads that are applied to the model. These loads do not change during the optimization iterations. An MBS extended scheme as introduced by (Häussler et al., 2001) can be employed to take the dynamic interaction between the FE model and the MBS system into account.

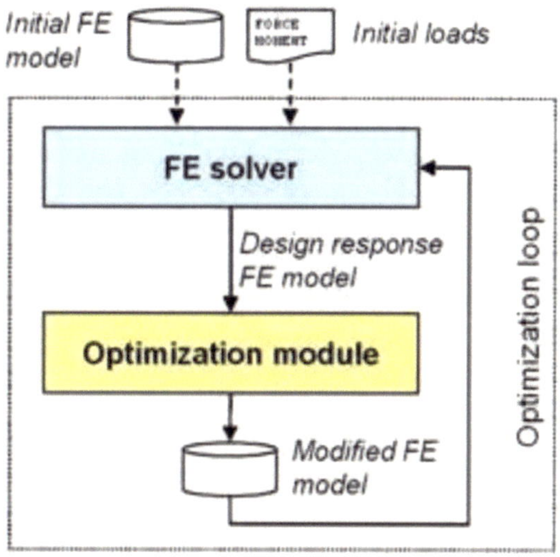

Fig. 4. Topology optimization scheme.

4.1 Topology optimization of robot thorax

The design of the central support structure of the upper body, the thorax, of the humanoid robot ARMAR III was determined with the help of topology optimization. The main functions of this element are the transmission of forces between arms, neck and torso joint and the integration of mechanical and electrical components, which must be accommodated for inside the robot's upper body. For instance four drive units for the elbows have to be integrated in the thorax to reduce the weight of the arms, electrical components like two PC-104s, four Universal Controller Modules (UCoM), A/D converters, DC/DC converters and force-moment controllers.

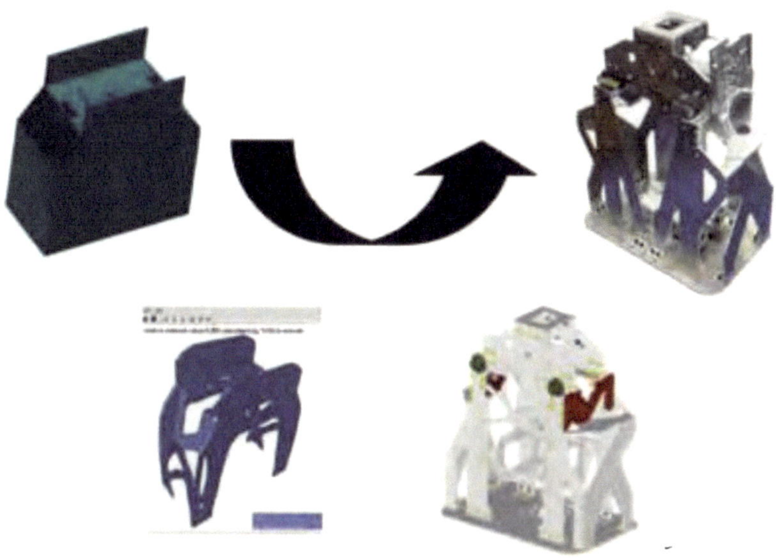

Fig. 5. Topology optimization of the thorax.

The left picture in figure 5 shows the initial FE model of the available design space including the geometric boundary conditions like the mechanical interfaces for the adjoining modules neck, arms and torso joint as well as the space reserved for important components like computers and controllers. Together with a set of static loads, this was the input for the optimization process. The bottom left picture shows the design as it was suggested by the optimization module after the final optimization loop. This design was then manually transferred into a 3d model in consideration of manufacturing restrictions. The picture on the right in Fig. 5 shows the assembled support structure made from high-strength aluminium plates. The result of the optimization is a stiff and lightweight structure with a total mass of 2.7 kg.

5. The upper body of ARMAR III

ARMAR III is s a full-size humanoid Robot which is the current demonstrator system of the collaborative research centre 588. It consists of a sensor head for visual and auditory perception of the environment, an upper body with two arms with a large range of motion for the manipulation of objects and a holonomic platform for omni-directional locomotion. ARMAR III has a modular design consisting of the following modules: head, neck joint, thorax, torso joint and two arms which are subdivided into shoulder, elbow, wrist and hands. The head and the holonomic platform were developed at the Research Center for Information Technologies (FZI), the hands were developed at the Institute for Applied Computer Science at the Forschungszentrum Karlsruhe (Beck et al, 2003; Schulz 2003). The modules for neck, torso and arms shown in the following figure were designed and manufactured at the Institute of Product Development (IPEK) at the University of Karlsruhe (TH).

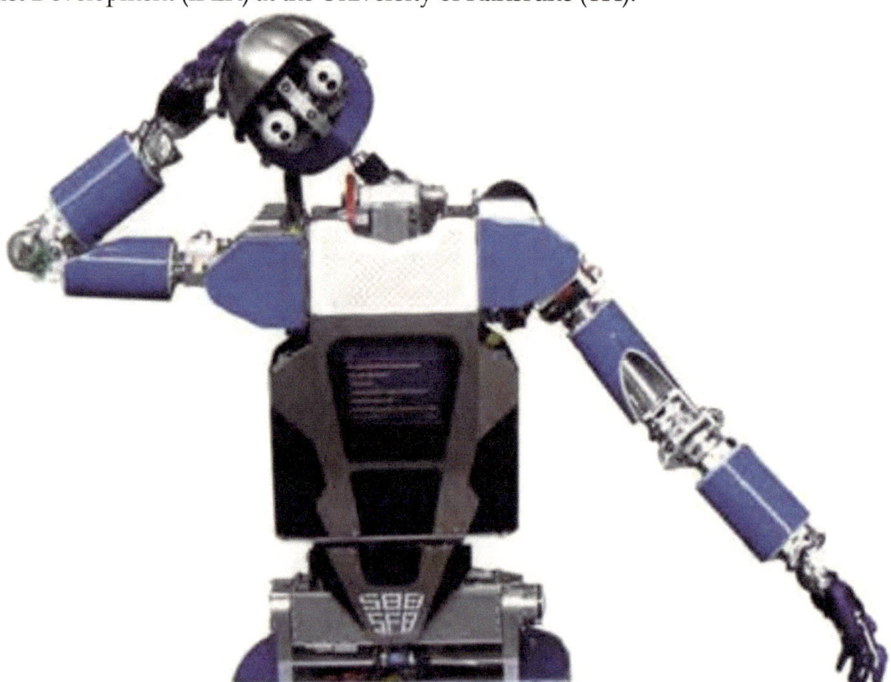

Fig. 6. The upper body of the humanoid robot ARMAR III.

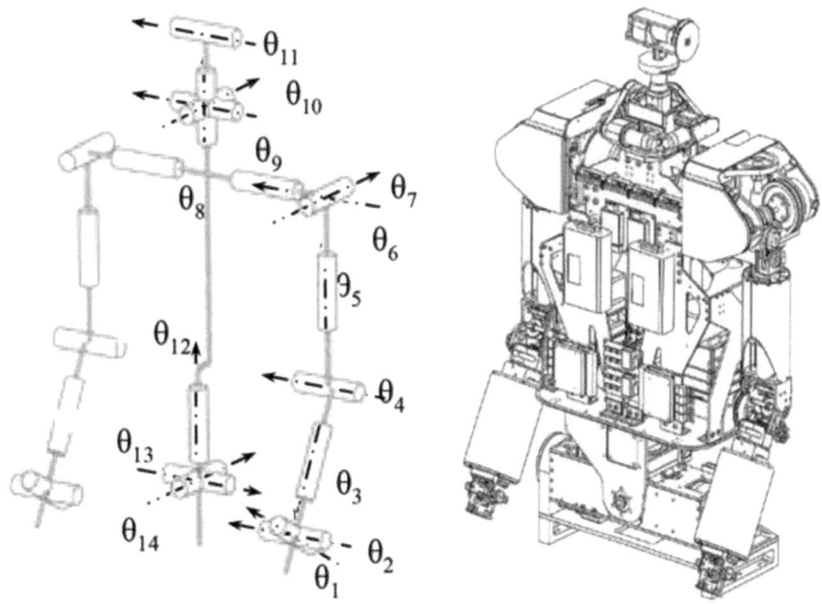

Fig. 7. Kinematics and CAD model of upper body of ARMAR III.

The size of the design space and the motion space of ARMAR III are similar to that of a human person with a height of approximately 175 cm. The main dimensions of the upper body can be seen in Fig. 8.

Table 1 gives an overview of the degrees of freedom and the motion range of all modules. Both arms have seven degrees of freedom. The three degrees of freedom in the shoulder provide a relatively wide range of motion. Together with two degrees of freedom in the elbow as well as in the wrist, the arm can be used for complex manipulation tasks that occur in the primary working environment of ARMAR III, the kitchen. Compared with other humanoid robots, the arm of ARMAR III provides large and humanlike ranges of motion. The neck joint with four degrees of freedom allows humanlike motion of the head.

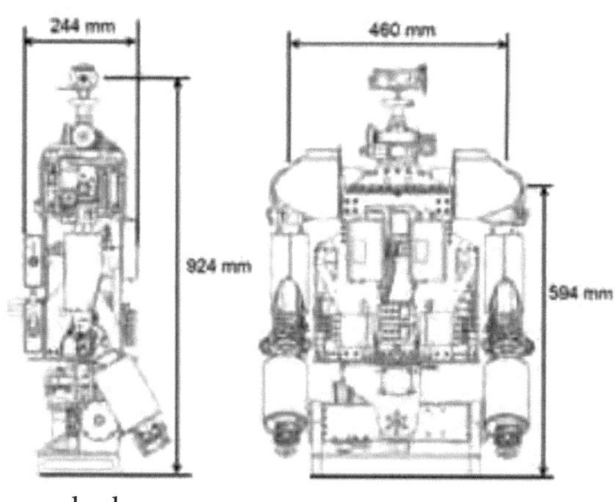

Fig. 8. Dimension of upper body.

	Part		D.O.F	amount	total
Degree of freedom	Wrist		2	2	4
	Elbow		2	2	4
	Shoulder		3	2	6
	Neck		4	1	4
	Torso		3	1	3
	Upper body				21
Range of motion	Wrist	θ_1	-30° to 30°		
		θ_2	-60° to 60°		
	Elbow	θ_3	-90° to 90°		
		θ_4	-10° to 150°		
	Shoulder	θ_5	-180° to 180°		
		θ_6	-45° to 180°		
		θ_7	-10° to 180°		
	Neck	θ_8	-180° to 180°		
		θ_9	-45° to 45°		
		θ_{10}	-45° to 45°		
		θ_{11}	-60° to 60°		
	Torso	θ_{12}	-180° to 180°		
		θ_{13}	-10° to 60°		
		θ_{14}	-20° to 20°		

Table 1. Degrees of freedom with range of motion.

5.1 Shoulder joint

The shoulder joint is the link between the arm and the torso. In addition to the realization of three degrees of freedom with intersecting axes in one point, the bowden cables for driving the elbow joint are guided through the shoulder joint from the elbow drive units in the torso to the elbow. The drive units of all joints are designed in a way, that their contributions to the inertia are as small as possible. Therefore the drive unit for panning the arm (Rot. 1), which has to provide the highest torque in the arm, is attached directly to the torso and does not contribute to the inertia of the moving part of the arm. The drive units for raising the arm (Rot. 2) and turning the arm around its longitudinal axis (Rot. 3) have been placed closely to the rotational axes to improve the dynamics of the shoulder joint. In order to achieve the required gear ratios in the very limited design space, Harmonic Drive transmissions, worm gear transmissions and toothed belt transmissions have been used.

These elements allow a compact design of the shoulder with a size similar to a human shoulder. As all degrees of freedom are realized directly in the shoulder, the design of the upper arm is slender. The integration of torque sensors in all three degrees of freedom is realized in two different ways. For the first degree of freedom strain gages are attached to a torsion shaft which is integrated in the drive train. The torque for raising and turning the arm is determined by force sensors that measure the axial forces in the worm gear shafts. In addition to the encoders, which are attached directly at the motors, angular sensors for all three degrees of freedom are integrated into the drive trains of the shoulder joints. The position sensors, which are located directly at the joints, allow quasi-absolute angular position measurement based on incremental optical sensors. A touch-sensitive artificial skin sensor, which can be used for collision detection or intuitive tactile communication, is attached to the front and rear part of the shoulder casing (Kerpa et al., 2003).

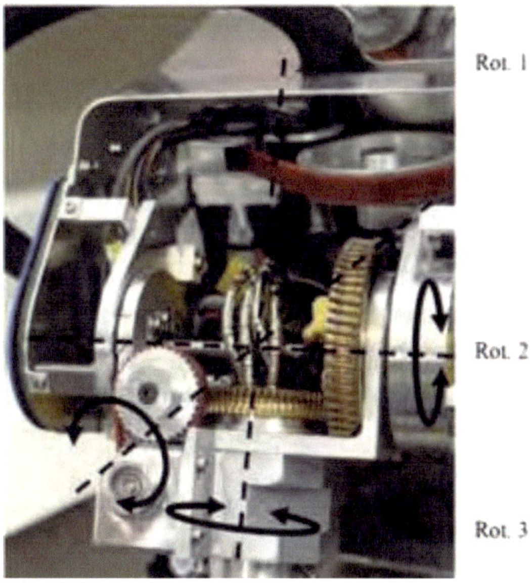

Fig. 9. Side view of the shoulder joint.

5.2 Elbow joint and upper arm

The elbow joint of ARMAR III has two degrees of freedom. These allow bending as well as rotating the forearm. The drive units, consisting of motor and Harmonic Drive transmissions, are not in the arm, but are located in the thorax of the robot. Thus the moving mass of the arm as well as the necessary design space are reduced, which leads to better dynamic characteristics and a slim form of the arm. The additional mass in the thorax contributes substantially less to the mass inertia compared to placing the drive units in the arm.

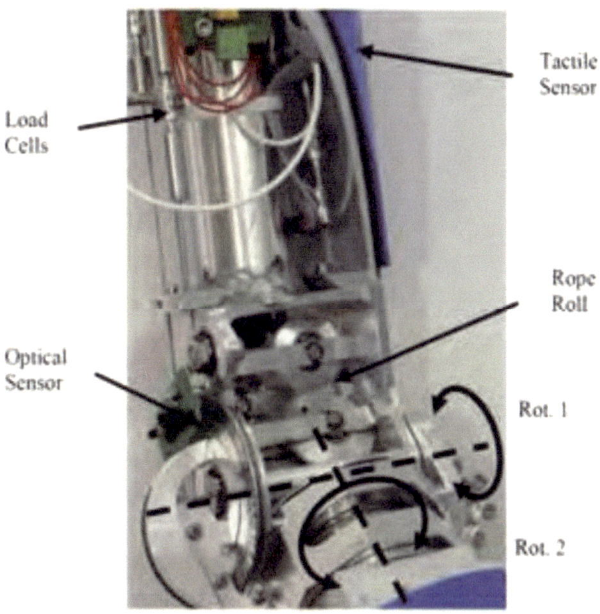

Fig. 10. Elbow joint.

Due to this concept, load transmission is implemented with the use of wire ropes, which are led from the torso through the shoulder to the elbow by rolls and bowden cables. In order to realize independent control of both degrees of freedom, the wire ropes for turning the forearm are led through the axis of rotation for bending the elbow. With altogether twelve rolls, this rope guide realizes the uncoupling of the motion of bending the elbow from rotating the forearm. In contrast to the previous version of the elbow, where the steel cables were guided by Bowden cables, this solution leads to smaller and constant friction losses which is advantageous for the control of this system.
Similar to the shoulder, the angular measurement is realized by encoders attached directly to the motors as well as optical sensors that are located directly at the joint for both degrees of freedom. In order to measure the drive torque, load cells are integrated in the wire ropes in the upper arm. As each degree of freedom in the elbow is driven by two wire ropes the measuring of force in the wire ropes can be done by differential measurements. Another possibility for measuring forces offers the tactile sensor skin, which is integrated in the cylindrical casing of the upper arm.
By placing the drive units in the thorax, there is sufficient design space left in the arm which can be used for electronic components that process sensor signals and which can be installed in direct proximity to the sensors in the upper arm.

5.3 Wrist joint and forearm

The wrist has two rotational degrees of freedom with both axes intersecting in one point. ARMAR III has the ability to move the wrist to the side as well as up and down. This was realized by a universal joint in very compact design. The lower arm is covered by a cylindrical casing with an outer diameter of 90 mm. The motors for both degrees of freedom are fixed at the support structure of the forearm. The gear ratio is obtained by a ball screw and a toothed belt or a wire rope respectively. The load transmission is almost free from backlash.

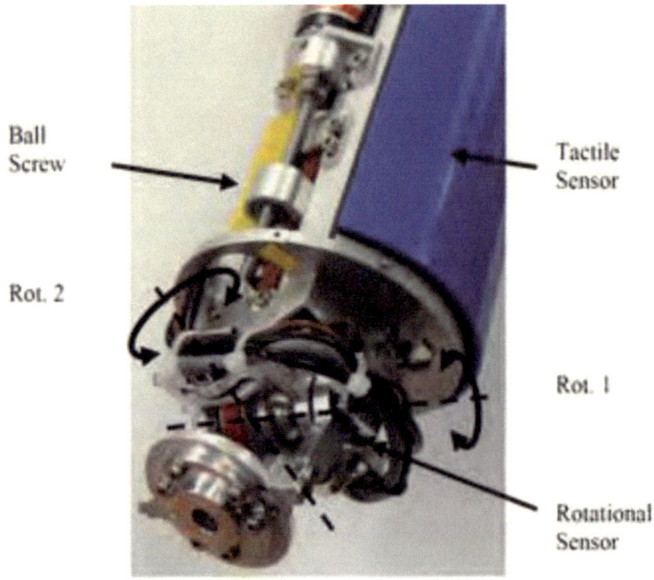

Fig. 11. Forearm with two degrees of freedom in the wrist.

By arranging the motors close to the elbow joint, the centre of mass of the forearm is shifted towards the body, which is an advantage for movements of the robot arm. Angular measurement in the wrist is realized by encoders at the motors and with quasi-absolute angular sensors directly at the joint. To measure the load on the hand, a 6-axis force and torque sensor is fitted between the wrist and the hand (Beck et al., 2003) (not shown in Fig. 11). The casing of the forearm is also equipped with a tactile sensor skin. The support structure of the forearm consists of a square aluminium profile. This rigid lightweight structure offers the possibility of cable routing on the inside and enough space for mounting electronic components on the flat surfaces of the exterior.

5.4 Neck joint

The complex kinematics of the human neck is defined by seven cervical vertebrae. Each connection between two vertebrae can be seen as a joint with three degrees of freedom. For this robot, the kinematics of the neck has been reduced to a serial kinematics with four rotational degrees of freedom. Three degrees of freedom were realized in the basis at the lower end of the neck. Two degrees of freedom allow the neck to lean forwards and backwards (1) and to both sides (2), another degree of freedom allows rotation around the longitudinal axis of the neck. At the upper end of the neck, a fourth degree of freedom allows nodding of the head. This degree of freedom allows more human-like movements of the head and improves the robots ability to look up and down and to detect objects directly in front of it.

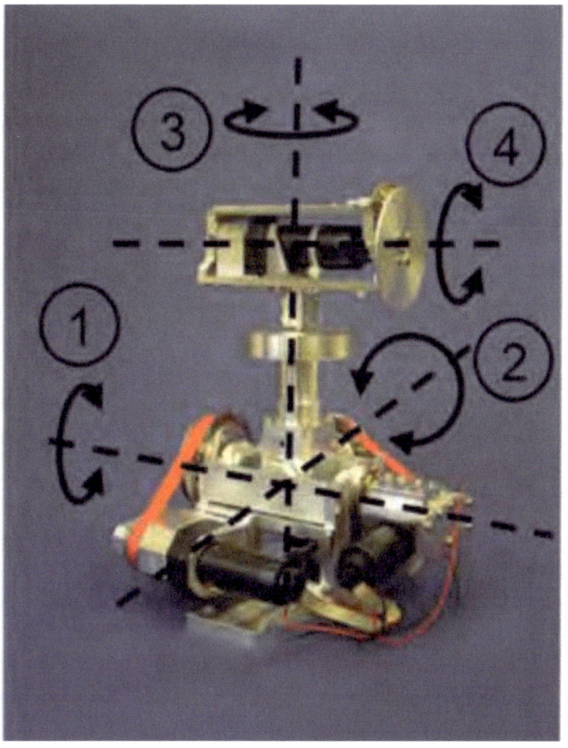

Fig. 12. Neck joint with four degrees of freedom.

For the conversion of torque and rotational speed, the drive train of each degree of freedom consists of Harmonic Drive transmissions either as only transmission element or, depending on the needed overall gear ratio, in combination with a toothed gear belt.

The drives for all degrees of freedom in the neck are practically free from backlash. The motors of all degrees of freedoms are placed as close as possible to the rotational axis in order to keep the moment of inertia small. The sensors for the angular position measurement in the neck consist of a combination of incremental encoders, which are attached directly to the motors, and quasi-absolute optical sensors, which are placed directly at the rotational axis. The neck module as depicted above weighs 1.6 kg.

5.5 Torso joint

The torso of the upper body of ARMAR III is divided into two parts, the thorax and the torso joint below it. The torso joint allows motion between the remaining upper body and the holonomic platform, similar to the functionality of the lower back and the hip joints in the human body. The kinematics of the torso joint does not exactly replicate the complex human kinematics of the hip joints and the lower back. The complexity was reduced in consideration of the functional requirements which result from the main application scenario of this robot in the kitchen. The torso joint has three rotational degrees of freedom with the axes intersecting in one point. The kinematics of this joint, as it is described in table 1 and Fig. 13, is sufficient to allow the robot to easily reach important points in the kitchen. For example in a narrow kitchen, the whole upper body can turn sideways or fully around without having to turn the platform. One special requirement for the torso joint is, that all cables for the electrical energy flow and information flow between the platform and the upper body need to go through the torso joint. All cables are to be led from the upper body to the torso joint in a hollow shaft with an inner bore diameter of 40 mm through the point of intersection of the three rotational axes. This significantly complicates the design of the joint, but the cable connections can be shorter and stresses on the cables due to compensational motions, that would be necessary if the cable routing was different, can be reduced. This simplifies the design of the interface between upper and lower body. For transportation of the robot, the upper and lower part of the body can be separated by loosening one bolted connection and unplugging a few central cable connections. Due to the special boundary conditions from the cable routing, all motors had to be placed away from the point of intersection of the three axes and the motor for the vertical degree of freedom Rot. 3 could not be positioned coaxially to the axis of rotation. The drive train for the degrees of freedom Rot. 1 and Rot. 3 consists of Harmonic Drive transmissions and toothed belt transmissions.

The drive train for the degree of freedom Rot.2 is different from most of the other drive trains in ARMAR III as it consists of a toothed belt transmission, a ball screw and a piston rod which transforms the translational motion of the ball screw into the rotational motion for moving the upper body sideways. This solution is suitable for the range of motion of 40°, it allows for a high gear ratio and the motor can be placed away from the driven axis and away from the point of intersection of the rotational axes.

In addition to the encoders, which are directly attached to the motors, two precision potentiometers and one quasi-absolute optical sensor are used for the angular position measurement.

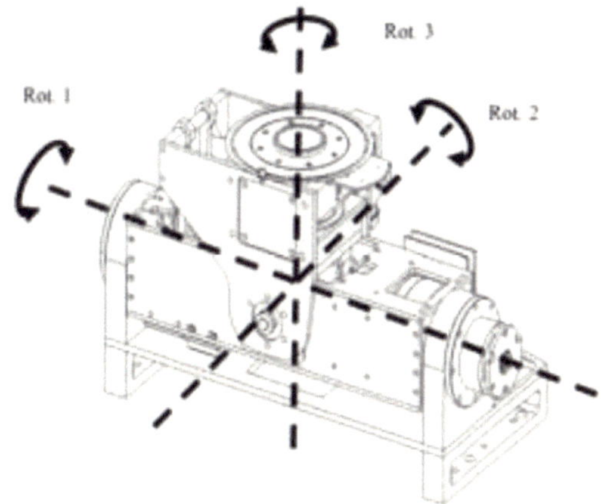

Fig. 13. Torso joint.

6. Conclusions and future work

Methods for the efficient development of modules for a humanoid robot were developed. Future work will be to create a database of system elements for humanoid robot components and the characterization for easier configuration of future humanoids. This database can then be used to generate consistent principle solutions for robot components more efficiently.
Topology optimization is a tool for designing and optimizing robot components which need to be light yet stiff. The thorax of ARMAR III was designed with the help of this method. For the simulation of mechatronic systems like humanoid robots, it is necessary to consider mechanical aspects as well as the behaviour of the control system. This is not yet realized in the previously described topology optimization process. The coupling between the mechanical system and the control system might influence the overall system's dynamic behaviour significantly. As a consequence, loads that act on a body in the system might be affected not only by the geometric changes due to optimization but also by the control system as well. The topology optimization scheme shown in Fig. 4 should be extended by means of integrating the dynamic system with a multi body simulation and the control system as depicted in Fig. 14.

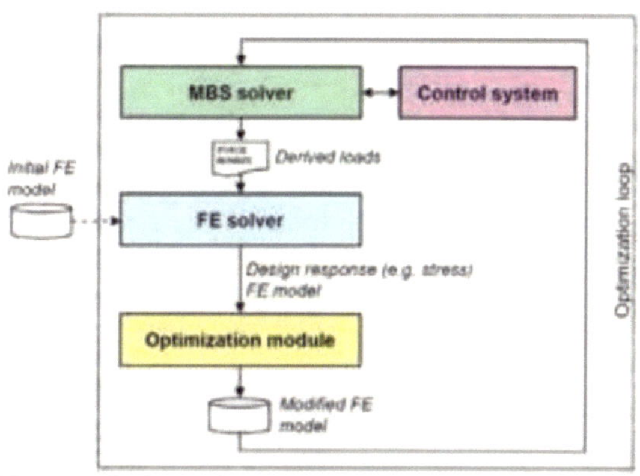

Fig. 14. Controlled MBS extended topology optimization.

The upper body of the humanoid robot ARMAR III was presented. The modules for neck, arms and torso were explained in detail. The main goals for the future work on ARMAR III are to further reduce the weight and to increase the energy efficiency, increase the payload and to design a closed casing for all robot joints.

7. Acknowledgement

The work presented in this chapter is funded by the German Research Foundation DFG in the collaborative research centre 588 "Humanoid robots - learning and cooperating multi- modal robots".

8. References

Asfour, T. (2003). Sensomotorische Bewegungskoordination zur Handlungsausführung eines humanoiden Roboters, Dissertation Faculty for Computer Science, University of Karlsruhe (TH)

Beck, S.; Lehmann, A.; Lotz, T.; Martin, J.; Keppler, R.; Mikut, R. (2003). Model-based adaptive control of a fluidic actuated robotic hand, Proc., GMA-Congress 2003

Bendsoe. M.; Sigmund, O. (2003) Topology Optimization - Theory, Methods, Application, Springer Verlag

Browning, T. R. (2001). Applying the Design Structure Matrix to System Decomposition and Integration Problems: A Review and New Directions, IEEE Transaction on Engineering Management, Vol. 48, No. 3

Häussler, P.; Emmrich ; D.; Müller, O.; Ilzhöfer, B.; Nowicki, L.; Albers, A. (2001). Automated Topology Optimization of Flexib-le Components in Hybrid Finite Element Multibody Systems using ADAMS/Flex and MSC.Construct, ADAMS European User's Conference, Berchtesgaden, Germany

Häussler, P. (2005). Ein neuer Prozess zur parameterfreien Formoptimierung dynamisch beanspruchter Bauteile in mechanischen Systemen auf Basis von Lebensdaueranalysen und hybriden Mehrkörpersystemen, dissertation Faculty for Mechanical Engineering, research reports of the Institute for Product Development, University of Karlsruhe, ISSN 1615-8113

Kerpa, O.; Weiss, K.; Wörn, H.; (2003). Development of Flexible Tactile Sensor System for a Humanoid Robot, Intelligent Robots and Systems IROS, Las Vegas USA

Minx, J.; Häussler, P.; Albers, A.; Emmrich D.; Allinger, P. (2004). Integration von FEM, MKS und Strukturoptimierung zur ganzheitlichen, virtuellen Entwicklung von dynamisch beanspruchten Bauteilen, NAFEMS seminar, analysis of multibody systems with FEM and MBS, October, 27th -28th, Wiesbaden

Ortiz, J.; Bir, G. (2006). Verification of New MSC.ADAMS Linearization Capability For Wind Turbine Applications, 44th AIAA Aerospace Sciences Meeting and Exhibit, Reno, Nevada

Pedersen, C.B.W.; Allinger, P. (2005). Recent Developments in the Commercial Implementation of Topology Optimization. TopoptSYMP2005 - IUTAM-Symposium- Topological design optimization of structures, machines and material - status and perspectives, Copenhagen, Denmark

Schäfer, C. (2000). Entwurf eines anthropomorphen Roboterarms: Kinematik, Arbeitsraumanalyse, Softwaremodellierung, dissertation Faculty for Computer Science, University of Karlsruhe (TH)

Schulz, S. (2003). Eine neue Adaptiv-Hand-Prothese auf der Basis flexibler Fluidaktoren, Dissertation, Faculty for Mechanical Engineering, University of Karlsruhe (TH) 2003

SysML Partners, (2005). Systems Modeling Language (SysML) specification version 1.0 alpha, 14, www.sysml.org

VDI Gesellschaft Entwicklung Konstruktion Vertrieb (Editor) (2004), VDI-Guideline 2206: Design methodology for mechatronic systems; Beuth Verlag GmbH, Berlin

2

Gait Transition from Quadrupedal to Bipedal Locomotion of an Oscillator-driven Biped Robot

Shinya Aoi and Kazuo Tsuchiya
Dept. of Aeronautics and Astronautics, Graduate School of Engineering, Kyoto University
Yoshida-honmachi, Sakyo-ku, Kyoto 606-8501, Japan

1. Introduction

Studies on biped robots have attracted interest due to such problems as inherent poor stability and the cooperation of a large degree of freedom. Furthermore, recent advanced technology, including hardware and software, allows these problems to be tackled, accelerating the interest. Actually, many sophisticated biped robots have already been developed that have successfully achieved such various motions as straight walking, turning, climbing slopes, rising motion, and running (Aoi & Tsuchiya, 2005; Aoi et al., 2004; Hirai et al., 1998; Kuniyoshi et al., 2004; Kuroki et al. 2003; Löffler et al., 2003; Nagasaki et al., 2004).

Steady gait for a biped robot implies a stable limit cycle in its state space. Therefore, different steady gait patterns have different limit cycles, and gait transition indicates that the state of the robot moves from one limit cycle to another. Even if the robot obtains steady gait patterns, their transition is not necessarily confirmed as completed. Thus, smooth transition between gait patterns remains difficult. To overcome such difficulty, many studies have concentrated on model-based approaches using inverse kinematics and kinetics. These approaches basically generate robot motions based on such criteria as zero moment point (Vukobratović et al., 1990) and manipulate robot joints using motors. However, they require accurate modeling of both the robot and the environment as well as complicated computations. The difficulty of achieving adaptability to various environments in the real world is often pointed out, which means that in these approaches the robot is too *rigid* to react appropriately to environmental changes. Therefore, the key issue in the control is to establish a *soft* robot by adequately changing the structure and response based on environmental changes.

In contrast to robots, millions of animal species adapt themselves to various environments by cooperatively manipulating their complicated and redundant musculoskeletal systems. Many studies have elucidated the mechanisms in their motion generation and control. In particular, neurophysiological studies have revealed that muscle tone control plays important roles in generating adaptive motions (Mori, 1987; Rossignol, 1996; Takakusaki et al., 2003), suggesting the importance of compliance in walking. Actually, many studies on robotics have demonstrated the essential roles of compliance. Specifically, by appropriately employing the mechanical compliance of robots, simple control systems have attained highly adaptive and robust motions, especially in hexapod (Altendorfer et al., 2001; Cham et al., 2004; Quinn et al., 2003; Saranli et al., 2001), quadruped (Fukuoka et al., 2003; Poulakakis et al., 2005), and biped robots (Takuma & Hosoda, 2006; Wisse et al., 2005). However, note that control systems using motors continue to have difficulty adequately manipulating compliance in robot joints.

On the other hand, neurophysiological studies have also clarified that animal walking is generated by central pattern generators (CPGs) that generate rhythmic signals to activate their limbs (Grillner, 1981, 1985; Orlovsky et al., 1999). CPGs modulate signal generation in response to sensory signals, resulting in adaptive motions. CPGs are widely modeled using nonlinear oscillators (Taga et al., 1991; Taga, 1995a,b), and based on such CPG models many walking robots and their control systems have been developed, in particular, for quadruped robots (Fukuoka et al., 2003; Lewis & Bekey, 2002; Tsujita et al., 2001), multi-legged robots (Akimoto et al., 1999; Inagaki et al., 2003), snake-like robots (Ijspeert et al., 2005; Inoue et al., 2004), and biped robots (Aoi & Tsuchiya, 2005; Aoi et al., 2004; Lewis et al., 2003; Nakanishi et al., 2004).

This paper deals with the transition from quadrupedal to bipedal locomotion of a biped robot while walking. These gait patterns originally have poor stability, and the transition requires drastic changes in robot posture, which aggravates the difficulty of establishing the transition without falling over. Our previous work developed a simple control system using nonlinear oscillators by focusing on CPG characteristics that are used for both quadruped and biped robots, revealing that they achieved steady and robust walking verified by numerical simulations and hardware experiments (Aoi & Tsuchiya, 2005; Aoi et al., 2004; Tsujita et al., 2001). In this paper, we use the same developed control system for both quadrupedal and bipedal locomotion of a biped robot and attempt to establish smooth gait transition. Specifically, we achieve stable limit cycles of these gait patterns and their transitions by moving the robot state from one limit cycle to another by cooperatively manipulating their physical kinematics through numerical simulations. This paper is organized as follows. Section 2 introduces the biped robot model considered in this paper. Section 3 explains the developed locomotion control system, and Section 4 addresses the approach of gait transition and numerical results. Section 5 describes the discussion and conclusion.

2. Biped robot model

Figure 1(a) shows the biped robot model considered in this paper. It consists of a trunk, a pair of arms composed of four links, and a pair of legs composed of six links. Each link is connected to the others through a single degree of freedom rotational joint. A motor is installed at each joint. Four touch sensors are attached to the sole of each foot, and one touch sensor is attached to the tip of the hand of each arm. The left and right legs are numbered Legs 1 and 2, respectively. The joints of the legs are also numbered Joints 1…6 from the side of the trunk, where Joints 1, 2, and 3 are yaw, roll, and pitch hip joints, respectively. Joint 4 is a pitch knee joint, and Joints 5 and 6 are pitch and roll ankle joints. The arms are also numbered in a similar manner. Joints 1 and 4 are pitch joints, Joint 2 is a roll joint, and Joint 3 is a yaw joint. To describe the configuration of the robot, we introduce angles $\theta^i_{\mathrm{A}j}$ and $\theta^i_{\mathrm{L}k}$ (i=1,2, j=1,…,4, k=1,…,6), which are rotation angles of Joint j of Arm i and Joint k of Leg i, respectively. The robot walks quadrupedally and bipedally, as shown in Figs. 1(b) and (c). Its physical parameters are shown in Table 1. The ground is modeled as a spring with a damper in numerical simulations.

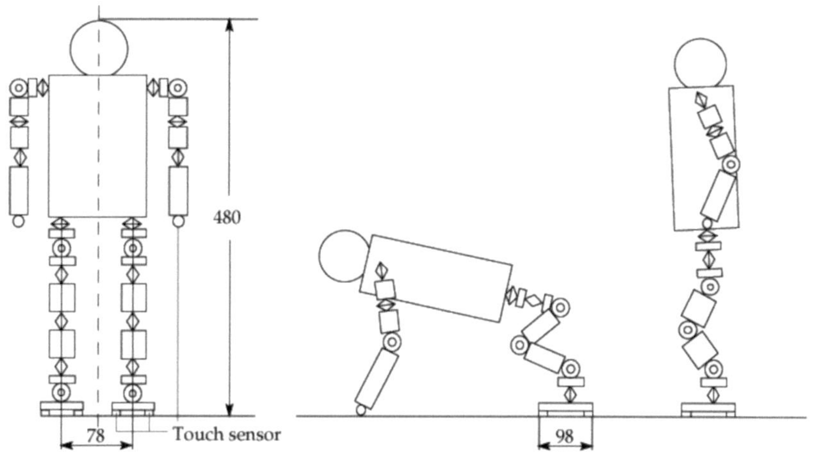

Fig. 1. Schematic model of a biped robot [mm].

Link	Mass [kg]	Length [m]
Trunk	2.34	0.20
Leg	1.32	0.28
Arm	0.43	0.25
Total	5.84	0.48

Table 1. Physical parameters of robot.

3. Locomotion control system

3.1 Concept of the control system

As described above, the crucial issue in controlling a biped robot is establishing a mechanism in which the robot adapts itself by changing its internal structure based on interactions between the robot's mechanical system and the environment. Neurophysiological studies have revealed that animal walking is generated by CPGs comprised of a set of neural oscillators present in the spinal cord. CPGs characteristically have the following properties:

1. CPGs generate inherent rhythmic signals that activate their limbs to generate rhythmic motions;
2. CPGs are sensitive to sensory signals from peripheral nerves and modulate signal generation in response to them.

Animals can immediately adapt to environmental changes and disturbances by virtue of these features and achieve robust walking.

We have designed a locomotion control system that has an internal structure that adapts to environmental changes, referring to CPG characteristics. In particular, we employed nonlinear oscillators as internal states that generate inherent rhythmic signals and adequately respond to sensory signals. Since the motor control of a biped robot generally uses local high-gain feedback control to manipulate the robot joints, we generated nominal joint motions using rhythmic signals from the oscillators. One of the most important factors in the dynamics of walking is the interaction between the robot and the external world, that is, dynamical interaction between the robot feet and the ground. The leg motion consists of swing and stance phases, and a harmonious balance must be achieved between these kinematical motions and dynamical interaction, which means that it is essential to adequately switch from one phase to another. Therefore, our developed control system focused on this point. Specifically, it modulated the signal generation of the oscillators and appropriately changed the leg motions from the swing to the stance phase based on touch sensors. Although we concisely describe the developed control system below, see our previous work (Aoi & Tsuchiya, 2005) for further details.

3.2 Developed locomotion control system

The locomotion control system consists of a motion generator and controller (see Fig. 2(a)). The former is composed of rhythm and trajectory generators. The rhythm generator has two types of oscillators: Motion and Inter (see Fig. 2(b)). As Motion oscillators, there are Leg 1, Leg 2, Arm 1, Arm 2, and Trunk oscillators. The oscillators follow phase dynamics in which they have interactions between themselves and receive sensory signals from touch sensors. The trajectory generator creates nominal trajectories of robot joints by phases of Motion oscillators, which means that it generates physical kinematics of the robot based on rhythmic signals from the oscillators. It receives outer commands and changes the physical kinematics to reflect the outer commands. The nominal trajectories are sent to the motion controller in which motor controllers manipulate the joint motions using the nominal trajectories as command signals. Note that physical kinematics is different between quadrupedal and bipedal locomotion, and except for the kinematics, throughout this paper we use the same control system regardless of gait patterns.

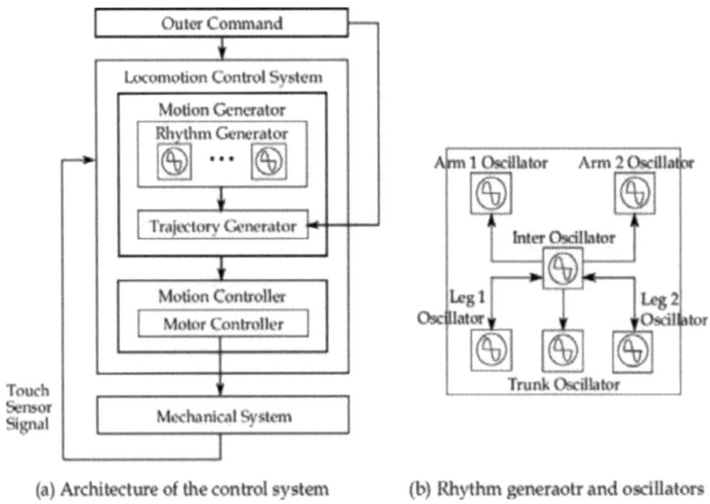

Fig. 2. Locomotion control system.

3.2.1 Trajectory generator

As mentioned above, the trajectory generator creates nominal trajectories of all joints based on the phases of the Motion oscillators. First, let φ_{L}^{i}, φ_{A}^{i}, φ_{T}, and φ_{I} (i=1,2) be the phases of Leg i, Arm i, Trunk, and Inter oscillators, respectively.

The nominal trajectories of the leg joints are determined by designing the nominal trajectory of the foot, specifically Joint 5, relative to the trunk in the pitch plane. The nominal foot trajectory consists of swing and stance phases (see Fig. 3). The former is composed of a simple closed curve that includes anterior extreme position (AEP) and posterior extreme position (PEP). This trajectory starts from point PEP and continues until the leg touches the ground. On the other hand, the latter consists of a straight line from the foot landing position (LP) to point PEP. Therefore, this trajectory depends on the timing of foot contact with the ground in each step cycle. Both in the swing and stance phases, nominal foot movement is designed to be parallel to the line that involves points AEP and PEP. The height and forward bias from the center of points AEP and PEP to Joint 3 of the leg are defined as parameters Δ_{L} and H_{L}, respectively. These two nominal foot trajectories provide nominal trajectories $\hat{\theta}^i_{\mathrm{L}j}$ (i=1,2, j=3,4,5) of Joint j (hip, knee, and ankle pitch joints) of Leg i by the functions of phase φ^i_{L} of Leg i oscillator written by $\hat{\theta}^i_{\mathrm{L}j}(\varphi^i_{\mathrm{L}})$, where we use $\varphi^i_{\mathrm{L}} = 0$ at point PEP and $\varphi^i_{\mathrm{L}} = \hat{\varphi}_{\mathrm{AEP}}$ at point AEP. Note that nominal stride $\hat{S}$ is given by the distance between points AEP and PEP, and duty factor $\hat{\beta}$ is given by the ratio between the nominal stance phase and step cycle durations.

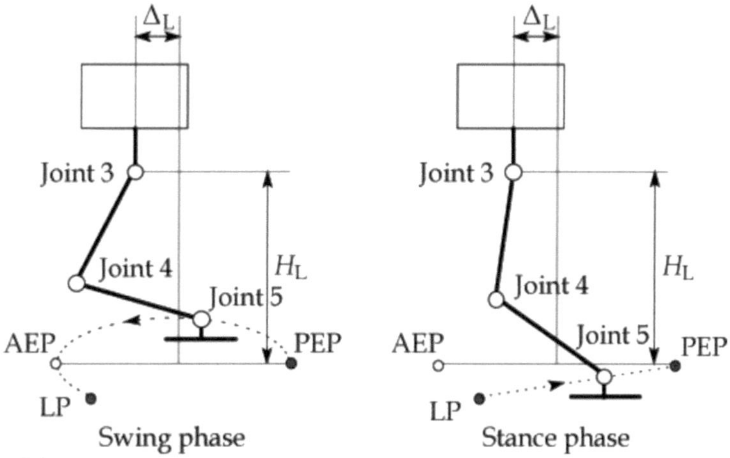

Fig. 3. Nominal foot trajectory.

The nominal trajectories of the arm joints are generated in a similar way to the leg joints described above except for the bend direction between Joint 4 of the arm and Joint 4 of the leg (see Fig. 4 below). Similar to the foot trajectory, the nominal trajectory of the hand, specifically the touch sensor at the tip of the arm, is designed relative to the trunk in the pitch plane, which consists of the swing and stance phases. Then, also from inverse kinematics, nominal trajectories $\hat{\theta}^i_{\mathrm{A}j}$ (i=1,2, j=1,…,4) of Joint j of Arm i are given by the functions of phase φ^i_{A} of Arm i oscillator. The nominal trajectories of the arm joints also have parameters Δ_{A} and H_{A}, similar to those of the leg joints, that use the same nominal stride $\hat{S}$ and duty ratio $\hat{\beta}$ as leg motions.

3.2.2 Rhythm generator and sensory signals

In the rhythm generator, Motion and Inter oscillators generate rhythmic behavior based on the following phase dynamics:

$$\begin{aligned}\dot{\varphi}_{\mathrm{I}} &= \hat{\omega} + g_{1\mathrm{I}} \\ \dot{\varphi}_{\mathrm{T}} &= \hat{\omega} + g_{1\mathrm{T}} \\ \dot{\varphi}_{\mathrm{A}}^{i} &= \hat{\omega} + g_{1\mathrm{A}}^{i} + g_{2\mathrm{A}}^{i} \quad i=1,2 \\ \dot{\varphi}_{\mathrm{L}}^{i} &= \hat{\omega} + g_{1\mathrm{L}}^{i} + g_{2\mathrm{L}}^{i} \quad i=1,2\end{aligned} \tag{1}$$

where $g_{1\mathrm{I}}$, $g_{1\mathrm{T}}$, $g_{1\mathrm{A}}^{i}$, and $g_{1\mathrm{L}}^{i}$ (i=1,2) are functions regarding the nominal phase relationship shown below, $g_{2\mathrm{A}}^{i}$ and $g_{2\mathrm{L}}^{i}$ (i=1,2) are functions arising from sensory signals given below, and $\hat{\omega}$ is the nominal angular velocity of each oscillator obtained by

$$\hat{\omega} = 2\pi \frac{1-\hat{\beta}}{\hat{T}_{\mathrm{sw}}} \tag{2}$$

where $\hat{T}_{\mathrm{sw}}$ is the nominal swing phase duration.

To establish stable walking, the essential problem is the coordination of joint motions. Interlimb coordination is the key. For example, both legs must move out of phase to prevent the robot from falling over during bipedal locomotion. Since the nominal joint trajectories for the limbs are designed by oscillator phases, interlimb coordination is given by the phase relation, that is, the phase differences between oscillators. Functions $g_{1\mathrm{I}}$, $g_{1\mathrm{T}}$, $g_{1\mathrm{A}}^{i}$, and $g_{1\mathrm{L}}^{i}$ in Eq. (1), which deal with interlimb coordination, are given by the phase differences between oscillators based on Inter oscillator, written by

$$\begin{aligned}g_{1\mathrm{I}} &= -\textstyle\sum_{i=1}^{2} K_{\mathrm{L}} \sin(\varphi_{\mathrm{I}} - \varphi_{\mathrm{L}}^{i} + (-1)^{i}\pi/2) \\ g_{1\mathrm{T}} &= -K_{\mathrm{T}} \sin(\varphi_{\mathrm{T}} - \varphi_{\mathrm{I}}) \\ g_{1\mathrm{A}}^{i} &= -K_{\mathrm{A}} \sin(\varphi_{\mathrm{A}}^{i} - \varphi_{\mathrm{I}} + (-1)^{i}\pi/2) \quad i=1,2 \\ g_{1\mathrm{L}}^{i} &= -K_{\mathrm{L}} \sin(\varphi_{\mathrm{L}}^{i} - \varphi_{\mathrm{I}} - (-1)^{i}\pi/2) \quad i=1,2\end{aligned} \tag{3}$$

where nominal phase relations are given so that both the arms and legs move out of phase and one arm and the contralateral leg move in phase and K_{L}, K_{A}, and K_{T} are gain constants.

In addition to physical kinematics and interlimb coordination, the modulation of walking rhythm is another important factor to generate walking. Functions $g_{2\mathrm{A}}^{i}$ and $g_{2\mathrm{L}}^{i}$ modulate walking rhythm through the modulation of the phases of oscillators based on sensory signals. Specifically, when the hand of Arm i (the foot of Leg i) lands on the ground, Arm i oscillator (Leg i oscillator) receives a sensory signal from the touch sensor (i=1,2). Instantly, phase φ_{A}^{i} of Arm i oscillator (phase φ_{L}^{i} of Leg i oscillator) is reset to nominal value $\hat{\varphi}_{\mathrm{AEP}}$ from value $\varphi_{\mathrm{land}}^{i}$ at the landing. Therefore, functions $g_{2\mathrm{A}}^{i}$ and $g_{2\mathrm{L}}^{i}$ are written by

$$\begin{aligned}g_{2\mathrm{A}}^{i} &= (\hat{\varphi}_{\mathrm{AEP}} - \varphi_{\mathrm{land}}^{i})\delta(t - t_{\mathrm{land}}^{i}) \quad i=1,2 \\ g_{2\mathrm{L}}^{i} &= (\hat{\varphi}_{\mathrm{AEP}} - \varphi_{\mathrm{land}}^{i})\delta(t - t_{\mathrm{land}}^{i}) \quad i=1,2\end{aligned} \tag{4}$$

where $\hat{\varphi}_{\mathrm{AEP}} = 2\pi(1-\hat{\beta})$, t_{land}^{i} is the time when the hand of Arm i (the foot of Leg i) lands on the ground (i=1,2) and $\delta(\cdot)$ denotes Dirac's delta function. Note that touch sensor signals not only modulate the walking rhythm but also switch the leg motions from the swing to stance phase, as described in Sec. 3.2.1.

4. Gait transition

This paper aims to achieve gait transition from quadrupedal to bipedal locomotion of a biped robot. As mentioned above, even if the robot establishes steady quadrupedal and bipedal locomotion, that is, each locomotion has a stable limit cycle in the state space of the robot, there is no guarantee that the robot can accomplish a transition from one limit cycle of quadrupedal locomotion to another of bipedal locomotion without falling over. Furthermore, these gait patterns originally have poor stability, and the transition requires drastic changes in robot posture, which aggravates the difficulty of preventing the robot from falling over during the transition.

4.1 Gait transition control

A biped robot generates quadrupedal and bipedal locomotion while manipulating many degrees of freedom in the joints. These gait patterns have different movements in many joints, which means that there are a million ways to achieve the transition. Therefore, the critical issue is designing gait transition.

In this paper, we generate both quadrupedal and bipedal locomotion of the robot based on physical kinematics. Figures 4(a) and (b) show schematics and parameters in quadrupedal and bipedal locomotion where COM indicates the center of mass of the trunk, l_{TA} and l_{TL} are the lengths from COM to Joint 1 of the arm and Joint 3 of the leg in the pitch plane, respectively, L_{A} and L_{L} are the forward biases from COM to the centers of the nominal foot and hand trajectories, respectively, ψ_{T} is the pitch angle of the trunk relative to the perpendicular line to the line that involves points AEP and PEP of the foot or the hand trajectory, and $*^{\mathrm{Q}}$ and $*^{\mathrm{B}}$ indicate the values of $*$ for quadrupedal and bipedal locomotion, respectively. Therefore, from a kinematical viewpoint, gait transition is achieved by changing parameters L_{A}, L_{L}, H_{A}, H_{L}, and ψ_{T} from values $L_{\mathrm{A}}^{\mathrm{Q}}$, $L_{\mathrm{L}}^{\mathrm{Q}}$, $H_{\mathrm{A}}^{\mathrm{Q}}$, $H_{\mathrm{L}}^{\mathrm{Q}}$, and $\psi_{\mathrm{T}}^{\mathrm{Q}}$ to values $L_{\mathrm{A}}^{\mathrm{B}}(=0)$, $L_{\mathrm{L}}^{\mathrm{B}}(=0)$, $H_{\mathrm{A}}^{\mathrm{B}}$, $H_{\mathrm{L}}^{\mathrm{B}}$, and $\psi_{\mathrm{T}}^{\mathrm{B}}$, while the robot walks. Note that from these figures, using parameters Δ_{A}, Δ_{L}, and ψ_{T}, parameters L_{A} and L_{L} are written as

$$\begin{aligned} L_{\mathrm{A}} &= l_{\mathrm{TA}} \sin\psi_{\mathrm{T}} + \Delta_{\mathrm{A}} \\ L_{\mathrm{L}} &= l_{\mathrm{TL}} \sin\psi_{\mathrm{T}} + \Delta_{\mathrm{L}} \end{aligned} \tag{5}$$

Animals generate various motions and smoothly change them by cooperatively manipulating their complicated and redundant musculoskeletal systems. To elucidate these mechanisms, many studies have investigated recorded electromyographic (EMG) activities, revealing that muscle activity patterns are expressed by the combination of several patterns, despite their complexity (d'Avella & Bizzi, 2005; Patla et al., 1985; Ivanenko et al., 2004, 2006). Furthermore, various motions have common muscle activity patterns, and different motions only have a few different specific patterns in combinations that express muscle activity patterns. This suggests that only a few patterns provide cooperation in different complex movements.

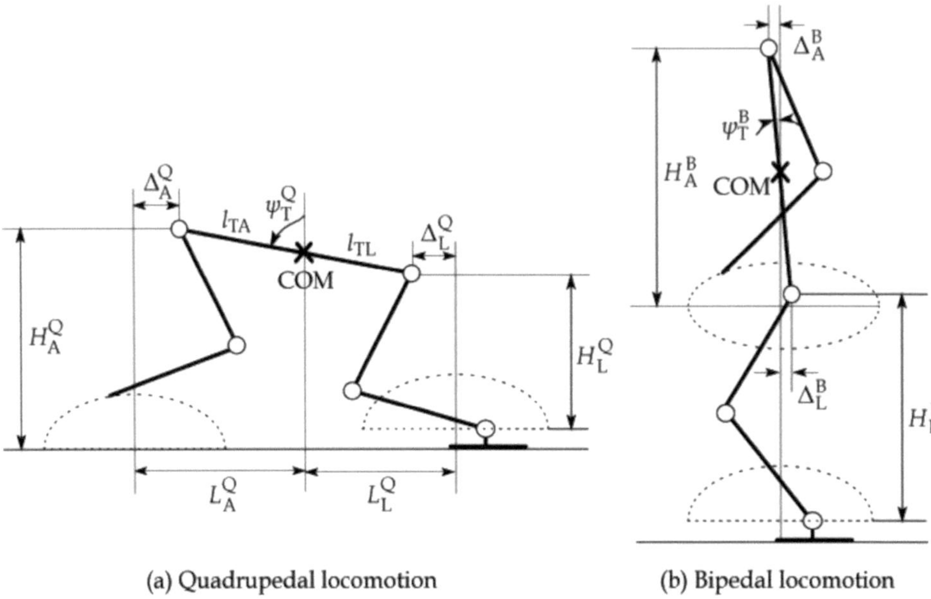

Fig. 4. Schematics and parameters in quadrupedal and bipedal locomotion.

Therefore, in this paper first we introduce a couple of parameters and then attempt to achieve gait transition by cooperatively changing the physical kinematics from quadrupedal to bipedal using the parameters. Specifically, two parameters, ξ_1 and ξ_2, are introduced, and parameters Δ_A, Δ_L, H_A, H_L, and ψ_T are designed as functions of parameters ξ_1 and ξ_2 by

$$\begin{aligned}
\Delta_A(\xi_1,\xi_2) &= \Delta_A^Q - \{l_{TA}\sin\psi_T(\xi_1,\xi_2) + \Delta_A^Q\}\xi_1 \\
\Delta_L(\xi_1,\xi_2) &= \Delta_L^Q - \{l_{TL}\sin\psi_T(\xi_1,\xi_2) + \Delta_L^Q\}\xi_1 \\
H_A(\xi_1,\xi_2) &= H_A^Q + (H_A^B - H_A^Q)\xi_2 \\
H_L(\xi_1,\xi_2) &= H_L^Q + (H_L^B - H_L^Q)\xi_2 \\
\psi_T(\xi_1,\xi_2) &= \psi_T^Q + (\psi_T^B - \psi_T^Q)\xi_2
\end{aligned} \tag{6}$$

This aims to use parameters ξ_1 and ξ_2 to change the distance between the foot and hand trajectories and the posture of the trunk, respectively. Using this controller, gait transition is achieved by simply changing introduced parameters (ξ_1, ξ_2) from (0, 0) to (1, 1), as shown in Fig. 5.

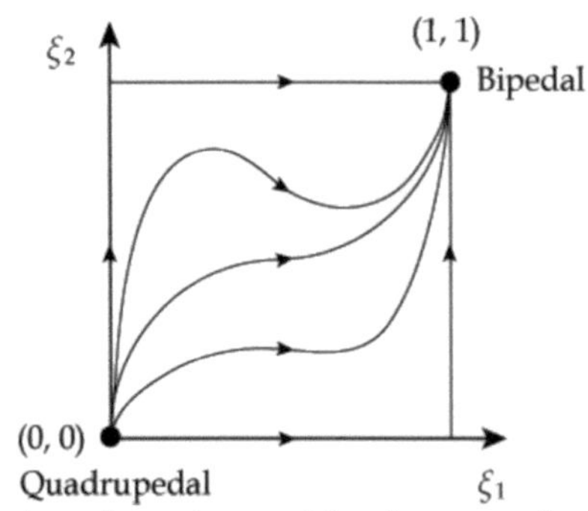

Fig. 5. Trajectories in $\xi_1 - \xi_2$ plane for gait transition from quadrupedal to bipedal locomotion.

4.2 Numerical results

In this section, we investigate whether the proposed control system establishes gait transition from quadrupedal to bipedal locomotion of the robot by numerical simulations. The following locomotion parameters were used: $\hat{S}$=5 cm, $\hat{\beta}$=0.5, $\hat{T}_{\text{sw}}$=0.3 s, K_{T}=10.0, K_{A}=2.0, K_{L}=2.0, l_{TA}=6.9 cm, and l_{TL}=7.6 cm; the remaining parameters for quadrupedal and bipedal locomotion are shown in Table 2. These parameters were decided so that the robot achieves steady quadrupedal and bipedal locomotion. That is, each locomotion has a stable limit cycle in the state space of the robot. Figures 6(a) and (b) show the roll motions of the robot relative to the ground during quadrupedal and bipedal locomotion, respectively, illustrating the limit cycles in these gait patterns. Note that due to setting $\hat{\beta}$=0.5 and the nominal phase differences of the oscillators as described in Sec. 3.2.2, a trot appears in the quadrupedal locomotion in which the robot is usually supported by one arm and the contralateral leg.

Parameter	Quadrupedal ($*^{\text{Q}}$)	Bipedal ($*^{\text{B}}$)
Δ_{A} [cm]	-3.0	1.4
Δ_{L} [cm]	4.0	-1.6
H_{A} [cm]	22.2	22.2
H_{L} [cm]	14.0	16.5
ψ_{T} [deg]	72	12

Table 2. Parameters for quadrupedal and bipedal locomotion.

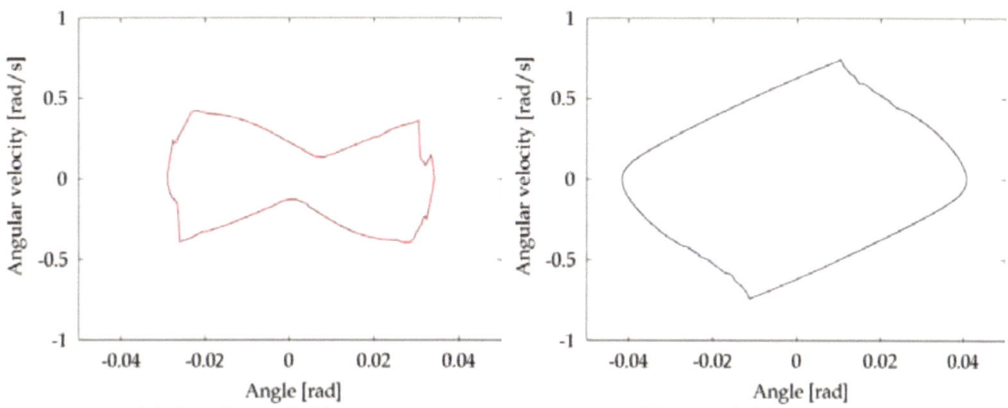

(a) Quadrupedal locomotion (b) Bipedal locomotion

Fig. 6. Roll motion relative to the ground.

To accomplish gait transition, parameters ξ_1 and ξ_2 are changed to reflect outer commands. Specifically, a trajectory in $\xi_1 - \xi_2$ plane is designed as the following two successive steps (see Fig. 7):

Step 1: while the robot walks quadrupedally, parameter ξ_1 increases from 0 to $\bar{\xi}_1$ during time interval T_1 s, where $0 \le \bar{\xi}_1 \le 1$.

Step 2: at the beginning of the swing phase of Arm 1, ξ_1 and ξ_2 increase from $\bar{\xi}_1$ to 1 and from 0 to 1, respectively, during time interval T_2 s.

Note that parameter $\xi_2 > 0$ geometrically indicates that the robot becomes supported only by its legs: that is, the appearance of bipedal locomotion.

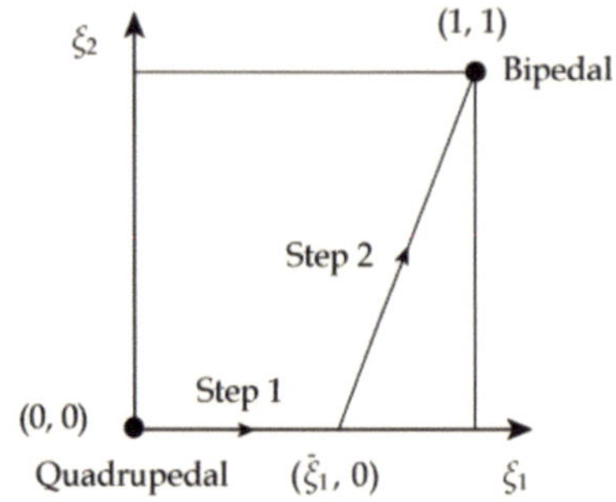

Fig. 7. Designed trajectory in $\xi_1 - \xi_2$ plane to change gait pattern from quadrupedal to bipedal locomotion.

Figures 8(a) and (b) show the roll motion of the robot relative to the ground during gait transition, by parameter $\bar{\xi}_1$=0.7. Specifically, in Fig. 8(a), time intervals T_1 and T_2 are both set at 20 s. Since the nominal step cycle is set at 0.6 s, the nominal kinematical trajectories of quadrupedal locomotion change slowly and gradually into bipedal locomotion, and it takes many steps to complete the change of gait patterns. Step 1 is from 10 to 30 s, and Step 2 is from 30.2 to 50.2 s. During Step 1, the foot and the hand positions come closer together, and the roll motion of the robot decreases. At the beginning of Step 2, since the robot becomes supported only by its legs, roll motion suddenly increases. However, during Step 2 roll motion gradually approaches a limit cycle of bipedal locomotion, and after Step 2 the motion converges to it. That is, the robot accomplishes gait transition from quadrupedal to bipedal locomotion. On the other hand, in Fig. 8(b), time intervals T_1 and T_2 are both set at 5 s. Step 1 is from 10 to 15 s, and Step 2 is from 15.1 to 20.1 s. In this case, although the nominal trajectories change relatively quickly, roll motion converges to a limit cycle of bipedal locomotion, and the robot achieves gait transition. Figure 9 displays the trajectory of the center of mass projected on the ground and the contact positions of the hands and the center of the feet on the ground during this gait transition. Figure 10 shows snapshots of this gait transition.

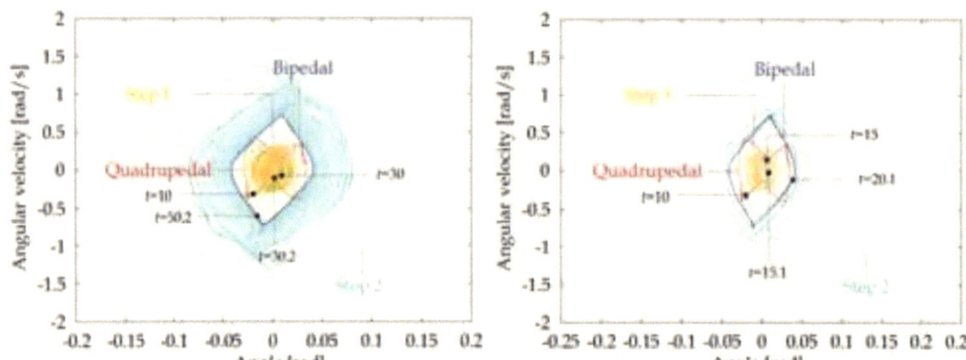

(a) Time intervals T_1, T_2=20 s (b) Time intervals T_1, T_2=5 s

Fig. 8. Roll motion relative to the ground during gait transition from quadrupedal to bipedal locomotion.

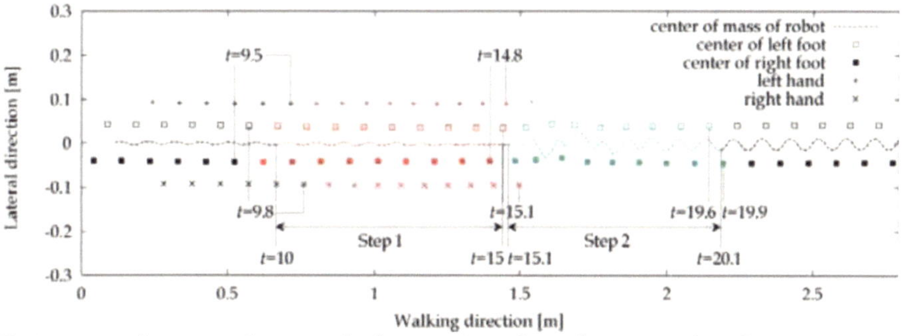

Fig. 9. Trajectory of center of mass of robot projected on the ground and contact positions of the hands and the center of feet on the ground. Time of feet and hands indicate when they land on the ground.

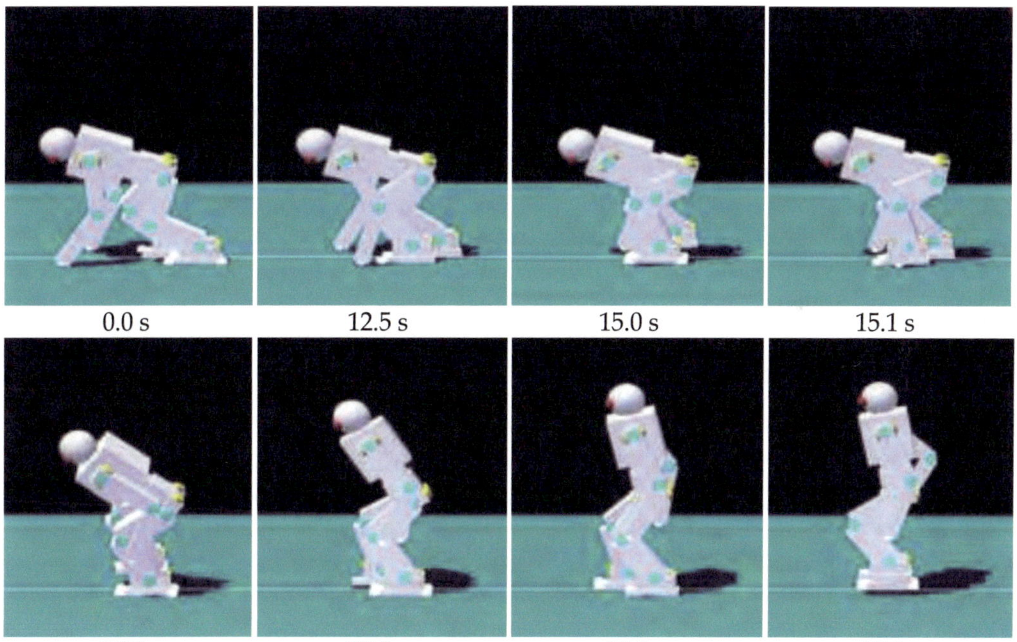

Fig. 10. Snapshots of gait transition from quadrupedal to bipedal locomotion.

5. Discussion

Kinematical and dynamical studies on biped robots are important for robot control. As described above, although model-based approaches using inverse kinematics and kinetics have generally been used, the difficulty of establishing adaptability to various environments as well as complicated computations has often been pointed out. In this paper, we employed an internal structure composed of nonlinear oscillators that generated robot kinematics and adequately responded based on environmental situations and achieved dynamically stable quadrupedal and bipedal locomotion and their transition in a biped robot. Specifically, we generated robot kinematical motions using rhythmic signals from internal oscillators. The oscillators appropriately responded to sensory signals from touch sensors and modulated the rhythmic signals and physical kinematics, resulting in dynamical stable walking of the robot. This means that a robot driven by this control system established dynamically stable and adaptive walking through the interaction between the dynamics of the mechanical system, the oscillators, and the environment. Furthermore, this control system needed neither accurate modeling of the robot and the environment nor complicated computations. It just relied on the timing of the touch sensor signals: it is a simple control system.

Since biped robots generate various motions manipulating many degrees of freedom, the key issue in control remains how to design their coordination. In this paper, we expressed two types of gait patterns using a set of several kinematical parameters and introduced two independent parameters that parameterized the kinematical parameters. Based on the introduced parameters, we changed the gait patterns and established gait transition. That is, we did not individually design the kinematical motion of all robot joints, but imposed kinematical restrictions on joint motions and contracted the degrees of freedom to achieve cooperative motions during the transition. Furthermore, we used the same control system between quadrupedal and bipedal locomotion except for the physical kinematics, which facilitated the design of such cooperative motions and established a smooth gait transition.

As mentioned above, the analysis of EMG patterns in animal motions clarified that common EMG patterns are embedded in the EMG patterns of different motions, despite generating such motions using complicated and redundant musculoskeletal systems (d'Avella & Bizzi, 2005; Patla et al., 1985; Ivanenko et al., 2004, 2006), suggesting an important coordination mechanism. In addition, kinematical studies revealed that covariation of the elevation angles of thigh, shank, and foot during walking displayed in three-dimensional space is approximately expressed on a plane (Lacquaniti et al., 1999), suggesting an important kinematical restriction for establishing cooperative motions. In designing a control system, adequate restrictions must be designed to achieve cooperative motions.

Physiological studies have investigated gait transition from quadrupedal to bipedal locomotion to elucidate the origin of bipedal locomotion. Mori et al. (1996), Mori (2003), and Nakajima et al. (2004) experimented on gait transition using monkeys and investigated the physiological differences in the control system. Animals generate highly coordinated and skillful motions as a result of the integration of nervous, sensory, and musculoskeletal systems. Such motions of animals and robots are both governed by dynamics. Studies on robotics are expected to contribute the elucidation of the mechanisms of animals and their motion generation and control from a dynamical viewpoint.

6. Acknowledgment

This paper is supported in part by Center of Excellence for Research and Education on Complex Functional Mechanical Systems (COE program of the Ministry of Education, Culture, Sports, Science and Technology, Japan) and by a Grant-in-Aid for Scientific Research on Priority Areas "Emergence of Adaptive Motor Function through Interaction between Body, Brain and Environment" from the Japanese Ministry of Education, Culture, Sports, Science and Technology.

7. References

Akimoto, K.; Watanabe, S. & Yano, M. (1999). An insect robot controlled by emergence of gait patterns. *Artificial Life and Robotics*, Vol. 3, 102–105.

Altendorfer, R.; Moore, N.; Komsuoglu, H.; Buehler, M.; Brown Jr., H.; McMordie, D.; Saranli, U.; Full, R. & Koditschek, D. (2001). RHex: A biologically inspired hexapod runner. *Autonomous Robots*, Vol. 11, No. 3, 207–213.

Aoi, S. & Tsuchiya, K. (2005). Locomotion control of a biped robot using nonlinear oscillators. *Autonomous Robots*, Vol. 19, No. 3, 219–232.

Aoi, S.; Tsuchiya, K. & Tsujita, K. (2004). Turning control of a biped locomotion robot using nonlinear oscillators. *Proc. IEEE Int. Conf. on Robotics and Automation*, pp. 3043-3048.

Cham, J.; Karpick, J. & Cutkosky, M. (2004). Stride period adaptation of a biomimetic running hexapod. *Int. J. Robotics Res.*, Vol. 23, No. 2, 141–153.

d'Avella, A. & Bizzi, E. (2005). Shared and specific muscle synergies in natural motor behaviors. *PNAS*, Vol. 102, No. 8, 3076–3081.

Fukuoka, Y.; Kimura, H. & Cohen, A. (2003). Adaptive dynamic walking of a quadruped robot on irregular terrain based on biological concepts. *Int. J. Robotics Res.*, Vol. 22, No. 3-4, 187–202.

Grillner, S. (1981). *Control of locomotion in bipeds, tetrapods and fish*. Handbook of Physiology, American Physiological Society, Bethesda, MD, pp. 1179–1236.

Grillner, S. (1985). Neurobiological bases of rhythmic motor acts in vertebrates. *Science*, Vol. 228, 143–149.

Hirai, K.; Hirose, M.; Haikawa, Y. & Takenaka, T. (1998). The development of the Honda humanoid robot. *Proc. IEEE Int. Conf. on Robotics and Automation*, pp. 1321-1326.

Ijspeert, A.; Crespi, A. & Cabelguen, J. (2005). Simulation and robotics studies of salamander locomotion. Applying neurobiological principles to the control of locomotion in robots. *Neuroinformatics*, Vol. 3, No. 3, 171–196.

Inagaki, S.; Yuasa, H. & Arai, T. (2003). CPG model for autonomous decentralized multi-legged robot system–generation and transition of oscillation patterns and dynamics of oscillators. *Robotics and Autonomous Systems*, Vol. 44, No. 3-4, 171–179.

Inoue, K.; Ma, S. & Jin, C. (2004). Neural oscillator network-based controller for meandering locomotion of snake-like robots. *Proc. IEEE Int. Conf. on Robotics and Automation*, pp. 5064–5069.

Ivanenko, Y.; Poppele, R. & Lacquaniti, F. (2004). Five basic muscle activation patterns account for muscle activity during human locomotion. *J. Physiol.*, Vol. 556, 267–282.

Ivanenko, Y.; Poppele, R. & Lacquaniti, F. (2006). Motor control programs and walking. *Neuroscientist*, Vol. 12, No. 4, 339–348.

Kuniyoshi, Y.; Ohmura, Y.; Terada, K.; Nagakubo, A.; Eitokua, S. & Yamamoto, T. (2004). Embodied basis of invariant features in execution and perception of whole-body dynamic actions—knacks and focuses of Roll-and-Rise motion. *Robotics and Autonomous Systems*, Vol. 48, No. 4, 189-201.

Kuroki, Y.; Fujita, M.; Ishida, T.; Nagasaka, K. & Yamaguchi, J. (2003). A small biped entertainment robot exploring attractive applications. *Proc. IEEE Int. Conf. on Robotics and Automation*, pp. 471-476.

Lacquaniti, F.; Grasso, R. & Zago, M. (1999). Motor patterns in walking. *News Physiol. Sci.*, Vol. 14, 168–174.

Lewis, M. & Bekey, G. (2002). Gait adaptation in a quadruped robot. *Autonomous Robots*, Vol. 12, No. 3, 301–312.

Lewis, M.; Etienne-Cummings, R.; Hartmann, M.; Xu, Z. & Cohen, A. (2003). An in silico central pattern generator: silicon oscillator, coupling, entrainment, and physical computation. *Biol. Cybern.*, Vol. 88, 137–151.

Löffler, K.; Gienger, M. & Pfeiffer, F. (2003). Sensors and control concept of walking ``Johnnie''. *Int. J. Robotics Res.*, Vol. 22, No. 3-4, 229–239.

Mori, S. (1987). Integration of posture and locomotion in acute decerebrate cats and in awake, free moving cats. *Prog. Neurobiol.*, Vol. 28, 161–196.

Mori, S. (2003). Higher nervous control of quadrupedal vs bipedal locomotion in non-human primates; Common and specific properties. *Proc. 2nd Int. Symp. on Adaptive Motion of Animals and Machines.*

Mori, S.; Miyashita, E.; Nakajima, K. & Asanome, M. (1996). Quadrupedal locomotor movements in monkeys (M. fuscata) on a treadmill: Kinematic analyses. *NeuroReport*, Vol. 7, 2277–2285.

Nagasaki, T.; Kajita, S.; Kaneko, K.; Yokoi, K. & Tanie, K. (2004). A running experiment of humanoid biped. *Proc. IEEE/RSJ Int. Conf. on Intelligent Robots and Systems*, pp. 136-141.

Nakajima, K.; Mori, F.; Takasu, C.; Mori, M.; Matsuyama, K. & Mori, S. (2004). Biomechanical constraints in hindlimb joints during the quadrupedal versus bipedal locomotion of M. fuscata. *Prog. Brain Res.*, Vol. 143, 183–190.

Nakanishi, J.; Morimoto, J.; Endo, G.; Cheng, G.; Schaal, S. & Kawato, M. (2004). Learning from demonstration and adaptation of biped locomotion. *Robotics and Autonomous Systems*, Vol. 47, No. 2-3, 79–91.

Orlovsky, G.; Deliagina, T. & Grillner, S. (1999). *Neuronal control of locomotion: from mollusc to man*. Oxford University Press.

Patla, A.; Calvert, T. & Stein, R. (1985). Model of a pattern generator for locomotion in mammals. *Am. J. Physiol.*, Vol. 248, 484–494.

Poulakakis, I.; Smith, J. & Buehler, M. (2005). Modeling and experiments of untethered quadrupedal running with a bounding gait: The Scout II robot. *Int. J. Robotics Res.*, Vol. 24, No. 4, 239–256.

Quinn, R.; Nelson, G.; Bachmann, R.; Kingsley, D.; Offi, J.; Allen, T. & Ritzmann, R. (2003). Parallel complementary strategies for implementing biological principles into mobile robots. *Int. J. Robotics Res.*, Vol. 22, No. 3, 169–186.

Rossignol, S. (1996). *Neural control of stereotypic limb movements*. Oxford University Press.

Saranli, U.; Buehler, M. & Koditschek, D. (2001). RHex: A simple and highly mobile hexapod robot. *Int. J. Robotics Res.*, Vol. 20, No. 7, 616–631.

Taga, G. (1995a). A model of the neuro-musculo-skeletal system for human locomotion I. Emergence of basic gait. *Biol. Cybern.*, Vol. 73, 97–111.

Taga, G. (1995b). A model of the neuro-musculo-skeletal system for human locomotion II. - Real-time adaptability under various constraints. *Biol. Cybern.*, Vol. 73, 113–121.

Taga, G.; Yamaguchi, Y. & Shimizu, H. (1991). Self-organized control of bipedal locomotion by neural oscillators in unpredictable environment. *Biol. Cybern.*, Vol. 65, 147–159.

Takakusaki, K.; Habaguchi, T.; Ohtinata-Sugimoto, J.; Saitoh, K. & Sakamoto, T. (2003). Basal ganglia efferents to the brainstem centers controlling postural muscle tone and locomotion: A new concept for understanding motor disorders in basal ganglia dysfunction. *Neuroscience*, Vol. 119, 293–308.

Takuma, T. & Hosoda, K. (2006). Controlling the walking period of a pneumatic muscle walker. *Int. J. Robotics Res.*, Vol. 25, No. 9, 861–866.

Tsujita, K.; Tsuchiya, K. & Onat, A. (2001). Adaptive gait pattern control of a quadruped locomotion robot. *Proc. IEEE/RSJ Int. Conf. on Intelligent Robots and Systems*, pp. 2318–2325.

Vukobratović, M.; Borovac, B.; Surla, D. & Stokić, D. (1990). *Biped locomotion—dynamics, stability, control and application*. Springer-Verlag.

Wisse, M.; Schwab, A.; van der Linde, R. & van der Helm, F. (2005). How to keep from falling forward: elementary swing leg action for passive dynamic walkers. *IEEE Trans. Robotics*, Vol. 21, No. 3, 393–401.

3

Estimation of the Absolute Orientation of a Five-link Walking Robot with Passive Feet

Yannick Aoustin, Gaëtan Garcia, Philippe Lemoine
Institut de Recherche en Communications et Cybernétique de Nantes (IRCCyN), École Centrale de Nantes, Université de Nantes, U.M.R. 6597, 1 rue de la Noë, BP 92101, 44321 Nantes Cedex 3, France.
e-mail: surname.name@irccyn.ec-nantes.fr

1. Introduction

One of the main objectives of current research on walking robots is to make their gaits more fluid by introducing imbalance phases. For example, for walking robots without actuated ankles, which are under actuated in single support, dynamically stable walking gaits have been designed with success (Aoustin & Formal'sky 2003; Chevallereau et al. 2003; Zonfrilli et al. 2002; Aoustin et al. 2006; Pratt et al. 2001; Westervelt et al. 2003). Both the design of reference motions and trajectories and the control of the mechanism along these trajectories usually require the knowledge of the whole state of the robot. This state contains internal variables which are easy to measure using encoders for example, and also the absolute orientation of the robot with respect to the horizontal plane. For robots with unilateral constraints with the ground, it may not be possible to derive the latter information from internal measurements, as in the case of the absolute orientation of a biped in single support. Of course, technological solutions exist such as accelerometers, gyrometers, inertial units... But the implementation of these solutions is expensive and difficult.

In order to overcome these difficulties, we propose to use a state observer which, based on the measurements of the joint variables and on a dynamic model of the robot, provides estimates of the absolute orientation of the walking robot during imbalance phases. This strategy was first validated in simulation for a three link biped without feet, using nonlinear high gain observers and a nonlinear observer based on sliding modes with a finite time convergence (Lebastard et al. 2006a) and (Lebastard et al. 2006b), for walking gaits composed of single support phases and impacts. The main drawback with this family of observers is that, when only some of the degrees of freedom are measured, a state coordinates transformation is necessary to design their canonical form (Gauthier & Bornard 1981; Krener & Respondek 1985; Bornard & Hammouri 1991; Plestan & Glumineau 1997).

In this chapter, the observer is an extended Kalman filter and it is applied to *SemiQuad*, a prototype walking robot built at our institute. *SemiQuad* is a five link mechanism with a platform and two double-link legs. It is designed to study quadruped gaits where both front legs (resp. rear legs) have identical movements. Its unique front leg (resp. rear leg) acts as the two front legs (resp. rear legs) of the quadruped, so that *SemiQuad* can be considered as an implementation of a virtual quadruped, hence its

name. The legs have passive (uncontrolled) feet that extend in the frontal plane. Due to this, the robot is stable in the frontal plane. Four motors located in haunches and knees drive the mechanism. The locomotion of the prototype is a planar curvet gait. In double support, our prototype is statically stable and over actuated. In single support, it is unstable and under actuated.
The reference walking gaits have been designed using an intuitive strategy which is such that the absolute orientation is not required. Still, it contains imbalance phases during which the observer can be used, and its results evaluated. The estimation results being correct, such an observer can be used for contexts where the absolute angle is necessary. Furthermore, the idea can be useful for other types of walking robots, such as bipeds with imbalance phases.
The organization of this chapter is the following. Section 2 is devoted to the model of *SemiQuad*. It also contains the data of its physical parameters. The intuitive gaits which were designed for *SemiQuad* are presented in section 3. The statement of the problem of estimation of the absolute orientation of *SemiQuad* is defined in Section 4. Simulation results and experimental results avec presented in section 5. Section 6 presents our conclusions and perspectives.

2. Presentation and dynamic models of *SemiQuad*

2.1 *SemiQuad*

The prototype (see figure 1) is composed of a platform and two identical double-link legs with knees. The legs have passive (uncontrolled) feet that extend in the frontal plane. Thus, the robot can only execute 2D motions in the sagittal plane, as considered here. There are four electrical DC motors with gearbox reducers to actuate the haunch joints between the platform and the thighs and the knee joints. Using a dynamic simulation, we have chosen parameters of the prototype (the sizes, masses, inertia moments of the links) and convenient actuators. The parameters of the four actuators with their gearbox reducers are specified in Table 1. The lengths, masses and inertia moments of each link of *SemiQuad* are specified in Table 2.

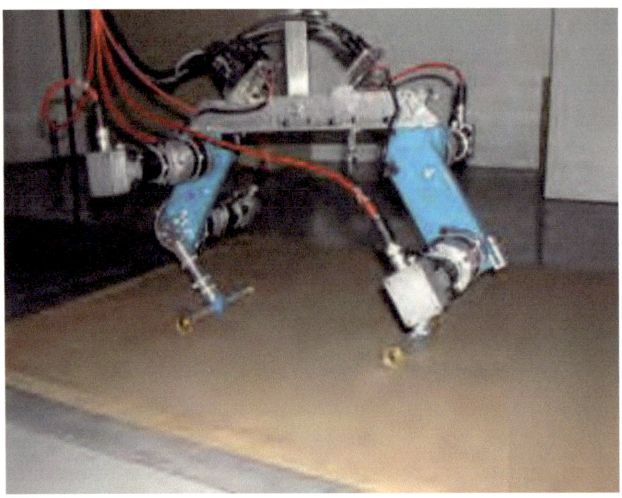

Fig. 1. Photography of *SemiQuad*.

DC motor +gearbox	length (m)	mass (kg)	gearbox ratio	Rotor inertia ($kg.m^2$)	Electromagnetical torque constant (N.m)/A
in haunch	0.23	2.82	50	$3.25\ 10^{-5}$	0.114
In knee	0.23	2.82	50	$2.26\ 10^{-5}$	0.086

Table 1. Parameters of actuators

physical parameters of each link	mass (kg)	length (m)	Center of mass locations (m)	Moment of inertia ($kg.m^2$) around the center of mass C_i (I=1,...,5)
links 1 and 5: shin	$m_1 = m_5 = 0.40$	$l_1 = l_5 = 0.15$	$s_1 = s_5 = 0.083$	$I_1 = I_5 = 0.0034$
link 3: platform +actuators in each haunch	$m_3 = 6.618$	$l_3 = 0.375$	$s_3 = 0.1875$	$I_3 = 0.3157$
links 2 and 4: thigh +actuators in each knee	$m_2 = m_4 = 3.21$	$l_2 = l_4 = 0.15$	$s_2 = s_4 = 0.139$	$I_2 = I_4 = 0.0043$

Table 2. Parameters of *SemiQuad*

The maximum value of the torque in the output shaft of each motor gearbox is 40 N.m . This saturation on the torques is taken into account to design the control law in simulation and in experiments. The total mass of the quadruped is approximately 14 kg. The four actuated joints of the robot are each equipped with one encoder to measure the angular position. The angular velocities are calculated using the angular positions. Currently the absolute platform orientation angle α (see figure 2) is not measured. As explained before, the proposed walking gait does not require this measurement. Each encoder has 2000 pts/rev and is attached directly to the motor shaft. The bandwidth of each joint of the robot is determined by the transfer function of the mechanical power train (motors, gearboxes) and the motor amplifiers that drive each motor. In the case of *SemiQuad,* we have approximately a 16 Hz bandwidth in the mechanical part of the joints and approximately 1.7 kHz for the amplifiers. The control computations are performed with a sample period of 5 ms (200 Hz). The software is developped in C language.

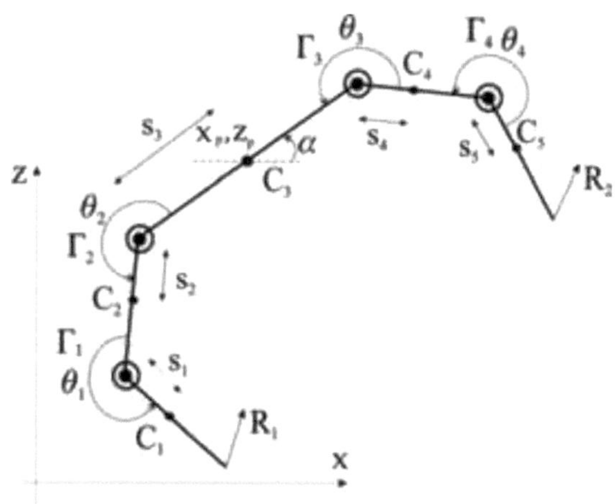

Fig. 2. *SemiQuad*'s diagram: generalized coordinates, torques, forces applied to the leg tips, locations of mass centers.

2.2 Dynamic model of *SemiQuad*

Figure 2 gives the notations of the torques, the ground reactions, the joint variables and the Cartesian position of the platform. Using the equations of Lagrange, the dynamic model of *SemiQuad* can be written:

$$D_e\ddot{q}_e + H_e\dot{q}_e + G_e = B_e\Gamma + D_{R_1}R_1 + D_{R_2}R_2 \tag{1}$$

with, $q_e := [q^T \ x_p \ z_p]^T$. The vector q is composed of the joint actuated variables and the absolute orientation of the platform, $q := [\theta_1 \ \theta_2 \ \theta_3 \ \theta_4 \ \alpha]^T$; (x_p, z_p) are the Cartesian coordinates of the platform center. The joint angle variables are positive for counter clockwise motion. $D_e(q)$ (7×7) is the inertia positive definite matrix. The matrix $H_e(q,\dot{q})$ (7×7) represents the Coriolis and centrifugal forces and $G_e(q)$ (7×1) is the vector of the gravity forces. B_e is a constant matrix composed of ones and zeros. Each matrix $D_{R_j}(q)$ (7×2) ($j=1,2$) is a jacobian matrix which represents the relation between feet Cartesian velocities and angular velocities. It allows to take into account the ground reaction R_j on each foot. If foot j is not in contact with the ground, then $R_j = [0,0]^T$. The term $\Gamma := [\Gamma_1 \ \Gamma_2 \ \Gamma_3 \ \Gamma_4]^T$ is the vector of four actuator torques. Let us assume that during the single support, the stance leg acts as a pivot joint. Our hypothesis is that the contact of the swing leg with the ground results in no rebound and no slipping of the swing leg. Then, in single support, equation (1) can be simplified and rewritten as:

$$D\ddot{q} + H\dot{q} + G = B\Gamma \tag{2}$$

As the kinetic energy $K = \frac{1}{2}\dot{q}^T D\dot{q}$ does not depend on the choice of the absolute frame (see (Spong, M. & Vidyasagar M. 1991)) and furthermore variable α defines the absolute orientation of *SemiQuad*, the inertia matrix $D(5\times5)$ does not depend on α, and then $D = D(\theta_1, \theta_2, \theta_3, \theta_4)$. As for the case of equation (1), the matrix $H(q,\dot{q})$ (5×5) represents the Coriolis and centrifugal forces and $G(q)$ (5×1) is the vector of gravity forces. B is a constant matrix composed of ones and zeros. Equation (2) can be written under the following state form:

$$\dot{x} = \begin{bmatrix} \dot{q} \\ D^{-1}\left(-H\dot{q} - G + B\Gamma\right) \end{bmatrix} = f(x) + g(q_{rel})\Gamma \tag{3}$$

With $x = [q^T \ \dot{q}^T]^T$ and the joint angle vector $q_{rel} = [\theta_1 \ \theta_2 \ \theta_3 \ \theta_4]^T$. The state space is defined as $x \in X \subset R^{10} = \{x = [q^T \ \dot{q}^T]^T \mid |\dot{q}| < \dot{q}_M < \infty; \ q \in]-\pi, \pi]^5\}$. From these definitions, it is clear that all state coordinates are bounded.

2.3 Discrete Model

Our objective is to design an extended Kalman filter with a sampling period Δ to observe the absolute orientation α. Then it is necessary to find a discrete model equivalent to (3). A possible solution is to write $\dot{x} \approx \frac{x_{k+1} - x_k}{\Delta}$ and (3) becomes:

$$x_{k+1} = x_k + \Delta(f(x_k) + g(q_{relk})\Gamma_k) \tag{4}$$

For *SemiQuad,* we have preferred to sample the generalized coordinates and their velocities using approximations to a different order by a Taylor development:

$$x(t+\Delta) = x(t) + \Delta\dot{x}(t) + \Delta^2\frac{\ddot{x}(t)}{2!} + ... + \Delta^n\frac{x^n(t)}{n!} + \varepsilon, \tag{5}$$

From (5), developments to second order and first order have been used for the generalized coordinates and their velocities, respectively. Limiting the order where possible limits the noise, but using second order developments for position variables introduces their second order derivative and allows to make use of the dynamic model.

$$\begin{pmatrix} q_{k+1} \\ \dot{q}_{k+1} \end{pmatrix} \approx \begin{pmatrix} q_k \\ \dot{q}_k \end{pmatrix} + \Delta\begin{pmatrix} \dot{q}_k \\ \ddot{q}_k \end{pmatrix} + \frac{\Delta^2}{2}\begin{pmatrix} \ddot{q}_k \\ 0_{5\times1} \end{pmatrix} \tag{6}$$

With this method, a discrete model of (3) is calculated. If we add the equation corresponding to the measurement of the joint angles, we obtain the following system:

$$\begin{aligned} x_{k+1} &= f_s(x_k, \Gamma_k) \\ y_k &= h(x_k) = (\theta_1 \quad \theta_2 \quad \theta_3 \quad \theta_4)^T = Cx_k \quad \text{with } C = I_{4x4} \end{aligned} \tag{7}$$

2.4 Impulsive impact equation

The impact occurs at the end of the imbalance phase, when the swing leg tip touches the ground, *i.e.* :

$x \in S = \{x \in X \mid q=q_f\}$ where q_f is the final configuration of *SemiQuad*. The instant of impact is denoted by T_i. Let us use the index 2 for the swing leg, and 1 for the stance leg during the imbalance phase in single support. We assume that:

- the impact is passive, absolutely inelastic;
- the swing leg does not slip when it touches the ground;
- the previous stance leg does not take off;
- the configuration of the robot does not change at the instant of impact.

Given these conditions, at the instant of an impact, the contact can be considered as impulsive forces and defined by Dirac delta-functions $R_j = I_{R_j}\Delta(t-T_i)$ (j=1, 2). Here $I_{R_j} = \begin{bmatrix} I_{R_{Nj}} & I_{R_{Tj}} \end{bmatrix}^T$, is the vector of the magnitudes of the impulsive reaction in leg j (Formal'sky 1974). The impact model is calculated by integrating (1) between T_i^- (just before impact) and T_i^+ (just after impact). The torques provided by the actuators at the joints, and Coriolis and gravity forces, have finite values, thus they do not influence the impact. Consequently the impact equation can be written in the following matrix form:

$$D_e(q)(\dot{q}_e^+ - q_e^-) = D_{R_1}(q)I_{R_1} + D_{R_2}(q)I_{R_2} \tag{8}$$

Here, q is the configuration of *SemiQuad* at $t = T_i$, $\dot{q}_e^-$ and $\dot{q}_e^+$ are the angular velocities just before and just after impact, respectively. Furthermore, the velocity of the stance leg tip before impact is null. Then we have:

$$D_{R_1}^T \dot{q}_e^+ = 0_{2\times1} \tag{9}$$

After impact, both legs are stance legs, and the velocity of their tip becomes null:

$$\begin{pmatrix} D_{R_1}^T \\ D_{R_2}^T \end{pmatrix} \dot{q}_e^+ = 0_{4\times1} \tag{10}$$

3. Walking gait strategy for *SemiQuad*

By analogy with animal gaits with no flight phase, *SemiQuad* must jump with front or back leg to realize a walking gait. There is no other way to avoid leg sliding. Then it is necessary to take into account that *SemiQuad* is under actuated in single support and over actuated in double support. Let us briefly present the adopted strategy to realize a walking gait for *SemiQuad*. It was tested first in simulation to study its feasibility and then experimentally (Aoustin et al. 2006). A video can be found on the web page *http://www.irccyn.ec-nantes.fr/irccyn/d/en/equipes/Robotique/Themes/Mobile*. Figure 4 represents a sequence of stick configurations for one step to illustrate the gait. Let us consider half step n as the current half step, which is composed of a double support, a single support on the rear leg and an impact on the ground. During the double support, *SemiQuad* has only three degrees of freedom. Then its movement can be completely prescribed with the four actuators. A reference movement is chosen to transfer the platform centre backwards. This is done by defining a polynomial of third order for both Cartesian coordinates x_p and z_p. The coefficients of these polynomials are calculated from a choice of initial and final positions, of the initial velocity and an intermediate position of the platform centre. The reference trajectories for the four actuated joint variables are calculated with an inverse geometric model. Then, by folding and immediately thrusting the front leg, the single support phase on the rear leg starts. During this imbalance phase, *SemiQuad* has five degrees of freedom and its rotation is free around its stance leg tip in the sagittal plane. Reference trajectories are specified with third order polynomial functions in time for the four actuated inter-link angles. However, the final time T_p of these polynomial functions is chosen smaller than the calculated duration T_{ss} of the single support phase, such that before impact the four desired inter-link angles (θ_{id} $i = 1,\dots,4$) become constant. Using this strategy, we obtain the desired final configuration of our prototype before the impact even if T_{ss} is not equal to the expected value.

$$\begin{aligned} \theta_{id} &= a_0 + a_1 t + a_2 t^2 + a_3 t^3 \quad \text{if} \quad t < T_p \\ \theta_{id} &= \theta_{ie}(T_p) \quad \text{if} \quad T_p \le t \le T_{ss} \end{aligned} \tag{11}$$

The coefficients of these polynomials are calculated from a choice of initial, intermediate and final configurations and of the initial velocity for each joint link. Just after impact, the next half step begins and a similar strategy is applied (figure 4). The tracking of the reference trajectories is achieved using a PD controller. The gains, which were adjusted using pole placement, were tested in simulation and in experiments. Figure 3 shows the evolutions of the absolute orientation variable $\alpha(t)$ and its velocity $\dot{\alpha}(t)$, obtained from the simulation of *SemiQuad* for five steps. These graphs clearly show that the dynamics of the absolute orientation cannot be neglected in the design of a control law based on a state feedback. The durations of the double support phase and the single support phase are 1.5 s and 0.4 s respectively.

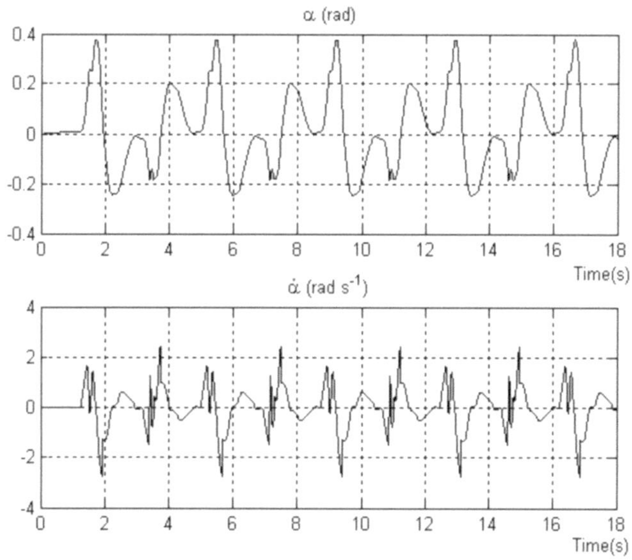

Fig. 3. Evolution of $\alpha(t)$ and $\dot{\alpha}(t)$ in simulation during the walking gait for five steps.

4. Observation of the absolute orientation of *SemiQuad*

System (3) is generically observable if the following matrix O has a full rank (see (Plestan & Glumineau 1997)), which is equal to 10, with ($k_1,...,k_p$) the p observability indexes.

$$O = \left(dh_1 \quad ... \quad dL_f^{k_1-1}h_1 \quad ... \quad dh_p \quad ... \quad dL_f^{k_p-1}h_p\right) \tag{12}$$

where dh is the gradient vector of function h (see system (7)) with respect to the state vector x, and $dL_f h$ is the Lie derivative of h along the vector field f. We have also checked that the equivalent property is satisfied by the discrete model. This means that, at each sampling time t_k, it is possible to find an observability matrix with 10 independent rows or columns.
Having checked system observability, we propose an extended Kalman filter to observe the absolute orientation. The design of this extended Kalman filter is now described.
In practice, due to uncertainty in the determination of parameters and to angular measurement errors, system (3), and of course system (7), are only approximations. A standard solution is to represent modelling and measurement errors as additive noises disturbing the system.
Let us consider the associated ``noisy'' system:

$$\begin{aligned} x_{k+1} &= f_s(x_k, \Gamma_k) + \alpha_k \\ y_k &= C(x_k) = (\theta_1 \quad \theta_2 \quad \theta_3 \quad \theta_4)^T + \beta_k \end{aligned} \tag{13}$$

In the case of a linear system, if α_k and β_k are white Gaussian noises, mutually independent and independent of x, the Kalman filter is an optimal estimator. When the system is not linear, it is possible to use the extended Kalman filter (EKF) by linearizing the evolution equation f_s and the observation equation (which is in our case linear) around the current state estimate. Although convergence and optimality are no longer guaranteed, the interest and the effectiveness of the extended Kalman filter have been proved in many experimental cases. The extended Kalman filter is very often used as a state observer (Bonnifait & Garcia 1998).

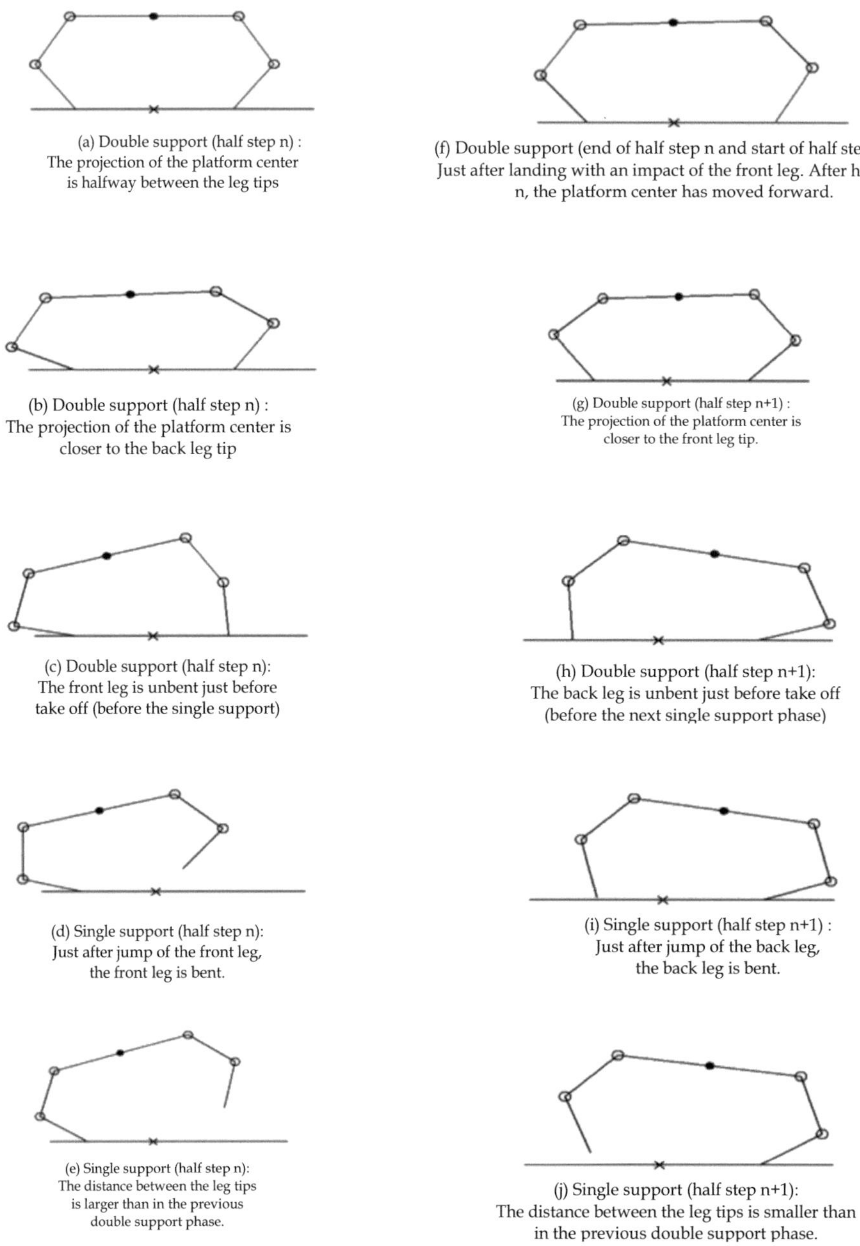

(a) Double support (half step n) : The projection of the platform center is halfway between the leg tips

(f) Double support (end of half step n and start of half step n+1): Just after landing with an impact of the front leg. After half step n, the platform center has moved forward.

(b) Double support (half step n) : The projection of the platform center is closer to the back leg tip

(g) Double support (half step n+1) : The projection of the platform center is closer to the front leg tip.

(c) Double support (half step n): The front leg is unbent just before take off (before the single support)

(h) Double support (half step n+1): The back leg is unbent just before take off (before the next single support phase)

(d) Single support (half step n): Just after jump of the front leg, the front leg is bent.

(i) Single support (half step n+1) : Just after jump of the back leg, the back leg is bent.

(e) Single support (half step n): The distance between the leg tips is larger than in the previous double support phase.

(j) Single support (half step n+1): The distance between the leg tips is smaller than in the previous double support phase.

Fig. 4. Plot of half steps n and n+1 of *SemiQuad*. as a sequence of stick figures.

In the case of our system, the equations of the extended Kalman filter are:

- *a priori* estimation: uses data available before t_{k+1}

$$\begin{aligned} \hat{x}_{k+1}^{-} &= f_s(\hat{x}_k, \Gamma_k) \\ P_{k+1}^{-} &= A_k P_k A_k^T + Q_\alpha \end{aligned} \tag{14}$$

- *a posteriori* estimation: uses data available at t_{k+1}

$$\begin{aligned} \hat{x}_{k+1} &= \hat{x}_{k+1}^{-} + K_{k+1}\left(y_{k+1} - C x_{k+1}^{-}\right) \\ P_{k+1} &= \left(I_{10\times10} - K_{k+1} C\right) P_{k+1}^{-} \end{aligned} \tag{15}$$

with:

$$A_k = \left(\frac{\partial f_s}{\partial x}\right)_{x=\hat{x}_k} \quad \text{and} \quad K_{k+1} = P_{k+1}^{-} C^T \left(C P_{k+1}^{-} C^T + Q_\beta\right)^{-1}$$

Here y_{k+1} are the measured joint variables, which are the first four components of vector x_k at time t_k, and $C\hat{x}_{k+1}^{-}$ is the *a priori* estimation of these joint variables. Q_α and Q_β are the covariance matrices of α_k and β_k, K is the Kalman gain and P the estimated covariance matrix of prediction (P^- at t_k) and estimation (P at t_k) errors.

5. Simulation and experimental results.

For the simulation and experimental tests, the physical parameters defined in section 2 are used. The sampling period Δ is equal to 5 ms. The incremental encoders having N=2000 points per revolution, the variance of measurement is taken equal to $\pi^2/\left(3\Delta N^2\right)$ =8.225 10^{-4}. The errors of incremental encoders being independent, we have chosen Q_β =8.225 10^{-4} $I_{4\times4}$. The components of Q_α for the configuration variables are determined by trial and error from simulation and experimental results. The components of Q_α for velocity variables are derived from the values for position variables, taking into account the sampling period, and are larger than those corresponding to position variables.

$$Q_\alpha = \begin{pmatrix} 3.0\ 10^{-10} I_{5x5} & 0_{5x5} \\ 0_{5x5} & 3.0\ 10^{-5} I_{5x5} \end{pmatrix}$$

The initialization of the covariance matrix P takes into account a determination of α less precise and fixes the variances on the velocities, as for Q_α, taking into account of the sampling period.

$$P_0 = \begin{pmatrix} 8\ 10^{-10} I_{4x4} & 0_{4x1} & 0_{4x5} \\ 0_{1x4} & 1.7\ 10^{-5} & 0_{1x5} \\ 0_{5x4} & 0_{5x1} & 5\ 10^{-2} I_{5x5} \end{pmatrix}$$

All the tests in simulations and in experiments were done following the scheme of figure 5. In simulations, the joint variables θ_i and their velocities $\dot{\theta}_i$ (i=1,2,3,4) obtained by integration of the direct dynamic model of *SemiQuad* and the control torques Γ_i (i=1,2,3,4) are the inputs of the extended Kalman filter. For the experimental tests, the joint variables θ_i (i=1,2,3,4) are measured and differentiated to obtain the velocities $\dot{\theta}_i$. These eight variables, together with the four experimental torques Γ_i (i=1,2,3,4) are the inputs of the extended Kalman filter. In both cases, the state vector $\begin{bmatrix}\theta_1 & \theta_2 & \theta_3 & \theta_4 & \alpha & \dot\theta_1 & \dot\theta_2 & \dot\theta_3 & \dot\theta_4 & \dot\alpha\end{bmatrix}^T$ is estimated.

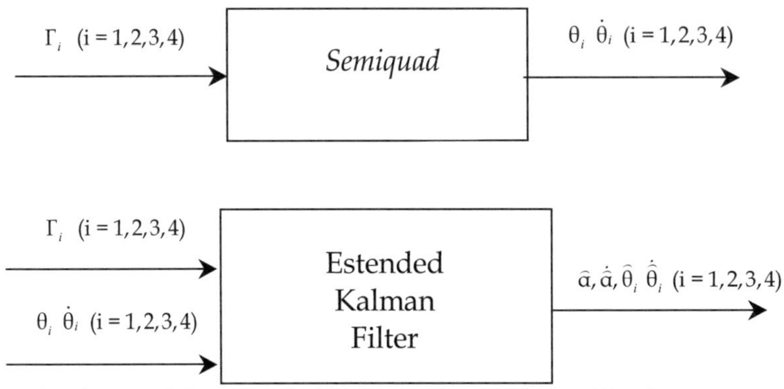

Fig. 5. Principle of tests of the extended Kalman filter with *SemiQuad*.

5.1 Simulation results

Figure 6 shows the evolution of the estimated and real orientations of the platform during a single support phase of the cyclic walking gait of *SemiQuad*. The initial error, which has been set to 0.0262 rad (1.5 degree), is rapidly reduced, andthe estimated orientation converges towards the real orientation. Let us notice that a maximum value of 1.5 degree for the initial error is reasonable because experimentally $\hat\alpha(0)$ is determined by the geometric model at the end of the previous double support.

In the model used by the observer, we do not consider any friction. We have performed robustness tests of our estimator by adding a viscous friction, $\Gamma_i = F_v\theta_i$ (i=1,2,3,4), and a Coulomb friction $\Gamma_i = F_s\dot\theta_i$ (i=1,2,3,4) to the simulation model. Figure 7 shows the estimated and real orientations of the platform of *SemiQuad* in the case when a viscous friction is added. The coefficient F_V equals to 0.1 N.m.s/rad. Similarly, figure 8 shows the estimated and the real orientations of the platform of *SemiQuad* in the case of a Coulomb friction, with a coefficient F_S equal to 0.2 N.m. Last robustness test (figure 9) presents the estimated and real orientations of the platform of *SemiQuad* with an inertia reduced by 5% for the platform in the simulator. In practice, 5% precision on inertia is feasible (see identification results in (Lydoire & Poignet 2003)).

From these robustness tests, we can conclude that we have no finite time convergence. However, the final relative errors of the estimated orientations of the platform of *SemiQuad* are small. Since it will be possible to update the initial condition of the estimator during the next double support phase, with the measurements of the encoders and the geometrical relations, such errors are not a problem.

5.2 Experimental results

At each step, the estimator is initialized with the configurations and the velocities at the end of the previous double support phase. At each sampling time, velocities are obtained by the usual differentiation operation $\dot{\theta}_i = \frac{\theta_i(k\Delta) - \theta_i((k-1)\Delta)}{\Delta}$ (i=1,2,3,4). No filtering is applied to smooth the measured joint variables θ_i, their velocities $\dot{\theta}_i$ and the four torques Γ_i (i=1,2,3,4).

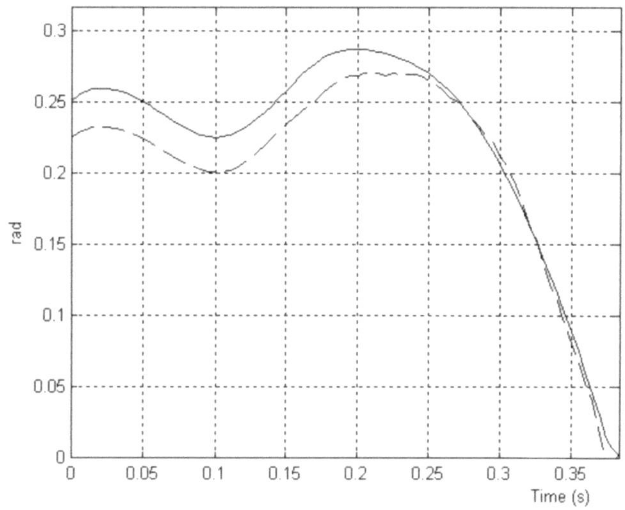

Fig. 6. Absolute orientation $\alpha(t)$ of the platform: real (solid) and estimated (dashed).

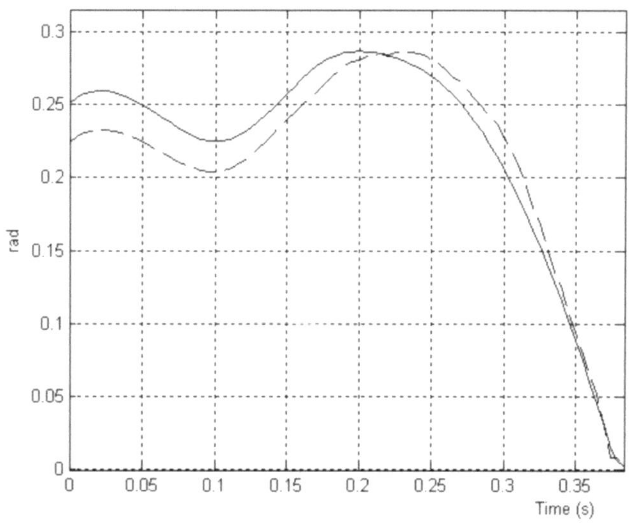

Fig. 7. Absolute orientation $\alpha(t)$ of the platform: real (solid) and estimated (dashed), with a supplement viscous friction.

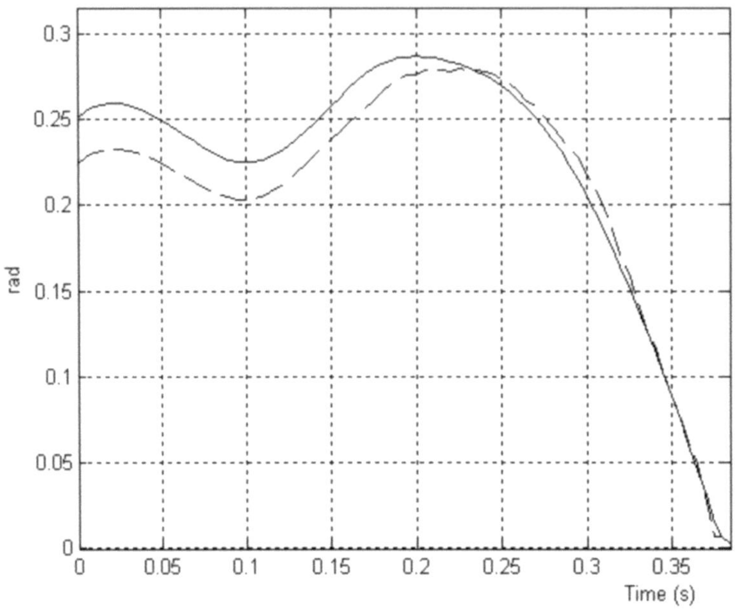

Fig. 8. Absolute orientation $\alpha(t)$ of the platform: real (solid) and estimated (dashed), with a supplement Coulomb friction.

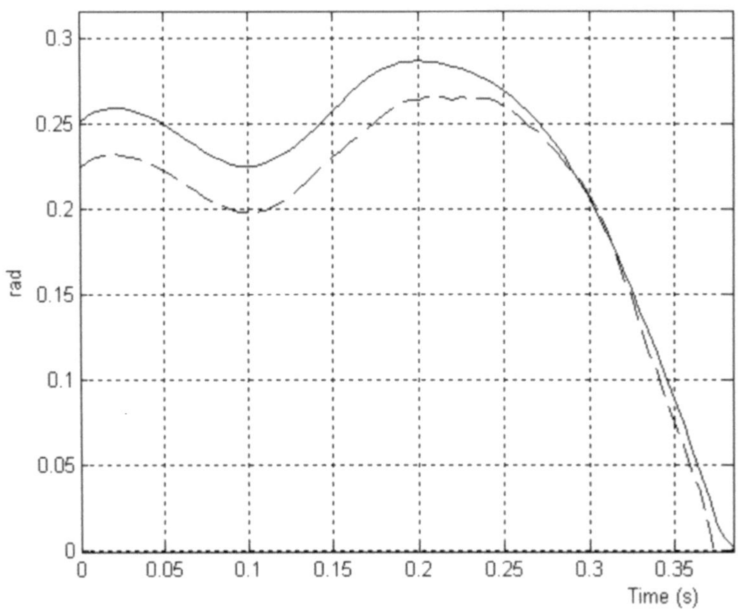

Fig. 9. Absolute orientation $\alpha(t)$ of the platform: real (solid) and estimated (dashed), with an error on the inertia of the platform.

Figure 10 shows the estimation of the absolute orientation of the platform $\alpha(t)$. The duration of the single support phase is 15% smaller than in simulation. It can probably be explained by the fact that our knowledge of the physical parameters of the robot is not perfect, and that we neglected effects such as friction in the joints.

Currently, there is no experimental measured data about the evolution of $\alpha(t)$ in single support, because *SemiQuad* is not equipped with a sensor such as a gyrometer or a gyroscope. However, in double support, using the geometric and kinematic models it is possible to calculate $\alpha(t)$ and $\dot{\alpha}(t)$. In addition to providing initial conditions for the observer, this also allows to calculate the orientation at the end of the single support phase, just after impact. The corresponding value is displayed as a star on the next graph, and is equal to 0.01 rad (0.57 degree). The difference between this value and the estimated value at the same instant is almost 3 degrees.

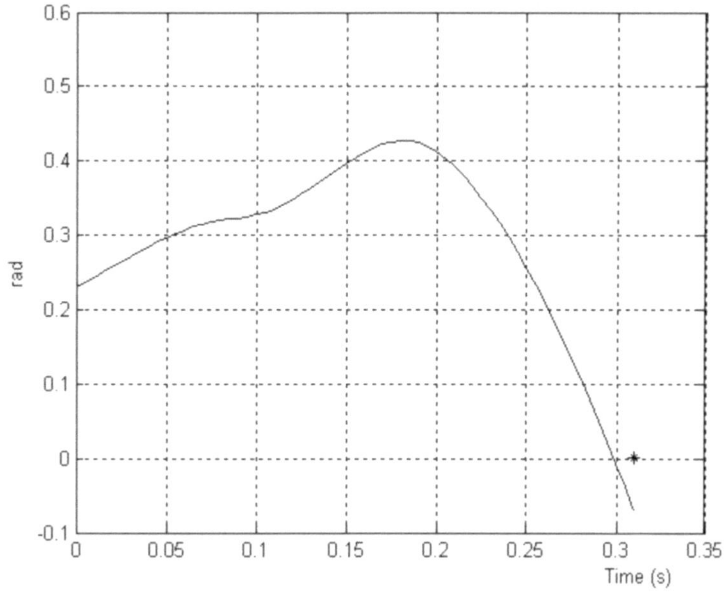

Fig. 10. Estimation of the absolute orientation $\alpha(t)$ of the platform using the experimental data.

6. Conclusion

An application of the digital extended Kalman filter has been presented for the problem of estimating the absolute orientation of *SemiQuad,* a 2D dynamically stable walking robot. There are some differences between simulations and experiments, but the results still prove the ability of our observer to yield a short term estimation of the orientation of the robot. The precision is sufficient for the estimation to be useful for calculating the dynamic model and monitoring the balance of the robot. This task is important for *SemiQuad,* and vital for bipeds, to which the idea is also applicable (see Lebastard, Aoustin, & Plestan 2006 for a different type of observer). Given the possibility to re-initialize the observer at each double support phase, even if very short, as it can be for bipeds, the observer can provide the absolute orientation over time without the use of any additional sensor, which was the goal of our work.

In a near future, we plan to equip *SemiQuad* with a gyrometer to fully evaluate the performance of our estimator over time. Our perspective is to use the estimated orientation in advanced feedback controllers.

7. References

Aoustin, Y. & Formal'sky, A. (2003). Control design for a biped: Reference trajectory based on driven angles as function of the undriven angle, *Journal of Computer and System Sciences International*, Vol. 42, No. 4, (July-August 2003) 159-176.

Chevallereau, C.; Abba, G.; Aoustin, Y.; Plestan, F. ; Westervelt, E. R.; Canudas de Wit, C. & Grizzle, J. W. (2003). Rabbit: A testbed for advanced control theory, *IEEE Control Systems Magazine*, Vol. 23, (October 2003) No. 5, 57-78.

Zonfrilli, F.; Oriolo, M. & Nardi, T. (2002). A biped locomotion strategy for the quadruped robot Sony ers-210, *Proceedings of IEEE Conference on Robotics and Automation*, pp. 2768-2774.

Aoustin, Y.; Chevallereau, C. & Formal'sky, A. (2006). Numerical and experimental study of the virtual quadruped-SemiQuad, *Multibody System Dynamics*, Vol. 16, 1-20.

Pratt, J.; Chew, C. M.; Torres, A.; Dilworth, P. & Pratt, G. (2001). Virtual model control: an intuitive approach for bipedal locomotion, *International Journal of Robotics Research,* Vol. 20, No. 2, 129-143.

Westervelt, E. R.; Grizzle, J. W. & Koditschek, D. E. (2003). Hybrid zero dynamics of planar biped walkers, *IEEE Transactions on Automatic Control*, Vol. 48, No. 1, (February 2003) 42-56.

Lebastard, V.; Aoustin, Y. & Plestan, F. (2006). Observer-based control of a walking biped robot without orientation measurement, *Robotica*, Vol. 24, 385-400.

Lebastard, V.; Aoustin, Y. & Plestan F. (2006) . Absolute orientation estimation for observer-based control of a five-link biped robot, *Springer Verlag Series Notes on Control and Information Sciences*, Editor: K. Kozlowski, Springer, vol. 335, Lecture Notes in Control and Information Sciences, 2006.

Gauthier, J.P. & Bornard, G. (1981). Observability for any u(t) of a class of nonlinear system. *IEEE Transactions on Automatic Control,* 26(4): 922-926.

Krener, A. & Respondek, W. (1985). Nonlinear observers with linearization error dynamics, *SIAM Journal Control Optimization*, Vol. 2, 197-216.

Bornard, G. & Hammouri, H. (1991). A high gain observer for a class of uniformly observable systems, *Proceedings of the 30th IEEE Conference on Decision and Control*, pp. 1494-1496.

Plestan, F. & Glumineau, A. (1997). Linearisation by generalized input-ouput injection, *Systems and Control Letters,* Vol. 31, 115-128.

Spong, M. & Vidyasagar M. (1991). Robot dynamics and control, *John Wiley*, New-York.

Formal'sky, A. M. (1982). Locomotion of Anthropomorphic Mechanisms, *Nauka, Moscow* [In Russian], 368 pages.

Bonnifait, P. & Garcia, G. (1998). Design and experimental validation of an odometric and goniometric localization system for outdoor robot vehicles, *IEEE Transactions on Robotics and Automation,* Vol. 14, No. 541-548.

Lydoire, F. & Poignet, Ph. (2003). Experimental dynamic parameters identification of a seven dof walking robot, *Proceedings of the CLAWAR'03 Conferencel*, pp. 477-484.

4

Teaching a Robotic Child - Machine Learning Strategies for a Humanoid Robot from Social Interactions

Artur Arsenio
Massachussets Institute of Technology[1]
USA

1. Introduction

Children love toys. Human caregivers often employ learning aids, such as books, educational videos, drawing boards, musical or textured toys, to teach a child. These social interactions provide a rich plethora of information to a child, and hence they should be extrapolated to a humanoid robot as well (Arsenio, 2004d).

Inspired in infant development, we aim at developing a humanoid robot's perceptual system through the use of learning aids, cognitive artifacts, and educational activities, so that a robot learns about the world according to a child's developmental phases (Arsenio, 2004c). Of course, the human caregiver plays a very important role on a robot's learning process (as it is so with children), performing educational and play activities with the robot (such as drawing, painting or playing with a toy train on a railway), facilitating robot's perception and learning. The goal is for the humanoid robot Cog (Figure 1) to see the world through the caregiver's eyes.

Fig. 1. The Humanoid Robot Cog.

This chapter will begin by addressing learning scenarios on section 2, which are employed to enable the robot to acquire input data in real-time to train learning algorithms. Tools developed to acquire such data, such as object / scene /texture segmentation (Arsenio,

[1] Research work was developed while the author was at MIT. The author is currently working at Siemens.

2004e), sound / face segmentation (Arsenio, 2004f), object / face / hand tracking (Arsenio, 2004e;f) or even robot actuation for active perception (Arsenio, 2003; Metta & Fitzpatrick, 2003) are described in the literature.

We do not treat children as machines, i.e., automatons. But this automaton view is still widely employed in industry to build robots. Building robots involves indeed the hardware setup of sensors, actuators, metal parts, cables, processing boards, as well as software development. Such engineering might be viewed as the robot genotype. But equally important in a child is the developmental acquisition of information in a social and cultural context (Vigotsky, 1962).

Therefore, for a humanoid robot to interact effectively in its surrounding world, it must be able to learn. Section 3 presents learning strategies applied to a diverse set of problems, so that the robot learns information about objects, scenes, people and actions. Training data for the algorithms is generated on-line, in real-time, while the robot is in operation. We will describe the development of these learning mechanisms, presenting statistics from a large plethora of experimental results.

We aim at introducing robots into our society and treating them as us, using child development as a metaphor for developmental learning of a humanoid robot

2. Children-like Social Interactions as Learning Scenarios for a Humanoid

An autonomous robot needs to be able to acquire and incrementally assimilate new information, to be capable of developing and adapting to its environment. The field of machine learning offers many powerful algorithms, but these often require off-line, manually inserted training data to operate. Infant development research suggests ways to acquire such training data from simple contexts, and use these experiences to bootstrap to more complex contexts. We need to identify situations that enable the robot to temporarily reach beyond its current perceptual abilities (Figure 2 shows a set of such situations), giving the opportunity for development to occur (Arsenio, 2004c;d; Metta & Fitzpatrick, 2003).

This led us to create children-like learning scenarios for teaching a humanoid robot. These learning experiments are used for transmitting information to the humanoid robot Cog to learn about objects' multiple visual and auditory representations from books, other learning aids, musical instruments and educational activities such as drawing and painting.

Our strategy relies heavily in human-robot interactions. For instance, it is essential to have a human in the loop to introduce objects from a book to the robot (as a human caregiver does to a child). A more rich, complete human-robot communication interface results from adding other aiding tools to the robot's portfolio (which facilitate as well the children' learning process).

This is achieved by selectively attending to the human actuator (hand or finger). Indeed, primates have specific brain areas to process the hand visual appearance (Perrett et al., 1990). Inspired by human development studies, emphasis will be placed on facilitating perception through the action of a human instructor, and on developing machine-learning strategies that receive input from these human-robot interactions (Arsenio, 2004a,d).

Multi-modal object properties are learned using these children educational tools, and inserted into several recognition schemes, which are then applied to developmentally acquire new object representations (Arsenio, 2004c).

Fig. 2. Cog learning from several social interactions.

2.1 Robot Skill Augmentation through Cognitive Artifacts

A human caregiver can introduce a robot to a rich world of visual information concerning objects' visual appearance and shape. But cognitive artifacts, which enhance perception, can also be applied to improve perception over other perceptual modalities, such as auditory processing.

We exploit repetition (rhythmic motion, repeated sounds) to achieve segmentation and recognition across multiple senses (Arsenio, 2004d;f). We are interested in detecting conditions that repeat with some roughly constant rate, where that rate is consistent with what a human can easily produce and perceive. This is not a very well defined range, but we will consider anything above 10Hz to be too fast, and anything below 0.1Hz to be too slow. Repetitive signals in this range are considered to be events in our system: waving a flag is an event, but the vibration of a violin string is not an event (too fast), and neither is the daily rise and fall of the sun (too slow). Such a restriction is related to the idea of natural kinds, where perception is based on the physical dimensions and practical interests of the observer.

Abrupt motions, such as a poking movement, which involve large variations of movement, are also used to extract percepts (Arsenio, 2003; Arsenio 2004c;d).

Teaching from Books

Learning aids are often used by human caregivers to introduce the child to a diverse set of (in)animate objects, exposing the latter to an outside world of colors, forms, shapes and contrasts, that otherwise might not be available to a child (such as images of whales and cows). Since these learning aids help to expand the child's knowledge of the world, they are a potentially useful tool for introducing new informative percepts to a robot.

Fig. 3. Object templates extracted from books.

Children's learning is hence aided by the use of audiovisuals, and especially books, during social interactions with their mother or caregiver. Indeed, humans often paint, draw or just read books to children during their childhood. Books are also a useful tool to teach robots different object representations and to communicate properties of unknown objects to them.

Figure 3 shows images of object templates extracted from books using an active object segmentation algorithm - active in the sense that a human points to the object in a book with his/her finger (Arsenio, 2004e). This human aided perceptual grouping algorithm extracts informative percepts from picture books (fabric, foam or cardboard books), by correlating such information with a periodically moving human actuator (finger), resulting on signal samples for objects' image templates. Whenever the interacting human makes repetitive sounds simultaneously, object sound signatures, as well as cross-modal signatures, are segmented as well (Fitzpatrick & Arsenio, 2004). This data is employed afterwards as inputs for learning (section 3).

Matching Geometric Patterns: Drawings, Paintings, Pictures ...

Object descriptions may come in different formats - drawings, paintings, photos, *et cetera*. Hence, the link between an object representation in a book and real objects recognized from the surrounding world can be established through object recognition. Objects will be recognized using geometric hashing (section 3), a widely used recognition technique. The algorithm operates on three different set of features: chrominance and luminance topological regions, and shape (determined by an object's edges), as shown by Figure 4. Except for a description contained in a book, which was previously segmented, the robot had no other knowledge concerning the visual appearance or shape of such object.

Additional possibilities include linking different object descriptions in a book, such as a drawing, as demonstrated by two samples of the experimental results presented in Figure 4. A sketch of an object contains salient features concerning its shape, and therefore there are advantages in learning, and linking, these different representations. This framework is also a useful tool for linking other object descriptions in a book, such as a photo, a painting, or a printing (Arsenio, 2004a;d).

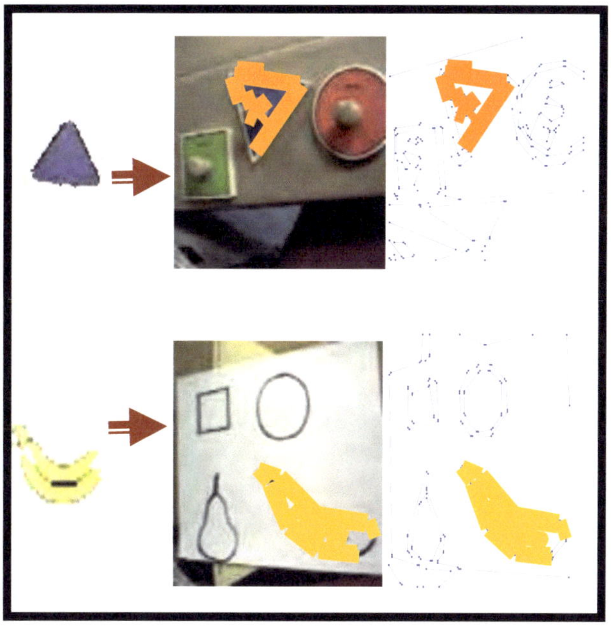

Fig. 4. Matching objects from books to real world objects and drawings.

Explorations into the World of Tools and Toys

A plethora of other educational tools and toys are widely used by educators to teach children, helping them to develop. Examples of such tools are toys (such as drawing boards), educational TV programs or educational videos. The Baby Einstein collection includes videos to introduce infants and toddlers to colors, music, literature and art. Famous painters and their artistic creations are displayed to children on the Baby Van Gogh video, from the mentioned collection. This inspired the design of learning experiments in which Cog is introduced to art using an artificial display (the computer monitor), as shown in Figure 5 (Arsenio, 2004d).

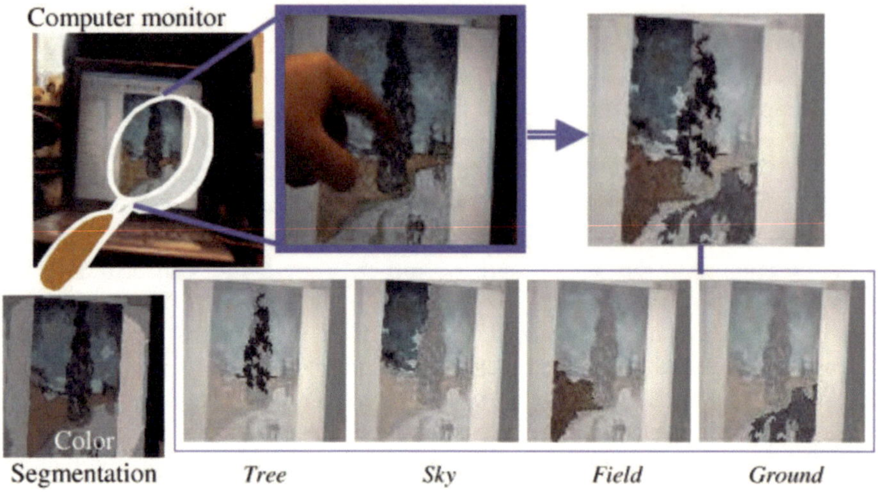

Fig. 5. The image of a painting by Vincent Van Gogh, Road with Cypress and Star, 1890 is displayed on a computer screen. Paintings are contextually different than pictures or photos, since the painter style changes the elements on the figure considerably. Van Gogh, a post-impressionist, painted with an aggressive use of brush strokes. But individual painting elements can still be grouped together by having a human actor tapping on their representation in the computer screen to group them together.

Learning First Words

Auditory processing is also integrated with visual processing to extract the name and properties of objects. However, hand visual trajectory properties and sound properties might be independent - while tapping on books, it is not the interacting human caregiver hand that generates sound, but the caregiver vocal system pronouncing sounds such as the object's name. Therefore, cross-modal events are associated together under a weak requirement: visual segmentations from periodic signals and sound segmentations are bound together if occurring temporally close (Fitzpatrick & Arsenio, 2004). This strategy is also well suited for sound patterns correlated with the hand visual trajectory (such as playing musical tones by shaking a rattle).

2.2 Educational, Learning Activities

A common pattern of early human-child interactive communication is through activities that stimulate the child's brain, such as drawing or painting. Children are able to extract information from such activities while they are being performed on-line. This capability motivated the implementation of three parallel processes which receive input data from three different sources: from an attentional tracker, which tracks the robot's attentional focus, and it is attracted to a new salient stimulus; from a multi-target tracking algorithm implemented to track simultaneously multiple targets; and from an algorithm that selectively attends to the human actuator (Arsenio, 2004d).

Learning Hand Gestures

Standard hand gesture recognition algorithms require an annotated database of hand gestures, built off-line. Common approaches, such as Space-Time Gestures (Darrel & Pentland, 1993), rely on dynamic programming. Others (Cutler & Turk, 1998) developed

systems for children to interact with lifelike characters and play virtual instruments by classifying optical flow measurements. Other classification techniques include state machines, dynamic time warping or Hidden Markov Models (HMMs).
We follow a fundamentally different approach, being periodic hand trajectories mapped into geometric descriptions of objects, to classify simple circular or triangular movements, for instance (Arsenio, 2004a). Figure 6b reports an experiment in which a human draws repetitively a geometric shape on a sheet of paper with a pen. The robot learns what was drawn by matching one period of the hand gesture to the previously learned shape (the hand gesture is recognized as circular in the Figure). Hence, the geometry of periodic hand trajectories is recognized in real-time to the geometry of objects in an object database, instead of being mapped to a database of annotated gestures.

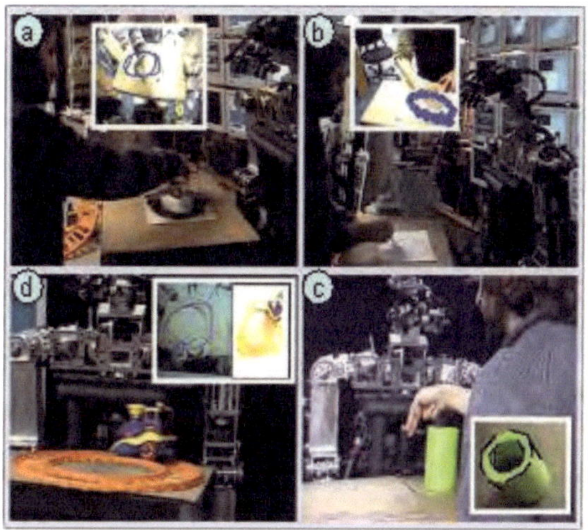

Fig. 6. Sample of experiments for object and shape recognition from hand gestures.

Object Recognition from Hand Gestures.
The problem of recognizing objects in a scene can be framed as the dual version of the hand gestures recognition problem. Instead of using previously learned object geometries to recognize hand gestures, hand gestures' trajectories are now applied to recover the geometric shape (set of lines computed by applying the Hough Transform) and appearance (given by an image template enclosing such lines) of a scene object (as seen by the robot). Visual geometries in a scene (such as circles) are recognized as such from hand gestures having the same geometry (as is the case of circular gestures). Figure 6a shows results for such task on an experiment in which an interacting human paints a circle. The robot learns what was painted (a circle) by matching the hand gesture to the shape defined by the ink on the paper. The algorithm is useful to identify shapes from drawing, painting or other educational activities (Arsenio, 2004d).

Shape from Human Cues
A very similar framework is applied to extract object boundaries from human cues. Indeed, human manipulation provides the robot with extra perceptual information concerning objects, by actively describing (using human arm/ hand/finger trajectories) object contours or the hollow parts of objects, such as a cup (see experiment with green cup in Figure 6c). Tactile perception of objects from the robot grasping activities has been actively pursued –

see for instance (Polana & Nelson, 1994; Rao et al., 1989). Although more precise, these techniques require hybrid position/ force control of the robot's manipulator end-effector so as not to damage or break objects.

Functional Constraints

Not only hand gestures can be used to detect interesting geometric shapes in the world as seen by the robot. For instance, certain toys, such as trains, move periodically on rail tracks, with a functional constraint fixed both in time and space. Therefore, one might obtain information concerning the rail tracks by observing the train's visual trajectory. To accomplish such goal, objects are visually tracked by an attentional tracker which is modulated by an attentional system (Arsenio, 2004d). The algorithm starts by masking the input world image to regions inside the moving object's visual trajectory (or outside but on a boundary neighborhood). Lines modeling the object's trajectory are then mapped into lines fitting the scene edges. The output is the geometry of the stationary object which is imposing the functional constraint on the moving object. Figure 6d shows as well an experiment for the specific case of extracting templates for rail tracks from the train's motion (which is constrained by the railway circular geometry).

2.3 Learning about People

Faces in cluttered scenes are located by a computationally efficient algorithm (Viola & Jones, 2001), which is applied to each video frame (acquired by a foveal camera). If a face is detected, the algorithm estimates a window containing that face, as shown in Figure 7. The novelty here consists on acquiring a large amount of samples of training data in real-time using a multi-object tracking algorithm (Arsenio, 2004d), which allows to group several image templates together - from different views of the same tracked face – into the same group.

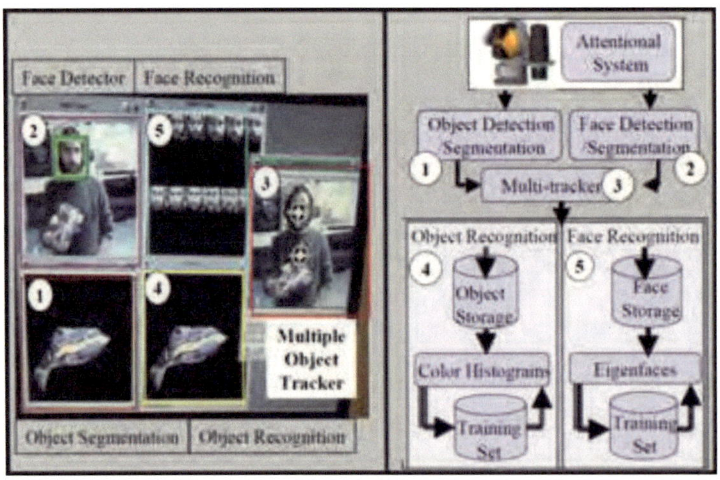

Fig. 7. Approach for segmenting and recognizing faces & objects. Training data for object/face recognition is extracted by keeping objects and others faces in memory for a while, generating a collection of training samples consisting of multiple segmentations of objects and faces. (left) on-line experiment on Cog (right) schematic organization. 1. Object segmentation 2. Face detection and segmentation 3. Multiple object tracking 4. Object Recognition 5. Face Recognition.

3. Machine Learning Algorithms

This section presents a collection of machine learning algorithms and methodologies implemented to emulate different cognitive capabilities on the humanoid robot Cog. This framework is effectively applied to a collection of AI, computer vision, and signal processing problems. It is shown to solve a broad spectrum of machine learning problems along a large categorical scope: actions, objects, scenes and people, using real-time data acquired from human-robot interactions as described by last section's learning scenarios.

Most of the learning algorithms introduced receive as training inputs information produced by other modules, such as object segmentation (Arsenio, 2004e). But all this training data is automatically annotated on-line, in real-time, instead of the standard off-line, manual annotation. This is accomplished by having human caregivers introducing new percepts to the robot (Arsenio, 2004a). Hence, we are motivated by cognitive development of human infants, which is bootstrapped by the helping hand that a human caregiver (and especially the infant's mother) provides to the infant (Arsenio, 2004c).

This chapter does not intend to propose new learning algorithms. Instead, the focus is placed on using existing learning algorithms to solve a wide range of problems.

One essential capability for a humanoid robot to achieve convincing levels of competency is object recognition. Therefore, a learning algorithm operating on object color histograms is first presented. A more robust algorithm employing geometric hashing techniques is also described. These algorithms are not however appropriate to tackle the face recognition problem, which is solved using the eigenfaces method. A similar eigenobjects based algorithm will be applied as well for sound recognition, using eigensounds.

Recognition of scenes is especially important for robot localization. An approach based on a contextual description of the scene envelope is also described. Contextual descriptions of a scene are modeled by a Mixture of Gaussians, being the parameters of such mixture estimated iteratively using the Expectation-Maximization algorithm.

The processing of cross-modal information, among different sensorial capabilities, leads to an innovative cross-modal object recognition scheme using a Dynamic Programming approach. Contextual features provide another source of information very useful to recognize objects - we apply a method similar to a mixture of experts: weighted cluster modeling. Another technique employed is Back-propagation Neural Networks for activities' identification (and for identifying the function of an object within an activity). This learning method is also shown to increase sound recognition rates compared to the eigensounds method. Both qualitative and quantitative experimental results are evaluated and discussed for each algorithm. Section 3 ends by referring briefly other learning strategies employed by the humanoid robot Cog, applied not only for perception but also for robot control by learning the underlying dynamic models.

3.1 Color Histograms

The object recognition algorithm needs to cluster object templates by classes according to their identity. Such task was implemented through color histograms - objects are classified based on the relative distribution of their color pixels. Since object masks are available from segmentation (Arsenio, 2004e), external global features do not affect recognition, and hence color histograms are appropriate. A multi-target tracking algorithm (Arsenio, 2004d) keeps track of object locations as the visual percepts change due to movement of the robot's active head. Ideally, a human actor should expose the robot to several views of the object being

tracked (if the object appearance is view-dependent), in order to link them to the same object. This way, a collection of object views becomes available as input.
Recognition works as follows. Quantization of each of the three color channels originates 8^3 groups G_i of similar colors. The number of image pixels n_{G_i} indexed to a group is stored as a percentage of the total number of pixels. The first 20 color histograms of an object category are saved into memory and updated thereafter. New object templates are classified according to their similarity with other object templates previously recognized for all object categories, by computing:

$$p = \sum_{i=1}^{8^3} \text{minimum}\,(n_{G_i}, n_{G'_i})$$

If $p < th$ (th set to 0,7) for all of the 20 histograms in an object category, then the object does not belong to that category. If this happens for all categories, then it is a new object. If $p \geq th$, then a match occurs, and the object is assigned to the category with maximum p.
Whenever an object is recognized into a given category, the average color histogram which originated a better match is updated. Given an average histogram which is the result of averaging m color histograms, the updating consists of computing the weighted average between this histogram (weight m) and the new color histograms (unit weight). This has the advantage that color histograms evolve as more samples are obtained to represent different views of an object.

Recog. objects	Big sofa	Small sofa	Table	Black chair	Green chair	Door	Blue door	Total
Errors%	4.35	18.2	5.56	0.0	0.0	3.57	14.29	5.5
Trials	23	11	18	18	4	28	7	109

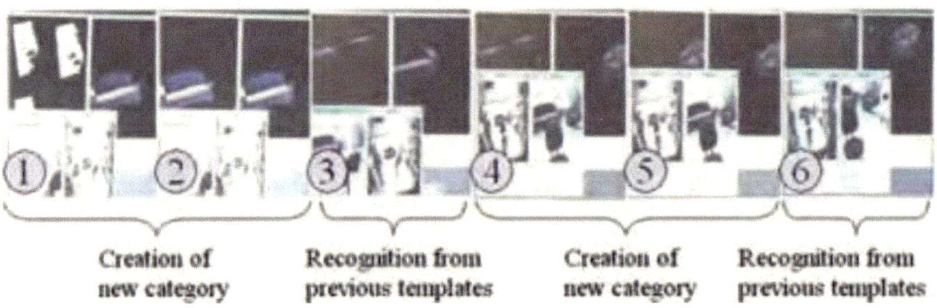

Fig. 8. (top) Recognition errors. Matches evaluated from a total of 11 scenes (objects are segmented and recognized more than once per scene). (bottom) Sequence from an on-line experiment of several minutes on the humanoid robot Cog: (1) The robot detects and segments a new object – a sofa; (2) New object is correctly assigned to a new category; (3) Object, not being tracked, is recognized from previous templates (as shown by the two sofa templates mapped to it); (4-5-6) Same sequence for a different, smaller sofa.

Experimental Results for Template Matching
Figure 8 presents quantitative performance statistics. It shows a sample of the system running on the humanoid robot Cog, while recognizing previously learned objects. Incorrect matches occurred due to color similarity among different objects (such as a big and a small

sofa). Errors arising from labeling an object in the database as a new object are chiefly due to drastic variations in light sources. Qualitative results from an on-line experiment of several minutes for object segmentation, tracking and recognition of new objects on the humanoid robot are also shown.
Out of around 100 samples from on-line experiments, recognition accuracy average was of 95%. Several experiments have shown however the algorithm not capable to differentiate among people's faces, although it differentiated correctly between faces and other objects.

3.2 Geometric Hashing

This object recognition algorithm consists of three independent algorithms. Each recognizer operates along orthogonal directions to the others over the input space (Arsenio, 2004b). This approach offers the possibility of priming specific information such as searching for a specific object feature (color, shape or luminance) independently of the others. The set of input features are:

- Color: groups of connected regions with similar color
- Luminance: groups of connected regions with similar luminance
- Shape. A Hough transform is applied to a contour image (from a Canny edge detector). Line orientation is determined using Sobel masks. Pairs of oriented lines are then used as input features

Geometric hashing (Wolfson & Rigoutsos, 1997) is a rather useful technique for high-speed performance. In this method, quasi-invariants are computed from training data in model images, and then stored in hash tables. Recognition consists of accessing and counting the contents of hash buckets. Recognition of objects has to occur over a variety of scene contexts. An adaptive Hash table (a hash table with variable-size buckets) algorithm was implemented to store affine color, luminance and shape invariants (which are view independent for small perspective deformations). Figure 4 displays two results from applying this algorithm using shape features (Arsenio, 2004d).

3.3 Eigenobjects – Principal Component Analysis

Component Analysis (PCA) is an efficient method to describe a collection of images. The corresponding eigenvectors are denominated eigenobjects.
Let the training set of M images from an object n be $\{\phi_1, \phi_2, \ldots \phi_M\}$ (see Figure 9). The average image of this set is defined by

$\psi = 1/M \sum_{i=1}^{M} \phi_i$. The covariance matrix for the set of training objects is thus given by (1):

$$C\phi_n = \sum_{i=1}^{M} \Gamma_i \Gamma_i^T = AA^T \qquad (1)$$

being $\Gamma_n = \phi_n - \psi$ the difference of each image from the mean, and $A = [\Gamma_1, \Gamma_2, \ldots, \Gamma_M]$.
Cropped faces are first rescaled to 128 × 128 images (size $S = 128^2$). Determining the eigenvectors and eigenvalues of the S^2 size covariance matrix C is untractable. However, C rank does not exceed $M - 1$. For $M < S^2$ there are only $M - 1$ eigenvectors associated to non-zero eigenvalues, rather than S^2. Let v_i be the eigenvectors of the M × M matrix $A^T A$. The eigenfaces μ_i are given by:

$$\mu_i = \sum_{k=1}^{M} v_{ik}\Gamma_k \quad i = 1,\dots M \tag{2}$$

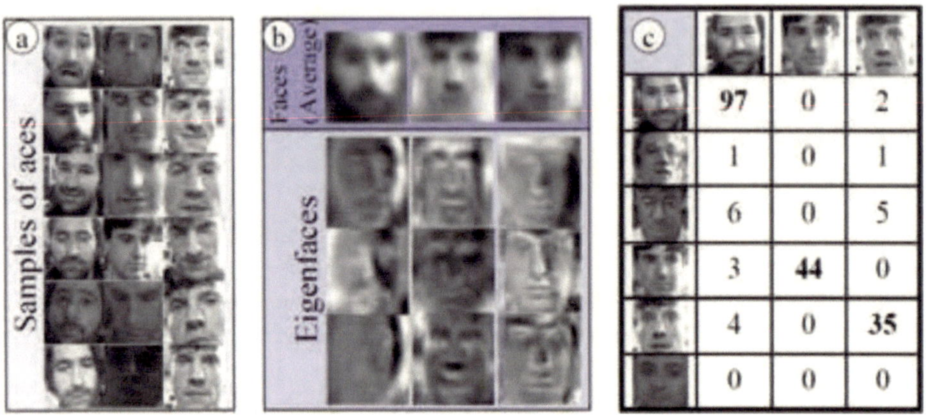

Fig. 9. a) Face image samples are shown for each of three people out of a database of six; b) Average face image for three people on the database, together with three eigenfaces for each one; c) Confusion table with face recognition results.

The number of basis functions is further reduced from M to M′ by selecting only the most meaningful M′ eigenvectors (with the largest associated eigenvalues), and ignoring all the others. Classification of the image of object φ consists of projecting it into the eigenobject components, by correlating the eigenvectors with it, for obtaining the coefficients $w_i=\mu_i(\phi-\psi)$, i= 1,..., M′ of this projection. The weights w_i form a vector $\Omega = \{w_1, w_2, \dots, w_{M'}\}$. An object is then classified by selecting the minimum L_2 distance to each object's coefficients in the database $\varepsilon_\phi = \left\| \Omega - \Omega_k \right\|$ where Ω_k describes the kth object class in the database. If ε_ϕ is below a threshold, then it corresponds to a new object.

Eigenfaces: Experimental Results

The eigenvectors are now denominated eigenfaces (Turk & Pentland, 1991), because they are face-like in appearance (see Figure 9 - the confusion table in this Figure presents results for recognizing three different people, being the average recognition accuracy of 88.9%). The training data set contains a lot of variation. Validation data corresponds to a random 20% of all the data.

Eigensounds for Sound Recognition

A collection of annotated acoustic signatures for each object are used as input data (see Figure 10) for a sound recognition algorithm by applying the eigenobjects method. A sound image is represented as a linear combination of base sound signatures (or *eigensounds*). Classification consists of projecting novel sounds to this space, determining the coefficients of this projection, computing the L_2 distance to each object's coefficients in the database, and selecting the class corresponding to the minimum distance.

Cross-modal information aids the acquisition and learning of unimodal percepts and consequent categorization in a child's early infancy. Similarly, visual data is employed here to guide the annotation of auditory data to implement a sound recognition algorithm. Training samples for the sound recognition algorithm are classified into different categories by the visual object recognition system or from information from the visual object tracking system. This enables the system, after training, to classify the sounds of objects not visible. The system was evaluated quantitatively by random selection of 10% of the segmented data for validation, and the remaining data for training. This process was randomly repeated three times. It is worth noticing that even samples received within a short time of each other often do not look too similar, due to background acoustic noise, noise on the segmentation process, other objects' sounds during experiments, and variability on how objects are moved and presented to the robot. For example, the car object is heard both alone and with a rattle (either visible or hidden). The recognition rate for the three runs averaged to 82% (86.7%, 80% and 80%). Recognition rates by object category were: 67% for the car, 91.7% for the cube rattle, 77.8% for the snake rattle and 83.3% for the hammer. Most errors arise from mismatches between (car and hammer) sounds.

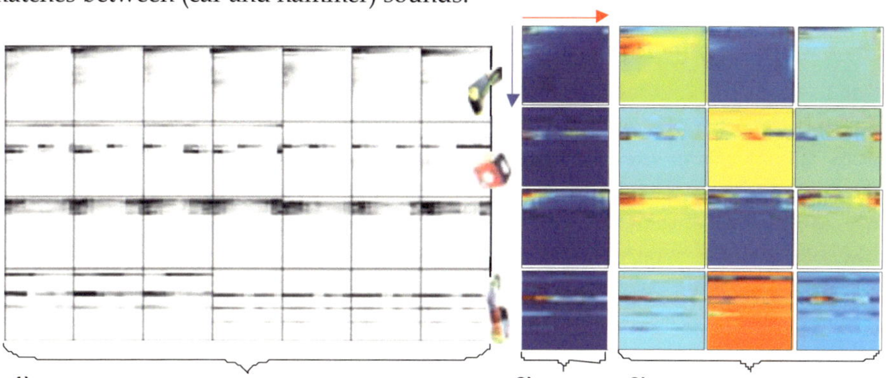

1) 7 random sound samples for each of 4 objects. From top to bottom: hammer, cube rattle, car and snake rattle, respectively **2)** Average sound images **3)** Eigenobjects corresponding to the three highest eigenvalues

Fig. 10. Sound recognition. Acoustic signatures for four objects are shown along the rows. (1) Seven sound segmentation samples are shown for each object, from a total of 28 (car), 49 (cube rattle), 23 (snake rattle) and 34 (hammer) samples. (2) Average acoustic signatures. The vertical axis corresponds to the frequency bands and the horizontal axis to time normalized by the period. (3) Eigensounds corresponding to the three highest eigenvalues. The repetitive nature of the sound generated by an object under periodic motion can be analyzed to extract an acoustic signature for that object. We search for repetition in a set of frequency bands, collecting those whose energies oscillate together with a similar period (Fitzpatrick & Arsenio, 2004).

3.4 Mixture of Gaussians for Scene Recognition

Wavelets (Strang & Nguyen, 1996) are employed to extract contextual features. Processing is applied iteratively through the low frequency branch of the wavelet transform over T=5 scales, while higher frequencies along the vertical, horizontal and diagonal orientations are stored (due to signal polarity, this corresponds to a compact representation of six orientations in three images). The input is thus represented by $v(x, y) = v(\vec{p}) = \{v_k(x, y), k=1,\ldots, N\}$, with N=3T=15. Each wavelet component at the ith level has dimensions $256/2^i \times 256/2^i$, and is down-sampled to an 8 × 8 image:

$$\bar{v}(x,y) = \sum_{i,j} v(i, j)h(i - x, j - y) \quad (3)$$

where h(x,y) is a Gaussian window. Thus, $\bar{v}(x,y)$ has dimension 960. Similarly to other approaches (Torralba, 2003), the dimensionality problem is reduced to become tractable by applying Principal Component Analysis (PCA). The image features $\bar{v}(\vec{p})$ are decomposed into the basis functions given by the PCA:

$$v(\vec{p}) = \sum_{i=1}^{D} c_i \varphi_k^i(\vec{p}) \;,\; c_i = \sum_{\vec{p},k} v_k(\vec{p}) \varphi_k^i(\vec{p}) \tag{4}$$

where the functions $\varphi_k^i(\vec{p})$ are the eigenfunctions of the covariance operator given by $v_k(\vec{p})$. These functions incorporate both spatial and spectral information. The decomposition coefficients are obtained by projecting the image features $v_k(\vec{p})$ into the principal components c_i, used hereafter as input context features.

The vector $\vec{c}$ ={c_i, i= 1, . . . , D} denotes the resulting D-dimensional input vector, with D=E_m, 2≤ D≤ Th_o, where m denotes a class, Th_o an upper threshold and E_m denotes the number of eigenvalues within 5% of the maximum eigenvalue. These features can be viewed as a scene's holistic (Oliva & Torralba, 2001) representation since all the regions of the image contribute to all the coefficients, as objects are not encoded individually. The effect of neglecting local features is reduced by mapping the foveal camera (which grabs data for the object recognition scheme based on local features) into the image from the peripheral view camera, where the weight of the local features $\vec{v}_I$ is strongly attenuated. The vector $\vec{p}$ is thus given in wide field of view retinal coordinates.

A collection of images is automatically annotated by the robot (Arsenio, 2004b;c) and used as training data. Mixture models are applied to find interesting places to put a bounded number of local kernels that can model large neighborhoods. In D-dimensions a mixture model is denoted by density factorization over multivariate Gaussians (spherical Gaussians were selected for faster processing times), for each object class n:

$$p(\vec{c} \mid o_n) = \sum_{m=1}^{M} b_m G(\vec{c}, \vec{\mu}_{m,n}, C_{m,n})$$

where G_m refers to the mth Gaussian with mean $\vec{\mu}_m$ and covariance matrix C_m, M is the number of Gaussian clusters, and b_m= p(G_m) are the weights of the local models. The estimation of the parameters will follow the EM algorithm (Gershenfeld, 1999):

- **E-step for** k-**iteration**: From the observed data $\vec{c}$, compute the a-posteriori cluster probability $e_{m,n}^k(l)$:

$$e_{m,n}^k(l) = p(c_{m,n} \mid \vec{c}) = \frac{b_{m,n}^k G(\vec{c}, \vec{\mu}_{m,n}^k, C_{m,n}^k)}{\sum_{m=1}^{M} b_{m,n}^k G(\vec{c}, \vec{\mu}_{m,n}^k, C_{m,n}^k)} \tag{5}$$

- **M-step for** k-**iteration**: cluster parameters are estimated according to the maximization of the join likelihood of the L training data samples

$$b_{m,n}^{k+1} = \sum_{l=1}^{L} e_{m,n}^{k} \;,\; \vec{\mu}_{m,n}^{k+1} = \frac{\sum_{l=1}^{L} e_{m,n}^{k}(l)\vec{c}_l}{\sum_{l=1}^{L} e_{m,n}^{k}(l)} \tag{6}$$

$$C_{m,n}^{k+1} = \frac{\sum_{l=1}^{L} e_{m,n}^{k}(l)(\vec{c}_l - \vec{\mu}_{m,n}^{k+1})(\vec{c}_l - \vec{\mu}_{m,n}^{k+1})^T}{\sum_{l=1}^{L} e_{m,n}^{k}(l)} \tag{7}$$

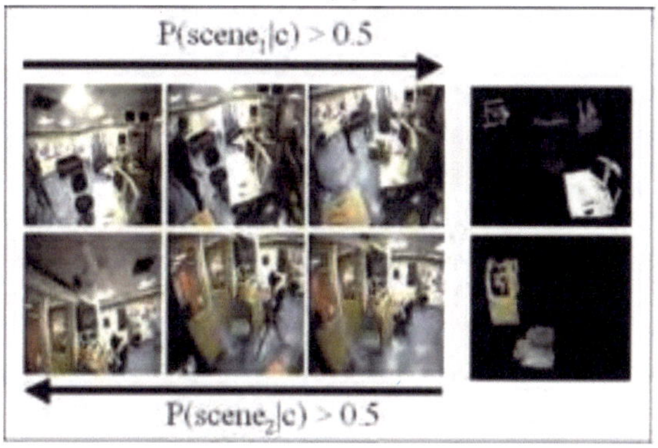

Fig. 11 Test images (wide field of view) organized with respect to $p(o_n \mid \vec{c})$. Top row: o_n=scene$_1$, $p(scene_1 \mid \vec{c}) > 0.5$; Bottom row: o_n=scene$_2$, $p(scene_2 \mid \vec{c}) > 0.5$. Scene descriptions shown in the right column are built on-line, automatically (Arsenio, 2004b;c).

The EM algorithm converges as soon as the cost gradient is small enough or a maximum number of iterations is reached. The probability density function (PDF) for an object n is then given by Bayes' rule $p(o_n \mid \vec{c}) = p(\vec{c} \mid o_n)p(o_n) / p(o_n)$ where $p(\vec{c}) = p(\vec{c} \mid o_n)p(o_n) + p(\vec{c} \mid \neg o_n)p(\neg o_n)$.

The same method applies for the out-of-class PDF $p(\vec{c} \mid \neg o_n)$ which represents the statistical feature distribution for the input data in which o_n is not present.

Finally, it is necessary to select the number M of gaussian clusters. This number can be selected as the one that maximizes the join likelihood of the data. An agglomerative clustering approach based on the Rissanen Minimum Description Length (MDL) order identification criterion (Rissanen, 1983) was implemented to automatically estimate M (Figure 11 shows algorithm results for classifying two scenes).

3.5 Dynamic Programming for Recognition from Cross-Modal Cues

Different objects have distinct acoustic-visual patterns which are a rich source of information for object recognition, if we can recover them. The relationship between object motion and the sound generated varies in an object-specific way. A hammer causes sound after striking an object. A toy truck causes sound while moving rapidly with wheels spinning; it is quiet when changing direction. These statements are truly cross-modal in nature. Features extracted from the visual and acoustic segmentations are what is needed to build an object recognition system (Fitzpatrick & Arsenio, 2004). The feature space for recognition consists of:

- Sound/Visual period ratios – the sound energy of a hammer peaks once per visual period, while the sound energy of a car peaks twice.
- Visual/Sound peak energy ratios – the hammer upon impact creates high peaks of sound energy relative to the amplitude of the visual trajectory.

Dynamic programming is applied to match the sound energy to the visual trajectory signal. Formally, let $S = (S_1, \ldots, S_n)$ and $V = (V_1, \ldots, V_m)$ be sequences of sound and visual trajectory energies segmented from n and m periods of the sound and visual trajectory signals, respectively. Due to noise, n may be different to m. If the estimated sound period is half the visual one, then V corresponds to energies segmented with 2m half periods (given by the distance between maximum and minimum peaks). A matching path $P = (P_1, \ldots, P_l)$ defines an alignment between S and M, where $\max(m, n) \leq l \leq m + n - 1$, and $P_k=(i, j)$, a match *k* between sound cluster *j* and visual cluster *i*. The matching constraints are set by:

- **The boundary conditions:** $P_1= (1, 1)$ and $P_l= (m, n)$.
- **Temporal continuity:** $P_{k+1} \in \{(i+1, j+1), (i+1, j), (i, j+1)\}$. Steps are adjacent elements of P.

The function cost $c_{i,j}$ is given by the square difference between V_i and S_j periods. The best matching path W can be found efficiently using dynamic programming, by incrementally building an m × n table caching the optimum cost at each table cell, together with the link corresponding to that optimum. The binding W will then result by tracing back through these links, as in the Viterbi algorithm.

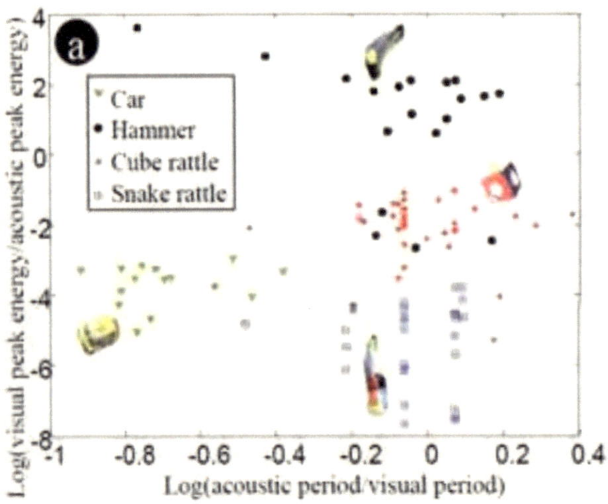

Confusion table

b	Car	Cube	Sna-ke	Ham-mer
Car	**30**	0	0	0
Cube	0	**52**	7	1
Snake	0	0	**45**	0
Hammer	0	5	0	**25**

Fig. 12. Object recognition from cross-modal clues. The feature space consists of period and peak energy ratios. The confusion matrix for a four-class recognition experiment is shown. The period ratio is enough to separate the cluster of the car object from all the others. Similarly, the snake rattle is very distinct, since it requires large visual trajectories for producing soft sounds. Errors for categorizing a hammer originated exclusively from erroneous matches with the cube rattle, because hammering is characterized by high energy ratios, and very soft bangs are hard to identify correctly. The cube rattle generates higher energy ratios than the snake rattle. False cube recognitions resulted mostly from samples with low energy ratios being mistaken for the snake.

Experimental Results: Figure 12 shows cross-modal features for a set of four objects. It would be hard to cluster automatically such data into groups for classification. But as in the sound recognition algorithm, training data is automatically annotated by visual recognition and tracking. After training, objects can be categorized from cross-modal cues alone. The system was evaluated by selecting randomly 10% of the data for validation, and the remaining data for training. This process was randomly repeated 15 times. The recognition rate averaged over all these runs were, by object category: 86.7%for the cube rattle, 100% for both the car and the snake rattle, and 83% for the hammer. The overall recognition rate was 92.1%. Such results demonstrate the potential for recognition using cross-modal cues.

3.6 Weighted Cluster Modeling

Objects in the world are situated, in the sense that they usually appear in specific places. Children are pretty good at learning the relative probability distribution of objects in a scene - for instance, chairs are most probable in front of desks, but not in a ceiling. The scene context puts a very important constraint on the type of places in which a certain object might be found. From a humanoid point of view, contextual selection of the attentional focus is very important both to constrain the search space for identifying or locating objects (optimizes computational resources) and also to determine common places on a scene to drop or store objects.

Therefore, it is important to develop a model for the contextual control of the attentional focus (location and orientation), scale selection and depth inference. The output space is defined by the 6-dimensional vector $\vec{x} = (\vec{p}, d, \vec{s}, \phi)$, where $\vec{p}$ is a 2D position vector, d is the object's depth (Arsenio, 2004b;c), $\vec{s}$ = (w, h) is a vector containing the principal components of the ellipse that models the 2D retinal size of the object, and ϕ is the orientation of such ellipse. Given the context $\vec{c}$, we need to evaluate the PDF $p(\vec{x} \mid o_n, \vec{c})$ from a mixture of (spherical) Gaussians (Gershenfeld, 1999),

$$p(\vec{x}, \vec{c} \mid o_n) = \sum_{m=1}^{M} b_{m,n} G(\vec{x}, \vec{\eta}_{m,n}, X_{m,n}) G(\vec{c}, \vec{\mu}_{m,n}^k, C_{m,n}^k) \tag{8}$$

The mean of the new Gaussian $G(\vec{x}, \vec{\eta}_{m,n}, X_{m,n})$ is now a function $\vec{\eta} = f(\vec{c}, \beta_{m,n})$, that depends on $\vec{c}$ and on a set of parameters $\beta_{m,n}$. A locally affine model was chosen for *f*, with $\{\beta_{m,n} = (\vec{a}_{m,n}, A_{i,n}) : \eta_{m,n} = \vec{a}_{m,n} + A^T \vec{c}\}$. The learning equations become now (Gershenfeld, 1999):

- **E-step for** k-**iteration**: From the observed data $\vec{c}$ and $\vec{x}$, compute the a-posteriori probabilities of the clusters:

$$e_{m,n}^k(l) = \frac{b_{m,n}^k G(\vec{x}, \vec{\eta}_{m,n}^k, X_{m,n}^k) G(\vec{c}, \vec{\mu}_{m,n}^k, C_{m,n}^k)}{\sum_{m=1}^{M} b_{m,n}^k G(\vec{x}, \vec{\eta}_{m,n}^k, X_{m,n}^k) G(\vec{c}, \vec{\mu}_{m,n}^k, C_{m,n}^k)}$$

- **M-step for** k-**iteration**: cluster parameters are estimated according to (where m indexes the M clusters, and l indexes the number L of samples):

$$C_{m,n}^{k+1} = \langle (\vec{c}_l - \vec{\mu}_{m,n}^{k+1})(\vec{c}_l - \vec{\mu}_{m,n}^{k+1})^T \rangle_m \tag{9}$$

$$A_{m,n}^{k+1} = (C_{m,n}^{k+1})^{-1} < (\vec{c} - \vec{\mu})(\vec{x} - \vec{\mu})^T >_m \quad (10)$$

$$a_{m,n}^{k+1} = < (\vec{x} - (A_{m,n}^{k+1})^T \vec{c}) >_m \quad (11)$$

$$X_{m,n}^{k+1} = < (\vec{x} - \vec{a}_{m,n}^{k+1} - (A_{m,n}^{k+1})^T \vec{c})(\vec{x} - \vec{a}_{m,n}^{k+1} - (A_{m,n}^{k+1})^T \vec{c})^T >_m \quad (12)$$

All vectors are column vectors and <>m in (9) represents the weighted average with respect to the posterior probabilities of cluster m.
The parameters $b_{m,n}^k$ and means $\vec{\mu}_{m,n}^{k+1}$ are estimated as before. The conditional probability follows then from the joint PDF of the presence of an object o_n, at the spatial location p, with pose ϕ, size $\vec{s}$ and depth d, given a set of contextual image measurements $\vec{c}$:

$$p(\vec{x} \mid o_n, \vec{c}) = \frac{b_{m,n}^k G(\vec{x}, \vec{\eta}_{m,n}^k, X_{m,n}^k) G(\vec{c}, \vec{\mu}_{m,n}^k, C_{m,n}^k)}{\sum_{m=1}^{M} b_{m,n}^k G(\vec{c}, \vec{\mu}_{m,n}^k, C_{m,n}^k)}$$

Object detection and recognition requires the evaluation of this PDF at different locations in the parameter space. The mixture of gaussians is used to learn spatial distributions of objects from the spatial distribution of frequencies in an image.
Figure 13 presents results for selection of the attentional focus for objects from the low-level cues given by the distribution of frequencies computed by wavelet decomposition.
Some furniture objects were not moved (such as the sofas), while others were moved in different degrees: the chair appeared in several positions during the experiment, while the table and door suffered mild displacements. Still, errors on the head gazing control added considerable location variability whenever a non-movable object was segmented and annotated. It demonstrates that, given an holistic characterization of a scene (by PCA on the image wavelet decomposition coefficients), one can estimate the appropriate places whether objects often appear, such as a chair in front of a table, even if no chair is visible at the time - which also informs that regions in front of tables are good candidates to place a chair. Object occlusions by people are not relevant, since local features are neglected, favoring contextual ones.

3.7 Back-propagation Neural Networks

Activity Identification

A feature vector for activity recognition was proposed by (Polana & Nelson, 1994) which accounts for 3-dimensional information: 2-dimensional spatial information plus temporal information. The feature vector is thus a temporal collection of 2D images. Each of these images is the sum of the normal flow magnitude (computed using a differential method) - discarding information concerning flow direction - over local patches, so that the final resolution is of 4×4. The normal flow accounts only for periodically moving pixels. Classification is then performed by a nearest centroid algorithm.
Our strategy reduces the dimensionality of the feature vector to 2-dimensional. This is done by constructing a 2D image which contains a description of an activity. Normalized length trajectories over one period of the motion are mapped to an image, in which the horizontal axis is given by the temporal scale, and the vertical axis by 6 elements describing position and 6 elements for velocities. The idea is to map trajectories into images. This is fundamentally different to the trajectory primal-sketch approach suggested in (Gould & Shah, 1989), which argues for compact representations involving motion discontinuities. We opt instead for using redundant information.

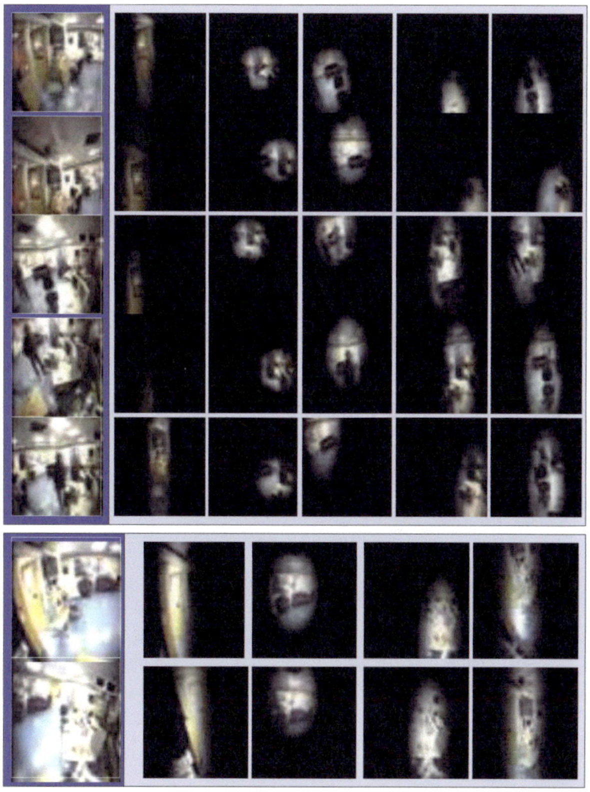

Fig. 13. Localizing and recognizing objects from contextual cues. (top) Samples of scene images are shown on the first column. The next five columns show probable locations based on context for finding a door, the smaller sofa, the bigger sofa, the table and the chair, respectively. Even if the object is not visible or present, the system estimates the places at which there is a high probability of finding such object. Two such examples are shown for the chair. Occlusion by humans do not change significantly the context. (bottom) Results in another day, with different lightning conditions.

Activities, identified as categories which include objects capable of similar motions, and the object's function in one activity, can then be learned by classifying 12 × 12 image patterns. One possibility would be the use of eigenobjects for classification (as described in this chapter for face and sound recognition). Eigenactivities would then be the corresponding eigenvectors. We opted instead for neural networks as the learning mechanism to recognize activities.

Target desired values, which are provided by the multiple object tracking algorithm, are used for the annotation of the training samples - all the training data is automatically generated and annotated, instead of the standard manual, offline annotation. An input feature vector is recognized into a category if the corresponding category output is higher than 0.5 (corresponding to a probability $p > 0{:}5$). Whenever this criterion fails for all categories, no match is assigned to the activity feature vector – since the activity is estimated as not yet in the database, it is labeled as a new activity.

We will consider the role of several objects in experiments taken for six different activities. Five of these activities involve periodic motion: cleaning the ground with a swiping brush; hammering a nail-like object with a hammer; sawing a piece of metal; moving a van toy; and playing with a swinging fish. Since more information is generated from periodic activities, they are used to generate both training and testing data. The remaining activity, poking a lego, is detected from the lego's discontinuous motion after poking. Figure 14 shows trajectories extracted for the positions of four objects from their sequences of images.

A three layer neural network is first randomly initialized. The input layer has 144 perceptron units (one for each input), the hidden layer has six units and the output layer has one perception unit per category to be trained (hence, five output units). Experiments are run with a set of (15, 12, 15, 6, 3, 1) feature vectors (the elements of the normalized activity images) for the swiping brush, hammer, saw, van toy, swinging fish and lego, respectively. A first group of experiments consists of selecting randomly 30% of these vectors as validation data, and the remaining as training data. The procedure is repeated six times, so that different sets of validation data are considered. The other two groups of experiments repeat this process for the random selection of 20% and 5% of feature vectors as validation data. The correspondent quantitative results are presented in figure 15.

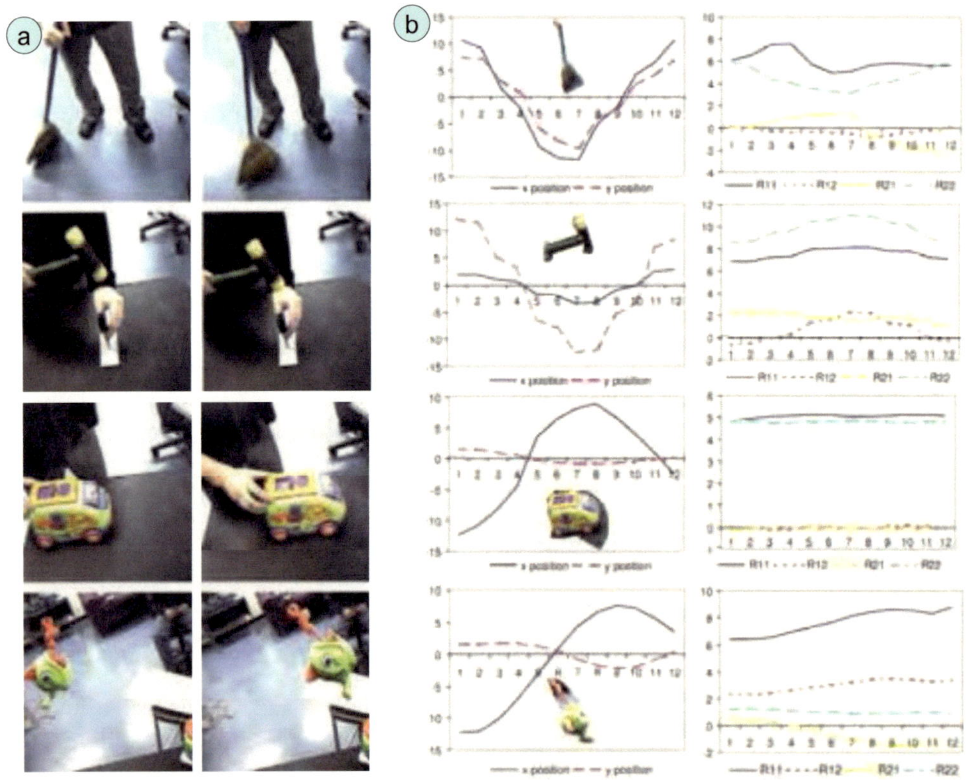

Fig. 14. a) Signals corresponding to one period segments of the object's trajectories normalized to temporal lengths of 12 points. From top to bottom: image sequence for a swiping brush, a hammer, a van toy and a swinging fish. b) Normalized centroid positions are shown in the left column, while the right column shows the (normalized and scaled) elements of the affine matrix R^i (where the indexes represents the position of the element on this matrix).

The lego activity, not represented in the training set, was correctly assigned as a new activity for 67% of the cases. The swinging fish was correctly recognized for just 17% of the cases, being the percentage of no matches equal to 57%. We believe that this poor result was due to the lack of a representative training set – this assumption is corroborated by the large number of times that the activity of swinging a fish was recognized as a new activity. The swiping brush was wrongly recognized for 3,7% of the total number of trials. The false recognitions occurred for experiments corresponding to 30% of the validation data. No recognition error was reported for smaller validation sets. All the other activities were correctly recognized for all the trials.

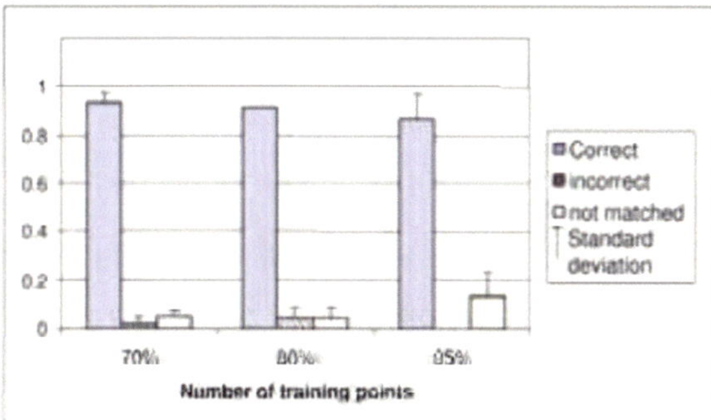

Confusion table (percentages)	Swiping Brush	Hammer	Saw	Van toy	fish	Not matched
Swiping Brush	**0.96**	0	0	0	0	0.037
Hammer	0	**1**	0	0	0	0
Saw	0	0	**1**	0	0	0
Van toy	0	0	0	**1**	0	0
fish	0.056	0	0	0.22	**0.17**	0.57
lego	0	0	0.056	0.056	0.22	**0.67**

Fig. 15. Experimental results for activity recognition (and the associated recognition of object function). Each experiment was ran six times for random initial conditions. Top graph) from left to right columns: 30%, 20% and 5% of the total set of 516 feature vectors are used as validation data. The total number of training and validation points, for each of the six trials (and for each of the 3 groups of experiments), is (15, 12, 15, 6, 3, 1) for the swiping brush, hammer, saw, van toy, swinging fish and lego, respectively. The three groups of columns show recognition, error and missed-match rates (as ratios over the total number of validation features). The bar on top of each column shows the standard deviation. Bottom table: Recognition results (as ratios over the total number of validation features). Row i and column j in the table show the rate at which object i was matched to object j (or to known, if j is the last column). Bold numbers indicate rates of correct recognitions.

Sound Recognition

An artificial neural network is applied off-line to the same data collected as before for sound recognition. The 32 × 32 sound images correspond to input vectors of dimension 1024. Hence, the neural network input layer contains 1024 perceptron units. The number of units in the hidden layer was set to six, while the output layer has four units corresponding to the four categories to be classified. The system is evaluated quantitatively by randomly selecting 40%, 30% and 5% of the segmented data for validation, and the remaining data for training. This process was randomly repeated six times. This approach achieves higher recognition rates when compared to eigensounds. The overall recognition rate is 96,5%, corresponding to a significant improvement in performance.

3.8 Other Learning Techniques

Other learning techniques exploited by Cog's cognitive system includes nearest-neighbor, locally linear receptive-field networks, and Markov models.

Locally Linear Receptive-field Networks

Controlling a robotic manipulator on the cartesian 3D space (eg. to reach out for objects) requires learning its kinematics - the mapping from joint space to cartesian space - as well as the inverse kinematics mapping. This is done through locally weighted regression and Receptive-field weighted regression, as proposed by (Schaal et al., 2000). This implementation on the humanoid robot Cog is described in detail by (Arsenio 2004c;d).

Markov Chains

Task descriptions can be modeled through a finite Markov Decision Process (MDP), defined by five sets <*S; A; P;R;O* >. Actions correspond to discrete, stochastic state-transitions $a \in A$={Periodicity, Contact, Release, Assembling, Invariant Set, Stationarity} from an environment's state $s_i \in S$ to the next state s_i+1, with probability $P^a_{s_i s_{i+1}} \in P$, where P is a set of transition probabilities $P^a_{ss'} = P_r\{s_{i+1} = s' \mid s, a\}$.

Task learning consists therefore on determining the states that characterize a task and mapping such states with probabilities of taking each possible action (Arsenio, 2003; Arsenio, 2004d).

4. Cognitive development of a Humanoid Robot

The work here described is part of a complex cognitive architecture developed for the humanoid robot Cog (Arsenio, 2004d), as shown in Figure 16. This chapter focused on a very important piece of this larger framework implemented on the robot. The overall framework places a special emphasis on incremental learning. A human tutor performs actions over objects while the robot learns from demonstration the underlying object structure as well as the actions' goals. This leads us to the object/scene recognition problem. Knowledge concerning an object is organized according to multiple sensorial percepts. After object shapes are learned, such knowledge enables learning of hand gestures. Objects are also categorized according to their functional role (if any) and their situatedness in the world. Learning *per si* is of diminished value without mechanisms to apply the learned knowledge. Hence, robot tasking deals with mapping learned knowledge to perceived information, for the robot to act on objects, using control frameworks such as neural oscillators and sliding-motion control (Arsenio, 2004).

Teaching a humanoid robot information concerning its surrounding world is a difficult task, which takes several years for a child, equipped with evolutionary mechanisms stored in its genes, to accomplish.

Learning aids such as books or educational, playful activities that stimulate a child's brain are important tools that caregivers extensively apply to communicate with children and to boost their cognitive development. And they also are important for human-robot interactions.

If in the future humanoid robots are to behave like humans, a promising venue to achieve this goal is by treating then as such, and initially as children - towards the goal of creating a 2-year-old-infant-like artificial creature.

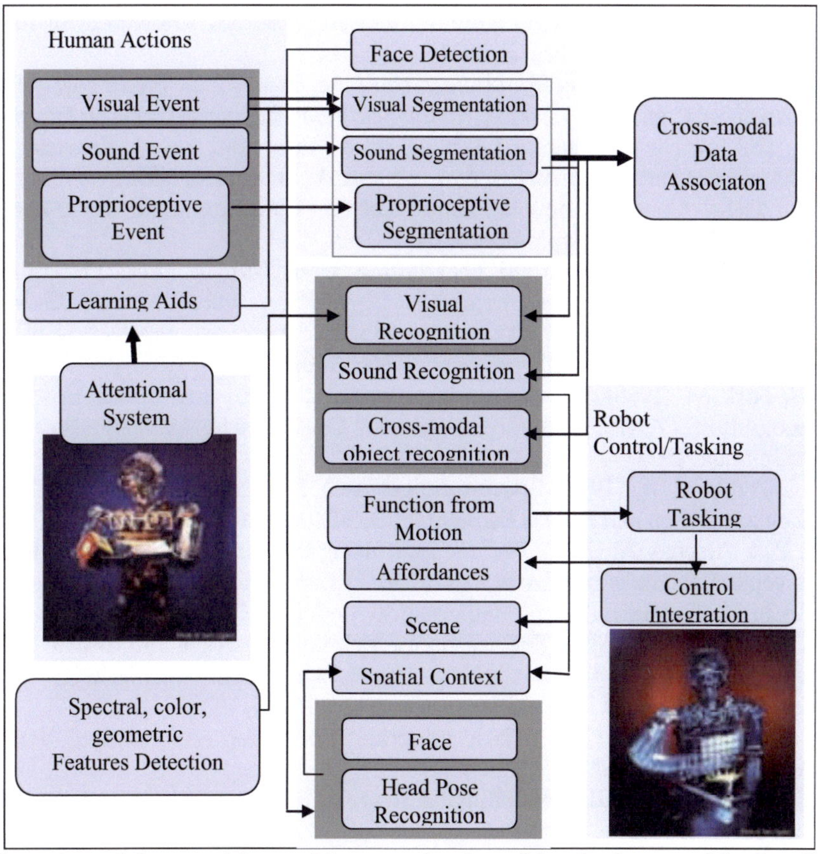

Fig. 16. Overview of the cognitive architecture developed for the humanoid robot Cog.

5. Conclusions

We proposed in this chapter the application of a collection of learning algorithms to solve a broad scope of problems. Several learning tools, such as Weighted-cluster modeling, Artificial Neural Networks, Nearest Neighbor, Hybrid Markov Chains, Geometric Hashing, Receptive Field Linear Networks and Principal Component Analysis, were extensively applied to acquire categorical information about actions, scenes, objects and people.
This is a new complex approach to object recognition. Objects might have various meanings in different contexts - a rod is labeled as a pendulum if oscillating with a fixed endpoint. From a visual image, a large piece of fabric on the floor is most often labeled as a tapestry, while it is most likely a bed sheet if it is found on a bed. But if a person is able to feel the fabric's material or texture, or the sound that it makes (or not) when grasped with other materials, then (s)he might determine easily the fabric's true function. Object recognition draws on many sensory modalities and the object's behavior, which inspired our approach.

6. References

Arsenio, A. (2003). Embodied vision - perceiving objects from actions. *Proceedings of IEEE International Workshop on Human-Robot Interactive Communication*, San-Francisco, 2003.

Arsenio, A. (2004a). Teaching a humanoid robot from books. *Proceedings of International Symposium on Robotics*, March 2004.

Arsenio, A. (2004b). Map building from human-computer interactions. *Proceedings of IEEE CVPR International Conference - Workshop on Real-time Vision for Human Computer Interaction*, 2004.

Arsenio, A. (2004c). Developmental Learning on a Humanoid Robot. *Proceedings of IEEE International Joint Conference on Neural Networks*, Budapest, 2004.

Arsenio, A. (2004d). Cognitive-developmental learning for a humanoid robot: A caregiver's gift, MIT PhD thesis, September 2004.

Arsenio, A. (2004e). Figure/ground segregation from human cues. *Proceedings of the IEEE/RSJ International Conference on Intelligent Robots and Systems (IROS-04)*, 2004.

Arsenio, A. (2004f). Object recognition from multiple percepts. *Proceedings of IEEE/RAS International Conference on Humanoid Robots*, 2004.

Cutler, R. & Turk, M. (1998). View-based interpretation of real-time optical flow for gesture recognition, *Proceedings of the International Conference on Automatic Face and Gesture Recognition*, 1998.

Darrel, T. & Pentland, A. (1993). Space-time gestures, *Proceedings of the IEEE Conference on Computer Vision and Pattern Recognition*, 335-340, New York, NY, 1993.

Fitzpatrick, P. & Arsenio, A. (2004). Feel the beat: using cross-modal rhythm to integrate robot perception. *Proceedings of Fourth International Workshop on Epigenetic Robotics*, Genova, 2004.

Gershenfeld, N. (1999). The nature of mathematical modeling. *Cambridge university press*, 1999.

Gould, K. & Shah, M. (1989). The trajectory primal sketch: A multi-scale scheme for representing motion characteristics. *Proceedings of IEEE International Conference on Computer Vision and Pattern Recognition*, pages 79–85, 1989.

Metta, G. & Fitzpatrick, P. (2003). Early integration of vision and manipulation. *Adaptive Behavior*, 11:2, pp. 109-128, June 2003.

Oliva, A. & Torralba, A. (2001). Modeling the shape of the scene: a holistic representation of the spatial envelope. *International Journal of Computer Vision*, pages 145–175, 2001.

Perrett, D.; Mistlin, A.; Harries, M. & Chitty, A. (1990). Understanding the visual appearance and consequence of hand action, *Vision and action: the control of grasping*, 163-180, Ablex, Norwood, NJ, 1990.

Polana, R. & Nelson, R.. (1994). Recognizing activities. *Proceedings of the 12th IAPR International Conference on Pattern Recognition*, October 1994.

Rao, K.; Medioni, G.& Liu, H. (1989). Shape description and grasping for robot hand-eye coordination, *IEEE Control Systems Magazine*, 9 (2) 22{29, 1989.

Rissanen, J. (1983). A universal prior for integers and estimation by minimum description length. *Annals of Statistics*, 11:417–431, 1983.

Schaal, S.; Atkeson, C. & Vijayakumar, S. (2000). Real-time robot learning with locally weighted statistical learning. *Proceedings of the International Conference on Robotics and Automation*, San Francisco, 2000.

Strang, G. & Nguyen, T. (1996). Wavelets and Filter Banks. *Wellesley-Cambridge Press*, 1996.

Torralba, A. (2003). Contextual priming for object detection. *International Journal of Computer Vision*, pages 153–167, 2003.

Turk, M. & Pentland, A. (1991). Eigenfaces for recognition. *Journal of Cognitive Neuroscience*, 3(1), 1991.

Vigotsky, L. (1962). Thought and language. *MIT Press*, Cambridge, MA, 1962.

Viola, P. & Jones, M. (2001). Robust real-time object detection. *Technical report, COMPAQ Cambridge Research Laboratory*, Cambridge, MA, 2001.

Wolfson, H. & Rigoutsos, I. (1997). Geometric hashing: an overview. *IEEE Computational Science and Engineering*, 4:10–21, 1997.

5

Biped Gait Generation and Control Based on Mechanical Energy Constraint

Fumihiko Asano[1], Masaki Yamakita[2,1], Norihiro Kamamichi[1] & Zhi-Wei Luo[3,1]
1. *Bio-Mimetic Control Research Center, RIKEN*
2. *Tokyo Institute of Technology*
3. *Kobe University*

1. Introduction

Realization of natural and energy-efficient dynamic walking has come to be one of the main subjects in the research area of robotic biped locomotion. Recently, many approaches considering the efficiency of gait have been proposed and McGeer's passive dynamic walking (McGeer, 1990) has been attracted as a clue to elucidate the mechanism of efficient dynamic walking. Passive dynamic walkers can walk down a gentle slope without any external actuation. Although the robot's mechanical energy is dissipated by heel-strike at the stance-leg exchange instant, the gravity potential automatically restores it during the single-support phase in the case of passive dynamic walking on a slope and thus the dynamic walking is continued. If we regard the passive dynamic walking as an active one on a level, it is found that the robot is propelled by the small gravity in the walking direction and the mechanical energy is monotonically restored by the virtual control inputs representing the small gravity effect. Restoration of the mechanical energy dissipated by heel-strike is a necessary condition common to dynamic gait generations from the mathematical point of view, and efficient active dynamic walking should be realized by reproducing this mechanism on a level. Mechanical systems satisfy a relation between the control inputs and the mechanical energy, the power-input for the system is equal to the time-derivative of mechanical energy, and we introduce a constraint condition so that the time-change rate of mechanical energy is kept positive constant. The dynamic gait generation is then specified by a simple redundant equation including the control inputs as the indeterminate variables and yields a problem of how to solve the equation in real-time. The ankle and the hip joint torques are determined according to the phases of cycle based on the pre-planned priority. The zero moment point (Vukobuatović & Stepanenko, 1972) can be easily manipulated by adjusting the ankle-joint torque, and the hip-joint torque in this case is secondly determined to satisfy the desired energy constraint condition with the pre-determined ankle-joint torque. Several solutions considering the zero moment point condition are proposed, and it is shown that a stable dynamic gait is easily generated without using any pre-designed desired trajectories. The typical gait is analyzed by numerical simulations, and an experimental case study using a simple machine is performed to show the validity of the proposed method.

2. Compass-like Biped Robot

In this chapter, a simplest planar 2-link full-actuated walking model, so-called *compass-like walker* (Goswami *et al.*, 1996), is chosen as the control object. Fig. 1 (left) shows the experimental walking machine and closed up of its foot which was designed as a nearly ideal compass-like biped model. This robot has three DC motors with encoders in the hip block to reduce the weight of the legs. The ankle joints are driven by the motors via timing belts. Table lists the values of the robot parameters. Fig. 1 (right) shows the simplest ideal compass-like biped model of the experimental machine, where m_H, m [kg] and $l = a + b$ [m] are the hip mass, leg mass and leg length, respectively. Its dynamic equation during the single-support phase is given by

$$\boldsymbol{M}(\boldsymbol{\theta})\ddot{\boldsymbol{\theta}} + \boldsymbol{C}(\boldsymbol{\theta},\dot{\boldsymbol{\theta}})\dot{\boldsymbol{\theta}} + \boldsymbol{g}(\boldsymbol{\theta}) = \boldsymbol{\tau}, \tag{1}$$

where $\boldsymbol{\theta} = \begin{bmatrix} \theta_1 & \theta_2 \end{bmatrix}^{\mathrm{T}}$ is the angle vector of the robot's configuration, and the details of the matrices are as follows:

$$\begin{aligned}
\boldsymbol{M}(\boldsymbol{\theta}) &= \begin{bmatrix} m_H l^2 + ma^2 + ml^2 & -mbl\cos(\theta_1 - \theta_2) \\ -mbl\cos(\theta_1 - \theta_2) & mb^2 \end{bmatrix}, \\
\boldsymbol{C}(\boldsymbol{\theta},\dot{\boldsymbol{\theta}}) &= \begin{bmatrix} 0 & -mbl\sin(\theta_1 - \theta_2)\dot{\theta}_2 \\ mbl\sin(\theta_1 - \theta_2)\dot{\theta}_1 & 0 \end{bmatrix}, \\
\boldsymbol{g}(\boldsymbol{\theta}) &= \begin{bmatrix} -(m_H l + ma + ml)\sin\theta_1 \\ mb\sin\theta_2 \end{bmatrix} g,
\end{aligned} \tag{2}$$

and the control torque input vector has the form of

$$\boldsymbol{\tau} = \boldsymbol{S}\boldsymbol{u} = \begin{bmatrix} 1 & 1 \\ 0 & -1 \end{bmatrix} \begin{bmatrix} u_1 \\ u_2 \end{bmatrix}. \tag{3}$$

The transition is assumed to be inelastic and without slipping. With the assumption and based on the law of conservation of angular momentum, we can derive the following compact equation between the pre-impact and post-impact angular velocities

$$\boldsymbol{Q}^{+}(\alpha)\dot{\boldsymbol{\theta}}^{+} = \boldsymbol{Q}^{-}(\alpha)\dot{\boldsymbol{\theta}}^{-}, \tag{4}$$

where

$$\begin{aligned}
\boldsymbol{Q}^{+}(\alpha) &= \begin{bmatrix} m_H l^2 + ma^2 + ml(l - b\cos(2\alpha)) & mb(b - l\cos(2\alpha)) \\ -mbl\cos(2\alpha) & mb^2 \end{bmatrix}, \\
\boldsymbol{Q}^{-}(\alpha) &= \begin{bmatrix} (m_H l^2 + 2mal)\cos(2\alpha) - mab & -mab \\ -mab & 0 \end{bmatrix},
\end{aligned} \tag{5}$$

and α [rad] is the half inter-leg angle at the heel-strike instant given by

$$\alpha = \frac{\dot{\theta}_1^{-} - \dot{\theta}_2^{-}}{2} = \frac{\dot{\theta}_2^{+} - \dot{\theta}_1^{+}}{2} > 0. \tag{6}$$

For further details of derivation, the authors should refer to the technical report by Goswami *et al.* This simplest walking model can walk down a gentle slope with suitable choices of physical parameters and initial condition. Goswami *et al.* discovered that this model exhibits period-doubling bifurcations and chaotic motion (Goswami *et al.*, 1996) when the slope angle increases. The nonlinear dynamics of passive walkers are very attractive but its mechanism has not been clarified yet.

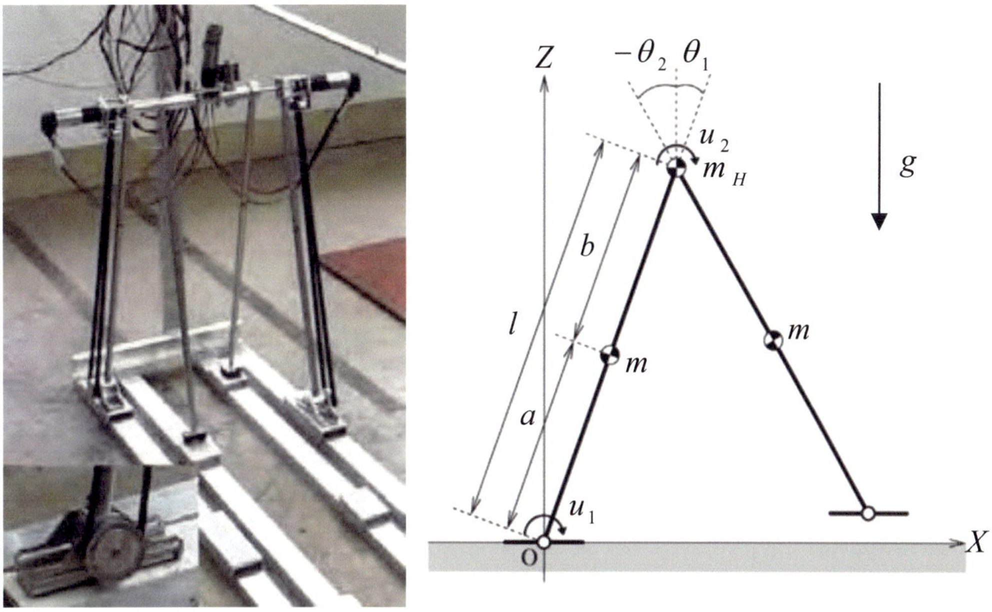

Fig. 1. Experimental walking machine and its foot mechanism (left), and its ideal model (right).

m_H		3.0	kg
m		0.4	kg
l	$= a + b$	0.680	m
a		0.215	m
b		0.465	m

Table 1. Physical parameters of the experimental machine.

3. Passive Dynamic Walking Mechanism Revisited

Passive dynamic walking has been considered as a clue to elucidate to clarify the essence of efficient dynamic walking, and the authors believe that it is worth investigating the automatic gait generation mechanism. The impulsive transition feature, non double-support phase, can be intuitively regarded as *vigor* for high-speed and energy-efficient walking. In order to get the *vigor*, the walking machine must restore the mechanical energy efficiently during the single-support phase, and the impulsive and inelastic collision with the ground dissipates it discontinuously. In the following, we describe it in detail.

The passive dynamic walker on a gentle slope can be considered to walk actively on a *virtual* level ground whose gravity is $g\cos\phi$ as shown in Fig. 2. The left robot in the figure is propelled forward by the small gravity element of $g\sin\phi$, and the right one walks by equivalent transformed torques. By representing this mechanism in the level walking, energy-efficient dynamic bipedal gait should be generated. The authors proposed virtual gravity concept for the level walking and called it "virtual passive dynamic walking." (Asano & Yamakita, 2001) The equivalent torques u_1 and u_2 are given by transforming the effect of the horizontal gravity element $g\sin\phi$ as shown in Fig. 2 left.

Let us define virtual total mechanical energy E_ϕ under the gravity condition of Fig. 2 as follows:

$$_\phi(\theta,\dot{\theta},\phi)=\frac{1}{2}\dot{\theta}\ M(\theta)\dot{\theta}\quad P(\theta,\phi) \tag{7}$$

where the *virtual* potential energy is given by

$$P(\theta,\phi)=\{(m_H l+ma+ml)\cos(\theta_1-\phi)-mb\cos(\theta_2-\phi)\}\,g\cos\phi\,. \tag{8}$$

In the case of passive dynamic walking on a slope, the total mechanical energy is kept constant during the single-support phase, whereas E_ϕ does not exhibit such behaviour. Fig. 3 shows the simulation results of passive dynamic walking on a gentle slope whose magnitude is 0.01 [rad]. (c) and (d) show the evolutions of the equivalent transformed torques and virtual energy E_ϕ, respectively. From (c), we can see that both u_1 and u_2 are almost *constant-like* and thus the ZMP should be kept within a narrow range. This property is effective in the virtual passive dynamic walking from the viewpoint of the stability of foot posture (Asano & Yamakita, 2001). It is apparent from (d) that the mechanical energy is dissipated at the transition instant and monotonically restored during the swing phase. Such energy behaviour can be considered as an indicator of efficient dynamic walking.

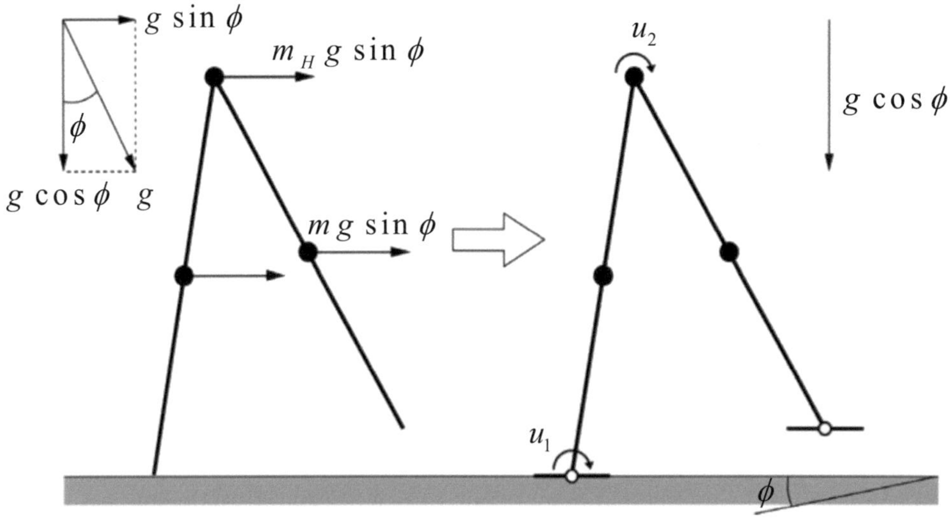

Fig. 2. Gravity acceleration mechanism of passive dynamic walking.

In general, we can state the following.

CH1) The total mechanical energy of the robot E_ϕ increases monotonically during the swing phase.

CH2) $\dot{\theta}_1>0$ always holds.

CH3) There exists an instant when $\theta_1-\theta_2=0$.

CH1 and CH2 always hold, regardless of physical and initial conditions, but CH3 does not always hold, as it depends on physical parameters and slope angle. We can confirm CH2 and CH3 from Fig. 3 (a) and (b). It is also clear that CH1 holds from Fig. 3 (d). From the results, the essence of a passive dynamic gait should be summarized as follows.

E1) The walking pattern is generated automatically, including impulsive transitions, and converges to a steady limit cycle.
E2) The total mechanical energy is restored during the single-support phase monotonically, and is dissipated at every transition instant impulsively by heel-strike with the floor.
E2 is considered to be an important characteristic for dynamic gait generation, and is the basic concept of our method. We will propose a simple method imitating the property in the next section.

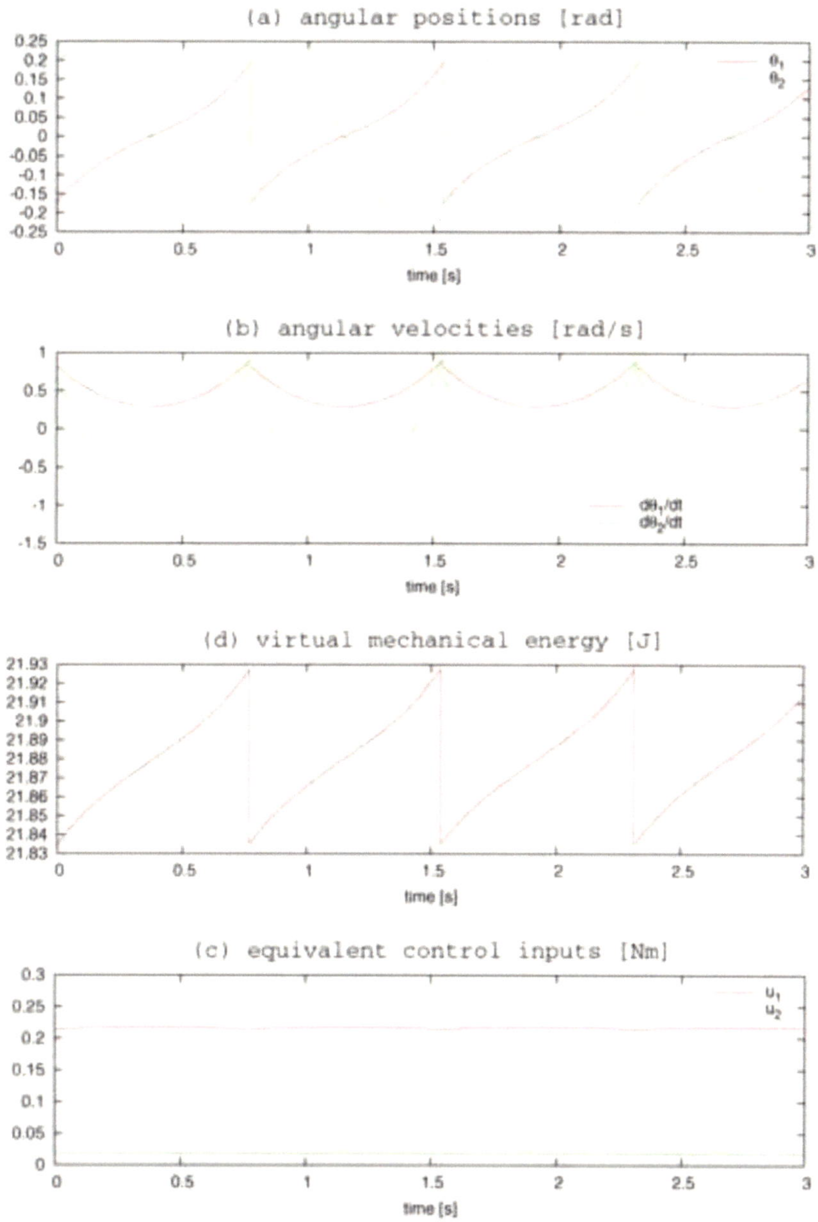

Fig. 3. Simulation results of passive dynamic walking on a slope where $\phi = 0.01$ [rad].

4.Energy Constraint Control

In our previous works, we have proposed virtual passive walking considering an artificial gravity condition called virtual gravity (Asano & Yamakita, 2001). This imitates the gravity acceleration mechanism of the original passive dynamic walking. A virtual gravity in the walking direction acts as a driving force for the robot and the stable limit cycle can be generated automatically without any gait design in advance. Determining a virtual gravity is, however, equivalent to that of control inputs, so there is no freedom to control other factors, for example, ZMP control. By imitating the property of monotonic energy restoration, however, we can formulate a simple method with a freedom of the control inputs.

4.1 The Control Law

The total mechanical energy of the robot can be expressed as

$$E\left(\boldsymbol{\theta},\dot{\boldsymbol{\theta}}\right)=\frac{1}{2}\dot{\boldsymbol{\theta}}^{\mathrm{T}}\boldsymbol{M}(\boldsymbol{\theta})\dot{\boldsymbol{\theta}}+P(\boldsymbol{\theta}), \tag{9}$$

where P is the potential energy. The power input to the system is the time-change rate of the total energy, that is

$$\dot{E}=\dot{\boldsymbol{\theta}}^{\mathrm{T}}\boldsymbol{\tau}=\dot{\boldsymbol{\theta}}^{\mathrm{T}}\boldsymbol{S}\boldsymbol{u}. \tag{10}$$

Suppose now that we use a simple control law imitating the characteristic CH1, monotonic energy restoration. Let $\lambda>0$ be a positive constant and consider the following condition:

$$\dot{E}=\lambda. \tag{11}$$

This means that the robot's mechanical energy increases monotonically with a constant rate of λ. We call this control or gait generation method "Energy Constraint Control (ECC)". In this method, the walking speed becomes faster w.r.t. the increase of λ, in other words, the magnitude of λ corresponds to the slope angle of virtual passive dynamic walking. Here let us consider the following output function:

$$H\left(\boldsymbol{\tau}\right)=\dot{E}-\lambda=\dot{\boldsymbol{\theta}}^{\mathrm{T}}\boldsymbol{\tau}-\lambda, \tag{12}$$

and the target constraint condition of Eq. (11) can be rewritten as $H\left(\boldsymbol{\tau}\right)=0$. Therefore, the ECC can be regarded in this sense as an output zeroing control.

Following Eqs. (10) and (11), the detailed target energy constraint condition is expressed as

$$\dot{E}=\dot{\boldsymbol{\theta}}^{\mathrm{T}}\boldsymbol{S}\boldsymbol{u}=\dot{\theta}_1u_1+\left(\dot{\theta}_1-\dot{\theta}_2\right)u_2=\lambda, \tag{13}$$

which is a redundant equation on the control inputs. The dynamic gait generation then yields a problem of how to solve the redundant equation for the control inputs u_1 and u_2 in real-time. The property of ECC strategy is that the control inputs can be easily determined by adjusting the feed-forward parameter λ, which can be determined by considering the magnitude of $\dot{E}$ of virtual passive dynamic walking.

4.2 Relation between ZMP and Reaction Moment

The actual walking machine has feet and a problem of reaction moment then arises. The geometrical specifications of the stance leg and its foot are shown in Fig. 4. In this chapter, the ZMP is calculated by the following approach. We assume:

1. The mass and volume of the feet can be ignored.
2. The sole always fits with the floor.

Under these assumptions, we can calculate the ZMP in the coordinate shown in Fig. 4 left as:

$$\text{ZMP} = -\frac{u_1}{R_n} \tag{14}$$

where u_1 [Nm] is the ankle torque acting not on the foot link but on the leg link and R_n [N] is the vertical element of the reaction force, respectively.

From Fig. 4, it is obvious that the ZMP is always shifted behind the ankle joint when driving the stance-leg forward, however, at the transition instant, the robot is critically affected by the reaction moment from the floor as shown in Fig. 4 right. Considering the reaction moment effect, we can reform the ZMP equation for the simplest model as follows:

$$\text{ZMP} = -\frac{u_1 + u_{\text{rm}}}{R_n} \tag{15}$$

where $u_{\text{rm}} > 0$ represents the equivalent torque of the reaction moment, and the ZMP is shifted backward furthermore. u_{rm} acts as a disturbance for the transition. Since the actual walking machines generally have feet with toe and heel, this problem arises. From the aforementioned point of view, we conclude that the ZMP should be shifted forward the ankle-joint just after the transition instant to cancel the reaction moment. Based on the observation, in the following, we consider an intuitive ZMP manipulation algorithm utilizing the freedom of the redundant equation of (13).

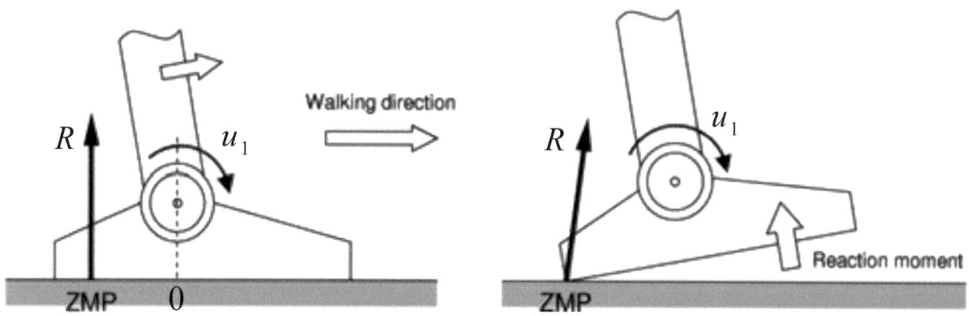

Fig. 4. Foot mechanism and reaction moment at the heel-strike instant.

4.3 Principal Ankle-joint Torque Control

From a practical point of view, as mentioned above, the two most important control factors of dynamic bipedal walking are mechanical energy restoration and ZMP control. To keep the energy constraint condition of Eq. (11), we should reconsider the solution algorithm. Firstly, we should consider mechanical energy restoration to generate a gait, and secondly, ZMP condition must be guaranteed without destroying the constraint condition. Based on the considerations, we first discuss the following solution approach:

1. Determine the value of λ.
2. Determine the ankle torque u_1.
3. By substituting λ and u_1 into Eq. (13), we can solve it for u_2.

In order to shift the ZMP, let us consider the following simple ankle-joint torque control:

$$u_1 = \begin{cases} u^- < 0 & s \le T \\ u^+ > 0 & \text{otherwise} \end{cases} \tag{16}$$

where s is a virtual time that is reset at every transition instant and $\dot{s} = 1.0$. This comes from the fact that u_1 must be negative to shift the ZMP forward the ankle-joint, and if $u_1 > 0$ the ZMP moves behind the ankle-joint. In this case, u_2 is obtained after u_1 as follows:

$$u_2 = \frac{\lambda - \dot{\theta}_1 u^{\pm}}{\dot{\theta}_1 - \dot{\theta}_2}. \tag{17}$$

Note that u_2 has a singularity at $\dot{\theta}_1 - \dot{\theta}_2 = 0$ which was mentioned before as CH3. This condition must be taken into account. We then propose a switching control law described later. Before it, we consider a more reasonable switching algorithm from u^- to u^+. In general, for most part of a cycle from the beginning, the condition $\dot{\theta}_1 - \dot{\theta}_2 > 0$ holds (See Fig. 3 (b)), and thus the sign of u_2 of Eq. (17) is identical with that of $\lambda - \dot{\theta}_1 u_1$. If $u_1 = u^-$, this sign is positive because of $\lambda, \dot{\theta}_1 > 0$ and $u^- < 0$. At the beginning of a cycle, $\lambda - \dot{\theta}_1 u^+$ increases monotonically because of $\ddot{\theta}_1 < 0$ (See Fig. 3 (b)). Therefore in general the condition

$$\frac{\mathrm{d}}{\mathrm{d}t}\left(\lambda - \dot{\theta}_1 u^+\right) = -\ddot{\theta}_1 u^+ > 0 \tag{18}$$

holds regardless of the system parameter choice. Therefore, if $\lambda < \dot{\theta}_1 u^+$ at the beginning, it is reasonable to switch when

$$\lambda - \dot{\theta}_1 u^+ = 0 \tag{19}$$

so as to keep u_2 of Eq. (17) always positive under the condition of $\dot{\theta}_1 - \dot{\theta}_2 > 0$. In addition, by this approach the hip-joint torque can always contribute the mechanical energy restoration. The switching algorithm of u_1 is summarized as follows:

$$u_1 = \begin{cases} u^- < 0 & \lambda \le \dot{\theta}_1 u^+ \\ u^+ > 0 & \text{otherwise} \end{cases}. \tag{20}$$

The value of u^+ must be determined empirically based on the simulation results of virtual passive dynamic walking, whereas u^- should be determined carefully so as not to destroy the original limit cycle or disturb the forward acceleration. Choosing the suitable combination between λ and u^+ is the most important for generating a stable limit cycle.

4.4 Principal Hip-joint Torque Control

As mentioned before, we must switch the controller to avoid the singularity of CH3 at the end of the single-support phase. As a new method, we propose the following new strategy:

1. Determine the value of λ.
2. Determine the hip torque u_2.
3. By substituting λ and u_2 into Eq. (13), we can solve it for u_1.

In this case, u_1 is determined by the following formula:

$$u_1 = \frac{\lambda - \left(\dot{\theta}_1 - \dot{\theta}_2\right)u_2}{\dot{\theta}_1}. \tag{21}$$

Note that here we use the assumption of CH2. In this paper, as a reasonable candidate of u_2, we consider the following form:

$$u_2 = \eta\left(\dot{\theta}_1 - \dot{\theta}_2\right). \tag{22}$$

Assuming $\eta > 0$, this leads the following inequality:

$$\left(\dot{\theta}_1 - \dot{\theta}_2\right)u_2 = \eta\left(\dot{\theta}_1 - \dot{\theta}_2\right)^2 \geq 0, \tag{23}$$

therefore it is found that this hip-joint torque u_2 also contributes the mechanical energy restoration.

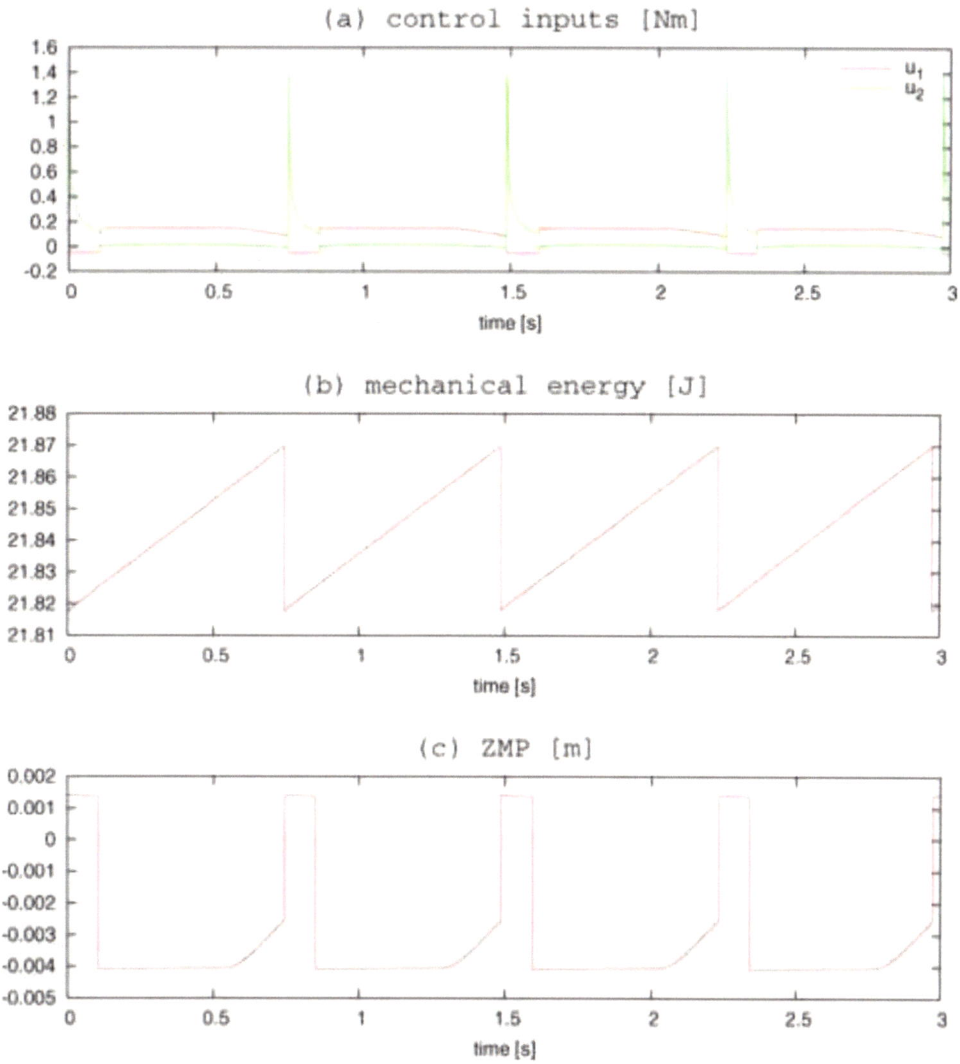

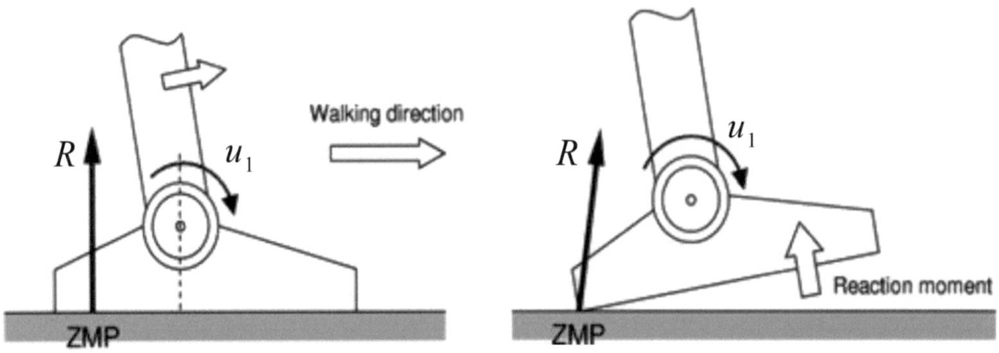

Fig. 5. Simulation results of dynamic walking by ECC considering ZMP control.

4.5 Switching Control

In order to manipulate the ZMP as well as to avoid the singularity, we must consider a switching algorithm from the principal ankle to hip-joint torque control. We here introduce the switching timing as $\theta_1 = \psi$ [rad]. At this instant, we reset η so that u_2 becomes continuous according to the following relationship:

$$u_2 = \eta\left(\dot{\theta}_1^{\text{sw}} - \dot{\theta}_2^{\text{sw}}\right) = \frac{\lambda - \dot{\theta}_1^{\text{sw}} u^+}{\dot{\theta}_1^{\text{sw}} - \dot{\theta}_2^{\text{sw}}}. \tag{24}$$

from which we can calculate η as follows:

$$\eta = \frac{\lambda - \dot{\theta}_1^{\text{sw}} u^+}{\left(\dot{\theta}_1^{\text{sw}} - \dot{\theta}_2^{\text{sw}}\right)^2} \tag{25}$$

where the superscript "sw" stands for the *switching* instant. The obtained η is used during its cycle and reset at every switching instant. Since u_2 is continuous, u_1 also becomes continuous.

Fig. 5 shows the simulation results of the dynamic walking by ECC with the proposed switching control. The control parameters are chosen as $\lambda = 0.07$ [J/s], $u^+ = 0.15$, $u^- = -0.05$ [Nm] and $\psi = 0.05$ [rad], respectively. By the effect of the principal ankle -joint torque control, the ZMP is shifted forward the ankle-joint without destroying the energy restoration condition. From Fig. 5 (b), we can see that the hip-joint torque becomes very large during the ZMP is shifted forward, but this does not affect the ZMP condition and the postural stability of foot is maintained.

4.6 Discussion

Here, we compare our method with approach proposed by Goswami *et al.* "energy tracking control." (Goswami *et al.*, 1997) Their approach is formulated as

$$\dot{E} = -\lambda_{\text{etc}}\left(E - E^*\right), \tag{26}$$

where E^* [J] (constant) is the reference energy and positive scalar λ_{etc} is the feedback gain. A solution of Eq. (13) by constant torque ratio $\mu > 0$ which gives the condition $u_1 = \mu u_2$ is obtained as

$$\boldsymbol{\tau} = \frac{-\lambda_{\text{etc}}\left(E - E^{*}\right)}{(\mu+1)\dot{\theta}_1 - \dot{\theta}_2}\begin{bmatrix} \mu+1 \\ -1 \end{bmatrix}, \tag{27}$$

and in our case, a solution by constant torque ratio is given by

$$\boldsymbol{\tau} = \frac{\lambda}{(\mu+1)\dot{\theta}_1 - \dot{\theta}_2}\begin{bmatrix} \mu+1 \\ -1 \end{bmatrix}. \tag{28}$$

Figs. 6 and 7 respectively show the simulation results of active dynamic walking on a level by the torques of Eqs. (27) and (28) without manipulating the ZMP actively. The two cases are equal in walking speed. From the simulation results, we can see that, in our approach, the maximum ankle-joint torque is about 3 times smaller than that of Goswami's approach and this yields better ZMP condition. In this sense, we should conclude that the mechanical energy must be restored efficiently but its time-change rate should be carefully chosen to guarantee the ZMP condition.

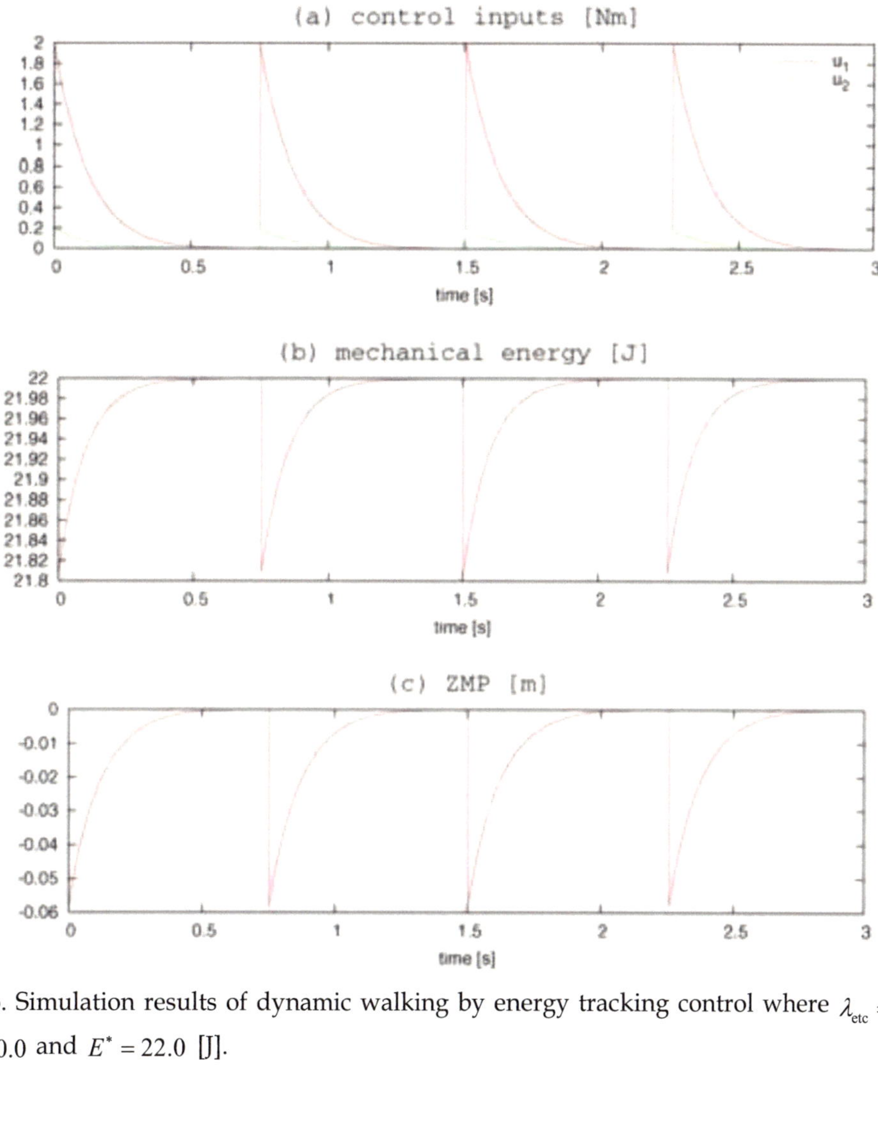

Fig. 6. Simulation results of dynamic walking by energy tracking control where $\lambda_{\text{etc}} = 10.0$, $\mu = 10.0$ and $E^{*} = 22.0$ [J].

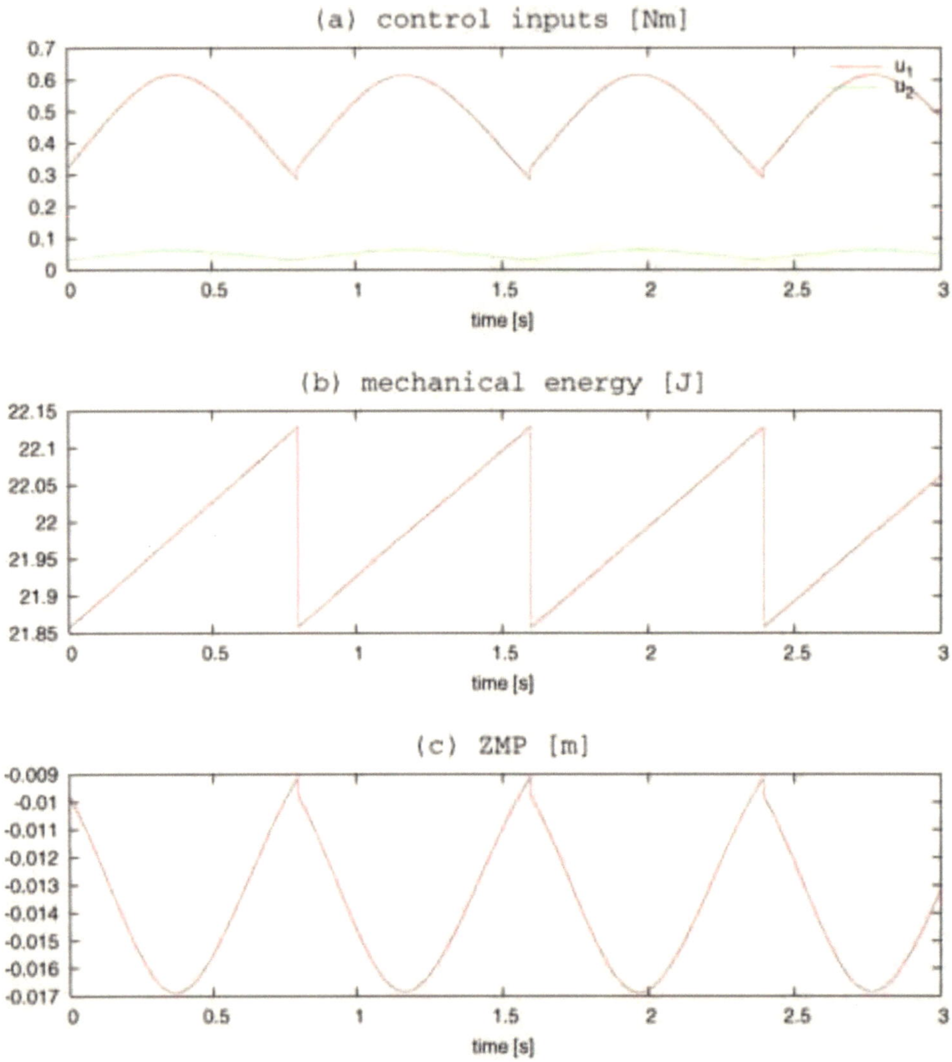

Fig. 7. Simulation results of dynamic walking by ECC where $\lambda = 0.34$ [J/s] and $\mu = 10.0$.

5. Experiments

In order to confirm the validity of the proposed method, we carried out actual walking experiment using our developed machine introduced in Fig.1. All encoders of the servomotors are interfaced to a computer (Pentium III 1.0 GHz) running Windows 98. To implement the control law, we used RtMaTX (Koga, 2000) for real-time computation with the sampling period 1.0 [ms].

Since the proposed methods are so called *model-matching control*, they are not robust for uncertainty. In this research, we use model following control of the motion generated by VIM (Virtual Internal Model) which is a reference model in computer. Every post-impact condition of VIM is reset to that of the actual machine. By using the VIM, the uncertainties of identification, which is crucial factor in the case of model matching control, can be compensated. The dynamics of VIM is given by

$$\hat{M}(\theta_d)\ddot{\theta}_d + \hat{C}(\theta_d, \dot{\theta}_d)\dot{\theta}_d + \hat{g}(\theta_d) = \tau_d, \quad (29)$$

where τ_d is the control input to drive the VIM and is determined by θ_d and $\dot{\theta}_d$. The control input for the actual robot is given by

$$\begin{aligned} \tau &= \hat{M}(\theta_d)u + \hat{C}(\theta_d, \dot{\theta}_d)\dot{\theta}_d + \hat{g}(\theta_d) \\ u &= \ddot{\theta}_d + K_D(\dot{\theta}_d - \dot{\theta}) + K_P(\theta_d - \theta) + K_I \int (\theta_d - \theta)\mathrm{d}t \end{aligned} \quad (30)$$

The virtual internal model started walking from the following initial condition:

$$\dot{\theta}(0) = \begin{bmatrix} 0.68 \\ 0.62 \end{bmatrix}, \theta(0) = \begin{bmatrix} -0.14 \\ 0.14 \end{bmatrix},$$

and its state was calculated and updated in real-time. At every transition instant, the angular positions of VIM were reset to that of the actual machine. PID controller drives the ankle-joint of the swing leg during the single-support phase so that the foot keeps the posture horizontal.

The experimental results are shown in Fig. 8. The adjustment parameters are chosen as $\lambda = 0.075$ [J/s], $u^+ = 0.15$, $u^- = -0.05$ [Nm] and $\psi = 0.05$ [rad] empirically. Fig. 8 (a) and (b) show the evolution of angular positions and velocities of the actual machine, respectively. The actual angular velocities are calculated by differentiation thorough a filter whose transfer function is $70/(s+70)$. A stable dynamic walking is experimentally realized based on ECC via model following control.

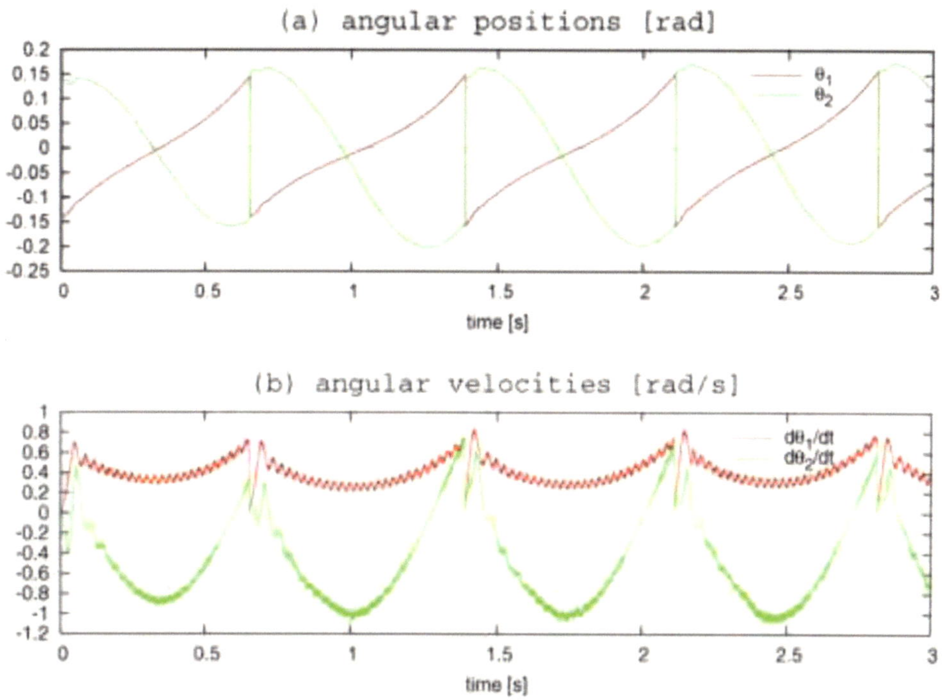

Fig. 8. Experimental results of dynamic walking by ECC.

6. Improving Robust Stability by Energy Feedback Control

Eq. (26) implies that the walking system becomes robust through the reference energy tracking. In other words, this control expands the basin of attraction of a limit cycle, however, our method Eq. (13) is so called the *feed-forward* control, which gives only energy change ratio without any information to attract the trajectories. Based on the observations, in this section, we firstly analyze the stability of the walking cycle and then consider an energy feedback control law in order to increase the robustness of the walking system.
Let us then consider an energy feedback control using a reference energy trajectory. Consider the following control

$$\dot{E} = \dot{\boldsymbol{\theta}}^{\mathrm{T}} \boldsymbol{S}\boldsymbol{u} = \dot{E}_{\mathrm{d}} - \zeta\left(E - E_{\mathrm{d}}\right), \tag{31}$$

which determines the control input so that the closed energy system yields

$$\frac{\mathrm{d}}{\mathrm{d}t}\left(E - E_{\mathrm{d}}\right) = -\zeta\left(E - E_{\mathrm{d}}\right) \tag{32}$$

where $\zeta > 0$ is the feedback gain. The original energy constraint control can be recognized as the case of $\dot{E}_{\mathrm{d}} = \lambda$ and $\zeta = 0$ in Eq. (31). By integrating Eq. (11) w.r.t. time, we can obtain the reference energy E_{d} using virtual time s as

$$E_{\mathrm{d}}(s) = E_0 + \lambda s \tag{33}$$

where E_0 [J] is the energy value when $s = 0$ [s]. A solution of Eq. (31) using constant torque ratio μ yields

$$\boldsymbol{S}\boldsymbol{u} = \frac{\dot{E}_{\mathrm{d}} - \zeta\left(E - E_{\mathrm{d}}\right)}{\left(\mu+1\right)\dot{\theta}_1 - \dot{\theta}_2} \begin{bmatrix} \mu+1 \\ -1 \end{bmatrix}. \tag{34}$$

Although *autonomy* of the walking system is destroyed by this method, we can improve the robustness of the walking system.
One way to examine the gait stability is Poincaré return map from a heel-strike collision to the next one. The Poincaré return map is denoted below as $\boldsymbol{F}$:

$$\boldsymbol{x}_{k+1} = \boldsymbol{F}\left(\boldsymbol{x}_k\right) \tag{35}$$

where the discrete state $\boldsymbol{x}_k$ is chosen as

$$\boldsymbol{x}_k = \begin{bmatrix} \theta_2^+[k] - \theta_1^+[k] \\ \dot{\theta}_1^+[k] \\ \dot{\theta}_2^+[k] \end{bmatrix}, \tag{36}$$

that is, relative hip joint angle and angular velocities just after k-th impact. The function $\boldsymbol{F}$ is determined based on Eqs. (1) and (3), but cannot be expressed analytically. Therefore, we must compute $\boldsymbol{F}$ by numerical simulation following an approximation algorithm.
In the case of steady walking, the relation $\boldsymbol{F}\left(\boldsymbol{x}^*\right) = \boldsymbol{x}^*$ holds and $\boldsymbol{x}^*$ is the equilibrium point of state at just after transition instant. For a small perturbation $\delta\boldsymbol{x}_k$ around the limit cycle, the mapping function $\boldsymbol{F}$ can be expressed in terms of Taylor series expansion as

$$\boldsymbol{F}\left(\boldsymbol{x}_k\right) = \boldsymbol{F}\left(\boldsymbol{x}^* + \delta\boldsymbol{x}_k\right) \approx \boldsymbol{x}^* + \nabla\boldsymbol{F} \cdot \delta\boldsymbol{x}_k \tag{37}$$

where

$$\nabla F \triangleq \left.\frac{\partial F(x)}{\partial x}\right|_{x=x*} \tag{38}$$

is the Jacobian (gradient) around x^*. By performing numerical simulations, ∇F can be calculated approximately. The all eigenvalues of ∇F are in the unit circle and the results are omitted. Although the robustness of the walking system is difficult to evaluate mathematically, the maximum singular value of ∇F should imply the convergence speed of gait; smaller the value is, faster the convergence to the steady gait is. Fig. 9 shows the analysis result of maximum singular value of ∇F w.r.t ζ in the Fig. 7 case with energy feedback control where $E_0 = 21.8575$ [J] and $\zeta = 10.0$. The maximum singular value monotonically decreases with the increase of ζ. The effect of improvement of the gait robustness by feedback control can be confirmed. Although applying this method destroys autonomy of the walking system, we can improve the robustness.

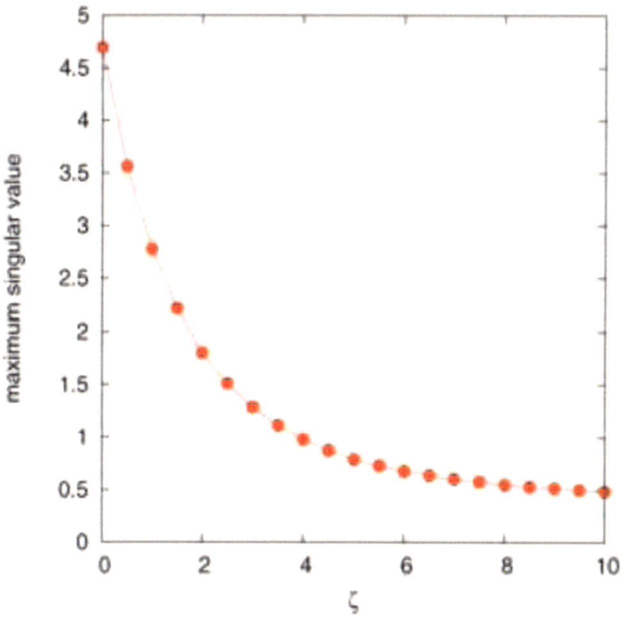

Fig. 9. Maximum singular value of ∇F w.r.t. the feedback gain ζ.

6. Extension to a Kneed Biped

This section considers an extension of ECC to a kneed biped model. We treat a simple planar kneed biped model shown in Fig. 10, and its dynamic equation is given by

$$\boldsymbol{M}(\boldsymbol{\theta})\ddot{\boldsymbol{\theta}} + \boldsymbol{C}(\boldsymbol{\theta},\dot{\boldsymbol{\theta}})\dot{\boldsymbol{\theta}} + \boldsymbol{g}(\boldsymbol{\theta}) = \boldsymbol{Su} = \begin{bmatrix} 1 & 1 \\ 0 & -1 \\ 0 & 0 \end{bmatrix} \begin{bmatrix} u_1 \\ u_2 \end{bmatrix} \tag{39}$$

We consider the following assumptions.

1. The knee-joint is passive.
2. It can be mechanically locked-on and off.

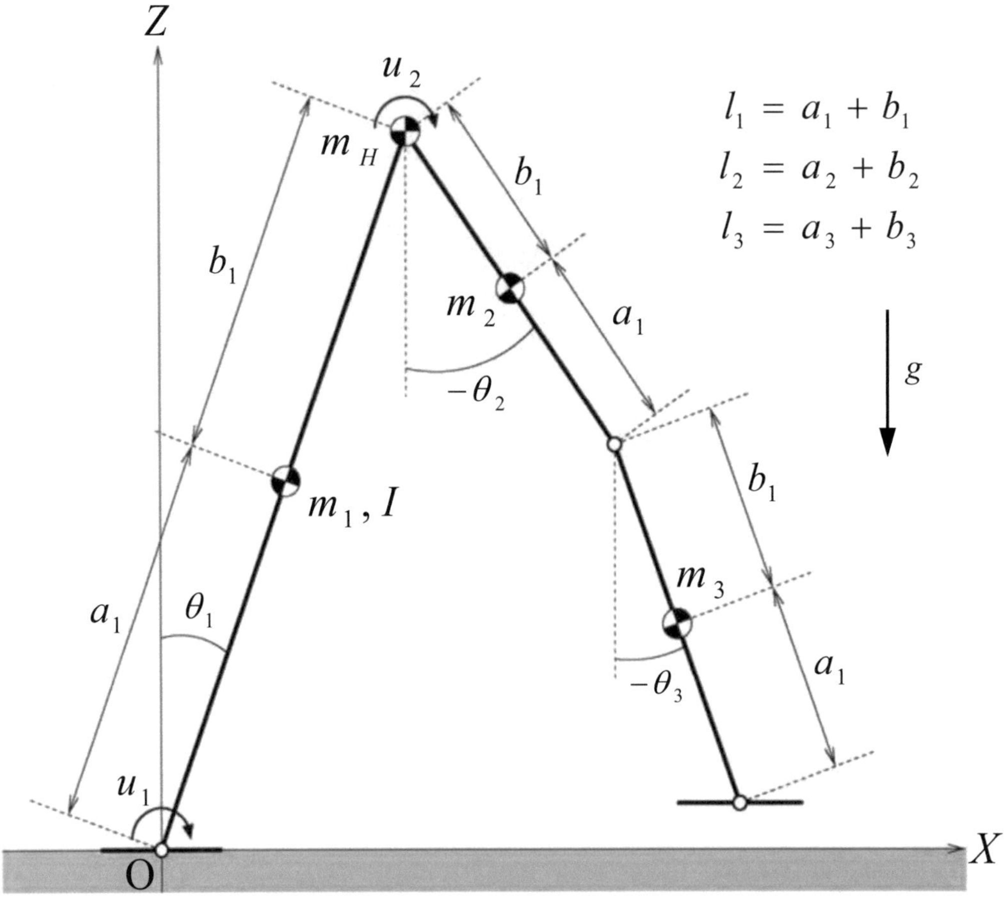

Fig. 10. Model of a planar underactuated biped robot.

The ECC then yields a problem of how to solve the following redundant equation:

$$\dot{E} = \dot{\theta}_1 u_1 + \left(\dot{\theta}_1 - \dot{\theta}_2\right) u_2 = \lambda \tag{40}$$

for the control inputs in real-time. Since the knee-joint is free, we can give the control input by applying the form of Eq. (28) as

$$\boldsymbol{S}\boldsymbol{u} = \begin{bmatrix} \dfrac{\lambda}{(\mu+1)\dot{\theta}_1 - \dot{\theta}_2} \begin{bmatrix} \mu+1 \\ -1 \end{bmatrix} \\ 0 \end{bmatrix}. \tag{40}$$

On the other hand, a kneed biped has a property of obstacle avoidance, in other words, guaranteeing the foot clearance by knee-bending. To improve the advantageous, we introduce an active knee-lock algorithm proposed in our previous work (Asano & Yamakita, 2001) in the following. The passive knee-strike occurs when $\theta_2 = \theta_3$ during the single-support phase, and its inelastic collision model is given by

$$\boldsymbol{M}(\boldsymbol{\theta})\dot{\boldsymbol{\theta}}^{+} = \boldsymbol{M}(\boldsymbol{\theta})\dot{\boldsymbol{\theta}}^{-} - \boldsymbol{J}_I^{\mathrm{T}} \lambda_I \tag{41}$$

where $\boldsymbol{J}_I = [0 \quad 1 \quad -1]^{\mathrm{T}}$ and λ_I is the Lagrange's indeterminate multiplier vector and means the impact force. We introduce an active knee-lock algorithm before the impact and mechanically lock the knee-joint at a suitable timing. Let us then consider the dissipated mechanical energy at this instant. Define the dissipated energy ΔE_{ks} as

$$\Delta E_{ks} \triangleq \frac{1}{2}(\dot{\theta}^{+})^{\mathrm{T}} M(\theta)\dot{\theta}^{+} - \frac{1}{2}(\dot{\theta}^{-})^{\mathrm{T}} M(\theta)\dot{\theta}^{-} \leq 0 \tag{42}$$

This can be rearranged by solving Eq. (41) as

$$\Delta E_{\mathrm{ks}} = -\frac{1}{2}\left(\dot{\theta}^{-}\right)^{\mathrm{T}} \boldsymbol{J}_I^{\mathrm{T}} \left(\boldsymbol{J}_I \boldsymbol{M}^{-1} \boldsymbol{J}_I^{\mathrm{T}}\right)^{-1} \boldsymbol{J}_I \dot{\theta}^{-} = -\frac{\left(\dot{\theta}_2^{-} - \dot{\theta}_3^{-}\right)^2}{2\boldsymbol{J}_I \boldsymbol{M}^{-1} \boldsymbol{J}_I^{\mathrm{T}}}. \tag{43}$$

This shows that the condition to minimize the energy dissipation is $\dot{\theta}_2^{-} = \dot{\theta}_3^{-}$, and this leads $\Delta E_{\mathrm{ks}} = 0$. In general, there exists the timing in the kneed gait. After locking-on the knee-joint, we should lock-off it and the timing should be chosen empirically following a certain trigger. In this section, we consider the trigger as $X_g = 0$ [m] where X_g is the *X*-position of the robot's center of mass. Fig. 11 shows the phase sequence of a cycle with the knee-lock algorithm, which consists of the following phases.

1. Start
2. 3-link phase I
3. Active knee-lock on
4. Virtual compass phase (2-link mode)
5. Active knee-lock off
6. 3-link phase II
7. Passive knee-strike
8. Compass phase (2-link mode)
9. Heel-strike

Fig. 12 shows the simulation results of dynamic walking by ECC where $\lambda = 5.0$ and $\mu = 4.0$. The physical parameters are chosen as Table 2. From Fig. 12 (b) and (d), it is confirmed that the passive knee-joint is suitably locked-on without energy-loss, and after that, active lock-off and passive knee-strike occur. Fig. 13 shows the stick diagram for one step. We can see that a stable dynamic bipedal gait is generated by ECC.

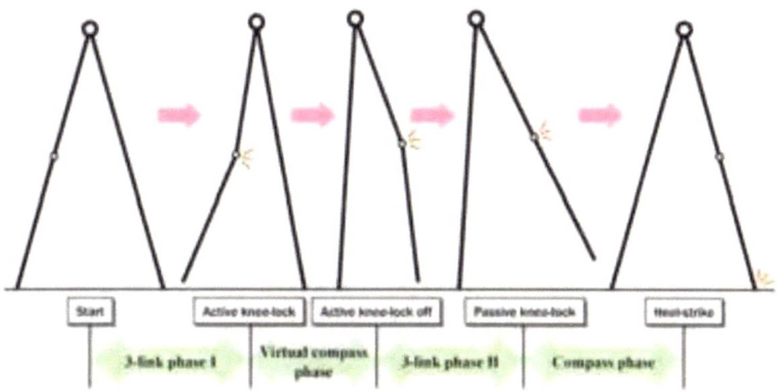

Fig. 11. Phase sequence of dynamic walking by ECC with active lock of free knee-joint.

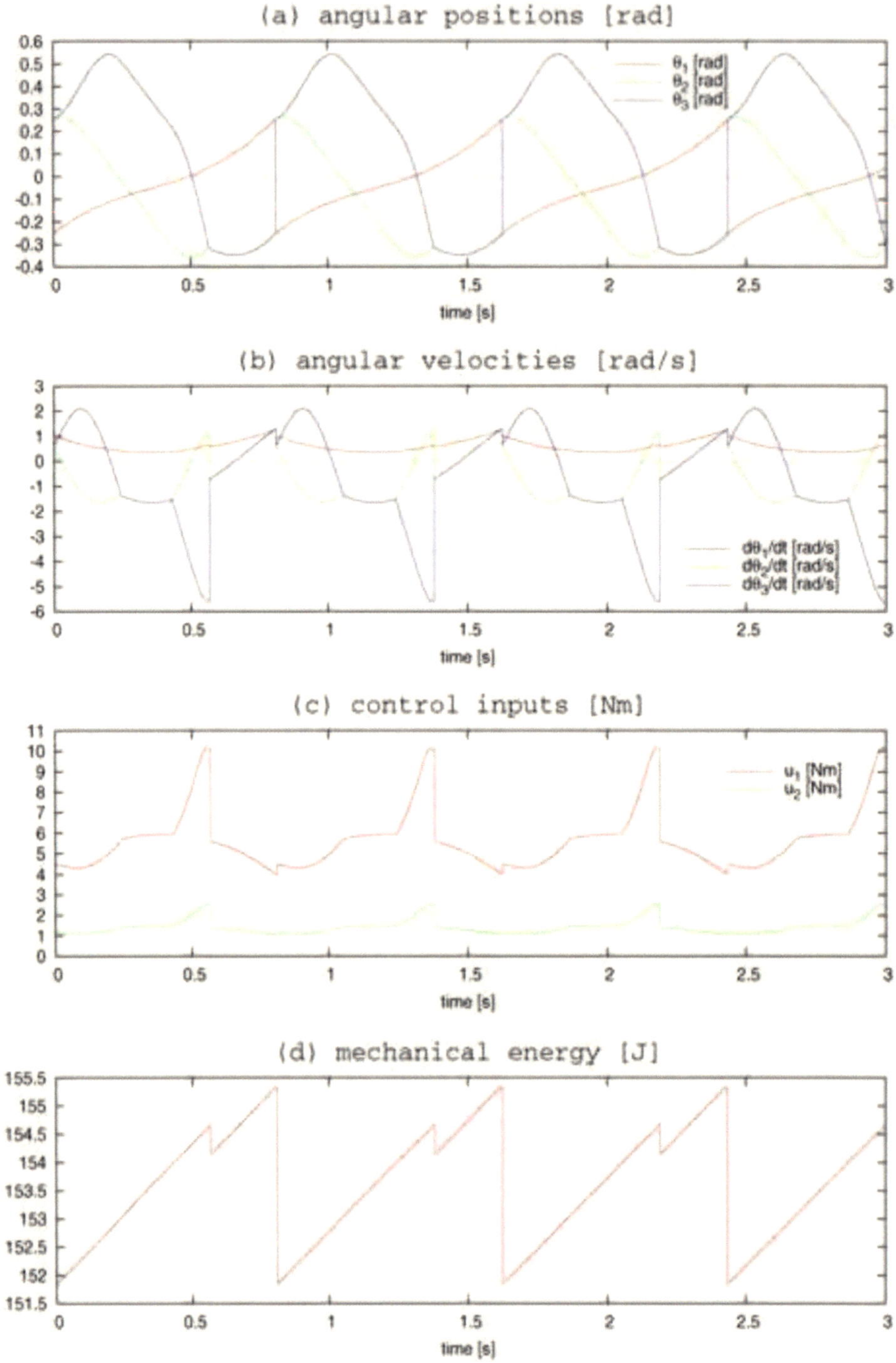

Fig. 12. Simulation results of dynamic walking of a kneed biped by ECC where $\lambda = 5.0$ [J/s] and $\mu = 4.0$.

m_1	$= m_2 + m_3$	5.0	kg
m_2		3.0	kg
m_3		2.0	kg
m_H		10.0	kg
I	$= m_2 m_3 (a_2 + b_3)^2 / m_1$	0.243	$\mathrm{kg \cdot m^2}$
a_1	$= (m_2 (l_3 + a_2) + m_3 a_3) / m_1$	0.52	m
b_1		0.48	m
a_2		0.20	m
b_2		0.30	m
a_3		0.25	m
b_3		0.25	m
l_1	$= a_1 + b_1$	1.00	m
l_2	$= a_2 + b_2$	0.50	m
l_3	$= a_3 + b_3$	0.50	m

Table 2. Parameters of the planar kneed biped.

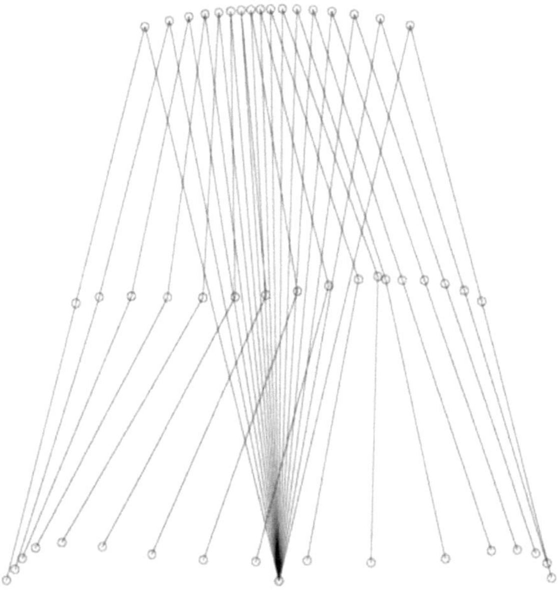

Fig. 16. Stick diagram of dynamic walking with free knee-joint by ECC

7. Conclusions and Future Work

In this chapter, we have proposed a simple dynamic gait generation method imitating the property of passive dynamic walking. The control design technique used in this study was shown to be effective to generate a stable dynamic gait, and numerical simulations and experiments have proved its validity.

The authors believe that an energy restoration is the most essential necessary condition of dynamic walking and its concept is worth to be taken into consideration to generate a natural and energy-efficient gait. In the future, extensions of our method to high-dof humanoid robots should be investigated.

8. References

Asano, F. & Yamakita, M. (2001). Virtual gravity and coupling control for robotic gait synthesis, *IEEE Trans. on Systems, Man and Cybernetics Part A*, Vol. 31, No. 6, pp. 737-745, Nov. 2001.

Goswami, A.; Thuilot, B. & Espiau, B. (1996). Compass-like biped robot Part I: Stability and bifurcations of passive gaits, Research Report INRIA 2613, 1996.

Goswami, A.; Espiau, B. & Keramane, A. (1997). Limit cycles in a passive compass gait biped and passivity-mimicking control laws, *Autonomous Robots*, Vol. 4, No. 3, pp. 273-286, Sept. 1997.

Koga, M. (2000). Numerical Computation with MaTX (In Japanese), Tokyo Denki Univ. Press, ISBN4-501-53110-X, 2000.

McGeer, T. (1990). Passive dynamic walking, *Int. J. of Robotics Research*, Vol. 9, No. 2, pp. 62-82, April 1990.

Vukobuatović, M. & Stepanenko, Y. (1972). On the stability of anthropomorphic systems, Mathematical Biosciences, Vol. 15, pp. 1-37, 1972.

6

Dynamic Simulation of Single and Combined Trajectory Path Generation and Control of A Seven Link Biped Robot

Ahmad Bagheri
Peiman Naserredin Musavi
Guilan University
Iran
bagheri@guilan.ac.ir

1. Introduction

Recently, numerous collaborations have been focused on biped robot walking pattern to trace the desired paths and perform the required tasks. In the current chapter, it has been focused on mathematical simulation of a seven link biped robot for two kinds of zero moment points (ZMP) including the Fixed and the Moving ZMP. In this method after determination of the breakpoints of the robot and with the aid of fitting a polynomial over the breakpoints, the trajectory paths of the robot will be generated and calculated. After calculation of the trajectory paths of the robot, the kinematic and dynamic parameters of the robot in Matlab environment and with respect to powerful mathematical functions of Matlab, will be obtained. The simulation process of the robot is included in the control process of the system. The control process contains Adaptive Method for known systems. The detailed relations and definitions can be found in the authors' published article [Musavi and Bagheri, 2007]. The simulation process will help to analyze the effects of drastic parameters of the robot over stability and optimum generation of the joint's driver actuator torques.

2. Kinematic of the robot

The kinematic of a seven link biped robot needs generation of trajectory paths of the robot with respect to certain times and locations in relevant with the assumed fixed coordinate system. In similarity of human and robot walking pattern, the process of path trajectory generation refers to determination of gait breakpoints. The breakpoints are determined and calculated with respect to system identity and conditions.

Afterward and in order to obtain comprehensive concept of the robot walking process, the following parameters and definitions will be used into the simulation process:

- Single Support phase: The robot is supported by one leg and the other is suspended in air
- Double support phase: The robot is supported by the both of its legs and the legs are in contact with the ground simultaneously

: Total traveling time including single and double support phase times. T_c -

T_c : Double support phase time which is regarded as 20% of T_d -

- T_m : The time which ankle joint has reached to its maximum height during walking cycle.
: Step number k -
Ankle joint maximum height: H_{ao} -
- L_{ao} : The horizontal traveled distance between ankle joint and start point when the ankle joint has reached to its maximum height.
Step length: D_s -
: Foot lift angle and contact angle with the level ground q_b, q_f -
- λ : Surface slope
- h_s : Stair level height-
- H_{st} : Foot maximum height from stair level
- x_{ed} : The horizontal distance between hip joint and the support foot (Fixed coordinate system) at the start of double support phase time.
- x_{sd} : The horizontal distance between hip joint and the support foot (Fixed coordinate system) at the end of double support phase time.
- F. C. S: The fixed coordinate system which would be supposed on support foot in each step.
-M.C : The mass centers of the links
- Saggital plane: The plane that divides the body into right and left sections.
- Frontal plane: The plane parallel to the long axis of the body and perpendicular to the saggital plane.

The saggital and frontal planes of the human body are shown in figure (1.1) where the transverse plane schematic and definition have been neglected due to out of range of our calculation domain.

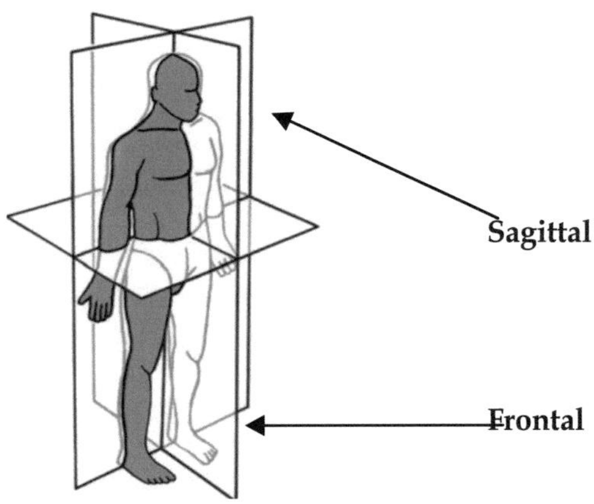

Fig. (1.1). The body configuration with respect to various surfaces.

The main role for the optimum trajectory path generation must be imagined upon hip and ankle joints of the robot. On the other hand, with creating smooth paths of the joints and with the aid of the breakpoints, the robot can move softly with its optimum movement parameters such as minimum actuator torques of joints (Shank) including integrity of the joints kinematic parameters.

The important parameters of the robot can be assumed as the listed above and are shown in figures (1.2) and (1.3).

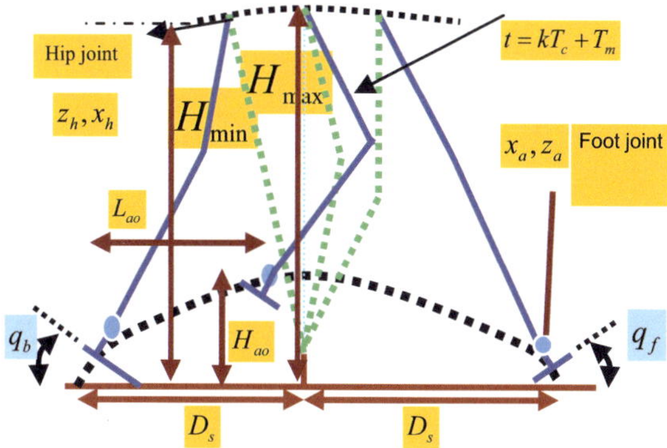

Fig. (1.2). The robot important parameters for calculation of trajectory path of a seven link biped robot.

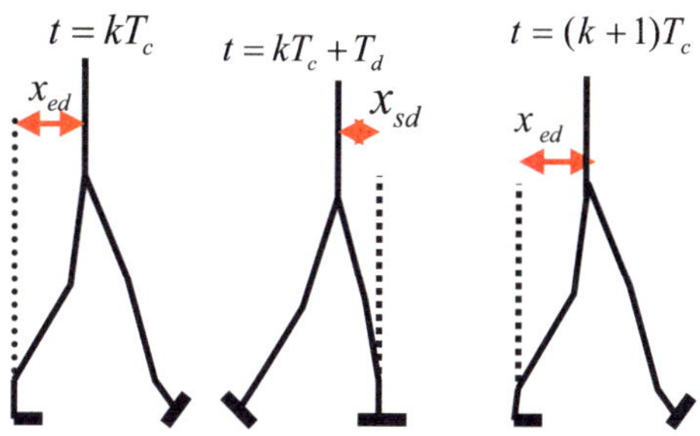

Fig. (1.3). The variables of hip: $x_{ed,}\ x_{sd}$.

With respect to saggital investigation of the robot, the most affecting parameters of the mentioned joints can be summarized as below:

1) **Hip joint**
2) **Ankle joint**

Obviously, the kinematic and dynamic attitude of shank joint will be under influence of the both of above mentioned joints. As can be seen from figure (1.2), the horizontal and vertical components of the joints play a great role in trajectory paths generation. This means that timing-process and location of the joints with respect to the fixed coordinate system which would be supposed on the support foot have considerable effects on the smooth paths and subsequently over stability of the robot. Regarding the above expressions and conditions, the vertical and horizontal components of the joints can be categorized and calculated as the following procedure. With respect to the conditions of the surfaces and figure (1.2) and (1.3), the components are classified as below:

2.1) Foot angle on horizontal surface or stair

$$\theta_a(t) = \begin{cases} -q_{gs} & t = kT_c \\ -q_b & t = kT_c + T_d \\ q_f & t = (k+1)T_c \\ q_{gf} & t = (k+1)T_c + T_d \end{cases} \tag{1}$$

2.2) Foot angle on declined surfaces

$$\theta_a(t) = \begin{cases} (\lambda + q_{gs}) & t = kT_c \\ \lambda - q_b & t = kT_c + T_d \\ \lambda + q_f & t = (k+1)T_c \\ \lambda + q_{gf} & t = (k+1)T_c + T_d \end{cases} \quad (2)$$

2.3) The displacements of Horizontal and Vertical Foot Traveling Over Horizontal Surface or Stair

With respect to figures (1.2) and (1.3), the horizontal and vertical components of ankle joint can be shown as below:

$$x_{ahor}(t) = \begin{cases} kD_s & t = kT_c \\ kD_s + l_{an}\sin q_b + \cdots & t = kT_c + T_d \\ l_{af}(1-\cos q_b) & \\ kD_s + L_{ao} & t = kT_c + T_m \\ (k+2)D_s - l_{an}\sin q_f - \cdots & t = (k+1)T_c \\ l_{ab}(1-\cos q_f) & \\ (k+2)D_s & t = (k+1)T_c + T_d \end{cases} \quad (3)$$

$$z_{ahor}(t) = \begin{cases} h_{gs} + l_{an} & t = kT_c \\ h_{gs} + l_{af}\sin q_b + l_{an}\cos q_b & t = kT_c + T_d \\ H_{ao} & t = kT_c + T_m \\ h_{ge} + l_{ab}\sin q_f + l_{an}\cos q_f & t = (k+1)T_c \\ h_{ge} + l_{an} & t = (k+1)T_c + T_d \end{cases} \quad (4)$$

$$z_{stair}(t) = \begin{cases} (k-1)h_{st} + l_{an} & t = kT_c \\ (k-1)h_{st} + l_{af}\sin q_b + \cdots & t = kT_c + T_d \\ l_{an}\cos q_b & \\ kh_{st} + H_{st} & t = kT_c + T_m \\ (k+1)h_{st} + l_{ab}\sin q_f + \cdots & t = (k+1)T_c \\ l_{an}\cos q_f & \\ (k+1)h_{st} + l_{an} & t = (k+1)T_c + T_d \end{cases} \quad (5)$$

2.4) The displacements of Horizontal and Vertical Foot Traveling Over declined Surface

$$x_{a,dec}(t) = \begin{cases} kD_s\cos\lambda - l_{an}\sin\lambda & t = kT_c \\ (kD_s + l_{af})\cos\lambda + \cdots & t = kT_c + T_d \\ l_{an}\sin(q_b - \lambda) - l_{af}\cos(q_b - \lambda) & \\ (kD_s + L_{ao})\cos\lambda & t = kT_c + T_m \\ ((k+2)D_s - l_{ab})\cos\lambda - \cdots & t = (k+1)T_c \\ l_{an}\sin(q_f + \lambda) + l_{ab}\cos(q_f + \lambda) & \\ (k+2)D_s\cos\lambda - l_{an}\sin\lambda & t = (k+1)T_c + T_d \end{cases} \quad (6)$$

$$z_{a,dec}(t)=\begin{cases} kD_s\sin\lambda+l_{an}\cos\lambda & t=kT_c \\ (kD_s+l_{af})\sin\lambda+\cdots & t=kT_c+T_d \\ +l_{an}\cos(q_b-\lambda)+l_{af}\sin(q_b-\lambda) & \\ (kD_s+L_{ao})\sin\lambda+H_{ao}\cos\lambda & t=kT_c+T_m \\ ((k+2)D_s-l_{ab})\sin\lambda+\cdots & t=(k+1)T_c \\ l_{an}\sin(q_f+\lambda)+l_{ab}\cos(\pi/2-(q_f+\lambda)) & \\ (k+2)D_s\sin\lambda+l_{an}\cos\lambda & t=(k+1)T_c+T_d \end{cases} \tag{7}$$

Assuming the above expressed breakpoints and also applying the following boundary condition of the robot during walking cycle, generation of the ankle joint trajectory path can be performed. The boundary conditions of the definite system are determined with respect to physical and geometrical specifications during movement of the system. As can be seen from figure (1.2) and (1.3), the linear and angular velocity of foot at the start and the end of double support phase equal to zero:

$$\begin{cases} \dot\theta_a(kT_c)=0 \\ \dot\theta_a((k+1)T_c+T_d)=0 \end{cases} \quad \begin{cases} \dot x_a(kT_c)=0 \\ \dot x_a((k+1)T_c+T_d)=0 \end{cases} \quad \begin{cases} \dot z_a(kT_c)=0 \\ \dot z_a((k+1)T_c+T_d)=0 \end{cases} \tag{8}$$

The best method for generation of path trajectories refers to mathematical interpolation. There are several cases for obtaining the paths with respect to various conditions of the movement such as number of breakpoints and boundary conditions of the system. Regarding the mentioned conditions of a seven link biped robot, Spline and Vandermonde Matrix methods seem more suitable than the other cases of interpolation process. The Vandermonde case is the simplest method with respect to calculation process while it will include calculation errors with increment of breakpoint numbers. The stated defect will not emerge on Spline method and it will fit the optimum curve over the breakpoints regardless the number of points and boundary conditions. With respect to low number of domain breakpoints and boundary conditions of a seven link biped robot, there are no considerable differences in calculation process of Vandermonde and Spline methods. For an example, with choosing one of the stated methods and for relations (7) and (8), a sixth-order polynomial or third-order spline can be fitted for generation of the vertical movement of ankle joint.

2.5) Hip Trajectory Interpolation for the Level Ground [Huang and et. Al, 2001] and Declined Surfaces

From figures (1.2) and (1.3), the vertical and horizontal displacements of hip joint can be written as below:

$$x_{h,Hor,\,stair}=\begin{cases} kD_s+x_{ed} & t=kT_c \\ (k+1)D_s-x_{sd} & t=kT_c+T_d \\ (k+1)D_s+x_{ed} & t=(k+1)T_c \end{cases} \tag{9}$$

$$x_{h,Dec} = \begin{cases} (kD_s + x_{ed})\cos\lambda & t = kT_c \\ ((k+1)D_s - x_{sd})\cos\lambda & t = kT_c + T_d \\ ((k+1)D_s + x_{ed})\cos\lambda & t = (k+1)T_c \end{cases} \tag{10}$$

$$z_{h,Hor.} = \begin{cases} H_{h\min} & t = kT_c \\ H_{h\max} & t = kT_c + .5(T_c - T_d) \\ H_{h\min} & t = (k+1)T_c \end{cases} \tag{11}$$

$$z_{h,Dec.} \quad \begin{matrix} H_{h\min}\cos\lambda \quad \cdots, t \quad kT_c \\ (kD_s \quad x_{ed})\sin\lambda \\ H_{h\max}\cos\lambda \quad \cdots, t \quad kT_c \quad .5(T_c \quad T_d) \\ (kD_s \quad x_{sd})\sin\lambda \\ H_{h\min}\cos\lambda \quad \cdots, \ t \quad (k \quad 1)T_c \\ ((k \quad 1)D_s \quad x_{ed})\sin\lambda, \end{matrix} \tag{12}$$

$$z_{stair} = \begin{cases} (k-1)h_s + H_{h\min} & t = kT_c \\ kh_s + H_{h\max} & t = kT_c + .5(T_c - T_d) \\ kh_s + H_{h\min} & t = (k+1)T_c \end{cases} \tag{13}$$

Where, in the above expressed relations, $H_{\min}$ and $H_{\max}$ indicate the minimum and maximum height of hip joint from the fixed coordinate system. Obviously and with respect to figure (1.2), the ankle and hip joint parameters including x_a, z_a and x_h, z_h play main role in optimum generation of the trajectory paths of the robot. With utilization of relations (1)-(13) and using the mathematical interpolation process, the trajectory paths of the robot will be completed.

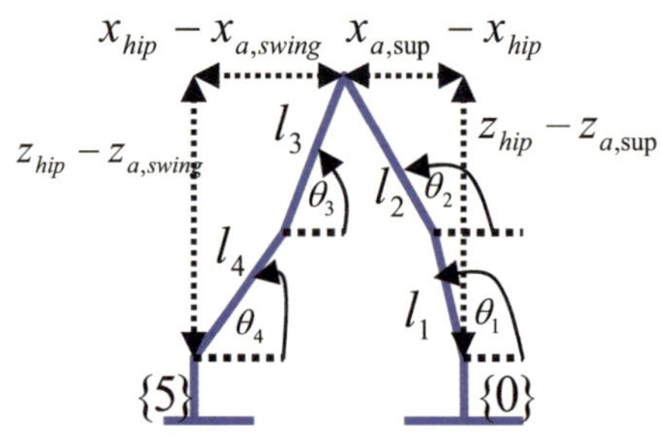

Fig. (1.4). The link's angles and configurations.

Regarding figure (1.4) and the trajectory paths generation of the robot based on four important parameters of the system (x_a, z_a and x_h, z_h), the first kinematic parameter of the robot can be obtained easily. On the other hand, with utilization of inverse kinematics, the link's angle of the robot will be calculated with respect to the domain nonlinear mathematical equations. As can be seen from figure (1.4), the equations can be written as below:

$$l_1 \cos(\pi-\theta_1)+l_2 \cos(\pi-\theta_2)=a$$
$$l_1 \sin(\pi-\theta_1)+l_2 \sin(\pi-\theta_2)=b \quad (14)$$

$$l_3 \cos(\theta_3)+l_4 \cos(\theta_4)=c \quad (15)$$
$$l_3 \sin(\theta_3)+l_4 \sin(\theta_4)=d$$

Where,

$a = x_{a,Sup} - x_{hip}$

$b = z_{hip} - z_{a,Sup}$

$c = x_{hip} - x_{a,swing}$

$d = z_{hip} - z_{a,swing}$

The all of conditions and the needed factors for solving of the relations (14) and (15) have been provided. The right hand of relations (14) and (15) are calculated from the interpolation process. For the stated purpose and with beginning to design program in MALAB environment and with utilization of strong commands such as **fsolve**, the angles of the links are calculated numerically. In follow and using kinematic chain of the robot links, the angular velocity and acceleration of the links and subsequently the linear velocities and accelerations are obtained. With respect to figure (1.5) and assuming the unit vectors parallel to the link's axis and then calculation of the link's position vectors relative to the assumed F.C.S, the following relations are obtained:

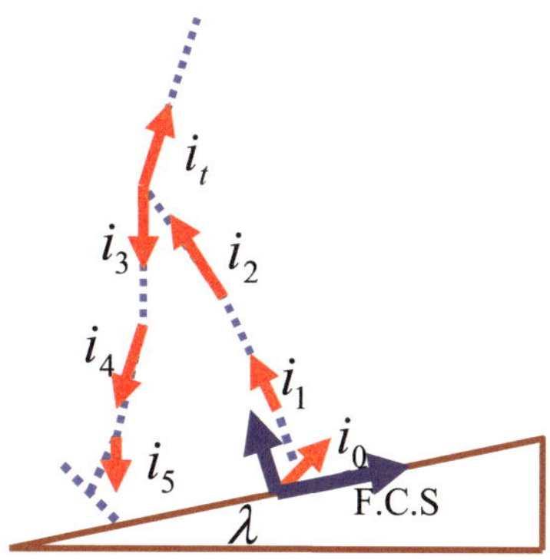

Fig. (1.5). The assumed unit vectors to obtain the position vectors.

$$\vec{r}_0 = l_{m.c}(\cos(\beta_{m.c}+q_f)I+\sin(\beta_{m.c}+q_f)K) \quad (16)$$

$$\vec{r}_1 = (l_0 \cos(\beta_{tot}+q_f)+l_{c1}\cos(\theta_1-\lambda))I + (l_{c1}\sin(\theta_1-\lambda)+l_0\sin(\beta_{tot}+q_f))K \quad (17)$$

$$\vec{r}_2 = (l_0 \cos(\beta_{tot}+q_f)+l_1\cos(\theta_1-\lambda) - l_{c2}\cos(\pi-\theta_2+\lambda))I + (l_1\sin(\theta_1-\lambda) + l_0\sin(\beta_{tot}+q_f)+l_{c2}\sin(\pi-\theta_2+\lambda))K \quad (18)$$

$$\begin{aligned}\vec{r}_3 &= (l_0\cos(\beta_{tot}+q_f)+l_1\cos(\theta_1-\lambda)\\ &-l_2\cos(\pi-\theta_2+\lambda)-l_{c3}\cos(\theta_3-\lambda))I\\ &+(l_1\sin(\theta_1-\lambda)+l_0\sin(\beta_{tot}+q_f)\\ &+l_2\sin(\pi-\theta_2+\lambda)-l_{c3}\sin(\theta_3-\lambda))K\end{aligned} \tag{19}$$

$$\begin{aligned}\vec{r}_4 &= (l_0\cos(\beta_{tot}+q_f)+l_1\cos(\theta_1-\lambda)\\ &-l_2\cos(\pi-\theta_2+\lambda)-l_3\cos(\theta_3-\lambda)\\ &-l_{c4}\cos(\theta_4-\lambda))I+(l_1\sin(\theta_1-\lambda)\\ &+l_0\sin(\beta_{tot}+q_f)+l_2\sin(\pi-\theta_2+\lambda)\\ &-l_3\sin(\theta_3-\lambda)-l_{c4}\sin(\theta_4-\lambda))K\end{aligned} \tag{20}$$

$$\begin{aligned}\vec{r}_5 &= (l_0\cos(\beta_{tot}+q_f)+l_1\cos(\theta_1-\lambda)\\ &-l_2\cos(\pi-\theta_2+\lambda)-l_3\cos(\theta_3-\lambda)\\ &-l_4\cos(\theta_4-\lambda)-l_{5m.c}\cos(\pi/2-\lambda+\beta_{s,m.s}-q_b))I\\ &+(l_1\sin(\theta_1-\lambda)+l_0\sin(\beta_{tot}+q_f)\\ &+l_2\sin(\pi-\theta_2+\lambda)-l_3\sin(\theta_3-\lambda)\\ &-l_4\sin(\theta_4-\lambda)-l_{5m.c}\sin(\pi/2-\lambda+\beta_{s,m.s}-q_b)))K\end{aligned} \tag{21}$$

$$\begin{aligned}\vec{r}_{tor} &= (l_0\cos(\beta_{tot}+q_f)+l_1\cos(\theta_1-\lambda)\\ &-l_2\cos(\pi-\theta_2+\lambda)+l_{tor}\cos(\pi/2-\theta_{tor}-\lambda))I\\ &+(l_1\sin(\theta_1-\lambda)+l_0\sin(\beta_{tot}+q_f)\\ &+l_2\sin(\pi-\theta_2+\lambda)+l_{tor}\sin(\pi/2-\theta_{tor}-\lambda))K\end{aligned} \tag{22}$$

As can be seen from relations (16)-(22), the all of position vectors have been calculated with respect to F.C.S for inserting into ZMP formula. The ZMP concept will be discussed in the next sub-section. Now, with the aid of first and second differentiating of relation (16)-(22), the linear velocities and accelerations of the link's mass centers can be calculated within relations (23)-(29).

$$\vec{v}_0 = l_{m.c}\vec{\omega}_0(-\sin(\beta_{m.c}+q_f)I+\cos(\beta_{m.c}+q_f)K) \tag{23}$$

$$\begin{aligned}\vec{v}_1 &= (-l_0\vec{\omega}_0\sin(\beta_{tot}+q_f)-l_{c1}\vec{\omega}_1\sin(\theta_1-\lambda))I\\ &+(l_{c1}\vec{\omega}_1\cos(\theta_1-\lambda)+l_0\cos(\beta_{tot}+q_f))K\end{aligned} \tag{24}$$

$$\begin{aligned}\vec{v}_2 &= (-l_0\vec{\omega}_0\sin(\beta_{tot}+q_f)-l_1\vec{\omega}_1\sin(\theta_1-\lambda)\\ &-l_{c2}\vec{\omega}_2\sin(\pi-\theta_2+\lambda))I+(l_1\vec{\omega}_1\cos(\theta_1-\lambda)\\ &+l_0\vec{\omega}_0\cos(\beta_{tot}+q_f)-l_{c2}\vec{\omega}_2\cos(\pi-\theta_2+\lambda))K\end{aligned} \tag{25}$$

$$\begin{aligned}\vec{v}_3 &= (-l_0\vec{\omega}_0\sin(\beta_{tot}+q_f)-l_1\vec{\omega}_1\sin(\theta_1-\lambda)\\ &-l_2\vec{\omega}_2\sin(\pi-\theta_2+\lambda)+l_{c3}\vec{\omega}_3\sin(\theta_3-\lambda))I\\ &+(l_1\vec{\omega}_1\cos(\theta_1-\lambda)+l_0\vec{\omega}_0\cos(\beta_{tot}+q_f)\\ &-l_2\vec{\omega}_2\cos(\pi-\theta_2+\lambda)-l_{c3}\vec{\omega}_3\cos(\theta_3-\lambda))K\end{aligned} \tag{26}$$

$$\begin{aligned}\vec{v}_4 &= (-l_0\vec{\omega}_0 \sin(\beta_{tot} + q_f) - l_1\vec{\omega}_1 \sin(\theta_1 - \lambda) \\ &- l_2\vec{\omega}_2 \sin(\pi - \theta_2 + \lambda) + l_3\vec{\omega}_3 \sin(\theta_3 - \lambda) \\ &+ l_{c4}\vec{\omega}_4 \sin(\theta_4 - \lambda))I + (l_1\vec{\omega}_1 \cos(\theta_1 - \lambda) \\ &+ l_0\vec{\omega}_0 \cos(\beta_{tot} + q_f) - l_2\vec{\omega}_2 \cos(\pi - \theta_2 + \lambda) - \\ &l_3\vec{\omega}_3 \cos(\theta_3 - \lambda) - l_{c4}\vec{\omega}_4 \cos(\theta_4 - \lambda))K\end{aligned} \tag{27}$$

$$\begin{aligned}\vec{v}_5 &= (-l_0\vec{\omega}_0 \sin(\beta_{tot} + q_f) - l_1\vec{\omega}_1 \sin(\theta_1 - \lambda) \\ &- l_2\vec{\omega}_2 \sin(\pi - \theta_2 + \lambda) + l_3\vec{\omega}_3 \sin(\theta_3 - \lambda) \\ &+ l_4\vec{\omega}_4 \sin(\theta_4 - \lambda) - l_{5m.c}\vec{\omega}_5 \sin(\pi/2 - \lambda + \beta_{s,m.c} - q_b))I \\ &+ (l_1\vec{\omega}_1 \cos(\theta_1 - \lambda) + l_0\vec{\omega}_0 \cos(\beta_{tot} + q_f) \\ &- l_2\vec{\omega}_2 \cos(\pi - \theta_2 + \lambda) - l_3\vec{\omega}_3 \cos(\theta_3 - \lambda) \\ &- l_4\vec{\omega}_4 \cos(\theta_4 - \lambda) + l_{5m.c}\vec{\omega}_5 \cos(\pi/2 - \lambda + \beta_{s,m.c} - q_b))K\end{aligned} \tag{28}$$

$$\begin{aligned}\vec{v}_{tor} &= (-l_0\vec{\omega}_0 \sin(\beta_{tot} + q_f) - l_1\vec{\omega}_1 \sin(\theta_1 - \lambda) - \\ &l_2\vec{\omega}_2 \sin(\pi - \theta_2 + \lambda) + l_{tor}\vec{\omega}_{tor} \sin(\pi/2 - \theta_{tor} - \lambda))I \\ &+ (\vec{\omega}_1 l_1 \cos(\theta_1 - \lambda) + l_0\vec{\omega}_0 \cos(\beta_{tot} + q_f) \\ &- l_2\vec{\omega}_2 \cos(\pi - \theta_2 + \lambda) - l_{tor}\vec{\omega}_{tor} \cos(\pi/2 - \theta_{tor} - \lambda))K\end{aligned} \tag{29}$$

Accordingly, the linear acceleration of the links can be calculated easily. After generation of the robot trajectory paths with the aid of interpolation process and with utilization of MATLAB commands, the simulation of the biped robot can be performed. Based on the all above expressed relations and the resulted parameters and subsequently with inserting the parameters into the program, the simulation of the robot are presented in simulation results.

3. Dynamic of the robot

In similarity of human and the biped robots, the most important parameter of stability of the robot refers to ZMP. The ZMP (Zero moment point) is a point on the ground whose sum of all moments around this point is equal to zero. Totally, the ZMP mathematical formulation can be presented as below:

$$x_{zmp} = \frac{\sum_{i=1}^{n} m_i(g\cos\lambda + \ddot{z}_i)x_i - \sum_{i=1}^{n} m_i(g\sin\lambda + \ddot{x}_i)z_i - \sum_{i=1}^{n} {}^{i}I_i\ddot{\theta}_i}{\sum_{i=1}^{n} m_i(g\cos\lambda + \ddot{z}_i)} \tag{30}$$

Where, $\ddot{x}_i$ and $\ddot{z}_i$ are horizontal and vertical acceleration of the link's mass center with respect to F.C.S where $\ddot{\theta}_i$ is the angular acceleration of the links calculated from the interpolation process. On the other hand, the stability of the robot is determined according to attitude of ZMP. This means that if the ZMP be within the convex hull of the robot, the stable movement of the robot will be obtained and there are no interruptions in kinematic parameters (Velocity of the links). The convex hull can be imagined as a projection of a pyramid with its heads on support and swing foots and also on the hip joint. Generally, the ZMP can be classified as the following cases:

1) **Moving ZMP**
2) **Fixed ZMP**

The moving type of the robot walking is similar to human gait. In the fixed type, the ZMP position is restricted through the support feet or the user's selected areas. Consequently, the significant torso's modified motion is required for stable walking of the robot. For the explained process, the program has been designed to find target angle of the torso for providing the fixed ZMP position automatically. In the designed program, q_{torso} shows the deflection angle of the torso determined by the user or calculated by auto detector mood of the program. Note, in the mood of auto detector, the torso needed motion for obtaining the mentioned fixed ZMP will be extracted with respect to the desired ranges. The desired ranges include the defined support feet area by the users or automatically by the designed program. Note, the most affecting parameters for obtaining the robot's stable walking are the hip's height and position. By varying the parameters with iterative method for x_{ed}, x_{sd} [Huang and et. Al, 2001] and choosing the optimum hip height, the robot control process with respect to the torso's modified angles and the mentioned parameters can be performed. To obtain the joint's actuator torques, the Lagrangian relation [Kraige, 1989] has been used at the single support phase as below:

$$\tau_i = H(q)\ddot{q} + C(q,\dot{q})\dot{q} + G(q_i) \tag{31}$$

where, $i = 0, 2, \cdots 6$ and H, C, G are mass inertia, coriolis and gravitational matrices of the system which can be written as following:

$$H(q)=\begin{bmatrix} h_{11} & h_{12} & h_{13} & h_{14} & h_{15} & h_{16} & h_{17} \\ h_{21} & h_{22} & h_{23} & h_{24} & h_{25} & h_{26} & h_{27} \\ h_{31} & h_{32} & h_{33} & h_{34} & h_{35} & h_{36} & h_{37} \\ h_{41} & h_{42} & h_{43} & h_{44} & h_{45} & h_{46} & h_{47} \\ h_{51} & h_{52} & h_{53} & h_{54} & h_{55} & h_{56} & h_{57} \\ h_{61} & h_{62} & h_{63} & h_{64} & h_{65} & h_{66} & h_{67} \end{bmatrix} \quad C(q,\dot{q})=\begin{bmatrix} c_{11} & c_{12} & c_{13} & c_{14} & c_{15} & c_{16} & c_{17} \\ c_{21} & c_{22} & c_{23} & c_{24} & c_{25} & c_{26} & c_{27} \\ c_{31} & c_{32} & c_{33} & c_{34} & c_{35} & c_{36} & c_{37} \\ c_{41} & c_{42} & c_{43} & c_{44} & c_{45} & c_{46} & c_{47} \\ c_{51} & c_{52} & c_{53} & c_{54} & c_{55} & c_{56} & c_{57} \\ c_{61} & c_{62} & c_{63} & c_{64} & c_{65} & c_{66} & c_{67} \end{bmatrix} \quad G(q)=\begin{bmatrix} G_1 \\ G_2 \\ G_3 \\ G_4 \\ G_5 \\ G_{tor} \end{bmatrix}$$

Obviously, the above expressed matrices show the double support phase of the movement of the robot where they are used for the single support phase of the movement. On the other hand, the relation (31) is used for the single support phase of the robot. Within the double support phase of the robot, due to the occurrence impact between the swing leg and the ground, the modified shape of relation (31) is used with respect to effects of the reaction forces of the ground [Lum and et. Al. 1999 and Westervelt, 2003, and Hon and et. Al., 1978]. For the explained process and in order to obtain the single support phase equations of the robot, the value of q_0 (as can be seen in figure (1.4)) must be put equal to zero. The calculation process of the above mentioned matrices components contain bulk mathematical relations. Here, for avoiding the aforesaid relations, just the simplified relations are presented:

$$
\begin{aligned}
& h_{11} = [m_1(l_{c1}^2 + l_{c1} l_e \cos(q_1 - \varphi))] + [m_2(l_1^2 + l_{c2}^2 + l_1 l_e \cos(q_1 - \varphi) + l_{c2} l_e \cos(q_2 - \varphi) + \cdots \\
& 2 l_1 l_{c2} \cos(q_2 - q_1))] + [m_3(l_1^2 + l_2^2 + l_{c3}^2 + l_1 l_e \cos(q_1 - \varphi) + l_2 l_e \cos(q_2 - \varphi) - l_{c3} l_e \cos(-q_3 + \varphi) + \cdots \\
& 2 l_1 l_2 \cos(q_2 - q_1) - 2 l_1 l_{c3} \cos(q_1 - q_3) - 2 l_2 l_{c3} \cos(q_2 - q_3))] + [m_4(l_1^2 + l_2^2 + l_3^2 + l_{c4}^2 + \cdots \\
& l_1 l_e \cos(q_1 - \varphi) + l_2 l_e \cos(q_2 - \varphi) - l_3 l_e \cos(q_3 - \varphi) - l_{c4} l_e \cos(-q_4 + \varphi) + 2 l_1 l_2 \cos(q_2 - q_1) - \cdots \\
& 2 l_1 l_3 \cos(q_3 - q_1) - 2 l_1 l_{c4} \cos(q_1 - q_4) - 2 l_2 l_3 \cos(q_3 - q_2) - 2 l_2 l_{c4} \cos(q_2 - q_4) + 2 l_{c4} l_3 \cos(q_3 - q_4))] + \\
& [m_5(l_1^2 + l_2^2 + l_3^2 + l_4^2 + l_{cfswing}^2 + l_1 l_e \cos(q_1 - \varphi) + l_2 l_e \cos(q_2 - \varphi) - l_3 l_e \cos(q_3 - \varphi) - \cdots \\
& l_4 l_e \cos(q_4 - \varphi) - l_{cfswing} l_e \cos(\varphi - (\pi/2) + q_{fswing} - \beta_{fswing}) + 2 l_1 l_2 \cos(q_2 - q_1) - 2 l_1 l_3 \cos(q_3 - q_1) - \cdots \\
& 2 l_1 l_4 \cos(q_4 - q_1) - 2 l_1 l_{cfswing} \cos(q_1 - (\pi/2) + q_{fswing} - \beta_{fswing}) - 2 l_2 l_3 \cos(q_3 - q_2) - \cdots \\
& 2 l_2 l_4 \cos(q_4 - q_2) - 2 l_2 l_{cfswing} \cos(q_2 - (\pi/2) + q_{fswing} - \beta_{fswing}) + 2 l_3 l_4 \cos(q_4 - q_3) + \cdots \\
& 2 l_4 l_{cfswing} \cos(q_4 - (\pi/2) + q_{fswing} - \beta_{fswing}) + 2 l_3 l_{cfswing} \cos(q_3 - (\pi/2) + q_{fswing} - \beta_{fswing}))] + \cdots \\
& [m_{tor}(l_1^2 + l_2^2 + l_{ctorso}^2 + l_1 l_e \cos(q_1 - \varphi) + l_2 l_e \cos(q_2 - \varphi) + l_e l_{ctorso} \cos(-q_{torso} - \varphi - (\pi/2)) + \cdots \\
& 2 l_1 l_2 \cos(q_2 - q_1) + 2 l_1 l_{ctorso} \cos(-q_{torso} - q_1 - (\pi/2)) + 2 l_2 l_{ctorso} \cos(q_{torso} + q_2 + (\pi/2))] + I_1 + I_2 + \cdots \\
& I_3 + I_4 + I_5 + I_{torso}
\end{aligned}
$$

$$
\begin{aligned}
& h_{12} = [m_1(l_{c1}^2)] + [m_2(l_1^2 + l_{c2}^2 + 2 l_1 l_{c2} \cos(q_2 - q_1))] + [m_3(l_1^2 + l_2^2 + l_{c3}^2 + 2 l_1 l_2 \cos(q_2 - q_1) - \cdots \\
& 2 l_1 l_{c3} \cos(q_1 - q_3) - 2 l_2 l_{c3} \cos(q_2 - q_3))] + [m_4(l_1^2 + l_2^2 + l_3^2 + l_{c4}^2 + 2 l_1 l_2 \cos(q_2 - q_1) - \cdots \\
& 2 l_1 l_3 \cos(q_3 - q_1) - 2 l_1 l_{c4} \cos(q_1 - q_4) - 2 l_2 l_3 \cos(q_3 - q_2) - 2 l_2 l_{c4} \cos(q_2 - q_4) + 2 l_{c4} l_3 \cos(q_3 - q_4))] + \\
& [m_5(l_1^2 + l_2^2 + l_3^2 + l_4^2 + l_{cfswing}^2 + 2 l_1 l_2 \cos(q_2 - q_1) - 2 l_1 l_3 \cos(q_3 - q_1) - 2 l_1 l_4 \cos(q_4 - q_1) - \cdots \\
& 2 l_1 l_{cfswing} \cos(q_1 - (\pi/2) + q_{fswing} - \beta_{fswing}) - 2 l_2 l_3 \cos(q_3 - q_2) - 2 l_2 l_4 \cos(q_4 - q_2) - \cdots \\
& 2 l_2 l_{cfswing} \cos(q_2 - (\pi/2) + q_{fswing} - \beta_{fswing}) + 2 l_3 l_4 \cos(q_4 - q_3) + \cdots \\
& 2 l_4 l_{cfswing} \cos(q_4 - (\pi/2) + q_{fswing} - \beta_{fswing}) + 2 l_3 l_{cfswing} \cos(q_3 - (\pi/2) + q_{fswing} - \beta_{fswing}))] + \cdots \\
& [m_{tor}(l_1^2 + l_2^2 + l_{ctorso}^2 + 2 l_1 l_2 \cos(q_2 - q_1) + 2 l_1 l_{ctorso} \cos(-q_{torso} - q_1 - (\pi/2)) + \cdots \\
& 2 l_2 l_{ctorso} \cos(q_{torso} + q_2 + (\pi/2))] + I_1 + I_2 + I_3 + I_4 + I_5 + I_{torso}
\end{aligned}
$$

$$\mathrm{h}_{13}=[m_2(l_{c2}^2+l_1l_{c2}\cos(q_2-q_1))]+[m_3(l_2^2+l_{c3}^2+l_1l_2\cos(q_2-q_1)-l_1l_{c3}\cos(q_1-q_3)-\cdots$$
$$2l_2l_{c3}\cos(q_2-q_3))]+[m_4(l_2^2+l_3^2+l_{c4}^2+l_1l_2\cos(q_2-q_1)-l_1l_3\cos(q_3-q_1)-\cdots$$
$$l_1l_{c4}\cos(q_1-q_4)-2l_2l_3\cos(q_3-q_2)-2l_2l_{c4}\cos(q_2-q_4)+2l_{c4}l_3\cos(q_3-q_4))]+\cdots$$
$$[m_5(l_2^2+l_3^2+l_4^2+l_{cfswing}^2+l_1l_2\cos(q_2-q_1)-l_1l_3\cos(q_3-q_1)-l_1l_4\cos(q_4-q_1)-\cdots$$
$$l_1l_{cfswing}\cos(q_1-(\pi/2)+q_{fswing}-\beta_{fswing})-2l_2l_3\cos(q_3-q_2)-2l_2l_4\cos(q_4-q_2)-\cdots$$
$$2l_2l_{cfswing}\cos(q_2-(\pi/2)+q_{fswing}-\beta_{fswing})+2l_3l_4\cos(q_4-q_3)+\cdots$$
$$2l_4l_{cfswing}\cos(q_4-(\pi/2)+q_{fswing}-\beta_{fswing})+2l_3l_{cfswing}\cos(q_3-(\pi/2)+q_{fswing}-\beta_{fswing}))]+\cdots$$
$$[m_{tor}(l_2^2+l_{ctorso}^2+l_1l_2\cos(q_2-q_1)+l_1l_{ctorso}\cos(-q_{torso}-q_1-(\pi/2))+2l_2l_{ctorso}\cos(q_{torso}+q_2+(\pi/2))]$$
$$+I_2+I_3+I_4+I_5+I_{torso}$$

$$\mathrm{h}_{14}=[m_3(l_{c3}^2-l_1l_{c3}\cos(q_1-q_3)-l_2l_{c3}\cos(q_2-q_3))]+[m_4(l_3^2+l_{c4}^2-l_1l_3\cos(q_3-q_1)-\cdots$$
$$l_1l_{c4}\cos(q_1-q_4)-l_2l_3\cos(q_3-q_2)-l_2l_{c4}\cos(q_2-q_4)+2l_{c4}l_3\cos(q_3-q_4))]+[m_5(l_3^2+l_4^2+\cdots$$
$$l_{cfswing}^2-l_1l_3\cos(q_3-q_1)-l_1l_4\cos(q_4-q_1)-l_1l_{cfswing}\cos(q_1-(\pi/2)+q_{fswing}-\beta_{fswing})-\cdots$$
$$l_2l_3\cos(q_3-q_2)-l_2l_4\cos(q_4-q_2)-l_2l_{cfswing}\cos(q_2-(\pi/2)+q_{fswing}-\beta_{fswing})+2l_3l_4\cos(q_4-q_3)+\cdots$$
$$2l_4l_{cfswing}\cos(q_4-(\pi/2)+q_{fswing}-\beta_{fswing})+2l_3l_{cfswing}\cos(q_3-(\pi/2)+q_{fswing}-\beta_{fswing}))]+I_3+I_4+I_5$$

$$\mathrm{h}_{15}=[m_4(l_{c4}^2-l_1l_{c4}\cos(q_1-q_4)-l_2l_{c4}\cos(q_2-q_4)+l_{c4}l_3\cos(q_3-q_4))]+[m_5(l_4^2+\cdots$$
$$l_{cfswing}^2-l_1l_4\cos(q_4-q_1)-l_1l_{cfswing}\cos(q_1-(\pi/2)+q_{fswing}-\beta_{fswing})-l_2l_4\cos(q_4-q_2)-\cdots$$
$$l_2l_{cfswing}\cos(q_2-(\pi/2)+q_{fswing}-\beta_{fswing})+l_3l_4\cos(q_4-q_3)+\cdots$$
$$2l_4l_{cfswing}\cos(q_4-(\pi/2)+q_{fswing}-\beta_{fswing})+l_3l_{cfswing}\cos(q_3-(\pi/2)+q_{fswing}-\beta_{fswing}))]+I_4+I_5$$

$$\mathrm{h}_{16}=[m_5(l_{cfswing}^2-l_1l_{cfswing}\cos(q_1-(\pi/2)+q_{fswing}-\beta_{fswing})-l_2l_{cfswing}\cos(q_2-(\pi/2)+q_{fswing}-\beta_{fswing})$$
$$+l_4l_{cfswing}\cos(q_4-(\pi/2)+q_{fswing}-\beta_{fswing})+l_3l_{cfswing}\cos(q_3-(\pi/2)+q_{fswing}-\beta_{fswing}))]+I_5$$

$$\mathrm{h}_{17}=[m_{torso}(l_{ctorso}^2+l_1l_{ctorso}\cos(-q_{torso}-(\pi/2)-q_1)+l_2l_{ctorso}\cos(q_{torso}+(\pi/2)+q_2))]+I_{toroso}$$

$$\mathrm{h}_{21}=[m_2(l_{c2}^2+l_{c2}l_e\cos(q_2-\varphi)+l_1l_{c2}\cos(q_2-q_1))]+[m_3(l_2^2+l_{c3}^2+l_2l_e\cos(q_2-\varphi)-\cdots$$
$$l_{c3}l_e\cos(-q_3+\varphi)+l_1l_2\cos(q_2-q_1)-l_1l_{c3}\cos(q_1-q_3)-2l_2l_{c3}\cos(q_2-q_3))]+[m_4(l_2^2+l_3^2+l_{c4}^2+\cdots$$
$$l_2l_e\cos(q_2-\varphi)-l_3l_e\cos(q_3-\varphi)-l_{c4}l_e\cos(-q_4+\varphi)+l_1l_2\cos(q_2-q_1)-l_1l_3\cos(q_3-q_1)-\cdots$$
$$l_1l_{c4}\cos(q_1-q_4)-2l_2l_3\cos(q_3-q_2)-2l_2l_{c4}\cos(q_2-q_4)+2l_{c4}l_3\cos(q_3-q_4))]+\cdots$$
$$[m_5(l_2^2+l_3^2+l_4^2+l_{cfswing}^2+l_2l_e\cos(q_2-\varphi)-l_3l_e\cos(q_3-\varphi)-l_4l_e\cos(q_4-\varphi)-\cdots$$
$$l_{cfswing}l_e\cos(\varphi-(\pi/2)+q_{fswing}-\beta_{fswing})+l_1l_2\cos(q_2-q_1)-l_1l_3\cos(q_3-q_1)-\cdots$$
$$l_1l_4\cos(q_4-q_1)-l_1l_{cfswing}\cos(q_1-(\pi/2)+q_{fswing}-\beta_{fswing})-2l_2l_3\cos(q_3-q_2)-\cdots$$
$$2l_2l_4\cos(q_4-q_2)-2l_2l_{cfswing}\cos(q_2-(\pi/2)+q_{fswing}-\beta_{fswing})+2l_3l_4\cos(q_4-q_3)+\cdots$$
$$2l_4l_{cfswing}\cos(q_4-(\pi/2)+q_{fswing}-\beta_{fswing})+2l_3l_{cfswing}\cos(q_3-(\pi/2)+q_{fswing}-\beta_{fswing}))]+\cdots$$
$$[m_{tor}(l_2^2+l_{ctorso}^2+l_2l_e\cos(q_2-\varphi)+l_el_{ctorso}\cos(-q_{torso}-\varphi-(\pi/2))+l_1l_2\cos(q_2-q_1)+\cdots$$
$$l_1l_{ctorso}\cos(-q_{torso}-q_1-(\pi/2))+2l_2l_{ctorso}\cos(q_{torso}+q_2+(\pi/2))]+I_2+I_3+I_4+I_5+I_{torso}$$

$$\mathrm{h}_{22}=[m_2(l_{c2}^2+l_1l_{c2}\cos(q_2-q_1))]+[m_3(l_2^2+l_{c3}^2+l_1l_2\cos(q_2-q_1)-l_1l_{c3}\cos(q_1-q_3)-\cdots$$
$$2l_2l_{c3}\cos(q_2-q_3))]+[m_4(l_2^2+l_3^2+l_{c4}^2+l_1l_2\cos(q_2-q_1)-l_1l_3\cos(q_3-q_1)-l_1l_{c4}\cos(q_1-q_4)-\cdots$$
$$2l_2l_3\cos(q_3-q_2)-2l_2l_{c4}\cos(q_2-q_4)+2l_{c4}l_3\cos(q_3-q_4))]+[m_5(l_2^2+l_3^2+l_4^2+l_{cfswing}^2+\cdots$$
$$l_1l_2\cos(q_2-q_1)-l_1l_3\cos(q_3-q_1)-l_1l_4\cos(q_4-q_1)-l_1l_{cfswing}\cos(q_1-(\pi/2)+q_{fswing}-\beta_{fswing})-2l_2l_3\cos(q_3-q_2)\cdots$$
$$-2l_2l_4\cos(q_4-q_2)-2l_2l_{cfswing}\cos(q_2-(\pi/2)+q_{fswing}-\beta_{fswing})+2l_3l_4\cos(q_4-q_3)+\cdots$$
$$2l_4l_{cfswing}\cos(q_4-(\pi/2)+q_{fswing}-\beta_{fswing})+2l_3l_{cfswing}\cos(q_3-(\pi/2)+q_{fswing}-\beta_{fswing}))]+\cdots$$
$$[m_{tor}(l_2^2+l_{ctorso}^2+l_1l_2\cos(q_2-q_1)+l_1l_{ctorso}\cos(-q_{torso}-q_1-(\pi/2))+2l_2l_{ctorso}\cos(q_{torso}+q_2+(\pi/2))]+\cdots$$
$$I_2+I_3+I_4+I_5+I_{torso}$$

$$\mathrm{h}_{23}=[m_2(l_{c2}^2)]+[m_3(l_2^2+l_{c3}^2-2l_2l_{c3}\cos(q_2-q_3))]+[m_4(l_2^2+l_3^2+l_{c4}^2-2l_2l_3\cos(q_3-q_2)-\cdots$$
$$2l_2l_{c4}\cos(q_2-q_4)+2l_{c4}l_3\cos(q_3-q_4))]+[m_5(l_2^2+l_3^2+l_4^2+l_{cfswing}^2-2l_2l_3\cos(q_3-q_2)-2l_2l_4\cos(q_4-q_2)-\cdots$$
$$2l_2l_{cfswing}\cos(q_2-(\pi/2)+q_{fswing}-\beta_{fswing})+2l_3l_4\cos(q_4-q_3)+2l_4l_{cfswing}\cos(q_4-(\pi/2)+q_{fswing}-\beta_{fswing})+\cdots$$
$$2l_3l_{cfswing}\cos(q_3-(\pi/2)+q_{fswing}-\beta_{fswing}))]+[m_{tor}(l_2^2+l_{ctorso}^2+2l_2l_{ctorso}\cos(q_{torso}+q_2+(\pi/2))]+\cdots$$
$$I_2+I_3+I_4+I_5+I_{torso}$$

$$h_{24}=[m_3(l_{c3}^2-l_2l_{c3}\cos(q_2-q_3))]+[m_4(l_3^2+l_{c4}^2-l_2l_3\cos(q_3-q_2)-l_2l_{c4}\cos(q_2-q_4)+2l_{c4}l_3\cos(q_3-q_4))]+\cdots$$
$$[m_5(l_3^2+l_4^2+l_{cfswing}^2-l_2l_3\cos(q_3-q_2)-l_2l_4\cos(q_4-q_2)-l_2l_{cfswing}\cos(q_2-(\pi/2)+q_{fswing}-\beta_{fswing})+\cdots$$
$$2l_3l_4\cos(q_4-q_3)+2l_4l_{cfswing}\cos(q_4-(\pi/2)+q_{fswing}-\beta_{fswing})+2l_3l_{cfswing}\cos(q_3-(\pi/2)+q_{fswing}-\beta_{fswing}))]+\cdots$$
$$I_3+I_4+I_5$$

$$h_{25}=[m_4(l_{c4}^2-l_2l_{c4}\cos(q_2-q_4)+l_{c4}l_3\cos(q_3-q_4))]+[m_5(l_4^2+l_{cfswing}^2-l_2l_4\cos(q_4-q_2)-\cdots$$
$$l_2l_{cfswing}\cos(q_2-(\pi/2)+q_{fswing}-\beta_{fswing})+l_3l_4\cos(q_4-q_3)+2l_4l_{cfswing}\cos(q_4-(\pi/2)+q_{fswing}-\beta_{fswing})+\cdots$$
$$l_3l_{cfswing}\cos(q_3-(\pi/2)+q_{fswing}-\beta_{fswing}))]+I_4+I_5$$

$$h_{26}=[m_5(l_{cfswing}^2-l_2l_{cfswing}\cos(q_2-(\pi/2)+q_{fswing}-\beta_{fswing})+l_4l_{cfswing}\cos(q_4-(\pi/2)+q_{fswing}-\beta_{fswing})+\cdots$$
$$l_3l_{cfswing}\cos(q_3-(\pi/2)+q_{fswing}-\beta_{fswing}))]+I_5$$

$$h_{27}=[m_{torso}(l_{ctorso}^2+l_2l_{ctorso}\cos(q_{torso}+(\pi/2)+q_2)]+I_{torso}$$

$$h_{31}=[m_3(l_{c3}^2-l_{c3}l_e\cos(-q_3+\varphi)-l_1l_{c3}\cos(q_1-q_3)-l_2l_{c3}\cos(q_2-q_3))]+[m_4(l_3^2+l_{c4}^2-\cdots$$
$$l_3l_e\cos(q_3-\varphi)-l_{c4}l_e\cos(-q_4+\varphi)-l_1l_3\cos(q_3-q_1)-l_1l_{c4}\cos(q_1-q_4)-l_2l_3\cos(q_3-q_2)-\cdots$$
$$l_2l_{c4}\cos(q_2-q_4)+2l_{c4}l_3\cos(q_3-q_4))]+[m_5(l_3^2+l_4^2+l_{cfswing}^2-l_3l_e\cos(q_3-\varphi)-\cdots$$
$$l_4l_e\cos(q_4-\varphi)-l_{cfswing}l_e\cos(\varphi-(\pi/2)+q_{fswing}-\beta_{fswing})-l_1l_3\cos(q_3-q_1)-\cdots$$
$$l_1l_4\cos(q_4-q_1)-l_1l_{cfswing}\cos(q_1-(\pi/2)+q_{fswing}-\beta_{fswing})-l_2l_3\cos(q_3-q_2)-\cdots$$
$$l_2l_4\cos(q_4-q_2)-l_2l_{cfswing}\cos(q_2-(\pi/2)+q_{fswing}-\beta_{fswing})+2l_3l_4\cos(q_4-q_3)+\cdots$$
$$2l_4l_{cfswing}\cos(q_4-(\pi/2)+q_{fswing}-\beta_{fswing})+2l_3l_{cfswing}\cos(q_3-(\pi/2)+q_{fswing}-\beta_{fswing}))]+I_3+I_4+I_5$$

$$\mathrm{h}_{32} = [m_3(l_{c3}^2 - l_1 l_{c3}\cos(q_1 - q_3) - l_2 l_{c3}\cos(q_2 - q_3))] + [m_4(l_3^2 + l_{c4}^2 - l_1 l_3 \cos(q_3 - q_1) - \cdots$$
$$l_1 l_{c4}\cos(q_1 - q_4) - l_2 l_3 \cos(q_3 - q_2) - l_2 l_{c4}\cos(q_2 - q_4) + 2l_{c4} l_3 \cos(q_3 - q_4))] + [m_5(l_3^2 + l_4^2 + \cdots$$
$$l_{cfswing}^2 - l_1 l_3 \cos(q_3 - q_1) - l_1 l_4 \cos(q_4 - q_1) - l_1 l_{cfswing}\cos(q_1 - (\pi/2) + q_{fswing} - \beta_{fswing}) - \cdots$$
$$l_2 l_3 \cos(q_3 - q_2) - l_2 l_4 \cos(q_4 - q_2) - l_2 l_{cfswing}\cos(q_2 - (\pi/2) + q_{fswing} - \beta_{fswing}) + 2l_3 l_4 \cos(q_4 - q_3) + \cdots$$
$$2l_4 l_{cfswing}\cos(q_4 - (\pi/2) + q_{fswing} - \beta_{fswing}) + 2l_3 l_{cfswing}\cos(q_3 - (\pi/2) + q_{fswing} - \beta_{fswing}))] + \cdots$$
$$I_3 + I_4 + I_5$$

$$\mathrm{h}_{33} = [m_3(l_{c3}^2 - l_2 l_{c3}\cos(q_2 - q_3))] + [m_4(l_3^2 + l_{c4}^2 - l_2 l_3 \cos(q_3 - q_2) - l_2 l_{c4}\cos(q_2 - q_4) + 2l_{c4} l_3 \cos(q_3 - q_4))] + \cdots$$
$$[m_5(l_3^2 + l_4^2 + l_{cfswing}^2 - l_2 l_3 \cos(q_3 - q_2) - l_2 l_4 \cos(q_4 - q_2) - l_2 l_{cfswing}\cos(q_2 - (\pi/2) + q_{fswing} - \beta_{fswing}) + \cdots$$
$$2l_3 l_4 \cos(q_4 - q_3) + 2l_4 l_{cfswing}\cos(q_4 - (\pi/2) + q_{fswing} - \beta_{fswing}) + 2l_3 l_{cfswing}\cos(q_3 - (\pi/2) + q_{fswing} - \beta_{fswing}))] + \cdots$$
$$+ I_3 + I_4 + I_5$$

$$\mathrm{h}_{34} = [m_3(l_{c3}^2)] + [m_4(l_3^2 + l_{c4}^2 + 2l_{c4} l_3 \cos(q_3 - q_4))] + [m_5(l_3^2 + l_4^2 + l_{cfswing}^2 + 2l_3 l_4 \cos(q_4 - q_3) + \cdots$$
$$2l_4 l_{cfswing}\cos(q_4 - (\pi/2) + q_{fswing} - \beta_{fswing}) + 2l_3 l_{cfswing}\cos(q_3 - (\pi/2) + q_{fswing} - \beta_{fswing}))] + \cdots$$
$$+ I_3 + I_4 + I_5$$

$$\mathrm{h}_{35} = [m_4(l_{c4}^2 + l_{c4} l_3 \cos(q_3 - q_4))] + [m_5(l_4^2 + l_{cfswing}^2 + l_3 l_4 \cos(q_4 - q_3) + \cdots$$
$$2l_4 l_{cfswing}\cos(q_4 - (\pi/2) + q_{fswing} - \beta_{fswing}) + l_3 l_{cfswing}\cos(q_3 - (\pi/2) + q_{fswing} - \beta_{fswing}))] + I_4 + I_5$$

$$\mathrm{h}_{36} = [m_5(l_{cfswing}^2 + l_4 l_{cfswing}\cos(q_4 - (\pi/2) + q_{fswing} - \beta_{fswing}) + l_3 l_{cfswing}\cos(q_3 - (\pi/2) + q_{fswing} - \beta_{fswing}))] + I_5$$

$$\mathrm{h}_{37} = 0$$

$$\mathrm{h}_{41} = [m_4(l_{c4}^2 - l_{c4}l_e\cos(-q_4+\varphi) - l_1l_{c4}\cos(q_1-q_4) - l_2l_{c4}\cos(q_2-q_4) + l_{c4}l_3\cos(q_3-q_4))] + \cdots$$
$$[m_5(l_4^2 + l_{cfswing}^2 - l_4l_e\cos(q_4-\varphi) - l_{cfswing}l_e\cos(\varphi-(\pi/2)+q_{fswing}-\beta_{fswing}) - l_1l_4\cos(q_4-q_1) - \cdots$$
$$l_1l_{cfswing}\cos(q_1-(\pi/2)+q_{fswing}-\beta_{fswing}) - l_2l_4\cos(q_4-q_2) - l_2l_{cfswing}\cos(q_2-(\pi/2)+q_{fswing}-\beta_{fswing}) + \cdots$$
$$l_3l_4\cos(q_4-q_3) + 2l_4l_{cfswing}\cos(q_4-(\pi/2)+q_{fswing}-\beta_{fswing}) + l_3l_{cfswing}\cos(q_3-(\pi/2)+q_{fswing}-\beta_{fswing}))] + \cdots$$
$$I_4 + I_5$$

$$\mathrm{h}_{42} = [m_4(l_{c4}^2 - l_1l_{c4}\cos(q_1-q_4) - l_2l_{c4}\cos(q_2-q_4) + l_{c4}l_3\cos(q_3-q_4))] + [m_5(l_4^2 + l_{cfswing}^2 - \cdots$$
$$l_1l_4\cos(q_4-q_1) - l_1l_{cfswing}\cos(q_1-(\pi/2)+q_{fswing}-\beta_{fswing}) - l_2l_4\cos(q_4-q_2) - \cdots$$
$$l_2l_{cfswing}\cos(q_2-(\pi/2)+q_{fswing}-\beta_{fswing}) + l_3l_4\cos(q_4-q_3) + 2l_4l_{cfswing}\cos(q_4-(\pi/2)+q_{fswing}-\beta_{fswing}) + \cdots$$
$$l_3l_{cfswing}\cos(q_3-(\pi/2)+q_{fswing}-\beta_{fswing}))] + I_4 + I_5$$

$$\mathrm{h}_{43} = [m_4(l_{c4}^2 - l_2l_{c4}\cos(q_2-q_4) + l_{c4}l_3\cos(q_3-q_4))] + [m_5(l_4^2 + l_{cfswing}^2 - l_2l_4\cos(q_4-q_2) - \cdots$$
$$l_2l_{cfswing}\cos(q_2-(\pi/2)+q_{fswing}-\beta_{fswing}) + l_3l_4\cos(q_4-q_3) + 2l_4l_{cfswing}\cos(q_4-(\pi/2)+q_{fswing}-\beta_{fswing}) + \cdots$$
$$l_3l_{cfswing}\cos(q_3-(\pi/2)+q_{fswing}-\beta_{fswing}))] + I_4 + I_5$$

$$\mathrm{h}_{44} = [m_4(l_{c4}^2 + l_{c4}l_3\cos(q_3-q_4))] + [m_5(l_4^2 + l_{cfswing}^2 - l_2l_4\cos(q_4-q_2) + l_3l_4\cos(q_4-q_3) + \cdots$$
$$2l_4l_{cfswing}\cos(q_4-(\pi/2)+q_{fswing}-\beta_{fswing}) + l_3l_{cfswing}\cos(q_3-(\pi/2)+q_{fswing}-\beta_{fswing}))] + I_4 + I_5$$

$$\mathrm{h}_{45} = [m_4(l_{c4}^2)] + [m_5(l_4^2 + l_{cfswing}^2 + 2l_4l_{cfswing}\cos(q_4-(\pi/2)+q_{fswing}-\beta_{fswing}))] + I_4 + I_5$$

$$\mathrm{h}_{46} = [m_5(l_{cfswing}^2 + l_4l_{cfswing}\cos(q_4-(\pi/2)+q_{fswing}-\beta_{fswing}))] + I_5$$

$$\mathrm{h}_{47} = 0$$

$$\mathrm{h}_{51}=[m_5(l_{cfswing}^2-l_{cfswing}l_e\cos(\varphi-(\pi/2)+q_{fswing}-\beta_{fswing})-l_1l_{cfswing}\cos(q_1-(\pi/2)+q_{fswing}-\beta_{fswing})-\cdots$$
$$l_2l_{cfswing}\cos(q_2-(\pi/2)+q_{fswing}-\beta_{fswing})+l_4l_{cfswing}\cos(q_4-(\pi/2)+q_{fswing}-\beta_{fswing})+\cdots$$
$$l_3l_{cfswing}\cos(q_3-(\pi/2)+q_{fswing}-\beta_{fswing}))]+I_5$$

$$\mathrm{h}_{52}=[m_5(l_{cfswing}^2-l_1l_{cfswing}\cos(q_1-(\pi/2)+q_{fswing}-\beta_{fswing})-l_2l_{cfswing}\cos(q_2-(\pi/2)+q_{fswing}-\beta_{fswing})+\cdots$$
$$l_4l_{cfswing}\cos(q_4-(\pi/2)+q_{fswing}-\beta_{fswing})+l_3l_{cfswing}\cos(q_3-(\pi/2)+q_{fswing}-\beta_{fswing}))]+I_5$$

$$\mathrm{h}_{53}=[m_5(l_{cfswing}^2-l_2l_{cfswing}\cos(q_2-(\pi/2)+q_{fswing}-\beta_{fswing})+\cdots$$
$$l_4l_{cfswing}\cos(q_4-(\pi/2)+q_{fswing}-\beta_{fswing})+l_3l_{cfswing}\cos(q_3-(\pi/2)+q_{fswing}-\beta_{fswing}))]+I_5$$

$$\mathrm{h}_{54}=[m_5(l_{cfswing}^2+l_4l_{cfswing}\cos(q_4-(\pi/2)+q_{fswing}-\beta_{fswing})+l_3l_{cfswing}\cos(q_3-(\pi/2)+q_{fswing}-\beta_{fswing}))]+I_5$$

$$\mathrm{h}_{55}=[m_5(l_{cfswing}^2+l_4l_{cfswing}\cos(q_4-(\pi/2)+q_{fswing}-\beta_{fswing}))]+I_5$$

$$\mathrm{h}_{56}=[m_5(l_{cfswing}^2)]+I_5$$

$$\mathrm{h}_{57}=0$$

$$h_{61}=[masstorso(l_{tor}^2+l_el_{tor}\cos(-q_{torso}-(\pi/2)-\varphi)+l_1l_{tor}\cos(-q_{torso}-(\pi/2)-q_1)+\cdots$$
$$l_2l_{tor}\cos(q_{torso}+(\pi/2)+q_2))]+I_{torso}$$

$$h_{62} = [masstorso(l_{tor}^2 + l_1 l_{tor}\cos(-q_{torso} - (\pi/2) - q_1) + l_2 l_{tor}\cos(q_{torso} + (\pi/2) + q_2))] + I_{torso}$$

●•••●

$$h_{63} = [masstorso(l_{tor}^2 + l_2 l_{tor}\cos(q_{torso} + (\pi/2) + q_2))] + I_{torso}$$

●•••●

$$h_{67} = [masstorso(l_{tor}^2)] + I_{torso}, h_{64,65,66} = 0$$

●•••●

In the all of above mentioned components, the following parameters have been used and they can be seen in figure (1.6) :

$$\phi = \beta_{m.c} + q_0, \varphi = \lambda + q_0 + \beta_{m.c}. \quad (32)$$

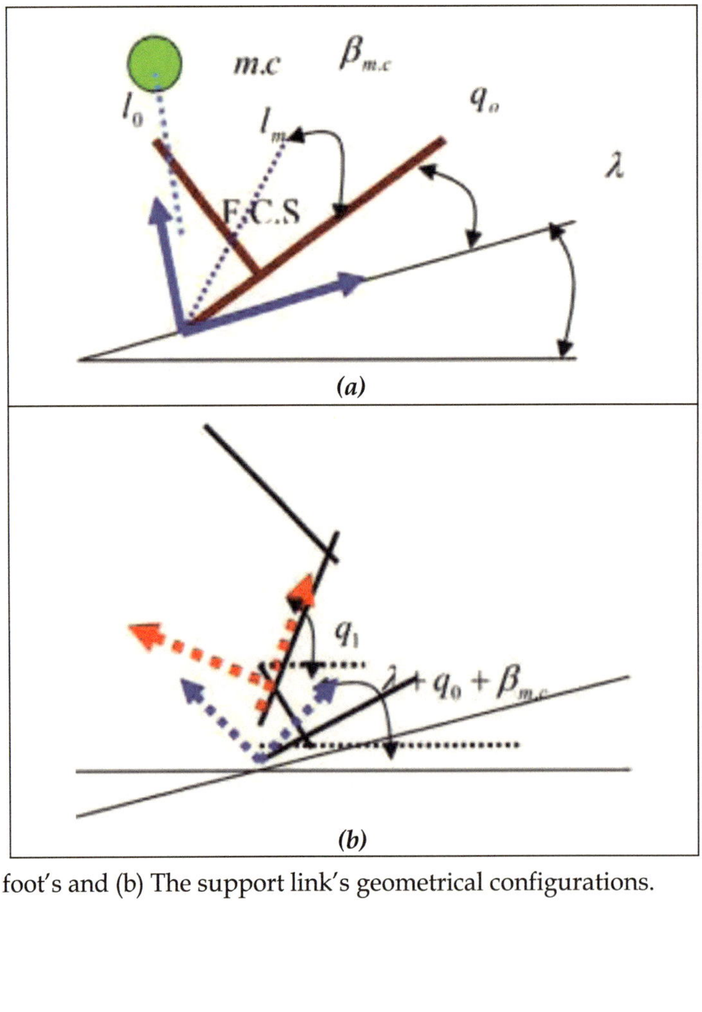

Fig. (1.6). (a) The foot's and (b) The support link's geometrical configurations.

Where, l_{fswing} and β_{fswing} refer to indexes of the swing leg with respect to geometrical configurations of the mass center of the swing leg as can be deducted from figure (1.6). The coriolis and gravitational components of the relation (30) can be calculated easily. After calculation of kinematic and dynamic parameters of the robot, the control process of the system will be emerged. In summery, the adaptive control procedure has been used for a known seven link biped robot. The more details and the related definitions such as the known and unknown system with respect to control aspect can be found in reference [Musavi and Bagheri, 2007 and Musavi, 2006, and Bagheri and Felezi, and et. Al., 2006]. For the simulation of the robot, the obtained parameters and relations are inserted into the designed program in Matlab environment. As can be seen in simulation results section, the most concerns refer to stability of the robot with respect to the important affecting parameters of the robot movements which indicate the ankle and hip joints parameters [Bagheri and Najafi and et. Al. 2006]. As can be seen from the simulations figures, the hip height and horizontal positions have considerable effects over the position of the ZMP and subsequently over the stability of the robot. To minimize the driver actuator torques of the joints, especially for the knee joint of the robot, the hip height which measured from the F.C.S has drastic role for diminution of the torques.

4. Simulation Results

In the designed program, the mentioned simulation processes for the two types of ZMP have been used for both of the nominal and un-nominal gait. For the un-nominal walking of the robot, the hip parameters (hip height) have been changed to consider the effects of the un-nominal motion upon the joint's actuator torques. The results are presented in figures (1.8) to (1.15) while the robot walks over declined surfaces for the single phase of the walking. Figure (1.15) shows combined path of the robot. The used specifications of the simulation of the robot are listed in table No. 1. Figures (1.8), (1.10) and (1.12) display the moving type of ZMP with the nominal walking of the robot. Figures (1.9), (1.11) and (1.13) show the same type of ZMP and also the un-nominal walking of the robot (with the changed hip height form the fixed coordinate system). Figure (1.14) shows the fixed ZMP upon descending surface. As can been seen from the table, the swing and support legs have the same geometrical and inertial values whereas in the designed program the users can choose different specifications. Note, the swing leg impact and the ground has been regarded in the designed program as given in references [Lum and et. Al. 1999 and Westervelt, 2003, and Hon and et. Al., 1978]. Below, the saggital movement and stability analysis of the seven link biped robot has been considered whereas the frontal considerations are neglected. For convenience, 3D simulations of the biped robot are presented. In table No. 1, l_{an}, l_{ab} and l_{af} present the foot profile which are displayed in figure (1.7). The program enables the user to compare the results as presented in figures where the paths for the single phase walking of the robot have been concerned. In the program with the aid of the given break points, either third-order spline or Vandermonde Matrix has been used for providing the different trajectory paths. With the aid of the designed program, the kinematic, dynamic and control parameters have been evaluated. Also, the two types of ZMP have been investigated. The presented simulations indicate the hip height effects over joint's actuator torques for minimizing energy consumption and especially obtaining fine stability margin. As can be seen in figures (9.h), (11.h) and (13.h), for the un-nominal walking of the robot with the lower hip height, the knee's actuator torque values is more than the obtained values as shown in figures (8.h), (10.h) and (12.h) (for the nominal gait with the higher hip height). This is due to the robot's need to bend its knee joint more at a low hip position. Therefore, the large knee joint torque is required to support the robot. Therefore, for reducing the load on the knee joint and consequently with respect to minimum energy consumption, it is essential to keep the hip at a high position. Finally, the trajectory path generation needs more precision with respect to the obtained kinematic relations to avoid the link's velocity discontinuities. The presented results have an acceptable consistency with the typical robot.

$l_{Sh.}$	$l_{Ti..}$	$l_{To.}$	l_{an}	l_{ab}	l_{af}
$0.3m$	$0.3m$	$0.3m$	$0.1m$	$0.1m$	$0.13m$
$m_{Sh.}$	$m_{Th.}$	$m_{To.}$	$m_{Fo.}$	D_s	T_c
$5.7kg$	$10kg$	$43kg$	$3.3kg$	$0.5m$	$0.9s$
T_d	T_m	H_{ao}	L_{ao}	x_{ed}	x_{sd}
$0.18s$	$0.4s$	$0.16m$	$0.4m$	$0.23m$	$0.23m$
g_{gs}	g_{gf}	H_{min}	H_{max}	h_s	H_s
0	0	$0.60m$	$0.62m$	$0.1m$	$0.15m$
I_{shank}		I_{tight}	I_{torso}	I_{foot}	
$0.02kgm^2$		$0.08kgm^2$	$1.4kgm^2$	$0.01kgm^2$	

Table 1. The simulated robot specifications.

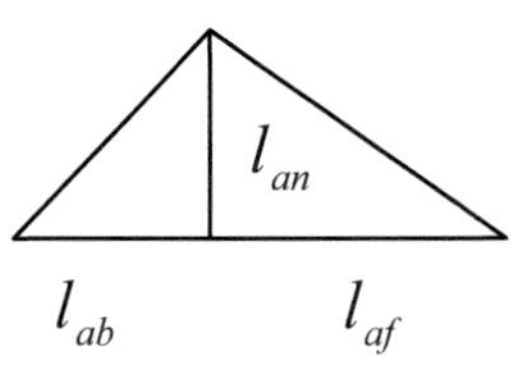

Fig. 1.7. The foot configuration.

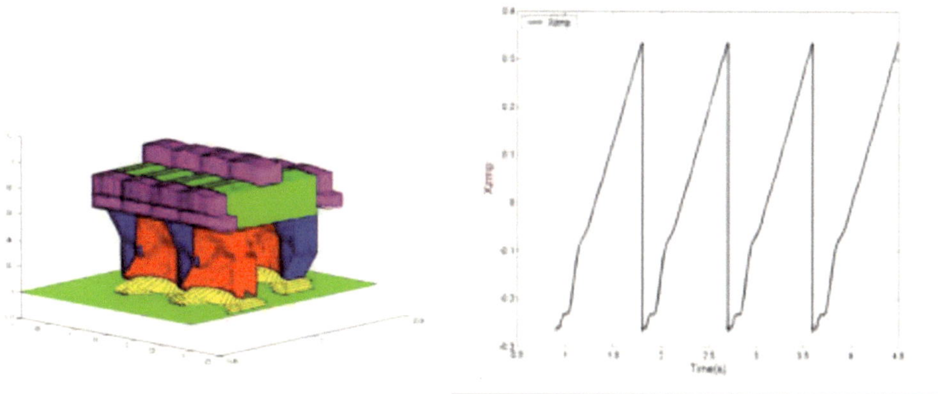

(a) Stick Diagram *(b) ZMP*

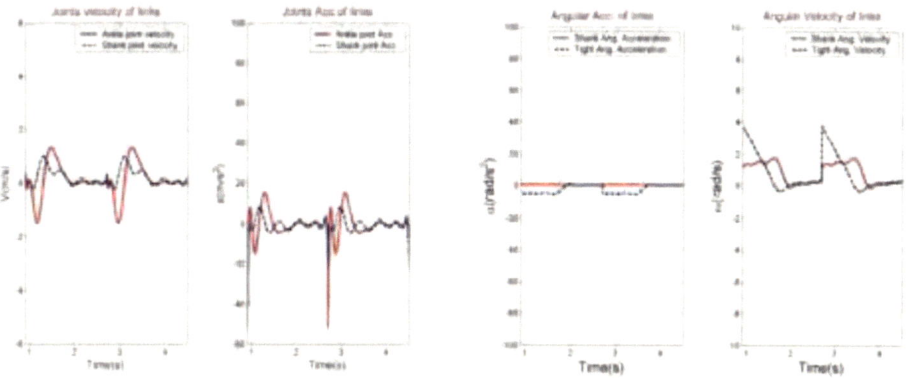

(c) Velocity *(d) Acceleration* *(e) Angular Vel.* *(f) Angular Acc.*

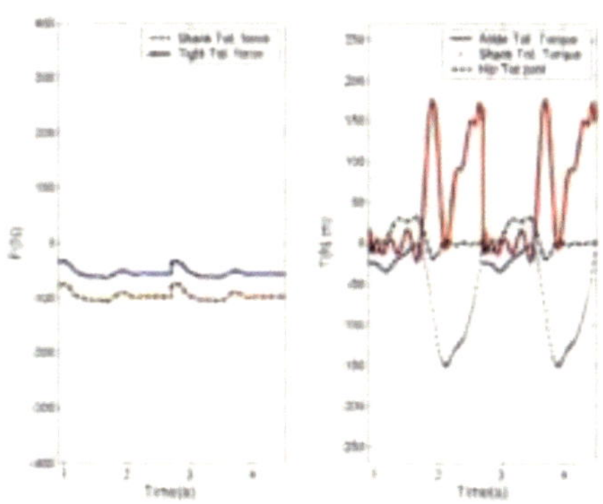

(j) Inertial Forces *(h) Driver Torques*

Fig. 1.8. *(a) The robot's stick diagram on* $\lambda = 0^\circ$ *, Moving ZMP,* $H_{\min}$ $0.60m, H_{\max}$ $0.62m$ *(b) The moving ZMP diagram in nominal gait which satisfies stability criteria (c) __: Shank M.C velocity, --: Tight M.C velocity (d)__: Shank M.C acceleration, --:Tight M.C acceleration (e) __: Shank angular velocity, --: Tight angular velocity (f) __: Shank angular acceleration, --: Tight angular acceleration (j) __: Shank M.C inertial force, --: Tight M.C inertial force (h) __: Ankle joint torque, --: Hip joint torque, …: Shank joint torque*

(a) Stick Daigram

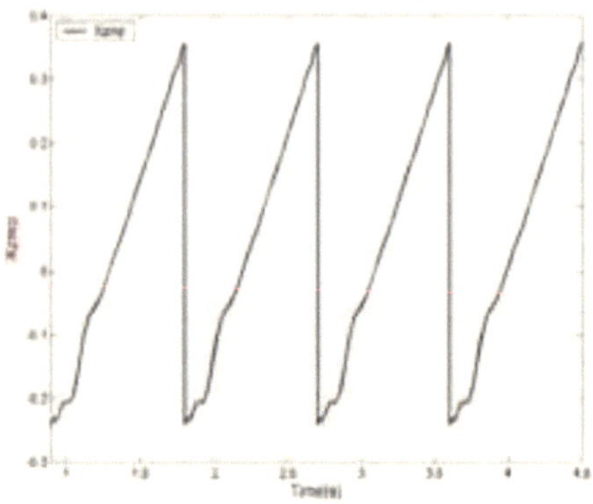

(b) ZMP

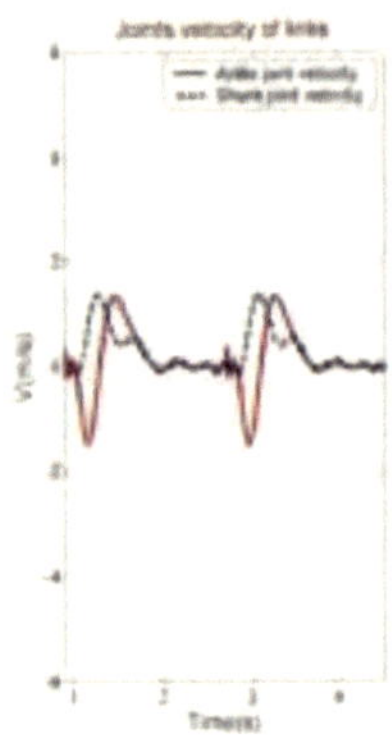

(c) Velocity

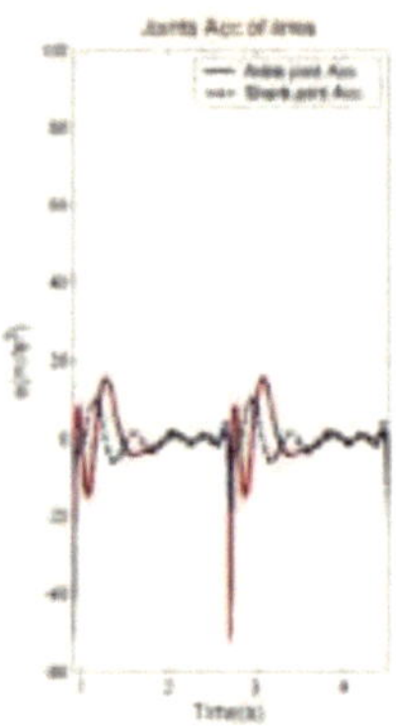

(d) Acceleration

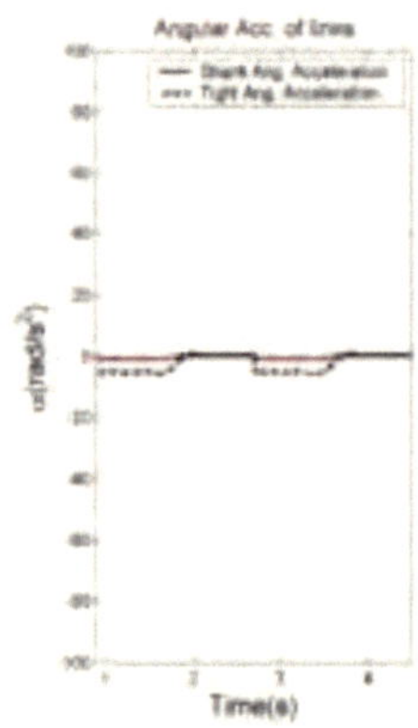

(e) Angular Vel.

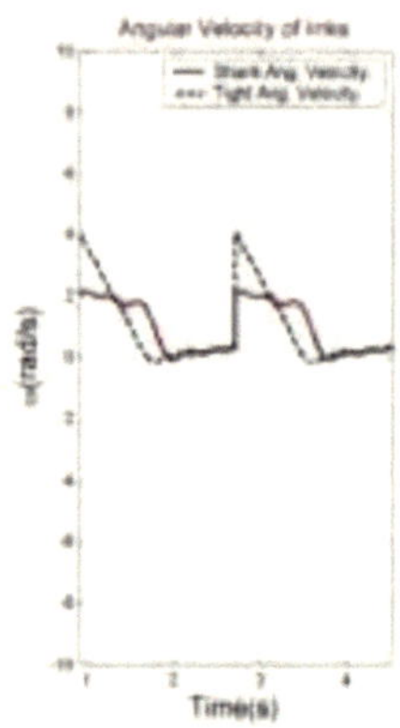

(f) Angular Acc.

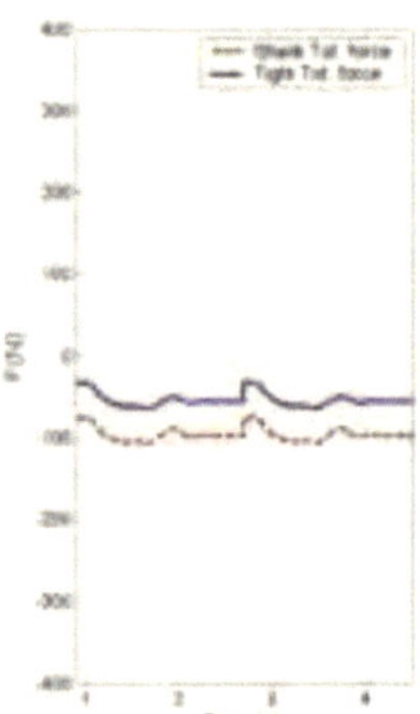

(j) Inertial Forces

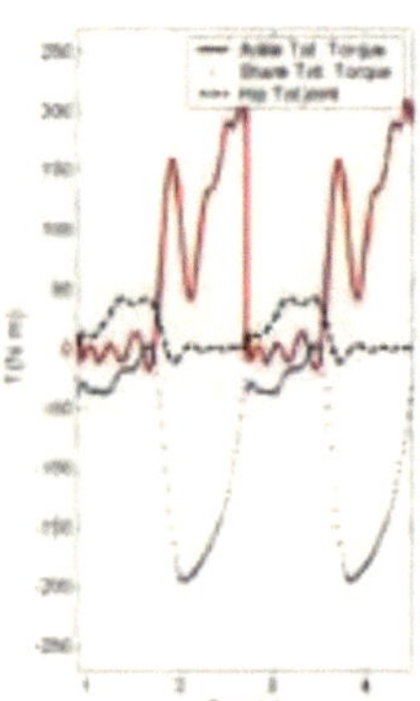

(h) Driver Torques

Fig. 1.9. *(a) The robot's stick diagram on* λ 0° *, Moving ZMP,* $H_{min} = 0.50m, H_{max} = 0.52m$ *(b) The moving ZMP diagram in nominal gait which satisfies stability criteria (c) __: Shank M.C velocity, --: Tight M.C velocity (d)__: Shank M.C acceleration, --:Tight M.C acceleration (e) __: Shank angular velocity, --: Tight angular velocity (f) __: Shank angular acceleration, --: Tight angular acceleration (j) __: Shank M.C inertial force, --: Tight M.C inertial force (h) __: Ankle joint torque, --: Hip joint torque, ...: Shank joint torque*

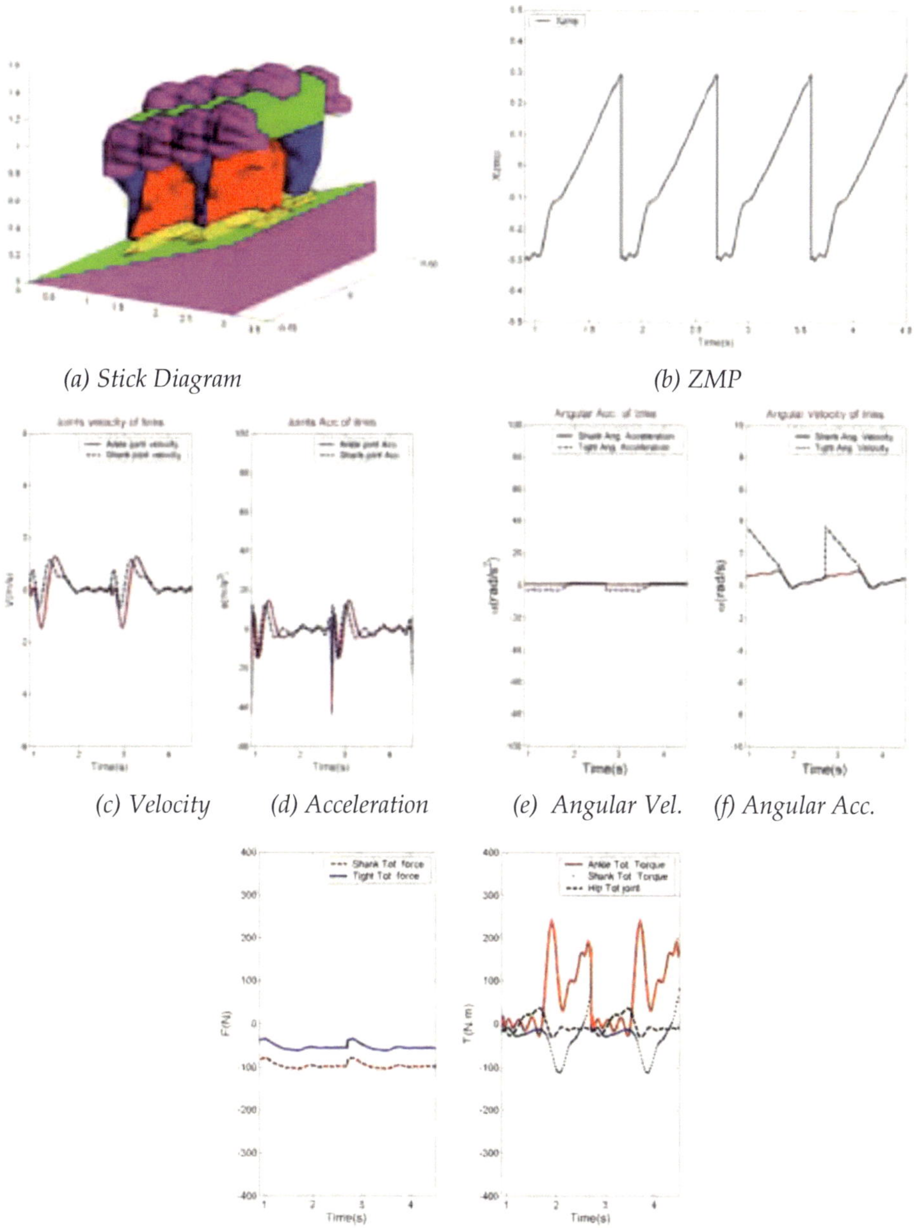

(a) Stick Diagram (b) ZMP

(c) Velocity (d) Acceleration (e) Angular Vel. (f) Angular Acc.

(j) Inertial Forces (h) Driver Torques

Fig. 1.10 (a) The robot's stick diagram on $\lambda = 10^\circ$ *, Moving ZMP,* $H_{\min} = 0.60m, H_{\max} = 0.62m$ *(b) The moving ZMP diagram in nominal gait which satisfies stability criteria (c) __: Shank M.C velocity, --: Tight M.C velocity (d)__: Shank M.C acceleration, --:Tight M.C acceleration (e) __: Shank angular velocity, --: Tight angular velocity (f) __: Shank angular acceleration, --: Tight angular acceleration (j) __: Shank M.C inertial force, --: Tight M.C inertial force (h) __: Ankle joint torque, --: Hip joint torque, …: Shank joint torque*

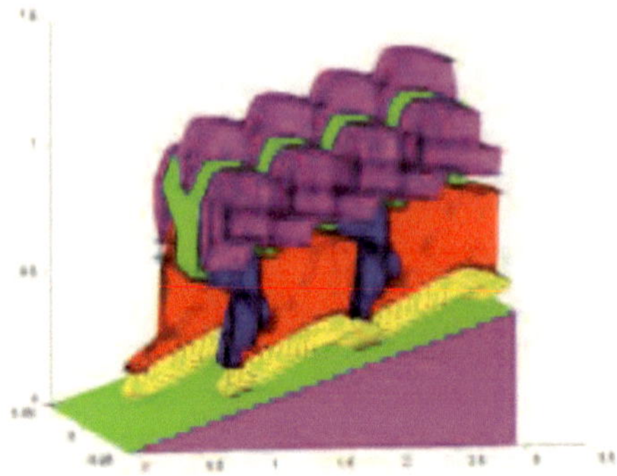

(a) Stick Diagram

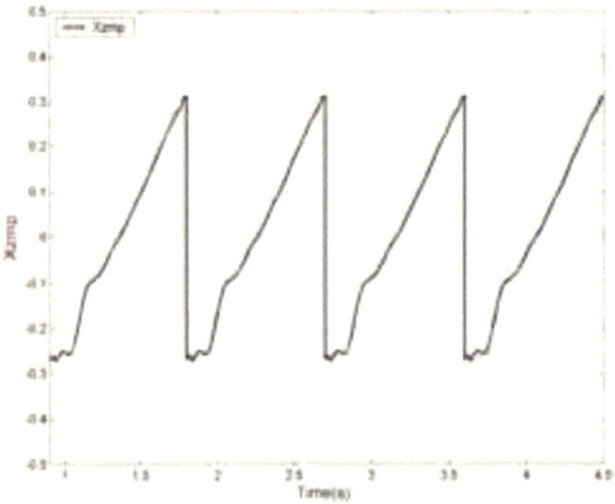

(b) ZMP

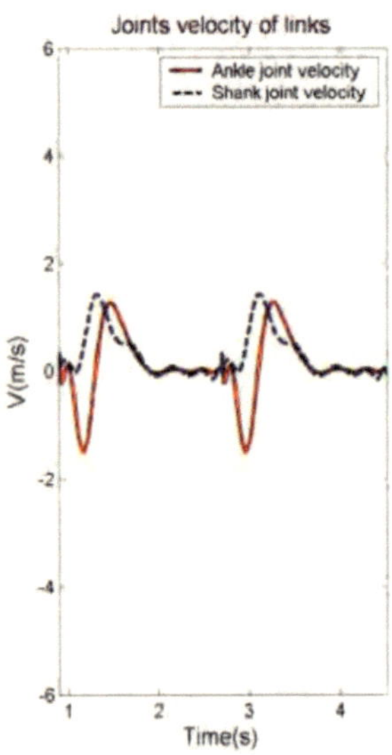

(c) Velocity

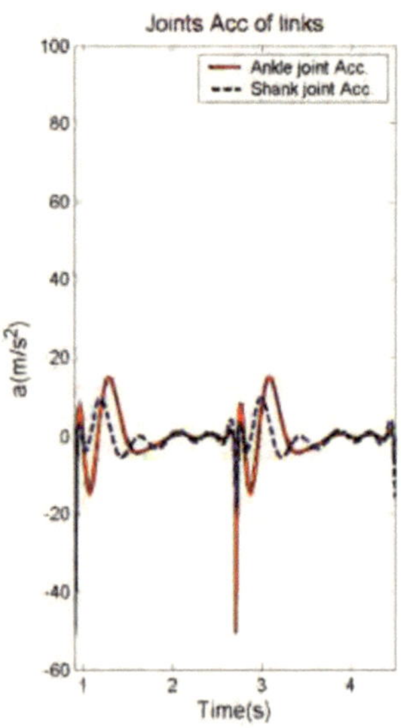

(d) Acceleration

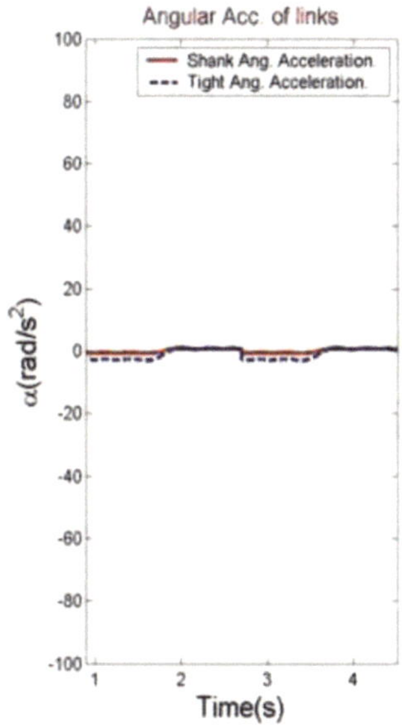

(e) Angular Vel.

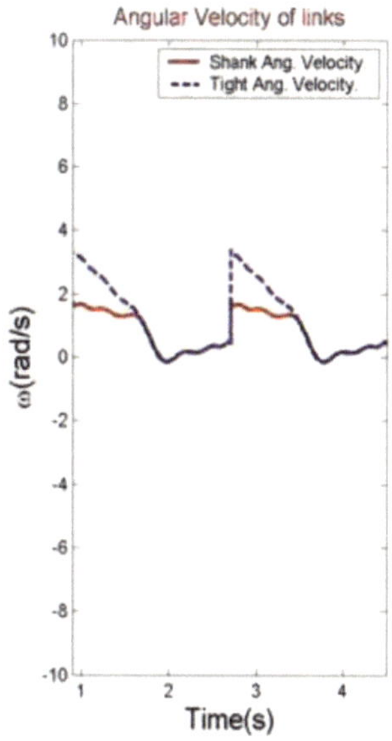

(f) Angular Acc.

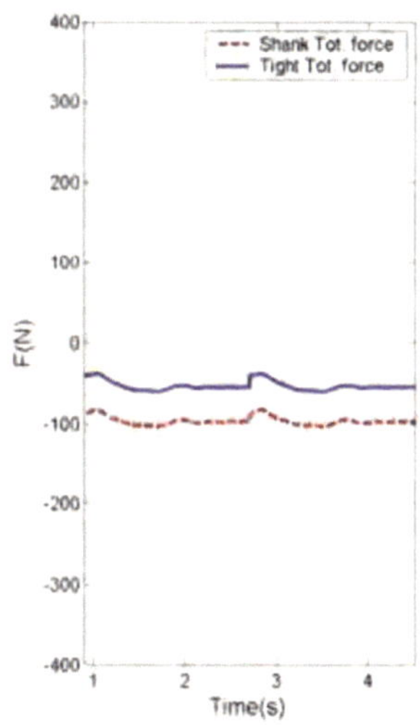

(j) Inertial Forces

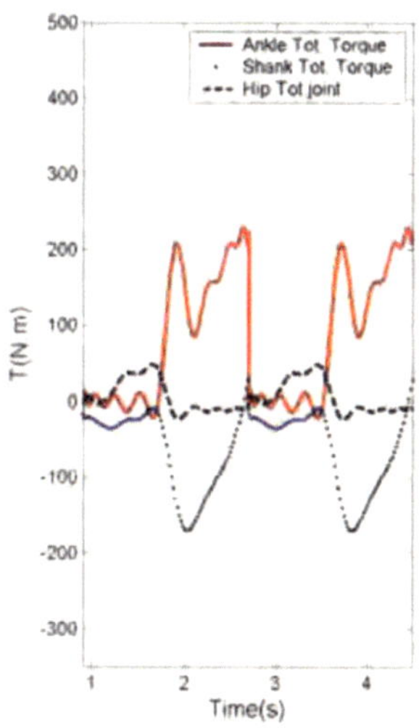

(h) Driver Torques

Fig. 1.11.

(a) The robot's stick diagram on $\lambda = 10^\circ$ *, Moving ZMP,* H_{min} $0.50m, H_{max}$ $0.52m$

(b) The moving ZMP diagram in nominal gait which satisfies stability criteria

(c) __: Shank M.C velocity, --: Tight M.C velocity

(d)__: Shank M.C acceleration, --:Tight M.C acceleration

(e) __: Shank angular velocity, --: Tight angular velocity

(f) __: Shank angular acceleration, --: Tight angular acceleration

(j) __: Shank M.C inertial force, --: Tight M.C inertial force

(h) __: Ankle joint torque, --: Hip joint torque, ...: Shank joint torque

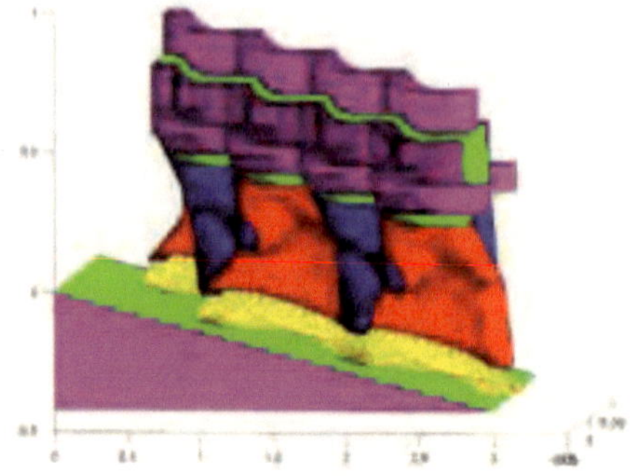

(a) Stick Diagram

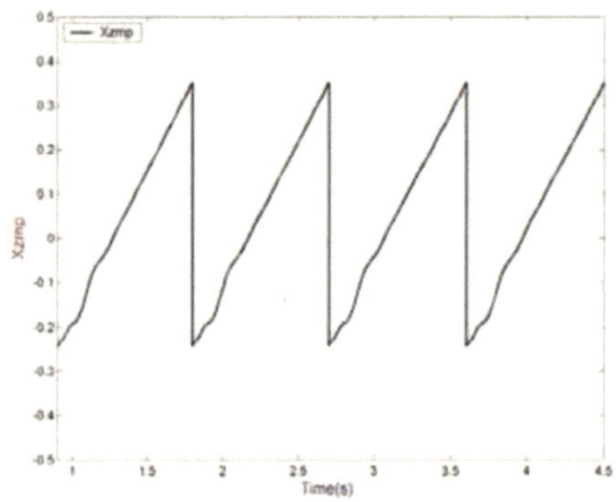

(b) ZMP

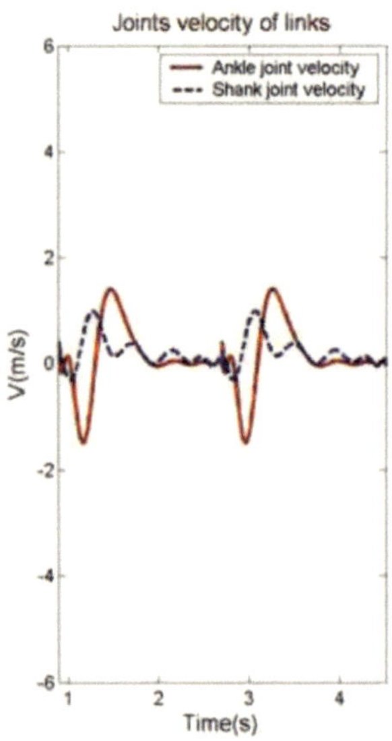

(c) Velocity

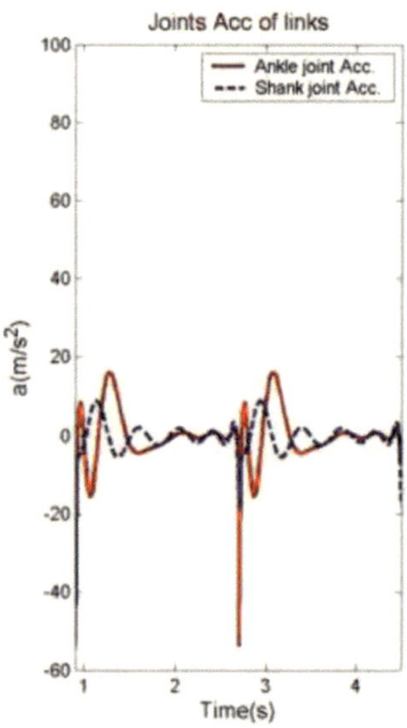

(d) Acceleration

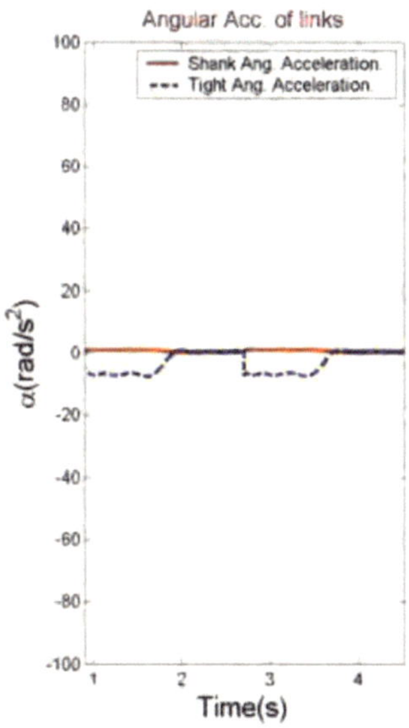

(e) Angular Vel.

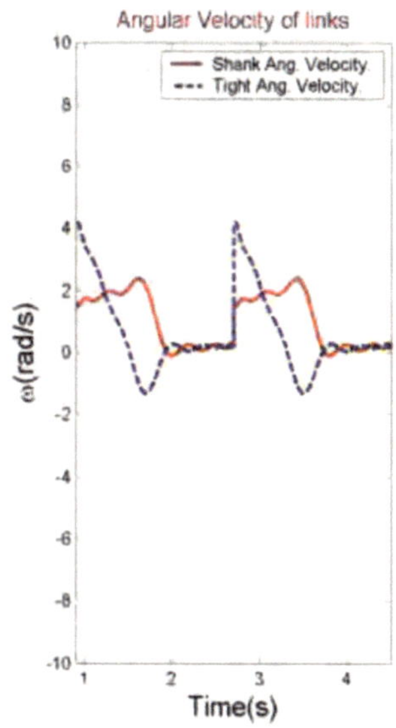

(f) Angular Acc.

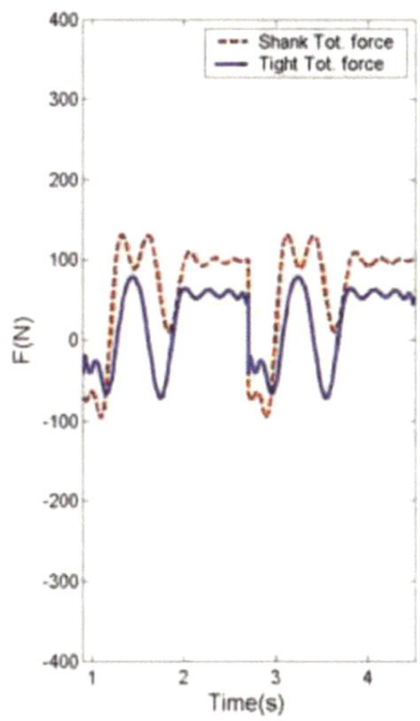

(j) Inertial Forces

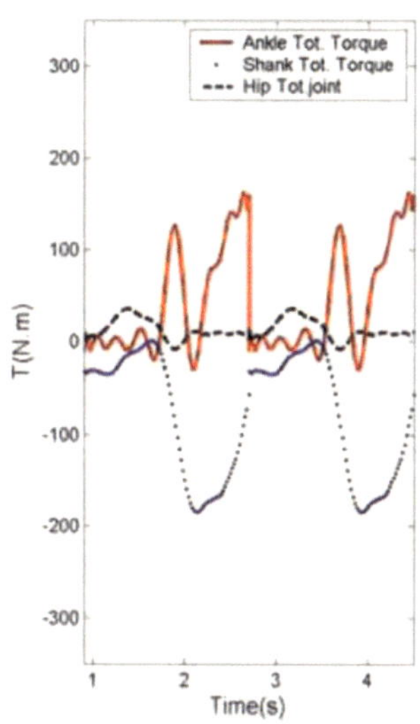

(h) Driver Torques

Fig. 1.12.

(a) The robot's stick diagram on $\lambda = -8^\circ$ *, Moving ZMP,* $H_{min} = 0.60m, H_{max} = 0.62m$

(b) The moving ZMP diagram in nominal gait which satisfies stability criteria

(c) __: Shank M.C velocity, --: Tight M.C velocity

(d)__: Shank M.C acceleration, --:Tight M.C acceleration

(e) __: Shank angular velocity, --: Tight angular velocity

(f) __: Shank angular acceleration, --: Tight angular acceleration

(j) __: Shank M.C inertial force, --: Tight M.C inertial force

(h) __: Ankle joint torque, --: Hip joint torque, …: Shank joint torque

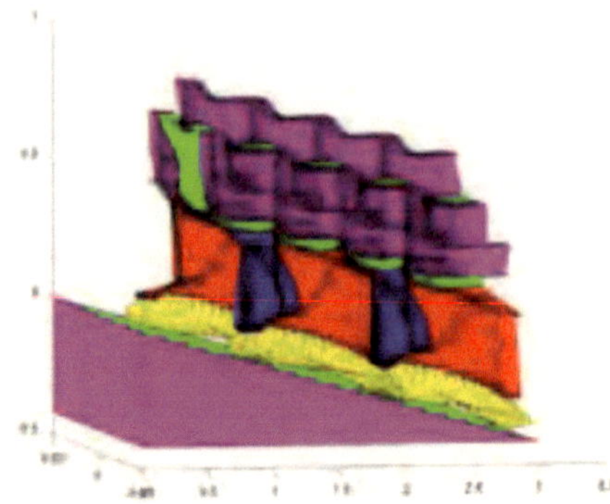

(a) Stick Diagram

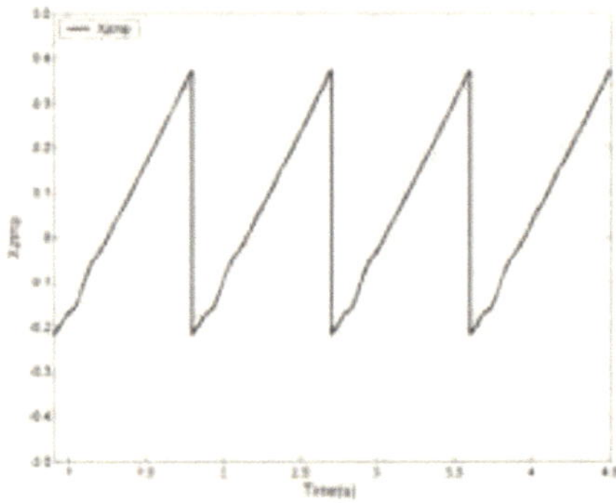

(b) ZMP

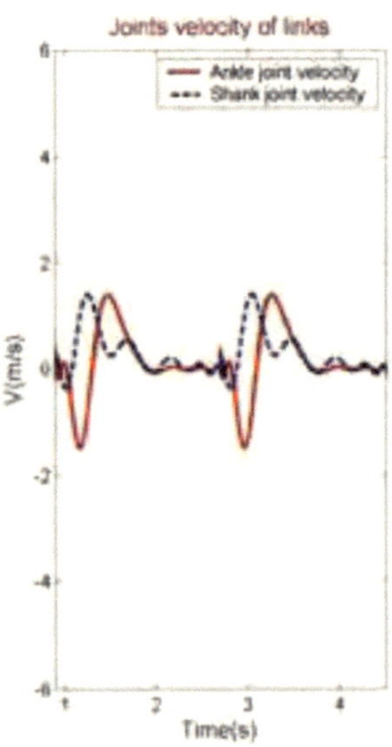

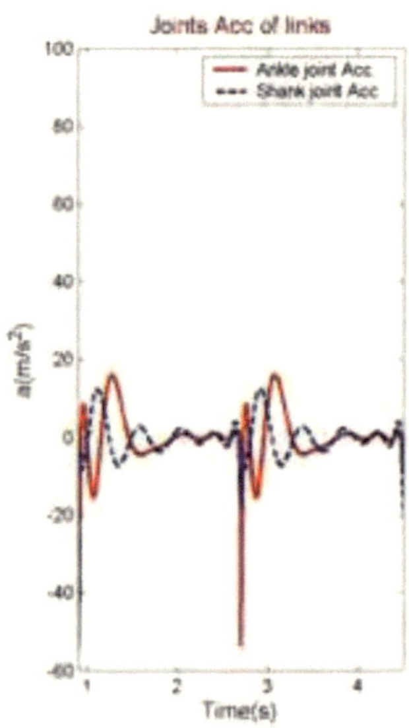

(c) Velocity *(d) Acceleration*

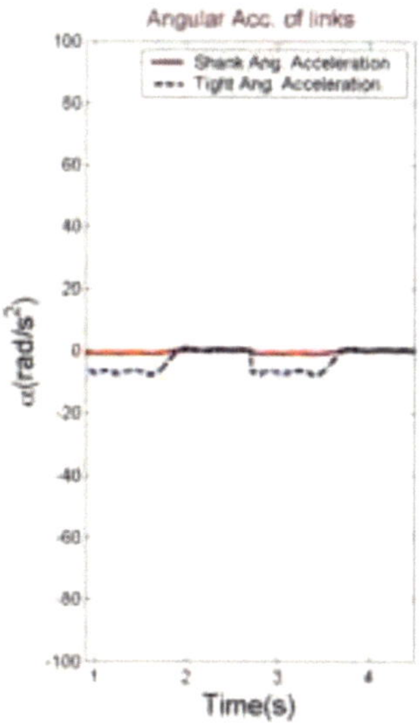

(e) Angular Vel.

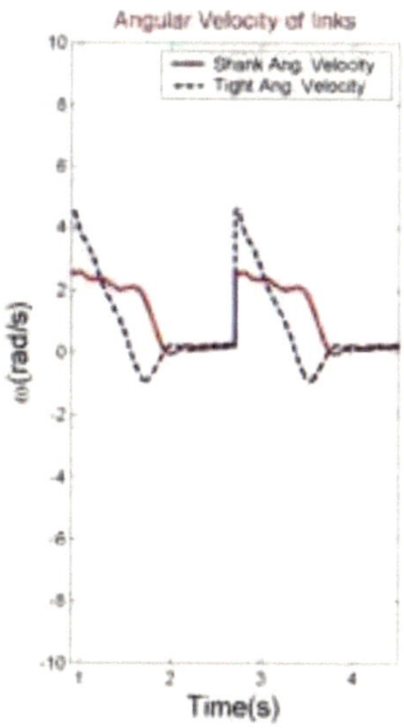

(f) Angular Acc.

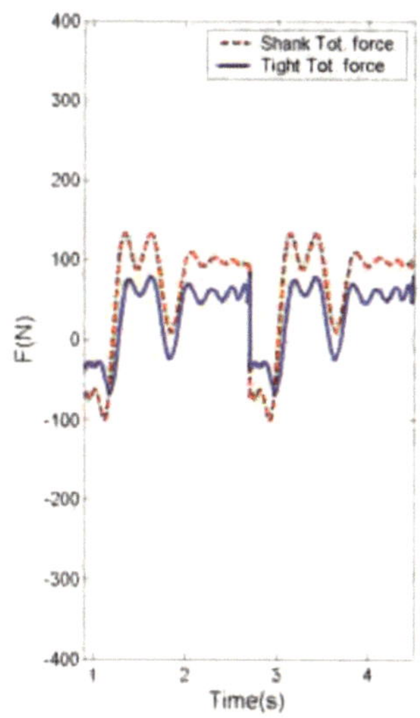

(j) Inertial Forces

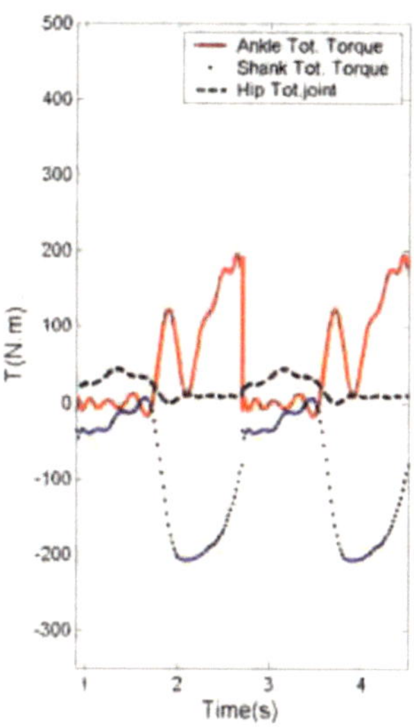

(h) Driver Torques

Fig. 1.13

(a) The robot's stick diagram on $\lambda = -8^\circ$ *, Moving ZMP,* $H_{min} = 0.50m, H_{max} = 0.52m$

(b) The moving ZMP diagram in nominal gait which satisfies stability criteria

(c) __: Shank M.C velocity, --: Tight M.C velocity

(d)__: Shank M.C acceleration, --:Tight M.C acceleration

(e) __: Shank angular velocity, --: Tight angular velocity

(f) __: Shank angular acceleration, --: Tight angular acceleration

(j) __: Shank M.C inertial force, --: Tight M.C inertial force

(h) __: Ankle joint torque, --: Hip joint torque, …: Shank joint torque

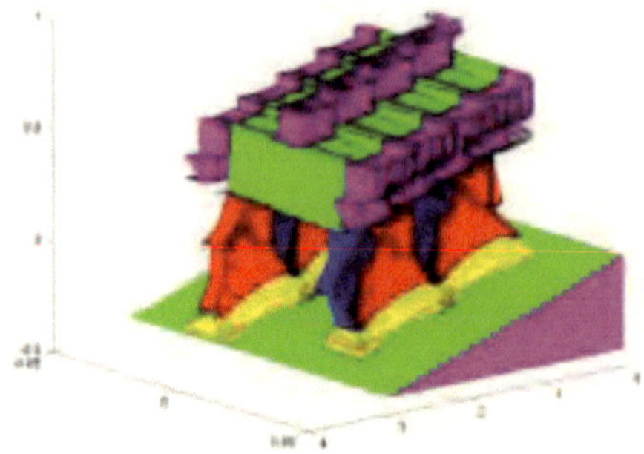

(a) Stick Diagram

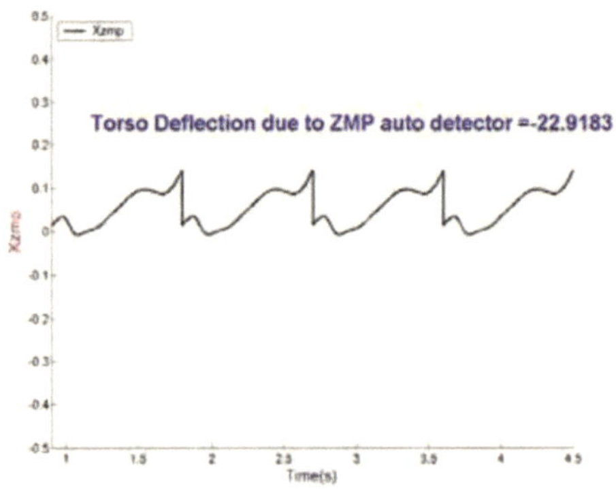

(b) ZMP

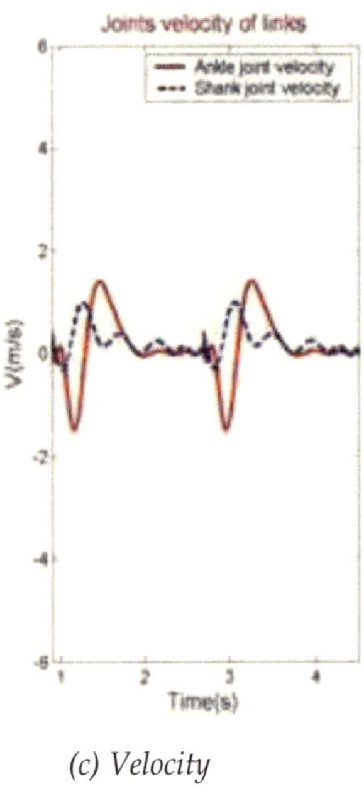

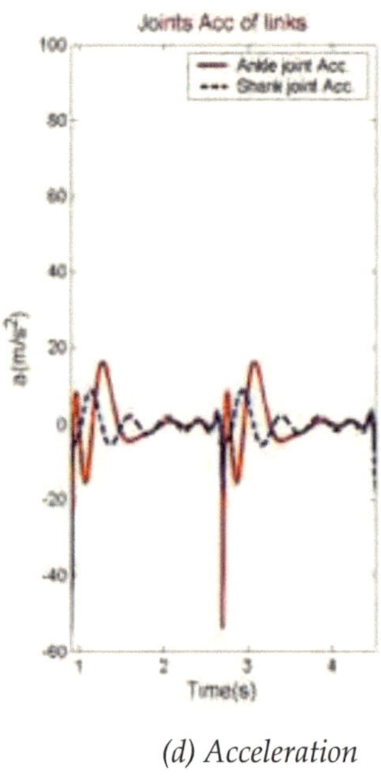

(c) Velocity

(d) Acceleration

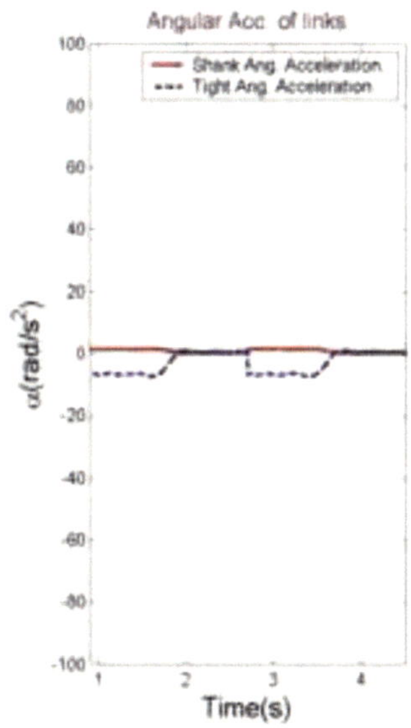

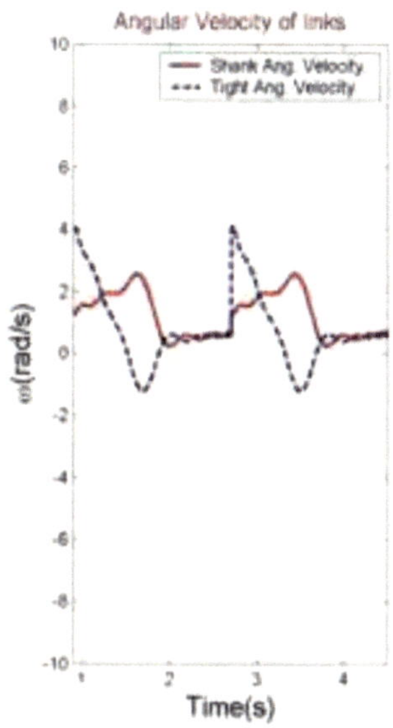

e) Angular Vel. *(f) Angular Acc.*

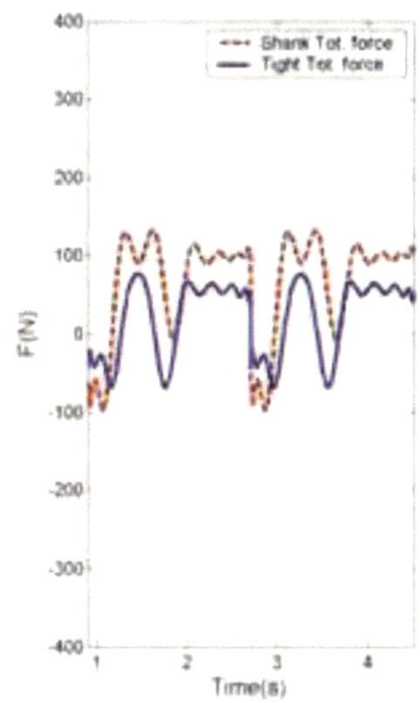

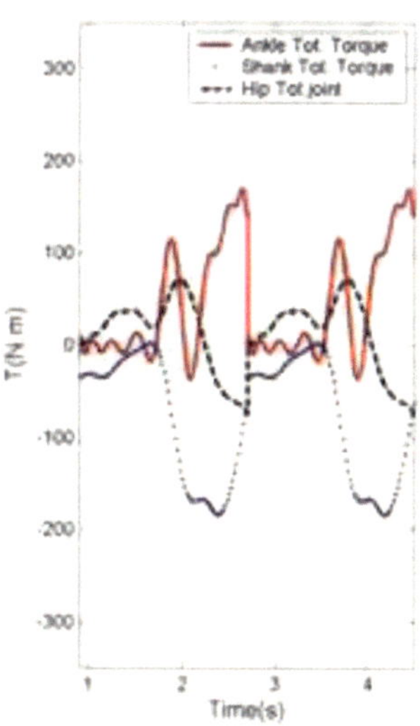

(j) Inertial Forces *(h) Driver Torques*

Fig. 1.14

(a) The robot's stick diagram on $\lambda = -8^\circ$ *, Fixed ZMP,* $H_{\min} = 0.60m, H_{\max} = 0.62m$

(b) The fixed ZMP diagram in nominal gait which satisfies stability criteria

(c) __: Shank M.C velocity, --: Tight M.C velocity

(d)__: Shank M.C acceleration, --:Tight M.C acceleration

(e) __: Shank angular velocity, --: Tight angular velocity

(f) __: Shank angular acceleration, --: Tight angular acceleration

(j) __: Shank M.C inertial force, --: Tight M.C inertial force

(h) __: Ankle joint torque, --: Hip joint torque, …: Shank joint torque

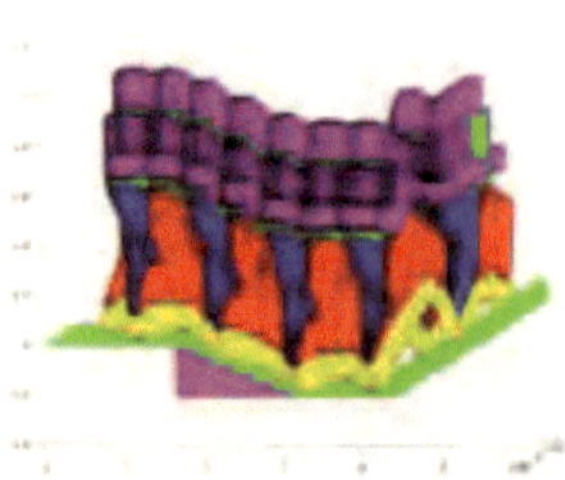

(a) Stick Diagram

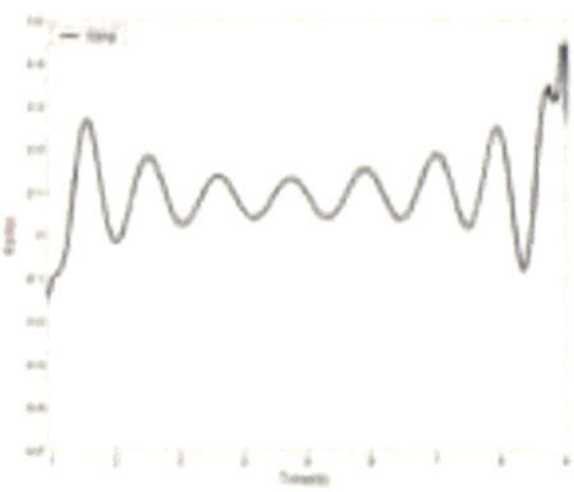

(b) ZMP

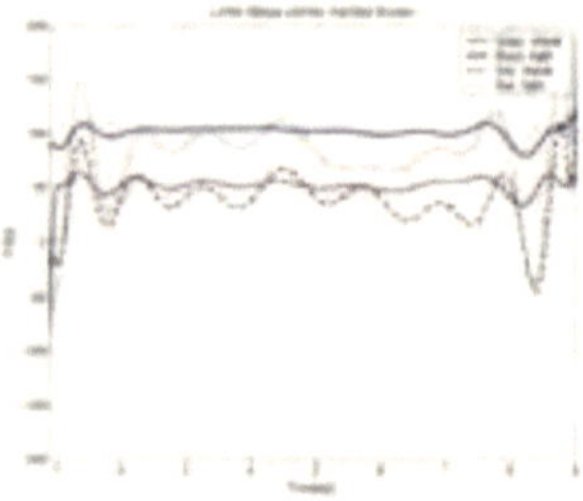

(c) Inertial Forces

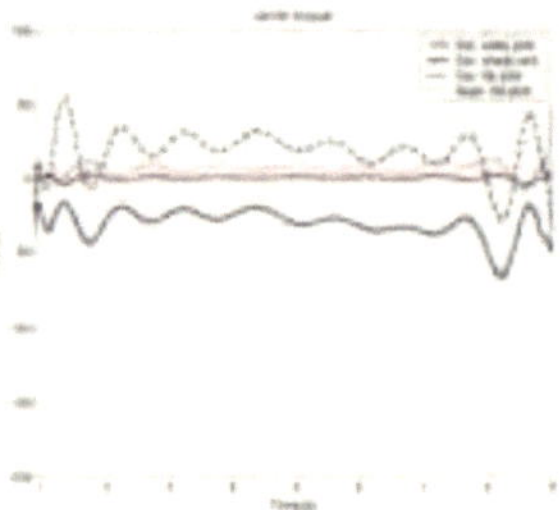

(d) Driver Torques

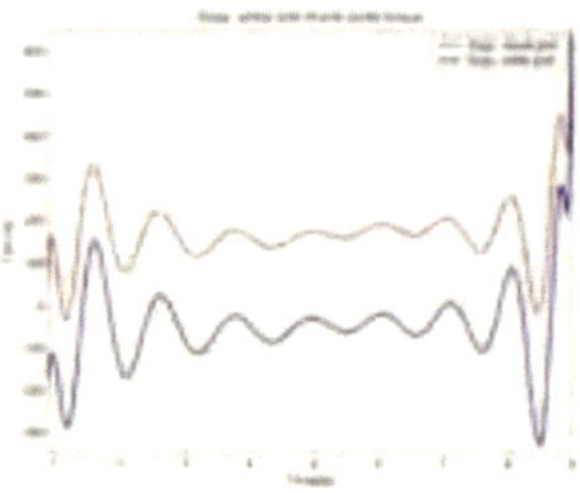

(e) Driver Torques

Fig. 1.15 *(a) The robot's stick diagram on combined surface, nominal motion, Moving ZMP,* $\lambda = -8^{\circ}$ *(b) The moving ZMP diagram in nominal gait which satisfies stability criteria (c) Inertial forces: __: Supp. tight, --: Supp. shank, ...: Swing tight, .-: Swing shank (d) Joint's torques: __: Swing shank joint, --: Swing ankle joint, ...: Supp. hip joint, .-: Swing hip joint (e) Joint's torques: __: Supp. ankle joint, --: Supp. shank joint*

5. References

Peiman Naseradin Mousavi, Ahmad Bagheri, "*Mathematical Simulation of a Seven Link Biped Robot and ZMP Considerations*", Applied Mathematical Modelling, Elsevier, 2007, Vol. 31/1.

Q. Huang, K. Yokoi, S. Kajita, K. Kaneko, H. Arai, N. Koyachi, K. Tanie, "Planning Walking Patterns For A Biped Robot", IEEE Transactions on Robotics and Automation, VOL 17, No. 3, June 2001.

John J. G, "Introduction to Robotics: Mechanics and Control", Addison-Wesley, 1989.

H. K. Lum, M. Zribi, Y. C. Soh, " Planning and Contact of a Biped Robot", International Journal of Engineering Science 37(1999), pp. 1319-1349

Eric R. Westervelt, " Toward a Coherent Framework for the Control of Plannar Biped Locomotion", A dissertation submitted in partial fulfillment of the requirements for the degree of Doctor of Philosophy, (Electrical Engineering Systems), In the University of Michigan, 2003.

H. Hon, T. Kim, and T.Park, "Tolerance Analysis of a Spur Gear Train," in Proc. 3rd DADS Korean user's Conf. 1978, pp. 61-81

Peiman Naseradin. Mousavi, "Adaptive Control of 5 DOF Biped Robot Moving On A Declined Surface", M.S Thesis, Guilan University, 2006.

A. Bagheri, M.E. Felezi, P. N. Mousavi, " Adaptivr Control and Simulation of a Seven Link Biped Robot for the Combined Trajectory Motion and Stability Investigations", WSEAS Transactions on Systems, Issue 5, Vol. 5, May 2006, pp: 1214-1222

A. Bagheri. F. Najafi, R. Farrokhi, R. Y. Moghaddam, and M. E. Felezi, "Design, Dynamic Modification, and Adaptive Control of a New Biped Walking Robot", International Journal of Humanoid Robotics, Vol. 3, Num.1, March 2006, pp 105-126

7

Analytical Criterions for the Generation of Highly Dynamic Gaits for Humanoid Robots: Dynamic Propulsion Criterion and Dynamic Propulsion Potential

Olivier Bruneau
Université de Versailles Saint-Quentin-en-Yvelines, Laboratoire d'Ingénierie des Systèmes de Versailles (LISV, EA 4048)
France

Anthony David
Université d'Orléans, Ecole Nationale Supérieure d'Ingénieurs de Bourges, Laboratoire Vision et Robotique (LVR, EA 2078)
France

1. Introduction

Many studies have been made to develop walking anthropomorphic robots able to move in environments well-adapted to human beings and able to cooperate with them. Among the more advanced projects of humanoid robots, one can mention : the Honda robots P2, P3 (Hirai, 1997) (Hirai et al., 1998) and Asimo (Sakagami et al., 2002), the HRP series developed by AIST (Kaneko et al., 1998) (Kajita et al., 2005) (Kaneko et al., 2004) (Morisawa et al., 2005), the small robot QRIO proposed by Sony (Nagasaka et al., 2004), the KHR series developed by KAIST (Kim et al., 2004) (Kim et al., 2005), the last robot of Waseda University having seven degrees of freedom per leg (Ogura et al., 2004), Johnnie (Lohmeier et al., 2004) and H7 (Kagami et al., 2005). These robots are namely able to climb stairs and to carry their own power supply during stable walking. The problem of dynamic locomotion and gait generation for biped robot has been studied theoretically and experimentally with quite different approaches. However, when searchers study the behavior or the design of dynamic walking robots, they inevitably meet a number of intrinsic difficulties related to these kinds of systems : a large number of parameters have to be optimized during the design process or have to be controlled during the locomotion task; the intrinsic stability of walking machines with dynamic behaviors is not robust; the coordination of the legs is a complex task. When human walks, it actively uses its own dynamic effects to ensure its propulsion. Today, some studies exploit the dynamic effects to generate walking gaits of robots. In this research field, four kinds of approaches are used. The first one uses pragmatic rules based on qualitative studies of human walking gaits (Pratt et al., 2001) (Sabourin et al., 2004). The second one focuses on the mechanical design of the robot in order to obtain natural passive dynamic gaits (Collins et al., 2005). The third one deals with theoretical studies of limit cycles (Westervelt et al., 2004) (Canudas-de-Wit et al., 2002). The fourth one creates various dynamic walking gaits with reference trajectories (Bruneau et al., 2001) (Chevallereau et al., 2003). However, all theses approaches are not really universal and do not allow online dynamic adaptation of the robot motions as function of the environment and of the real capabilities of the system.

Our objective is to carry out a more adaptive and universal approach based on the dynamic equations in order to generate walking gaits with a high dynamic behavior. Moreover, we do not wish to impose exact temporal desired trajectories. On the contrary, the capabilities of the robot, in term of intrinsic dynamics and actuator torques, are constantly evaluated and exploited as well as possible with an aim of reaching a desired average walking rate of the system. In order to do this, we propose a strategy based on two dynamic criterions, which takes into account, at every moment, the intrinsic dynamics of the system and the capabilities of the actuators torques in order to modify the dynamics of the robot directly. These criterions called the "dynamic propulsion criterion" and the "dynamic propulsion potential" permit to produce analytically the motions of the legs during all phases and all transition phases without specification of the events to perform. Even if this method is universal, we illustrate it for a planar under-actuated robot without foot : RABBIT (Buche, 2006). The particularity of this robot is that it can not be statically stable because the contact area between the legs and the ground is very small. Thus, a control strategy based on dynamic considerations is required.

The organization of this chapter is as follows. In the second part the presentation of the robot Rabbit and its characteristics are given. In part three, dynamics of the biped robot Rabbit are introduced. Then, in part four, the two dynamic criterions are expressed : the "dynamic propulsion criterion" and the "dynamic propulsion potential". In part five, we describe the control strategy, based on these criterions, to generate highly dynamic walking gaits. The sixth part gives the simulation results obtained with this analytical approach.

2. The bipedal robot RABBIT

The robot Rabbit (fig. 1.) is an experimentation of seven French laboratories (IRCCYN Nantes, LAG Grenoble, LMS Poitiers, LVR Bourges, LGIPM Metz, LIRMM Montpellier, LRV (new name : LISV) Versailles) concerning the control of a biped robot for walking and running gaits generation within the framework of a CNRS project ROBEA (Sabourin et al., 2006) (Chevallereau et al., 2003).

This robot is composed of two legs and a trunk and has not foot. The joints are located at the hip and at the knee. This robot has four motors : one for each hip, one for each knee. Rabbit weighs 32 kilograms and measures 1 meter. Its main masses and lengths are given in table 1. Its motions are included in sagittal planes, by using a radial bar link fixed at a central column that allows to guide the direction of progression of the robot around a circle. As Rabbit moves only in the sagittal plane, it can be modeled in two dimensions. This robot represents the minimal system able to generate walking and running gaits. Since the contact between the robot and the ground is just one point (passive dof (degree of freedom)), the robot is under-actuated during the single support phase: there are only two actuators, at the hip and at the knee of the stance leg, to control three parameters :

-the trunk evolution along the moving direction $\vec{x}$ at the reference point

-the trunk evolution along the gravity direction $\vec{y}$ at the reference point

-the trunk angle evolution around the rotation direction $\vec{z}$

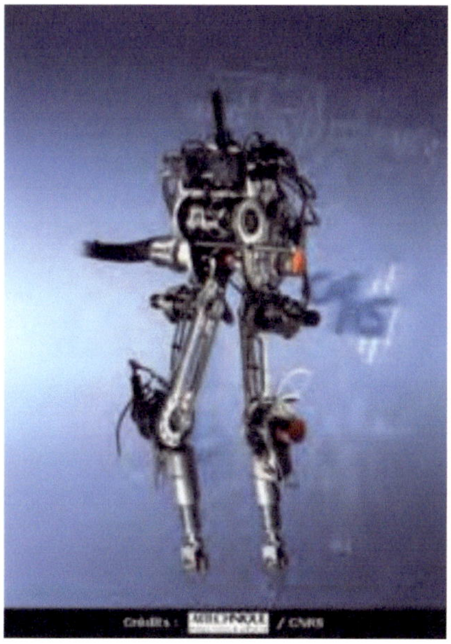

Fig. 1. Robot Rabbit.

	Weight (Kg)	Length (m)
Trunk	12	0,2
Thigh	6,8	0,4
Shin	3,2	0,4

Table 1. Characteristics of Rabbit.

The configuration of Rabbit is described by the parameters given in fig.2. The reference's point R_p is selected at the hip of the robot for the definition of the position, velocity and acceleration of the trunk.

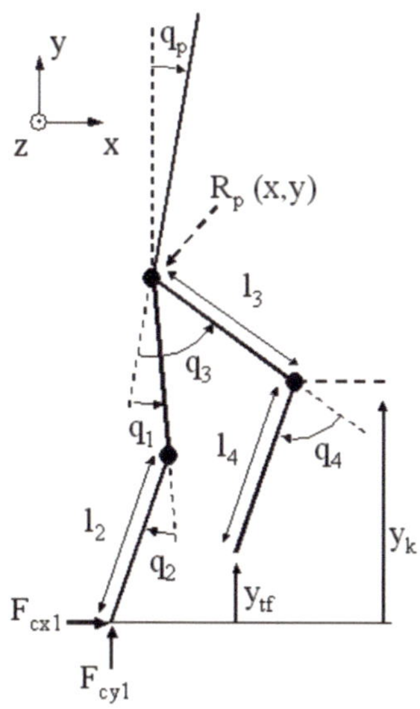

Fig. 2. Model parameters of Rabbit.

3. Analytical dynamics of Rabbit

The approach is based on the dynamic equations with the lagrangian form for the robot Rabbit at the reference point R_p.

3.1 Lagrangian Dynamic Equations

The lagrangian motion equations, applied at R_p, are written in two subsystems:

$$\overline{\overline{M}}_p\overline{a}_p+\overline{\overline{H}}_k\overline{a}_k+\overline{C}_p+\overline{G}_p=\overline{\overline{D}}_{1p}^{T}\overline{F}_{c1}+\overline{\overline{D}}_{2p}^{T}\overline{F}_{c2} \tag{1}$$

$$\overline{\overline{H}}_k^{T}\overline{a}_p+\overline{\overline{M}}_k\overline{a}_k+\overline{C}_k+\overline{G}_k=\overline{\overline{D}}_{1k}^{T}\overline{F}_{c1}+\overline{\overline{D}}_{2k}^{T}\overline{F}_{c2}+\overline{\Gamma} \tag{2}$$

where $\overline{a}_p=\left[\ddot{x},\ddot{y},\ddot{q}_p\right]^T$ is the trunk accelerations vector at R_p, $\overline{a}_k=\left[\ddot{q}_1,\ddot{q}_2,\ddot{q}_3,\ddot{q}_4\right]^T$ the joint accelerations vector, $\overline{\overline{F}}_{cj}$ (j=1,2) the contacting forces applied to the feet and $\overline{\Gamma}$ the joint torques vector. The index "p" refers to the trunk and the index "k" to the legs. The subsystem (1), coming from the first three motion equations, is related to the trunk. This subsystem is said passive. The subsystem (2), coming from the last four motion equations, is related to the legs. This subsystem is said active.

3.2 Trunk Dynamics and External Generalized Force Expression

The trunk dynamics and the external generalized forces expression, at the reference point, are obtained by isolating $\overline{a}_k$ in (2) and by injecting it in (1). The terms which are functions of the trunk configuration uniquely are isolated in the inertia matrix ${}^p\overline{\overline{M}}_p$, centrifugal and Coriolis vector ${}^p\overline{C}_p$ and gravity vector ${}^p\overline{G}_p$. The other terms related with legs or coupled between legs and trunk are grouped in the matrix ${}^k\overline{\overline{M}}_p$ and vectors ${}^k\overline{C}_p$ and ${}^k\overline{G}_p$. Thus we can write :

$$\begin{aligned}\overline{\overline{M}}_p&={}^p\overline{\overline{M}}_p+{}^k\overline{\overline{M}}_p\\ \overline{C}_p&={}^p\overline{C}_p+{}^k\overline{C}_p \qquad \text{with } {}^p\overline{C}_p=\overline{0}\\ \overline{G}_p&={}^p\overline{G}_p+{}^k\overline{G}_p\end{aligned} \tag{3}$$

The following equations system is also obtained:

$$\begin{aligned}{}^p\overline{\overline{M}}_p\overline{a}_p+{}^p\overline{C}_p+{}^p\overline{G}_p=&-\left[{}^k\overline{\overline{M}}_p-\overline{\overline{H}}_k\overline{\overline{M}}_k^{-1}\overline{\overline{H}}_k^{T}\right]\overline{a}_p\\ &+\overline{\overline{H}}_k\overline{\overline{M}}_k^{-1}\left[\overline{C}_k+\overline{G}_k-\overline{\Gamma}\right]-{}^k\overline{C}_p-{}^k\overline{G}_p\\ &+\left[\sum_{i=1}^{2}\overline{\overline{D}}_{ip}^{T}\overline{F}_{ci}-\overline{\overline{H}}_k\overline{\overline{M}}_k^{-1}\left(\overline{\overline{D}}_{1k}^{T}\overline{F}_{c1}+\overline{\overline{D}}_{2k}^{T}\overline{F}_{c2}\right)\right]\end{aligned} \tag{4}$$

The left term of (4) represents trunk dynamics whereas the right term is the external generalized forces $\overline{F}_r$ applied to the trunk at R_p.

3.3 Computed Torque Method

The actuator torques can be used on one hand to control the evolution of the trunk and on the other hand to satisfy constraints related to the locomotion (no-contact between the swing leg and the ground and landing without impact). In the second case, the actuator torques will be function of joint trajectories. The method used to calculate these desired actuator torques is the computed torque method using non linear decoupling of the dynamics. For that, we express the actuator torques $\overline{\Gamma}$ as function of the joint accelerations $\overline{a}_k$ by isolating $\overline{a}_p$ in (1) and by injecting it in (2):

$$\begin{aligned}\overline{\Gamma} &= \left[\overline{\overline{M}}_k - \overline{\overline{H}}_k^T \overline{\overline{M}}_p^{-1} \overline{\overline{H}}_k\right]\overline{a}_k + \left[\overline{C}_k - \overline{\overline{H}}_k^T \overline{\overline{M}}_p^{-1} \overline{C}_p\right] + \left[\overline{G}_k - \overline{\overline{H}}_k^T \overline{\overline{M}}_p^{-1} \overline{G}_p\right] \\ &- \left[\overline{\overline{D_{1k}}}^T \overline{F_{c1}} + \overline{\overline{D_{2k}}}^T \overline{F_{c2}} - \overline{\overline{H}}_k^T \overline{\overline{M}}_p^{-1}\left(\overline{\overline{D_{1p}}}^T \overline{F_{c1}} + \overline{\overline{D_{2p}}}^T \overline{F_{c2}}\right)\right]\end{aligned} \tag{5}$$

Then, we separate (5) in two subsystems. A subsystem, named A, composed of the equations of the actuator torques used to generate a desired joint trajectory. A subsystem, named B, composed of the other equations. In the subsystem B, we isolate the joint acceleration, associated to the actuator torques not used to imposed a desired joint trajectory. In this case, if we inject their expressions in the subsystem A, this subsystem A express the actuator torques used to impose joint trajectories as function of the desired joint acceleration and of the other actuator torques. So, knowing the configuration of the robot and the desired joint accelerations, given by the desired joint trajectories, we can calculated the actuator torques.

3.4 Close-Loop Constraint

During the double contact phase, we have to respect the close-loop constraint along the $\vec{y}$ direction when we calculate the actuator torques, in order to avoid a no desired take off of a leg. At first, this geometric constraint is expressed as follows:

$$y_{p2} - y_{p1} = g(q_p, q_1, q_2, q_3, q_4) \tag{6}$$

where y_{p1} and y_{p2} are respectively the foot position of the leg 1 and the leg 2 along the $\vec{y}$ direction. Then, we have to express the close-loop constraint in terms of actuator torques constraint. For that, we derive (6) twice:

$$\ddot{y}_{p2} - \ddot{y}_{p1} = f(\ddot{q}_p, \ddot{q}_1, \ddot{q}_2, \ddot{q}_3, \ddot{q}_4, \dot{q}_p, \dot{q}_1, \dot{q}_2, \dot{q}_3, \dot{q}_4, q_p, q_1, q_2, q_3, q_4) \tag{7}$$

where $\ddot{y}_{p2} - \ddot{y}_{p1}$ is equal to zero. Then, we replace the joint acceleration vector by its expression obtained with $f = \ddot{g}$ (5):

$$\overline{a}_k = \left[\overline{\overline{M}}_k - \overline{\overline{H}}_k^T \overline{\overline{M}}_p^{-1} \overline{\overline{H}}_k\right]^{-1} \begin{bmatrix} \overline{\Gamma} - \left[\overline{C}_k - \overline{\overline{H}}_k^T \overline{\overline{M}}_p^{-1} \overline{C}_p\right] - \left[\overline{G}_k - \overline{\overline{H}}_k^T \overline{\overline{M}}_p^{-1} \overline{G}_p\right] \\ + \begin{bmatrix} \overline{\overline{D_{1k}}}^T \overline{F_{c1}} + \overline{\overline{D_{2k}}}^T \overline{F_{c2}} \\ - \overline{\overline{H}}_k^T \overline{\overline{M}}_p^{-1}\left(\overline{\overline{D_{1p}}}^T \overline{F_{c1}} + \overline{\overline{D_{2p}}}^T \overline{F_{c2}}\right) \end{bmatrix} \end{bmatrix} \tag{8}$$

Thus, we obtain the close-loop constraint as function of the joint torques:

$$0 = f(\overline{\Gamma}, \ddot{q}_p, \dot{q}_p, \dot{q}_1, \dot{q}_2, \dot{q}_3, \dot{q}_4, q_p, q_1, q_2, q_3, q_4) \tag{9}$$

4. Dynamic criterions

Based on the previous equations, two dynamic criterions are proposed : the dynamic propulsion criterion and the dynamic propulsion potential.

4.1 Dynamic Propulsion Criterion

To start, we define the dynamic propulsion criterion. The idea of this criterion is to quantify exactly the acceleration required instantaneously to the robot to generate the desired trajectory, for the three dof. We note that the desired trajectory can be expressed just as an average desired speed of the robot at R_p. Then, knowing the configuration of the robot, the aim is to calculate the joint torques to generate these desired accelerations.

To do this, we define two generalized forces. Let $\overline{F}_r$ be the external generalized force applied to the trunk at R_p by the two legs and generating the real trajectory. Let $\overline{F}_i$ be the external generalized force which should be applied to the trunk by the two legs at R_p to generate the desired trajectory. Thus, the dynamic propulsion of the robot will be ensured if $\overline{F}_r$ is equal to $\overline{F}_i$. Finally, the difference $\overline{F}_i$–$\overline{F}_r$ represents the dynamic propulsion criterion that we have to minimize. Knowing the configuration of the robot, $\overline{F}_i$ and the expression of the real generalized forces (right term of eq. 4), we can calculate the desired joint torques $\overline{\Gamma}^d$ to generate the desired trajectory:

$$\begin{aligned} \overline{F}_i = & -\left[{}^k\overline{\overline{M}}_p - \overline{\overline{H}}_k \overline{\overline{M}}_k^{-1} \overline{\overline{H}}_k^T\right]\overline{a}_p \\ & + \overline{\overline{H}}_k \overline{\overline{M}}_k^{-1}\left[\overline{C}_k + \overline{G}_k - \overline{\Gamma}^d\right] - {}^k\overline{C}_p - {}^k\overline{G}_p \\ & + \left[\sum_{i=1}^{2} \overline{\overline{D}}_{ip}^{\,T}\, \overline{F}_{ci} - \overline{\overline{H}}_k \overline{\overline{M}}_k^{-1}\left(\overline{\overline{D}}_{1k}^{\,T}\, \overline{F}_{c1} + \overline{\overline{D}}_{2k}^{\,T}\, \overline{F}_{c2}\right)\right] \end{aligned} \tag{10}$$

To calculate $\overline{F}_i$, we use the expression of the trunk dynamics obtained in left term of (4), where we replace the real values $\ddot{x}$, $\dot{x}$, x, $\ddot{y}$, $\dot{y}$, y, $\ddot{q}_p$, $\dot{q}_p$ and q_p of the trajectory by their desired values $\ddot{x}^d$, $\dot{x}^d$, x^d, $\ddot{y}^d$, $\dot{y}^d$, y^d, $\ddot{q}_p^d$, $\dot{q}_p^d$ and q_p^d (in paragraph 5.1, we will see how obtain these desired values). The desired trajectory being defined in acceleration terms as well as in velocity and position terms, we replace:

$$\overline{a}_p^d = \left[\ddot{x}^d \;\; \ddot{y}^d \;\; \ddot{q}_p^d\right]^T \tag{11}$$

by:

$$\overline{a}_p^d = \begin{bmatrix} \ddot{x}^d + K_v\left(\dot{x}^d - \dot{x}\right) + K_p\left(x^d - x\right) \\ \ddot{y}^d + K_v\left(\dot{y}^d - \dot{y}\right) + K_p\left(y^d - y\right) \\ \ddot{q}_p^d + K_v\left(\dot{q}_p^d - \dot{q}_p\right) + K_p\left(q_p^d - q_p\right) \end{bmatrix} \tag{12}$$

what gives us the following expression of $\overline{F}_i$:

$$\overline{F}_i = {}^p\overline{\overline{M}}_p \overline{a}_p^d + {}^p\overline{C}_p^d + {}^p\overline{G}_p^d \tag{13}$$

4.2 Dynamic Propulsion Potential

With the dynamic propulsion criterion, we know the required acceleration to generate the desired trajectory and the actuator torques we must apply. However, does the robot have the capability to generate this required acceleration?

To answer this question, we introduce the dynamic propulsion potential $\widetilde{P}$. This criterion represents the capability of a system to generate a desired generalized force, i.e. the minimum and the maximum generalized force that it can generate.

For a planar system (Oxy plane), $\widetilde{P}$ is written as follows :

$$\widetilde{P}=\begin{pmatrix}P_{Fx}\\ P_{Fy}\\ P_{Mz}\end{pmatrix}=\begin{pmatrix}\left[F_{rx}^{\min};F_{rx}^{\max}\right]\\ \left[F_{ry}^{\min};F_{ry}^{\max}\right]\\ \left[M_{rz}^{\min};M_{rz}^{\max}\right]\end{pmatrix} \tag{14}$$

This criterion integrates the technology used as well as the constraints related to the system's design. In the case of Rabbit, we define three criterions, one for each leg ($\widetilde{P}_1,\widetilde{P}_2$) and one for the robot ($\widetilde{P}$).

The expressions of $\widetilde{P}_1$ and $\widetilde{P}_2$ are based on the expression of the external generalized force generated by each leg and the field of variation of the actuator torques. The expression of the external generalized force performed by each leg is obtained with the decomposition of (4):

$$\begin{aligned}\overline{F}_{rj}=&-\left\{{}^{kj}\overline{\overline{M}}_p-\overline{\overline{H}}_{kj}\overline{\overline{M}}_{kj}^{-1}\overline{\overline{H}}_{kj}^{T}\right\}\overline{a}_p+\overline{\overline{H}}_{kj}\overline{\overline{M}}_{kj}^{-1}\left[\overline{C}_{kj}+\overline{G}_{kj}-\begin{pmatrix}\Gamma_{hj}\\ \Gamma_{gj}\end{pmatrix}\right]-{}^{kj}\overline{C}_p-{}^{kj}\overline{G}_p\\ &+\left[\overline{\overline{D}}_{jp}^{T}-\overline{\overline{H}}_{kj}\overline{\overline{M}}_{kj}^{-1}\overline{\overline{D}}_{jk}^{T}\right]\overline{F}_{cj}\end{aligned} \tag{15}$$

where Γ_{hj} and Γ_{gj} are respectively the hip and knee actuator torque of the leg j.

Then, we have to determine the field of variation of the actuator torques. For that, the idea is to limit the field of variation of the actuator torques, given by the actuator technology, such as the joint actuators have the capability to stop the motion before reaching the joint limits. To determinate the limits of the field of variation of each actuator torque, we use three relations. The first relation (16) integrates the actuator technology. The joint velocity and the actuator torque give the maximum and the minimum actuator torque that the joint actuator can generate for the next sample time :

$$(\dot{q}_i,\Gamma_i)\Rightarrow\left(\Gamma_i^{\max},\Gamma_i^{\min}\right) \tag{16}$$

The second relation (17), based on (8), gives the expression of the joint acceleration as function of the actuator torque:

$$\Gamma_i\Rightarrow\ddot{q}_i \tag{17}$$

The third relation (18) gives the new joint velocity ($\dot{q}_i^{t+\delta t}$) and position ($q_i^{t+\delta t}$), knowing the new joint acceleration ($\ddot{q}_i^{t+\delta t}$) and the old angular velocity ($\dot{q}_i^{t}$) and position (q_i^{t}) :

$$\begin{aligned}\dot{q}_i^{t+\delta t}&=\dot{q}_i^{t}+\ddot{q}_i^{t+\delta t}\,\delta t\\ q_i^{t+\delta t}&=q_i^{t}+\dot{q}_i^{t}\,\delta t+\ddot{q}_i^{t+\delta t}\,\frac{\delta t^2}{2}\end{aligned} \tag{18}$$

where δt is the sample time.

With these three relations and the maximum and minimum joint limits $\left(q_i^{\min},q_i^{\max}\right)$, we limit the field of variation of the actuator torque. Here is explained the method to determinate the maximum torque that the joint actuator can generate, but the principle is the same for the minimum torque. This method is given by fig. 3.

With (16), we determine the maximum torque $\Gamma_i^{\max}$ that the technology of the joint actuator allows to generate. Then, we check if the joint actuator has the capability to stop the joint motion before reaching the joint limit $q_i^{\max}$, if we apply $\Gamma_i^{\max}$. In order to perform that, we calculate, with (17) and (18), the new joint configuration that we would have obtained if we had applied $\Gamma_i^{\max}$. Then, in order to stop the joint motion, we apply the minimum torque that the joint actuator can generate in this new joint configuration. We continue this loop as long as the joint velocity is positive. And the minimum torque is recalculated at each step of time as function of the new joint configuration with (16). Then, we compare the joint position obtained with the joint limit $q_i^{\max}$. If it is lower than $q_i^{\max}$, $\Gamma_i^{\max}$ can be applied because the joint actuator has the capability to stop the joint motion before reaching the joint limit. Otherwise, $\Gamma_i^{\max}$ is reduced until obtaining the maximum value allowing to stop the joint motion before reaching the joint limit.

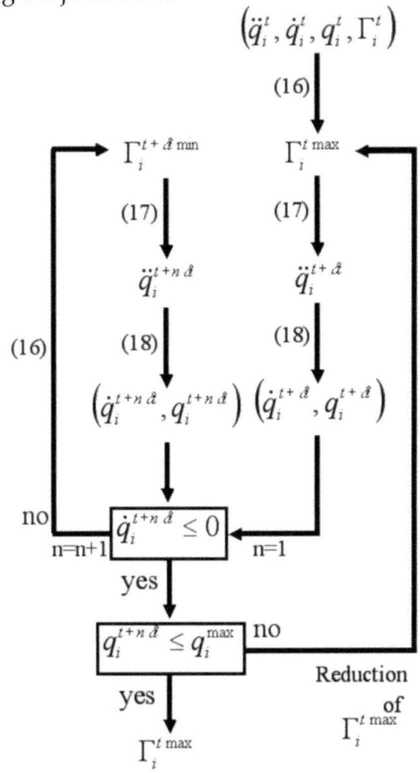

Fig. 3. Determination of the real maximum torque.

To summarize, the algorithm of fig. 3. permits to calculate the lower and upper limit of the effective torque that we can really apply, because it takes into account the configuration of the robot, the actuator technology and the joint limits:

$$\tilde{\Gamma}_i = \left[\Gamma_{ieff}^{\min}; \Gamma_{ieff}^{\max}\right] \tag{19}$$

So, knowing the field of variation of the torque that the joint actuators can generate (19) and using (15), the dynamic propulsion potential of each leg and the dynamic propulsion potential of the robot are computed.
Related to the design of Rabbit, one joint actuator per leg has an important influence on the propulsion along the moving direction. It is the knee joint for the stance leg and the hip joint for the swing leg. So, we decide to consider only the field of variation of one actuator torque

per leg to calculate its dynamic propulsion potential. Consequently, the dynamic propulsion potential $\widetilde{P}_i$ of the leg i is:

$$\widetilde{P}_i = \begin{pmatrix} \min\{F_{rx}(\Gamma_{hi}^{\min};\Gamma_{gi}), F_{rx}(\Gamma_{hi}^{\max};\Gamma_{gi})\}; \max\{F_{rx}(\Gamma_{hi}^{\min};\Gamma_{gi}), F_{rx}(\Gamma_{hi}^{\max};\Gamma_{gi})\} \\ \min\{F_{ry}(\Gamma_{hi}^{\min};\Gamma_{gi}), F_{ry}(\Gamma_{hi}^{\max};\Gamma_{gi})\}; \max\{F_{ry}(\Gamma_{hi}^{\min};\Gamma_{gi}), F_{ry}(\Gamma_{hi}^{\max};\Gamma_{gi})\} \\ \min\{M_{rz}(\Gamma_{hi}^{\min};\Gamma_{gi}), M_{rz}(\Gamma_{hi}^{\max};\Gamma_{gi})\}; \max\{M_{rz}(\Gamma_{hi}^{\min};\Gamma_{gi}), M_{rz}(\Gamma_{hi}^{\max};\Gamma_{gi})\} \end{pmatrix} \quad (20)$$

If the leg i is is the swing leg, and :

$$\widetilde{P}_i = \begin{pmatrix} \min\{F_{rx}(\Gamma_{hi};\Gamma_{gi}^{\min}), F_{rx}(\Gamma_{hi};\Gamma_{gi}^{\max})\}; \max\{F_{rx}(\Gamma_{hi};\Gamma_{gi}^{\min}), F_{rx}(\Gamma_{hi};\Gamma_{gi}^{\max})\} \\ \min\{F_{ry}(\Gamma_{hi};\Gamma_{gi}^{\min}), F_{ry}(\Gamma_{hi};\Gamma_{gi}^{\max})\}; \max\{F_{ry}(\Gamma_{hi};\Gamma_{gi}^{\min}), F_{ry}(\Gamma_{hi};\Gamma_{gi}^{\max})\} \\ \min\{M_{rz}(\Gamma_{hi};\Gamma_{gi}^{\min}), M_{rz}(\Gamma_{hi};\Gamma_{gi}^{\max})\}; \max\{M_{rz}(\Gamma_{hi};\Gamma_{gi}^{\min}), M_{rz}(\Gamma_{hi};\Gamma_{gi}^{\max})\} \end{pmatrix} \quad (21)$$

If the leg i is the stance leg.

Knowing the potential of each leg, we calculate the dynamic propulsion potential $\widetilde{P}$ of the robot. During the single contact phase, with no close-loop constraints to respect we have directly:

$$\widetilde{P} = \widetilde{P}_1 + \widetilde{P}_2 \quad (22)$$

During the double contact phase, we have to respect the close-loop constraint along the $\vec{y}$ direction. So, we limit again the field of variation of the actuator torques at the knee joint by taking into account this constraint. Equation (9) establishing a direct relation between the two actuator torques, we recalculate the minimum and the maximum torque of each joint actuator compatible with the minimum and the maximum torque of the other joint torque. This gives the new field of variation of the torque in (19) in the case of the double contact phase. Then, we recalculate with (21) the dynamic propulsion potential of each leg and, in the same way as during the single contact phase, the potential of the robot with (22).

5. Control Strategy

Based on the dynamic propulsion of the two legs ($\widetilde{P}_1, \widetilde{P}_2$) and of the robot ($\widetilde{P}$), we are able to produce for a desired average walking speed (v_a) the motion during the different locomotion phases and to detect the necessity to perform a transition between one phase to another. To perform this, we have to define on one hand the trajectory of R_p as function of v_a and on the other hand we have to define the control strategy to generate this desired trajectory.

5.1 Trajectory definition

The desired trajectory is defined by v_a and the desired trunk angle, is fixed equal to zero. With these two parameters, the nine parameters $\ddot{x}^d$, $\dot{x}^d$, x^d, $\ddot{y}^d$, $\dot{y}^d$, y^d, $\ddot{q}_p^d$, $\dot{q}_p^d$ and q_p^d are computed:

$$\begin{aligned} & x^d = x + v_a \delta t \;\; ; \;\; \dot{x}^d = v_a \;\; ; \;\; \ddot{x}^d = \frac{v_a - \dot{x}}{\delta t} \;\; ; \\ & y^d = y \;\; ; \;\; \dot{y}^d = \dot{y} \;\; ; \;\; \ddot{y}^d = \ddot{y} \;\; ; \\ & q_p^d = 0 \;\; ; \;\; \dot{q}_p^d = \frac{q_p^d - q_p}{\delta t} \;\; ; \;\; \ddot{q}_p^d = \frac{\dot{q}_p^d - \dot{q}_p}{\delta t} \;\; ; \end{aligned} \quad (23)$$

where δt is the sample time. We note that the evolution of the height y^d is free.

5.2 Control Strategy based on generalized forces

We wish to generate the desired trajectory along the $\vec{x}$ direction and around the $\vec{z}$ direction. For that, with (13), we determine F_{ix} and M_{iz} that the robot has to perform. Then, we verify if the robot has the capability to generate F_{ix} and M_{iz}. In order to do that, we compare F_{ix} to the lower and upper bounds of P_{Fx}. If F_{ix} is included in P_{Fx}, the robot has the capability to generate F_{ix}. Otherwise, F_{ix} is limited to the maximum force that the robot can generate along the $\vec{x}$ direction if F_{ix} is larger than the upper limit of P_{Fx} and limited to the minimum force if F_{ix} is lower than the lower limit of P_{Fx}. We use the same algorithm for M_{iz}. We note that the direction with the most important constraint is along the $\vec{x}$ direction. Thus, with the dynamic propulsion criterion (10) along the $\vec{x}$ direction and around the $\vec{z}$ direction, we calculate the actuator torques we must apply to generate F_{ix} and M_{iz}:

$$F_{ix} = f(\overline{\Gamma}) = f(\Gamma_{h1}, \Gamma_{g1}, \Gamma_{h2}, \Gamma_{g2}) \tag{24}$$

$$M_{iz} = f(\overline{\Gamma}) = f(\Gamma_{h1}, \Gamma_{g1}, \Gamma_{h2}, \Gamma_{g2}) \tag{25}$$

However, the constraints related to the Rabbit's design and to the generation of walking gaits have to be taken into account. So, we complete the calculation of the desired actuator torques by taking into account these constraints as function of the different phases of the walking gaits.

1) Double Contact Phase: During the double contact phase, we have to respect the close-loop constraint (9). As we have three equations ((24), (25) and (9)) to check, we use a generalized inverse to calculate the four desired joint torques.

2) Transition from the Double to the Single Contact Phase: In order to not exceed the capability of the legs, we initiate the single contact phase when the dynamic propulsion potential of one leg is weak, i.e. when the force generated by the leg along or around a direction reaches the maximum or the minimum forces that it can generate. In the case of walking gaits, the dynamic propulsion potential concerned is that of the back leg along the $\vec{x}$ direction.

During this phase, we have to ensure the take off the back leg. For that, we define a desired height for the tip of the back leg, then a trajectory to joint it. The time to reach this desired height is not fixed arbitrarily. We determine the shortest time to reach this height such as the capability of the knee actuator of the back leg is not exceeded. So, this time and consequently the trajectory is recalculated at each step of time. We decide to use the knee actuator of the back leg because it has a very weak influence on the propulsion of the robot.

With inverse kinematics, we calculate the desired joint acceleration at the knee joint and, with the computed torque method, the desired actuator torque at the knee joint.

With the three other joints, we perform the dynamic propulsion. As the swing leg has a weak influence on the propulsion around $\vec{z}$, we decide to distribute F_{ix} on each leg, as function of their dynamic propulsion potential along the $\vec{x}$ direction, that gives us the following decomposition for (24):

$$F_{ix1} = f(\Gamma_{h1}, \Gamma_{g1}) \tag{26}$$

$$F_{ix2} = f(\Gamma_{h2}) \tag{27}$$

if the leg 1 is the stance leg and the leg 2 the swing leg. So, as we have three equations ((26), (27) and (25)), we can calculate the last three desired actuator torques directly.

3) Single Contact Phase: During this phase, we use the same strategy as during the transition phase. The only difference is that we use the knee actuator of the swing leg to preserve the desired height for the tip of the swing leg and not to join it.
4) Transition from the Single to the Double Contact Phase: In order to not exceed the capability of the legs, we initiate the return to the double contact phase when the dynamic propulsion potential of one leg is weak.
During this transition, we have to ensure the landing of the swing leg, which has to avoid a strong impact. So, we must control the trajectory of the tip of the swing leg along the $\vec{x}$ and $\vec{y}$ directions. In order to carry out this, we use the same strategy as this one used to control the height of the swing leg's tip during the single contact phase. However, we have to control the trajectory along two directions. Thus, we decide to use the two joint actuators of the swing leg.
With the two other joint actuators we perform the dynamic propulsion. As we do not use the swing leg to ensure the dynamic propulsion, we compare the force and the torque, that the stance leg has to perform to generate F_{ix} and M_{iz} respectively, with its dynamic propulsion potential. If the stance leg does not have the capability to generate them, we limit them. So, as we have two equations ((24) and (25)), we can calculate the last two desired actuator torques directly.

6. Simulation results

With these criterions, we develop a strategy to control RABBIT and we validate it with the simulation of the first step of a walking gait. To start, we define the desired average velocity at R_p. As an illustration, linear velocity $\dot{x}$ is imposed and the desired trunk angle q_p is equal to zero. The evolution of the height y is free. So, we ensure the dynamic propulsion along the x direction and around the z direction only. The next idea is to modify, via the joint torques, the dynamic effects generated by the legs in order to make $\overline{F}_r$ converge towards $\overline{F}_i$. With the equations along the x direction and around the z direction, we can calculate $\overline{\Gamma}^d$ directly. However, in this case, we do not take into account the constraints related to the robot's design and to the generation of walking gaits. These constraints are introduced in the next section as functions of locomotion phases.

6.1 Double support phase

During this phase, we have four joint actuators to ensure the dynamic propulsion and the closed loop constraint (Fig. 2. with y_{tf}=0). As we have three equations, we use a generalized inverse to calculate $\overline{\Gamma}^d$. In order to not exceed the capability of the robot, we use the dynamic propulsion potential of each leg. We compare the force generated by the legs to their dynamic propulsion potential along the x direction. If a leg reaches its maximum capability, we decide to start its take off to perform the single support phase.
We see that the robot generates the desired moving velocity at R_p (Fig. 4.), without exceeding its capability nor the capability of each leg along the x direction (Fig.5). After 1,3 seconds, the maximum capability of the back leg along the x direction reduces. Indeed, the knee joint of the back leg approaches its joint stop (Fig.6). So, when the back leg reaches its maximum capability, the take off starts in order to perform the single support phase.

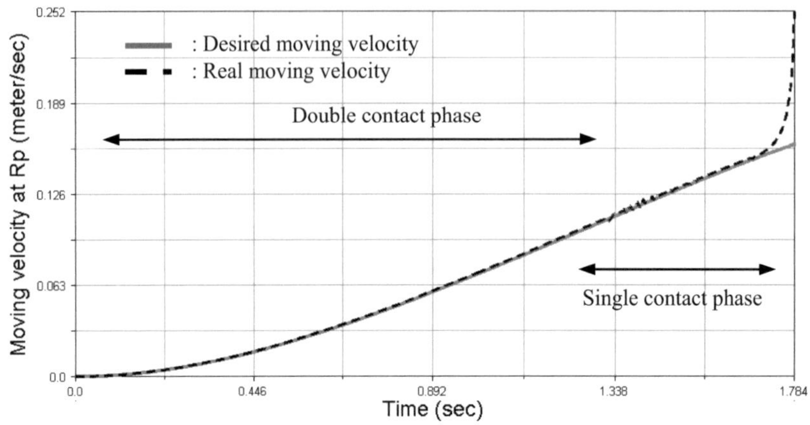

Fig. 4. Desired and real Moving velocity at R_p.

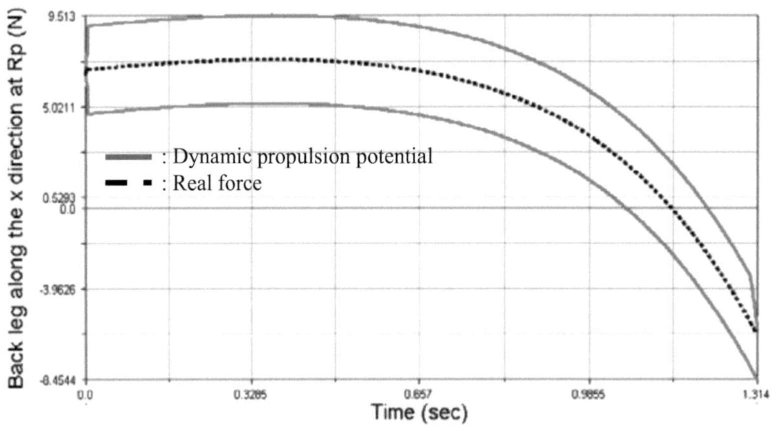

Fig. 5. Dynamic propulsion potential and real force of the back leg along the x direction at R_p.

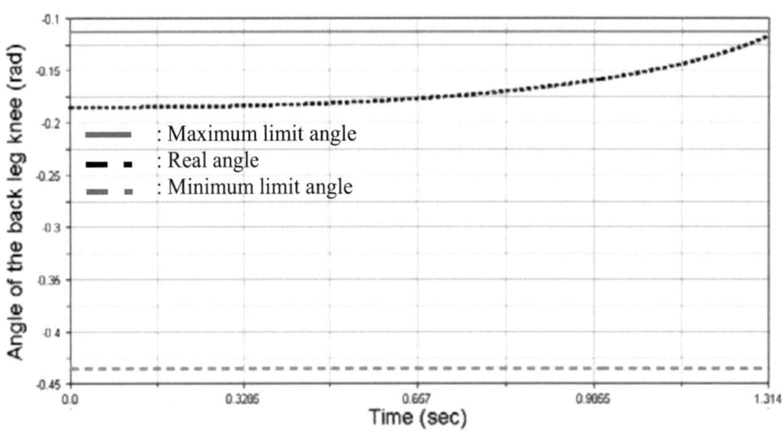

Fig. 6. Angle and limit angle of the back leg knee.

6.2 Single support phase

During this phase, we have four joint actuators to ensure the dynamic propulsion and to avoid the contact between the tip of the swing leg and the ground. In order to do this, the swing leg's knee is used to avoid the contact while the three other joints perform the dynamic propulsion of the robot.

The joint torque of the swing leg' knee is calculated with a computed torque method using non linear decoupling of the dynamics. The desired joint acceleration, velocity and position of the swing leg knee are calculated with inverse kinematics. We express the joint torque of the swing leg' knee as function of the three other joint torques and of the desired control vector components of the swing leg's knee.

With the three other joints we perform the dynamic propulsion. It is possible that the robot does not have the capability to propel itself along the x direction dynamically. In this case, we limit the desired force with its dynamic propulsion potential. Then, we distribute this desired force with the dynamic propulsion potential of each leg. In order to keep a maximum capability for each leg, the desired force generated by each leg is chosen to be as further as possible from the joint actuators limits. In this case, we have three equations, one for the desired force along the x direction for each leg and one for the desired force around the z direction, to calculate the three joint torques. The joint torque of the swing leg's knee is replaced by its expression in function of the three other joint torques. So, we calculate the three joints torque performing the dynamic propulsion, then the joint torque avoiding the contact between the tip of the swing leg and the ground.

We see that the robot can ensure the propulsion along the x direction (Fig.7) and generates the desired moving velocity (single contact phase in Fig. 4.). Moreover, the control strategy involves naturally the forward motion of the swing leg (Fig.8). After 1,675 seconds, the robot can not ensure exactly the desired propulsion along the x direction. Indeed, the swing leg is passed in front of the stance leg and the system is just like an inverse pendulum submitted to the gravity and for which rotational velocity increases quickly.

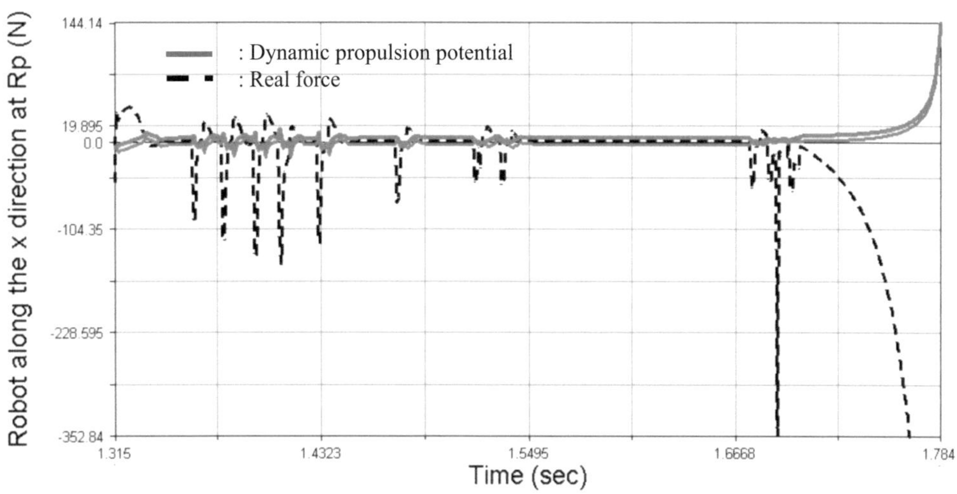

Fig. 7. Dynamic propulsion potential and real force of the robot along the x direction at R_{p}..

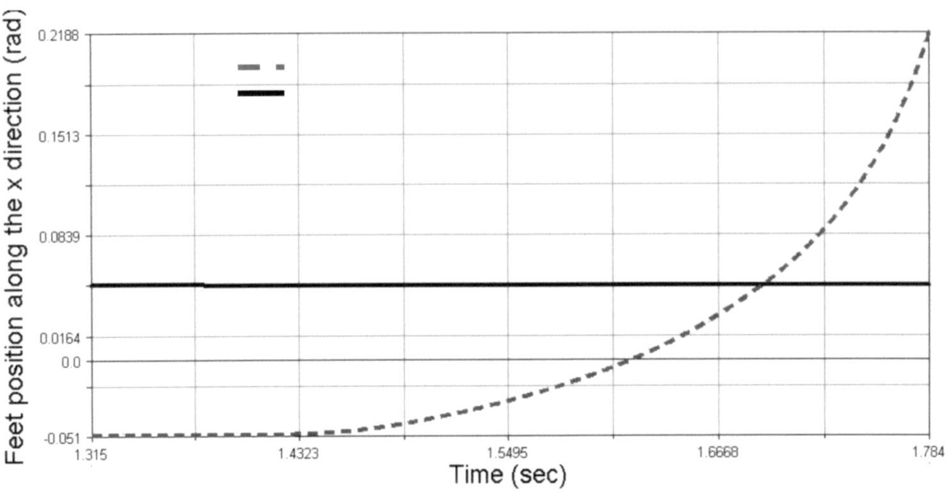

Fig. 8. Feet position along the x direction.

7. Conclusions and future work

In this paper, we presented an analytical approach for the generation of walking gaits with a high dynamic behavior. This approach, using the dynamic equations, is based on two dynamic criterions: the dynamic propulsion criterion and the dynamic propulsion potential. Furthermore, in this method, the intrinsic dynamics of the robot as well as the capability of its joint torques are taken into account at each sample time.

In this paper, in order to satisfy the locomotion constraints, for instance the no-contact between the tip of the swing's leg and the ground, we selected joint actuators, for instance the knee to avoid this contact during the single contact phase. Our future work will consist in determining the optimal contribution of each joint actuator by using the concept of dynamic generalized potential in order to satisfy at the same time the dynamic propulsion and the locomotion constraints. In this case, just one desired parameter will be given to the robot: the average speed. Thus, all the decisions concerning the transitions between two phases or the motions during each phase will be fully analytically defined.

8. References

Bruneau, O.; Ouezdou, F.-B. & Fontaine, J.-G. (2001). Dynamic Walk of a Bipedal Robot Having Flexible Feet, *Proceedings of IEEE International Conference On Intelligent Robots and Systems*, 2001.

Buche., G. (2006). *ROBEA Home Page*, http://robot-rabbit.lag.ensieg.inpg.fr/English/, 2006.

Canudas-de-Wit, C.; Espiau, B. & Urrea, C. (2002). Orbital Stabilization of Underactuated Mechanical System, *Proceedings of IFAC*, 2002.

Chevallereau, C. & Djoudi, D. (2003). Underactuated Planar Robot Controlled via a Set of Reference Trajectories, *Proceedings of International Conference On Climbing and Walking Robots*, September 2003.

Collins, S. H. & Ruina, A. (2005). A Bipedal Walking Robot with Efficient and Human-Like Gait, *Proceedings of IEEE International Conference On Robotics and Automation*, Barcelona, April 2005, Spain.

Hirai, K. (1997). Current and Future Perspective of Honda Humanoid Robot, *Proceedings of IEEE International Conference On Intelligent Robots and Systems*, PP. 500-508, 1997.

Hirai, K.; Hirose, M. ; Haikawa, Y. & Takenaka, T. (1998). The Development of Honda Humanoid Robot, *Proceedings of IEEE International Conference On Robotics and Automation*, PP. 1321-1326, 1998.

Kagami, S.; Nishiwaki, K.; Kuffner Jr, J.-J.; Kuniyoshi, Y.; Inaba, M. & Inoue, H. (2005). Online 3D Vision, Motion Planning and Biped Locomotion Control Coupling System of Humanoid Robot : H7, *Proceedings of IEEE International Conference On Intelligent Robots and Systems*, PP. 2557-2562, 2005.

Kajita, S.; Nagasaki, T.; Kaneko, K.; Yokoi, K. & Tanie, K. (2005). A Running Controller of Humanoid Biped HRP-2LR, *Proceedings of IEEE International Conference On Robotics and Automation*, PP. 618-624, Barcelona, April 2005, Spain.

Kaneko, K.; Kanehiro, F.; Kajita, S.; Yokoyama, K.; Akachi, K.; Kawasaki, T.; Ota, S. & Isozumi, T. (1998). Design of Prototype Humanoid Robotics Plateform for HRP, *Proceedings of IEEE International Conference On Intelligent Robots and Systems*, PP. 2431-2436, 1998.

Kaneko, K.; Kanehiro, F. & Kajita, S. (2004). Humanoid Robot HRP-2, *Proceedings of IEEE International Conference On Robotics and Automation*, PP. 1083-1090, 2004.

Kim, J.-Y.; Park, I.-W.; Lee, J.; Kim, M.-S.; Cho, B.-K. & Oh, J.-H. (2005). System Design and Dynamic Walking of Humanoid Robot KHR-2, *Proceedings of IEEE International Conference On Robotics and Automation*, PP. 1443-1448, Barcelona, April 2005, Spain.

Kim, J.-Y. & Oh, J.-H. (2004). Walking Control of the Humanoid Platform KHR-1 based on Torque Feedback Control, *Proceedings of IEEE International Conference On Robotics and Automation*, PP. 623-628, 2004.

Lohmeier, S.; Löffler, K. ; Gienger, M.; Ulbrich, H. & Pfeiffer, F. (2004). Computer System and Control of Biped "Johnnie", *Proceedings of IEEE International Conference On Robotics and Automation*, PP 4222-4227, 2004.

Morisawa, M.; Kajita, S.; Kaneko, K.; Harada, K.; Kanehiro, F.; Fujiwara, K. & H. Hirukawa, H. (2005). Pattern Generation of Biped Walking Constrained on Parametric Surface, *Proceedings of IEEE International Conference On Robotics and Automation*, PP. 2416-2421, Barcelona, April 2005, Spain.

Nagasaka, K.; Kuroki, Y.; Suzuki, S.; Itoh,Y. and Yamaguchi, J. (2004). Integrated Motion Control for Walking, Jumping and Running on a Small Bipedal Entertainment Robot, *Proceedings of IEEE International Conference On Intelligent Robots and Systems*, PP. 3189-3194, 2004.

Pratt, J. E.; Chew, C.-M.; Torres, A.; Dilworth, P. & Pratt, G. (2001). Virtual Model Control : an Intuitive Approach for Bipedal Locomotion, *International Journal of Robotics Research*, vol. 20, pp. 129-143, 2001.

Sabourin, C.; Bruneau, O. & Fontaine, J.-G. (2004). Pragmatic Rules for Real-Time Control of the Dynamic Walking of an Under-Actuated Biped Robot, *Proceedings of IEEE International Conference On Robotics and Automation*, April 2004.

Sabourin, C.; Bruneau, O. & Buche, G. (2006). Control strategy for the robust dynamic walk of a biped robot, *International Journal of Robotics Research*, Vol.25, N°9, pp. 843--860.

Sakagami,Y.; Watanabe,R.; Aoyama, C.; Matsunaga, S.; Higaki, N. & Fujimura, K. (2002). The Intelligent ASIMO: System Overview and Integration, *Proceedings of IEEE International Conference On Intelligent Robots and Systems*, PP. 2478-2483, 2002.

Westervelt, E. R.; Buche, G. & Grizzle J. W. (2004). Experimental Validation of a Framework for the Design of Controllers that Induce Stable Walking in Planar Bipeds, *International Journal of Robotics Research*, vol. 23, no. 6, June 2004.

8

Design of a Humanoid Robot Eye

Giorgio Cannata*, Marco Maggiali**
**University of Genova*
Italy
*** Italian Institute of Technology*
Italy

1. Introduction

This chapter addresses the design of a robot eye featuring the mechanics and motion characteristics of a human one. In particular the goal is to provide guidelines for the implementation of a tendon driven robot capable to emulate saccadic motions.
In the first part of this chapter the physiological and mechanical characteristics of the eye-plant[1] in humans and primates will be reviewed. Then, the fundamental motion strategies used by humans during saccadic motions will be discussed, and the mathematical formulation of the relevant *Listing's Law* and *Half-Angle Rule*, which specify the geometric and kinematic characteristics of ocular saccadic motions, will be introduced.
From this standpoint a simple model of the eye-plant will be described. In particular it will be shown that this model is a good candidate for the implementation of Listing's Law on a purely mechanical basis, as many physiologists believe to happen in humans. Therefore, the proposed eye-plant model can be used as a reference for the implementation of a robot emulating the actual mechanics and actuation characteristics of the human eye.
The second part of this chapter will focus on the description of a first prototype of fully embedded robot eye designed following the guidelines provided by the eye-plant model. Many eye-head robots have been proposed in the past few years, and several of these systems have been designed to support and rotate one or more cameras about independent or coupled *pan-tilt* axes. However, little attention has been paid to emulate the actual mechanics of the eye, although theoretical investigations in the area of modeling and control of human-like eye movements have been presented in the literature (Lockwood et al., 1999; Polpitiya & Ghosh, 2002; Polpitiya & Ghosh, 2003; Polpitiya et al., 2004).
Recent works have focused on the design of embedded mechatronic robot eye systems (Gu et al., 2000; Albers et al., 2003; Pongas et al., 2004). In (Gu et al., 2000), a prosthetic implantable robot eye concept has been proposed, featuring a single degree-of-freedom. Pongas et al., (Pongas et al., 2004) have developed a mechanism which actuates a CMOS micro-camera embedded in a spherical support. The system has a single degree-of-freedom, and the spherical shape of the eye is a purely aesthetical detail; however, the mechatronic approach adopted has addressed many important engineering issues and led to a very

[1] By eye-plant we mean the eye-ball and all the mechanical structure required for its actuation and support.

interesting system. In the prototype developed by Albers et al., (Albers et al., 2003) the design is more *humanoid*. The robot consists of a sphere supported by *slide bearings* and moved by a stud constrained by two gimbals. The relevance of this design is that it actually exploits the spherical shape of the eye; however, the types of ocular motions which could be generated using this system have not been discussed.
In the following sections the basic mechanics of the eye-plant in humans will be described and a quantitative geometric model introduced. Then, a first prototype of a tendon driven robot formed by a sphere hold by a low friction support will be discussed. The second part of the chapter will described some of the relevant issues faced during the robot design.

2. The human eye

The human eye has an almost spherical shape and is hosted within a cavity called *orbit*; it has an average diameter ranging between *23 mm* and *23.6 mm*, and weighs between *7 g* and *9 g*. The eye is actuated by a set of six *extra-ocular muscles* which allow the eye to rotate about its centre with negligible translations (Miller & Robinson, 1984; Robinson, 1991).
The rotation range of the eye can be approximated by a cone, formed by the admissible directions of fixation, with an average width of about 76 *deg* (Miller & Robinson, 1984). The action of the extra-ocular muscles is capable of producing accelerations up to *20.000 deg* sec^{-2} allowing to reach angular velocities up to *800 deg* sec^{-1} (Sparks, 2002).
The extra-ocular muscles are coupled in *agonostic/antagonistic* pairs, and classified in two groups: *recti (medial/lateral* and *superior/inferior*), and *obliqui* (*superior/inferior*). The four recti muscles have a common origin in the bottom of the orbit (*annulus of Zinn*); they diverge and run along the eye-ball up to their *insertion points* on the *sclera* (the eye-ball surface). The insertion points form an angle of about *55 deg* with respect to the optical axis and are placed symmetrically (Miller & Robinson, 1984; Koene & Erkelens, 2004). (Fig. 1, gives a qualitative idea of the placement of the four recti muscles.) The obliqui muscles have a more complex path within the orbit: they produce actions almost orthogonal to those generated by the recti, and are mainly responsible for the torsion of the eye about its optical axis. The superior oblique has its origin from the annulus of Zinn and is routed through a connective sleeve called *troclea*; the inferior oblique starts from the side of the orbit and is routed across the orbit to the eye ball.

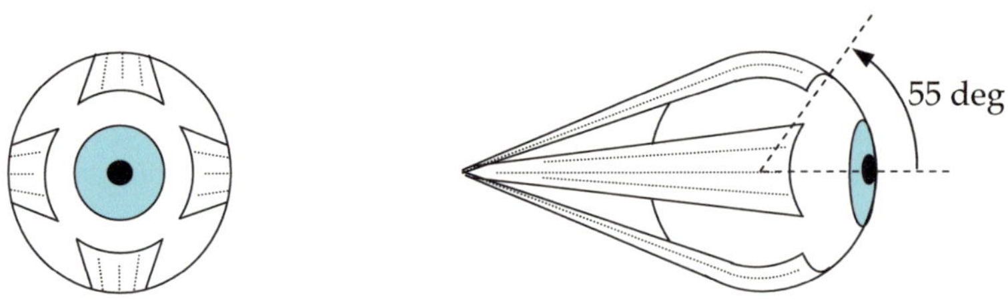

Fig. 1. Frontal and side view of the eye: qualitative placement of recti muscles.

Recent anatomical and physiological studies have suggested that the four recti have an important role for the implementation of saccadic motions which obey to the so called *Listing's Law*. In fact, it has been found that the path of the recti muscles within the orbit is constrained by soft connective tissue (Koornneef, 1974; Miller, 1989, Demer et al., 1995, Clark et al. 2000, Demer et al., 2000), named *soft-pulleys*. The role of the soft-pulleys to generate ocular motions compatible with Listing's Law in humans and primates is still debated (Hepp, 1994; Raphan, 1998; Porrill et al., 2000; Wong et al., 2002; Koene & Erkelens 2004; Angelaki, 2004); however, analytical and simulation studies suggest that the implementation of Listing's Law on a mechanical basis is feasible (Polpitiya, 2002; Polpitiya, 2003; Cannata et al., 2006; Cannata & Maggiali, 2006).

3. Saccadic motions and Listing's Law

The main goal of the section is to introduce saccades and provide a mathematical formulation of the geometry and kinematics of saccadic motions, which represent the starting point for the development of models for their implementation.
Saccadic motions consist of rapid and sudden movements changing the direction of fixation of the eye. Saccades have duration of the order of a few hundred milliseconds, and their high speed implies that these movements are open loop with respect to visual feedback (Becker, 1991); therefore, the control of the rotation of the eye during a saccade must depend only on the mechanical and actuation characteristics of the eye-plant. Furthermore, the lack of any stretch or proprioceptive receptor in extra-ocular muscles (Robinson, 1991), and the unclear role of other sensory feedback originated within the orbit (Miller & Robinson, 1984), suggest that the implementation of Listing's Law should have a strong mechanical basis.
Although saccades are apparently controlled in open-loop, experimental tests show that they correspond to regular eye orientations. In fact, during saccades the eye orientation is determined by a basic principle known as Listing's Law, which establishes the amount of eye torsion for each direction of fixation. Listing's Law has been formulated in the mid of the 19th century, but it has been experimentally verified on humans and primates only during the last 20 years (Tweed & Vilis, 1987; Tweed & Vilis, 1988; Tweed & Vilis, 1990; Furman & Schor, 2003).
Listing's Law states that there exists a specific orientation of the eye (with respect to a head fixed reference frame $<h> = \{h_1,h_2,h_3\}$), called *primary position*. During saccades any physiological orientation of the eye (described by the frame $<e> = \{e_1,e_2,e_3\}$), with respect to the primary position, can be expressed by a unit quaternion q whose (unit) rotation axis, v, always belongs to a head fixed plane, $\mathcal{L}$. The normal to plane $\mathcal{L}$ is the eye's direction of fixation at the primary position. Without loss of generality we can assume that e_3 is the fixation axis of the eye, and that $<h> \equiv <e>$ at the primary position: then, $\mathcal{L} = span\{h_1, h_2\}$. Fig. 2 shows the geometry of Listing compatible rotations.
In order to ensure that $v \in \mathcal{L}$ at any time, the eye's angular velocity ω, must belong to a plane $\mathcal{P}_\omega$, passing through v, whose normal, n_ω, forms an angle of $\theta/2$ with the direction of fixation at the primary position, see Fig. 3. This property, directly implied by Listing's Law, is usually called *Half Angle Rule*, (Haslwanter, 1995). During a generic saccade the plane $\mathcal{P}_\omega$ is rotating with respect to both the head and the eye due to its dependency from v and θ. This fact poses important questions related to the control mechanisms required to implement the Listing's Law, also in view of the fact that there is no evidence of sensors in the eye-plant capable to detect how $\mathcal{P}_\omega$ is oriented. Whether Listing's Law is implemented in humans and primates on a mechanical basis, or it requires an active feedback control action, processed by the brain, has been debated among neuro-physiologists in the past few years. The evidence of the so called soft pulleys, within the orbit, constraining the extra ocular muscles, has

suggested that the mechanics of the eye plant could have a significant role in the implementation of Half Angle Rule and Listing's Law (Quaia & Optican, 1998; Raphan 1998; Porril et al., 2000; Koene & Erkelens, 2004), although counterexamples have been presented in the literature (Hepp, 1994; Wong et al., 2002).

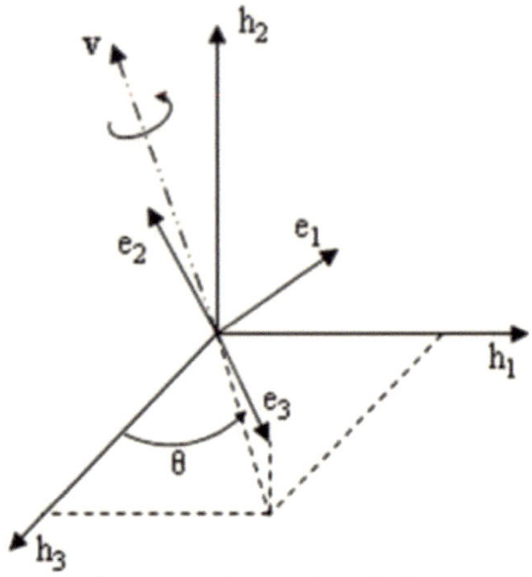

Fig. 2 Geometry of Listing compatible rotations. The finite rotation of the eye fixed frame <*e*>, with respect to <*h*> is described by a vector v always orthogonal to h_3.

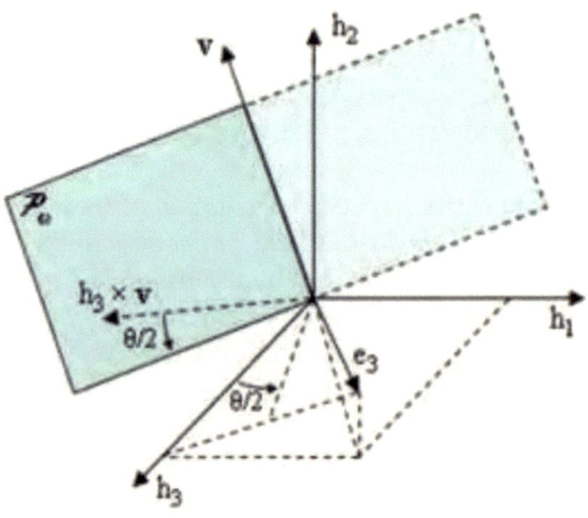

Fig. 3. *Half Angle Rule* geometry. The eye's angular velocity must belong to the plane $\mathscr{P}_\omega$ passing through axis v.

4. Eye Model

The eye in humans has an almost spherical shape and is actuated by six extra-ocular muscles. Each extra-ocular muscle has an insertion point on the sclera, and is connected with the bottom of the orbit at the other end. Accordingly to the rationale proposed in (Haslwanter, 2002; Koene & Erkelens, 2004), only the four rectii extra-ocular muscles play a significant role during saccadic movements. In (Lockwood et al., 1989), a complete 3D model of the eye plant including a non linear dynamics description of the extra-ocular muscles has been proposed. This model has been extended in (Polpitiya & Ghosh, 2002; Polpitiya & Ghosh, 2003), including also a description of the soft pulleys as elastic suspensions (springs). However, this model requires that the elastic suspensions perform particular movements in order to ensure that Listing's Law is fulfilled. The model proposed in (Cannata et al., 2006; Cannata & Maggiali, 2006), and described in this section, is slightly simpler than the previous ones. In fact, it does not include the dynamics of extra-ocular muscles, since it can be shown that it has no role in implementing Listing's Law, and models soft pulleys as fixed pointwise pulleys. As it will be shown in the following, the proposed model, for its simplicity, can also be used as a guideline for the design of humanoid tendon driven robot eyes.

4.1 Geometric Model of the Eye

The eye-ball is assumed to be modeled as a homogeneous sphere of radius R, having 3 rotational degrees of freedom about its center. Extra-ocular muscles are modeled as non-elastic thin wires (Koene & Erkelens, 2004), connected to pulling force generators (Polpitiya & Ghosh, 2002). Starting from the insertion points placed on the eye-ball, the extra-ocular muscles are routed through head fixed pointwise pulleys, emulating the soft-pulley tissue. The pointwise pulleys are located on the rear of the eye-ball, and it will be shown that appropriate placement of the pointwise pulleys and of the insertion points has a fundamental role to implement the Listing's Law on a purely mechanical basis.
Let O be the center of the eye-ball, then the position of the pointwise pulleys can be described by vectors $\boldsymbol{p}_i$, while, at the primary position insertion points can be described by vectors $\boldsymbol{c}_i$, obviously assuming that $|\boldsymbol{c}_i| = R$. When the eye is rotated about a generic axis v by an angle θ, the position of the insertion points can be expressed as:

$$r_i = R(v, \theta)\, c_i \quad \forall i = 1 \ldots 4 \tag{1}$$

where $R(v, \theta)$ is the rotation operator from the eye to the head coordinate systems.
Each extra-ocular muscle is assumed to follow the shortest path from each insertion point to the corresponding pulley, (Demer et al., 1995); then, the path of the each extra-ocular muscle, for any eye orientation, belongs to the plane defined by vectors $\boldsymbol{r}_i$ and $\boldsymbol{p}_i$. Therefore, the torque applied to the eye by the pulling action $\tau_i \geq 0$, of each extra-ocular muscle, can be expressed by the following formula:

$$m_i = \tau_i \frac{r_i \times p_i}{|r_i \times p_i|} \quad \forall i = 1 \ldots 4 \tag{2}$$

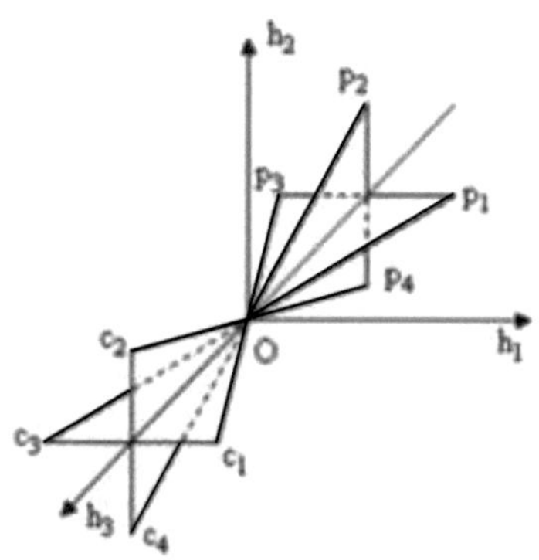

Fig. 4. The relative position of pulleys and insertion points when the eye is in the *primary position*.

From expression (2), it is clear that $|p_i|$ does not affect the direction or the magnitude of m_i so we can assume in the following that $|p_i| = |c_i|$. Instead, the orientation of the vectors p_i, called *principal directions*, are extremely important. In fact, it is assumed that p_i and c_i are symmetric with respect to the plane $\mathcal{L}$; this condition implies:

$$(v \cdot c_i) = (v \cdot p_i) \quad \forall i = 1 \ldots 4, \forall v \in \mathcal{L} \tag{3}$$

Finally, it is assumed that insertion points are symmetric with respect to the fixation axis:

$$(h_3 \cdot c_i) = (h_3 \cdot c_j) \quad \forall i, j = 1 \ldots 4 \tag{4}$$

and

$$(c_3 - c_1) \cdot (c_4 - c_2) = 0 \quad \forall i, j = 1 \ldots 4 \tag{5}$$

4.2 Properties of the Eye Model

In this section we review the most relevant properties of the proposed model. First, it is possible to show that, for any eye orientation compatible with Listing's Law, all the torques m_i produced by the four rectii extra-ocular muscles belong to a common plane passing through the finite rotation axis $v \in \mathcal{L}$, see (Cannata et al., 2006) for proof.

Theorem 1: Let $v \in \mathcal{L}$ be the finite rotation axis for a generic eye orientation, then there exists a plane $\mathcal{M}$, passing through v such that:

$$m_i \in \mathcal{M} \quad \forall i = 1 \ldots 4$$

A second important result is that, at any Listing compatible eye's orientation, the relative positions of the insertion points and pointwise pulleys form a set of parallel vectors, as stated by the following theorem, see (Cannata et al., 2006) for proof.

Theorem 2: Let $v \in \mathcal{L}$ be the finite rotation axis for a generic eye orientation, then:

$$(r_i - p_i) \times (r_j - p_j) = 0 \quad \forall i, j = 1 \ldots 4$$

Finally, it is possible to show that planes $\mathcal{M}$ and $\mathcal{P}_\omega$ are coincident, see (Cannata et al., 2006) for proof.

Theorem 3: Let $\mathbf{v} \in \mathcal{L}$ be the finite rotation axis for a generic eye orientation, then:

$$\mathbf{m}_i \in \mathcal{P}_\omega \quad \forall i = 1 \ldots 4$$

Remark 1: Theorem (3) has in practice the following significant interpretation. For any Listing compatible eye orientation any possible torque applied to the eye, and generated using only the four *rectii* extra-ocular muscles, must lay on plane $\mathcal{P}_\omega$.

The problem now is to show, according to formula (2), when arbitrary torques $m_i \in \mathcal{P}_\omega$ can be generated using only pulling forces. Theorem 2 and theorem 3 imply that m_i are all orthogonal to the vector n_ω, normal to plane $\mathcal{P}_\omega$. Therefore, formula (2) can be rewritten as:

$$\boldsymbol{\tau} = -n_\omega \times \left(\sum_{i=1}^{4} \gamma_i r_i\right) \tag{6}$$

where:

$$\gamma_i = \frac{\tau_i}{|n_\omega \times r_i|} \geq 0 \quad \forall i = 1 \ldots 4 \tag{7}$$

take into account the actual pulling forces generated by the extra-ocular muscles. From formula (6), it is clear that τ is orthogonal to a *convex* linear combination of vectors r_i. Then, it is possible to generate any torque vector laying on plane $\mathscr{P}_\omega$, as long as $\boldsymbol{n}_\omega$ belongs to the convex hull of vectors $\boldsymbol{r}_i$ as shown in Fig. 5.

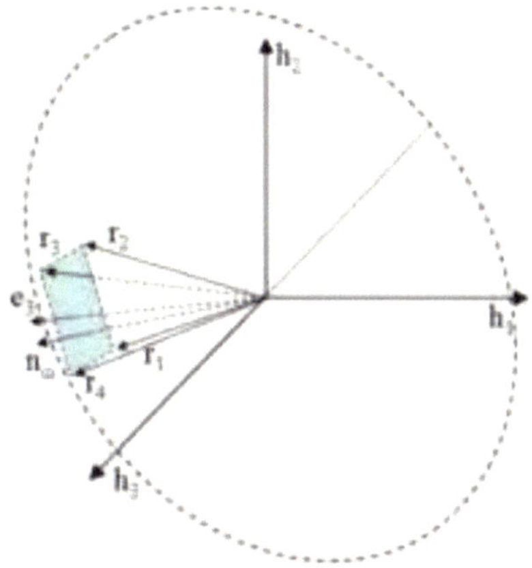

Fig. 5. When vector $\boldsymbol{n}_\omega$ belongs to the convex hull of vectors $\boldsymbol{r}_i$ then rectii extra-ocular muscles can generate any admissible torque on P_ω.

Remark 2: The discussion above shows that the placement of the insertion points affects the range of admissible motions. According to the previous discussion when the eye is in its primary position any torque belonging to plane $\mathcal{L}$ can be assigned. The angle β formed by the insertion points with the optical axis determines the actual eye workspace. For an angle $\beta = 55\ deg$ the eye can rotate of about $45\ deg$ in all directions with respect to the direction of fixation at the primary position.

Assume now that, under the assumptions made in section 3, a simplified dynamic model of the eye could be expressed as:

$$I\dot{\omega} = \boldsymbol{\tau} \tag{8}$$

where I is a scalar describing the momentum of inertia of the eye-ball, while $\dot{\omega}$ is its angular acceleration of the eye. Let us assume at time 0 the eye to be in the primary position, with zero angular velocity (zero state). Then, the extra-ocular muscles can generate a resulting torque of the form:

$$\boldsymbol{\tau} = v\,\theta(t) \tag{9}$$

where $v \in \mathcal{L}$ is a constant vector and $\theta(t)$ a scalar control signal. Therefore, ω and $\dot{\omega}$ are parallel to v; then, it is possible to reach any Listing compatible orientation, and also, during the rotation, the Half Angle Rule is satisfied. Similar reasoning can be applied to control the eye orientation to the primary position starting from any Listing compatible orientation and zero angular velocity.

The above analysis proves that saccadic motions from the primary position to arbitrary secondary positions can be implemented on a mechanical basis. However, simulative examples, discussed in (Cannata & Maggiali, 2006), show that also generic saccadic motions can be implemented adopting the proposed model. Theoretical investigations on the model properties are currently ongoing to obtain a formal proof of the evidence provided by the simulative tests.

5. Robot Eye Design

In this section we will give a short overview of a prototype of humanoid robot eye designed following the guidelines provided by the model discussed in the previous section, while in the next sections we will discuss the most relevant design details related to the various modules or subsystems.

5.1 Characteristics of the Robot

Our goal has been the design of a prototype of a robot eye emulating the mechanical structure of the human eye and with a comparable working range. Therefore, the first and major requirement has been that of designing a spherical shape structure for the eye-ball and to adopt a tendon based actuation mechanism to drive the ocular motions. The model discussed in the previous sections allowed to establish the appropriate quantitative specifications for the detailed mechanical design of the system.
At system level we tried to develop a fairly integrated device, keeping also into account the possible miniaturization of the prototype to human scale. The current robot eye prototype has a cylindrical shape with a diameter of about *50 mm* and an overall length of about *100 mm*, Fig. 6; the actual eye-ball has a diameter of *38.1 mm* (i.e. about 50% more than the human eye). These dimensions have been due to various trade-offs during the selection of the components available off-the-shelf (e.g. the eye-ball, motors, on board camera etc.), and budget constraints.

5.2 Components and Subsystems.

The eye robot prototype consists of various components and subsystems. The most relevant, discussed in detail in the next sections, are: the eye-ball, the eye-ball support, the pointwise pulleys implementation, the actuation and sensing system, and the control system architecture.

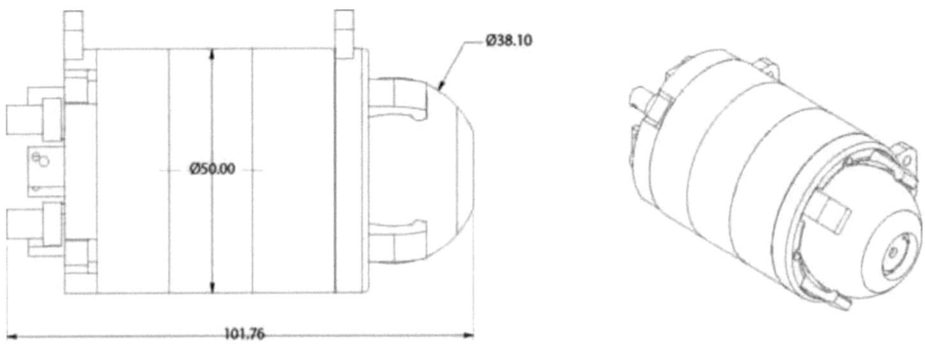

Fig. 6. Outline of the robot eye.

The design of the eye-ball and its support structure, has been inspired by existing ball transfer units. To support the eye-ball it has been considered the possibility of using thrust-bearings, however, this solution has been dropped since small and light components for a miniature implementation (human sized eye), were not available. The final design has been based on the implementation of a low friction (PTFE) slide bearing, which could be easily scaled to smaller size.
The actuation is performed by tendons, i.e. thin stiff wires, pulled by *force generators*. The actuators must provide a linear motion of the tendons with a fairly small stroke (about *30*

mm, in the current implementation, and less than *20 mm* for an eye of human size), and limited pulling force. In fact, a pulling force of *2.5 N* would generate a nominal angular acceleration of about $6250\ rad\ sec^{-2}$, for an eye-ball with a mass *50 g* and radius of *20 mm*, and about $58000\ rad\ sec^{-2}$ in the case of an eye of human size with a mass of *9 g* and radius of *12 mm*. The actuators used in the current design are standard miniature DC servo motors, with integrated optical encoder, however, various alternative candidate solutions have been taken into account including: shape memory alloys and artificial muscles. According to recent advances, (Carpi et al., 2005; Cho & Asada, 2005), these technologies seem very promising as alternative solutions to DC motors mostly in terms of size and mass (currently the mass of the motors is about *160 g*, *i.e.* over *50%* of the total mass of the system, without including electronics). However, presently both shape memory alloys and artificial muscles require significant engineering to achieve operational devices, and therefore have not be adopted for the first prototype implementation.
In the following the major components and subsystems developed are reviewed.

6. The Eye-Ball

The eye ball is a precision PTFE sphere having a diameter of 38.1 *mm* (1.5*in*). The sphere has been CNC machined to host a commercial CMOS camera, a suspension spring, and to route the power supply and video signal cables to the external electronics. A frontal flange is used to allow the connection of the tendons at the specified insertion points, and to support miniature screws required to calibrate the position of the camera within the eye ball. On the flange it is eventually placed a spherical cover purely for aesthetical reasons. Fig. 7 and Fig. 8 show the exploded view and the actual eye-ball.

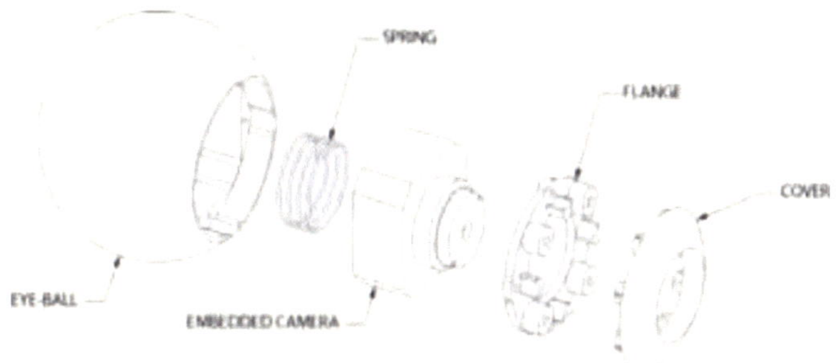

Fig. 7. Exploded view of the eye-ball.

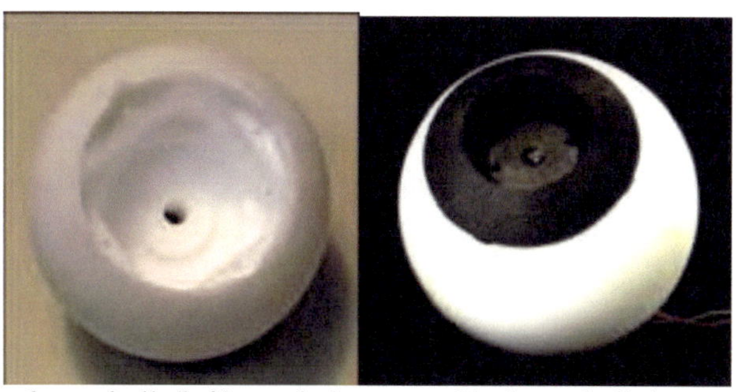

Fig. 8. The machined eye-ball (left), and the assembled eye-ball (camera cables shown in background).

The insertion points form an angle of *55 deg*, with respect to the (geometric) optical axis of the eye, therefore the eye-ball can rotate of about *45 deg* in all directions. The tendons used are monofiber nylon coated wires having a nominal diameter of *0.25 mm*, well approximating the geometric model proposed for the extra-ocular muscles.

7. Supporting Structure

The structure designed to support the eye ball is formed by two distinct parts: a low friction support, designed to hold the eye ball, Fig. 9, and a rigid flange used to implement the pointwise pulleys, and providing appropriate routing of the actuation tendons) required to ensure the correct mechanical implementation of Listing's Law, Fig. 13.

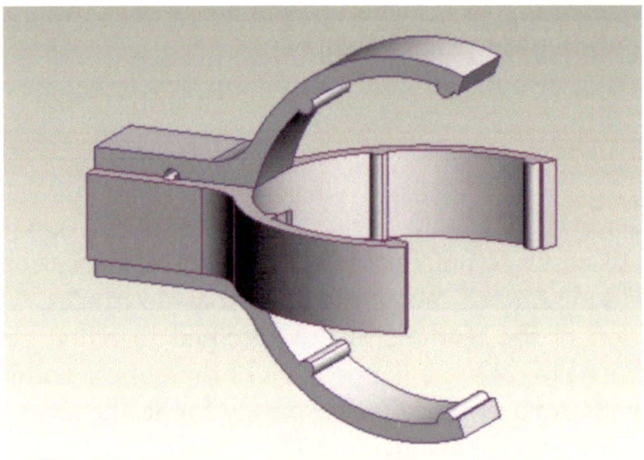

Fig. 9. CAD model of the eye-ball support (first concept).

7.1 The Eye-Ball Support

The eye-ball support is made of two C-shaped PTFE parts mated together, Fig. 10. The rear part of the support is drilled to allow the routing of the power supply and video signal cables to and from the on board camera.
The eight *bumps* on the C-shaped parts are the actual points of contact with the eye-ball. The placement of the contact points has been analysed by simulation in order to avoid interference with the tendons. Fig. 11 shows the path of one insertion point when the eye is rotated along the boundary of its workspace (i.e. the fixation axis is rotated to form a cone with amplitude of *45 deg*). The red marker is the position of the insertion point at the primary position while the green markers represent the position of two frontal contact points. The *north pole* in the figure represents the direction of axis h_3. The frontal bumps form an angle of *15 deg* with respect to the *equatorial* plane. The position of the rear bumps is constrained by the motion of the camera cables coming out from the eye-ball. To avoid interferences the rear bumps form an angle of *35 deg* with respect to *equatorial* plane of the eye.

7.2 The Pointwise Pulleys

The rigid flange, holding the eye-ball support, has the major function of implementing the pointwise pulleys. The pulleys have the role of constraining the path of the tendons so that, at every eye orientation, each tendon passes through a given head fixed point belonging to the principal direction associated with the corresponding pointwise pulley.

Let us assume the eye in a Listing compatible position *A*, then we may assume that a generic tendon is routed as sketched in Fig. 12. The pulley shown in the figure is tangent to the principal direction at point $\boldsymbol{p}_i$, and it allows the tendon to pass through $\boldsymbol{p}_i$. Assume now to rotate the eye to another Listing compatible position *B*; if the pulley could tilt about the principal axis, during the eye rotation, the tangential point $\boldsymbol{p}_i$ would remain the same so that the tendon is still routed through point $\boldsymbol{p}_i$. Therefore, the idea of the pulley tilting (about the principal axis) and possibly rotating (about its center), fully meets the specifications of the pointwise pulleys as defined for the eye model.

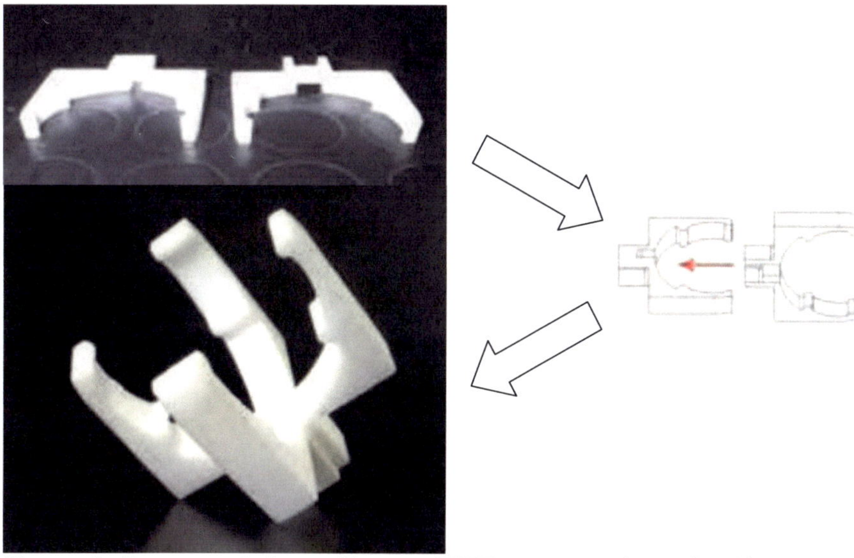

Fig. 10. The eye-ball support is formed by two PTFE parts mated together (final design).

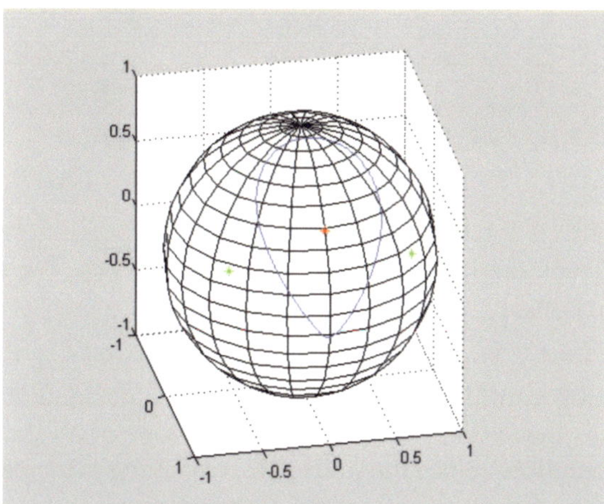

Fig. 11. During Listing compatible eye motions, the insertion points move within the region internal to the blue curve. The red marker represents the position of the insertion point at the primary position, while the green markers are the positions of (two of) the frontal contact points on the eye-ball support. The fixation axis at the primary position, $\boldsymbol{h}_3$, points upward.

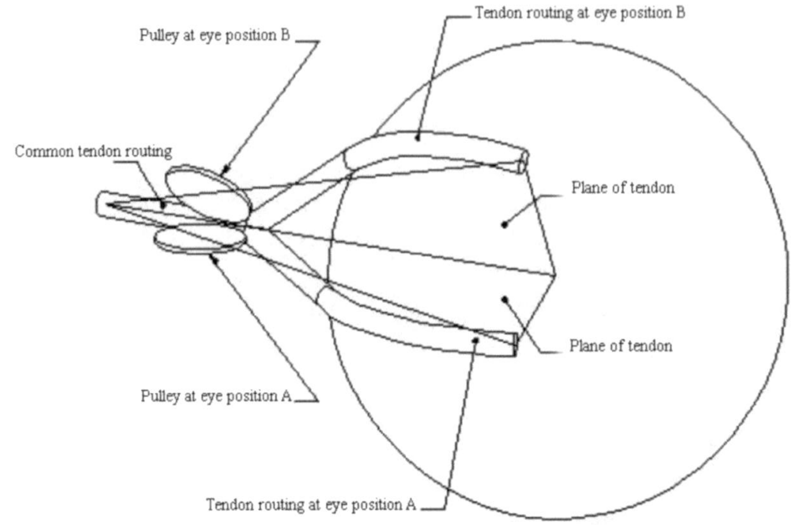

Fig. 12. Sketch of the tendon's paths, showing the tilting of the routing pulley when the eye is rotated from position *A* to position *B* (tendons and pulleys diameters not to scale).

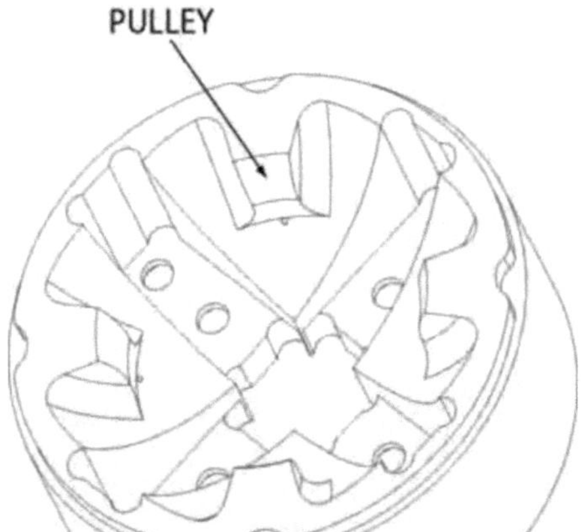

Fig. 13. Detail of the flange implementing the pointwise pulleys. The tendon slides along a section of a toroidal surface.

A component featuring these characteristics could be implemented, but its miniaturization and integration has been considered too complex, so we decided to implement a *virtual pulley*, to be intended as the surface formed by the envelope of all the tilting pulleys for all the admissible eye orientations. Since the pulley tilts about the principal axis at point p_i, then the envelope is a section of a torus with inner diameter equal to the radius of the tendon, and external radius equal to the radius selected for the pulley. Then, the implementation of the virtual pulley has been obtained by machining a section of a torus on the supporting flange as shown in Fig. 13.
The assembly of the eye-ball and its supporting structure is shown in Fig. 14.

8. Sensing and Actuation

The robot-eye is actuated by four DC servo-motors, producing a maximum output torque of *12 mNm,* and pulling four independent tendons routed to the eye as discussed in the previous section. The actuators are integrated within the structure supporting the eye, as shown in Fig. 15 and Fig. 16.

Fig. 14. The eye-ball and its supporting parts.

Fig. 15. Four DC motor actuate the tendons; pulleys route the tendons towards the eye-ball

The servo-motors are equipped with optical encoders providing the main feedback for the control of the ocular movements. A second set of sensors for measuring the mechanical tension of the tendons is integrated in the robot. In fact, as the tendons can only apply pulling forces, control of the ocular movements can be properly obtained only if slackness of the tendons or their excessive loading is avoided. The tension sensors are custom made and integrated within the supporting structure of the eye, Fig. 17.
Each sensor is formed by an infrared led/photodiode couple separated by a mobile *shutter*, preloaded with a phosphore-bronze spring. As shown in Fig. 18, the tension of the tendon counter-balances the pre-load force thus varying the amount of IR radiation received. The sensor output is the current generated by the photodiode according to the following equation:

$$I_p = k_p \gamma(f) E0 \tag{10}$$

where I_p is the current generated by the photodiode, k_p is the characteristic parameter of the photodiode, $E0$ is the IR radiation emitted by the led, and $\gamma(f)$ is a monotonic non-linear function of the tendon's tension depending on the system geometry. Each sensor is calibrated and a look-up table is used to map its current to tension characteristic.

Fig. 16. The body of the MAC-EYE robot. All the motors are integrated within the body; furthermore, all the cables from and to the eye-ball and the embedded optical tension sensors are routed inside of the structure.

Fig. 17. Implementation of the embedded sensors for measuring the mechanical tension of the tendons. The picture shows the *shutter* and its preloading spring, cables and tendons.

9. Robot Control System

9.1 The control architecture

The control architecture is implemented as a two level hierarchical system. At low level are implemented two control loops for each actuator. In the first loop a P-type control action regulates the tension of the *i-th* tendon at some constant reference value f_i^*, while in the second loop a PI-type action controls the motor velocity as specified by signal $\dot{\bar{q}}_i^*$, see Fig. 19. The tension feedback control loop makes the eye-ball backdrivable, so the eye can be positioned *by hand*. Both the reference signals $\dot{\bar{q}}_i^*$ and f_i^* are generated by the higher level control modules which implement a position based PI-type controller.

Currently, the major task implemented is that of controlling the eye position during emulated saccadic movements. Therefore, coordinated signals q_i^* and $\dot{q}_i^*$ must be generated for all the actuators. In order to compute the appropriate motor commands a geometric and a kinematic eye model are implemented as described below.

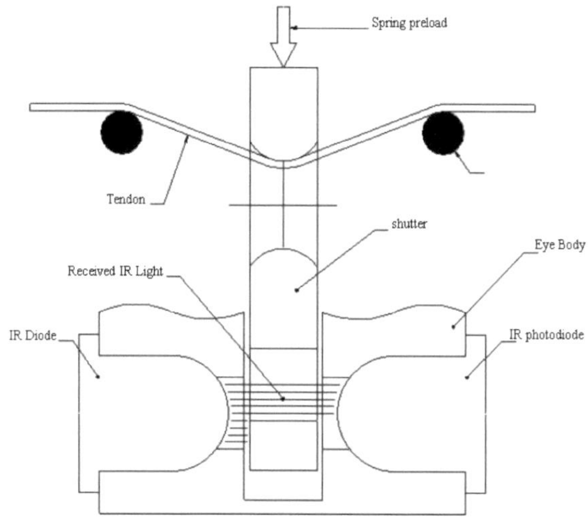

Fig. 18. Sketch of the tension sensor (not to scale).

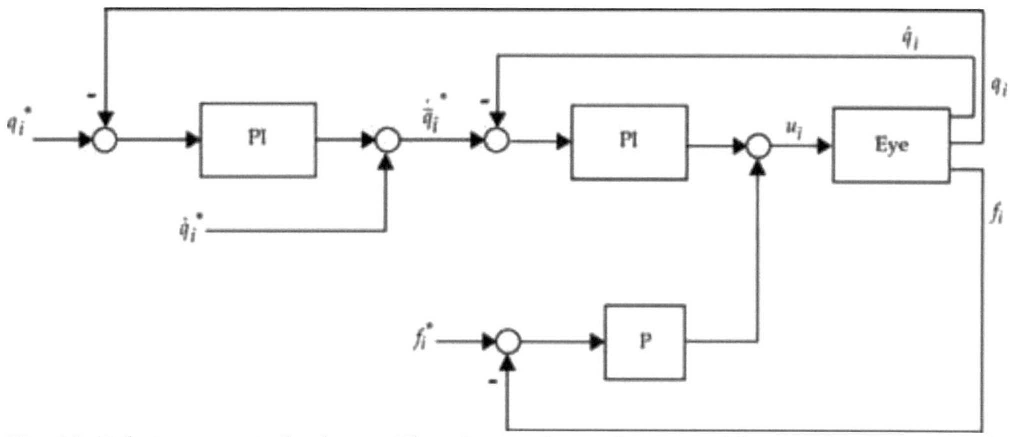

Fig. 19. Robot eye control scheme. The picture shows the control loops related to a single actuator.

Assume that a given reference trajectory for the eye is given and expressed by a rotation matrix $R^*(t)$, and an angular velocity $\omega^*(t)$. Then, an algebraic and a differential mapping, relating the eye orientation to the displacement of the tendons, can be easily computed. In fact, as shown in Fig. 20, for a given eye orientation the tendon starting from the insertion point $\boldsymbol{r}_i$ remains in contact with the eye-ball up to a point $\boldsymbol{t}_i$ (point of tangency). From $\boldsymbol{t}_i$ the tendon reaches the pointwise pulley and then moves towards the motor which provides the actuation force. For a given position of the eye there exists a set of displacements of the *free end* of the tendons corresponding to the amount of rotation to be commanded to the motors. According to Fig. 20, for each tendon this amount can be computed (with respect to a reference eye position, e.g. the primary position), using the formula:

$$x_i = R(\phi_i - \phi_{0i}) \tag{11}$$

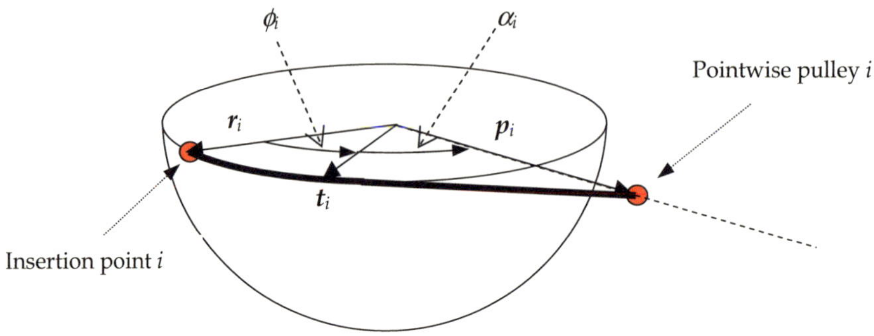

Fig. 20. The plane of tendon i for a generic eye orientation. Angle α_i is constant for any eye orientation.

where x_i is the amount of displacement of the *free end* of the tendon, while ϕ_i and ϕ_{0i} are the angles (with positive sign), formed by vectors $\boldsymbol{r}_i$ and $\boldsymbol{t}_i$ at a generic eye orientation, and at the primary position, respectively. Signs are chosen so that if $x_i < 0$ then the corresponding tendon must be pulled in order to orient the eye as specified by matrix $R^*(t)$. In order to compute x_i the angle ϕ_i can be determined, as follows. According to Fig. 20, the angle α_i must be constant for any eye orientation and can be expressed as:

$$\alpha_i = \cos^{-1}\left(\frac{R}{d_i}\right) \tag{12}$$

If the eye orientation is known with respect to frame <h>, then $\boldsymbol{r}_i$ is known, hence:

$$Rd_i \cos(\alpha_i + \phi_i) = \boldsymbol{r}_i \cdot \boldsymbol{p}_i \tag{13}$$

and finally, from equations (12) and (13), we obtain:

$$\phi_i = \cos^{-1}\left(\frac{\boldsymbol{r}_i \cdot \boldsymbol{p}_i}{Rd_i}\right) - \cos^{-1}\left(\frac{R}{d_i}\right) \tag{14}$$

The time derivative of ϕ_i can be computed by observing that $\dot{\boldsymbol{r}}_i = \omega \times \boldsymbol{r}_i$, where ω is the angular velocity of the eye, then, the time derivative of equations (13) can be written as:

$$\frac{d}{dt}(\boldsymbol{r}_i \cdot \boldsymbol{p}_i) = -Rd_i \sin(\alpha_i + \phi_i)\dot{\phi}_i \tag{15}$$

Therefore, we obtain the following equality:

$$(\omega \times \boldsymbol{r}_i) \cdot p_i = -Rd_i \sin(\alpha_i + \phi_i)\dot{\phi}_i \tag{16}$$

Then, by observing that $(\omega \times \boldsymbol{r}_i) \cdot p_i = (\boldsymbol{r}_i \times p_i) \cdot \omega$, we have:

$$\dot{\phi}_i = -\frac{1}{|\boldsymbol{r}_i \times p_i|}(\boldsymbol{r}_i \times p_i) \cdot \omega \tag{17}$$

Then, if $R^*(t)$ and $\omega^*(t)$ are the desired trajectory and angular velocity of the eye, the reference motor angles and velocities can be computed using formulas (11), (14) and (17) as:

$$
\begin{aligned}
q_i^*(t) &= \frac{R}{R_m}\left[\cos^{-1}\left(\frac{\boldsymbol{r}_i^* \cdot \boldsymbol{p}_i}{Rd_i}\right) - \cos^{-1}\left(\frac{R}{d_i}\right) - \phi_{i0}\right] \\
\dot{q}_i^*(t) &= -\frac{R}{R_m}\frac{1}{\left|r_i^* \times p_i\right|}\left(r_i^* \times p_i\right)\cdot \omega^*
\end{aligned}
\qquad (18)
$$

where R_m is the radius of the motor pulley.

9.2 The control architecture

The computer architecture of the robot eye is sketched in Fig. 23. A PC based host computer implements the high level position based control loop and the motor planning algorithm (18). Currently the high level control loop runs at a rate of *125 Hz.*

The low level control algorithms are implemented on a multi-controller custom board, Fig. 21, featuring four slave micro-controllers (one for each motor), operating in parallel and coordinated by a master one managing the communications through CAN bus with higher level control modules or other sensing modules (e.g. artificial vestibular system). The master and slave microcontrollers communicates using a multiplexed high speed (*10 Mbits*) serial link and operate at a rate of 1.25*KHz*.

Fig. 21. The prototype custom embedded real-time controller. The board has dimensions of 69 × 85mm^2.

Fig. 22. Complete stereoscopic robot system

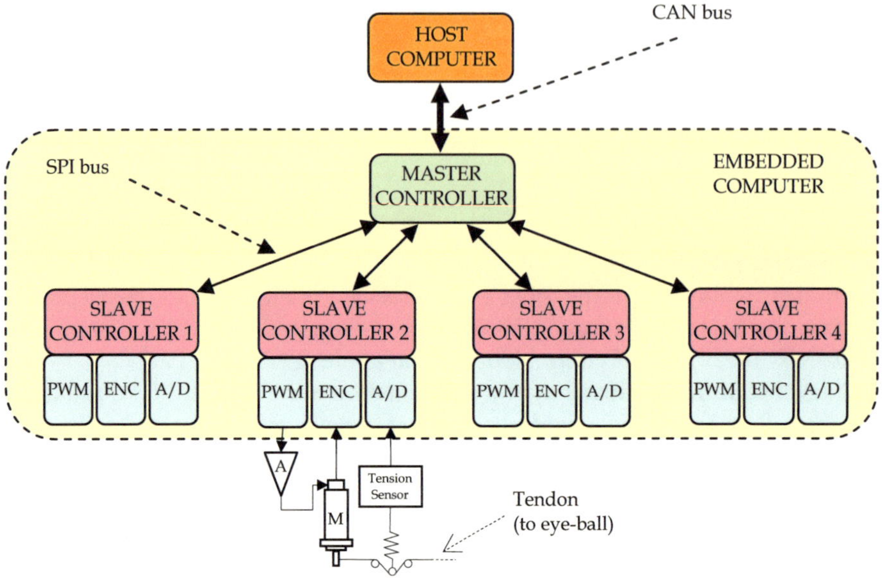

Fig. 23. Sketch of the computer control architecture.

10. Conclusions

This chapter has shown the feasibility of a tendon driven robot eye featuring the implementation of Listing's Law on a mechanical basis. The design of the robot is based on a model which is strongly inspired on assumptions derived from physiological evidence.

The achievements discussed in this chapter represent the starting point for the development of a more complex humanoid robot eye. From the mechanical point of view the most important advancement is represented by the possibility of extending the actuation to six tendons in order to enable the implementation of more complex ocular motions as the *vestibulo-ocular reflex*. From a theoretical point of view a more complete analysis of the properties of the proposed eye model could provide further understanding of the dynamics of saccadic motions. Furthermore, other issues such as eye miniaturization and embedding of image processing algorithms modules for direct visual feedback represent important challenges for the development of fully operational devices.

11. References

Albers, A. ; Brudniok, S. & Burger, W. (2003). The Mechanics of a Humanoid, *Proceedings of Humanoids 2003*, ISBN 3-00-012047-5; Karlsruhe, September 2003, IEEE, Germany.

Angelaki, D. E. & Hess, B. J. M., (2004). Control of eye orientation: where does the brain's role ends and the muscles begins, *European Journal of Neurosciences*, vol. 19, pp. 1-10, 2004, ISSN 0270-6474.

Becker, W. (1991). Eye Movements, *Carpenter, R.H.S. ed., Macmillan*, pp. 95-137,1991, ISSN 0301-4738.

Biamino, D. & Piazza, A. (2005). Studio Progetto e Realizzazione di una Coppia di Occhi Robotici con Sistema di Controllo Embedded, *Master Degree Thesis*, Faculty of Engineering, University of Genova, 2005, no ISBN or ISSN.

Cannata, G. ; D'Andrea, M.; Monti, F. & Maggiali, M. (2006). Implementation of Listing's Law for a Robot Eye, *Proceedings of Sy.Ro.Co 2006*, Bologna (Italy), Sept. 6-8, no ISBN or ISSN, 2006.

Cannata, G. & Maggiali, M. (2006). Implementation of Listing's Law for a Tendon Driven Robot Eye, *Proceedings of. IEEE Conf. on Intelligent Robots and Systems, IROS 2006*, pp. 3940-3945, ISBN 1-4244-0259-X, Beijing, Oct. 9-15, 2006, IEEE, China.

Clark, R. A.; Miller, J.M. & Demer, J. L. (2000). Three-dimensional Location of Human Rectus Pulleys by Path Inflection in Secondary Gaze Positions, *Investigative Ophthalmology and Visual Science*, vol. 41, pp. 3787-3797, 2000, ISSN 0146-0404.

Carpi F.; Migliore A.; Serra G. & De Rossi, D. (2005). Elical Dielectric Elastomer Actuators, *Smart Material and Structures*, vol. 14, pp. 1210-1216, 2005, ISSN 0964-1726.

Cho, K. & Asada, H. H. (2005). Segmentation theory for design of a multi-axis actuator array using segmented binary control. *Proceedings of the American Control Conference*, vol. 3, pp. 1969-1974, ISBN 0-7803-649, *ACC*, 2005.

Demer, J. L.; Miller J. M.; Poukens V.; Vinters, H. V. & Glasgow B.J. (1995). Evidence for fibromuscular pulleys of the recti extraocular muscles, *Investigative Ophthalmology and Visual Science*, vol. 36, pp. 1125-1136, 1995, ISSN 0030-3747.

Demer, J. L.; Ho, S. Y. & Pokens, V. (2000). Evidence for Active Control of Rectus Extraocular Muscle Pulleys, *Invest. Ophtalmol. Visual Sci.*, vol. 41, pp. 1280-1290, 2000, ISSN 0042-6989.

Furman, J. M. & Schor, R. (2003). Orientation of Listing's plane during static tilt in young and older human subjects, *Vision Res.*, vol. 43, pp. 67-76, 2003, ISSN 0042-6989.

Gu, J.; Meng, M., Cook, A. & Faulkner, M. G. (2000). A study of natural movement of artificial eye plant. *Robotics and Autonomous System*, vol. 32, pp. 153-161 , 2000, ISSN 0921-8890.

Haslwanter, T. (1995). Mathematics of Three-dimensional Eye Rotations, *Vision Res.*, vol. 35, pp. 1727-1739, 1995, ISSN 0042-6989.

Haslwanter, T. (2002). Mechanics of Eye Movements Implications of the Orbital Revolution, *Ann. N. Y. Acad. Sci.*, vol. 956, pp. 33-41, ,2002, ISSN 0077-8923.

Hepp, K. (1994). Oculomotor control: Listing's law and all that, *Current Opinion in Neurobiology*, vol. 4, pp. 862-868, 1994 ISSN 0959-4388.

Koene, A. R. & Erkelens, C.J. (2004). Properties of 3D rotations and their relation to eye movement control, *Biol. Cybern.*, vol. 90, pp. 410-417, Jul. 2004, ISSN 0340-1200.

Koornneef, L. (1974). The first results of a new anatomical method of approach to the human orbit following a clinical enquiry, *Acta Morphol Neerl Scand*, vol. 12, n. 4, pp. 259-282, 1974, ISSN 0001-6225.

Lockwood-Cooke, P.; Martin, C. F. & Schovanec L. (1999). A Dynamic 3-d Model of Ocular Motion, *Proceedings of 38th Conference of Decision and Control*, ISBN 0-7803-5253-X, Phoenix, December,1999.

Miller, J. M. & Robinson, David A. (1984). A Model of the Mechanics of Binocular Alignment. *Computer and Biomedical Research*, vol.17, pp. 436-470, 1984, ISSN 0010-4809.

Miller, J. M. (1989). Functional anatomy of normal human rectus muscles, *Vision Res.*, pp. 223-240, vol. 29, 1989, ISSN 0042-6989.

Polpitiya, A. D. & Ghosh, B. K. (2002). Modelling and Control of Eye-Movements with Muscolotendon Dynamics, *Proceedings of American Control Conference*, pp. 2313-2318, ISBN 1902815513, Anchorage, May 2002.

Polpitiya, A. D. & Ghosh, B. K. (2003). Modeling the Dynamics of Oculomotor System in Three Dimensions, *Proceedings of Conference on Decision and Control*, pp. 6418-6422, ISBN 0-7803-7925-X, Maui, Dec. 2003.

Polpitiya, A. D.; Ghosh, B. K., Martin, C. F. & Dayawansa, W. P. (2004). Mechanics of the Eye Movement Geometry of the Listing Space, *Proceedings of American Control Conference*, ISBN 0-444-81933-9, 2004.

Pongas, D., Guenter, F., Guignard, A. & Billard, A. (2004). Development of a Miniature Pair of Eyes With Camera for the Humanoid Robot Robota, *Proceedings of IEEE-RAS/RSJ International Conference on Humanoid Robots - Humanoids 2004*, vol. 2, pp. 899- 911, ISBN 0-7803-8863-1,Santa Monica, Los Angeles, 2004.

Porrill, J.; Warren, P. A. & Dean, P., (2000). A simple control law generates Listing's positions in a detailed model of the extraocular muscle system, *Vision Res.*, vol. 40, pp. 3743-3758, 2000, ISSN 0042-6989.

Quaia, C. & Optican, L. M. (1998). Commutative Saccadic Generator Is Sufficient to Control a 3D Ocular Plant With Pulleys, *The Journal of Neurophysiology*, pp. 3197-3215, vol. 79, 1998, ISSN 0022-3077.

Raphan, T. (1998). Modeling Control of Eye Orientation in Three Dimensions. I. Role of Muscle Pulleys in Determining Saccadic Trajectory, *The Journal of Neurophysiology*, vol. 79, pp. 2653-2667, 1998, ISSN 0022-3077.

Robinson, D. A., (1991). Overview, In: *Carpenter, R.H.S. ed., Macmillan* ,pp. 320-332,1991, ISSN 0301-4738.

Straumann D.; Zee. D. S.; Solomon D. & Kramer P. D., (1996). Validity of Listing's law during fixations, saccades, smooth pursuit eye movements, and blinks, *Exp. Brain Res.*, vol. 112, pp. 135-146, 1996, ISSN 0014-4819.

Sparks D. L. (2002). The brainstem control of saccadic eye movements. *Nature Rev.*, vol. 3, pp. 952-964, December, 2002, ISSN 1471-003X.

Tweed, D. & Vilis, T. (1987). Implications of Rotational Kinematics for the Oculomotor System in Three dimensions, *The Journal of Neurophysiology*, vol. 58, no.4, pp. 832-849, Oct. 1987, ISSN 0022-3077.

Tweed D. & Vilis T., (1988). Rotation Axes of Saccades, *Ann. N. Y. Acad. Sci.*, vol. 545, pp. 128-139, 1988, ISSN 0077-8923.

Tweed D. & Vilis T., (1990). Geometric relations of eye position and velocity vectors during saccades, *Vision. Res.*, vol. 30, n. 1, pp. 111-127, 1990, ISSN: 0042-6989.

Wong, A. M. F.; Tweed D. & Sharpe, J. A., (2002). Adaptive Neural Mechanism for Listing's Law Revealed in Patients with Sixth Nerve Palsy, *Investigative Ophtalmology and Visual Science*, vol. 43, n. 1, pp. 112-118, Jan. 2002, ISSN 0146-0404.

9

Multicriteria Optimal Humanoid Robot Motion Generation

Genci Capi[1], Yasuo Nasu[2], Mitsuhiro Yamano[2], Kazuhisa Mitobe[2]
University of Toyama[1]
Yamagata University[2]
Japan

1. Introduction

Humanoid robots, because of their similar structure with humans, are expected to operate in hazardous and emergency environments. In order to operate in such environments, the humanoid robot must be highly autonomous, have a long operation time and take decisions based on the environment conditions. Therefore, algorithms for generating in real time the humanoid robot gait are central for development of humanoid robot.

In the early works, the humanoid robot gait is generated based on the data taken from human motion (Vukobratovic et al. 1990). Most of the recent works (Roussel 1998, Silva & Machado 1998, Channon 1996) consider minimum consumed energy as a criterion for humanoid robot gait generation. Roussel (1998) considered the minimum consumed energy gait synthesis during walking. The body mass is concentrated on the hip of the biped robot. Silva & Machado (1998) considered the body link restricted to the vertical position and the body forward velocity to be constant. The consumed energy, related to the walking velocity and step length, is analyzed by Channon (1996). The distribution functions of input torque are obtained by minimizing the joint torques.

In our previous works, we considered the humanoid robot gait generation during walking and going up-stairs (Capi et al. 2001) and a real time gait generation (Capi et al. 2003). In addition of minimum consumed energy (MCE) criteria, minimum torque change (MTC) (Uno et al. 1989, Nakano et al. 1999) was also considered. The results showed that MCE and MTC gaits have different advantages. Humanoid robot motion generated based on MCE criterion was very similar with that of humans. Another advantage of MCE criterion is the long operation time when the robot is actuated by a battery. On the other hand, MTC humanoid robot motion was more stable due to smooth change of torque and link accelerations.

Motivated from these observations, it will be advantageous to generate the humanoid robot motion such that different criteria are satisfied. This belongs to a multiobjective optimization problem. In a multiobjective optimization problem there may not exist one solution that is the best with respect to all objectives. Usually, the aim is to determine the tradeoff surface, which is a set of nondominated solution points, known as Pareto-optimal or noninferior solutions.

The multiobjective problem is almost always solved by combining the multiple objectives into one scalar objective using the weighting coefficients. Therefore, to combine different objectives in a single fitness function, an a-priori decision is needed about the relative importance of the objectives, emphasizing a particular type of solution. These techniques often require some problem-specific information, such as total range each objective covers. In complex problems, such as humanoid robot gait generation, this information is rarely known in advance, making the selection of single objective weighting parameters difficult. In addition, there is no rational basis of determining adequate weights and the objective function so formed may lose significance due to combining non-commensurable objectives. To avoid this difficulty, the e-constraint method for multiobjective optimization was presented (Becerra & Coello). This method is based on optimization of the most preferred objective and considering the other objectives as constraints bounded by some allowable levels. These levels are then altered to generate the entire Pareto-optimat set. The most obvious weaknesses of this approach are that it is time-consuming and tends to find weakly nondominated solutions.

In this paper, we present a multiobjective evolutionary algorithm (MOEA) (Coello 1999, Herrera et al. 1998) technique for humanoid robot gait synthesis. The main advantage of the proposed algorithm is that in a single run of evolutionary algorithm, humanoid robot gaits with completely different characteristics are generated. Therefore, the humanoid robot can switch between different gaits based on the environment conditions. In out method, the basic idea is to encode the humanoid robot gait parameters in the genome and take the parameters of the non-dominated optimal gaits in the next generation. The specific questions we ask in this study are: 1) whether MOEA can successfully generate the humanoid robot gait that satisfies different objective functions in a certain degree, 2) whether the humanoid robot gait optimized by MOEA in simulation can indeed be helpful in hardware implementation.

In order to answer these questions, we considered the MCE and MTC cost functions as criteria for "Bonten-Maru" humanoid robot gait synthesis. We employed a real number MOEA. Simulation and experimental results show a good performance of the proposed method. The non-dominated optimal Pareto optimal solutions have a good distribution and humanoid robot gait varies from satisfying each of both considered objectives to satisfying both of them. Therefore, as a specific contribution of proposed method is that in a single run of MOEA are generated humanoid robot gaits with completely different characteristics, making it possible to select the appropriate gait based on our preferences. In order to further verify how the optimized gait will perform on real hardware, we implemented the optimal gait using the "Bonten-Maru" humanoid robot. The results show that in addition of energy consumption, the optimized gait was stable and with a small impact due to the smooth change of the joint torques.

2. Multiobjective Evolutionary Algorithm

2.1 Multiobjective Optimization Problem

In multiobjective optimization problems there are many (possibly conflicting) objectives to be optimized, simultaneously. Therefore, there is no longer a single optimal solution but rather a whole set of possible solutions of equivalent quality. In contrast to fully ordered scalar search spaces, multidimensional search spaces are only partially ordered, i.e. two different solutions are related to each other in two possible ways: either one dominates the other or none of them is dominated. Consider without loss of generality the following multiobjective maximization problem with m decision variables x (parameters) and n objectives:

$$y = f(x) = (f_1(x_1, \ldots\ldots x_m), \ldots, f_n(x_1, \ldots\ldots x_m)) \quad (1)$$

where $x = (x_1, \ldots\ldots x_m) \in X$, $y = (y_1, \ldots\ldots y_n) \in Y$ and where x is called decision (parameter) vector, X parameter space, y objective vector and Y objective space. A decision vector $a \in X$ is said to dominate a decision vector $b \in X$ (also written as a >b) if and only if:

$$\forall i \in \{1, \ldots, n\} : f_i(a) \geq f_i(b) \wedge \exists j \in \{1, \ldots, n\} : f_j(a) > f_j(b) \quad (2)$$

The decision vector a is called Pareto-optimal if and only if a is nondominated regarding the whole parameter space X. Pareto-optimal parameter vectors cannot be improved in any objective without causing degradation in at least one of the other objectives. They represent in that sense globally optimal solutions. Note that a Pareto-optimal set does not necessarily contain all Pareto optimal solutions in X. The set of objective vectors corresponding to a set of Pareto-optimal parameter vectors is called "Pareto-optimal front".

In extending the ideas of SOEAs to multiobjective cases, two major problems must be addressed: -- How to accomplish fitness assignment and selection in order to guide the search towards the Pareto-optimal set? -- How to maintain a diverse population in order to prevent premature convergence and achieve a well distributed, wide spread trade-off front? Note that the objective function itself no longer qualifies as fitness function since it is a vector valued and fitness has to be a scalar value. Different approaches to relate the fitness function to the objective function can be classified with regard to the first issue. The second problem is usually solved by introducing elitism and intermediate recombination. Elitism is a way to ensure that good individuals do not get lost (by mutation or set reduction), simply by storing them away in an external set, which only participates in selection. Intermediate recombination, on the other hand, averages the parameter vectors of two parents in order to generate one offspring.

2.2 Nondominated Sorting Genetic Algorithm

NSGA was employed to evolve the neural controller where the weight connections are encoded as real numbers. Dias & Vasconcelos (2002) compared the NSGA with four others multiobjective evolutionary algorithms using two test problems. The NSGA performed better than the others did, showing that it can be successfully used to find multiple Pareto-optimal solutions. In NSGA, before selection is performed, the population is ranked on the basis of domination using Pareto ranking, as shown in Fig. 1. All nondominated individuals are classified in one category with a dummy fitness value, which is proportional to the population size (Srivinas, & Deb 1995). After this, the selection, crossover, and mutation usual operators are performed.

In the ranking procedure, the nondominated individuals in the current population are first identified. Then, these individuals are assumed to constitute the first nondominated front with a large dummy fitness value (Srivinas, & Deb 1995). The same fitness value is assigned to all of them. In order to maintain diversity in the population, a sharing method is then applied. Afterwards, the individuals of the first front are ignored temporarily and the rest of population is processed in the same way to identify individuals for the second nondominated front. A dummy fitness value that is kept smaller than the minimum shared dummy fitness of the previous front is assigned to all individuals belonging to the new front. This process continues until the whole population is classified into nondominated fronts. Since the nondominated fronts are defined, the population is then reproduced according to the dummy fitness values.

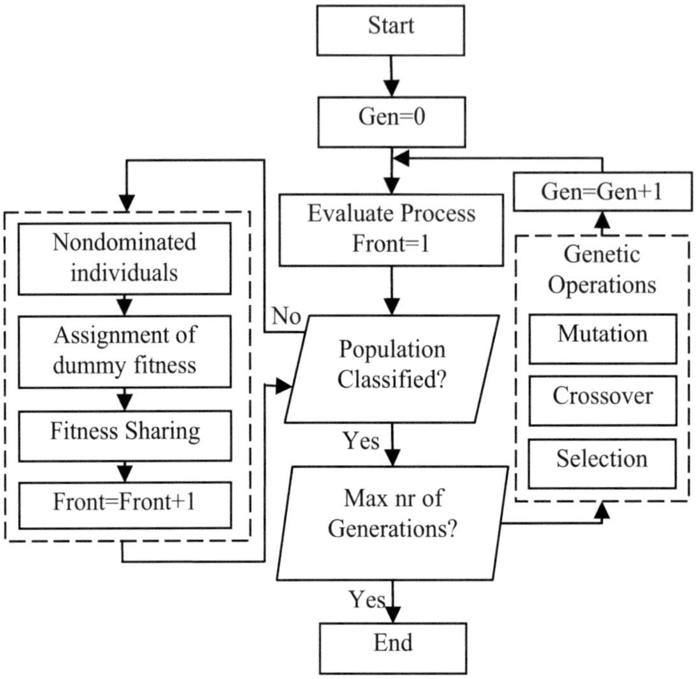

Fig. 1. Flowchart of NSGA.

Fitness Sharing: In genetic algorithms, sharing techniques aim at encouraging the formation and maintenance of stable subpopulations or niches (Zitzler et al. 2000). This is achieved by degrading the fitness value of points belonging to a same niche in some space. Consequently, points that are very close to, with respect to some space (decision space X in this paper), will have their dummy fitness function value more degraded. The fitness value degradation of near individuals can be executed using (3) and (4), where the parameter d_{ij} is the variable distance between two individuals i and j, and σ_{shared} is the maximum distance allowed between any two individuals to become members of a same niche. In addition, df_i is the dummy fitness value assigned to individual i in the current front and df'_i is its corresponding shared value. N_{pop} is the number of individuals in the population. The sharing function (Sh) measures the similarity level between two individuals. The effect of this scheme is to encourage search in unexplored regions. For details about niching techniques, see Sareni et al. (1998).

$$Sh(d_{ij}) = \begin{cases} 1 - \left(\dfrac{d_{ij}}{\sigma_{shared}} \right)^2, & \text{if } d_{ij} < \sigma_{shared} \\ 0, & \text{if } d_{ij} \geq \sigma_{shared} \end{cases} \tag{3}$$

$$df_i' = df_i \left[\sum_{j=1}^{Npop} Sh(d_{ij}) \right]^{-1} \tag{4}$$

3. Optimal Gait Generation

During motion, the arms of the humanoid robot will be fixed on the chest. Therefore, it can be considered as a five-link biped robot in the saggital plane, as shown in Fig. 2. The motion of the biped robot is considered to be composed from a single support phase and an instantaneous double support phase. The friction force between the robot's feet and the ground is considered to be great enough to prevent sliding. During the single support phase, the ZMP must be within the sole length, so the contact between the foot and the ground will remain. In our work, we calculate the ZMP by considering the link mass concentrated at one point. To have a stable periodic walking motion, when the swing foot touches the ground, the ZMP must jump in its sole. This is realized by accelerating the body link. To have an easier relative motion of the body, the coordinate system from the ankle joint of the supporting leg is moved transitionally to the waist of the robot ($O_1X_1Z_1$). Referring to the new coordinate system, the ZMP position is written as follows:

$$\overline{X}_{ZMP} = \frac{\sum_{i=1}^{5} m_i(\ddot{\overline{z}}_i + \ddot{z}_w + g_z)\overline{x}_i - \sum_{i=1}^{5} m_i(\ddot{\overline{x}}_i + \ddot{x}_w)(\overline{z}_i + z_w)}{\sum_{i=1}^{5} m_i(\ddot{\overline{z}}_i + \ddot{z}_w + g_z)}, \quad (5)$$

where m_i is mass of the particle "*i*", x_w and z_w are the coordinates of the waist with respect to the coordinate system at the ankle joint of supporting leg, $\overline{x}_i$ and $\overline{z}_i$ are the coordinates of the mass particle "*i*" with respect to the $O_1X_1Z_1$ coordinate system, $\ddot{\overline{x}}_i$ and $\ddot{\overline{z}}_i$ are the acceleration of the mass particle "*i*" with respect to the $O_1X_1Z_1$ coordinate system.

Based on the formula (3), if the position, $\overline{x}_i, \overline{z}_i$, and acceleration, $\ddot{\overline{x}}_i, \ddot{\overline{z}}_i$, of the leg part (i=1,2,4,5), the body angle, θ_3, and body angular velocity, $\dot{\theta}_3$, are known, then because $\ddot{\overline{x}}_3, \ddot{\overline{z}}_3$ are functions of $l_3, \theta_3, \dot{\theta}_3, \ddot{\theta}_3$, it is easy to calculate the body angular acceleration based on the ZMP position. Let (0) and (f) be the indexes at the beginning and at the end of the step, respectively. At the beginning of the step, $\ddot{\theta}_{30}$ causes the ZMP to be in the position ZMP_{jump}. At the end of the step, the angular acceleration $\ddot{\theta}_{3f}$ is calculated in order to have the ZMP at the position ZMP_f, so that the difference between $\ddot{\theta}_{3f}$ and $\ddot{\theta}_{30}$ is minimal. Therefore, the torque necessary to change the acceleration of the body link will also be minimal.

3.1 Objective Functions

The gait synthesis problem, with respect to walking or going up-stairs, consists on finding the joint angle trajectories, to connect the first and last posture of the biped robot for which the consumed energy and torque change are minimal. For the MCE cost function, it can be assumed that the energy to control the position of the robot is proportional to the integration of the square of the torque with respect to time, because the joint torque is proportional with current. Therefore, minimizing the joint torque can solve the MCE problem (Capi 2002).

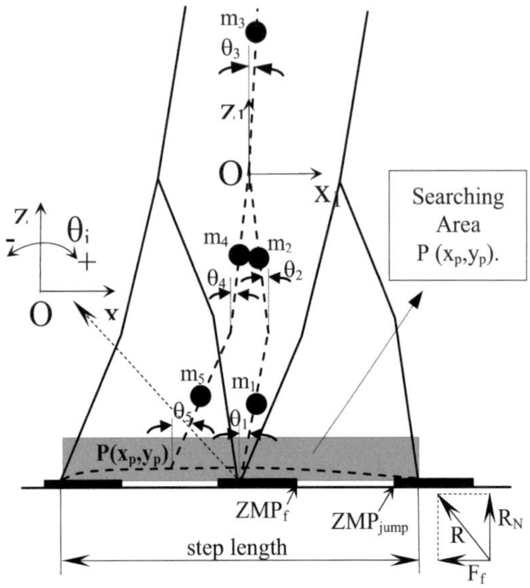

Fig. 2. Five-link humanoid robot.

The cost function J, which is a quantity proportional to the energy required for the motion, is defined as follows:

$$J = \frac{1}{2}\left(\int_0^{t_f} \tau^T \tau dt + \Delta\tau_{jump}^2 \Delta t + \int_0^{t_f} C dt\right), \tag{6}$$

where: t_f is the step time, τ is the torque vector, $\Delta\tau_{jump}$ and Δt are the addition torque applied to the body link to cause the ZMP to jump and its duration time, and C is the constraint function, given as follows:

$$C = \begin{cases} 0 \text{ - if the constraints are satisfied,} \\ c_i \text{ - if the constraints are not satisfied,} \end{cases}$$

c denotes the penalty function vector. We consider the following constraints for our system.

1) The walking to be stable or the ZMP to be within the sole length.
2) The distance between the hip and ankle joint of the swing leg must not be longer then the length of the extended leg.
3) The swing foot must not touch the ground prematurely.

The torque vector is calculated from the inverse dynamics of the five-link biped robot as :

$$J(\theta)\ddot{\theta} + X(\theta)\dot{\theta}^2 + Y\dot{\theta} + Z(\theta) = \tau \,. \tag{7}$$

where $J(\theta)$ is the mass matrix (5x5), $X(\theta)$ is the matrix of centrifugal coefficients (5x5), Y is the matrix of Coriolis coefficients (5x5), $Z(\theta)$ is the vector of gravity terms (5x1), τ is the generalized torque vector (5x1), and $\theta, \dot{\theta}, \ddot{\theta}$ are 5x1 vectors of joint variables, joint angular velocities and joint angular accelerations, respectively.

The MTC model (Uno 1989, Nakano 1999) is based on smoothness at the torque level. The cost is the integrated squared torque change summed over the joints and the movement. In the MTC, the objective function to be minimized is expressed by:

$$J_{torquechange} = \frac{1}{2}\left(\int_0^{t_f} \left(\frac{d\tau}{dt}\right)^T \left(\frac{d\tau}{dt}\right) dt + \left(\frac{\Delta\tau}{\Delta t}\right)^2 + \int_0^{t_f} C dt\right). \tag{8}$$

4. Boundary Conditions and GA Variables

To have a continuous periodic motion, the humanoid robot posture has to be the same at the beginning and at the end of the step. Therefore, the following relations must be satisfied:

$$\theta_{10} = \theta_{5f},\ \theta_{20} = \theta_{4f},\ \theta_{1f} = \theta_{50},\ \theta_{2f} = \theta_{40},\ \theta_{30} = \theta_{3f}. \tag{9}$$

In order to find the best posture, the optimum value of θ_{10}, θ_{20} and θ_{30} must be determined by GA. For a given step length, it is easy to calculate θ_{40} and θ_{50}. When referring to Figure 2, it is clear that links 1, 2, 4 at the beginning of the step and links 2, 4, 5 at the end of the step, change the direction of rotation. Therefore, we can write:

$$\dot{\theta}_{10} = \dot{\theta}_{20} = \dot{\theta}_{40} = \dot{\theta}_{2f} = \dot{\theta}_{4f} = \dot{\theta}_{5f} = 0. \tag{10}$$

The angular velocity of link 1 at the end of the step and link 5 at the beginning of the step is considered to be the same. In order to find the best value of angular velocity, we consider it as one variable of GA, because the rotation direction of these links does not change. GA will determine the optimal value of the angular velocity of the body link, which is considered to be the same at the beginning and at the end of the step. The following relations are considered for the angular acceleration:

$$\ddot{\theta}_{10} = \ddot{\theta}_{5f},\ \ddot{\theta}_{20} = \ddot{\theta}_{4f},\ \ddot{\theta}_{1f} = \ddot{\theta}_{50},\ \ddot{\theta}_{2f} = \ddot{\theta}_{40}. \tag{11}$$

In this way, during the instantaneous double support phase, we don't need to apply an extra torque to change the angular acceleration of the links. To find the upper body angle trajectory, an intermediate angle θ_{3p} and its passing time t_3 are considered as GA variables. To determine the angle trajectories of the swing leg, the coordinates of an intermediate point $P(x_p,z_p)$ and their passing time t_p, are also considered as GA variables. The searching area for this point is shown in Figure 2. Based on the number of constraints, the degree of the time polynomial for $\theta_1, \theta_2, \theta_3, \theta_4$ and θ_5 are 3, 3, 7,6 and 6, respectively.

	Body	Lower leg	Upper leg	Lower leg + foot
Mass [kg]	12	2.93	3.89	4.09
Inertia [kg m^2]	0.19	0.014	0.002	0.017
Length [m]	0.3	0.2	0.204	0.284
CoM dist.[m]	0.3	0.09	0.1	0.136

Table 1. "Bonten-Maru" humanoid robot link parameters.

5. Results

5.1 "Bonten-Maru" Humanoid Robot

In the simulations and experiments, we use the the "Bonten-Maru" humanoid robot (Nasu et al. 2002, Takeda et al. 2001). The parameter values are presented in Table 1 and the robot is shown in Fig. 3(a). The "Bonten-Maru I" humanoid robot is 1.2 m high and weights 32 kg, like an 8 years old child. The "Bonten-Maru I" is a research prototype, and as such has undergone some refinement as different research direction are considered. During the

design process, some predefined degree of stiffness, accuracy, repeatability, and other design factors have been taken into consideration. The link dimensions are determined such that to mimic as much as possible the humans. In the "Bonten-Maru" humanoid robot, a DC motor actuates each joint. The rotation motion is transmitted by a timing belt and harmonic drive reduction system. Under each foot are four force sensors, two at the toe and two across the heel. These provide a good indication of both contact with the ground, and the ZMP position. The head unit has two CCD cameras (542x492 pixels, Monochrome), which are connected to the PC by video capture board. A Celeron based microcomputer (PC/AT compatible) is used to control the system.
The dof are presented in Fig. 3(b). The high number of dof gives the "Bonten-Maru I" humanoid robot the possibility to realize complex motions. The hip is a ball-joint, permitting three dof; the knee joint one dof; the ankle is a double-axis design, permitting two. The shoulder has two dof, the elbow and wrist one dof. The DC servomotors act across the three joints of the head, where is mounted the eye system, enabling a total of three dof. The distribution of dof is similar with the dof in human limbs.

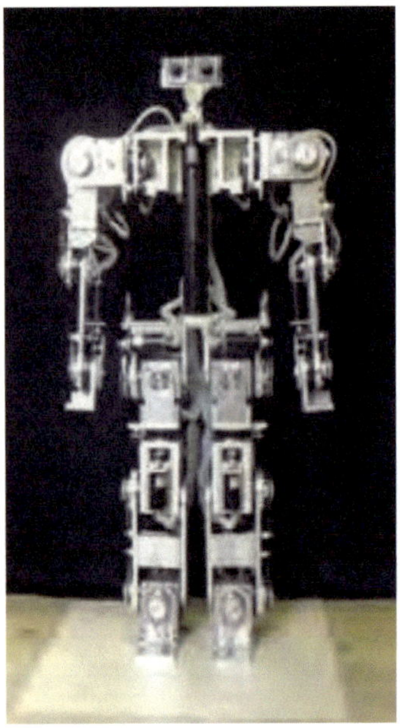
(a)

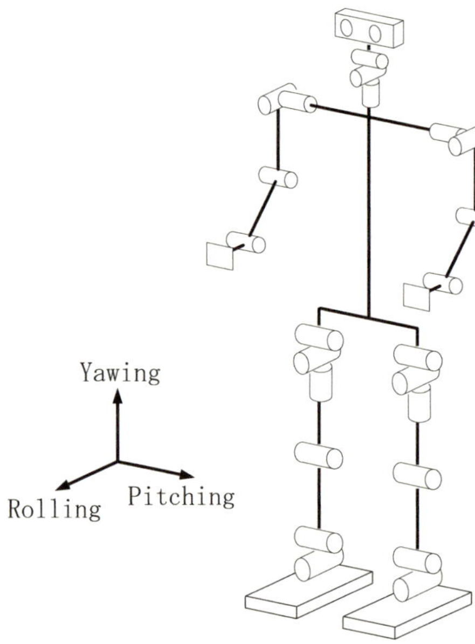

(b)

Fig. 3. "Bonten-Maru" humanoid robot.

5.2 Simulation and Experimental Results

Due to difficulties of binary representation when dealing with continuous search space with large dimension, real coded GA (Herrera 1998) is used in this study. The decision variables are represented by real numbers within their lower and upper limits. We employed a standard crossover operator and the non-uniform mutation. In all optimization runs, crossover and mutation probabilities were chosen as 0.9 and 0.3, respectively. On all optimization runs, the population size was selected as 50 individuals and the optimization terminated after 100 generations. The maximum size of the Pareto-optimal set was chosen as 50 solutions.

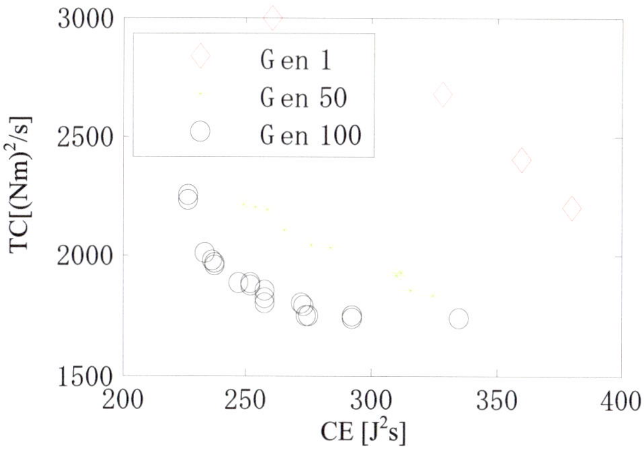

Fig. 4. Pareto optimal solution for different generations.

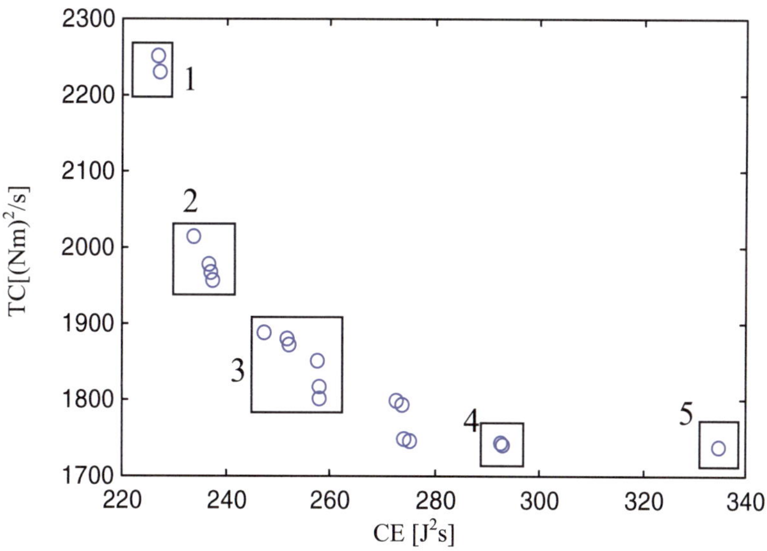

Fig. 5. Pareto front of nondominated solutions after 100 generations.

Based on the parameters of the "Bonten-Maru" humanoid robot the step length used in the simulations varies from 0.2m to 0.55m. The bounds, within which the solution is sought, change according to the step length and step time. In the following, we present the results for the step length 0.42m and step time 1.2s.

Fig. 4 shows the Pareto optimal front for generations 1, 50 and 100. During the first 50 generations there is a great improvement on the quality and distribution of Pareto optimal solutions. From this figure, it can be deduced that the MOEA is equally capable of finding the best solution for each objective when two conflicting objectives are considered simultaneously.

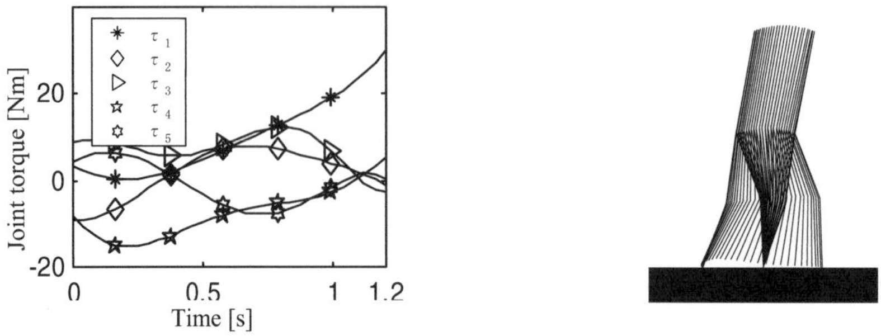

(a) Box 1 result (MCE).

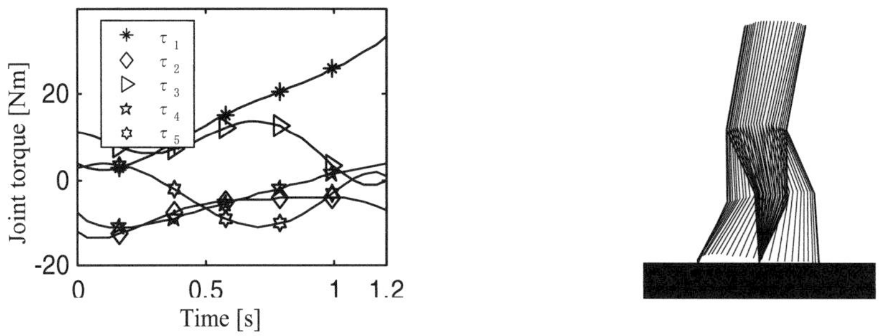

(b) Box 5 result (MTC).

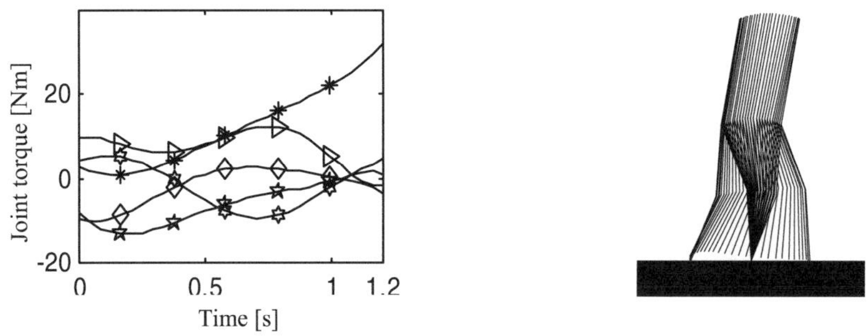

(c) Box 3 results.

Fig. 6. Different results from Pareto-front solutions.

Fig. 5 shows the Pareto-optimal trade-off front after 100 generations. We can observe the existence of a clear tradeoff between the two objectives. In addition, the obtained reference solution set has a good distribution (similar to uniform distribution). One of the interesting features of the resulting Pareto front is the almost exponential relation between the MCE and MTC cost functions. Results in Box 1 and Box 5 are at the extreme ends of the Pareto front. Box1 represents Pareto solutions with high value of MTC function, but low energy consumption. Based on the Pareto-optimal solutions, we can choose whether to go for minimal CE (Box 1 in Fig. 4) at the expense of a less smoothens in the torque or choose some intermediate result. If we are interested for a low consumed energy humanoid robot gait, without neglecting the smoothness in the torque change, the results shown in Boxes 2, 3 are the most important. The results in Box 2, show that by a small increase in the energy consumption (2.2%), we can decrease the MTC fitness function by around 12.1%. Also, the energy can be reduced by 14.5% for a small increase in the MTC cost function (Box 4).

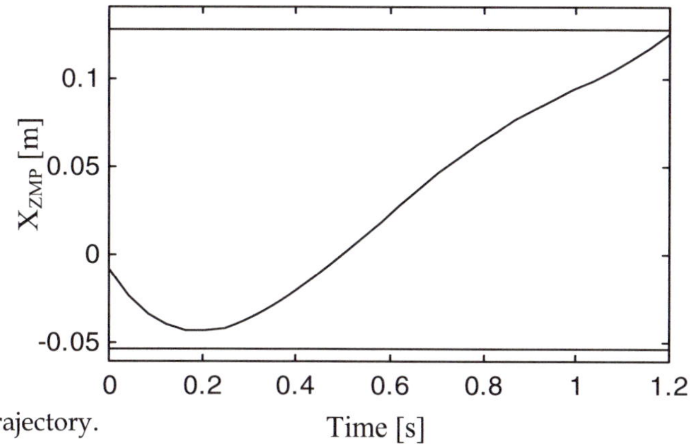

Fig. 7. ZMP trajectory.

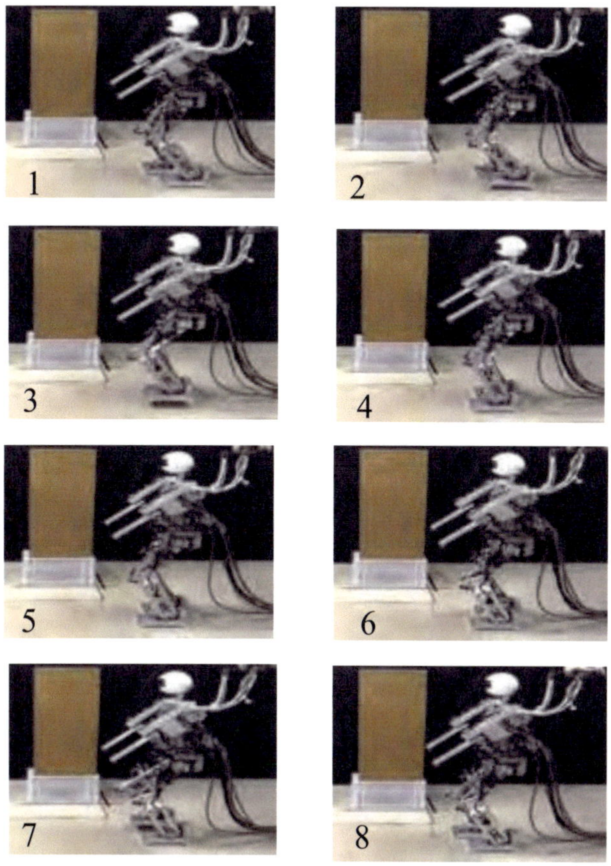

Fig. 8. Video capture of robot motion.

The torque vector (τ_i) and the optimal gaits for different results of Pareto front solutions are shown in Fig. 6. The robot posture is straighter, similar to humans, for MCE cost function (Fig.6(a)). Torque value is low for MCE gait and the torques change smoothly for MTC gait (Fig. 6(b)). The optimal gait generated by Box 3 solutions satisfies both objective functions. The energy consumption is increased by 9% but on the other hand the value of MTC cost function is decreased by 19.2%.
The ZMP trajectory is presented in Fig. 7 for humanoid robot gait generated by Box 3 result. The ZMP is always between the dotted lines, which present the length of the foot. At the end of the step, the ZMP is at the position ZMP_f, as shown in Fig. 2. At the beginning of the step, the ZMP is not exactly at the position ZMP_{jump} because of the foot's mass. It should be noted that the mass of the lower leg is different when it is in supporting leg or swing leg.
In order to investigate how the optimized gaits in simulation will perform in real hardware, we transferred the optimal gaits that satisfy both objective functions on the "Bonten-Maru" humanoid robot (Fig. 8). The experimental results show that in addition of reduction in energy consumption, the humanoid robot gait generated by Box 3 solutions was stable. The impact of the foot with the ground was small.

6. Conclusion

This paper proposed a new method for humanoid robot gait generation based on several objective functions. The proposed method is based on multiobjective evolutionary algorithm. In our work, we considered two competing objective functions: MCE and MTC. Based on simulation and experimental results, we conclude:

- Multiobjective evolution is efficient because optimal humanoid robot gaits with completely different characteristics can be found in one simulation run.
- The nondominated solutions in the obtained Pareto-optimal set are well distributed and have satisfactory diversity characteristics.
- The optimal gaits generated by simulation gave good performance when they were tested in the real hardware of "Bonten-Maru" humanoid robot.
- The optimal gait reduces the energy consumption and increases the stability during the robot motion.

In the future, it will be interesting to investigate if the robot can learn in real time to switch between different gaits based on the environment conditions. In uneven terrains MTC gaits will be more

7. References

Becerra, L. B., and Coello, C. A. (2006). Solving Hard Multiobjective Optimization Problems Using e-Constraint with Cultured Differential Evolution, in Thomas Philip Runarsson, Hans-Georg Beyer, Edmund Burke, Juan J. Merelo-Guervós, L. Darrell Whitley and Xin Yao (editors), Parallel Problem Solving from Nature - PPSN IX, 9th International Conference, pp. 543--552, Springer. Lecture Notes in Computer Science Vol. 4193.

Capi, G., Nasu, Y., Barolli, L., Mitobe, K., and Takeda, K. (2001). Application of genetic algorithms for biped robot gait synthesis optimization during walking and going up-stairs, *Advanced Robotics Journal*, Vol. 15, No. 6, 675-695.

Capi, G., Nasu, Y., Barolli, L., Mitobe, K., Yamano, .M., and Takeda, K. (2002) A new gait optimization approach based on genetic algorithm for walking biped robots and a neural network implementation, *Information Processing Society of Japan (IPSJ)*, Vol. 43, No. 4, 1039-1049.

Capi, G., Nasu, Y., Barolli, L., and Mitobe, K. (2003). Real time gait generation for autonomous humanoid robots: a case study for walking, *Robotics and Autonomous Systems*, Vol. 42, No. 2, 107-116.

Channon, P. H, Pham, D. T., and Hopkins, S. H. (1996). A variational approach to the optimization of gait for a bipedal robot, *Journal of Mechanical Engineering Science*, Vol. 210, 177-186.

Coello, C. A. C. (1999) A comprehensive survey of evolutionary based multiobjective optimization techniques, *Knowledge and Information Systems*, Vol. 1, No. 3, pp. 269-308.

Dias, A. H. F. & De Vasconcelos, J. A. (2002). Multiobjective genetic algorithms applied to solve optimization problems, *IEEE Transactions on Magnetic*, Vol. 38, No. 2, 1133-1136.

Herrera, F., Lozano, M. and Verdegay, J. L. (1998). Tackling real-coded genetic algorithms: operators and tools for behavioral analysis, *Artificial Intelligence Review*, Vol. 12, No. 4, 265-319.

Nakano, E., Imamizu, H., Osu, R., Uno, Y. Gomi, H., Yoshioka, T., and Kawato, M. (1999) Quantitative examinations of internal representations for arm trajectory planning: minimum commanded torque change model, *The Journal of Neurophysiology*, Vol. 81, No. 5, 2140-2155.

Nasu, Y., Capi, G., Yamano, M. (2002). "Bonten-Maru I": Development and Perspectives of a Humanoid Robot Project, *Proc. of Pacific Conference on Manufacturing (PCM2002)*, pp. 240-245.

Roussel, L., Canudas-de-Wit, C., and Goswami, A. (1998). Generation of energy optimal complete gait cycles for biped robots, *Proc. IEEE Int. Conf. on Robotics and Automation*, pp. 2036-2041.

Sareni, B., Krähenbühl, L. and Nicolas, A. (1998). Niching genetic algorithms for optimization in electromagnetics-I Fundamentals, *IEEE Transactions on Magnetic*, Vol. 34, 2984–2987.

Silva, F. M. and Machado, J. A. T. (1999). Energy analysis during biped walking, *Proc. IEEE Int. Conf. On Robotics and Automation*, pp. 59-64.

Srivinas, N. & Deb, K. (1995). Multiobjective optimization using non-dominated sorting in genetic algorithms, *Evolutionary Computation*, Vol. 2, No. 3, 279-285.

Takeda, K., Nasu, Y., Capi, G., Yamano, M., Barolli, L., Mitobe, K. (2001). A CORBA-Based approach for humanoid robot control, *Industrial Robot-an International Journal*, Vol.28, No.3, 242-250.

Uno, Y., Kawato, M., and Suzuki, R. (1989). Formulation and control of optimal trajectory in human multijoint arm movement, *Biol. Cybernet.* Vol. 61, pp. 89-101.

Vukobratovic, M., Borovac, B., Surla, D. and D. Stokic. (1990). *Biped Locomotion, Dynamics, Stability, Control and Application*. Springer-Verlag Berlin.

Zitzler, E., Deb, K., and Thiele, L. (2000). Comparison of multiobjective evolutionary algorithms: empirical results, *Evolutionary Computation*, Vol. 8, No. 2, 173-195.

10

An Incremental Fuzzy Algorithm for The Balance of Humanoid Robots

Erik Cuevas[1,2], Daniel Zaldivar[1,2], Ernesto Tapia[2] and Raul Rojas[2]
[1]*Universidad de Guadalajara, CUCEI,* [2]*Freie Universität Berlin, Institut für Informatik*
Mexico, Germany

1. Introduction

Humanoid robots base their appearance on the human body (Goddard et al., 1992; Kanehira et al., 2002; Konno et al., 2000). Minimalist constructions have at least a torso with a head, arms or legs, while more elaborated ones include devices that assemble, for example, human face parts, such as eyes, mouth, and nose, or even include materials similar to skin. Humanoid robots are systems with a very high complexity, because they aim to look like humans and to behave as they do.

Mechanical control, sensing, and adaptive behaviour are the constituting *logical parts* of the robot that allow it to "behave" like a human being. Normally, researchers study these components by modelling only a *mechanical part* of the humanoid robot. For example, artificial intelligence and cognitive science researches consider the robot from the waist up, because its visual sensing is located in its head, and its behavior with gestures normally uses its face or arms. Some engineers are mostly interested in the autonomy of the robot and consider it from the waist down. They develop mathematical models that control the balance of the robot and the movement of its legs (Miller, 1994; Yamaguchi et al., 1999; Taga et al., 1991), allowing the robot to walk, one of the fundamental behaviours that characterizes human beings.

Examples of such mathematical models are *static* and *dynamic walking*. The static walking model controls the robot to maintain its *center of gravity* (*COG*) inside a stable support region, while the dynamic walking model maintains the *zero moment point* (*ZMP*) inside the support region. Kajita et al. (1992) designed and developed an almost ideal 2-D model of a biped robot. He supposed, for simplicity, that the robot's COG moves horizontally and he developed a control law for initiation, continuation and termination of the walking process. Zhen and Shen (1990) proposed a scheme to enable robot climbing on inclined surfaces. Force sensors placed in the robot's feet detect transitions of the terrain type, and motor movements correspondingly compensate the inclination of robot.

The models mentioned above can be, however, computationally very expensive, and prohibitive for its implementation in microcontrollers. Control algorithms for a stable walking must be sufficiently robust and smooth, to accomplish a balance correction without putting in risk the mechanical stability of the robot. This could be resolved by using a controller that modifies its parameters according to a mathematical model, which considers certain performance degree required to offer the enough smoothness.

Fuzzy Logic is especially advantageous for problems which cannot be easily represented by a fully mathematical model, because the process is mathematically too complex and computationally expensive, or some data is either unavailable or incomplete. The real-world language used in Fuzzy Control also enables the incorporation of approximate human logic into computers. It allows, for example, partial truths or multi-value truths within the model. Using linguistic modeling, as opposed to mathematical modeling, greatly simplifies system design and modification. It generally leads to quicker development cycles, easy programming, and fairly accurate control. It is important, however, to underline that fuzzy logic solutions are usually not aimed at achieving the computational precision of traditional techniques, but aims at finding acceptable solutions in shorter time.

The Incremental Fuzzy Control algorithm fulfils the robustness and smoothness requirements mentioned above, even its implementation in microcontrollers. Such an algorithm is relatively simple and computationally more efficient than other adaptive control algorithms, because it consists of only four fuzzy rules. The algorithm demonstrates a smooth balance control response between the walking algorithm and the lateral plane control: one adaptive gain varies incrementally depending on the required performance degree.

The objective of this chapter is to describe the incremental fuzzy algorithm, used to control the balance of lateral plane movements of humanoid robots. This fuzzy control algorithm is computationally economic and allows a condensed implementation. The algorithm was implemented in a PICF873 microcontroller. We begin on the next section with the analysis of the balance problem, and follow later with the description of the controller structure. Afterwards, we explain important considerations about the modification of its parameters. Finally, we present experimental results of algorithm, used on a real humanoid robot, "Dany walker", developed at the Institut für Informatik of the Freie Universität Berlin.

2. Robot Structure

The humanoid robot "Dany Walker" used in our research was built only from the waist down. It consists of 10 low-density aluminium links. They are rotational on the pitch axis at the hip, knee and ankle. Each link consists of a modular structure. The links form a biped robot with 10 degrees of freedom, see Fig. 1.

The robot structure and its mass distribution affect directly the dynamic of the humanoid (Cuevas et al., 2004), therefore, the movement of the *Center of Masses* (*COM*) has a significant influence on the robot stability. In order to achieve *static* stability, we placed the COM as low as possible. To such purpose, our design uses short legs, see Fig. 2

To compensate the disturbances during walking, our construction enables lateral movements of the robot. Thus, it was possible to control the lateral balance of the robot by swaying the waist using four motors in the lateral plane: two at the waist and two at the ankles, see Fig. 3.

Fig. 1 The biped robot "Dany Walker"

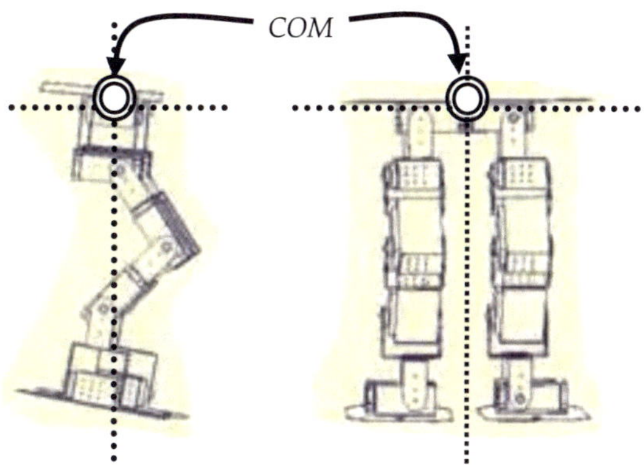

Fig. 2. Dany Walker's *COM* location.

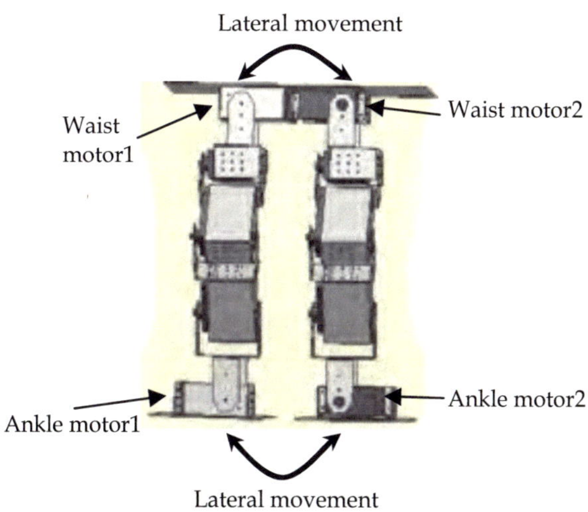

Fig. 3. Lateral balance of the motors.

3. Balance control criterion

We used the dynamic walking model to define our balance criterion. It consists of maintaining the *Zero Moment Point* (*ZMP*) inside the support region (Vukobratovic & Juricic, 1969; Vukobratovic, 1973). We implemented a feedback-force system to calculate the *ZMP*, and feed it in to the fuzzy PD incremental controller to calculate the *ZMP* error. Then, the controller adjusts the lateral robot's positions to maintain the *ZMP* point inside of the support region.
To achieve stable dynamic walking, the change between simple supports phase and double supports phase should be smooth. In the beginning of the double supports phase, the foot returns from the air and impacts against the floor, generating strong forces that affect the walking balance (Cuevas et al., 2005). The intensity of these forces is controlled by imposing velocity and acceleration conditions on saggital motion trajectories. This is achieved by using smooth cubic interpolation to describe the trajectories. In this chapter, we only discuss the control of the lateral motion (balance).

3.1 Zero Moment Point (ZMP)

The *ZMP* is the point on the ground where the sum of all momentums is zero. Using this principle, the *ZMP* is computed as follows:

$$x_{ZMP} = \frac{\sum_i m_i(\ddot{z}+g)x_i - \sum_i m_i \ddot{x} z_i - \sum_i I_{iy} \ddot{\theta}_{iy}}{\sum_i m_i(\ddot{z}+g)} \tag{1}$$

$$y_{ZMP} = \frac{\sum_i m_i(\ddot{z}+g)y_i - \sum_i m_i \ddot{x} z_i - \sum_i I_{ix} \ddot{\theta}_{ix}}{\sum_i m_i(\ddot{z}+g)}, \tag{2}$$

where $(x_{ZMP}, y_{ZMP}, 0)$ are the *ZMP* coordinates, (x_i, y_i, z_i) is the mass center of the link i in the coordinate system, m_i is the mass of the link i, and g is the gravitational acceleration. I_{ix} and I_{iy} are the inertia moment components, θ_{iy} and θ_{ix} are the angular velocity around the axes x and y, taken as a point from the mass center of the link i. The force sensor values are directly used to calculate the *ZM*. For the lateral control, it is only necessary to know the *ZMP* value for one axis. Thus, the *ZMP* calculus is simplified using the formula

$$P_{ZMP} = \frac{\sum_{i=1}^{3} f_i \mathbf{r}_i}{\sum_{i=1}^{3} f_i}, \tag{3}$$

where f_i represents the force at the i sensor, and r_i represents the distance between the coordinate origin and the point where the sensor is located. Figure 4 shows the distribution of sensors (marked with tree circles) used for each robot's foot.
The total *ZMP* is obtained by the difference between the ZMPs at each foot:

$$Total\_P_{ZMP} = P_{ZMP1} - P_{ZMP2}, \tag{4}$$

where P_{ZMP1} is the *ZMP* for one foot and P_{ZMP2} is the ZMP for the other.

Figure 5 shows the *ZMP* point (black point) for two robot's standing cases, one before to give a step (left), and other after give a step (right). The pointed line represents the support polygon.

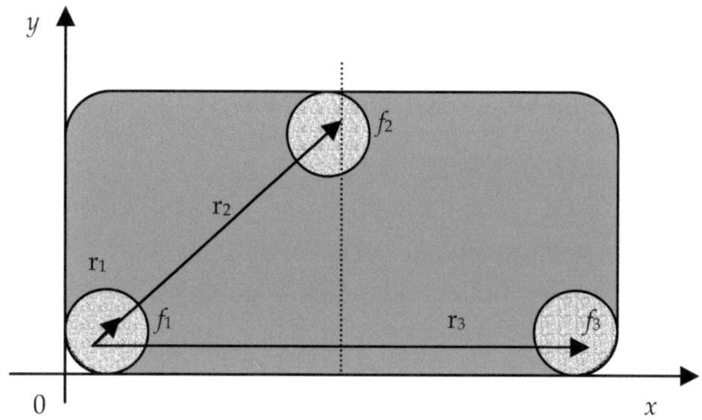

Fig. 4. Sensors distribution at the robot's foot.

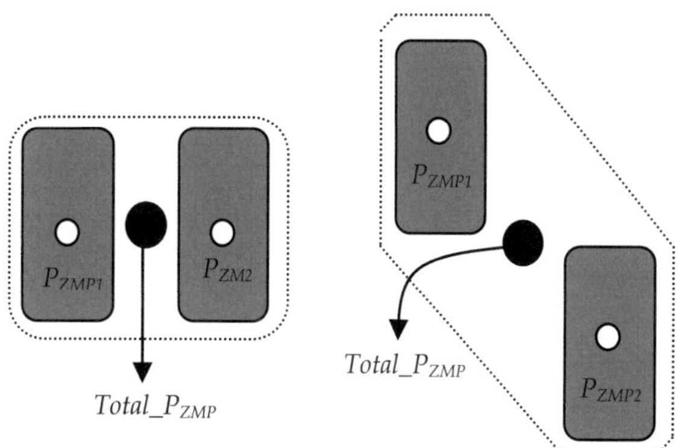

Fig. 5. The black point represents the Robot ZMP before giving a step (left), and after giving a step (right).

4. The Fuzzy PD incremental algorithm.

We propose the fuzzy PD incremental control algorithm, as a variant for the fuzzy PD controller (Sanchez et al., 1998), to implement the biped balance control. The fuzzy PD incremental control algorithm consists of only four rules and has the structure illustrated in Fig. 6.

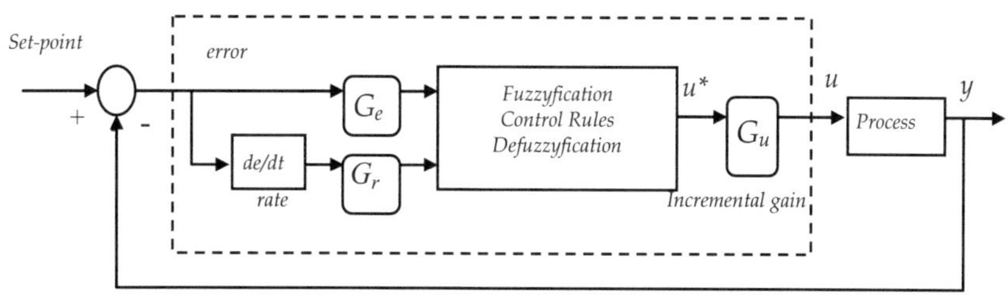

Fig. 6. Fuzzy PD incremental algorithm structure.

The gains G_u, G_e and G_r are determined by tuning and they correspond respectively to the output gains, the error (*ZMP error*) and error rate (*ZMP rate*) gains. The value u^* is the defuzzyficated output or "crisp" output. The value u is

$$u = \begin{cases} G_u * (u^*) & if\,|e| < \theta \;\; (for\;\; t = 0,\;\; G_{inc} = 0) \\ \\ G_{inc} + Inc & if\,|e| > \theta \end{cases} \tag{5}$$

Where e is the error (error*G_e), θ is an error boundary selected by tuning, and G_{inc} is the incremental gain obtained adding the increment "*Inc*".
Figure 7 shows the flow diagram for the incremental gain of u.

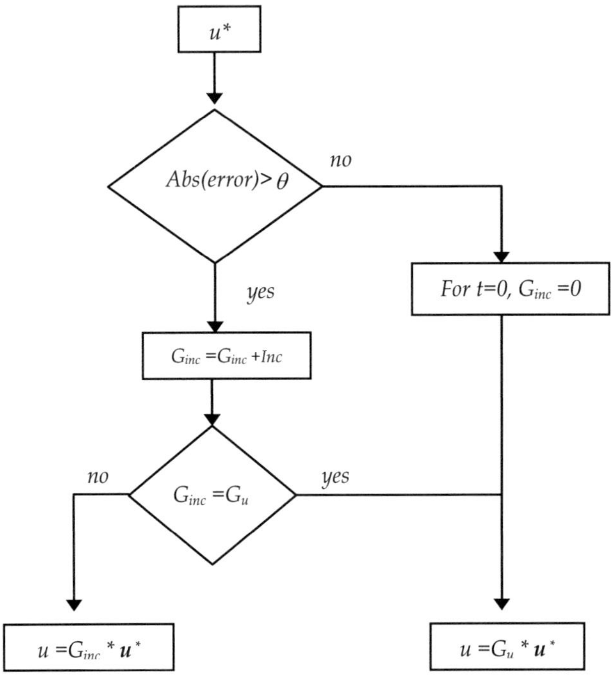

Fig. 7. Flow diagram for the incremental gain of G_u.

Figure 8 shows the area where the absolute error is evaluated and the controller output is incremental (u=G_{inc}+*Inc*).

4.1 Fuzzyfication

As is shown in figure 9, there are two inputs to the controller: error and rate. The error is defined as:

$$error = setpoint - y \tag{6}$$

Rate it is defined as it follows:

$$rate = (ce - pe) / sp \tag{7}$$

Where *ce* is the current error, *pe* is the previous error and *sp* is the sampling period. Current and previous error, are referred to an error without gain. The fuzzy controller has a single incremental output, which is used to control the process

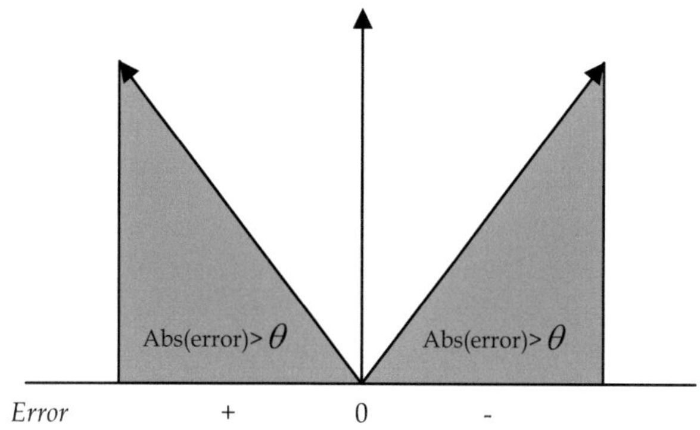

Fig. 8. Fuzzy PD incremental absolute error area.

The input an output membership functions for the fuzzy controller are shown in Fig. 9 and Fig. 10, respectively. Fig. 9 shows that each input has two linguistic terms. For the error input are: *Ge* negative error* (*en*) and *Ge* positive error* (*ep*) and for the rate input are: *Gr*negative rate* (*rn*) and *Gr * positive rate* (*rp*), while the output fuzzy terms are shown in Fig. 10 and they are: *Negative output* (*on*), *Zero output* (*oz*) and *Positive output* (*op*).
As shown in Fig. 9, the same function is applied for *error* and *rate* but with different scaling factors: *Ge* and *Gr* respectively.
H and L are two positive constants to be determined. For convenience we will take H=L to reduce the number of control parameters to be determined. The membership functions for the input variables, error and rate, are defined by:

$$\mu_{ep} = \frac{L + (G_e * error)}{2L}$$
$$\mu_{en} = \frac{L - (G_e * error)}{2L}$$
$$\mu_{rp} = \frac{L + (Gr * rate)}{2L}$$
$$\mu_{rn} = \frac{L - (Gr * rate)}{2L} \quad (8)$$

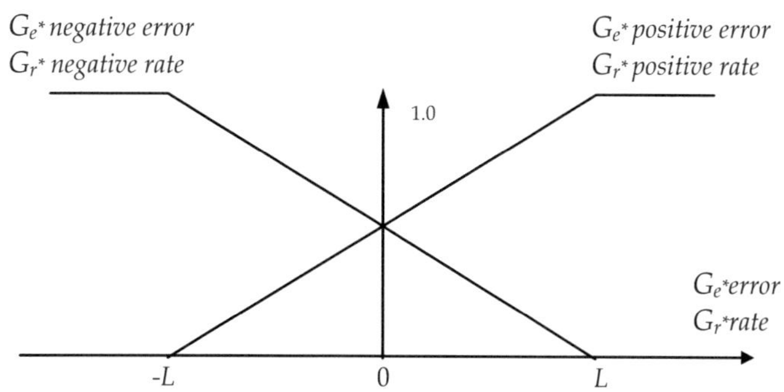

Fig. 9. Input membership functions.

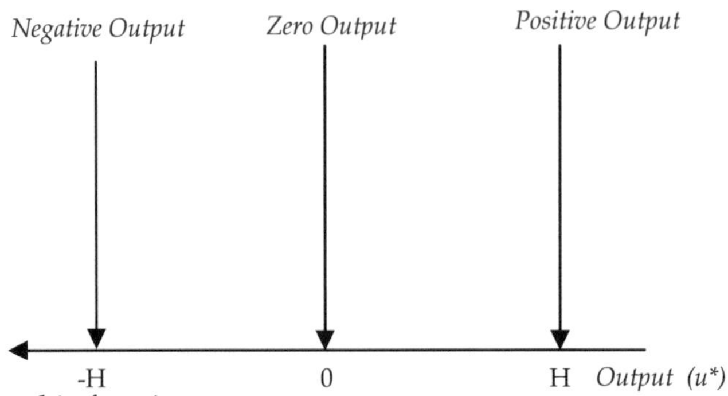

Fig. 10. Input membership functions.

4.2 Fuzzy rules and defuzzyfication.

There exist four rules to evaluate the fuzzy PD incremental controller (Miller 1994):

R1. If *error* is *ep* and *rate* is *rp* then output is *op*

R2. If *error* is *ep* and *rate* is *rn* then output is *oz*

R3. If *error* is *en* and *rate* is *rp* then output is *oz*

R4. If *error* is *en* and *rate* is *rn* then output is *on*

The determination of these rules can be accomplished easily if the system evolution is analyzed in the different operation points. For example, when the *error* and the *rate* increase (rule 1), it means that the system response decreases and moves away from the *setpoint,* for this reason it is necessary to apply a *positive* stimulus that allows to increase the system output. The figure 11 shows the determination of the different rules based on the system response.

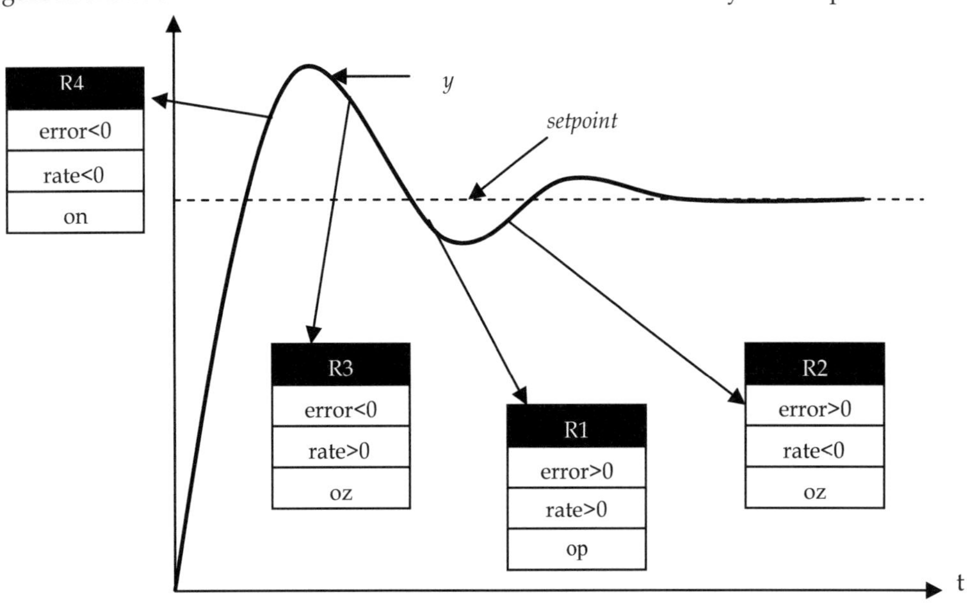

Fig. 11. Determination of the different rules based on the system response (see Text).

For the fuzzy condensed controller proposed, the input error and rate values ranges can be represented in 20 input regions (IC), like is shown in Fig. 12. If the membership functions are evaluated, the 4 control rules, the simplification H=L, and the defuzzyfication is applied in each one of the 20 inputs combinations, then 9 equations can be obtained (Sanchez et al., 1998), which can determine the control signal u that should be applied, depending on the region it is located. In other words, to implement the fuzzy condensed controller, only will be necessary to know the region in which the inputs variables are located and later evaluate the corresponding equation for this region. For example, the first equation acts in regions IC1, IC2, IC5, IC6. The figure 13 shows the control surface of the fuzzy condensed controller considering H=L=1.

Finally, the defuzzyfication method used is the gravity center, in this case is represented by:

$$u = \frac{-H(\mu_{R4(x)}) + 0(\mu_{R2(x)+}\mu_{R3(x)}) + H(\mu_{R1(x)})}{\mu_{R4(x)} + (\mu_{R2(x)+}\mu_{R3(x)}) + \mu_{R1(x)}} \tag{9}$$

5. Real time results

The first test applied to the balance controller was an x-direction impulse. The robot was standing in balance and then a push in the lateral direction was applied to the biped robot. Figure 14 shows the evolution of the *ZMP* value in centimeters during 10 seconds.

Figure 15 shows the error and rate evolution during the application of the x-direction impulse. The incremental PD fuzzy controller's parameters were: *Ge*=2, *Gr*=2, and *Gu*=1.

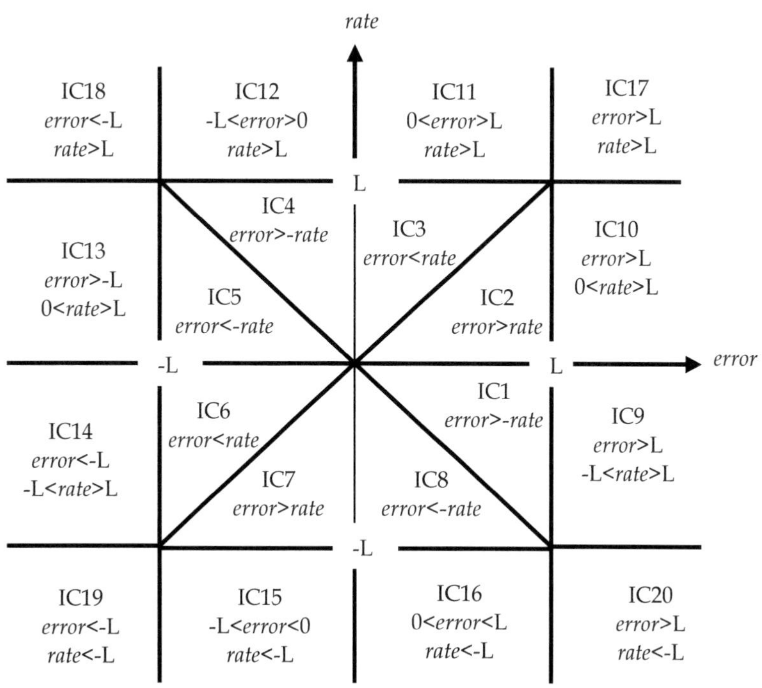

Fig. 12. Input regions.

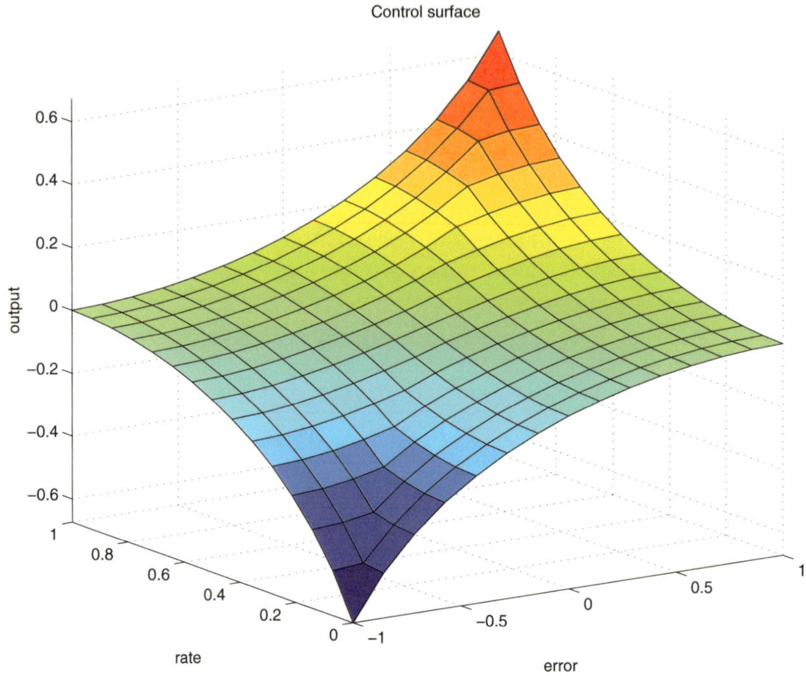

Fig. 13. Control surface of the fuzzy condensed controller considering H=L=1.

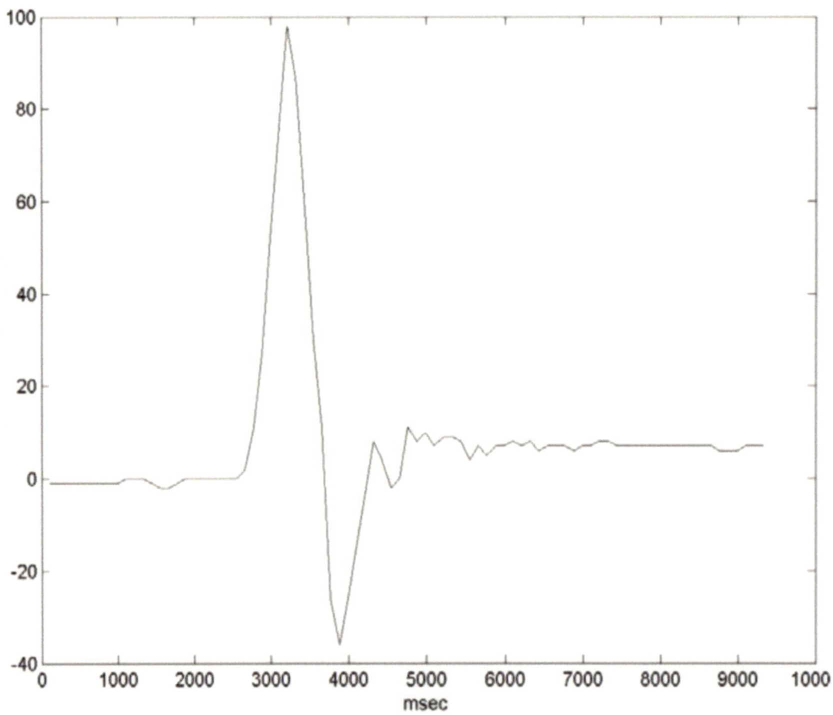

Fig. 14. ZMP value evolution achieved by the controller, for a x-direction impulse.

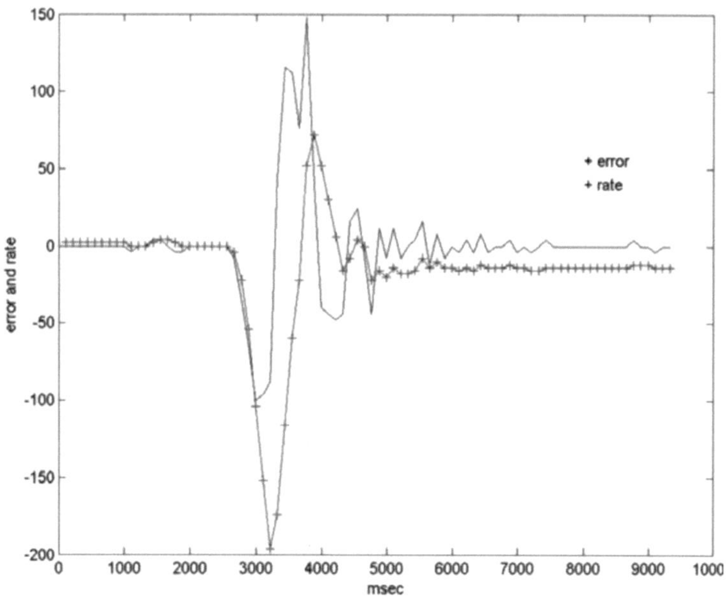

Fig. 15. Error and rate evolution for a x-direction impulse.

Finally, a walking test balance was applied to the robot. The robot gave some steps in a flat surface, while the balance controller compensated the ZMP value. Figure 16 shows the evolution of the ZMP value (in centimeters) achieved by the controller during 15 seconds. Figure 17 shows error and rate evolution during the robots walk. The incremental PD fuzzy controller's parameters were: *Ge*=2, *Gr*=2, and *Gu* =1.

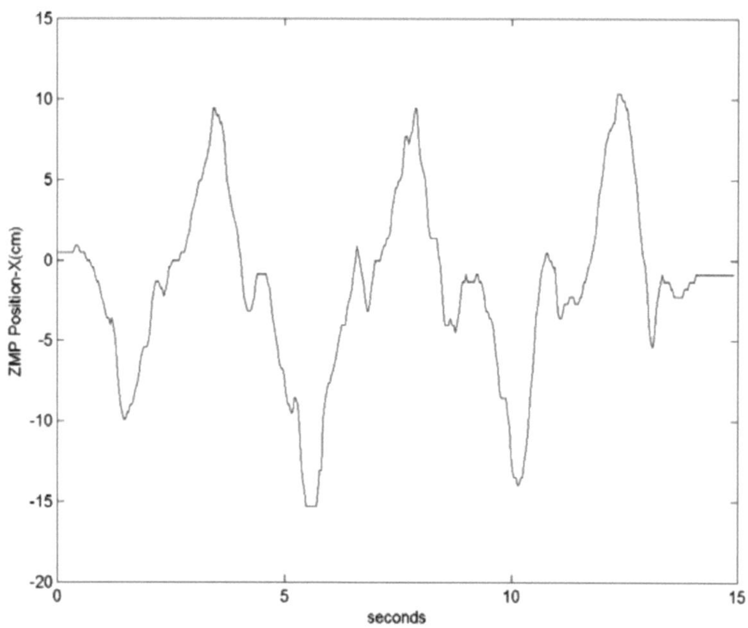

Fig. 16. ZMP value evolution achieved by the controller during the robot walk.

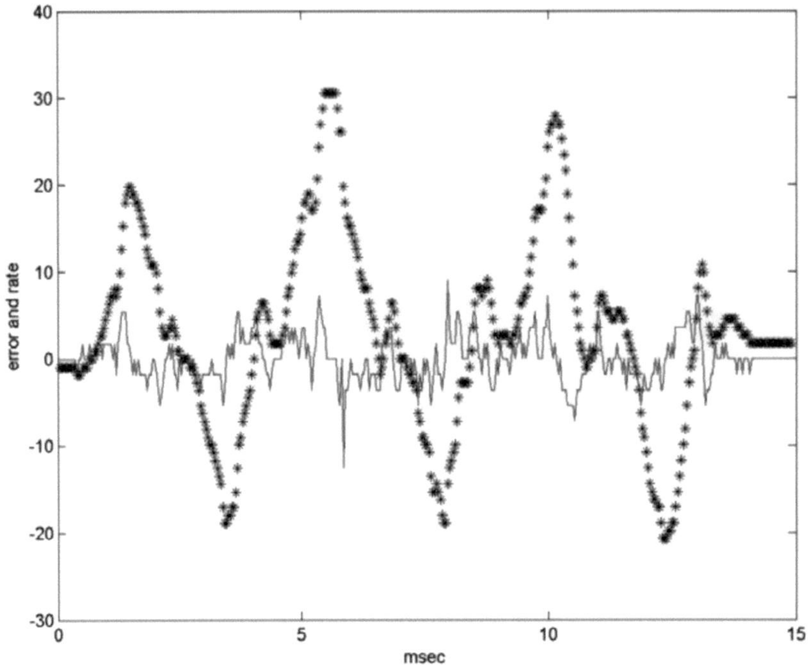

Fig. 17. Error and rate evolution during the robot walk.

The motor's position output was computed by the controller during the walk. It is showed at Figure 18. This motor's position value is only just for one waist motor.

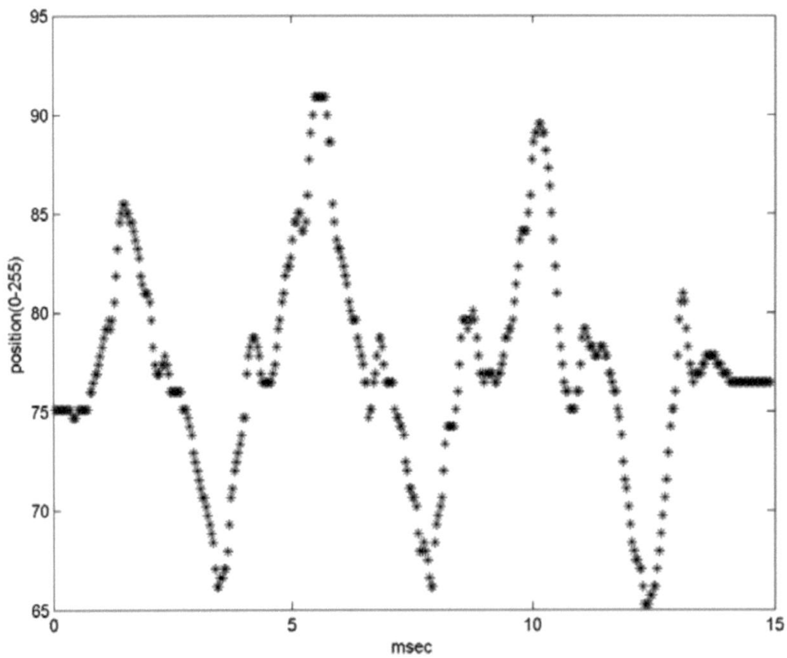

Fig. 18. Motor's position output during the robot walk..

6. Conclusion

Figure 11 shows the controller's performance and response for an *x*-direction impulse. The response was fast, approximately 2 seconds, until the controller reached a *ZMP* value near to zero. This feature allows the biped robot to gain stability even during walking (figure 13), maintaining the *ZMP* always inside the support polygon.

Our experiments with the fuzzy PD incremental controller algorithm demonstrated that it is computationally economic, all running in a PIC microcontroller, and appropriated for balance control. The algorithm was successfully used in real-time with the biped robot "Dany walker" (videos available at http://www.inf.fu-berlin.de/~zaldivar).

The algorithm proposed in this chapter could be also used in other robots structures with a different dynamic, and even with other degrees of freedom. It would be only necessary to adjust the controller's gain parameters to the particular structure.

We plan to use the information given by an inclinometer along with the *ZMP* value. In this case, the goal of the bipedal balance robot control will be to achieve an inclination value of cero and to maintain the *ZMP* at the center, or inside of the support polygon.

The bipedal robot used in this work is part of a project that is being developed at the Freie Universität Berlin.

7. References

Goddard, R.; Zheng, Y. & Hemami, H. (1992). Control of the heel-off to toe-off motion of a dynamic biped gait, *IEEE Trans. Syst. Man. Cayb.*, 1992, *vol.22*, no. 1

Kanehira, N.; Kawasaki, T.; Ohta, S.; Isozumi, T.; Kawada, T.; Kanehiro, F.; Kajita, S. & Kaneko, K. (2002). Design and experiments of advanced module (HRPL-2L) for humanoid robot (HRP-2) development, Proc. 2002, IEEE-RSJ *Int. Conf. Intell. Rob. Sys. EPFL*, Lausanne, Switzerland, 2002, 2455-2460.

Konno, A.; Kato, N.; Shirata, S.; Furuta, T. & Uchiyama, M. (2000). Development of a light-weight biped humanoid robot in Proc. 2000 *IEEE-RSJ Int. Con. Intell. Rob. Sys.*, 2000, 1565-1570.

Miller, W. (1994). Real-time neural network control of a biped walking robot, *IEEE contr. Syst., vol. 14*, 1994, 41-48.

Yamaguchi, J.; Soga, E.; Inoue, S. & Takanishi, A. (1999). Development of a biped robot-control method of a whole body cooperative dynamic biped walking, in Proc. 1999 *IEEE Int. Conf. Robot. And Aut.*, Detroit, Michigan, 1999, 368-374.

Taga, G.; Yamaguchi Y. & Shimizu, H. (1991). Self-organized control of bipedal locomotion by neural oscilators in unpredictable environment, *Biol. Cyb. 65* 1991, 147-165.

Kajita, S.; Yamaura, T. & Kobayashi, A. (1992). Dynamic walking control of a biped robot along a potiential energy conserving orbit, *IEEE Trans. Robot Automat. Vol. 8*, no. 4, 1992. pp 437-438.

Zheng, Y. & Shen, J. (1990). Gait synthesis for the SD-2 biped robot to climb sloped surface, *IEEE, Trans. Robot Automat. Vol 6*, no 1. , 1990, 86-96.

Cuevas, E.; Zaldívar, D. & Rojas, R. (2004). Bipedal robot description, *Technical Report B-03-19, Freie Universität Berlin, Fachbereich Mathematik und Informatik*, Berlin, Germany, 2004.

Vukobratovic, M. & Juricic, D. (1969). Contribution to the synthesis of biped gait, *IEEE Trans. Bio-Med. Eng., vol. BME-16*, no. 1 , 1969, 1-6.

Vukobratovic, M. (1973). How to control artificial anthropomorphic system, *IEEE Trans. Syst. Man. Cyb., vol. SMC-3,* no. 5, 1973, 497-507.

Cuevas, E.; Zaldívar, D. & Rojas, R. (2005). Dynamic Control Algorithm for a Biped Robot, *The 7th IASTED International Conference on CONTROL AND APPLICATIONS,Cancún, México, CA 2005.*

Sanchez, E.; Nuno, L.; Hsu, Y. & Guanrong, C. (1998). Real Time Fuzzy Swing-up Control for an Underactuated Robot, *JCIS '98 Proceedings,Vol 1* , N.C., USA, 1998.

11

Spoken Language and Vision for Adaptive Human-Robot Cooperation

Peter Ford Dominey
CNRS – University of Lyon
France

1. Introduction

In order for humans and robots to cooperate in an effective manner, it must be possible for them to communicate. Spoken language is an obvious candidate for providing a means of communication. In previous research, we developed an integrated platform that combined visual scene interpretation with speech processing to provide input to a language learning model. The system was demonstrated to learn a rich set of sentence-meaning mappings that could allow it to construct the appropriate meanings for new sentences in a generalization task. We subsequently extended the system not only to understand what it hears, but also to describe what it sees and to interact with the human user. This is a natural extension of the knowledge of sentence-to-meaning mappings that is now applied in the inverse scene-to-sentence sense (Dominey & Boucher 2005). The current chapter extends this work to analyse how the spoken language can be used by human users to communicate with a Khepera navigator, a Lynxmotion 6DOF manipulator arm, and the Kawada Industries HRP-2 Humanoid, to program the robots' behavior in cooperative tasks, such as working together to perform an object transportation task, or to assemble a piece of furniture. In this framework, a system for Spoken Language Programming (SLP) is presented. The objectives of the system are to 1. Allow the human to impart knowledge of how to accomplish a cooperative task to the robot, in the form of a sensory-motor action plan. 2. To allow the user to test and modify the learned plans. 3. To do this in a semi-natural and real-time manner using spoken language and visual observation/demonstration. 4. When possible, to exploit knowledge from studies of cognitive development in making implementation choices. With respect to cognitive development, in addition to the construction grammar model, we also exploit the concept of "shared intentions" from developmental cognition as goal-directed action plans that will be shared by the human and robot during cooperative activities.
Results from several experiments with the SLP system employed on the different platforms are presented. The SLP is evaluated in terms of the changes in efficiency as revealed by task completion time and number of command operations required to accomplish the tasks. Finally, in addition to language, we investingate how vision can be used by the robot as well to observe human activity in order to able to take part in the observed activities. At the interface of cognitive development and robotics, the results are interesting in that they (1) provide concrete demonstration of how cognitive science can contribute to human-robot interaction fidelity, and (2) they suggest how robots might be used to experiment with theories on the implementation of cognition in the developing human.

2. Language and Meaning

Crangle and Suppes (1994) stated: "(1) the user should not have to become a programmer, or rely on a programmer, to alter the robot's behavior, and (2) the user should not have to learn specialized technical vocabularies to request action from a robot." Spoken language provides a very rich and direct means of communication between cooperating humans (Pickering & Garrod 2004). Language essentially provides a vector for the transmission of meaning between agents, and should thus be well adapted for allowing humans to transmit meaning to robots. This raises technical issues of how to extract meaning from language.
Construction grammar (CxG) provides a linguistic formalism for achieving the required link from language to meaning (Goldberg 2003). Indeed, grammatical constructions define the direct mapping from sentences to meaning. Meaning of a sentence such as (1) is represented in a predicate-argument (PA) structure as in (2), based on generalized abstract structures as in (3). The power of these constructions is that they employ abstract argument "variables" that can take an open set of values.

(1) John put the ball on the table.
(2) Transport(John, Ball, Table)
(3) Event(Agent, Object, Recipient)

We previously developed a system that generates PA representations (i.e. meanings) from video event sequences. When humans performed events and described what they were doing, the resulting <sentence, meaning> input pairs allowed a separate learning system to acquire a set of grammatical constructions defining the sentences. The resulting system could describe new events and answer questions with the resulting set of learned grammatical constructions (Dominey & Boucher 2005).
PA representations can be applied to commanding actions as well as describing them. Hence the CxG framework for mapping between sentences and their PA meaning can be applied to action commands as well. In either case, the richness of the language employed will be constrained by the richness of the perceptual and action PA representations of the target robot system. In the current research we examine how simple commands and grammatical constructions can be used for action command in HRI. Part of the challenge is to define an intermediate layer of language-commanded robot actions that are well adapted to a class of HRI cooperation tasks.
This is similar to the language-based task analysis in Lauria et al. (2002). An essential part of the analysis we perform concerns examining a given task scenario and determining the set of action/command primitives that satisfy two requirements. 1. They should allow a logical decomposition of the set of tasks into units that are neither too small (i.e. move a single joint) nor too large (perform the whole task). 2. They should be of general utility so that different tasks can be performed with the same set of primitives.

3. Spoken Language Programming

For some tasks (e.g. navigating with a map) the sequence of human and robot actions required to achieve the task can be easily determined before beginning execution. Other tasks may require active exploration of the task space during execution in order to find a solution. In the first case, the user can tell the robot what to do before beginning execution, while in the second, instruction will take place during the actual execution of the task.

In this context, (Lauria et al. 2002) asked naïve subjects to provide verbal instructions to a robot in a miniature-town navigation task. Based on an analysis of the resulting speech corpora, they identified a set of verbal action chunks that could map onto robot control primitives. More recently, they demonstrated the effectiveness of navigation instructions translated into these primitive procedures for actual robot navigation (Kyriacou et al. 2005). This research indicates the importance of implementing the mapping between language and behavioural primitives for high-level natural language instruction or programming. The current study extends these results by introducing a conditional (e.g. if-then) component to the programming.

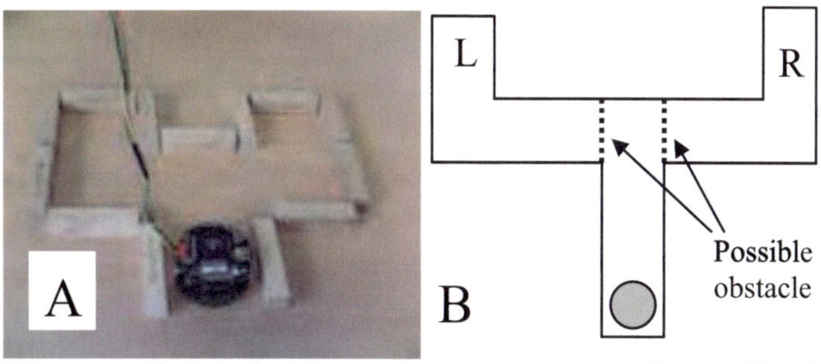

Fig. 1. Khepera robot and object transportation scenario. A. Physical set up. B. Labeled schematic representation.

Figure 1 illustrates a given scenario for HRI. In Figure 1A we see the Khepera robot on a table top, and in Figure 1B a schematic of the robot (represented as the gray disc) in a two arm maze. Consider a task in which User1 sends the Khepera robot to User2 who gives it an object to transport back to User1. Getting to User2 requires a conditional choice between two paths based on the location of an obstacle that is not known in advance, at the locations indicated by the dotted lines. Once the robot determines the location of the obstacle, it should choose the path that is not blocked, and make its way to the end of the arm. There it should turn around, and wait for User2, to place an object on its carrying surface. When User2 has performed this, User1 will say "continue" and the robot should then return to the beginning of the maze arm where User1 will take the transported block, and the process then continues.

On-Line Commanding: The simplest solution for controlling the robot in this task, which involves no learning, is for User1 simply to tell the robot what to do, step by step. Depending on whether the obstacle is at the left or right arm position, the human User1 would decide whether to tell the robot to take the right or left pathway. Likewise, once at the end point, User1 would wait until the User2 placed the object before commanding the robot to turn around and come back.

Programming a Pre-Specified Problem: Again, in many cases, the problem is known to the user in advance (possibly because the user has now "walked" the robot through the problem as just described) and can be specified prior to execution.

The objective of SLP is to provide a framework in which non-specialist humans can explain such a task to a robot, using simple spoken language. While the task described above remains relatively simple, explaining it to the robot already levies several interesting requirements on the system, and meeting these requirements will provide a fairly general and robust capability for SLP. In particular, this task requires the following: (1) The user should be able to provide a sequence of primitive commands that will be executed in a particular order. (2) The user should be able to specify conditional execution, based on the values of robot sensors. (3) The user should be able to tell the robot that at a certain point it should wait for some sensory event to occur before proceeding with its sequence execution., as demonstrated below.

In addition, for tasks that become increasingly long and complex, the user may make mistakes in his/her initial specification of the task, so the SLP system should allow the user to test, "debug" and modify programs. That is, once the user has specified the program, he should be able to execute it and - if there is a problem - modify it in a simple and straightforward manner.

On-line Commanding with Repetitive Sub-Tasks: On-line commanding allows the user to be responsive to new situations, and to learn him/herself by taking the robot through a given task or tasks. On the other hand, for tasks that are well defined, the user can program the robot as defined above. In between these two conditions there may arise situations in which during the course of solving a cooperative problem with the robot, the user comes to see that despite the "open endedness" of a given problem set, there may repetitive subtasks that occur in a larger context. In this type of situation, the human user may want to teach the robot about the repetitive part so this can be executed as an autonomous subroutine or "macro" while the user still remains in the execution loop for the components that require his/her decision.

Fig. 2. Human-Robot cooperation in a furniture construction task. A. Robot takes a table leg from User2. B. Robot hands the leg to User1. C. Robot holds the table steady while User1 attaches the leg.

Figure 2 illustrates such an HRI scenario that involves two humans and the HRP-2 cooperating in the construction of a small table. The construction task will involve fixing the legs to the surface of the table with wood screws. User1 on the left interacts with the robot and with User2 on the right via spoken language.

User1 will command the robot to prepare to receive one of the table legs that will be passed to it by User2. The robot waits until it receives a "continue" signal from User1, and will then pass the leg to User1 who will take the leg, and then ask the robot to hold the table top steady allowing User1 to attach the leg to the table. User1 then tells the robot to release the table. At this point, the first leg has been fixed to the table, and the "get the leg and attach it" sequence can be repeated. This task thus has a repeating subsequence that can be applied to each of the four legs of the table. Experiments below address this task. These experiments identify the utility of a more complex command structure based on grammatical constructions.

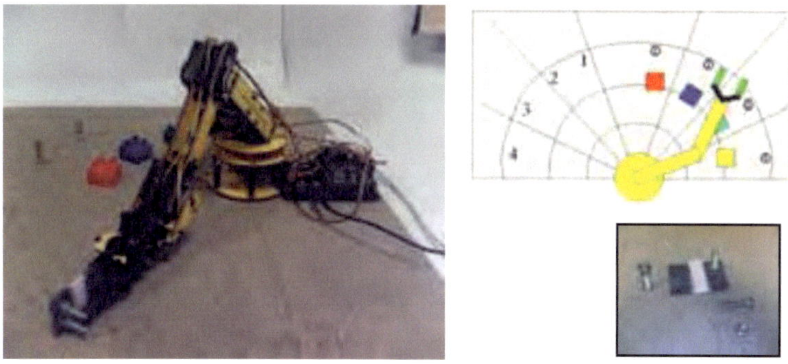

Fig. 3. Cooperative manipulation task scenario. User asks for screws from locations identified by color. While robot holds the plastic "table" (as illustrated) the user can insert the screws into the table (see inset).

Influence of Robot Command Structure on Language: Part of the central principal of construction grammar (and the related domain of cognitive linguistics) is that human language is made up of a structured inventory of grammatical constructions (Goldberg 2003). These constructions reflect the meaning of prototypical events that are basic to human experience. Thus for example the "ditransitive" construction involves one agent transferring an object to another, as in "John gave Mary the ball." From a robotics perspective, the action predicate-argument Move(Object, Location) can thus be considered as a template for a motor program that allows the use of constructions of the form "Give the X to Y". This robot PA involves localizing X and Y, and then grasping X and transporting the effector to Y to transfer X to Y. To the extent that such an action PA (APA) is built into the repertoire of the robot (or to the extent that it can be built up from primitives like localize(X), transport-to(X) etc.) rich grammatical constructions can be used to generate and interpret sentences like "give me the red block," or "put the blue block next to the red block" for robot control.

Figure 3 illustrates a simplified version of the table construction scenario that uses the Lynx6 arm for cooperative manipulation in a construction task. Experiments below examine cooperative construction with the Lynx6 based on lessons learned from the HRP-2 experiments, including more complex commands based on grammatical constructions.

4. Implementation of Spoken Language Programming

Because of the potential influence that the robot itself will have on spoken language interaction, we chose to develop a core SLP system, and then to adapt the system to three different robot platforms the K-Team Khepera, the LynxMotion Lynx 6 DOF manipulator arm, and the Kawada Industries HRP-2 Humanoid. This allows us to explore the navigation vs. manipulation dimension, and within the manipulation dimension the Lynx allows for rapid prototyping while the HRP-2 allows for a much more robust and human-scale human-robot interaction.

Spoken Language Programming Core: The core SLP system is presented in Figures 4 and 5, and Table 1. The core system provides the command, conditional control, and programming/teaching capabilities based on spoken language.

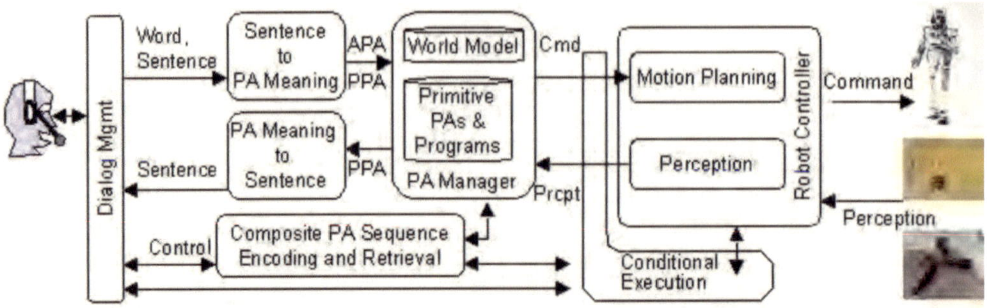

Fig. 4. Spoken Language Programming Architecture. The fundamental unit of meaning is the predicate-argument (PA) representation. APA - Action Predicate-Argument representation. PPA - Perceptual Predicate-Argument representation. Spoken language commands are interpreted as individual words or grammatical constructions, and the command and possible arguments are extracted in Sentence to PA Meaning. The robot level commands (CMD) are extracted by the PA manager. World Model includes specification of object locations, and related state information. Robot commands are then issued to the Controller to effect the commands. Conditional commands including IF *CONDITION*.. OTHERWISE, and WAIT are implemented as part of the Conditional Execution. Learning-related commands LEARN, OK/SAVE-MACRO, RUN MACRO are handled by the Sequence Encoding and Retrieval function. Event perception and description, provided by PA Meaning to Sentence are implemented in Dominey & Boucher (2005).

Dialog management and spoken language processing (voice recognition, and synthesis) is provided by an "off the shelf" public domain product, the CSLU Rapid Appication Development (RAD) Toolkit (http://cslu.cse.ogi.edu/toolkit/). RAD provides a state-based dialog system capability, in which the passage from one state to another occurs as a function of recognition of spoken words or phrases; or evaluation of Boolean expressions. Figure 5 illustrates the major dialog states.

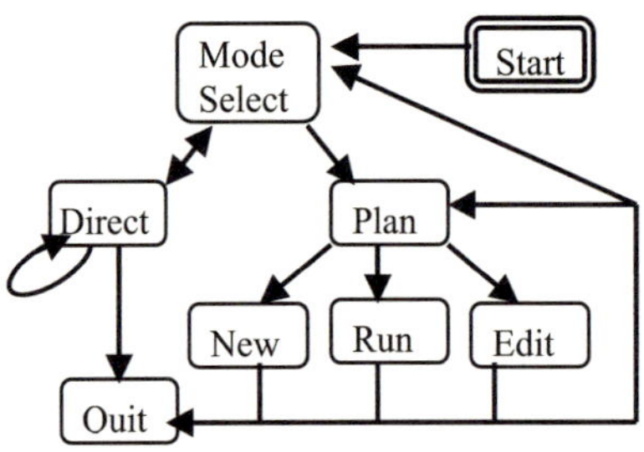

Fig. 5. Major Dialog States

In Figure 5, when the system starts in the "Mode Select" state it asks the user "*Shall we learn by planning or by demonstration?*"

Learning by Demonstration: If the user replies "demonstration" (note: In all dialogs, human responses will be indicated by *italics*) then the system transitions to the "Direct" state, and interrogates the subject with "*Now what?*" The user can reply with one of the robot-specific motor commands, which the robot will execute and then again prompt "*Now what?*" In the course of this direct interaction, if the user determines that there is a repetitive component in the task, as illustrated in the discussion of Figure 2, above, the user can invoke the "Learn" command. This indicates to the system to begin to store the subsequent commands. When the repetitive subsequence has been performed, the user then issues the command "OK" and the subsequence is stored, and can then be executed in-line with the command "Macro". This learning, in which the user "demonstrates" the performance by directing the robot through the execution is illustrated in Exps 3-5.

Learning by Planning: In Figure 5, when the system starts in the "Mode Select" state it asks the user "*Shall we learn by planning or by demonstration?*" If the user replies "planning" then the system transitions to the "Plan" state, and interrogates the subject with "*Shall we make a new plan, run the old plan, edit it or quit ?*"

If the user replies "New plan" then the system continuously interrogates the user with "*now what?,*" with the user responding each time with the next motor command or conditional command until the user replies "stop." In this manner, command by command without execution, the user specifies the planned program using spoken language. After "stop", the plan is stored, and the system returns to the Plan state. Here, the user can now run the plan to see how it works. If the user detects that there is a problem with one of the elements in the stored plan, at the "Plan" state, he has the option of saying that he wants to "edit" the plan.

Editing Learned Programs: When the system enters the Edit state, it displays the plan graphically as a list of commands and corresponding conditions, and then ask the user what he wants to modify. The user can then respond with a two word phrase in which the first word is "action" or "condition", and the second word is the number of the action or condition in question. The system then asks the user for the new value for that component, and then again asks what the user wants to modify. When the editing is finished, the system then returns to the Plan state where the user can either execute the modified plan, or do something else.

Conditional Execution: In addition to the learning-related commands, we have also identified the requirement for commands related to conditional behavior execution and control. These commands are specified in Table 1. Perhaps the most obvious of these the "if -then-else" construction.

Conditional Commands	**Effects / Correspondence**
If *condition*	Begin a conditional segment
Otherwise	End a conditional segment
Wait *condition*	Interrupt execution until *condition* is met
Continue	User command to proceed when waiting This is one of the conditions for wait.
Learning-Related Commands	
Learn	Begin macro learning
Ok/Stop	End macro learning
Macro	Run stored macro

Table 1. Conditional and Learning-Related Commands.

Indeed this is the first concrete example of the use of a multi-word construction. When the "if *condition*" construciton is issued, the user specifies the "if" and the corresponding condition in the same utternence (e.g. "if left clear"), and the "if" and the "left clear" condition are paired together in the stored sequence plan. The user then continues to specify the succession of commands that should logically follow in case the "if" condition is satisfied. When the user terminates that logical section, he indicates this with the conditional command "otherwise". The subsequent succession of commands corresponds to those that should be executed in case the condition of the "if" fails. During sequence execution in the Run state, when the system encounters an "if", it tests the corresponding condition. When the condition is true, the system executes up to the "otherwise" statement. When false, it skips to the otherwise, and then resumes execution of the subsequent commands.

The behavioral scenarios above also identified the need for a conditional wait, in which the execution of a stored sequence waits for a sensory condition to become true. Like the "if", the "wait *condition*" construction uses the identifier "wait" followed by a condition, which is paired with the wait command. Then, during execution of the plan in the Run state, when the system encounters a "wait" action, it tests the corresponding condition, and waits until that condition is true before proceeding. The "continue" condition (indicated by the user saying "continue") is the single robot-independent condition.

Robot Specific Command: Given the robot-independent component of the SLP architecture, we now explain how this is adapted to the specific robots. The final behavioral result of a robot-specific action command that is issued either directly or as part of a learned plan is the execution of the corresponding action on the robot. We new specify the specific commands for each of the three robot platforms we used.

K-Team Khepera: The Khepera (K-Team) is a small two-wheeled robot equipped with proximity sensors that has been extensively explored in sensory-motor robotics. Based on the requirements for the cooperative task described above, a set of primitive actions, and senor states was identified for the Khepera. These robot-specific sensory-motor functions are identified in Table 2.

The Khepera is connected to the SLP system PC via the RS232 serial port. Commands are issued to the Khepera via this connection, and the values of the forward, left and right position sensors are read from this connection as well.

Motor Commands	**Resulting Actions**
Explore	Move forward until obstacle is encountered
Left	Turn 90° left
Right	Turn 90° right
Specifiable Conditions	**Corresponding check**
Left clear	Check left proximity sensor
Right clear	Check right proximity sensor
Obstacle removed	Check forward proximity sensor

Table 2. Khepera Action Commands and Sensory Conditions.

Kawada Industries HRP-2 Humanoid: Based on the preliminary analysis of the table-building scenario above, a set of primitive actions was identified for the Kawada Industries HRP-2 humanoid robot under the control of the OpenHRP controller (Dominey et al. 2007). The HRP-2 has 30 controlled degrees of freedom, 8 of which are used in this study. Each of the functions in Table 3 corresponds to a particular posture that is specified as the angles for a subset of the 30 DOFs. These actions have been implemented by handwritten python scripts that specify final, hardcoded, joint angles and motion durations for the given postures. The python script execution is triggered remotely by the CSLU toolkit, and communicates directly with the low-level OpenHRP framework. The motion is achieved by linearly interpolating joint angles between the starting and final configurations, for each specific action. We have chosen these simple actions in order to demonstrate the feasibility of the overall approach in the table-building scenario, and more complex functions are currently under development.

Motor Command	**Resulting Actions**
Prepare	Move both arms to neutral position, rotate chest to center, elevate left arm, avoiding contact with the work surface (5 DOF)
Left open	Open left hand (1 DOF)
Left close	Close left hand (1 DOF)
Give it to me	Rotate hip to pass the object in left hand to user on the right (1 DOF)
Hold	Center hip, raise right arm preparing to hold table top (5 DOF)
Right open	Open right hand (1 DOF)
Right close	Close right hand (1 DOF)

Table 3. HRP-2 Action Commands.

Lynx 6 DOF Manipulator: We also performed SLP experiments with a 6DOF Lynx 6 robot manipulator arm. The robot controller is connected to the Pentium PC running RAD via the RS232 serial port, and provides access to the 6 joint angles which can be controlled individually and in parallel with a time-to-completion parameter. Based on task requirements for object manipulation and the miniature task building tasks, we pre-localized 8 locations in the workspace along a fixed radius that correspond to the color and number locations indicated in Fig. 3. We then manually determined the joint angles for taking the manipulator to these regions, above the workspace, and for lowering the manipulator to be able to grasp objects at these locations. These postures were then used to create the action commands in Table 4.

Action Command	**Resulting Actions**
Put the *Color* screw at location *number*.	Move to location marked by *color*, pick up the screw, place it at location marked by *number*. Return to neutral position.
Give me the *color* screw.	Move to location marked by *color*, pick up the screw, place it at location three. Return to neutral position.
Open gripper	Open gripper
Close gripper	Close gripper
Hold the table	Move gripper to location 3, preparing to hold the table

Table 4. Lynx 6 Action Commands

Note that in Table 4 that the "put" and "give me" commands are multiword constructions that take arguments. These constructions correspond to the ditransitive construction from section II.B.

5. Experimental Results Part 1

We now provide results from a series of experiments that were performed based on the scenarios described in Section 3. These experiments were performed with human users who were familiar with the system.

5.1 Programming Conditional Navigation with Khepera

The current experiment tests whether the user can use spoken language to explain a complex conditional task to the robot, based on the scenario described in section 3 and Fig1 in which User1 sends the Khepera robot to User2 who gives it an object to transport back to User1. Getting to User2 requires a conditional choice between two paths based on the location of an obstacle that is not known in advance.

In the following dialog, robot language is specified in *italics*, and human responses in normal text. Technical and explanatory annotation is in <>'s.

<u>Dialog for Cooperative Navigation and Transportation:</u>

Shall we learn by planning or by demonstration ? Planning. *Shall we make a new plan, run the old plan, edit it or quit ?* New plan. *Now what?* <*user begins to specify the plan*>

Explore. *Now what?* If left clear. <the user specifies that if the left is clear, the following should be performed >

Now what? Left. *Now what?* Explore *Now what?* Right *Now what?* Explore. *Now what?* Left. *Now what?* Left. <the user has now specified the plan segment that takes the robot to location L in Fig 1, and turns it around, ready for the return trip>

Now what? Wait until I say continue. <user specifies that before continuing execution of the return trip the robot should wait for a verbal "continue" signal. During this period the second human team member places the Lego block on the robot. Speech pattern match on "wait *any continue" where *any matches anything.>

<The following plan segment takes the robot back to the starting location, where the principal team member can decide where to place the block.>

Now what? Explore. *Now what?* Left. *Now what?* Explore. *Now what?* Right. *Now what?* Explore. *Now what?* Left? *Now what?* Left. *Now what?*

< The planning for the conditional execution of the path to the left is finished. This plan segment now specifies the "else' - what should happen if the left arm is blocked. It is similar to what happens for the left arm case.>

Otherwise. *Now what?* Right. *Now what?* Explore. *Now what?* Left. *Now what?* Explore. *Now what?* Left. *Now what?* Left. *Now what?* Wait until I say continue. <Robot now at the R, ready to come back after User2 places the block.> *Now what?* Explore. *Now what?* Left. <This is an error that will turn the robot into the wall> *Now what?* Explore. *Now what?* Right. *Now what?* Explore. *Now what?* Left *Now what?* Left. *Now what?* Stop. <Stop indicates that the program should be stored, and then executed>

When the stored plan executed we observed that when the obstacle was placed to the right, the robot correctly detected that the left was clear, and proceeded along the left arm to the location marked L. It then announced that a conditional wait was encountered, and that it was waiting for the spoken command "continue". This allowed User1 to verify that User2 had placed a Lego on the robot, so that he could then say "continue" and the robot continued back to the starting location, where User1 could recover the block. We then placed the obstacle at the left location. The conditional execution of the "if" determined that the test condition failed, and execution skipped to the "otherwise" section of the plan. In execution of this plan segment, an error in the plan was detected. After the robot started to return to User1 (after the "continue" command had been issued) the plan incorrectly specified that the robot turn left (into the wall) rather than right (towards the center of the maze).

That is, during the specification of a plan of this length, it is possible that the user makes an error in a command, particularly in left vs. right distinctions when the robot is moving towards the human, in which the human and robot reference frames are reversed. The editing capability described above thus allowed the user to listen to the program and see it on the screen and then modify the errors by indicating the number of the action in the program, and what it should be changed to.

5.2 Programming Cooperative Manipulation with HRP-2

We now move from the navigation domain to object manipulation. The current experiment exploits the bimanual manipulation capabilities of the HRP-2 humanoid for a human-robot cooperation task that involves constructing a piece of furniture (Dominey et al. 2007).

In the following paragraphs we provide the literal transcription of the human-robot dialog, annotated with explanation about the corresponding human and robot behavior and key system functions in <> brackets. Again, robot language will be indicated in *italics*, and human language in plain text.

Phase 1. Spoken language commanding.

Users 1 and 2 (Fig 3) begin their first interaction with the robot. User1 controls the task via spoken language. Because the task is new to him, he proceeds for the moment only by commanding the robot, with no programming.

Dialog: <Robot initiates the interaction with the prompt> *I am ready.* <the human responsd with the command> Prepare. *You said prepare?* Yes. <Because of possible errors in speech recognition, we choose to have an explicit validation on each command as illustrated here>. *Preparing to grasp.* <Robot raises left hand > *I am ready.* Left open. *Opening left hand.* <Robot opens left hand> *I am ready.* <User2 places the table leg in the robot's left hand (See Fig 2)> Left close. *Closing left hand.* <Robot closes left hand to grasp the table leg> *I am ready.* Give it to me. *Giving you the object.* <Robot turns to the right, holding the table leg out to User1> *I am ready.* Left open. Yes. *Opening left hand.* <As the Robot opens the hand, User1 reaches for and grasps the table leg> *I am ready.* Hold. <the first of several primitives required for the robot to hold the table while User1 attaches the leg.> *You said give it to me?* No. <This is an example of a speech recognition error.> *I am ready.* Hold. Yes. *Getting ready to hold.* <The robot orients to the work surface and raises its right hand. *I am ready.* Right open. *Opening right hand.* <With the robot's right hand open, User1 places the table top within the robot's grasp>. *I am ready.* Right close. *Closing right hand. I am ready.* Wait. *Waiting for your signal* <During this wait period, User1 attaches the leg to the table, and then gives the "continue" command when ready.> Continue. *I am ready.* Right open. *Opening right hand.* <User1 takes the table from the robot and prepares for the next leg.> *I am ready.*

Phase 2: Programming the Robot

At this point, User1 has used spoken language tele-operation to enlist the robot's assistance in attaching the first leg to the table. Given the knowledge of the task User1 has gained, he is now ready to "program" the robot for the repetitive component of the task by using the "learn" command.

Dialog: Learn. *You said learn?* Yes. *I'll start learning now. I am ready.* Prepare. *You said prepare?* Yes. *Preparing to grasp....*(same sequence as for the first leg) ... Right open. *You said right open?* Yes. *Opening right hand.* <User1 takes the table>. *I am ready.* OK. *You said OK?* Yes. *OK we will store this plan. I am ready.*

Execution of Stored Program Macro
Now the stored program can be used via the "macro" command.
Dialog: Macro. *You said macro?* Yes. *Running the macro. Preparing to grasp. Waiting for your signal.* Continue. *Closing left hand.* <Robot takes leg from User2> *Giving you the object. Opening left hand.* <User1 takes leg>. *Getting ready to hold. Waiting for your signal.* <User1 place table in robot's right hand> Continue. *Closing right hand. Waiting for your signal* <User1 attaches the table leg>. Continue. *Opening right hand. In line macro finished. I am ready.* <Third leg is now attached. User1 takes the table. Same procedure with macro for attaching fourth leg.>
Performance Analysis: During the assembly of the first two legs, User1 issued 10 distinct commands for each of the two legs. Installing legs 1 and 2 took 3min 25sec and 3min 24sec. Once the program was learned, for legs 3 and 4, a single command initiated the program, and the user was only required to issue 3 "continue" commands in order to indicate to the robot that he was ready to proceed. Execution time was reduced to 2:11 and 2:33 respectively. We performed a statistical analysis of the variance (ANOVA) in indivisual action completion times examining the effects of Repetition (i.e. first and second leg in either the CMD or PRG mode), and Programming condition (i.e. CMD vs PRG). Only the Programming condition had a significant effect on the completion times (ANOVA, Programming Effect: $F(1,6) = 109$, $p < 0.0001$).
We performed a second experiment in which the same primitives were used, with absolutely no change to the software, to disassemble the table. The use of the programming capability for the third and fourth leg (executed in 2:51 and 2:51 respectively) yielded significant reductions in execution time as compared with the first two legs (executed in 3:57 and 4:11 respectively). To compare performance in the two experiments we performed a 3 way ANOVA with the factors Experiment (assembly, disassembly), Programming (with vs. without, i.e. PRG vs CMD), and Repetition (First vs. second repetition in each condition). For the completion times were elevated for the CMD vs PRG conditions, i.e. action execution was slower when programming was not used. The ANOVA reveled that only the Programming effect was significant ($F(1,6) = 277$, $p < 0.0001$).

5.3 Grammatical Constructions for Manipulation with Lynx6 Arm

The programming by command sequencing illustrated in the previous experiment clearly leads to improvements in performance and facilitation of the task for the human. Further progress can be made by the implementation of higher level action functions that respect the natural interaction categories that can arise in these cooperative manipulation tasks. In particular, the user who is building the table repeatedly interacts with the robot so that the robot will pass him the next leg of the table. The action semantics can be captured in the construction "Give me X", or more generally "Move the object at location X to location Y" where X specifies a target object location and Y a destination location where that object is to be placed.
We hypothesized that the availability of this kind of construction would simplify the task for the user, and also speed up the execution process. We test this hypothesis with the Lynx

6 robot, as a prototype platform before subsequent implementation on the HRP-2. The table construction task with the Lynx employed a plastic table surface and four 3cm threaded bolts with nuts (see insert Fig.3). These served as legs, and the robot was used to fetch the legs and hold the table while the user fixed the legs to the table, as in the previous experiment. As illustrated in Fig. 3 we designated 4 areas indicated by color for the four legs to be placed initially, and four numbered locations where these legs could be placed using a construction of the form: Put the *X* leg at location *Y*, where X was one of the four color designations and Y one of the four numbered designations (see Fig 3). We also implemented a construction of the form : Give me the *X* leg, which commanded the robot to fetch the leg at location color, and place it at location 3 which was just in front of the user.

Dialog:
Command an action. Give me the red screw. *You want the red screw?* Yes. Giving the red screw <robot takes screw at red location> to you <and passes it to the user>. *Command an action.* Start learning. <this initiates memory of the subsequent command sequence> *Command an action.* Hold. *Preparing to hold. Command an action.* Close gripper. *Closing gripper. Command an action.* Wait. *Waiting for your signal.* <User insertes and tightens screw while robot holds the grey plastic "table". See Fig 3> Continue. *Command an action.* Open. *Opening gripper. Command an action.* Save the macro. *You said save the macro?* Yes.
Command an action. Give me the blue screw. *You want the blue screw?* Yes. *Giving the blue screw... to you. Command an action.* Run the macro. *You said run the macro?* Yes. *Preparing to hold. Closing gripper. Waiting for your signal.* <User attaches screw while robot holds the table steady>. Continue. *Opening gripper.*
Command an action. Give me the green screw. *You want the green screw?* Yes. *Giving the green screw... to you. Command an action.* Run the macro. *You said run the macro?* Yes. *Preparing to hold. Closing gripper. Waiting for your signal.* Continue. *Opening gripper.*
Command an action. Give me the yellow screw. *You want the yellow screw?* Yes. *Giving the yellow screw... to you. Command an action.* Run the macro. *You said run the macro?* Yes. *Preparing to hold. Closing gripper. Waiting for your signal.* Continue. *Opening gripper.*
The total execution time for the assembly was less than five minutes. With the macro use, the user required five spoken commands per leg, and three of these were confirmations of commands and the continue signal. The two main commands are "Give me.." and "run the macro." With respect to the HRP-2, the availability of the "Give me the X screw" PA command eliminated the need to sequence a number of lower level postural commands into the macro, which instead was limited to implementing the "hold the table" function.

6. Programming Shared Behavior by Observation and Imitation

If spoken language is one of the richest ways to transmit knowledge about what to do, demonstration and imitation of observed behavior are clearly another. One of the current open challenges in cognitive computational neuroscience is to understand the neural basis of the human ability to observe and imitate action. The results from such an endeavor can then be implemented and tested in robotic systems. Recent results from human and non-human primate behavior, neuroanatomy and neurophysiology provide a rich set of observations that allow us to constrain the problem of how imitation is achieved. The current research identifies and exploits constraints in these three domains in order to develop a system for goal directed action perception and imitation.

One of the recurrent findings across studies of human imitation is that in the context of goal directed action, it is the goal itself that tends to take precedence in defining what is to be imitated, rather than the means (Bekkering et al. 2000, Tomasello et al. 2005). Of course in some situations it is the details (e.g. kinematics) of the movement itself that are to be imitated (see discussion inCarpetner and Call 2007, Cuijpers et al. 2006), but the current research focuses on goal based imitation. This body of research helped to formulate questions concerning what could be the neurophysiological substrates for goal based imitation. In 1992 di Pellegrino (et al.) in the Rizzolatti lab published the first results on "mirror" neurons, whose action potentials reflected both the production of specific goal-directed action, and the perception of the same action being carried by the experimenter. Since then, the premotor and parietal mirror system has been studied in detail in monkey (by single unit recording) and in man (by PET and fMRI) reviewed in Rizzolatti & Craighero (2004).

In the context of understanding imitation, the discovery of the mirror system had an immense theoretical impact, as it provided justification for a common code for action production and perception. In recent years a significant research activity has used simulation and robotic platforms to attempt to link imitation behavior to the underlying neurophysiology at different levels of detail (see Oztop et al. (2006) for a recent and thorough review, edited volume (Nehaniv & Dautenhahn 2007), and a dedicated special issue of Neural Networks (Billard & Schaal 2006)). Such research must directly address the question of how to determine what to imitate. Carpenter and Call (2007) distinguish three aspects of the demonstration to copy: the physical action, the resulting change in physical state, and the inferred goal - the internal representation of the desired state. Here we concentrate on imitation of the goal, with the advantage of eliminating the difficulties of mapping detailed movement trajectories across the actor and imitator (Cuijpers et al. 2006).

Part of the novelty of the current research is that it will explore imitation in the context of cooperative activity in which two agents act in a form of turn-taking sequence, with the actions of each one folding into an interleaved and coordinated intentional action plan. With respect to constraints derived from behavioral studies, we choose to examine child development studies, because such studies provide well-specified protocols that test behavior that is both relatively simple, and pertinent. The expectation is that a system that can account for this behavior should extend readily to more complex behavior, as demonstrated below.

Looking to the developmental data, Warneken, Chen and Tomasello (2006) engaged 18-24 month children goal-oriented tasks and social games which required cooperation. In one of the social games, the experiment began with a demonstration where one participant sent a wooden block sliding down an inclined tube and the other participant caught the block in a tin cup that made a rattling sound. This can be considered more generally as a task in which one participant moves an object so that the second participant can then in turn manipulate the object. This represents a minimal case of a coordinated action sequence. After the demonstration, in Trials 1 and 2 the experimenter sent the block down one of the tubes three times, and then switched to the other, and the child was required to choose the same tube as the partner. In Trials 3 and 4 during the game, the experimenter interrupted the behavior for 15 seconds and then resumed.

Behaviorally, children successfully participated in the game in Trials 1 and 2. In the interruption Trials 3 and 4 they displayed two particularly interesting types of response that were (a) to attempt to perform the role of the experimenter themselves, and/or (b) to reengage the experimenter with a communicative act. This indicates that the children had a clear awareness both of their role and that of the adult in the shared coordinated activity. This research thus identifies a set of behavioral objectives for robot behavior in the perception and execution of cooperative intentional action. Such behavior could, however, be achieved in a number of possible architectures.

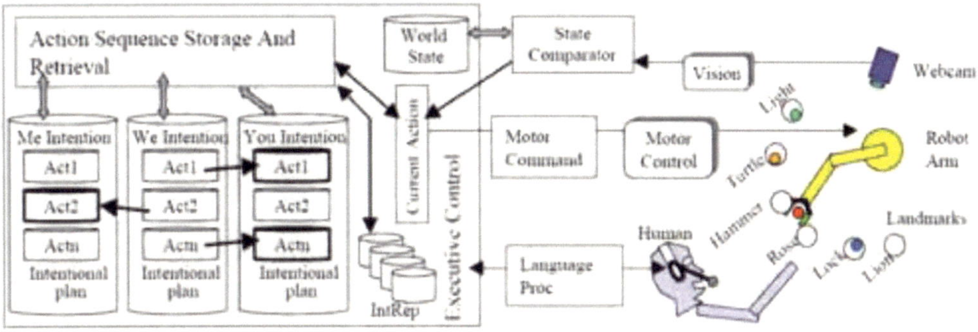

Fig. 6. Cooperation System. In a shared work-space, human and robot manipulate objects (green, yellow, read and blue circles corresponding to dog, horse, pig and duck), placing them next to the fixed landmarks (light, turtle, hammer, etc.). *Action*: Spoken commands interpreted as individual words or grammatical constructions, and the command and possible arguments are extracted using grammatical constructions in Language Proc. The resulting Action(Agent, Object, Recipient) representation is the Current Action. This is converted into robot command primitives (Motor Command) and joint angles (Motor Control) for the robot. *Perception*: Vision provides object location input, allowing action to be perceived as changes in World State (State Comparator). Resulting Current Action used for action description, imitation, and cooperative action sequences. *Imitation*: The user performed action is perceived and encoded in Current Action, which is then used to control the robot under the supervision of Executive Control. *Cooperative Games*. During observations, individual actions are perceived, and attributed to the agent or the other player (Me or You). The action sequence is stored in the We Intention structure, that can then be used to separately represent self vs. other actions.

7. Implementing Shared Planning

In a comment on Tomasello et al (2005) on understanding and sharing intention, Dominey (2005) analyses how a set of initial capabilities can be used to provide the basis for shared intentions. This includes capabilities to : 1. perceive the physical states of objects, 2. perceive (and perform) actions that change these states, 3. distinguish between self and other, 4. perceive emotional/evaluation responses in others, and 5. learn sequences of predicate-argument representations.

The goal is to demonstrate how these 5 properties can be implemented within the constraints of the neurophysiology data reviewed above in order to provide the basis for performing these cooperative tasks. In the current experiments the human and robot cooperate by moving physical objects to different positions in a shared work-space as illustrated in Figures 6 and 7. The 4 moveable objects are pieces of a wooden puzzle, representing a dog, a pig, a duck and a cow. These pieces can be moved by the robot and the user in the context of cooperative activity. Each has fixed to it a vertically protruding metal screw, which provides an easy grasping target both for the robot and for humans. In addition there are 6 images that are fixed to the table and serve as landmarks for placing the moveable objects, and correspond to a light, a turtle, a hammer, a rose, a lock and a lion, as partially illustrated in Figures 6 & 7. In the interactions, human and robot are required to place objects in zones next to the different landmarks, so that the robot can more easily determine where objects are, and where to grasp them. Figure 6 provides an overview of the architecture, and Figure 7, which corresponds to Experiment 6 provides an overview of how the system operates.

Representation: The structure of the internal representations is a central factor determining how the system will function, and how it will generalize to new conditions. Based on the neurophysiology reviewed above, we use a common representation of action for both perception and production. Actions are identified by the agent, the object, and the target location to move that object to. As illustrated in Figure 6, by taking the short loop from vision, via Current Action Representation, to Motor Command, the system is thus configured for a form of goal-centered action imitation. This will be expanded upon below.

A central feature of the system is the World Model that represents the physical state of the world, and can be accessed and updated by vision, motor control, and language, similar to the Grounded Situation Model of Mavridis & Roy (2006). The World Model encodes the physical locations of objects that is updated by vision and proprioception (i.e. robot action updates World Model with new object location). Changes in the World Model in terms of an object being moved allows the system to detect actions in terms these object movements. Actions are represented in terms of the agent, the object and the goal of the action, in the form MOVE(object, goal location, agent). These representations can be used for commanding action, for describing recognized action, and thus for action imitation and narration, as seen below. In order to allow for more elaborate cooperative activity, the system must be able to store and retrieve actions in a sequential structure.

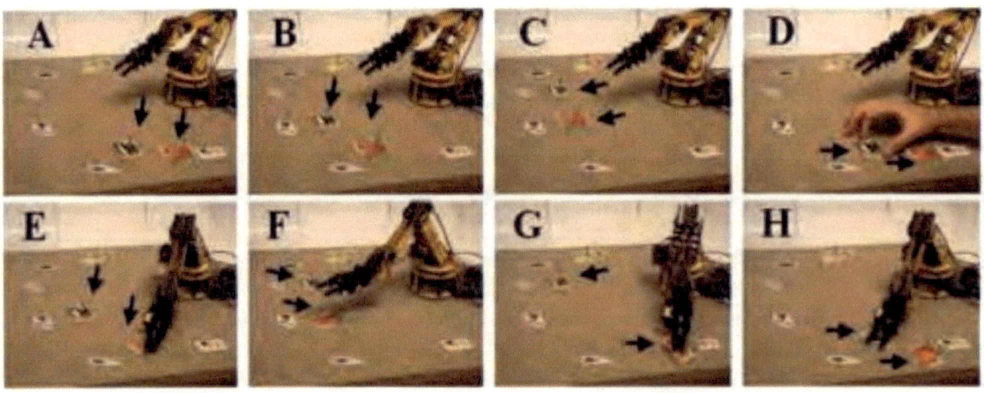

Fig. 7. Cooperative task of Exp 5-6. Robot arm, with 6 landmarks (Light, turtle, hammer, rose, lock and lion from top to bottom). Moveable objects include Dog and Horse. In A-D, human demonstrates a "horse chase the dog" game, and successively moves the Dog then Horse, indicating that in the game, the user then the robot are agents, respectively. After demonstration, human and robot "play the game". In each of E – F user moves Dog, and robot follows with Horse. In G robot moves horse, then in H robot detects that the user is having trouble and so "helps" the user with the final move of the dog. See Exp 5 & 6.

Visual perception: Visual perception is a challenging technical problem. To simplify, standard lighting conditions and a small set (n = 10) of visual object to recognize are employed (4 moveable objects and 6 location landmarks). A VGA webcam is positioned at 1.25 meters above the robot workspace. Vision processing is provided by the Spikenet Vision System (http://www.spikenet-technology.com/). Three recognition models for each object at different orientations (see Fig. 8.I) were built with an offline model builder. During real-time vision processing, the models are recognized, and their (x, y) location in camera coordinates are provided. Our vision post-processing eliminates spurious detections and returns the reliable (x, y) coordinates of each moveable object in a file. The nearest landmark is then calculated.

Motor Control & Visual-Motor Coordination: While visual-motor coordination is not the focus of the current work, it was necessary to provide some primitive functions to allow goal directed action. All of the robot actions, whether generated in a context of imitation, spoken command or cooperative interaction will be of the form *move(x to y)* where *x* is a member of a set of visually perceivable objects, and *y* is a member of the set of fixed locations on the work plan.

Robot motor control for transport and object manipulation with a two finger gripper is provided by the 6DOF Lynx6 arm (www.lynxmotion.com). The 6 motors of the arm are coordinated by a parallel controller connected to a PC computer that provides transmission of robot commands over the RS232 serial port.

Human users (and the robot) are constrained when they move an object, to place it in one of the zones designated next to each of the six landmarks (see Fig 3). This way, when the nearest landmark for an object has been determined, this is sufficient for the robot to grasp that object at the prespecified zone.

I.

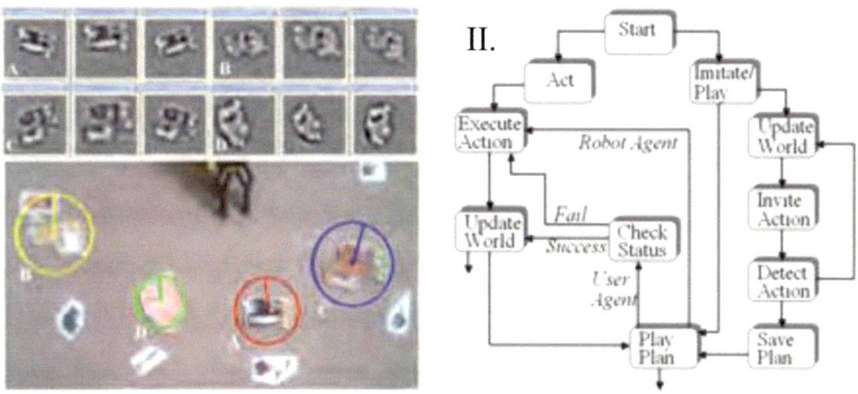

Figure 8. I. Vision processing. Above: A. – D. Three templates each for the Dog, Duck, Horse and Pig objects at three different orientations. Below, encompassing circles indicate template recognition for the four different objects near different fixed landmarks, as seen from the camera over the robot workspace. II. Dialog flow of Control

In a calibration phase, a target point is marked next to each of the 6 fixed landmark locations, such that they are all on an arc that is equidistant to the center of rotation of the robot arm base. For each, the rotation angle of Joint 0 (the rotating shoulder base) necessary to align the arm with that point is then determined, along with a common set of joint angles for Joints 1 – 5 that position the gripper to seize any of the objects. Angles for Joint 6 that controls the closing and opening of the gripper to grasp and release an object were then identified. Finally a neutral position to which the arm could be returned in between movements was defined. The system was thus equipped with a set of primitives that could be combined to position the robot at any of the 6 grasping locations, grasp the corresponding object, move to a new position, and place the object there.

Cooperation Control Architecture: The spoken language control architecture illustrated in Fig 8.II is implemented with the CSLU Rapid Application Development toolkit (http://cslu.cse.ogi.edu/toolkit/). This system provides a state-based dialog management system that allows interaction with the robot (via the serial port controller) and with the vision processing system (via file i/o). It also provides the spoken language interface that allows the user to determine what mode of operation he and the robot will work in, and to manage the interaction via spoken words and sentences.

Figure 8.II illustrates the flow of control of the interaction management. In the Start state the system first visually observes where all of the objects are currently located. From the start state, the system allows the user to specify if he wants to ask the robot to perform actions (Act), to imitate the user, or to play (Imitate/Play). In the Act state, the user can specify actions of the form "Put the dog next to the rose" and a grammatical construction template is used to extract the action that the robot then performs. In the Imitate state, the robot first verifies the current state (Update World) and then invites the user to demonstrate an action (Invite Action). The user shows the robot one action. The robot re-observes the world and detects the action based on changes detected (Detect Action). This action is then saved and transmitted (via Play the Plan with Robot as Agent) to execution (Execute action). A predicate(argument) representation of the form Move(object, landmark) is used both for action observation and execution.

Imitation is thus a minimal case of Playing in which the "game" is a single action executed by the robot. In the more general case, the user can demonstrate multiple successive actions, and indicate the agent (by saying "You/I do this") for each action. The resulting intentional plan specifies what is to be done by whom. When the user specifies that the plan is finished, the system moves to the Save Plan, and then to the Play Plan states. For each action, the system recalls whether it is to be executed by the robot or the user. Robot execution takes the standard Execute Action pathway. User execution performs a check (based on user response) concerning whether the action was correctly performed or not. If the user action is not performed, then the robot communicates with the user, and performs the action itself. Thus, "helping" was implemented by combining an evaluation of the user action, with the existing capability to perform a stored action representation.

8. Experimental Results Part 2

For each of the 6 following experiments, equivalent variants were repeated at least ten times to demonstrate the generalized capability and robustness of the system. In less than 5 percent of the trials, errors of two types were observed to occur. Speech errors resulted from a failure in the voice recognition, and were recovered from by the command validation check (Robot: "Did you say …?"). Visual image recognition errors occurred when the objects were rotated beyond 20° from their upright position. These errors were identified when the user detected that an object that should be seen was not reported as visible by the system, and were corrected by the user re-placing the object and asking the system to "look again". At the beginning of each trial the system first queries the vision system, and updates the

World Model with the position of all visible objects. It then informs the user of the locations of the different objects, for example "The dog is next to the lock, the horse is next to the lion." It then asks the user "Do you want me to act, imitate, play or look again?", and the user responds with one of the action-related options, or with "look again if the scene is not described correctly.

Validation of Sensorimotor Control: In this experiment, the user says that he wants the "Act" state (Fig 8.II), and then uses spoken commands such as "Put the horse next to the hammer". Recall that the horse is among the moveable objects, and hammer is among the fixed landmarks. The robot requests confirmation and then extracts the predicate-argument representation - *Move(X to Y)* - of the sentence based on grammatical construction templates. In the Execute Action state, the action *Move(X to Y)* is decomposed into two components of *Get(X)*, and *Place-At(Y)*. *Get(X)* queries the World Model in order to localize X with respect to the different landmarks, and then performs a grasp at the corresponding landmark target location. Likewise, *Place-At(Y)* simply performs a transport to target location Y and releases the object. Decomposing the *get* and *place* functions allows the composition of all possible combinations in the *Move(X to Y)* space. Ten trials were performed moving the four object to and from different landmark locations. Experiment 1 thus demonstrates (1) the ability to transform a spoken sentence into a Move(X to Y) command, (2) the ability to perform visual localization of the target object, and (3) the sensory-motor ability to grasp the object and put it at the specified location. In ten experimental runs, the system performed correctly.

Imitation: In this experiment the user chooses the "imitate" state. As stated above, imitation is centered on the achieved ends - in terms of observed changes in state - rather than the means towards these ends. Before the user performs the demonstration of the action to be imitated, the robot queries the vision system, and updates the World Model (Update World in Fig 8.II) and then invites the user to demonstrate an action. The robot pauses, and then again queries the vision system and continues to query until it detects a difference between the currently perceived world state and the previously stored World Model (in State Comparator of Fig 1, and Detect Action in Fig 8.II), corresponding to an object displacement. Extracting the identity of the displaced object, and its new location (with respect to the nearest landmark) allows the formation of an *Move(object, location)* action representation. Before imitating, the robot operates on this representation with a meaning-to-sentence construction in order to verify the action to the user, as in "Did you put the dog next to the rose?" It then asks the user to put things back as they were so that it can perform the imitation. At this point, the action is executed (Execute Action in Fig 8.II). In ten experimental runs the system performed correctly. This demonstrates (1) the ability of the system to detect the goals of user-generated actions based on visually perceived state changes, and (2) the utility of a common representation of action for perception, description and execution.

A Cooperative Game: The cooperative game is similar to imitation, except that there is a sequence of actions (rather than just one), and the actions can be effected by either the user or the robot in a cooperative manner. In this experiment, the user responds to the system request and enters the "play" state. In what corresponds to the demonstration in Warneken et al. (2006) the robot invites the user to start showing how the game works. The user then begins to perform a sequence of actions. For each action, the user specifies who does the action, i.e. either "you do this" or "I do this". The intentional plan is thus stored as a sequence of action-agent pairs, where each action is the movement of an object to a particular target location. In Fig 6, the resulting interleaved sequence is stored as the "We

intention", i.e. an action sequence in which there are different agents for different actions. When the user is finished he says "play the game". The robot then begins to execute the stored intentional plan. During the execution, the "We intention" is decomposed into the components for the robot (Me Intention) and the human (You intention).
In one run, during the demonstration, the user said "I do this" and moved the horse from the lock location to the rose location. He then said "you do this" and moved the horse back to the lock location. After each move, the robot asks "Another move, or shall we play the game?". When the user is finished demonstrating the game, he replies "Play the game." During the playing of this game, the robot announced "Now user puts the horse by the rose". The user then performed this movement. The robot then asked the user "Is it OK?" to which the user replied "Yes". The robot then announced "Now robot puts the horse by the lock" and performed the action. In two experimental runs of different demonstrations, and 5 runs each of the two demonstrated games, the system performed correctly. This demonstrates that the system can learn a simple intentional plan as a stored action sequence in which the human and the robot are agents in the respective actions.
Interrupting a Cooperative Game: In this experiment, everything proceeds as in the previous experiment, except that after one correct repetition of the game, in the next repetition, when the robot announced "Now user puts the horse by the rose" the user did nothing. The robot asked "Is it OK" and during a 15 second delay, the user replied "no". The robot then said "Let me help you" and executed the move of the horse to the rose. Play then continued for the remaining move of the robot. This illustrates how the robot's stored representation of the action that was to be performed by the user allowed the robot to "help" the user.
A More Complex Game: In order to more explicitly test the intentional sequencing capability of the system, this experiment replicates the Cooperative Game experiment but with a more complex task, illustrated in Figure 7. In this game (Table 5), the user starts by moving0 the dog, and after each move the robot "chases" the dog with the horse, till they both return to their starting places.

Action	User identifies agent	User Demonstrates Action	Ref in Figure 7
1.	I do this	Move dog from the lock to the rose	B
2.	You do this	Move the horse from the lion to the lock	B
3.	I do this	Move the dog from the rose to the hammer	C
4.	You do this	Move the horse from the lock to the rose	C
5.	You do this	Move the horse from the rose to the lion	D
6.	I do this	Move the dog from the hammer to the lock	D

Table 5. Cooperative "horse chase the dog" game specified by the user in terms of who does the action (indicated by saying) and what the action is (indicated by demonstration). Illustrated in Figure 7.

As in the simplified cooperative game, the successive actions are visually recognized and stored in the shared "We Intention" representation. Once the user says "Play the game", the final sequence is stored, and then during the execution, the shared sequence is decomposed into the robot and user components based on the agent associated with each action. When the user is the agent, the system invites the user to make the next move, and verifies (by asking) if the move was ok. When the system is the agent, the robot executes the movement. After each move the World Model is updated. As before two different complex games were learned, and each one "played" 5 times. This illustrates the learning by demonstration (Zollner et al. 2004) of a complex intentional plan in which the human and the robot are agents in a coordinated and cooperative activity.

Interrupting the Complex Game: As in Experiment 4, the objective was to verify that the robot would take over if the human had a problem. In the current experiment this capability is verified in a more complex setting. Thus, when the user is making the final movement of the dog back to the "lock" location, he fails to perform correctly, and indicates this to the robot. When the robot detects failure, it reengages the user with spoken language, and then offers to fill in for the user. This is illustrated in Figure 7H. This demonstrates the generalized ability to help that can occur whenever the robot detects the user is in trouble. These results were presented in Dominey (2007).

9. Discussion

This beginning of the 21st century marks a period where humanoid robot mechatronics and the study of human and artificial cognitive systems come in parallel to a level of maturity sufficient for significant progress to be made in making these robots more human-like in there interactions. In this context, two domains of interaction that humans exploit with great fidelity are spoken language, and the visual ability to observe and understand intentional action. A good deal of research effort has been dedicated to the specification and implementation of spoken language systems for human-robot interaction (Crangle & Suppes 1994, Lauria et al. 2002, Severinson-Eklund 2003, Kyriacou et al. 2005, Mavrides & Roy 2006). The research described in the current chapter extends these approaches with a Spoken Language Programming system that allows a more detailed specification of conditional execution, and by using language as a compliment to vision-based action perception as a mechanism for indicating how things are to be done, in the context of cooperative, turn-taking behavior.

The abilities to observe an action, determine its goal and attribute this to another agent are all clearly important aspects of the human ability to cooperate with others. Recent research in robot imitation (Oztop et al. 2006, Nehaniv & Dautenhahn 2007, Billard & Schaal 2006) and programming by demonstration (Zollner et al. 2004) begins to address these issues. Such research must directly address the question of how to determine what to imitate. Carpenter and Call (2007) The current research demonstrates how these capabilities can contribute to the "social" behavior of learning to play a cooperative game, playing the game, and helping another player who has gotten stuck in the game, as displayed in 18-24 month children (Werneken et al. 2006, Werneken & Tomasello 2006). While the primitive bases of such behavior is visible in chimps, its full expression is uniquely human. As such, it can be considered a crucial component of human-like behavior for robots (Carpenter & Call 2007).

The current research is part of an ongoing effort to understand aspects of human social cognition by bridging the gap between cognitive neuroscience, simulation and robotics (Dominey 2003, 2005, et al. 2004, 2006, 2007; Dominey & Boucher 2005), with a focus on the role of language. The experiments presented here indicate that functional requirements derived from human child behavior and neurophysiological constraints can be used to define a system that displays some interesting capabilities for cooperative behavior in the context of spoken language and imitation. Likewise, they indicate that evaluation of another's progress, combined with a representation of his/her failed goal provides the basis for the human characteristic of "helping." This may be of interest to developmental scientists, and the potential collaboration between these two fields of cognitive robotics and human cognitive development is promising. The developmental cognition literature lays out a virtual roadmap for robot cognitive development (Dominey 2005, Werneken et al. 2006). In this context, we are currently investigating the development of hierarchical means-end action sequences. At each step, the objective will be to identify the behavior characteristic and to implement it in the most economic manner in this continuously developing system for human-robot cooperation.

At least two natural extensions to the current system can be considered. The first involves the possibility for changes in perspective. In the experiments of Warneken et al. the child watched two adults perform a coordinated task (one adult launching the block down the tube, and the other catching the block). At 24 months, the child can thus observe the two roles being played out, and then step into either role. This indicates a "bird's eye view" representation of the cooperation, in which rather than assigning "me" and "other" agent roles from the outset, the child represents the two distinct agents A and B for each action in the cooperative sequence. Then, once the perspective shift is established (by the adult taking one of the roles, or letting the child choose one) the roles A and B are assigned to me and you (or vice versa) as appropriate.

This actually represents a minimal change to our current system. First, rather than assigning the "you" "me" roles in the We Intention at the outset, these should be assigned as A and B. Then, once the decision is made as to the mapping of A and B onto robot and user, these agent values will then be assigned accordingly. Second, rather than having the user tell the robot "you do this" and "I do this" the vision system can be modified to recognize different agents who can be identified by saying their name as they act, or via visually identified cues on their acting hands.

The second issue has to do with inferring intentions. The current research addresses one cooperative activity at a time, but nothing prevents the system from storing multiple such intentional plans in a repertory (IntRep in Fig 6). In this case, as the user begins to perform a sequence of actions involving himself and the robot, the robot can compare this ongoing sequence to the initial subsequences of all stored sequences in the IntRep. In case of a match, the robot can retrieve the matching sequence, and infer that it is this that the user wants to perform. This can be confirmed with the user and thus provides the basis for a potentially useful form of learning for cooperative activity.

In conclusion, the current research has attempted to build and test a robotic system for interaction with humans, based on behavioral and neurophysiological requirements derived from the respective literatures. The interaction involves spoken language and the performance and observation of actions in the context of cooperative action. The experimental results demonstrate a rich set of capabilities for robot perception and subsequent use of cooperative action plans in the context of human-robot cooperation. This work thus extends the imitation paradigm into that of sequential behavior, in which the learned intentional action sequences are made up of interlaced action sequences performed in cooperative alternation by the human and robot. While many technical aspects of robotics (including visuomotor coordination and vision) have been simplified, it is hoped that the contribution to the study of imitation and cooperative activity is of some value.

Acknowledgements: I thank Jean-Paul Laumond, Eiichi Yoshida and Anthony Mallet from the LAAS Toulouse for cooperation with the HRP-2 as part of the French-Japanese Joint

Robotics Laboratory (AIST-Japan, CNRS-France). I thank Mike Tomasello, Felix Warneken, Malinda Carpenter and Elena Lieven for useful discussions during a visit to the MPI EVA in Leipzig concerning shared intentions; and Giacomo Rizzolatti for insightful discussion concerning the neurophysiology of sequence imitation at the IEEE Humanoids meeting in Genoa 2006. This research is supported in part by the French Minister of Research under grant ACI-TTT, and by the LAFMI.

10. References

Bekkering H, WohlschlagerA , Gattis M (2000) Imitation of Gestures in Children is Goal-directed, The Quarterly Journal of Experimental Psychology: Section A, 53, 153-164

Billard A, Schaal (2006) Special Issue: The Brain Mechanisms of Imitation Learning, Neural Networks, 19(1) 251-338

Boucher J-D, Dominey PF (2006) Programming by Cooperation: Perceptual-Motor Sequence Learning via Human-Robot Interaction, Proc. Simulation of Adaptive Behavior, Rome 2006.

Calinon S, Guenter F, Billard A (2006) On Learning the Statistical Representation of a Task and Generalizing it to Various Contexts. Proc IEEE/ICRA 2006.

Carpenter M, Call Josep (2007) The question of 'what to imitate': inferring goals and intentions from demonstrations, in Chrystopher L. Nehaniv and Kerstin Dautenhahn Eds, *Imitation and Social Learning in Robots, Human sand Animals,* Cambridge Univerity Press, Cambridge.

Crangle C. & Suppes P. (1994) Language and Learning for Robots, CSLI lecture notes: no. 41, Stanford.

Cuijpers RH, van Schie HT, Koppen M, Erlhagen W, Bekkering H (2006) Goals and means in action observation: A computational approach, *Neural Networks* 19, 311-322,

di Pellegrino G, Fadiga L, Fogassi L, Gallese V, Rizzolatti G (1992) Understanding motor events: a neurophysiological study. *Exp Brain Res.*;91(1):176-80.

Dominey PF (2005) Toward a construction-based account of shared intentions in social cognition. Comment on Tomasello et al. 2005, *Beh Brain Sci.* 28:5, p. 696.

Dominey PF, (2003) Learning grammatical constructions from narrated video events for human–robot interaction. *Proceedings IEEE Humanoid Robotics Conference,* Karlsruhe, Germany

Dominey PF, Alvarez M, Gao B, Jeambrun M, Weitzenfeld A, Medrano A (2005) Robot Command, Interrogation and Teaching via Social Interaction, Proc. IEEE Conf. On Humanoid Robotics 2005.

Dominey PF, Boucher (2005) Learning To Talk About Events From Narrated Video in the Construction Grammar Framework, *Artificial Intelligence,* 167 (2005) 31–61

Dominey PF, Boucher, J. D., & Inui, T. (2004). Building an adaptive spoken language interface for perceptually grounded human–robot interaction. In *Proceedings of the IEEE-RAS/RSJ international conference on humanoid robots.*

Dominey PF, Hoen M, Inui T. (2006) A neurolinguistic model of grammatical construction processing. *Journal of Cognitive Neuroscience.*18(12):2088-107.

Dominey PF, Mallet A, Yoshida E (2007) Progress in Spoken Language Programming of the HRP-2 Humanoid, *Proc. ICRA 2007, Rome*

Dominey PF (2007) Sharing Intentional Plans for Imitation and Cooperation: Integrating Clues from Child Developments and Neurophysiology into Robotics, Proceedings of the AISB 2007 Workshop on Imitation.

Fong T, Nourbakhsh I, Dautenhaln K (2003) A survey of socially interactive robots. *Robotics and Autonomous Systems*, 42 3-4, 143-166.

Goga, I., Billard, A. (2005), Development of goal-directed imitation, object manipulation and language in humans and robots. In M. A. Arbib (ed.), *Action to Language via the Mirror Neuron System*, Cambridge University Press (in press).

Goldberg A. Constructions: A new theoretical approach to language. Trends in Cognitive Sciences 2003; 7: 219–24.

Kozima H., Yano H. (2001) A robot that learns to communicate with human caregivers, in: *Proceedings of the International Workshop on Epigenetic Robotics*,.

Kyriacou T, Bugmann G, Lauria S (2005) Vision-based urban navigation procedures for verbally instructed robots. Robotics and Autonomous Systems, (51) 69-80

Lauria S, Buggmann G, Kyriacou T, Klein E (2002) Mobile robot programming using natural language. Robotics and Autonomous Systems 38(3-4) 171-181

Mavridis N, Roy D (2006). Grounded Situation Models for Robots: Where Words and Percepts Meet. Proceedings of the IEEE/RSJ International Conference on Intelligent Robots and Systems (IROS)

Nehaniv CL, Dautenhahn K eds. (2007) *Imitation and Social Learing in Robots, Humans and Animals*, Cambridge University Press, Cambridge.

Nicolescu M.N., Mataric M.J. : Learning and Interacting in Human-Robot Domains, IEEE Trans. Sys. Man Cybernetics B, 31(5) 419-430.

Oztop E, Kawato M, Arbib M (2006) Mirror neurons and imitation: A computationally guided review. *Neural Networks*, (19) 254-271

Pickering MJ, Garrod S. (2004) Toward a mechanistic psychology of dialogue.Behav Brain Sci. Apr;27(2):169-90

Rizzolatti G, Craighero L (2004) The Mirror-Neuron system, *Annu. Rev. Neuroscience* (27) 169-192

Severinson-Eklund K., Green A., Hüttenrauch H., Social and collaborative aspects of interaction with a service robot, Robotics and Autonomous Systems 42 (2003) 223–234

Sommerville A, Woodward AL (2005) Pulling out the intentional structure of action: the relation between action processing and action production in infancy. *Cognition*, 95, 1-30.

Tomasello M, Carpenter M, Cal J, Behne T, Moll HY (2005) Understanding and sharing intentions: The origins of cultural cognition, *Beh. Brain Sc;*. 28; 675-735.

Torrey C, Powers A, Marge M, Fussel SR, Kiesker S (2005) Effects of Adaptive Robot Dialogue on Information Exchange and Social Relations, Proceedings HRI 2005.

Warneken F, Chen F, Tomasello M (2006) Cooperative Activities in Young Children and Chimpanzees, *Child Development*, 77(3) 640-663.

Warneken F, Tomasello M (2006) Altruistic helping in human infants and young chimpanzees, *Science*, 311, 1301-1303

Zöllner R., Asfour T., Dillman R.: Programming by Demonstration: Dual-Arm Manipulation Tasks for Humanoid Robots. Proc IEEE/RSJ Intern. Conf on Intelligent Robots and systems (IROS 2004).

12

Collision-Free Humanoid Reaching: Past, Present, and Future

Evan Dmmwright and Maja Mataric
University of Southern California
United States

1. Abstract

Most recent humanoid research has focused on balance and locomotion. This concentration is certainly important, but one of the great promises of humanoid robots is their potential for effective interaction with human environments through manipulation. Such interaction has received comparatively little attention, in part because of the difficulty of this task. One of the greatest obstacles to autonomous manipulation by humanoids is the lack of efficient collision-free methods for reaching. Though the problem of reaching and its relative, pick-and-place, have been discussed frequently in the manipulator robotics literature- e.g., (Lozano-Pérez et al., 1989); (Alami et al., 1989); (Burridge et al., 1995)- researchers in humanoid robotics have made few forays into these domains. Numerous subproblems must be successfully addressed to yield significant progress in humanoid reaching. In particular, there exist several open problems in the areas of algorithms, perception for modeling, and control and execution. This chapter discusses these problems, presents recent progress, and examines future prospects.

2. Introduction

Reaching is the one of the most important tasks for humanoid robots, endowing them with the ability to manipulate objects in their environment. Unfortunately, getting humanoids to reach efficiently and safely, without collision, is a complex problem that requires solving open subproblems in the areas of algorithms, perception for modeling, and control and execution. The algorithmic problem requires the synthesis of collision-free joint-space trajectories in the presence of moving obstacles. The perceptual problem, with respect to modeling, is comprised of acquiring sufficiently accurate information for constructing a geometric model of the environment. Problems of control and execution are concerned with correcting deviation from reference trajectories and dynamically modifying these trajectories during execution to avoid unexpected obstacles. This chapter delves into the relevant subproblems above in detail, describes the progress that has been made in solving them, and outlines the work remaining to be done in order to enable humanoids to perform safe reaching in dynamic environments.

3. Problem statement

The problem of reaching is formally cast as follows. Given:

1. a world $= \mathbb{R}^3$
2. the current time t_o; T is then defined as the interval $[t_o, \infty]$
3. a robot
4. a smooth manifold $\subset \mathbb{R}^n$ called the *state space* of; let $\kappa :\ \longrightarrow$ be a function that maps state-space to the robot's configuration space
5. the state transition equation is $\dot{x} = f(x, u, t)$, where $x \in$ and $u : T \rightarrow$ generates a vector of control inputs ($\subset \mathbb{R}^m$) as a function of time
6. a nonstationary obstacle region $(t) \subset\ , \forall t \in T;\ _{obs}(t), \forall t \in T$ is then the projection of obstacles in the robot's configuration space into state-space (i.e., $\kappa^{-1}(\ _{obs}(t)) =\ _{obs}(t)$ and $_{free}(t) = \ -\ _{obs}(t)$).
7. $\subset \mathbb{R}^6$ is the reachable workspace[1] of
8. a direct kinematics function, $F : \rightarrow$ that transforms robot states to operational-space configurations of one of the robot's end effectors
9. a set of feasible operational-space goal functions of time, G such that $\forall g \in G, g : T \rightarrow$
10. a feasible state-space Boolean function $\delta : T \times g \times \ \rightarrow \ 0,1$ where $g \in G$
11. $x_0 \in\ _{free}$, the state of the robot at t_0

generate the control vector function $u(.)$ from time $t > t_0$ such that $x_t \in\ _{free}(t)$ where $x_t = x_0 + \int_{t_0}^{t} f(x, u(t), t)\, dt$, for $t > t_0$ and there exists a time t_j for which $\min_{g \in G} F(x(t_i)) - g(t_i) < \epsilon$ and $\delta(t_i, g(t_i), x(t_i)) = 1$ for all $t_i > t_j$, or correctly report that such a function $u(.)$ does not exist.

Informally, the above states that to solve the reaching problem, the commands sent to the robot must cause it to remain collision-free and, at some point in the future, cause both the operational space distance from the end-effector to one of the goals to remain below a given threshold ε and the state-space of the robot to remain in an admissable region.

The implications of the above formal definition are:

- The state transition function $f(.)$ should accurately reflect the dynamics of the robot. Unfortunately, due to limitations in mechanical modeling and the inherent uncertainty of how the environment might affect the robot, $f(.)$ will only approximate the true dynamics. Section 4.3 discusses the ramifications of this approximation.
- The robot must have an accurate model of its environment. This assumption will only be true if the environment is instrumented or stationary. The environments in which humanoids are expected to operate are dynamic (see #6 above), and this chapter will assume that the environment is not instrumented. Constructing an accurate model of the environment will be discussed in Section 4.2.
- The goals toward which the robot is reaching may change over time; for example, the robot may refine its target as the robot moves nearer to it. Thus, even if the target itself is stationary, the goals may change given additional information. It is also possible that the target is moving (e.g., a part moving on an assembly line). The issue of changing targets will be addressed in Section 4.1.
- Manipulation is not explicitly considered. It is assumed that a separate process can grasp or release an object, given the operational-space target for the hand and the desired configuration for the fingers (the Boolean function $\delta(.)$ is used to ensure

---

[1] The *reachable workspace* is defined by Sciavicco & Siciliano (2000) to be the region of operational-space that the robot can reach with at least one orientation.

that this latter condition is satisfied). This assumption is discussed further in the next section.

3 Related work

A considerable body of work relates to the problem defined in the previous section yet does not solve this problem. In some cases, researchers have investigated similar problems, such as developing models of human reaching. In other cases, researchers have attempted to address both reaching and manipulation. This section provides an overview of these alternate lines of research, though exhaustive surveys of these areas are outside of the scope of this chapter. Humanoids have yet to autonomously reach via locomotion to arbitrary objects in known, static environments, much less reach to objects without collision in dynamic environments. However, significant progress has been made toward solving this problems recently. This section concludes with a brief survey of methods that are directly applicable toward solving the reaching problem.

3.1 Models of reaching in neuroscience

A line of research in neuroscience has been devoted to developing models of human reaching; efficient, human-like reaching for humanoids has been one of the motivations for this research. Flash & Hogan (1985), Bullock et al. (1993), Flanagan et al. (1993), Crowe et al. (1998) and Thor-oughman & Shadmehr (2000) represent a small sample of work in this domain. The majority of neuroscience research into reaching has ignored obstacle avoidance, so the applicability of this work toward safe humanoid reaching has not been established. Additionally, neuroscience often considers the problem of *pregrasping,* defined by Arbib et al. (1985) as a configuration of the fingers of a hand before grasping such that the position and orientation of the fingers with respect to the palm's coordinate system satisfies *a priori* knowledge of the object and task requirements. In contrast to the neuroscience approach, this chapter attempts to analyze the problem of humanoid reaching from existing subfields in robotics and computer science. Recent results in the domains of motion planning, robot mapping, and robot control architectures are used to identify remaining work in getting humanoids to reach safely and efficiently. This chapter is unconcerned with generating motion that is natural in appearance by using pregrasping and human models of reaching, for example.

3.2 Manipulation planning

Alami et al. (1997), Gupta et al. (1998), Mason (2001), and Okada et al. (2004) have considered the problem of *manipulation planning,* which entails planning the movement of a workpiece to a specified location in the world without stipulating how the manipulator is to accomplish the task. Manipulation planning requires reaching to be solved as a subproblem, even if the dependence is not explicitly stated. As noted in LaValle (2006), existing research in manipulation planning has focused on the geometric aspects of the task while greatly simplifying the issues of grasping, stability, friction, mechanics, and uncertainty. The reaching problem is unconcerned with grasping (and thereby friction) by presuming that reaching and grasping can be performed independently. The definition provided in Section 2 allows for treatment of mechanics (via $f(.)$, the state transition function) and stability and uncertainty (by stating the solution to the problem in terms of the observed effects rather than the desired commands). Additionally, the problem of reaching encompasses more

tasks than those used in pick-and-place; for example, both pointing and touching can be considered as instances of reaching.
A late development by Stilman & Kuffner, Jr. (2007) addresses manipulation planning amongst movable obstacles. The reaching problem as defined in Section 2 permits the obstacles to be movable by the humanoid. Many of the issues described in this chapter (e.g., constructing a model of the environment, monitoring execution, etc.) need to be resolved to fully explore this avenue, but the ability to move obstacles as necessary will certainly admit new solutions to some instances of the reaching problem.

3.3 Recent work directly applicable to humanoid reaching

Recent work in humanoid robotics and virtual humans is directly applicable toward efficient, safe humanoid reaching. For example, work by Brock (2000) permits reaching in dynamic environments by combining a planned path with an obstacle avoidance behavior. More recent work by Kagami et al. (2003) uses stereo vision to construct a geometric model of a static environment and motion planning and inverse kinematics to reach to and grasp a bottle with a stationary humanoid robot. Liu & Badler (2003); Kallmann et al. (2003); Bertram et al. (2006) and Drumwright & Ng-Thow-Hing (2006) focused on developing algorithms for humanoid reaching; the algorithms in the latter two works are *probabilistically complete,* while the algorithms in (Liu & Badler, 2003) and (Kallmann et al., 2003) are not complete in any sense. All four works assumed static environments, perfect control and holonomic constraints.

4. Outstanding issues

This section discusses the issues that remain to solve the reaching problem, as follows:

1. Constructing an accurate model of the environment
2. Planning collision-free motions in dynamic environments
3. Correcting deviation from the desired trajectory due to imperfect control
4. Avoiding both fixed and moving obstacles during trajectory execution

The first item has received the least research attention to date and therefore includes the majority of open problems in collision-free humanoid reaching. Section 4.2 discusses progress and prospects in this area. Planning collision-free motions, at least in static environments, has received considerable attention; Section 4.1 discusses why this problem is challenging from an algorithmic standpoint and addresses extensions to dynamic environments. Finally, correcting deviation from the planned trajectory and avoiding obstacles during trajectory execution are key to reach the target in a safe manner. Section 4.3 discusses these two issues.

4.1 Algorithmic issues

The best studied aspect of the reaching problem is the algorithmic component, which is an extension to the general motion planning introduced below. Section 4.1.1 formally relates the problem of reaching to the general motion planning problem, and analyzes the complexity of the latter. Section 4.1.2 introduces sample-based motion planning, a paradigm for circumventing the intractability of motion planning; the following section discusses the extension of a popular sample-based motion planner to respect differential constraints. Finally, Sections 4.1.4–4.1.8 discuss motion planning issues highly relevant to humanoid reaching, namely planning under uncertainty, potential incompleteness resulting from multiple inverse kinematics solutions, planning to nonstationary goals, and planning in dynamic environments.

Researchers have investigated ways for planning collision-free motions from one configuration to another since the introduction of the *Piano Mover's Problem,* also known as the *Static Mover's Problem* Reif (1979). The Piano Mover's problem can be stated as follows (excerpted from LaValle (2006)).

Given:

1. a world , where $= \mathbb{R}^2$ or $= \mathbb{R}^3$
2. a semi-algebraic obstacle region $\subset$
3. a collection of m semi-algebraic links of a robot, 1, 2, … m
4. the configuration space of the robot; $_{\text{free}}$ can then be defined as the subset of configuration space which does not cause the robot's geometry to intersect with any obstacles
5. the initial configuration of the robot, $\mathbf{q}_I$
6. the goal configuration of the robot, $\mathbf{q}_G$

generate a continuous path $\tau : [0,1] \rightarrow _{free}$ such that $\tau(0) = \mathbf{q}_I$ and $\tau(1) = \mathbf{q}_G$ or correctly report that such a path does not exist. The term *semi-algebraic* refers to a geometric representation that is composed by Boolean operations on implicit functions.

4.1.1 Complexity of motion planning

The definition of the Piano Mover's Problem is quite similar to the problem formulation for reaching at the beginning of this chapter. Indeed, an instance of the reaching problem can be transformed into an instance Piano Mover's Problem given the following constraints:

- the obstacle region, is stationary (i.e., $(t_i) = (t_j), \forall t_i, t_j \in T$)
- $=$ from which follows $\kappa(x) \triangleq x, \forall x \in$ and $\mathbf{x}_0 = \mathbf{q}_I$
- $=$
- G consists of a single element, g, which is nonstationary, and there exists only one robot configuration $\mathbf{q}_G$ that results in g (i.e., $F^{-1}(g) = \mathbf{q}_G$)
- $\int_{t_a}^{t_b} f(x, u(t), t)\, dt = u(t_b) - x(t_a)$ (implies that the command is the new robot state)

Reif (1979) showed that the Piano Mover's Problem is PSPACE-complete, implying NP-hardness. Additionally, the best known algorithm for solving the Piano Mover's problem (complexity-wise) is Canny's Roadmap Algorithm (Canny, 1993), which exhibits running-time exponential in the configuration space; aside from being intractable, the algorithm is reportedly quite difficult to implement (LaValle, 2006). Later work by Reif & Sharir (1994) proved that planning motions for a robot with fixed degrees-of-freedom and velocity constraints in the presence of moving obstacles with known trajectories is PSPACE-hard; thus, the constraints that were imposed transforming the reaching problem into the Piano Mover's Problem are unlikely to make the former problem easier.

4.1.2 Sample-based motion planning

The paradigm of *sample-based motion planning* was introduced with the advent of the *randomized potential field* (Barraquand & Latombe, 1991). Sample-based algorithms trade algorithmic completeness for excellent average-case performance and ease of implementation. In fact, completeness was not cast aside; rather, it was relaxed to lesser constraints, *probabilistic completeness* and *resolution completeness.* It is said that an algorithm is

probabilistically complete if the probability of finding a solution, if it exists, tends to unity as the number of samples increase. Similarly, a motion planning algorithm is resolution complete if a solution, if it exists, will be found in finite time given sufficiently dense sampling resolution over the domain of configuration space. Note that neither weaker form of completeness permits this class of algorithm to definitively state that a path does not exist. Finally, the underlying approach of sample-based planning is quite different from adapting classical search algorithms (i.e., A*, best-first, etc.) to a discretized grid; such an approach is generally intractable due to the combinatorial explosion of configuration space.
The most popular sample-based algorithm is currently the rapidly-exploring random tree (RRT) LaValle (1998). RRTs are popular due to their inherent bias toward the largest Voronoi region of configuration space (i.e., the largest unexplored region) during exploration. Efficient exploration is critical for sample-based algorithms because their running times generally increase linearly with the number of samples.

4.1.3 Motion planning under differential constraints

In addition to the advantage of efficient exploration, RRTs allow for planning under differential (e.g., nonholonomic) constraints, through *kinodynamic planning*. Kinodynamic planning plans in the control space, rather than in configuration space, and is therefore able to respect dynamic constraints. Kinodynamic planning theoretically permits motion to be planned for hu-manoids, which generally use bipedal locomotion and are nonholonomically constrained. As might be expected, kinodynamic planning is harder computationally than in the unconstrained case; additionally, kinodynamic planning requires a model of the system's dynamics and a control method for solving a *two-point boundary value problem*[2].
Planning directly in the state-actuator space of the humanoid is infeasible: the motion planner would not only have to avoid obstacles but also balance the humanoid and provide locomotion. An accurate model of the robot's dynamics would be required as well. Alternatively, planning could occur over the robot's configuration space augmented with planar position and orientation of the base. Constraints would be enforced kinematically rather than dynamically, and a trajectory rescaling mechanism could be used to enforce the dynamic constraints after planning. For example, kinematic constraints could be used to allow the base to translate or rotate, but not translate and rotate simultaneously. Planning in this augmented configuration space would require a locomotion controller that translates desired differential position and orientation of the base into motor commands. Once a plan were constructed in the augmented configuration space, the locomotion controller and joint-space controllers would generate the proper motor commands for a given configuration space trajectory.
Finally, it should be noted that if humanoids moved on holonomic bases, simpler methods could potentially be employed for humanoid reaching. For example, Maciejewski & Klein (1985) combines inverse kinematics with obstacle avoidance for redundant manipulators under holonomic constraints; some mobile manipulators fit into this category. Though the method of Maciejewski and Klein can cause the robot to become trapped in local minima, it presents minimal computational requirements.

---

[2] In the context of this work, the two-point boundary problem can be considered to be the problem of getting from a given state to a desired state under nonholonomic constraints. For example, the problem of parallel parking a car can be considered a two-point boundary problem: a series of actions is required to move the car into a parking space, even if only a simple translation is required (e.g., the car initially is aligned laterally with the parking space).

4.1.4 Motion planning for humanoid reaching

The RRT has been applied successfully to reaching for humanoids in virtual environments in (Kuffner, Jr., 1998); (Liu & Badler, 2003); (Kallmann et al, 2003); (Bertram et al, 2006) and (Drumwright & Ng-Thow-Hing, 2006), among others. Additionally, the RRT has been applied to reaching for an embodied humanoid by Kagami et al. (2003), although, as stated in Section 2, the environment was considered to be stationary and locomotion was not utilized.

Unfortunately, several assumptions of current, effective motion planning algorithms are unsuitable for humanoid reaching, as follows:

1. An accurate model of the environment and a precise mechanism for control are required.
2. A single goal configuration is assumed.
3. Targets for reaching do not change over time.
4. The environment is static.

These issues are explored further in the remainder of this section.

4.1.5 Planning under uncertainty

This chapter generally assumes that a sufficiently accurate model of the environment can be constructed and that the robot can be controlled with some degree of accuracy. However, these assumptions are often unrealistic in real-world robotics. It would be advantageous to construct plans that would maximize the distance from the robot to obstacles, for example, to minimize deleterious effects of uncertainty. LaValle & Hutchinson (1998) have explored the issue of planning under sensing and control uncertainty; however, their use of dynamic programming (Bellman, 1957) has restricted applications to planar robots. As an alternative, Lazanas & Latombe (1995) proposed an approach based on *landmarks,* regions of the state space where sensing and control are accurate. The assumption that such regions exist is significant. The limited application of these two methods illustrates the difficulty of planning while maximizing objectives, which is known as *optimal planning* (LaValle, 2006). Finally, Baumann (2001) proposes a planning method that iteratively modifies a trajectory; a fitness function judges the quality of the modified trajectory versus the original, in part based on distance to obstacles. However, this work has yet to be subjected to peer-review.

4.1.6 Incompleteness resulting from multiple IK solutions

A single operational-space goal generally corresponds to an infinite number of robot configurations, given the hyper-redundant degrees-of-freedom of most humanoids[3]. It is possible that collision-free paths exist only for a subset of the space of collision-free inverse kinematics solutions. Drumwright & Ng-Thow-Hing (2006) addressed that problem by maintaining a list of goal configurations that is continually grown using inverse kinematics; the motion planner frequently attempts to reach arbitrary goals in the list. Thus, the motion planner can avoid becoming stuck in planning to unreachable goals by not committing to any particular goal.

4.1.7 Motion planning to nonstationary goals

Planning trajectories to nonstationary goals has received little attention in the motion planning community; however, two concerns are evident. The goals may change with

---

[3] We assume that all available degrees-of-freedom are used for reaching, rather than for achieving secondary tasks like singularity avoidance, e.g., (Tanner & Kyriakopoulos, 2000).

increasing sensory data, leading to new estimates of the target's position and orientation. Additionally, the target itself may be moving, perhaps with a predictable trajectory.
The first issue is readily solvable using existing methods. If already in the process of planning, the goal can be replaced without ill effects: the results from sample-based planning methods will not be invalidated. If a plan has been determined, replanning can be performed using the results already determined (i.e., the roadmap, tree, etc.) to speed computation. Alternatively, inverse kinematics can be utilized to modify the generated plan slightly, somewhat in the manner used for motion retargeting by Choi & Ko (2000).
Moving targets are far more difficult to manage. Current sample-based planning methods have not focused on this problem, and the complexity of adapting existing methods to this purpose is unknown. Again, it is imaginable that existing plans could be modified using inverse kinematics, though replanning may be required if the target is moving quickly. Alternatively, Brock's method (discussed in Section 4.3) could potentially be utilized with some modifications toward this purpose.

4.1.8 Motion planning in dynamic environments

A simple means to address motion planning in dynamic environments adds time as an extra dimension of configuration-space. As LaValle (2006) notes, the planning process must constrain this extra dimension to move forward only. This approach of augmenting the configuration space with time can fail because the dynamics of the robot are not considered: it may not be possible or advisable for a robot to generate sufficient forces to follow the determined trajectory. An alternative to this approach is to use kinodynamic planning, as described in Section 4.1.3 to plan in the control space of the robot. However, kinodynamic planning does not appear to be applicable to configuration spaces above twelve dimensions, in addition to the difficulties with this approach described in Section 4.1.3.
Dynamic replanning, which refers to fast replanning as needed, is an alternative to methods which plan around dynamic obstacles. Dynamic replanning does not require the trajectories of dynamic obstacles to be known (dynamic obstacles are treated as stationary), and thus avoids the additional complexity of planning around these obstacles. Dynamic replanning may be the best option for the high-dimensional configuration spaces of humanoids. Kallmann & Mataric (2004) has explored online modification of existing sample-based roadmaps for virtual humanoids; unfortunately, that work indicated that the modification is likely no faster than generating a new plan. However, newer algorithms by Ferguson & Stentz (2006) and Zucker et al. (2007), also based on RRTs, have proven adept at replanning in nonstationary environments with large configuration spaces by modifying existing trees dynamically. Additionally, Drumwright & Ng-Thow-Hing (2006) have indicated that even planning anew in real-time for humanoids using a slightly modified RRT algorithm is nearly performable using current computers.
Recent work by Jaillet & Siméon (2004) and van den Berg & Overmars (2005) has taken a two-phase approach to motion planning in dynamic environments. The first phase entails constructing a probabilistic roadmap (Kavraki et al., 1996) over the persistent parts of the environment offline. In the online phase, a graph search algorithm finds a feasible path around dynamic obstacles. These methods are compelling because the bulk of computation, constructing the roadmap, is performed offline. Nevertheless, further research is required to determine efficient ways to update the roadmap as the environment changes (e.g., doors are opened, furniture is moved, etc.) before these methods can be used for humanoid reaching,

Regardless of the method employed, a significant concern is that the process of planning can itself cause possible solutions to disappear, because planning occurs in real-time. An additional significant concern is the requirement of many methods that the trajectories of the obstacles to be known *a priori;* filtering techniques– e.g., Kalman filters Kalman & Bucy (1961)- might permit short-term predictions, but long-term predictions will be problematic unless the obstacles follow balistic trajectories (which will prove difficult for the robot to avoid anyhow).

4.1.9 Summary

Substantial progress has been made in motion planning in the past decade, leading to tractable solutions of high-dimensional planning problems. Indeed, in the areas of planning to nonstationary goals and planning in dynamic environments, researchers are on the cusp of solutions that are viable for humanoid reaching. However, considerable work still remains in the area of planning under uncertainty.

4.2 Perception for modeling issues

We ignore the problems of object recognition and object pose determination, which are beyond the scope of this paperm, and focus on the perceptual issues related to modeling the environment for purposes of collision detection[4], assuming that the humanoid is equipped with a directional range finder. The problem of simultaneous localization and mapping (SLAM) is well studied in the mobile robotics community, where significant success has been achieved at developing methods to construct maps of $2\frac{1}{2}$D environments; some success has been achieved building three-dimensional maps as well. The natural inclination is to see whether SLAM techniques for mobile robots can be adapted to the problem of environment modeling toward humanoid reaching. Human environments are dynamic, humanoids manipulate objects (thus changing the world), and environment modeling is conducted with respect to planning; these challenges make current SLAM methods for mobile robotics difficult to utilize for humanoid reaching. Indeed, significant obstacles remain before humanoids can construct environment models suitable for motion planning. The remainder of this section discusses the relevant issues toward this purpose, namely representation, localization, exploration, and nonstationary environments.

4.2.1 Representation of environment model

There exist several possible representations for modeling the environment, including volumetric point sets (Thrun et al$_v$ 2003); occupancy grids (Moravec & Elfes, 1985), (Elfes, 1989); 3D models (Teller, 1998), (Kagami et al., 2003) and feature-based maps (Kuipers et al, 1993). However, the application of reaching presents several requirements. First, the representation must allow for fast intersection testing. Second, the representation must be able to efficiently manage the prodigious amounts of data that 3D range scans generate. These first two requirements exclude volumetric point sets. Fast updates of the representation from sensory data are also required. This stipulation excludes the use of 3D models, which require considerable post-processing including iso-surface extraction via the

[4] Alfhough the goal functions for the reaching problem are given in the global coordinate frame, it is quite natural to use ego-centric frames instead. As a result, localization is likely not required to perform reaching, except for effectively constructing the environment model.

Marching Cubes algorithm (Lorensen & Cline, 1987) or stitching (Turk & Levoy, 1994) and aligning and registering (Mayer & Bajcsy 1993); (Pito, 1996). In addition to the requirements listed above, the ability to readily extract high-level features from the representation for use with localization, described in Section 4.2.2, would be advantageous.

Drumwright et al. (2006) used an octree, which may be considered as an extension of occupancy grids, for modeling environments for reaching. The octree representation results in efficient storage, permits fast updates to the representation from range data, and is capable of performing fast intersection testing with the robot model. Drumwright et al. (2006) assumed perfect localization, and we are unaware of work that extracts features from octrees (or their two-dimensional equivalent, quadtrees) for the purpose of localization. However, there is precedent for feature extraction from octrees, e.g., (Sung et al., 2002).

An alternative to the octree representation would use high-level features (e.g., landmarks, objects, or shapes) as the base representation. Such features would serve well for input to one of the popular methods for simultaneous localization and mapping (SLAM) such as (Smith & Cheeseman, 1985); (Smith et al., 1990) or (Fox et al., 1999). Recognition of objects in the context of mapping and localization has been limited generally to those objects which are of interest to mobile robots, including doors (Avots et al., 2002), furniture (Hähnel et al., 2003), and walls (Martin & Thrun, 2002). Additionally, representations used typically involve geometric primitives, which fail to realize the potential benefit of using identified objects to infer occluded features. The difficulty of performing object recognition in unstructured environments makes the near-term realization of this benefit unlikely.

4.2.2 Localization

The humanoid must be able to localize itself (i.e., know its planar position and orientation) with respect to the environment, if not globally. Recent work by Fox et al. (1999) indicates that localization can be performed successfully with range data even in highly dynamic environments. This work, as well as other probabilistic approaches, typically can provide measures of certainty of their localization estimates. The estimated variance can be used to filter updates of the environment model; the environment model will be updated only if the certainty of the prediction is above a given threshold. Alternatively, localization accuracy can be quite high, e.g., on the order of centimeters Yamamoto et al. (2005), if the environment is modestly instrumented.

4.2.3 Nonstationary environments

Modeling nonstationary environments requires the management of three types of obstacles, which we call *dynamic, movable,* and *static.* Dynamic obstacles are capable of moving on their own (e.g., humans, cars, etc.), while movable obstacles (e.g., cups, furniture, books, etc.) must be moved by a dynamic obstacle. Static obstacles, such as walls, are incapable of movement. The three cases would ideally be managed separately, but a possible solution could treat movable and static obstacles identically yet allow for gradual changes to the environment.

Three recent approaches have made strides toward modeling nonstationary environments. The first approach, introduced independently by Wang & Thorpe (2002) and Hähnel et al. (2002), attempts to identify dynamic objects and filter them from the sensory data. The second approach, (Hähnel et al., 2003), uses an expectation-maximization (Dempster et al., 1977) based algorithm to repeatedly process the sensor readings and predict whether a given feature belongs to a dynamic obstacle; dynamic obstacles are not incorporated into the constructed map. Finally, Biswas et al. (2002) attempt to model dynamic obstacles as movable rigid bodies using an occupancy grid and an expectation-maximization based algorithm. Unfortunately, the first two approaches described above are ill-suited for movable obstacles, while the second approach is unable to handle dynamic obstacles.

4.2.4 Exploration

Exploration of the environment to facilitate reaching requires both directed locomotion and directed/active sensing. Both exploration and active sensing have been studied extensively in the context of mobile robot mapping, it is unlikely that this research is fully applicable toward our problem. Exploration in the mapping context is conducted in a greedy manner: movement is directed to build a complete map of the environment with respect to some criterion, e.g., minimum time (Yamauchi, 1997), minimum energy expenditure (Mei et al., 2006), maximum safety (González-Baños & Latombe, 2002), etc. In contrast, the reaching problem requires a balance between exploration and exploitation: the environment must be sufficiently explored, but not at the expense of finding a valid collision-free path. The balance between exploration and exploitation precludes the use of active sensing frameworks, such as that described by Mihaylova et al. (2002), to guide exploration. Finally, moving obstacles must be avoided during exploration, perhaps by using reactive approaches like VFH Borenstein & Koren (1989) or potential fields Khatib (1986).

Choset & Burdick (2000) refer to environmental exploration in the context of motion planning *sensor based motion planning,* and introduced a new data structure, the *hierarchical generalized voronoi graph,* to solve this problem. Unfortunately, their work was targeted toward mobile robots, and it appears highly unlikely to scale to humanoid robots with high-dimensional configuration spaces.

Researchers have yet to propose methods to perform directed exploration with respect to motion planning for humanoid robots. One possibility is to adapt the concepts employed for mobile robot exploration to humanoids. For example,*frontier-based exploration* Yamauchi (1997), which directs the robot to move to the area between open space and uncharted territory (i.e., the frontier), could be combined with a heuristic to select frontiers nearest the target.

4.2.5 Summary

This section covered the relevant issues necessary to construct a model of the environment for reaching. The issues of localization and selecting a proper representation for modeling the environment seem manageable through existing research. However, considerable work remains in the areas of modeling nonstationary environments and exploration for motion planning.

4.3 Control and Execution issues

Ideally, kinematic commands could be issued to a humanoid, the robot would execute those commands with perfect precision, and obstacles would not move. Those assumptions do not hold in the real-world: humanoids are typically holonomically constrained, controllers are imperfect, robots exhibit some degree of inaccuracy due to mechanical tolerances, the model of the environment may be flawed, and obstacles can appear suddenly.

The process of getting a manipulator to reach to an operational space target generally consists of the following sequence of steps:

1. A motion planner generates kinematic commands; the environment is often considered to be static for manipulators, so motion planning may consist of using cubic or quinticsplines to form a trajectory with via points.
2. The kinematic commands are input to a mechanism for rescaling the trajectory. The trajectory is rescaled to a longer duration, if necessary, to permit the robot to execute the trajectory without deviating from the prescribed path and prevent possible damage to the robot.
3. The commands are continually fed to a controller, which translates kinematic commands into actuator commands.

This process runs counter to best practices in mobile robotics because it provides no means to monitor the execution of the trajectory and react to exigent events, such as humans getting in the way. For humanoids, the process above must provide a means to supervise the robot's execution of the plan to avoid obstacles and correct deviation from the plan.
This section discusses the causes, effects, and remedies for controller deviation and examines execution monitoring, which attempts to avoid obstacles while driving the robot to its target.

4.3.1 Controller deviation

Robot controllers that minimize command error generally incorporate a model of the robot's dynamics; in particular, an inverse dynamics model of the robot is utilized to map desired accelerations to actuator torques. Controllers that use inverse dynamics models will still exhibit some deviation from the desired trajectory due to approximations of the robot's true dynamics (e.g., Coulombic friction, infinitely rigid links, etc.) Of far more concern, however, is that an algorithm does not exist for computing the inverse dynamics for robots with floating-bases experiencing contact constraints (e.g., for humanoids standing on a floor). As a result, humanoids are frequently controlled using feedback controllers (e.g., PD, PID, $PIDD^2$, etc.) These controllers necessarily result in deviation from the commanded trajectory because their generated motor commands are solely a function of command error.
For systems with holonomic differential constraints, deviation from the reference trajectory is the primary concern. For systems that are nonholonomically constrained, such as bipeds, deviation results in an additional difficulty: correcting deviation may require solving a two-point boundary value problem. As Section 4.1.3 noted, determining a general solution to this problem for humanoids is currently not practical. Thus, the question remains: given that the robot has deviated from its trajectory and is not in danger of colliding with obstacles, how is the deviation corrected? As in Section 4.1.3, we assume that the humanoid has a locomotion controller with inputs of desired planar position and orientation differentials for the base. We make no assumptions about the types of controllers employed for motor control. If the robot deviates from its planned trajectory for either the base or one of the joints, feedback is incorporated into the next commands sent to the controller. If the locomotion controller is unable to accomodate the desired differentials due to kinematic constraints[5], it reports this problem, and a new path for the base is planned that respects these constraints. The advantage of this approach is that the planning method can remain ignorant of the robot's dynamics model and controller internals. The most significant issue remaining is to determine whether the humanoid is capable of becoming stuck in a particular location due to kinematic constraints; however, this issue is entirely dependent upon the robot's mechanics and controllers and is thus robot-specific.

[5] As in Section 4.1.3, we assume that the trajectory has been rescaled to respect dynamic constraints.

4.3.2 Global methods for execution monitoring

The execution of the planned trajectory must be monitored to avoid both dynamic obstacles and static obstacles that were circumlocuted during planning but, due to unintentional deviation from the trajectory, that the robot is in imminent danger of contacting. Ideally, a policy would exist that could issue avoidance commands to the robot with expediency. Such a policy, denoted as $\mathbf{u} \leftarrow \pi(q, \dot{q})$, would issue commands (i.e., $\mathbf{u}$) as a function of the current state of the robot $(q, \dot{q})$ and the current obstacle region, . This policy theoretically could be constructed offline through reinforcement learning or dynamic programming; practically, however, the high-dimensionality of the robot's state-action spaces and the infinite number of configurations of the environment make this approach intractable (Bellman, 1957). Alternatively, it is conceivable that a policy could be determined nearly instantaneously using reinforcement learning as the environmental model changes (i.e., $\pi(.)$ would be a function of the robot's configuration only). Again, such an approach is currently infeasible given the robot's high-dimensional state-action spaces, and dynamic replanning would likely prove to be a better alternative. Yang & LaValle (2000) provide a method for *feedback motion planning,* which entails constructing motion plans that are viable from entire regions of state-space. Perhaps because objective criteria, upon which dynamic programming is based, do not have to be considered, their method seems to be slightly more tractable than dynamic programming. However, this method has only been applied to five-dimensional configuration spaces, and it appears unlikely to scale much further.

The methods described above are known as "global" methods, because good actions (i.e, those that lead to the goal and away from obstacles) are always generated, regardless of the configurations of the robot and environment. As noted above, the combined state-action spaces are immense, thus generally preventing global methods from being used for humanoids. "Local methods", in contrast to global methods, are generally tractable, but may result in failure to reach the goal; fortunately, these methods typically work well at avoiding obstacles.

4.3.3 Local methods for execution monitoring

One local method, by Brock & Khatib (2002), currently permits reaching with fast obstacle avoidance. This method is based on *elastic strips,* an extension of Khatib's operational space framework Khatib (1986) that employs virtual rubber bands. These virtual rubber bands warp planned trajectories away from obstacles. The most significant drawback to using elastic strips lies within its use of the robot's joint-space inertia matrix; this matrix is frequently unknown-it requires precise knowledge of the robot's inertial properties- and changes if the robot carries objects or interacts with the environment. In addition, like all local methods, elastic strips can cause the robot to become stuck in local minima. However, Brock did implement this method on both a mobile manipulator in an environment populated with moving obstacles and a simulated humanoid.

4.3.4 Summary

Control and locomotion are both significant problems in humanoid robotics, and assessing future prospects in these areas is difficult. However, the assumption of a planar controller for locomotion seems reasonable; it would prove difficult to have the roboticist juggle balance, locomotion, and task performance simultaneously. Indeed, many current humanoid designs prohibit direct access to actuators.

For execution monitoring, only elastic strips has proven successful for real-time obstacle avoidance on a mobile manipulator that followed a given trajectory. Depending on how seriously inaccuracies in the joint-space inertia matrix affect elastic strips, this method might prove extendible to humanoids. Viable alternatives include dynamic replanning and the offline-online probabilistic roadmap based methods of Jaillet & Siméon (2004) and van den Berg & Overmars (2005) described in Section 4.1, using dense roadmaps of configuration space. Regardless of the method chosen for execution monitoring, dynamic obstacles still require representation in the robot's perceptual model of the environment, as described in Section 4.2.

5 Conclusion

The past decade has seen remarkable progress in the types of motion planning problems that have been solved. However, many of these successes have been in application domains far removed from robotics. Additionally, humanoids have yet to perform tasks autonomously in human environments. This chapter formally defined one of the most important problems for humanoids, efficient, safe reaching in dynamic, unknown environments. Progress in the three areas critical areas to solving this problem- algorithms, perception for modeling, and execution-was surveyed and future prospects were examined. Table 1 summarizes the key issues in these areas as well as progress and prospects.

Issue	Progress and prospects
Planning under uncertainty	Difficult for humanoids, possible solution is dynamic replanning
Incompleteness resulting from multiple IK solutions	Solved by (Bertram et al., 2006) and (Drumwright & Ng-Thow-Hing, 2006)
Planning to nonstationary goals	Has received little focus, possible solution is modification of existing plans using inverse kinematics
Planning in dynamic environments	Dynamic replanning and offline-online probabilistic roadmaps seem to be promising directions; effectiveness toward humanoid reaching needs to be assessed
Representation of environment model	Octree approach works well, high-level features might work better in the future
Localization	Highly active area of research; (Fox et al., 1999) indicates good localization feasible even in nonstationary environments
Modeling nonstationary environments	Considerable work remaining; current work fails to address movable obstacles and dynamic obstacles simultaneously
Exploration	Considerable work remaining; exploration for planning humanoid motions yet to be addressed
Controller deviation	Better controllers will address this problem; can be mitigated by assuming differential locomotion controller
Execution monitoring	(Brock & Khatib, 2002) is one possible solution; dynamic replanning is another

Table 1. The core issues with respect to humanoid reaching discussed in this chapter with brief summaries of the findings.

With respect to algorithms, control, and execution monitoring, dynamic replanning seems to be one of the best solutions for humanoid reaching. As computing power continues to grow, dynamic replanning becomes increasingly viable for planning reaching in nonstationary environments. For modeling the robot's environment, computational power seems to be less of an issue. Instead, new algorithms are required to balance the tradeoff between exploration and exploitation and to perceive and classify fully dynamic and semi dynamic obstacles. These three areas critical to humanoid reaching are currently the focus of intensive research, and results applicable toward humanoid reaching are on the horizon.

6. References

Alami, R., Laumond, J.-P, & Siméon, T.(1997). Two manipulation planning algorithms. In J.-P. Laumond & M. Overmars (Eds.), *Algorithms for robot motion and manipulation.* Wellesley MA: A.K. Peters.

Alami, R., Siméon, T., & Laumond, J.-P. (1989). A geometric approach to planning manipulation tasks. In *Proc. intl. symp. on robotics research (ISRR)* (pp. 113-119).

Arbib, M. A., Iberall, T., & Lyons, D. (1985). Coordinated control programs for movements of the hand. *Experimental Brain Research,* 111-129.

Avots, D., Lim, E., Thibaux, R., & Thrun, S. (2002). A probabilistic technique for simultaneous localization and door state estimation with mobile robots in dynamic environments. In *Proc. of the IEEE conf. on intelligent robots and systems (IROS).* Lausanne, Switzerland.

Barraquand, J., & Latombe, J.-C. (1991, Dec). Robot motion planning: a distributed representation approach. *Intl. J. of Robotics Research, 10(6),* 628-649.

Baumann, A. (2001). *Robot motion planning in time-varying environments.* Unpublished doctoral dissertation.

Bellman, R. E.(1957). *Dynamic programming.* Dover Publications.

Bertram, D., Kuffner, Jr., J. J., Dillmann, R., & Asfour, T. (2006, May). An integrated approach to inverse kinematics and path planning for redundant manipulators. In *Proc. of the IEEE intl. conf. on robotics and automation (ICRA).* Orlando, FL, USA.

Biswas, R., Limketkai, B., Sanner, S., & Thrun, S. (2002). Towards object mapping in dynamic environments with mobile robots. In *Proc. of the IEEE conf. on intelligent robots and systems (IROS).* Lausanne, Switzerland.

Borenstein, J., & Koren, Y. (1989, Sept-Oct). Real-time obstacle avoidance for fast mobile robots. *IEEE Trans, on Systems, Man, and Cybernetics, 19(5),* 1179-1187.

Brock, O. (2000). *Generating robot motion: the integration of planning and execution.* Unpublished doctoral dissertation, Stanford University.

Brock, O., & Khatib, O. (2002, Dec). Elastic strips: A framework for motion generation in human environments. *Intl. J. of Robotics Research, 21(12),* 1031-1052.

Bullock, D., Grossberg, S., & Guenther, F.(1993). A self-organizing neural model of motor equivalent reaching and tool use by a multijoint arm. */. of Cognitive Neuroscience,* 5, 408-435.

Burridge, R. R., Rizzi, A. A., & Koditschek, D. E. (1995, Aug). Toward a dynamical pick and place. In *Proc. of the 1995IEEE/RS] intl. conf. on intelligent robots and systems* (Vol. 2, pp. 292-297).

Canny, J. (1993). Computing roadmaps of general semi-algebraic sets. *The Computer Journal, 36(5),* 504-514.

Choi, K.-J., & Ko, H.-S. (2000). On-line motion retargeting. *Journal of Visualization and Computer Animation, 11,* 223-235.

Choset, H., & Burdick, J. (2000). Sensor based motion planning: The hierarchical generalized voronoi graph. *Intl. J. of Robotics Research, 19(2),* 96-125.

Crowe, A., Porrill, J., & Prescott, T. (1998, Sept). Kinematic coordination of reach and balance. *Journal of Motor Behavior, 30(3),* 217-233.

Dempster, A. P., Laird, A. N., & Rubin, D. B. (1977). Maximum likelihood from incomplete data via the EM algorithm. *J. of the Royal Statistical Society, Series B,* 39(1), 1-38.

Drumwright, E., & Ng-Thow-Hing, V.(2006, Oct). Toward interactive reaching in static environments for humanoid robots. In *Proc. of IEEE/RSJ intl. conf. on intelligent robots and systems (IROS).* Beijing.

Drumwright, E., Ng-Thow-Hing, V., & Matarić, M.(2006, Dec). The task matrix framework for platform independent humanoid programming. In *Proc. of IEEE intl. conf. on humanoid robotics (humanoids).* Genova, Italy.

Elfes, A. (1989, June). Using occupancy grids for mobile robot perception and navigation. *Computer,* 22(6), 46-57.

Ferguson, D., & Stentz, A. (2006, May). Replanning with RRTs. In *Proc. of the IEEE intl. conf. on robotics and automation (ICRA).* Orlando, FL.

Flanagan, J. R., Feldman, A. G., & Ostry, D. J.(1993). Control of trajectory modifications in target-directed reaching. *Journal of Motor Behavior,* 25(3), 140-152.

Flash, T., & Hogan, N. (1985, July). The coordination of arm movements: An experimentally confirmed mathematical model. *The Journal ofNeuroscience, 5(7),* 1688-1703.

Fox, D., Burgard, W., & Thrun, S. (1999). Markov localization for mobile robots in dynamic environments. *Journal of Artificial Intelligence Research, 11,* 391-427.

González-Baños, H., & Latombe, J.-C. (2002, Oct-Nov). Navigation strategies for exploring indoor environments. *Intl. J. of Robotics Research,* 22(10-11), 829-848.

Gupta, K., Ahuactzin, J., & Mazer, E. (1998). Manipulation planning for redundant robots: A practical approach. *Intl. J. of Robotics Research,* 17(7).

Hähnel, D., Schulz, D., & Burgard, W. (2002, Sept-Oct). Mapping with mobile robots in populated environments. In *Proc. of the IEEE/RSJ intl. conf. on intelligent robots and systems (IROS).* Lausanne, Switzerland.

Hähnel, D., Triebel, R., Burgard, W., & Thrun, S. (2003, Sept). Map building with mobile robots in dynamic environments. In *Proc. of the IEEE intl. conf. on robotics and automation.* Taipei, Taiwan.

Jaillet, L., & Siméon, T.(2004, Sept-Oct). A PRM-based motion planner for dynamically changing environments. In *Proc. of the IEEE conf. on intelligent robots and systems (IROS)* (Vol. 2, pp. 1606-1611). Sendai, Japan.

Kagami, S., Kuffner, Jr., J. J., Nishiwaki, K., Inaba, M., & Inoue, H. (2003, April). Humanoid arm motion planning using stereo vision and RRT search. *Journal of Robotics and Mechatronics, 15(2),* 200-207.

Kallmann, M., Aubel, A., Abaci, T., & Thalmann, D. (2003). Planning collision-free reaching motions for interactive object manipulation and grasping. In *Proc. of euro-graphics* (p. 313-322). Grenada, Spain.

Kallmann, M., & Matarić, M.(2004, April). Motion planning using dynamic roadmaps. In *Proc. of IEEE intl. conf. on robotics and automation (ICRA)* (p. 4399-4404). New Orleans, LA.

Kalman, R., & Bucy, R. (1961). New results in linear filtering and prediction theory. *Trans, of the ASME-Journal of Basic Engineering, 83,* 95-107.

Kavraki, L. E., Svestka, P., Latombe, J.-C, & Overmars, M. (1996). Probabilistic roadmaps for path planning in high dimensional configuration spaces. *IEEE Trans, on Robotics and Automation,* 22(4), 566-580.

Khatib, O. (1986). Real-time obstacle avoidance for manipulators and mobile robots. *The Intl. J. of Robotics Research,* 5(1), 90-98.

Kuffner, Jr., J. J. (1998). Goal-directed navigation for animated characters using realtime path planning and control. *Lecture Notes in Computer Science, 1537,*171-186.

Kuipers, B., Froom, R., Lee, W. K., & Pierce, D.(1993). The semantic hierarchy in robot learning. In J. Connell & S. Mahadevan (Eds.), *Robot learning.* Boston, MA: Kluwer Academic Publishers.

LaValle, S. (1998, October). *Rapidly-exploring random trees: a new tool for path planning* (Tech. Rep. No. TR 98-11). Computer Science Dept., Iowa State University.

LaValle, S.(2006). *Planning algorithms.* Cambridge University Press.

LaValle, S., & Hutchinson, S. (1998, Jan). An objective-based framwork for motion planning under sensing and control uncertainties. *Intl. J. of Robotics Research, 17(1),* 19-42.

Lazanas, A., & Latombe, J. C. (1995). Motion planning with uncertainty: A landmark approach. *Artificial Intelligence,* 76(1-2), 285-317.

Liu, Y., & Badler, N. I. (2003, May). Real-time reach planning for animated characters using hardware acceleration. In *Proceedings of computer animation and social agents* (p. 86-93). New Brunswick, NJ: IEEE Computer Society.

Lorensen, W. E., & Cline, H. E. (1987). Marching cubes: a high resolution 3D surface construction algorithm. *Computer Graphics,* 22(4), 163-169.

Lozano-Pérez, T., Jones, J. L., Mazer, E., & O'Donnell, P. A. (1989). Task-level planning of pick-and-place robot motions. *Computer,* 22(3), 21-29.

Maciejewski, A. A., & Klein, C. A. (1985). Obstacle avoidance for kinematically redundant manipulators in dynamically varying environments. *Intl. J. of Robotics Research,* 4(3), 109-117.

Martin, C, &Thrun, S.(2002). Real-time acquisition of compact volumetric maps with mobile robots. In *IEEE intl. conf. on robotics and automation (ICRA).* Washington, DC.

Mason, M. T.(2001). *Mechanics of robotic manipulation.* Cambridge, MA: MIT Press.

Mayer, J., & Bajcsy R. (1993, May). Occlusions as a guide for planning the next view. *IEEE Trans, on Pattern Analysis and Machine Intelligence, 15(5),* 417-433.

Mei, Y., Lu, Y.-H., Hu, Y. C, & Lee, C. S. G. (2006). Energy-efficient mobile robot exploration. In *Proc. of the IEEE intl. conf. on robotics and automation (ICRA).* Orlando, FL.

Mihaylova, L., Lefebvre,T., Bruyninckx, H., Gadeyne, K., & De Schutter, J.(2002). Active sensing for robotics- a survey. In *Intl. conf. on numerical methods and applications.*

Moravec, H., & Elfes, A. (1985, March). High-resolution maps from wide-angle sonar. In *Proc. of IEEE intl. conf. on robotics and automation (ICRA).* St. Louis, MO.

Okada, K., Haneda, A., Nakai, H., Inaba, M., & Inoue, H. (2004, Sept-Oct). Environment manipulation planner for humanoid robots using task graph that generates action sequence. In *Proc. of the IEEE/RS] intl. conf. on intelligent robots and systems (IROS).* Sendai, Japan.

Pito, R.(1996). A sensor based solution to the next best view problem. In 941-945 (Ed.), *Proc. IEEE intl. conf. on pattern recognition* (Vol. 1).

Reif, J. H. (1979). Complexity of the mover's problem and generalizations. In *IEEE symposium on foundations of computer science* (pp. 421-427). San Juan, Puerto Rico.

Reif, J. H., & Sharir, M. (1994). Motion planning in the presence of moving obstacles. *Journal of the Association of Computing Machinery, 41,* 764-790.

Sciavicco, L., & Siciliano, B. (2000). *Modeling and control of robot manipulators, 2nd ed.* London: Springer-Verlag.

Smith, R., & Cheeseman, P. (1985). *On the representation and estimation of spatial uncertainty* (Tech. Rep. Nos. TR 4760, TR 7239). SRI.

Smith, R., Self, M., & Cheeseman, P. (1990). Estimating uncertain spatial relationships in robotics. In I. Cox & G. Wilfong (Eds.), *Autonomous robot vehicles* (pp. 167-193). Springer-Verlag.

Stilman, M., & Kuffner, Jr., J. J. (2007, April). Manipulation planning among movable obstacles. In *To appear in proc. of the IEEE intl. conf. on robotics and automation (ICRA).* Rome.

Sung, R., Corney, J., & Clark, D.(2002). Octree based assembly sequence generation. *J. of Computing and Information Science in Engineering.*

Tanner, H. G., & Kyriakopoulos, K. J. (2000). Nonholonomic motion planning for mobile manipulators. In *Proc. of the IEEE intl. conf. on robotics and automation (ICRA)* (pp. 1233-1238). San Francisco, CA.

Teller, S. (1998). Automated urban model acquisition: Project rationale and status. In *Proc. of the image understanding workshop* (pp. 455-462).

Thoroughman, K. A., & Shadmehr, R.(2000, Oct). Learning of action through adaptive combination of motor primitives. *Nature, 407,* 742-747.

Thrun, S., Hähnel, D., Ferguson, D., Montemerlo, M., Triebel, R., Burgard, W., et al. (2003). A system for volumetric robotic mapping of abandoned mines. In *Proc. of the IEEE intl. conf. on robotics and automation (ICRA).*

Turk, G., & Levoy M. (1994, July). Zippered polygon meshes from range images. *Computer Graphics (Proc. of ACM SIGGRAPH),* 311-318.

van den Berg, J. R, & Overmars, M. H. (2005, Oct). Roadmap-based motion planning in dynamic environments. *IEEE Trans, on Robotics, 21(5).*

Wang, C.-C., & Thorpe, C. (2002, May). Simultaneous localization and mapping with detection and tracking of moving objects. In *Proc. of the IEEE intl. conf. on robotics and automation (ICRA).* Washington, D.C.

Yamamoto, Y., Pirjanian, P., Munich, M., DiBernardo, E., Goncalves, L., Ostrowski, J., et al. (2005). Optical sensing for robot perception and localization. In *Proc. of IEEE workshop on advanced robotics and its social impacts* (pp. 14-17).

Yamauchi, B. (1997, July). A frontier based approach for autonomous exploration. In *Proc. of IEEE intl. symp. on computational intelligence in robotics and automation (CIRA).* Monterey, CA.

Yang, X. S., & LaValle, S. (2000, April). A framework for planning feedback motion strategies based on a random neighborhood graph. In *Proc. of the IEEE intl. conf. on robotics and automation (ICRA)* (Vol. 1, pp. 544-549). San Francisco, CA.

Zucker, M., Kuffner, Jr., J. J., & Branicky, M. (2007, April). Multipartite RRTs for rapid replanning in dynamic environments. In *Proc. of the IEEE intl. conf. on robotics and automation (ICRA).* Rome, Italy.

13

Minimum Energy Trajectory Planning for Biped Robots

Yasutaka Fujimoto
Yokohama National University
Japan

1. Introduction

Biped robots have potential ability to elevate mobility of robotic system and they attract general attention in the last decade. Due to their form, it is easy to introduce them into our living space without preparing special infrastructure. There are many theoretical and experimental studies on the biped robots. Recently, several autonomous biped humanoid robots have been developed. In 1996, the first autonomous bipedal humanoid robot with battery power supply was developed by Honda Motor Co., Ltd. (Hirai, et al., 1998). Takanishi and his co-workers at Waseda University built a human-size bipedal humanoid robot and proposed a basic control method of whole body cooperative dynamic biped walking (Yamaguchi et al., 1999). Same group also developed a biped robot having two Stewart platforms as legs (Sugahara, et al., 2003). Kaneko and his colleagues at the National Institute of Advanced Industrial Science and Technology (AIST) also developed a bipedal humanoid robot of 1.54m height and 58kg weight mainly for the purpose of investigating fundamental techniques and applications (Kaneko et al., 2004). Same group also proposed a method of a biped walking pattern generation by using a preview control of the zero-moment point(ZMP) (Kajita, et al., 2003). The ZMP and its extension (FRI; foot-rotation indicator) are basic stability criteria for biped robots (Vukobratovic, et al, 2001, Goswami, 1999). Nishiwaki and his co-researchers at University of Tokyo studied a humanoid walking control system that generates body trajectories to follow a given desired motion online (Nishiwaki, et al. 2003). Loffler and his colleagues at Technical University of Munich also designed a biped robot to achieve a dynamically stable gait pattern (Loffler, et al., 2003).

Generally, careful design is required for development of a bipedal robot. Selection of gears and actuators, taking their rate, power, torque, and weight into account, is especially important. Developments of power conversion technology based on semiconductor switching devices and rare-earth permanent magnet materials Nd-Fe-B in combination with optimal design of the electromagnetic field by the finite element method enable improvement of power/weight ratio of actuators. They contribute to the realization of such autonomous biped robots. However, they are still underpowered to achieve fast walking and running motions as the same that human does. There are several researches on running control of biped model (Raibert, 1986, Hodgins, 1996, Kajita, et al., 2002). The first human-size running robot was developed at AIST (Nagasaki, et al., 2004). In December 2005, Honda Motor Co., Ltd. announced their new bipedal humanoid robot that could run at 6km/h. Kajita, et al. proposed a method to generate a running pattern and reported that it requires at least 28 to 56 times more powerful actuators than that of their actual humanoid robot HRP1, and also the consumption power is estimated ten times bigger than the human runner (Kajita, et al., 2002)

In this chapter, a method to generate a trajectory of a running motion with minimum energy consumption is introduced. It is useful to know the lower bound of the consumption energy when we design the biped robot and select actuators. The generation of low-energy trajectories for biped robots remained an open problem. Usually, it is formulated as an optimal control problem. Since symbolic expression of motion equation of robots becomes extremely complicated in the case that the number of links increases, only specific simple type of a structure of robots was investigated and simplified assumptions such as ignoring the effects of centripetal forces were made in the past works (Roussel, et al., 1998). In this chapter, exact and general formulation of optimal control for biped robots based on recursive representation of motion equation is introduced. As an illustrative numerical example, the method is applied to a five link planar biped robot of 1.2m height and 60kg weight. The robot with the generated minimum energy running pattern with 1m/sec speed consumes only 45W on average. It is found that big peak power and torque is required for the knee joints but its consumption power is negative small and the main work is done by the hip joints.

The rest of this chapter is organized as follows. In Section 2, the problem definition is introduced where the formulation of the biped running robot is given. The minimization of consumption energy is explained in the Section 3, and the computational scheme is proposed in the Section 4. The numerical study of a five link planar biped robot is provided in Section 5, and conclusions are outlined in the Section 6.

2. Model of Biped Running Robot

2.1 Basic Model of a Biped Robot

Consider a three dimensional bipedal robot with open-chain mechanism consisting of N joints and N+1 rigid links. Since the robot is not fixed on the ground, it is modelled as a free-fall manipulator which has N+6 motion-degree-of-freedom (Fujimoto & Kawamura, 1995, 1998). The model in general form is given by

$$H\ddot{x} + C\dot{x} + g = u + J^T f_E \tag{1}$$

where $H \in \Re^{(N+6)\times(N+6)}$ is an inertia matrix, $C \in \Re^{(N+6)\times(N+6)}$ specifies centrifugal and Coriolis effects, $g \in \Re^{N+6}$ specifies gravity effects, $x = (\theta, \varphi, p) \in \Re^{N+6}$ specifies displacements of joints, and posture and position of a base link. $u = (n, 0, 0) \in \Re^{N+6}$ specifies input generalized forces. $f_E \in \Re^6$ is external force and $J \in \Re^{6\times(N+6)}$ is a Jacobian matrix for tip of the support leg. Also $\theta, n \in \Re^N$ specify joint angles and input joint torques, respectively. $\varphi, p \in \Re^{N+3}$ specify posture and position of a base link, respectively. An example of the coordinates for the case of planner biped is shown in Fig. 1.

2.2 Model of Support Phase

The motion of running is decomposed to two phases; single-leg support phase and flight phase.

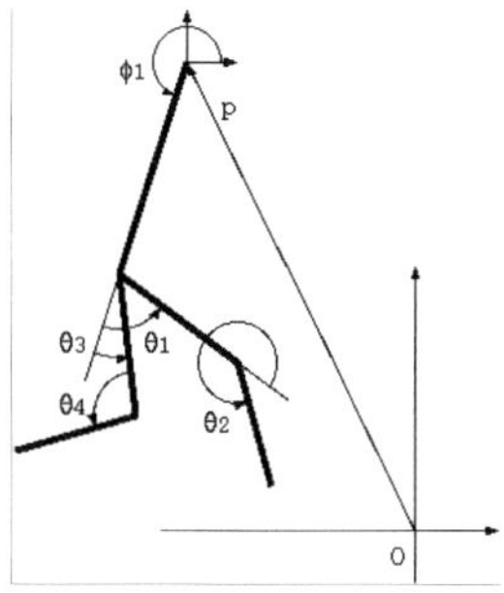

Fig. 1. A five link biped robot.

Let the tip position of the support leg with respect to the origin of the base link system be $h(\theta,\varphi)$. Then its position with respect to the origin of the world coordinate system is represented by

$$y = h(\theta,\varphi) + p \tag{2}$$

Since the foot of the support leg is fixed on the ground during single support phase, it is subject to the following conditions.

$$\dot{y} = J\dot{\theta} + R\dot{\varphi} + \dot{p} = 0 \tag{3}$$

$$\ddot{y} = J\ddot{\theta} + \dot{J}\dot{\theta} + R\ddot{\varphi} + \dot{R}\dot{\varphi} + \ddot{p} = 0 \tag{4}$$

where $J = \partial h / \partial \theta^T$ and $R = \partial h / \partial \varphi^T$ are Jacobian matrices.
Eliminating $\dot{p}$, $\ddot{p}$, and f_E from (1) using (3) and (4), the following dynamics is obtained.

$$H_s \ddot{x}_s + C_s \dot{x}_s + g_s = u_s \tag{5}$$

where $x_s = (\theta_s, \varphi_s)$ and $u_s = (n_s, 0)$. The subscript s represents variables during support phase. The state equation is given by

$$\dot{w}_s = \begin{bmatrix} \dot{x}_s \\ H_s^{-1}(u_s - C_s \dot{x}_s - g_s) \end{bmatrix} \tag{6}$$

where $w_s = (x_s, \dot{x}_s) = (\theta_s, \varphi_s, \dot{\theta}_s, \dot{\varphi}_s) \in \Re^{2N+6}$ is a state vector.

2.3 Model of Flight Phase

During flight phase, the external force f_E in the motion equation (1) is zero. The conservation law of angular momentum is already included equivalently, which corresponds to no existence of external forces. Therefore the dynamics becomes

$$H_f \ddot{x}_f + C_f \dot{x}_f + g_f = u_f \tag{7}$$

where $x_f = (\theta_f, \varphi_f, p_f)$ and $u_f = (n_f, 0, 0)$. The subscript f represents variables during flight phase. The state equation is given by

$$\dot{w}_f = \begin{bmatrix} \dot{x}_f \\ H_f^{-1}(u_f - C_f \dot{x}_f - g_f) \end{bmatrix} \tag{8}$$

where $w_f = (x_f, \dot{x}_f) = (\theta_f, \varphi_f, p, \dot{\theta}_f, \dot{\varphi}_f, \dot{p}) \in \Re^{2(N+6)}$ is a state vector.

3. Generation of Minimum Energy Consumption Gait

It is useful to know the lower bound of the consumption energy when we design the bipedal robot and select actuators. An ideal actuator is assumed in this chapter to investigate the consumption energy, although real actuators with high-ratio gears such as harmonic gears have large frictions and roughly 70% of energy efficiency. The energy regeneration is considered. The problem is to find input joint torques and initial posture that minimize input energy during running motion under the condition that the robot takes completely periodic and symmetric motion, given the step period and the stride. The problem is described as follow.

$$\text{minimize} \quad E = \int_0^T \dot{\theta}^T n dt \tag{9}$$

$$\text{subject to} \quad \begin{cases} \theta(T) = K\theta(0) \\ \varphi(T) = \varphi(0) \\ p(T) = p(0) + S \\ H\ddot{x} + C\dot{x} + g = u + J^T f_E \end{cases} \tag{10}$$

where T is a period for one step. S is the stride. K is a coordinate conversion matrix;

$$K = \begin{bmatrix} I_b & 0 & 0 \\ 0 & 0 & I_\ell \\ 0 & I_\ell & 0 \end{bmatrix} \tag{11}$$

where I_b and I_ℓ are identity matrices whose dimensions are same as number of joints in body and one leg, respectively.
Since the structure of dynamics varies depending on the phase as shown in the previous section, a reflection of time axis is introduced. A new time axis is given by

$$\tau = \begin{cases} \dfrac{t}{\alpha T} & \text{for} \quad 0 \le t \le \alpha T \\ \dfrac{T-t}{(1-\alpha)T} & \text{for} \quad \alpha T \le t \le T \end{cases} \tag{12}$$

The timing chart of events are shown in Fig. 2. All variables in the rest of this section are functions of τ.

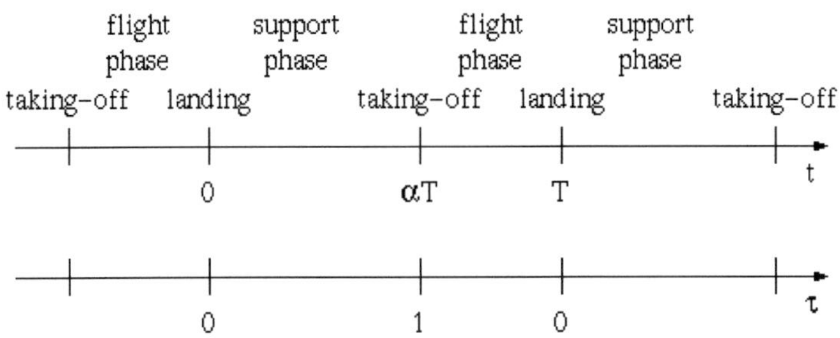

Fig. 2. Timing chart of events.

The objective function (9) is represented by

$$\begin{aligned} E &= \int_0^{\alpha T} \dot{\theta}^T n\, dt + \int_{\alpha T}^{T} \dot{\theta}^T n\, dt \\ &= \int_0^1 \left(\alpha \dot{\theta}_s^T n_s + (1-\alpha)\dot{\theta}_f^T n_f\right) T d\tau \end{aligned} \tag{13}$$

The state equations (6) for $0 \le t \le \alpha T$ and (8) for $\alpha T \le t \le T$ are transformed onto τ -axis as follows.

$$\frac{dw_s}{d\tau} = \alpha T \begin{bmatrix} \dot{x}_s \\ H_s^{-1}(u_s - C_s \dot{x}_s - g_s) \end{bmatrix} \tag{14}$$

$$\frac{dw_f}{d\tau} = (\alpha - 1)T \begin{bmatrix} \dot{x}_f \\ H_f^{-1}(u_f - C_f \dot{x}_f - g_f) \end{bmatrix} \tag{15}$$

State variables should include the support phase ratio α in order to find its optimal value as well. The support phase ratio is constant. Then, the following differential equation is introduced.

$$\frac{d\alpha}{d\tau} = 0 \tag{16}$$

Finally, the problem (9)(10) is transformed into a Bolza problem.
minimize

$$\widetilde{E} = g(z_1) + \int_0^1 \left(f_0(z(\tau), v(\tau)) + \lambda(\tau)^T \left(\frac{dz(\tau)}{d\tau} - f(z(\tau), v(\tau)) \right) \right) d\tau \tag{17}$$

where $z = (w_s, w_f, \alpha) = (x_s, \dot{x}_s, x_f, \dot{x}_f, \alpha) = (\theta_s, \varphi_s, \dot{\theta}_s, \dot{\varphi}_s, \theta_f, \varphi_f, p_f, \dot{\theta}_f, \dot{\varphi}_f, \dot{p}_f, \alpha) \in \Re^{4N+19}$ is a state vector, $v = (n_s, n_f) \in \Re^{2N}$ is a input vector, $\lambda \in \Re^{4N+19}$ is a Lagrange multiplier, $z_1 = z(1)$ is state at the terminal period, and

$$f_0(z, v) = \left(\alpha \dot{\theta}_s^T n_s + (1 - \alpha)\dot{\theta}_f^T n_f\right)T \tag{18}$$

$$f(z, v) = \begin{bmatrix} \alpha T \dot{x}_s \\ \alpha T H_s^{-1}(u_s - C_s \dot{x}_s - g_s) \\ (\alpha - 1)T \dot{x}_f \\ (\alpha - 1)T H_f^{-1}(u_f - C_f \dot{x}_f - g_f) \\ 0 \end{bmatrix} \tag{19}$$

The function $g(z_1)$ represents penalty for the terminal condition that is introduced to guarantee continuity of the state variable at the instant of the taking-off.

$$\begin{aligned} g(z_1) = W\Big(&\|\theta_s(1) - \theta_f(1)\|^2 + \|\dot{\theta}_s(1) - \dot{\theta}_f(1)\|^2 + \|\varphi_s(1) - \varphi_f(1)\|^2 \\ &+ \|\dot{\varphi}_s(1) - \dot{\varphi}_f(1)\|^2 + \|p_s(1) - p_f(1)\|^2 + \|\dot{p}_s(1) - \dot{p}_f(1)\|^2 \Big) \end{aligned} \tag{20}$$

where W is an weight coefficient. The variables p_s and $\dot{p}_s$ are implicitly defined, which are represented by functions of w_s. Variation of the extended objective function $\widetilde{E}$ is given by

$$\begin{aligned} \delta\widetilde{E} = &-\lambda_0^T \frac{\partial z_0}{\partial z_0'^T} \delta z_0' + \left(\frac{\partial g}{\partial z_1} + \lambda_1 \right)^T \delta z_1 + \int_0^1 \Bigg(\left(\frac{\partial f_0}{\partial z} - \frac{\partial f^T}{\partial z} \lambda - \frac{d\lambda}{d\tau} \right)^T \delta z \\ &+ \left(\frac{\partial f_0}{\partial v} - \frac{\partial f^T}{\partial v} \lambda \right)^T \delta v + \left(\frac{dz}{d\tau} - f \right)^T \delta\lambda \Bigg) d\tau \end{aligned} \tag{21}$$

where $\lambda_0 = \lambda(0)$ and $\lambda_1 = \lambda(1)$. In this equation, we assume that the initial state z_0 is a function of certain variables which consist of partial set of the state, namely, a part of the initial state is independent and the other depends on it. Let the independent initial state variables be $z_0' = (\theta_s(0), \varphi_s(0), \dot{\theta}_f(0), \dot{\varphi}_f(0), \dot{p}_f(0), \alpha(0))$. The rest of the initial state are decided by

$$\theta_f(0) = K\theta_s(0) \tag{22}$$

$$\varphi_f(0) = \varphi_s(0) \tag{23}$$

$$p_f(0) = -h(\theta_s(0), \varphi_s(0)) + S \tag{24}$$

$$\begin{bmatrix} \dot{\theta}_s(0) \\ \dot{\varphi}_s(0) \end{bmatrix} = \begin{bmatrix} K & 0 & 0 \\ 0 & I & 0 \end{bmatrix} \left(I - H_f^{-1} \tilde{J}_f^T \left(\tilde{J}_f H_f^{-1} \tilde{J}_f^T \right)^{-1} \tilde{J}_f \right) \begin{bmatrix} \dot{\theta}_f(0) \\ \dot{\varphi}_f(0) \\ \dot{p}_f(0) \end{bmatrix} \tag{25}$$

The first three equations are coordinate conversion at the instant of landing and the last is the condition of perfectly inelastic collision at the instant of landing. Let the impulsive external force at the foot of support leg be δf. The impact force δf is inflicted at the instant of the landing and the generalized velocity changes discontinuously. From (1), the generalized momentum after the collision is given by

$$H\dot{x}_+ = H\dot{x}_- + \tilde{J}^T \delta f \tag{26}$$

where $\dot{x}_+$ and $\dot{x}_-$ denote the generalized velocities after and before collision, respectively. $\tilde{J} = [J, R, I]$ is an extended Jacobian. Since it is support phase after the collision, the condition (3) holds, namely,

$$\tilde{J}\dot{x}_+ = 0 \tag{27}$$

Describing (26) and (27) for $\dot{x}_+$ and δf, the following equation is obtained.

$$\begin{bmatrix} H & -\tilde{J}^T \\ -\tilde{J} & 0 \end{bmatrix} \begin{bmatrix} \dot{x}_+ \\ \delta f \end{bmatrix} = \begin{bmatrix} H\dot{x}_- \\ 0 \end{bmatrix} \tag{28}$$

Eliminating δf from (28), we have

$$\dot{x}_+ = \left(I - H^{-1} \tilde{J}^T \left(\tilde{J} H^{-1} \tilde{J}^T \right)^{-1} \tilde{J} \right) \dot{x}_- \tag{29}$$

Here, $\dot{x}_+$ corresponds to $\dot{x}_s(0)$ which is the generalized velocity of the support phase at $\tau = 0$, and $\dot{x}_-$ corresponds to $\dot{x}_f(0)$ which is the generalized velocity of the flight phase at $\tau = 0$. Taking into account the coordinate conversion between left and right leg, (29) is transformed into the form of (25).

From (21), the following conditions are obtained.

$$\lambda_1 = -\frac{\partial g}{\partial z_1} \tag{30}$$

$$\frac{d\lambda}{d\tau} = \frac{\partial f_0}{\partial z} - \frac{\partial f^T}{\partial z} \lambda \tag{31}$$

$$\frac{dz}{d\tau} = f \tag{32}$$

Also the gradients are given by

$$\frac{\partial \widetilde{E}}{\partial z_0'} = -\frac{\partial z_0^T}{\partial z_0'}\lambda_0 \tag{33}$$

$$\frac{\partial \widetilde{E}}{\partial v} = \frac{\partial f_0}{\partial v} - \frac{\partial f^T}{\partial v}\lambda \tag{34}$$

To find the optimal solution, the conjugate gradient method in infinite dimensional space (Hilbert space) is applied to this problem. The procedures of the algorithm are as follows.

1) The initial solution $(z_0', v(\tau))$ is given.
2) The initial state z_0 is computed by (22)-(25).
3) The differential equation (32) is solved using z_0 as the initial condition.
4) λ_1 is computed by (30) using the final value z_1.
5) The differential equation (31) is backwardly solved using z_1.
6) The gradients for z_0' and $v(\tau)$ are computed by (33)(34) using $z(\tau)$, $\lambda(\tau)$, and $v(\tau)$.
7) The temporary solution $(z_0', v(\tau))$ is updated toward the direction of the conjugate gradient.
8) If the gradient is not small enough, return to 2.

Finally, the input joint torques $n(t)$, the joint angles $\theta(t)$, the posture and position (of the base link) $\varphi(t)$, $p(t)$, their derivatives $\dot{\theta}(t)$, $\dot{\varphi}(t)$, and the support phase ratio α are obtained. A general method to compute the partial derivatives in (30)-(34) is proposed in the next section.

4. Computational Scheme for Partial Derivative

It is difficult to calculate the partial derivatives in (30)-(34) symbolically, because basically it costs very much to obtain a symbolic expression of the equation of motion (1). In this section, a computational scheme for the partial derivatives based on numerical representation of motion equation is proposed. Finally we can compute them easily by using forward-backward recursive Newton-Eular formulation.

Each partial derivative appeared in (30)-(34) is represented as follows.

$$\frac{\partial f}{\partial z^T} = T \times \begin{bmatrix} 0 & \alpha I & 0 & 0 & \dot{x}_s \\ \alpha\frac{\partial f_s}{\partial x_s^T} & \alpha\frac{\partial f_s}{\partial \dot{x}_s^T} & 0 & 0 & f_s \\ 0 & 0 & 0 & (\alpha-1)I & \dot{x}_f \\ 0 & 0 & (\alpha-1)\frac{\partial f_f}{\partial x_f^T} & (\alpha-1)\frac{\partial f_f}{\partial \dot{x}_f^T} & f_f \\ 0 & 0 & 0 & 0 & 0 \end{bmatrix} \tag{35}$$

where

$$f_s = H_s^{-1}\left(u_s - C_s \dot{x}_s - g_s\right) \tag{36}$$

$$f_f = H_f^{-1}\left(u_f - C_f \dot{x}_f - g_f\right) \tag{37}$$

And then,

$$\frac{\partial f_s}{\partial x_{si}} = -H_s^{-1}\left(\frac{\partial C_s}{\partial x_{si}}\dot{x}_s + \frac{\partial g_s}{\partial x_{si}}\right) - H_s^{-1}\frac{\partial H_s}{\partial x_{si}} f_s \tag{38}$$

$$\frac{\partial f_f}{\partial x_{fi}} = -H_f^{-1}\left(\frac{\partial C_f}{\partial x_{fi}}\dot{x}_f + \frac{\partial g_f}{\partial x_{fi}}\right) - H_f^{-1}\frac{\partial H_f}{\partial x_{fi}} f_f \tag{39}$$

$$\frac{\partial f}{\partial v^T} = \begin{bmatrix} 0 & 0 \\ \alpha T H_s^{-1} P_s & 0 \\ 0 & 0 \\ 0 & (\alpha - 1) T H_f^{-1} P_f \\ 0 & 0 \end{bmatrix} \tag{40}$$

where $P_s = [I,0]^T \in \Re^{(N+3)\times N}$, $P_f = [I,0,0]^T \in \Re^{(N+6)\times N}$ are selection matrices. And also,

$$\frac{\partial f_0}{\partial z} = \begin{bmatrix} 0 \\ 0 \\ \alpha T n_s \\ 0 \\ 0 \\ 0 \\ 0 \\ (1-\alpha) T n_f \\ 0 \\ 0 \\ T\left(\dot{\theta}_s^T n_s - \dot{\theta}_f^T n_f\right) \end{bmatrix} \tag{41}$$

$$\frac{\partial f_0}{\partial v} = \begin{bmatrix} \alpha T \dot{\theta}_s \\ (1-\alpha) T \dot{\theta}_f \end{bmatrix} \tag{42}$$

$$\frac{\partial z_0}{\partial z_0'^T} = \begin{bmatrix} I & 0 & 0 \\ \Gamma & \Lambda & 0 \\ \widetilde{K} & 0 & 0 \\ 0 & I & 0 \\ 0 & 0 & 1 \end{bmatrix} \tag{43}$$

$$\frac{\partial g}{\partial z_1} = 2W \times \begin{bmatrix} \theta_{s1} - \theta_{f1} + J_s^T\left(h_s + p_{f1}\right) + \left[\dot{\theta}_{s1}^T \frac{\partial J_s^T}{\partial \theta_{s1i}}\left(J_s\dot{\theta}_{si} + \dot{p}_{f1}\right)\right] \\ \varphi_{s1} - \varphi_{f1} + R_s^T\left(h_s + p_{f1}\right) + \left[\dot{\varphi}_{s1} \frac{\partial R_s^T}{\partial \varphi_{s1i}}\left(R_s\dot{\varphi}_{s1} + \dot{p}_{f1}\right)\right] \\ \dot{\theta}_{s1} - \dot{\theta}_{f1} + J_s^T\left(J_s\dot{\theta}_{s1} + \dot{p}_{f1}\right) \\ \dot{\varphi}_{s1}^T - \dot{\varphi}_{f1} + R_s^T\left(R_s\dot{\varphi}_{s1} + \dot{p}_{f1}\right) \\ -\theta_{s1} + \theta_{f1} \\ -\varphi_{s1} + \varphi_{f1} \\ h_s + p_{f1} \\ -\dot{\theta}_{s1} + \dot{\theta}_{f1} \\ -\dot{\varphi}_{s1} + \dot{\varphi}_{f1} \\ J_s\dot{\theta}_{s1} + \dot{p}_{f1} \\ 0 \end{bmatrix} \tag{44}$$

where the subscript 1 corresponds to the value at $\tau = 1$, and

$$\widetilde{K} = \begin{bmatrix} K & 0 \\ 0 & I \\ -J_s & -R_s \end{bmatrix} \tag{45}$$

$$\Lambda = \widetilde{K}_l\left(I - H_f^{-1}\widetilde{J}_f^T\left(\widetilde{J}_f H_f^{-1}\widetilde{J}_f^T\right)^{-1}\widetilde{J}_f\right) \tag{46}$$

$$\widetilde{K}_l = \begin{bmatrix} K & 0 & 0 \\ 0 & I & 0 \end{bmatrix} \tag{47}$$

$$\Gamma = \left[\frac{\partial \Lambda}{\partial x_{si}}\dot{x}_f\right] \tag{48}$$

$$\frac{\partial \Lambda}{\partial x_{si}} = \widetilde{K}_l H_f^{-1}\frac{\partial H_f}{\partial x_{fi}}H_f^{-1}\widetilde{J}_f^T\Omega\widetilde{J}_f - H_f^{-1}\frac{\partial \widetilde{J}_f^T}{\partial x_{fi}}\Omega\widetilde{J}_f - H_f^{-1}\widetilde{J}_f^T\Omega\frac{\partial \widetilde{J}_f^T}{\partial x_{fi}}$$
$$+ H_f^{-1}\widetilde{J}_f^T\Omega\left(\frac{\partial \widetilde{J}_f}{\partial x_{fi}}H_f^{-1}\widetilde{J}_f^T - \widetilde{J}_f H_f^{-1}\frac{\partial H_f}{\partial x_{fi}}H_f^{-1}\widetilde{J}_f^T + \widetilde{J}_f H_f^{-1}\frac{\partial \widetilde{J}_f^T}{\partial x_{fi}}\right)\Omega\widetilde{J}_f\widetilde{K}_r \tag{49}$$

$$\Omega = \left(\widetilde{J}_f H_f^{-1}\widetilde{J}^T\right)^{-1} \tag{50}$$

$$\widetilde{K}_r = \begin{bmatrix} K & 0 \\ 0 & I \\ 0 & 0 \end{bmatrix} \tag{51}$$

The partial derivatives appeared in (38), (39), (44), (48), and (49) are computed by using modified Newton-Euler formulations.

5. Numerical Study of Five-link Planar Biped

The proposed method is applied to a five-link planar biped robot. The specification of the robot and the control parameters are shown in Table 1. The robot is 1.2 [m] height and 60 [kg] weight. The coordinates are taken as shown in Fig. 1.
The optimal trajectories are computed as shown in Fig. 3-Fig. 6. Snapshots of the running motion are also shown in Fig. 7. The solid lines and the dashed lines show the trajectories for the right leg and the left leg, respectively. In Fig. 4, there are some discontinuous velocities due to the impact force at the instant of the landing. In Fig. 5, the peak torque for the hip joints appears at the beginnings of the swinging. For the knee joints, the torque becomes maximal at the instant of the landing and keeps over 100N.m. In Fig. 6, the positive peak power for the hip and knee joints appears during kicking motion. For the hip joints, the power becomes negative peak value at the beginning of the flight phase. This means that the hip joints absorb the energy of the kicking motion. The negative peak power for the knee joints appears at the instant of the landing. Namely, knee joints absorb the impact power between the foot and the ground.

body length and weight	0.6m, 20kg
thigh length and weight	0.3m, 10kg
shin length and weight	0.3m, 10kg
total height & weight	1.2m, 60kg
stride S	0.5m
period of one step T	0.5s
running speed T/S	1m/s

Table 1. Specifications of robot and control parameters.

	hip	knee
peak angular velocity [rad/s]	14.5	11.6
peak torque [N.m]	48.2	117.6
peak power (positive) [W]	355	1265
peak power (negative) [W]	-699	-849
consumption power [W]	27.9	-5.37
total consumption power [W]	45.1	

Table 2. Actuator requirement.

Table 2 shows requirements for the actuators based on this result. It is found that very big power is required for knee joints. However, its total consumption power has small negative value. Therefore, the main work is done by the hip joints. Since the negative power is also big, for real robots, the introduction of the energy regeneration mechanism such as elastic actuators or combination of high back-drivable actuators and bidirectional power converters is effective to reduce the total consumption power.

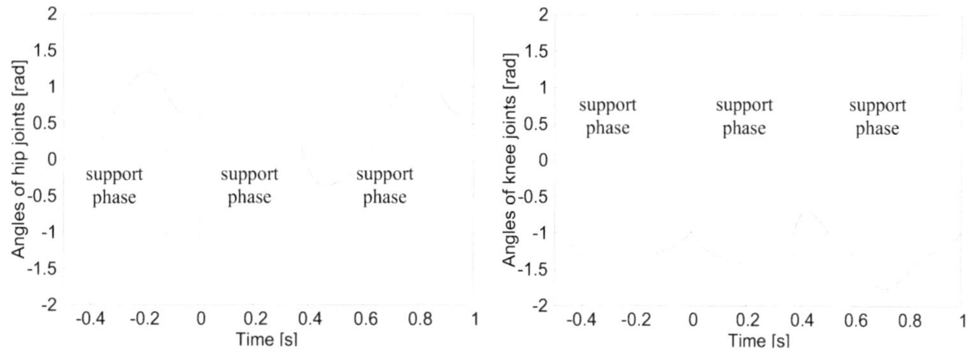

(a) hip joint (b) knee joint.

Fig. 3. Joint angles (solid line: right leg, dashed line: left leg).

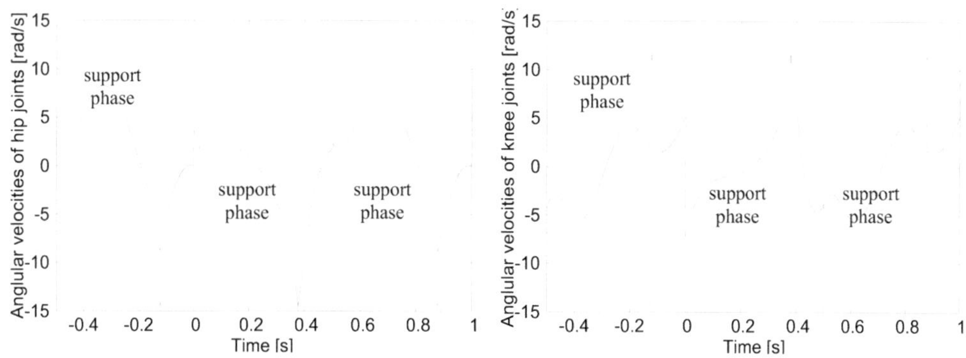

(a) hip joint (b) knee joint.

Fig. 4. Angular velocities of joints (solid line: right leg, dashed line: left leg).

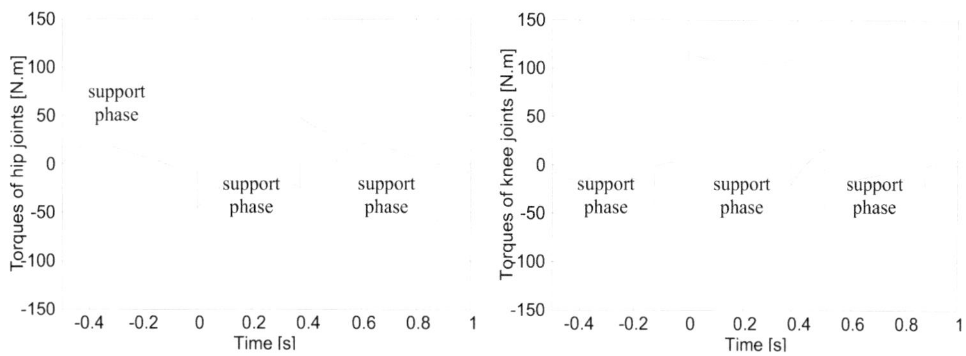

(a) hip joint (b) knee joint.

Fig. 5. Joint torques (solid line: right leg, dashed line: left leg).

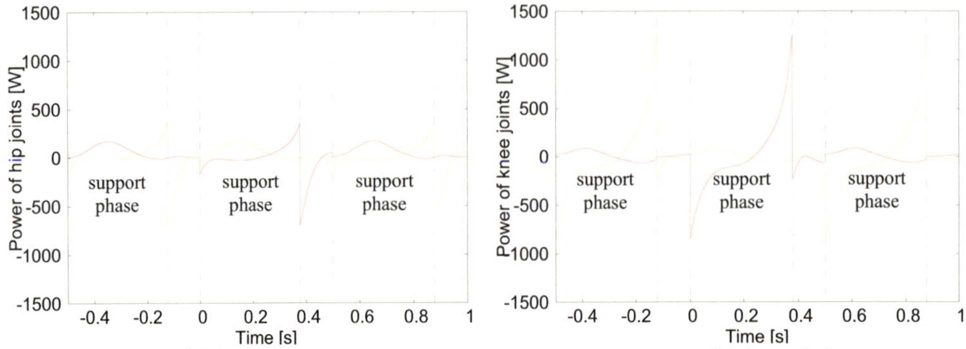

(a) hip joint (b) knee joint.

Fig. 6. Joint powers (solid line: right leg, dashed line: left leg).

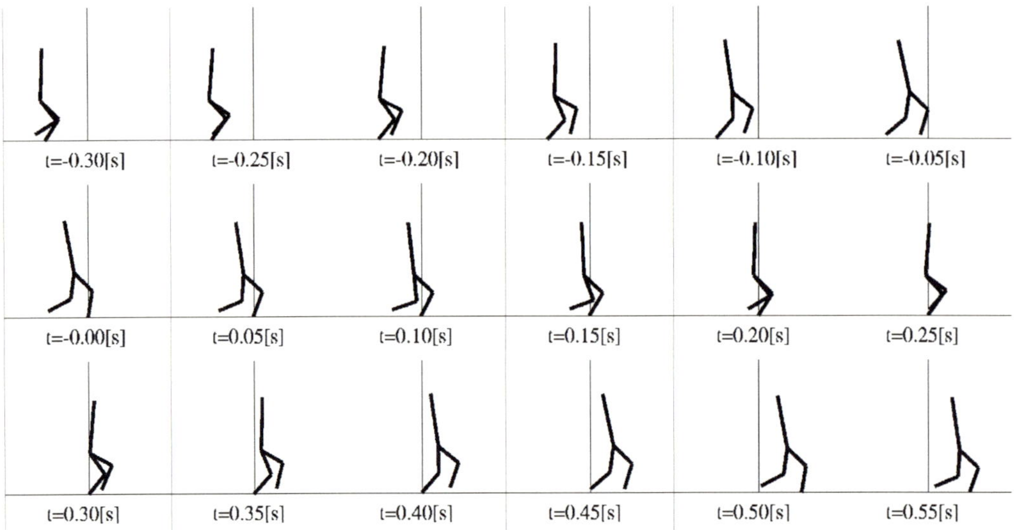

Fig. 7. Snapshots of running trajectory.

6. Conclusion

In this chapter, the method to generate a trajectory of a running motion with minimum energy consumption is proposed. It is useful to know the lower bound of the consumption energy when we design the bipedal robot and select actuators. The exact and general formulation of optimal control for biped robots based on numerical representation of motion equation is proposed to solve exactly the minimum energy consumption trajectories. Through the numerical study of a five link planar biped robot, it is found that big peak power and torque is required for the knee joints but its consumption power is small and the main work is done by the hip joints.

8. References

Fujimoto, Y. & Kawamura, A. (1995). Three Dimensional Digital Simulation and Autonomous Walking Control for Eight-axis Biped Robot, *Proceedings of IEEE*

International Conference on Robotics and Automation, pp. 2877-2884, 0-7803-1965-6, Nagoya, May 1995, IEEE, New York

Fujimoto, Y. & Kawamura, A. (1998). Simulation of an Autonomous Biped Walking Robot Including Environmental Force Interaction, *IEEE Robotics and Automation Magazine*, Vol. 5, No. 2, June 1998, pp. 33–42, 1070-9932

Goswami, A. (1999). Foot-Rotation Indicator (FRI) Point: A New Gait Planning Tool to Evaluate Postural Stability of Biped Robot, *Proceedings of IEEE International Conference on Robotics and Automation*, pp. 47-52, 0-7803-5180-0, Detroit, May 1999, IEEE, New York

Hirai, K.; Hirose, M.; Haikawa, Y. & Takenaka, T. (1998). TheDevelopment of Honda Humanoid Robot, *Proceedings of IEEE International Conference on Robotics and Automation*, pp. 1321-1326, 0-7803-4300-X, Leuven, May 1998, IEEE, New York

Hodgins, J. K. (1996). Three-Dimensional Human Running, *Proceedings of IEEE International Conference on Robotics and Automation*, pp. 3271-3277, 0-7803-2988-0, Minneapolis, April 1996, IEEE, New York

Kajita, S.; Nagasaki, T.; Yokoi, K.; Kaneko, K. & Tanie, K. (2002). Running Pattern Generation for a Humanoid Robot, *Proceedings of IEEE International Conference on Robotics and Automation*, pp. 2755-2761, 0-7803-7272-7, Washington DC, May 2002, IEEE, New York

Kajita, S.; Kanehiro, F.; Kaneko, K.; Fujiwara, K.; Harada, K.; Yokoi, K. & Hirukawa, H. (2003). Biped Walking Pattern Generation by using Preview Control of Zero-Moment Point, *Proceedings of IEEE International Conference on Robotics and Automation*, pp. 1620–1626, 0-7803-7736-2, Taipei, May 2003, IEEE, New York

Kaneko, K.; Kanehiro, F.; Kajita, S.; Hirukaka, H.; Kawasaki, T.; Hirata, M.; Akachi, K. & Isozumi, T. (2004). Humanoid Robot HRP-2, *Proceedings of IEEE International Conference on Robotics and Automation*, pp. 1083-1090, 0-7803-8232-3, New Orleans, April 2004, IEEE, New York

Loffler, K.; Gienger, M. & Pfeiffer, F. (2003). Sensor and Control Design of a Dynamically Stable Biped Robot, *Proceedings of IEEE International Conference on Robotics and Automation*, pp. 484-490, 0-7803-7736-2, Taipei, May 2003, IEEE, New York

Nagasaki, T.; Kajita, S.; Kaneko, K.; Yokoi, K. & Tanie, K. (2004). A Running Experiment of Humanoid Biped, *Proceedings of IEEE/RSJ International Conference on Intelligent Robots and Systems*, pp. 136-141, 0-7803-8463-6, Sendai, September 2004, IEEE, New York

Nishiwaki, K.; Kagami, S.; Kuffner J. J.; Inaba, M. & Inoue, H. (2003). Online Humanoid Walking Control System and a Moving Goal Tracking Experiment, *Proceedings of IEEE International Conference on Robotics and Automation*, pp. 911-916, 0-7803-7736-2, Taipei, May 2003, IEEE, New York

Raibert, M., H. (1986). *Legged Robots That Balance*, MIT Press, 0-262-18117-7, Cambridge

Roussel, L.; Canudas-de-Wit, C. & Goswami, A. (1998). Generation of Energy Optimal Complete Gait Cycles for Biped Robots, *Proceedings of IEEE International Conference on Robotics and Automation*, pp. 2036–2041, 0-7803-4300-X, Leuven, May 1998, IEEE, New York

Sugahara, Y.; Endo, T.; Lim, H. & Takanishi, A. (2003). Control and Experiments of a Multi-purpose Bipedal Locomotor with Parallel Mechanism, *Proceedings of IEEE*

International Conference on Robotics and Automation, pp. 4342-4347, 0-7803-7736-2, Taipei, May 2003, IEEE, New York

Vukobratovic, M.; Borovac, B. & Surdilovic, D. (2001). Zero-Moment Point - Proper Interpretation and New Applications, *Proceedings of International Conference on Humanoids Robots*, pp. 237-244, 4-9901025-0-9, Tokyo, November 2001, IEEE, New York

Yamaguchi, J.; Soga, E.; Inoue, S. & Takanishi, A. (1999). Development of a Bipedal Humanoid Robot—Control Method of Whole Body Cooperative Dynamic Biped Walking, *Proceedings of IEEE International Conference on Robotics and Automation*, pp. 368-374, 0-7803-5180-0, Detroit, May 1999, IEEE, New York

14

Real-time Vision Based Mouth Tracking and Parameterization for a Humanoid Imitation Task

Sabri Gurbuz[a,b], Naomi Inoue[a,b] and Gordon Cheng[c,d]
[a]*NICT Cognitive Information Science Laboratories, Kyoto, Japan*
[b]*ATR Cognitive Information Science Laboratories, Kyoto, Japan*
[c]*ATR-CNS Humanoid Robotics and Computational Neuroscience, Kyoto, Japan*
[d]*JST-ICORP Computational Brain Project, Kawaguchi, Saitama, Japan.*

1. Introduction

Robust real-time stereo facial feature tracking is one of the important research topics for a variety multimodal Human-Computer, and human robot Interface applications, including telepresence, face recognition, multimodal voice recognition, and perceptual user interfaces (Moghaddam et al., 1996; Moghaddam et al., 1998; Yehia et al., 1988). Since the motion of a person's facial features and the direction of the gaze is largely related to person's intention and attention, detection of such motions with their 3D real measurement values can be utilized as a natural way of communication for human robot interaction. For example, addition of visual speech information to robot's speech recognizer unit clearly meets at least two practicable criteria: It mimics human visual perception of speech recognition, and it may contain information that is not always present in the acoustic domain (Gurbuz et al., 2001). Another application example is enhancing the social interaction between humans and humanoid agents with robots learning human-like mouth movements from human trainers during speech (Gurbuz et al., 2004; Gurbuz et al., 2005).
The motivation of this research is to develop an algorithm to track the facial features using stereo vision system in real world conditions without using prior training data. We also demonstrate the stereo tracking system through a human to humanoid robot mouth mimicking task. Videre stereo vision hardware and SVS software system are used for implementing the algorithm.
This work is organized as follows. In section 2, related earlier works are described. Section 3 discusses face RIO localization. Section 4 presents the 2D lip contour tracking and its extention to 3D. Experimental results and discussions are presented in Section 5. Conclusion is given in Section 6. Finally, future extention is described in Section 7.

2. Related Work

Most previous approaches to facial feature tracking utilize skin tone based segmentation from single camera exclusively (Yang & Waibel, 1996; Wu et al., 1999; Hsu et al., 2002; Terrillon & Akamatsu, 1999; Chai & Ngan, 1999). However, color information is very sensitive to lighting conditions, and it is very difficult to adapt the skin tone model to a dynamically changing environment in real-time.

Kawato and Tetsutani (2004) proposed a mono camera based eye tracking technique based on six-segmented filter (SSR) which operates on integral images (Viola & Jones, 2001). Support vector machine (SVM) classification is employed to verify pattern between the eyes passed from the SSR filter. This approach is very attractive and fast. However, it doesn't benefit from stereo depth information. Also SVM verification fails when the eyebrows are covered by the hair or when the lighting conditions are significantly different than the SVM training conditions.

Newman et al., (2000) and Matsumoto et al., (19990) proposed to use 3D model fitting technique based on virtual spring for 3D facial feature tracking. In the 3D feature tracking stage each facial feature is assumed to have a small motion between the current frame and the previous one, and the 2D position in the previous frame is utilized to determine the search area in the current frame. The feature images stored in the 3D facial model are used as templates, and the right image is used as a search area firstly. Then this matched image in 2D feature tracking is used as a template in left image. Thus, as a result, 3D coordinates of each facial feature are calculated. This approach requires 3D facial model beforehand. For example, error in selection of a 3D facial model for the user may cause inaccurate tracking results.

Russakoff and Herman (2000) proposed to use stereo vision system for foreground and background segmentation for head tracking. Then, they fit a torso model to the segmented foreground data at each image frame. In this approach, background needs to be modeled first, and then the algorithm selects the largest connected component in the foreground for head tracking.

Although all approaches reported success under broad conditions, the prior knowledge about the user model or requirement of modeling the background creates disadvantage for many practical usages. The proposed work extends these efforts to a universal 3D facial feature tracking system by adopting the six-segmented filter approach Kawato and Tetsutani (2004) for locating the eye candidates in the left image and utilizing the stereo information for verification. The 3D measurements data from the stereo system allows verifying universal properties of the facial features such as convex curvature shape of the nose explicitly while such information is not present in the 2D image data directly. Thus, stereo tracking not only makes tracking possible in 3D, but also makes tracking more robust. We will also describe an online lip color learning algorithm which doesn't require prior knowledge about the user for mouth outer contour tracking in 3D.

3. Face ROI Localization

In general, face tracking approaches are either image based or direct feature search based methods. Image based (top-down) approaches utilize statistical models of skin color pixels to find the face region first, accordingly pre-stored face templates or feature search algorithms are used to match the candidate face regions as in Chiang et al. (2003). Feature based approaches use specialized filters directly such as templates or Gabor filter of different frequencies and orientations to locate the facial features.

Our work falls into the latter category. That is, first we find the eye candidate locations employing the integral image technique and the six segmented rectangular filter (SSR) method with SVM. Then, the similarities of all eye candidates are verified using the stereo system. The convex curvature shape of the nose and first and second derivatives around the nose tip are utilized for the verification. The nose tip is then utilized as a reference for the selection of the mouth ROI. At the current implementation, the system tracks the person closest to the camera only, but it can be easily extended to a multiple face tracking algorithm.

3.1 Eye Tracking

The pattern of the between the eyes are detected and tracked with updated pattern matching. To cope with scales of faces, various scale down images are considered for the detection, and an appropriate scale is selected according to the distance between the eyes (Kawato and Tetsutani, 2004). The algorithm calculates the intermediate representation of the input image called "Integral image", described in Viola & Jones (2001). Then, a SSR filter is used for fast filtering of bright-dark relations of the eye region in the image. Resulting face candidates around the eyes are further verified by perpendicular relationship of nose curvature shape as well as the physical distance between the eyes, and eye level and nose tip.

3.2 Nose Bridge and Nose Tip Tracking

The human nose has a convex curvature shape and the ridge of the nose from the eye level to the tip of the nose lies on a line as depicted in Fig. 1. Our system utilizes the information in the integral intensity profile of convex curvature shape. The peak of the profile of a segment that satisfies Eqn. 1 using the filter shown in Fig.2 is the convex hull point. A convolution filter with three segments traces the ridge with the center segment greater than the side segments, and the sum of the intensities in all three segments gives a maximum value on the convex hull point. Fig.2 shows an example filter with three segments that traces the convex hull pattern starting from the eye line. The criteria for finding the convex hull point on an integral intensity profile of a row segment is as follows,

$$S_1 < S_2 < S_3 \quad and \quad \arg\{\max_j(S_1 + 3S_2 + S_3)\}, \tag{1}$$

where S_i denotes the integral value of the intensity of a segment in the maximum filter shown in Fig. 2, and j is the center location of the filter in the current integral intensity profile. The filter is convolved with the integral intensity profile of every row segment. A row segment typically extends over 5 to 10 rows of the face ROI image, and a face ROI image typically contains 20 row segments. Integral intensity profiles of row segments are processed to find their hull points (see Fig.1 using Equation 1 until either the end of the face ROI is reached or until Eqn. 1 is no longer satisfied. For the refinement process, we found that the first derivative of the 3D surface data as well as the first derivative of the intensity at the nose tip are maximum, and the second derivative is zero at the nostril level (Gurbuz etal., 2004a).

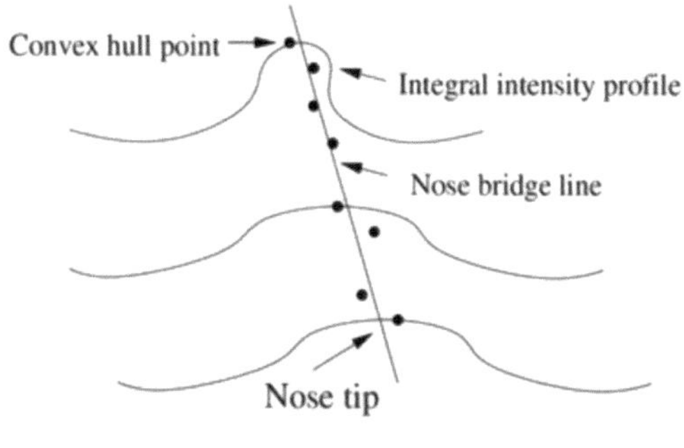

Fig. 1. Nose bridge line using its convex hull points from integral intensity projections.

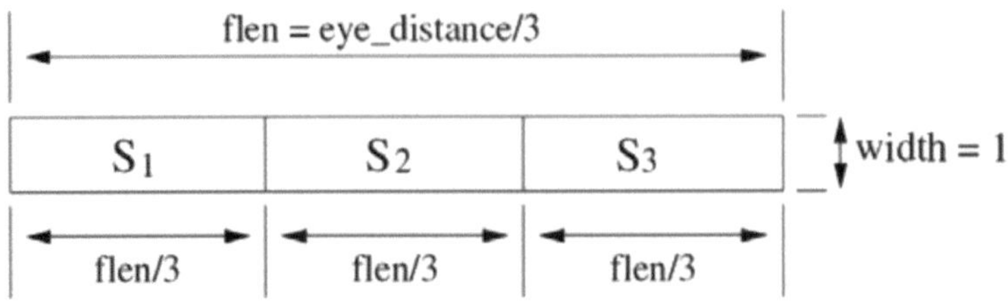

Fig. 2. A three-segment filter for nose bridge tracing.

4. Lip Tracking

The nose tip location is then utilized for the initial mouth ROI selection. Human mouth has dynamic behavior and even dynamic colors as well as presence or absence of tongue and teeth. Therefore, at this stage, maximum-likelihood estimation of class conditional densities for subsets of lip (w_1) and non-lip (w_2) classes are formed in real-time for the Bayes decision rule from the left camera image. That is, multivariate class conditional Gaussian density parameters are estimated for every image frame using an unsupervised maximum-likelihood estimation method.

4.1 Online Learning and Extraction of Lip and Non-lip Data Samples

In order to alleviate the influence of ambient lighting on the sample class data, chromatic color transformation is adopted for color representation (Chiang et al., 2003; Yang et al., 1998). It was pointed out (Yang et al., 1998) that human skin colors are less variant in the chromatic color space than the RGB color space. Although in general the skin-color distribution of each individual may be modeled by a multivariate normal distribution, the parameters of the distribution for different people and different lighting conditions are significantly different. Therefore, online learning and sample data extraction are important keys for handling different skin-tone colors and lighting changes. To solve these two issues, the authors proposed an adaptation approach to transform the previous developed color model into the new environment by combination of known parameters from the previous frames. This approach has two drawbacks in general. First, it requires an initial model to start, and second, it may fail in the case of a different user with completely different skin-tone color starts using the system.

We propose an online learning approach to extract sample data for lip and non-lip classes to estimate their distribution in real time. Chiang et al. (2003) in their work provides hints for this approach. They pointed out that lip colors are distributed at the lower range of green channel in the (r,g) plane. Fig. 4 shows an example distribution of lip and non-lip colors in the normalized (r,g) space.

Utilizing the nose tip, time dependent (r,g) spaces for lip and non-lip are estimated for every fame by allowing ε % (typically 10%) of the non-lip points stay within the lip (r,g) space as shown in Fig. 4. Then, using the obtained (r,g) space information in the initial classification, the pixels below the nostril line that falls within the lip space are considered as lip pixels, and the other pixels are considered as non-lip pixels in the sample data set extraction process, and RGB color values of pixels are stored as class attributes, respectively.

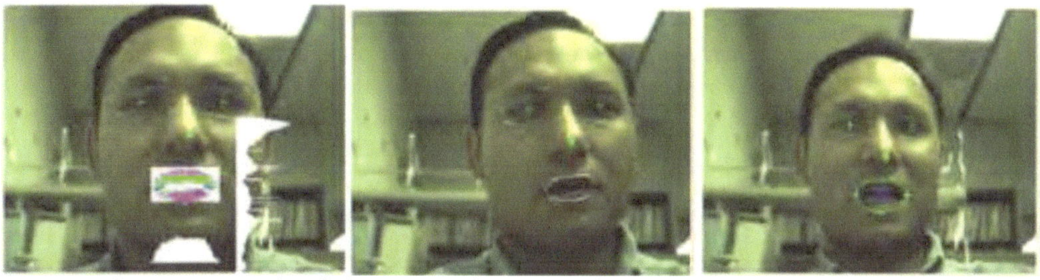

Fig. 3. Left image: result of the Bayes decision rule, its vertical projection (bottom) and integral projection of intensity plane between nose and chin (right). Middle image: estimated outer lip contour using the result of the Bayes rule. Right image: a parameterized outer lip contour.

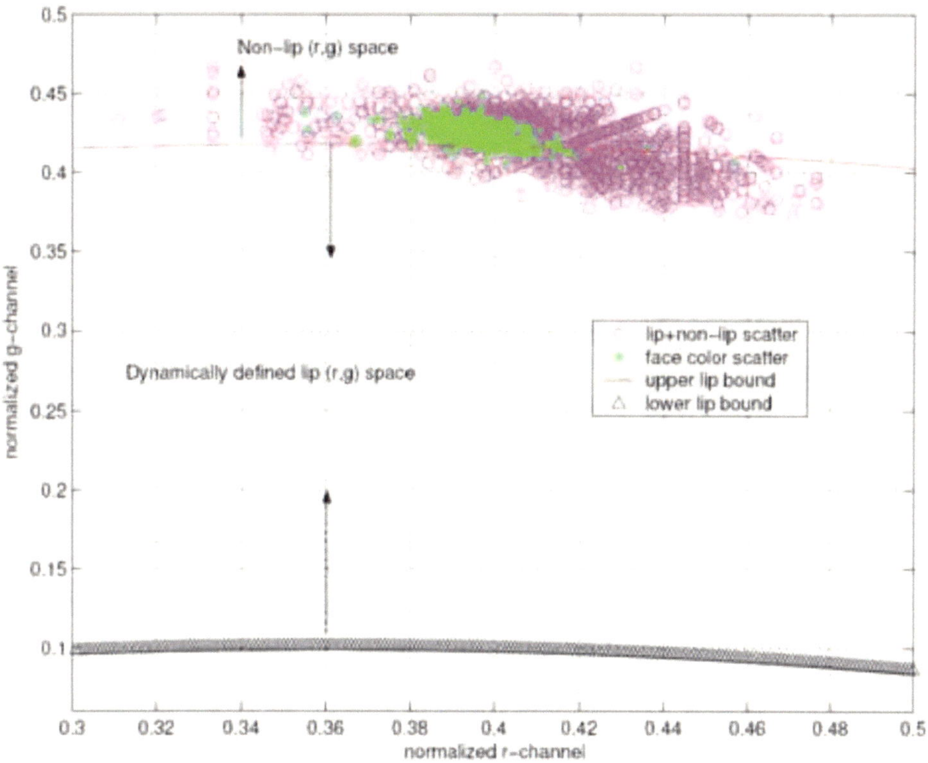

Fig. 4. Dynamically defined lip and non-lip (r,g) spaces.

In most cases, sample data contains high variance and it is preferable to separate the data into subsets according to its time dependent intensity average. Let avg_L and D_k be the intensity average and k[th] subset of the lip class, respectively. The subsets of the lip class are separated according to lip class' intensity average as

$$\begin{cases} \text{assign to } D_1 & \text{if } x_{intensity} < avg_L, \\ \text{assign to } D_2 & \text{if } avg_L/2 < x_{intensity} < 3avg_L/2, \\ \text{assign to } D_3 & \text{if } x_{intensity} > avg_L. \end{cases} \quad (2)$$

Using the same concept in Eqn. 2, we also separate the non-lip data samples into subsets according to intensity average of the non-lip class. Fig. 5 depicts simplified conditional density plots in 1D for the subsets of an assumed non-lip class.

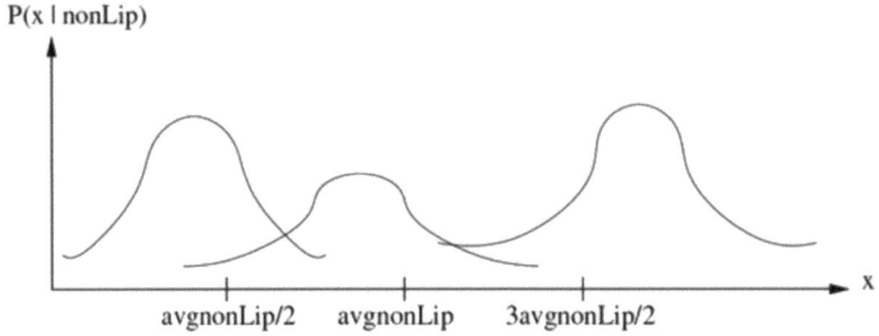

Fig. 5. Example class conditional densities for subsets of non-lip class.

4.2 Maximum-Likelihood Estimation of Class Conditional Multivariate Normal Densities

The mean vector and covariance matrix are the sufficient statistics to completely describe a distribution of the normal density. We utilize a maximum-likelihood estimation method for the estimation of a class conditional multivariate normal density described by

$$p(\mathbf{x}|i) = \frac{1}{\sqrt{(2\pi)^n \|\Sigma_i\|}} exp\{-\frac{1}{2}(\mathbf{x} - \mu_i)\Sigma_i^{-1}(\mathbf{x} - \mu_i)^T\}, \tag{3}$$

where i may be w_1, or w_2, or subset of a class. $\mu_i = E[x]$ is the mean value of the ith class. $\sum_i$ is the n x n (in this work, n is number of color attributes so n = 3) covariance matrix defined as

$$\Sigma_i = E[(\mathbf{x} - \mu_i)(\mathbf{x} - \mu_i)^T] \tag{4}$$

where $\|\cdot\|$ represents the determinant operation, and $E[\cdot]$ represents the expected value of a random variable. Unbiased estimates of the parameters μ_i and $\sum_i$ are estimated by using the sample mean and sample covariance matrix.

4.3 Bayes Decision Rule

Let **x** be an observation vector formed from RGB attributes of a pixel location in an image frame. Our goal is to design a Bayes classifier to determine whether **x** belongs to w_1 or w_2 in two class classification problem. The Bayes test using a posteriori probabilities may be written as follows:

$$p(w_1|\mathbf{x}) \mathop{\gtrless}_{w_2}^{w_1} p(w_2|\mathbf{x}), \tag{5}$$

where $p(w_i \mid x)$ is the a posteriori probability of w_i given x. Equation (5) shows that if the probability of w_1 given x is larger than the probability of w_2, then x is declared belonging to w_1, and vice versa. Since direct calculation of $p(w_i \mid x)$ is not practical, we can re-write the a posteriori probability of w_i using Bayes' Theorem in terms of a priori probability and the conditional density function $p(x \mid w_i)$, as

$$p(w_i|\mathbf{x}) = \frac{p(\mathbf{x}|w_i)p(w_i)}{p(\mathbf{x})} \tag{6}$$

where $p(x)$ is the density function and is positive constant for all classes. Then, re-arranging both sides, we obtain

$$L(\mathbf{x}) = \frac{p(\mathbf{x}|w_1)}{p(\mathbf{x}|w_2)} \underset{w_2}{\overset{w_1}{\gtrless}} \frac{p(w_2)}{p(w_1)} \tag{7}$$

where $L(x)$ is called the likelihood ratio, and $p(w_1)/p(w_2)$ is called the threshold value} of the likelihood ratio for the decision. Because of the exponential form of the densities involved in Equation (7), it is preferable to work with the monotonic discriminant functions obtained by taking the logarithm as follows.

$$q_{w_i}(\mathbf{x}) = ln(p(\mathbf{x}|w_i)p(w_i)), \tag{8}$$

thus, by re-arranging Equation (8), we get

$$q_{w_i}(\mathbf{x}) = -\frac{1}{2}(\mathbf{x} - \mu_i)\Sigma_i^{-1}(\mathbf{x} - \mu_i)^T + c_i \tag{9}$$

Where

$$c_i = \ln p(w_i) - (1/2)\ln 2\pi - (1/2)\|\Sigma_i\|$$

is a constant for this image frame. In general, Equation (9) has only nonlinear quadratic form and a summation, and using this equation, the Bayes rule can be implemented for real-time lip tracking as follows.

$$q^*_{w_1}(\mathbf{x}) \underset{w_2}{\overset{w_1}{\gtrless}} q^*_{w_2}(\mathbf{x}), \tag{10}$$

Where

$$q^*_i(\mathbf{x}) = \max\{q_i^{(1)}(\mathbf{x}), q_i^{(2)}(\mathbf{x}), q_i^{(3)}(\mathbf{x})\}$$

for $i = \{w_1, w_2\}$ and referring to Fig. 5. Threshold value of the likelihood ratio as shown in Eqn. (7) is based on a priori class probabilities. In our implementation, equally likely a priori class probabilities are assumed.

4.4 Mouth Outer Contour Parameterization in 2D

After mouth tracking algorithm locates the mouth region, outer lip contours of the speaker's lips in left camera image are detected (see Fig. 3). Then, the outer contour as a whole is parameterized by a generalized ellipse shape which is obtained using the estimated outer contour data. A parametric contour is found that corresponds to the general quadratic equation $a_1x^2 + a_2xy + a_3y^2 + a_4x + a_5y + a_6 = 0$, where a_is are constants, and a_1 and a_3 are non-zero. Let us denote the 2D positions over the traced outer lip contour as

$$\begin{bmatrix} x_1 & x_2 & x_3 & \cdots & x_N \\ y_1 & y_2 & y_3 & \cdots & y_N \end{bmatrix}. \tag{10}$$

The basic form used in the elliptical parameter estimation in matrix notation is $Ma = 0$ where $a = [a_1 \cdots a_6]^T$ The dimensionality of *M* is the number of points, *N*, in the segment multiplied by 6 (that is, *N x* 6). Each row of *M* corresponds to one point in the segment. The parameters of each contour are then solved using the least-squares method to find a_i s, where $i = 1,2,\cdots,6$.

Using the estimated parameters, parametric lip contour data can be re-generated for each image frame. Five points are sufficient to represent a general elliptical shape, leading to a significant data reduction and representation.

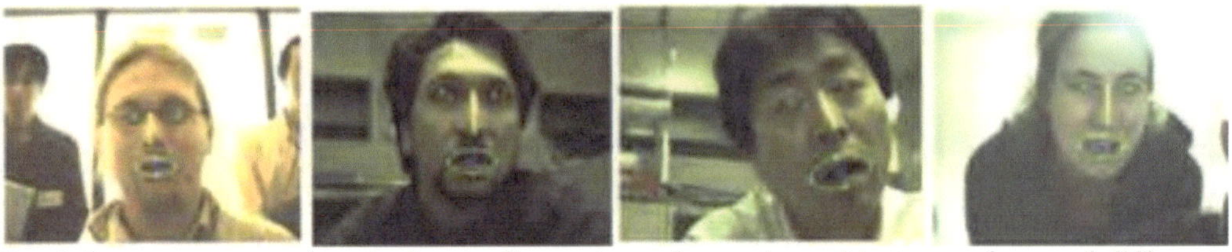

Fig. 6. Screen capture of tracked outer lip contours for various skin tone colors and different lighting conditions.

4.5 Estimation of 3D Mouth Outer Contour

Once the outer lip contour points of the speaker's lips in left camera image are found then their stereo disparity values from the right image can be calculated utilizing the previously found horopter information. Fig. 7 shows stereo and disparity images. Knowing a pixel location (x,y) in the left camera image and its disparity in the right camera image, we can calculate its 3D (X,Y,Z) coordinates with respect to the camera coordinate system as shown in Fig. 8.

Fig. 7. Screen capture of the left and right camera images, and their disparity map around the face region of interest.

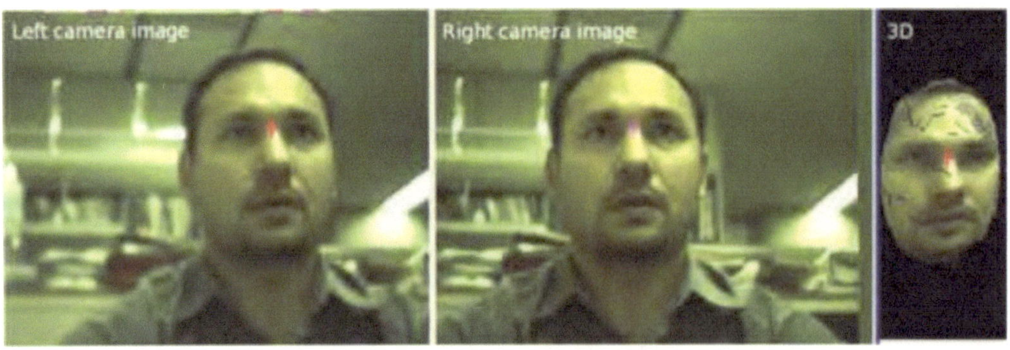

Fig. 8. Screen capture of the left and right camera images, and OpenGL plot of texture mapped 3D face reconstructed by the stereo algorithm.

5. Experimental Results and Discussion

In this paper, our work focused on a real-time stereo facial feature tracking algorithm. Intensity information is used for initial eye candidates in the left image. Relationship of the eyes and the nose are verified using 3D data. Then, the nose tip is utilized as a reference point for mouth ROI. RGB color attributes of lip pixels are used for the lip tracking.

The proposed stereo facial feature tracking algorithm has been tested on live data from various users without using any special markers or paintings. Fig. 6 shows tracked lip contour results for various users under various lighting conditions. The stereo facial feature tracking algorithm which utilizes Videre stereo hardware works around 20 frames per second (un-optimized) on a 2 GHz notebook PC under Windows platform. We also demonstrate the developed stereo tracking system through a human to humanoid robot mouth mimicking task. The humanoid head (Infanoid) is controlled via serial communication connected to a PC. Commands to the mouth is send every 50 ms (20 hz), at the same rate as the processing of facial features tracking - thus, allowing real-time interaction. The communication between the vision system and the control PC is via a 1Gbit Ethernet network. The opening and closing of the person's mouth is directly mapped to the mouth of the humanoid's mouth, with a simple geometric transform. The Infanoid robot head is shown in Fig. 10 which was developed by Kozima (2000).

6. Conclusions

A new method for stereo facial feature tracking of individuals in real world conditions is described. Specifically, stereo face verification and an unsupervised online parameter estimation algorithm for the Bayesian rule is proposed. That is, a speaker's lip colors are learned from the current image frame using the nose tip as a reference point. Vertical and horizontal integral projections are utilized to guide the algorithm in picking out the correct lip contour. In the final stage, estimated outer contour data for every image frame of the left camera is parameterized as a generalized ellipse. Then, utilizing the contour pixel locations in the left camera image and their disparity of the right camera image, we calculate their 3D (X,Y,Z) coordinates with respect to the camera coordinate system. Future work for vision includes extraction of 3D coordinates of other facial points such as eyebrows, chin and cheek, and further extending the work to the multiple face tracking algorithm.

7. Humanoid Robotics Future Extension

Future extensions of this work include developing a machine learning method for smooth mouth movement behavior to enable humanoids to learn visual articulatory motor tapes for any language with minimal human intervention. Fig. 9 shows the flow diagram of the system. Such a system should extract and store motor tapes by analyzing a human speaker's audio-visual speech data recorded from predetermined phonetically-balanced spoken text to create a mapping between the sound units and the time series of the mouth movement parameters representing the mouth movement trajectories. These motor tapes can then be executed with the same time index of the audio, yielding biologically valid mouth movements during audio synthesis.

We call the system the *text-to-visual speech (TTVS) synthesis* system. It can be combined with a concatinative speech synthesis system, such as Festival (Black & Taylor, 1997; Sethy & Narayanan, 2002; Chang et al., 2000) to create a text-to-audiovisual speech synthesis system for humanoids.

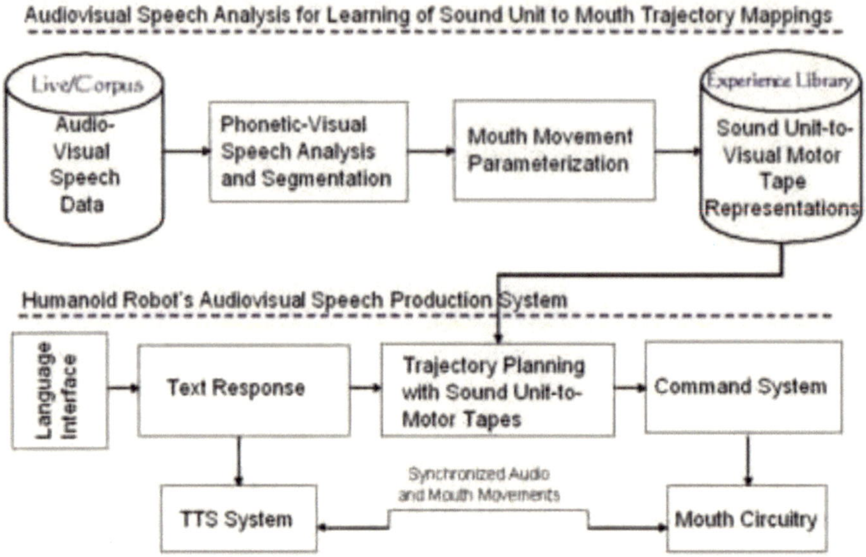

Fig. 9. Future extension to a TTS based speech articulation system for Humanoids.

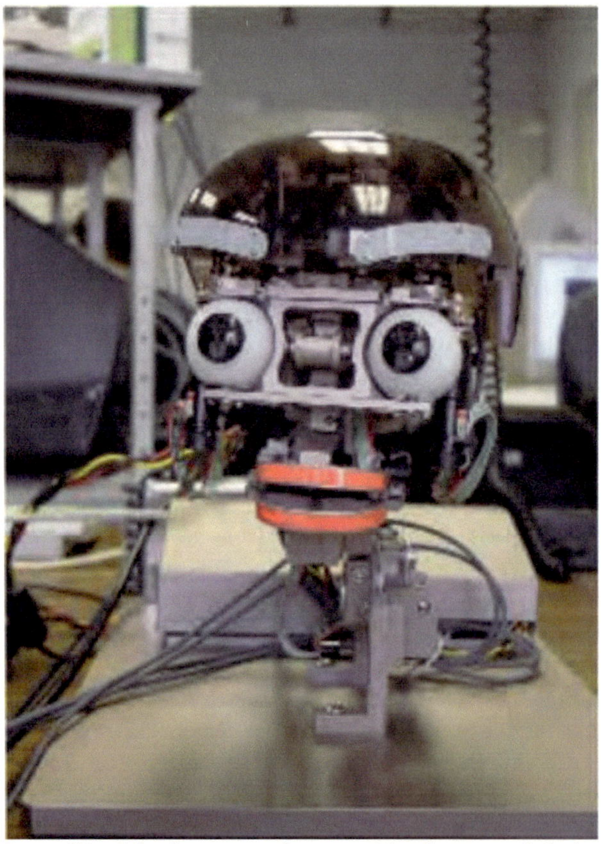

Fig. 10. Infanoid robot utilized for human to humanoid robot mouth imitation task.

A concatenative synthesis system creates indexed waveforms by concatenating parts (diphones) of natural speech recorded from humans. Using the same concatenative synthesis concept, the proposed TTVS system can concatenate corresponding mouth movement primitives. Thus, the system is capable of generating sequences of entirely novel visual speech parameters that represent the mouth movement trajectories of the spoken text. A humanoid agent equipped with TTS and TTVS systems can produce novel utterances, and so is not limited to those recorded in the original audio-visual speech corpus. With these capabilities, the humanoid robot can robustly emulate a person's audiovisual speech. A detailed explanation of this extension is described in Gurbuz et al. (2004b). Also, we will extend the work to include imitation of other facial movements, as the vision system expands to stereo and track additional features such as eyebrows, and perform perception studies to ascertain the effect of more accurate speech and face movement cues on naturalness and perceptibility in humanoids.

8. Acknowledgment

This research was conducted as part of "Research on Human Communication" with funding from the National Institute of Information and Communications Technology (NiCT), Japan. Thanks are due to Dr. Hideki Kozima for the use of his Infanoid robot and Shinjiro Kawato for the original eye tracking work extended in this paper.

9. References

Black, A., & Taylor, P. (1997). *The Festival Speech Synthesis System*. University of Edinburgh.

Chai, D. & Ngan, K. N., (1999). Face segmentation using skin-color map in videophone applications. IEEE Trans. on Circuits and Systems for Video Technology 9 (4), 551{564.

Chang, S., Shari, L. & Greenberg, S. (2000). Automatic phonetic transcription of spontaneous speech (American English), *International Conference on Spoken Language Processing, Beijing, China.*

Chiang, C. C., Tai, W. K., Yang, M. T., Huang, Y. T. & Huang, C. J., (2003). A novel method for detecting lips, eyes and faces in real-time. Real-Time Imaging 9, 277-287.

Gurbuz, S., Shimizu, T. & Cheng, G. (2005). Real-time stereo facial feature tracking: Mimicking human mouth movement on a humanoid robot head. IEEE-RAS/RSJ International Conference on Humanoid Robots (Humanoids 2005).

Gurbuz, S., Kinoshita, K., & Kawato, S., (2004a). Real-time human nose bridge tracking in presence of geometry and illumination changes. Second International Workshop on Man-Machine Symbiotic Systems, Kyoto, Japan.

Gurbuz, S., Kinoshita, K., Riley, M. & Yano, S., (2004b). Biologically valid jaw movements for talking humanoid robots. IEEE-RAS/RSJ International Conference on Humanoid Robots (Humanoids 2004), Los Angeles, CA, USA.

Gurbuz, S., Tufekci, Z., Patterson, E. & Gowdy, J., (2001). Application of affine invariant Fourier descriptors to lipreading for audio-visual speech recognition. Proceedings of ICASSP.

Hsu, R. L., Abdel-Mottaleb, M. & Jain, A. K., (2002). Face detection in color images. IEEE Trans. on PAMI 24 (5), 696-706.

Kawato, S. & Tetsutani, N., (2004). Scale adaptive face detection and tracking in real time with SSR filter and support vector machine. Proc. of ACCV, vol. 1.

Kozima, H., (2000). NICT infanoid: An experimental tool for developmental psycho-robotics. International Workshop on Developmental Study, Tokyo.

Matsumoto, Y. & Zelinsky, A., (1999). Real-time face tracking system for human robot interaction. Proceedings of 1999 IEEE International Conference on Systems, Man, and Cybernetics (SMC'99). pp. 830-835.

Moghaddam, B., Nastar, C. & Pentland, A., (1996). Bayesian face recognition using deformable intensity surfaces. In: IEEE Conf. on Computer Vision and Pattern Recognition.

Moghaddam, B., Wahid, W. & Pentland, A., (1998). Beyond eigenfaces: Probabilistic matching for face recognition. In: International Conference on Automatic Face and Gesture Recognition.

Newman, R., Matsumoto, Y., Rougeaux, S. & Zelinsky, A., (2000). Real-time stereo tracking for head pose and gaze estimation. Proceedings. Fourth IEEE International Conference on Automatic Face and Gesture Recognition.

NICT-Japan, Infanoid project. http://www2.nict.go.jp/jt/a134/infanoid/robot-eng.html.

Russako_, D. & Herman, M., (2000). Head tracking using stereo. Fifth IEEE Workshop on Applications of Computer Vision.

Sethy, A. & Narayanan, S. (2002). Rifined speech segmentation for concatanative speech synthesis, *International Conference on Spoken Language Processing, Denver, Colorado.*

Terrillon, J. C. & Akamatsu, S., (1999). Comparative performance of different chrominance spaces for color segmentation and detection of human faces in complex scene images. Proc. 12th Conf. on Vision Interface. pp. 180-187.

Viola, P. & Jones, M., (2001). Robust real-time object detection. Second International Workshop on Statistical and Computational Theories of Vision-Modeling, Learning, Computing, and Sampling, Vancouver, Canada,.

Wu, H., Cheng, Q. & Yachida, M., (1999). Face detection from color images using a fuzzy pattern matching method. IEEE Trans. on PAMI 21 (6), 557-563.

Yang, J., Stiefelhagen, R., Meier, U. & Waibel, A., (1998). Visual tracking for multimodal human computer interaction. Proceedings of the SIGCHI conference on Human factors in computing systems.

Yang, J. & Waibel, A., (1996). A real-time face tracker. In: Proc. 3rd IEEE Workshop on Application of Computer Vision. pp. 142-147.

Yehia, H., Rubin, P.E., & Vatikiotis-Bateson, E. (1998). "Quantitative association of vocal tract and facial behavior," *Speech Communication*, no. 26, pp. 23–44.

15

Clustered Regression Control of a Biped Robot Model

Olli Haavisto and Heikki Hyötyniemi
Helsinki University of Technology, Control Engineering Laboratory
Finland

1. Introduction

Controlling of a biped walking mechanism is a very challenging multivariable problem, the system being highly nonlinear, high-dimensional, and inherently unstable. In almost any realistic case the exact dynamic equations of a walking robot are too complicated to be utilized in the control solutions, or even impossible to write in closed form.

Data-based modelling methods try to form a model of the system using only observation data collected from the system inputs and outputs. Traditionally, the data oriented methods are used to construct a *global* black-box model of the system explaining the whole sample data within one single function structure. Feedforward neural networks, as presented in (Haykin, 1999), for example, typically map the input to the output with very complicated and multilayered grid of neurons and the analysis of the whole net is hardly possible. Local learning methods (Atkeson *et al.*, 1997), on the other hand, offer a more structured approach to the problem. The overall mapping is formed using several *local* models, which have a simple internal structure but are alone valid only in small regions of the input-output space. Typically, the local models used are *linear*, which ensures the *scalability* of the model structure: Simple systems can be modelled, as well as more complex ones, using the same structure, only the number of the local models varies.

In robotics, local modelling has been used quite successfully to form inverse dynamics or kinematic mappings that have then been applied as a part of the actual controller (Vijayakumar *et al.*, 2002). However, when trying to cover the whole high-dimensional input-output space, the number of local models increases rapidly. Additionally, external reference signals are needed for the controller to get the system function as desired.

To evaluate the assumption of simple local models, a feedback structure based on linear local models, *clustered regression*, is used here to implement the gait of a biped walking robot model. The local models are based on *principal component analysis* (see Basilevsky, 1994) of the local data. Instead of mapping the complete inverse dynamics of the biped, only one gait trajectory is considered here. This means that the walking behaviour is stored in the model structure. Given the current state of the system, the model output estimate is directly used as the next control signal value and no additional control solutions or reference signals are needed. The walking cycle can become *automated*, so that no higher-level control is needed.

This text summarizes and extends the presentation in (Haavisto & Hyötyniemi, 2005).

2. Biped model

2.1 Structure of the mechanism

The biped model used in the simulations is a two-dimensional, five-link system which has a torso and two identical legs with knees. To ensure the possibility of mathematical simulation, the biped model was chosen to be quite simple. It can, however, describe the walking motion rather well and is therefore applied in slightly different forms by many researchers (see e.g. Chevallereau *et al.*, 2003; Hardt *et al.*, 1999; Juang, 2000).
Fig. 1 shows the biped and the coordinates used. Each link of the robot is assumed to be a rigid body with uniformly distributed mass so that the centre of mass is located in the middle of the link. The interaction with the walking surface is modelled by adding external forces F to the leg tips. As the leg touches the ground, the corresponding forces are switched on to support the leg. The actual control of the biped gait is carried out using the joint moments M, which are applied to both thighs and knees.

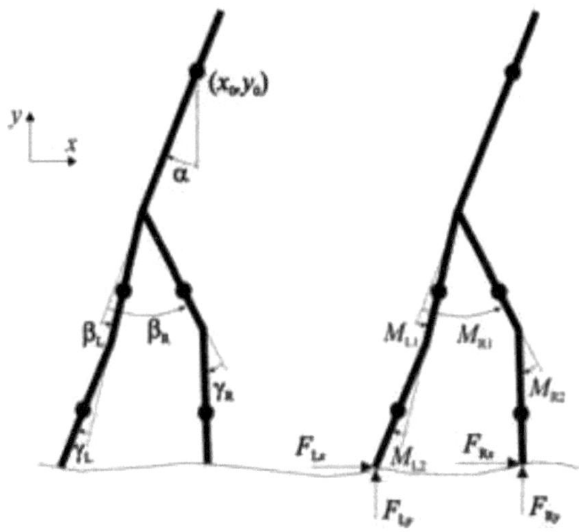

Fig. 1. The coordinates used (left) and the external forces (right).

As such, the system has seven degrees of freedom, and seven coordinates must be chosen to describe the configuration of the biped. The coordinate vector now is

$$q = \begin{pmatrix} x_0 & y_0 & \alpha & \beta_L & \beta_R & \gamma_L & \gamma_R \end{pmatrix}^T . \tag{1}$$

Here the subindices L and R refer to the "left" and "right" leg of the biped, respectively. The dynamic equations of the system can be derived using *Lagrangian mechanics* (Wells, 1967) and the result has the following form:

$$A(q)\,\ddot{q} = b(q, \dot{q}, M, F) , \tag{2}$$

where the "dotted variables" denote time derivatives of the coordinate vector. The exact formulas of the *inertia matrix* $A(q)$ and the right hand side vector are quite complex and therefore omitted here (see Haavisto & Hyötyniemi, 2004). The minimum dimension n of the state vector representing the system state is 14 (the 7 coordinates and their time derivatives); however, because of the efficient compression carried out by the proposed data-based modelling technique, the dimension of the "extended state" could be higher, including measurements that are not necessarily of any use in modelling (redundancy among the measurements being utilized for elimination of noise).

2.2 Simulator

To simulate the biped walking on a surface, a Matlab/Simulink (MathWorks, 1994-2007) block was developed. It handles internally the calculation of the support forces F using separate PD controllers for each force component: The force opposing the movement is the stronger the farther the foot tip has gone beneath the ground surface (P term), and the faster the foot tip is moving downward (D term). By adjusting these controller parameters, different kinds of ground surface properties can be simulated (very hard surfaces, however, resulting in high force peaks, so that short simulation step sizes have to be applied). Also the knee angles are automatically limited to a range defined by the user. To obtain this, the simulation block adds an additional moment to the knee joints if the angle is exceeding or going under the given limits. This limitation can be used for example to prevent the knee turning to the "wrong" direction, that is, to keep the angles γ_L and γ_R positive.

The input of the simulation block is the four dimensional moment vector

$$M = \begin{pmatrix} M_{\mathrm{L1}} & M_{\mathrm{R1}} & M_{\mathrm{L2}} & M_{\mathrm{R2}} \end{pmatrix}^T, \tag{3}$$

and the output the 14-dimensional state of the system augmented with leg tip touch sensor values s_L and s_R is

$$\begin{pmatrix} q^T & \dot{q}^T & s_{\mathrm{L}} & s_{\mathrm{R}} \end{pmatrix}^T. \tag{4}$$

If the leg is touching the ground, the sensor value equals to 1, otherwise 0.

The biped dynamics, ground support forces and knee angle limiters are simulated in continuous time, but the input and output of the block are discretized using zero order hold. This allows the use of discrete-time control methods. A more detailed description of the used biped model and simulation tool is presented in the documentation (Haavisto & Hyötyniemi, 2004).

2.3 Data collection

In general, the gait movement of a biped walker can be divided into two phases according to the number of legs touching the ground. In the *double support phase* (DSP) both legs are in contact with the ground and the weight is moving from the rear leg to the front leg. In the *single support phase* (SSP) only the *stance leg* is touching the ground, while the *swing leg* is swinging forward. Because the biped is symmetrical regarding the left and right leg, the whole gait can be described with one DSP and one SSP. In every other step the stance and swing leg signals are switched to model the left and right leg swing in turns. This means that it is necessary to model only one DSP and one SSP with, say, left leg acting as the rearmost or swing leg and right leg as the foremost or stance leg. Accordingly, the data collected during this work were transformed to cover only these DSP and SSP cases.

In order to collect sample data from the gait of the walking biped, a rough PD control scheme was developed. It was based on predetermined reference signals and four separate and discrete PD controllers, which produced the moment inputs for the biped. Fig. 2 shows the closed loop simulation model of the controlled biped. The *PD controller* block includes the separate PD controllers, whereas the *Biped model* block simulates the actual walking system and the surface interaction.

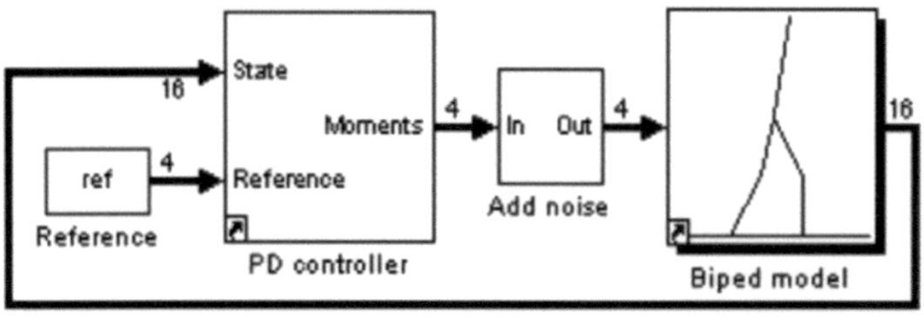

Fig. 2. The sample data were produced by simulating the PD controlled biped

The PD controller was able to control the biped so that a cyclic gait movement was attained. To reach more varying data, independent white noise was added to each moment signal component. The input and output data of the biped system were collected of a 40 seconds walk, during which the biped took over 40 steps. The sampling interval was equal to the discretization period $h = 0.005$ sec.

The data were mean-zeroed and scaled to unit variance in order to make the different components equally significant. Also the x_0 coordinate of the state vector was omitted because it is monotonically increasing during the walk instead of being cyclic. As a result, the modelling data had 8000 samples of the 15 dimensional (reduced) state vector and the corresponding four dimensional moment vectors. The points were located around a nominal trajectory in the state and moment space.

In the following, the *state vector* at time kh is denoted by $u(k)$ and the *moment vector* by $y(k)$ since these are the *input* and the *output* of the *controller*. Additionally, the state vector without the touch sensor values is denoted by $\tilde{u}(k)$.

3. Clustered Regression

The data-based model structure, *clustered regression*, that is applied here is formed purely by the statistical properties of the sample data. The main idea is to divide the data into *clusters* and set an *operating point* in the centre of each cluster. Every operating point, additionally, has its own local linear regression model which determines the local input-output mapping in the cluster region. A scheme of the model structure is presented in Fig. 3.

Fig. 3. The local models cover the whole data along the trajectory

3.1 Local model structure

Each local model calculates a linear *principal component regression* (PCR) estimate of the given regression input. Principal components show the orthogonal directions of the highest variances in the input data. It is assumed here that variance is carrying information and the most important principal components are therefore relevant in the data, whereas the smaller ones can be omitted as noise. The idea of the PCR is first to map the input data to the lower dimensional principal component subspace and then use multivariable linear regression to get the output estimate.

Let us denote the input signal for the local model p at the time instant kh as $u^p(k)$, and the corresponding output as $y^p(k)$. When the input data are mapped onto the principal components, one gets the latent variable signal $x^p(k)$. To simplify the regression structure the latent variable data are locally scaled to unit variance before the regression mapping.

Now the whole regression structure of the local model p can be stored in the following statistics, which are calculated using the data in the corresponding cluster:

C_u^p Expectation value of the input vector u^p.

C_y^p Expectation value of the output vector y^p.

P_{xx}^p Inverse of the latent variable data x^p covariance matrix.

$R_{x\tilde{u}}^p$ Cross-covariance of the latent and reduced input data.

R_{yx}^p Cross-covariance of the ouput and latent variable data.

R_{uu}^p Covariance of the input data.

Note that in the following the statistics are calculated using the collected data, which gives merely the *estimates* of the real quantities. It is assumed, however, that the estimates are accurate enough and no difference is made between them and the real values.

The regression calculation itself is based on a Hebbian and anti-Hebbian learning structure (see 6.1). Assuming that the statistics correspond to the real properties of the data, the output estimate of the model p given an arbitrary input vector can be expressed in mathematical terms as

$$\hat{y}^p(k) = R_{yx}^p \left(P_{xx}^p\right)^2 R_{x\tilde{u}}^p \left(\tilde{u}(k) - C_{\tilde{u}}^p\right) + C_y^p . \tag{5}$$

Here $C_{\tilde{u}}^p$ is the expectation value of the reduced state vector $\tilde{u}^p$. In (5) the input is first transformed to the operating point p centred coordinates by removing the input mean value; then the estimate is calculated and the result is shifted back to the original coordinates in the output space by adding the output mean vector.

A cost value for the estimate made in the unit p should be evaluated to measure the error of the result. The cost can depend for example on the distance between the cluster centre and the estimation point:

$$J^p\left(u(k)\right) = \frac{1}{2}\left(u(k) - C_u^p\right)^T H^p \left(u(k) - C_u^p\right), \tag{6}$$

where H^p is a constant weighting matrix.

3.2 Combination of the local models

The overall estimate of the clustered regression structure is calculated as a weighted average of all the local models. Assuming that the number of the operating points is N, one has

$$\hat{y}(k) = \frac{\sum_{p=1}^{N} K^p(k)\hat{y}^p(k)}{\sum_{p=1}^{N} K^p(k)} . \tag{7}$$

The weights should naturally be chosen so that the currently best matching local models affect the final results the most, whereas the further ones are practically neglected. Let us choose

$$H^p = \frac{1}{\sigma}\left(R_{uu}^p\right)^{-1}, \tag{8}$$

that is, the weighting matrix for each local model cost calculation equals the scaled inverse of the input data covariance matrix, σ being the scaling factor; assuming that the data in each local model are normally distributed, the *maximum likelihood* estimate for the combined output value is obtained when the weights in (7) are chosen as

$$K^p(k) = \frac{1}{\sqrt{(2\pi)^n \left|\left(H^p\right)^{-1}\right|}} \exp\left(-J^p\left(u(k)\right)\right), \tag{9}$$

and $\sigma = 1$. The simulations, however, showed that a more robust walking behaviour is reached with larger scaling parameter values. This increases the effect of averaging in the combination of the local model estimates.

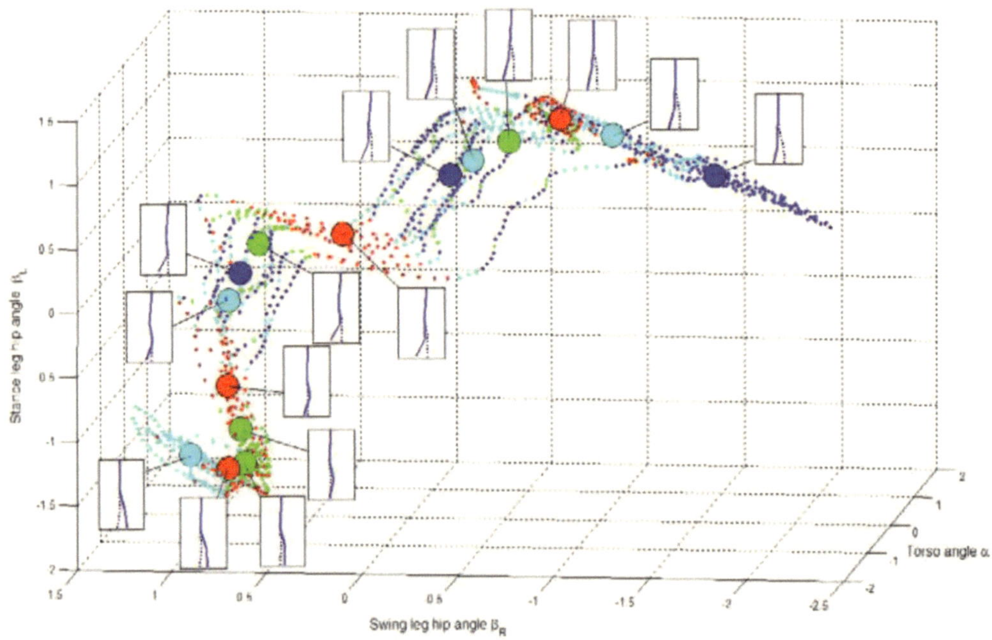

Fig. 4. Clustered teaching data and the operating points

4. Clustered Regression Control

4.1 Teaching the model

The clustered regression model was formed using the sample data collected from the PD controlled gait. The model input was the state vector $u(k)$ and output the corresponding control signal $y(k)$. The data were first divided into N = 15 clusters located along the nominal trajectory in the state and output space. Based on the data belonging to each cluster the estimates of the statistics listed in 3.1 were calculated with eight principal component directions in each model.

Fig. 4 shows part of the clustered teaching data projected to three input variables. The operating point centres are shown as large circles with the same shade as the data belonging to the corresponding clusters. Also the state of the system in the operating point centres is drawn. Clearly the start of a new DSP is located in the right part of the figure. Following the nominal trajectory the state of the biped then changes to a SSP after some four operating points. Finally, the step ends just before the beginning of a new DSP in the lower left corner of the figure, where the state of the system is almost identical with the original state, only the legs are transposed.

4.2 Biped behavior

To test the attained model, a Simulink block was developed to realize the estimation of the control related to the measured current state of the system. The block could now be used to control the biped instead of the PD controller module applied in the data collection. It appeared that the taught clustered regression model was able to keep the biped walking and produced a very similar gait as the original PD controller. Fig. 5 shows two steps of the both gaits.

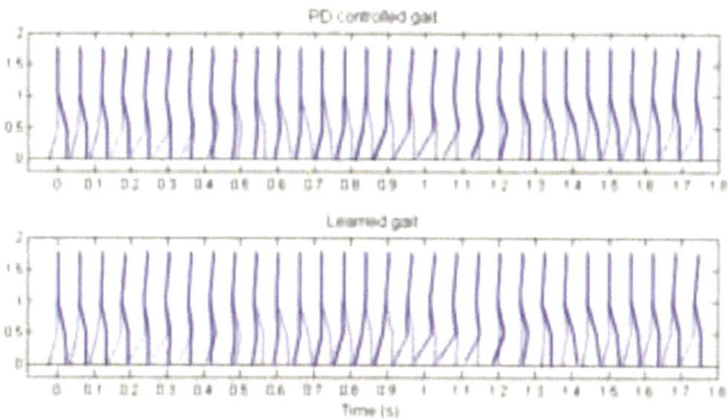

Fig. 5. The learned gait was qualitatively quite similar to the original one

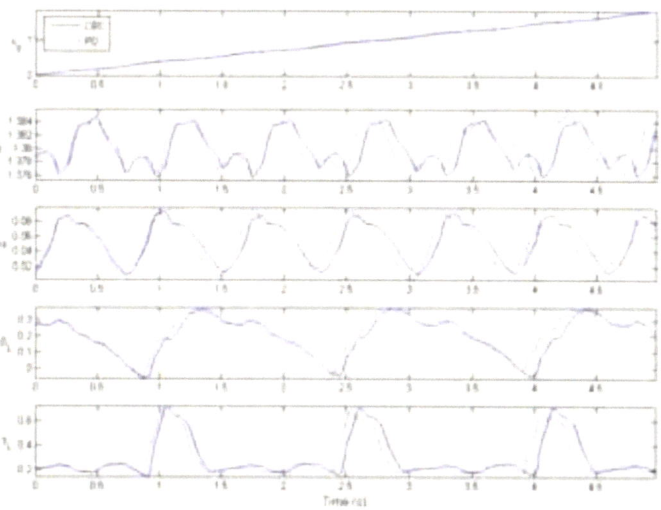

Fig. 6. The clustered regression controlled system (CRC) is functioning a bit slower than the PD controlled one.

The biggest difference between the two gaits is that the clustered regression controlled one is proceeding a little slower. When comparing the coordinate value changes, the lag of the clustered regression controlled system is clearly seen (Fig. 6). The variations in the PD controlled signals are due to the added noise in the teaching data generation. However, the clustered regression controller can reproduce the original behaviour very accurately.

5. Optimization

5.1 Dynamic programming

It was shown that the clustered regression controller is able to repeat the unoptimized behaviour used in the teaching. It would be very beneficial, however, to reach a more optimal control scheme so that the biped could, e.g., walk faster with smaller energy consumption; this would also be a strong evidence of the validity of the selected model structure when explaining biological systems. It would be interesting to see if the biped would learn new *modes* of moving: For example, if the speed were emphasized, would it finally learn to run?

The optimization procedure of the clustered regression structure could be compared with *dynamic programming* (Bellman, 1957) that is based on the *principle of optimality*. In this case this idea can be formulated as follows:

> *An optimal control sequence has the property that whatever the initial state and initial control are, the remaining controls must constitute an optimal control sequence with regard to the state resulting from the first control.*

In general, a dynamic optimization problem can be solved calculating the control from the end to the beginning. Starting from the final state, the optimal control leading to that state is determined. Then the same procedure is performed in the *previous* state and so on. Finally, when the initial state is reached, the whole optimal control sequence is attained.

This means that one can form the *global* optimized behaviour using *local* optimization. When compared with the clustered regression structure, the local optimization should now be done inside every cluster or operating point. When the system enters the operating point region, the controller is assumedly able to guide it optimally to the next region.

Another fact from the theory of optimal control states that for a quadratic (infinite-time) optimization problem for a linear (affine) system, the optimal control law is also a linear (affine) function of the state. This all means that, assuming that the model is locally linearizable, a globally (sub)optimal control can be implemented in the proposed clustered regression framework.

5.2 Optimization principle

As the walking motion is cyclic, one cannot choose the "first" or "last" operating point. Instead, the optimization can be carried out gradually in each cluster. A trial-and-error based optimization scheme which was used successfully in the previous work (Hyötyniemi, 2002) is presented in the following.

As a starting point for the optimization an unoptimized but easily attained behaviour is taught to the clustered regression controller (the original PD control). Then the current cost of the control for one cycle is calculated. The cost criterion can be chosen rather freely so that in the minimum of the criterion the behaviour reaches the desired optimum - for example, low energy consumption and high walking speed.

When a stable and cyclic motion is obtained the control is slightly varied at one time instant, and the new data point is adapted to the clustered regression model. If the new model gives a better control, that is, the cost is now smaller, it is accepted. On the other hand, if the cost is higher, the new model is rejected. After that, a new change can be made and similarly evaluated. Repeating these steps the controller can hopefully learn a more optimal behaviour, although the process may be quite slow and nothing prevents it ending to a local minimum of the cost function. This random search is an example of *reinforcement learning*; now, because of the mathematically simple underlying model structure, the adaptation of the model can still be relatively efficient.

5.3 Adaptive learning

In order to optimize the control, it should be possible to update the clustered regression structure using new data. This adaptation, however, turned out to be quite hard to realize because of the properties of the data clusters.

It was detected during the PD controlled gait reproduction that at least eight principal component directions need to be considered in each local PCR model to reach an accurate enough control estimate. The relative importance of each principal component can be described by the variance of the corresponding latent variable component. Fig. 7 shows the averages of the variances of the local models.

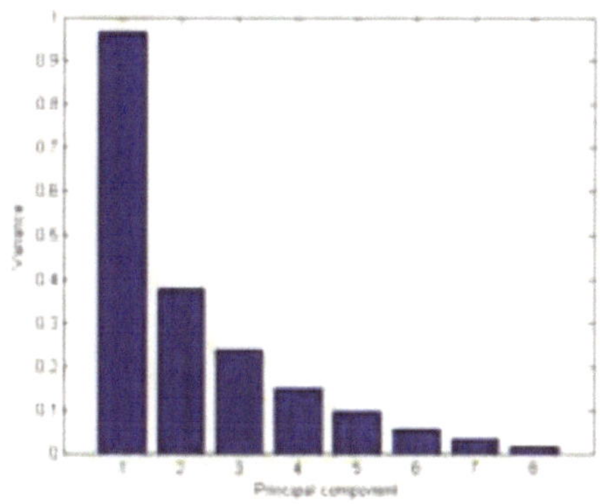

Fig. 7. Averages of the principal component variances in the local models

Clearly, the ratio of the first and last principal component variances is huge, which means that it is very hard to iteratively determine the last principal component directions. In practice, this means that when the training is repeated, always using the earlier model for construction of data for the next model, the model sooner or later degenerates. The tiny traces of relevant control information are outweighted by measurement noise.

6. Discussion

The presented scheme was just a mathematical model applying engineering intuition for reaching good control, and the main goal when constructing the model was simplicity. However, it seems that there is a connection to real neuronal systems.

It has been observed (Haykin, 1999; Hyötyniemi, 2004) that simple Hebbian/anti-Hebbian learning in neuronal grids results in a *principal subspace model* of the input data. This Hebbian learning is the principle that is implemented by the simple neurons (Hebb, 1949).

The principal subspace captures the principal components, and it is a mathematically valid basis for implementations of principal component regressions.
In (Hyötyniemi, 2006) the Hebbian learning principle is extended by applying *feedback through environment*. It turns out that when the nonideality is taken into account - exploiting signals also exhausts them - there is a negative feedback in the system; one can reach stabilization of the Hebbian adaptation without additional nonlinearities, and emergence of the principal subspace without complicated hierarchies among neurons. There can exist *self-regulation* and *self-organization* in the neuronal system, meaning that the adaptation of the global model can be based solely on the local interactions between individual neurons.
But as the biped structure is highly nonlinear, one needs to extend the linear model; here this was accomplished by introducing the clustered structure with submodels. How could such extension be motivated in a neuronal system, as it now seems that some kind of central coordination is necessary to select among the submodels and to master their adaptation? Again, as studied in (Hyötyniemi, 2006), there emerges *sparse coding* in a system with the Hebbian feedback learning. It can be claimed that in sparse coding the basis vectors are rotated to better match the physically relevant features in data - such behaviour has been detected, for example, in visual cortex (Földiák, 2002). In Hebbian feedback learning the correlating structures become separated, and they compete for activation; without any centralized control structures, the signals become distributed among the best matching substructures. As seen from outside, the net effect is to have "virtual" clusters, with smooth interpolation between them.

7. Conclusions

In this work, a clustered regression structure was used to model and control a walking biped robot model. It was shown that the purely data-based model is accurate enough to control the biped. The control structures can also be motivated from the physiological point of view.
The main problem is that to successfully reproduce the walking gait, the clustered regression controller should learn to keep the system well near the nominal trajectory. If the state of the system drifts too far from the learned behaviour, the validity of the local models strongly weakens and the system collapses. As a result, the robustness of the controller is dependent on the amount of source data variation obtained by additional noise. However, the clustered regression structure was unable to control the biped with the required noise level present in the PD controlled simulations. This complicates the iterative optimization process.
It was also shown that the management of no less than eight principal components is necessary; the "visibility ratio" between these principal components, or the ratio between variances, is over three decades. This also dictates the necessary signal-to-noise ratio. It seems that such accuracy cannot be achieved in biological systems; to construct a biologically plausible control scheme, the model structure has to be modified.
There are various directions to go. Principal components are always oriented towards the input data only, neglecting the output, or the outcome of the actions. This problem should somehow be attacked. One possibility would be to explicitly control the adaptation based on the observed outputs, so that rather than doing standard principal component analysis, some kind of *partial least squares* (PLS) (Hyötyniemi, 2001) approach would be implemented. Also, the data could be rescaled to emphasize the less visible PC directions.

It is also important to point out that the physical model of the biped robot is only a model, and some phenomena present in the data may be caused by the simulator. Especially the leg contacts with the ground may introduce some oscillations which probably would not appear in data collected from a real robot. This means that some of the less important principal component directions may as well describe these irrelevant effects thus making the modelling problem harder. In the future work, it would be interesting to analyze the principal component directions in each local model one by one and try to find out which of them are really connected to the actual walking motion.

8. References

Atkeson, C. G.; Moore, A. W. & Schaal, S. (1997). Locally Weighted Learning. *Artificial Intelligence Review*, vol. 11, pp. 11-73.

Basilevsky, A.(1994). *Statistical Factor Analysis and Related Methods*. John Wiley & Sons, New York.

Bellman, R. (1957). *Dynamic Programming*. Princeton University Press, Princeton, New Jersey.

Chevallereau, C. & *et al.* (2003). RABBIT: A Testbed for Advanced Control Theory. *IEEE Control Systems Magazine*, vol. 23, no. 5, pp. 57-79.

Földiák, P. (2002). Sparse coding in the primate cortex. In Arbib M.A. (ed.). *The Handbook of Brain Theory and Neural Networks*. MIT Press, Cambridge, Massachusetts (second edition).

Haavisto, O. & Hyötyniemi, H. (2004). *Simulation Tool of a Biped Walking Robot Model*. Technical Report 138, Helsinki University of Technology, Control Engineering Laboratory.

Haavisto, O. & Hyötyniemi, H. (2005). Data-based modeling and control of a biped robot. *Proceedings of The 6th IEEE International Symposium on Computational Intelligence in Robotics and Automation - CIRA2005*, Espoo, Finland, June 27-30, pp. 427-432.

Hardt, M.; Kreutz-Delgado, K. & Helton, J. W. (1999). Optimal Biped Walking with a Complete Dynamic Model. *Proceedings of The 38th Conference on Decision and Control*, Phoenix, Arizona, pp. 2999-3004.

Haykin, S. (1999). *Neural Networks, a Comprehensive Foundation*. Prentice Hall, Upper Saddle River, New Jersey (second edition).

Hebb, D.O. (1949): *The Organization of Behavior: A Neuropsychological Theory*. John Wiley & Sons, New York.

Hyötyniemi, H. (2001). *Multivariate Regression: Techniques and Tools*. Technical Report 125, Helsinki University of Technology, Control Engineering Laboratory.

Hyötyniemi, H. (2002). Life-like Control. *Proceedings of STeP 2002 - Intelligence, the Art of Natural and Artificial*, Oulu, Finland, December 15-17, pp. 124-139.

Hyötyniemi, H. (2004). *Hebbian Neuron Grids: System Theoretic Approach*. Technical Report 144, Helsinki University of Technology, Control Engineering Laboratory.

Hyötyniemi, H. (2006). *Neocybernetics in biological systems*. Technical Report 151, Helsinki University of Technology, Control Engineering Laboratory.

Juang, J. (2000). Fuzzy Neural Network Approaches for Robotic Gait Synthesis. *IEEE Transactions on Systems, Man, and Cybernetics - Part B: Cybernetics*, vol. 30, no. 4, pp. 594-601.

Wells, D. A. (1967). *Theory and Problems of Lagrangian Dynamics with a Treatment of Euler's Equations of Motion, Hamilton's Equations and Hamilton's Principle.* Schaum publishing, New York.

The MathWorks, Inc., *Matlab - The Language of Technical Computing*. See the WWW pages at www.mathworks.com.

Vijayakumar, S.; Shibata, T.; Conradt, J. & Schaal, S. (2002). Statistical Learning for Humanoid Robots. *Autonomous Robots*, vol. 12, no. 1, pp. 55-69.

16

Sticky Hands

Joshua G. Hale[1] & Frank E. Pollick[2]
[1]*JST ICORP Computational Brain Project, Japan*
[2]*Dept. of Psychology, University of Glasgow, Scotland*

1. Introduction

Sticky Hands is a unique physically-cooperative exercise that was implemented with a full-size humanoid robot. This involved the development of a novel biologically-inspired learning algorithm capable of generalizing observed motions and responding intuitively. Since Sticky Hands has thus far been performed as an exercise for the mutual development of graceful relaxed motion and comfortable physical intimacy between two humans, performing the exercise with a humanoid robot represents a conceptual advance in the role of humanoid robots to that of partners for human self-development.

Engendering a sense of comfortable physical intimacy between human and robot is a valuable achievement: humans must be able to interact naturally with humanoid robots, and appreciate their physical capabilities and requirements in given situations in order to cooperate physically. Humans are attuned to performing such assessments with regard to other humans based on numerous physical and social cues that it will be essential to comprehend and replicate in order to facilitate intuitive interaction.

The chapter expands these issues in more detail, drawing attention to the relevance of established research in motion control for computer graphics and robotics, human motion performance, and perceptual psychology. In particular, attention is paid to the relevance of theories of human motion production, which inspire biomimetic motion performance algorithms, and experimental results shedding light on the acuity of human perception of motion and motion features. We address the following questions: How do we interact naturally, instinctively, intuitively, and effectively with the most adaptable of multipurpose machines, *i.e.*, humanoid robots? How do we make interactions natural, graceful, and aesthetically pleasing? How do we encourage human attribution in the perception of humanoid robots? How can we expand the concept of humanoid robots through novel applications? How can we draw on knowledge already existing in other fields to inspire developments in humanoid robotics?

It is with this perspective, and by way of reply to these inquiries, that we begin this chapter by describing the implementation of the Sticky Hands system, its intuitive learning and generalization system, and hardware. We illustrate how a study of biological processes inspired various aspects of the system, such as the plastic memory of its learning system. We present experimental results and evaluations of the Sticky Hands system.

We conclude the chapter by returning to the main philosophical themes presented in the paper, and describe a fruitful future direction for humanoid robotics research: we conclude that there is a synergistic relationship between neuroscience and humanoid robotics with the following four benefits: (i) neuroscience inspires engineering solutions *e.g.*, as we have illustrated, through human motor production and motion perception research (although the breadth of relevant knowledge is much wider than the scope of these two examples); (ii) engineering implementations yield empirical evaluations of neuroscientific theories of human capabilities; (iii) engineering implementations inspire neuroscientific hypotheses; (iv) engineering solutions facilitate new neuroscientific experimental paradigms, *e.g.*, through the recording, analysis and replication of motion for psychovisual experiments.

We believe that exploiting this synergy will yield rapid advances in both fields, and therefore advocate parallel neuroscientific experimentation and engineering development in humanoid robotics. Regarding research building on the Sticky Hands system, experiments involving physical interaction will thus further the development of humanoids capable of interacting with humans, and indeed humans capable of interacting with humanoids.

2. The Sticky Hands game

'Sticky Hands' was drawn from Tai Chi practice, which often includes physical contact exercises with a partner. The game involves forming contact with a partner gently and maintaining it while moving in a previously undetermined pattern. The goal is to develop an ability to perform relaxed and graceful motion, through learning to be sensitive to the forces transmitted via contact and intuitively predict one's partner's movements. When one partner yields the other must push, and vice versa. Prolonged practice reveals the development of intuition and conditioned responses so that the contact may be preserved with a very slight force throughout complex spontaneous sequences of motion. Aspects of the game include becoming comfortable with physical contact, a mutual goal of personal development, and a fulfilling and calming influence. Some people therefore regard it as a form of spiritual development. It is possible for an expert to educate a beginner by encouraging them to perform graceful and rewarding movements, and breaking down the tension in the student's motion.

Our goal was to have a humanoid robot take the role of one partner and play Sticky Hands with a human. In order to rationalise the interaction, we defined a specific variant of the game involving contact using one hand. Partners stand and face each other, raise one hand to meet their partner's and begin making slow circling motions. The path of the contact point may then diverge into a spontaneously developing trajectory as the partners explore their range of physical expression. The robot DB may be seen playing the game in Fig. 1.

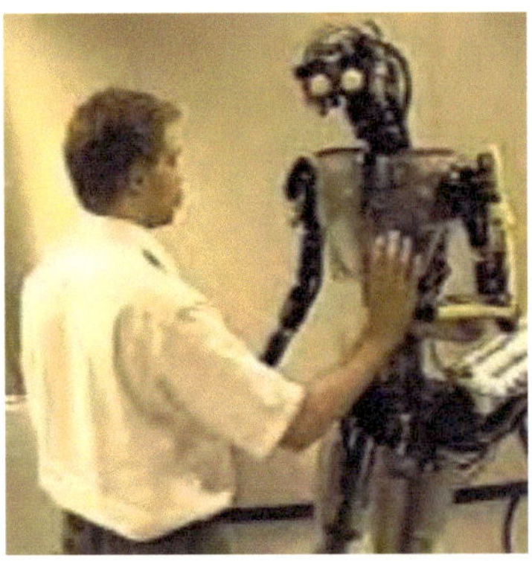

Fig. 1. Playing Sticky Hands with humanoid robot DB.

Sticky Hands, being a novel interaction, presents several interesting challenges for the design of an intelligent system, including complex motion planning that mimics the ineffable quality of human intuition. As a basis the robot must be capable of moving while remaining compliant to contact forces from a human, and refining this behaviour, be able to learn from and mimic the motion patterns of humans playing the game. By imitating the types of pattern produced by humans, the robot may reflect creativity and encourage people to explore their own range of motion. To accomplish this while simultaneously experiencing new motion patterns that may develop unpredictably it is necessary to provide a learning algorithm capable of generalising motion it has seen to new conditions, maintaining a suitably evolving internal state.

This work was motivated by a desire to explore physical interaction between humans and robots. This desire grew from the observation that the arising problem for the design of appropriate intelligent systems is how to communicate and cooperate with people (Takeda et al.1997). Using the principle that physical interaction is a familiar and reliable method for most humans, Sticky Hands broaches the area of physical cooperation and the subtle but significant issue of communication (Coppin et al. 2000) through physical movement (Adams et al. 2000; Hikiji 2000). Our work therefore considers human imitation (Scassellati 2000), which we consider to be of particular relevance for the future -as computer science and robotics develop it becomes clear that humans and robots will cooperate with a wide range of tasks. Moreover, we explored the use of a humanoid robot as a playmate facilitating a human's self-development. In this context the robot assumes a new social role involving a physically intimate and cooperative interaction. We can also hope that through the interaction, people will be encouraged to attribute the robot an anthropomorphic identity rather than considering it as a mechanical entity. Such a shift of perspective heralds ways of making human and robot interactions more natural.

Humanoid robotics embodies a certain fascination with creating a mechanical entity analogous to our human selves. There are of course many other valid motivations for creating humanoids (Bergener et al. 1997), not least among which is the nature of our environment - being highly adapted to human sensory and motor capabilities it begs for artificial agents with analogous capabilities that can usefully coexist in our own living and working spaces. Having an anthropomorphic shape and organic motion also makes working with such robots aesthetically and emotionally more pleasing. The production of human-like, and emotionally expressive styles of movement are of particular relevance. The Sticky Hands interaction embodies human-like motion and autonomy. The target of maintaining minimal contact force is tangible. The creative, anticipatory aspect however, is enhanced by initiative. The challenges posed by these problems motivated the development of a highly generalized learning algorithm, and a theoretical investigation of expressive motion styles.

We continue this section with a system overview describing the relationship between robot control and learning in the Sticky Hands system. We then outline the robot control issues, describing additional sensing technology and explaining how we achieved hand placement compliant to external forces. We then discuss the learning algorithm which observes trajectories of the hand throughout interaction with a human and predicts the development of a current trajectory.

2.1. System overview

The control system for robotic Sticky Hands was treated as three components which are shown in Fig. 2. The *robot motor controller* is responsible for positioning the hand, and obeys a planned trajectory supplied by the learning algorithm. It also estimates the contact force between the human's and robot's hands, adjusting the trajectory plan to compensate for contact force discrepancies. The actual hand position data was smoothed to remove noise, and sent back to the learning algorithm.

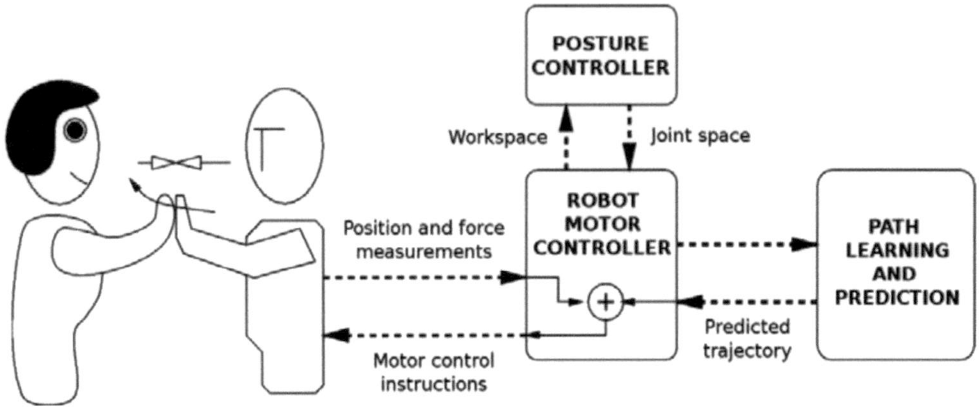

Fig. 2. System breakdown.

The *path learning and prediction* component is a learning algorithm that outputs a predicted hand trajectory and processes the actual observed trajectories supplied by the motor controller. The learning algorithm observes the evolution of the hand trajectory continuously. It learns motion patterns, and generalises them to predict future developments in the hand trajectory. The input and output are both sequences of position vectors. The robot controller makes use of the *posture controller*. The posture controller uses a straightforward inverse kinematics routine to generate joint configurations satisfying Cartesian hand placement targets.

2.2 Robot control

The Sticky Hands exercise was performed by a 30 DOF SARCOS anthropomorphic robot (Atkeson et al. 2000) that may be seen in Fig. 3. Each joint is powered hydraulically, and has angle and load sensors. Joints are servoed independently by the low-level controller using torques proportional to the angular offset between the measured and target angles, and negatively proportional to the angular velocity at each joint. This yields proportional gains/spring-damper control at each joint, where the torque at a given joint is calculated as:

$$\tau = k_s(\theta_t - \theta) + k_d(\dot{\theta}_t - \dot{\theta}) \tag{1}$$

θ_t and $\dot{\theta}_t$ are the target angle and angular velocity (usually $\dot{\theta}_t = 0$), θ and $\dot{\theta}$ are the current angle and angular velocity. k_s and k_d are the spring stiffness and damping parameters respectively.

Oscillations caused by this proportional gains controller were avoided by employing an inverse dynamics algorithm to estimate the torques necessary to hold a position. Since the robot is anchored off the ground by its pelvis, standing and balancing did not constitute problems. The Sticky Hands exercise involves only one hand and the chain of joints from the anchor point to the robot's hand encompasses 10 DOFs. The chain is kinematically redundant, so an iterative inverse kinematics algorithm was used (Tevatia & Schaal 2000). The 20 unused DOFs were resolved according to a default posture.

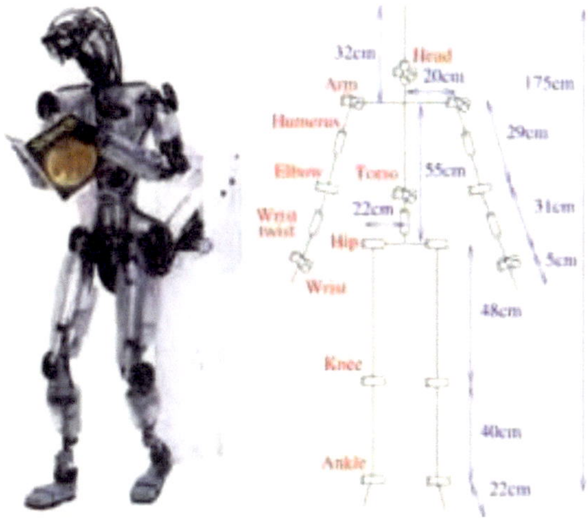

Fig. 3. DB kinematics.

The robot was required to balance a force applied by the human player against its hand. Trajectories were performed while incorporating any adjustments needed to balance the contact force, thus making the hand *actively compliant* to changes of the contact force. The simplest method of determining the contact force is direct measurement, *i.e.*, by means of a force transducer between the robot's and human's hands. Fig. 4 shows the attachment of a force transducer between the robot's hand and a polystyrene hemisphere intended to facilitate an ergonomic surface for the human to contact. The transducer measured forces in the X, Y and Z directions. The target position of the hand was translated to compensate if the contact force exceeded a 5N threshold. In order to establish a balance point against the force applied by the human, we subtracted a small quantity in the Z (forward-backward) direction from the measured force prior to thresholding. We also implemented a method of responding to the contact force using only the sensors internal to the SARCOS robot, *i.e.*, joint load and angle. In both cases we assumed that the interactions were not so forceful as to necessitate more than a constant adjustment in hand position, *i.e.*, continuous pressure would result in a yielding movement with a constant velocity.

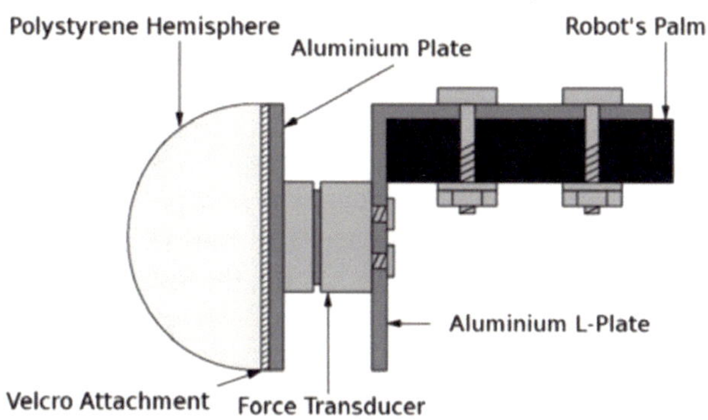

Fig. 4. Force transducer attachment.

It is possible to calculate the externally applied force using the internal sensors of DB. This involves measuring the joint angles, estimating the torques necessary to hold the position using inverse dynamics, and subtracting the measured loads. Any discrepancy should be torques due to external loads other than gravity, because gravity is the only external force accounted for in the inverse dynamics calculation. The method relies on having a very accurate inverse dynamics model, and unfortunately inaccuracies in the dynamic model and load sensors necessitated a large contact force to facilitate suitable thresholding of the results.

We therefore measured the positional offset between target and actual hand positions instead, assuming any large *positional* discrepancy was caused by external non-gravity forces. We were able to threshold by 2cm (yielding an effective force threshold of about 12N). When the low-level controller was required to follow a suggested trajectory, the hand was directed to 5cm further forward from the suggested position. The human was assumed to supply a force sufficient to maintain the hand in the actual target position specified by the suggested trajectory. This method did not require readings from the joint load sensors to respond to external forces because physical interactions were registered purely through the position of the robot's hand -which requires only joint angle readings to compute. If the human reduced or increased the contact force the robot's hand would move beyond the 2cm threshold and the suggested trajectory consequently adjusted to compensate. While the contact force was entirely in the Z direction, perturbations in the X and Y as well as Z directions were monitored, and accommodated by translating the target position when the 2cm threshold was reached. This indirect kinematic method facilitated the use of a significantly lighter contact force than with inverse dynamics based force estimation.

Any reasonable positioning and compliance strategy is in fact compatible with the high-level Sticky Hands control system. Ideally, the force threshold is minimal, since the threshold determines the minimum contact force. The larger the contact force the less relaxed the human's motion may be. The way the redundancy in robot hand placement is resolved does not significantly affect the contact force but on the other hand may have an effect on the appearance of the robot's motion. We discuss this point later. We also compare this kinematic hand positioning technique with the force transducer method and present traces of the forces measured during interaction with a human.

The trajectories supplied by the learning algorithm were described using piecewise linear splines. The robot controller ran at 500Hz and the learning algorithm ran at 10Hz. The sequence of predictions output by the learning algorithm was interpreted as the advancing end-point of a spline. The knots of this spline were translated to compensate for any discrepancy in the contact force detected by the motor controller. This translation was accomplished smoothly, in order to prevent the hand from jerking in response to contact forces: after translating the knots, a negative translation vector was initialized to bring the knots back to their original position. This negative translation was gradually decayed to zero during each cycle of the motor controller. The sum of the translated knots and the negative translation vector thus interpolated from the original knot positions and the translated knot positions.

2.3 'Prototype set' learning algorithm

3D point samples describing the robot's hand trajectory were fed as input into the learning algorithm. A vector predicting the progression of the trajectory was output in return for each sample supplied. The learning algorithm fulfilled the following properties, which are required by the nature of the Sticky Hands exercise.

- **Generalise** observed trajectories for prediction of similar new trajectories with different orientation, scale, curvature, position or velocity
- **Extrapolate** properties of a new trajectory for prediction in the absence of similar observed trajectories
- Maintain a **fluid internal state** in order to cope with the evolving nature of trajectories through continuous update, replacement and 'forgetting' of recorded information
- **Handle branch points** where similar observed trajectories diverge
- Tolerate **noise** causing inaccuracy in position samples
- Facilitate a **parameterisable time bound,** ensuring real time operation
- Facilitate a **parameterisable memory bound,** in order to fully exploit the host architecture

We refer to the structure used to record the instantaneous properties of the input trajectory as a 'prototype'. This paradigm may be compared to the work of Stokes et al. (1999) who presented a method for identifying cyclic patterns and their significance in *space-line* samples. Our process focuses rather on the immediate instant of a trajectory. Salient features are recorded for efficient retrieval but no internal classifications of higher level structures such as cycles are made. This is the essence of the generalisation and branch point handling properties of our algorithm since the recorded properties of any instant of an observed trajectory may be used to predict the development of any new trajectory. Moreover, no information about correlations between trajectories is maintained in an explicit form by the prediction process.

The 'prototype' is defined mathematically below and the creation of prototypes from raw geometrical information is presented. The utilisation of prototypes for *prediction* and *extrapolation* is also demonstrated. Then the issue of how to select the most appropriate prototype for predicting a given trajectory from a memory bank of prototypes is addressed. The memory bank is maintained according to a reinforcement principle designed to ensure an efficient use of memory by 'forgetting' prototypes that are not necessary. This search procedure requires a distance metric between prototypes and has been optimised. The reader may find it useful to refer to Fig. 5 throughout this prototype learning section.

2.3.1 Prediction using prototypes

Given a sequence of input position samples, $\{p_k : k \in \mathrm{N}\}$. The prototype P_i corresponding to p_i is defined using p_{i-1}, p_i and p_{i+1}:

$$P_i = (p_i, v_i, a_i, T_i) \tag{2}$$

$$v_i = p_{i+1} - p_i \in \mathrm{R}^3 \tag{3}$$

$$a_i = \frac{|p_{i+1} - p_i|}{|p_i - p_{i-1}|} \in \mathrm{R} \tag{4}$$

$$T_i = \left(\cos\left(\frac{\theta_i}{2}\right), \sin\left(\frac{\theta_i}{2}\right)(p_{i+1} - p_i) \times (p_i - p_{i-1}) \right) \tag{5}$$

$$\theta_i = \cos^{-1}\left(\frac{(p_i - p_{i-1}).(p_{i+1} - p_i)}{|p_i - p_{i-1}||p_{i+1} - p_i|} \right) \in \mathrm{R} \tag{6}$$

v_i is the velocity out of p_i (scaled by an arbitrary factor) and a_i is a scalar indicating the magnitude of the acceleration. The *direction* of the acceleration is deducible from T_i, which is a quaternion describing the change in direction between v_i and v_{i-1} as a rotation through their mutually orthogonal axis.

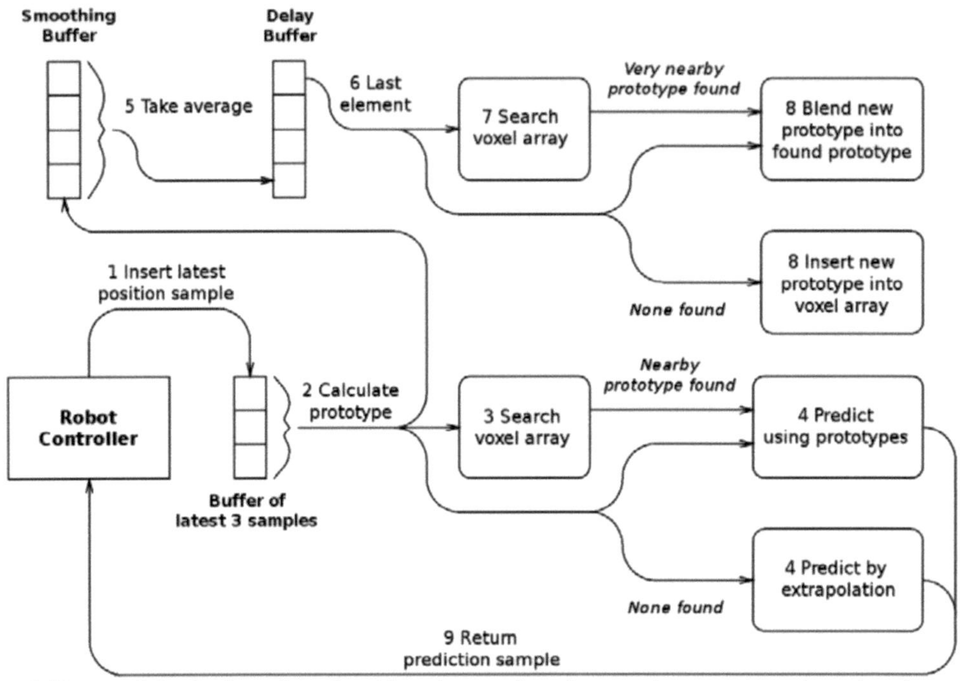

Fig. 5. Datapath in the learning algorithm (arrows) and execution sequence (numbers).

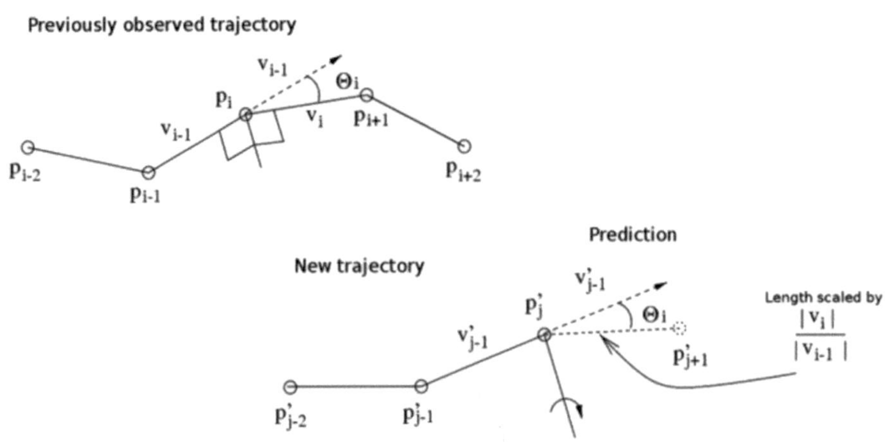

Fig. 6. Trajectory prediction using a prototype.

The progression of a trajectory $\{p'_k : k \in \mathrm{N}\}$ at a given instant may be predicted using a prototype. Suppose that for a particular trajectory sample p'_j, it is known that P_i

corresponds best to p'_j, then $p'_j + a_i T_i (p'_j - p'_{j-1})$ is an estimate for p'_{j+1}. Pre-multiplication of a 3-vector by T_i denotes quaternion rotation in the usual way. This formula applies the bend and acceleration occurring at p_i to predict the position of p'_j. We also linearly blend the position of p_i into the prediction, and the magnitude of the velocity so that p'_j combines the actual position and velocity of p_i with the prediction duplicating the bending and accelerating characteristics of p_i (see Fig. 6):

$$p'_{j+1} = p'_j + s_j T_i . \frac{p'_j - p'_{j-1}}{| p'_j - p'_{j-1} |} + g_p (p_i - p'_j) \tag{7}$$

$$s_j = (1 - g_v) a_i \, | p'_j - p'_{j-1} | + g_v \, | v_i | \tag{8}$$

g_p and g_v are blending ratios used to manage the extent to which predictions are entirely general, or repeat previously observed trajectories, *i.e.*, how much the robot wants to repeat what it has observed. We chose values of g_p and g_v in the range [0.1, 0.001] through empirical estimation. g_p describes the tendency of predictions to gravitate spatially towards recorded motions, and g_v has the corresponding effect on velocity.

In the absence of a corresponding prototype we can calculate P'_{j-1}, and use it to estimate p'_{j+1}, thus extrapolating the current characteristics of the trajectory. Repeated extrapolations lie in a single plane determined by p_{i-2}, p_{i-1} and p_i, and maintain the trajectory curvature (rotation in the plane) measured at p'_j. We must set $g_p = 0$ since positional blending makes no sense when extrapolating, and would cause the trajectory to slow to a halt, *i.e.*, the prediction should be based on an extrapolation of the immediate velocity and turning of the trajectory and not averaged with its current position since there is no established trajectory to gravitate towards.

2.3.2 Storage and retrieval

Ideally, when predicting p'_{j+1}, an observed trajectory with similar characteristics to those at p'_j is available. Typically a large set of recorded prototypes is available, and it is necessary to find the closest matching prototype P_i or confirm that no suitably similar prototype exists. The prototype P'_{j-1} which is generated from the current trajectory can be used as a basis for identifying similar prototypes corresponding to similar, previously observed trajectories. We define a distance metric relating prototypes in order to characterise the closest match.

$$d(P_i, P_j) = 1 - \cos(\theta') + \frac{| p_i - p_j |}{M_p} \tag{9}$$

Where,

$$\begin{aligned} &\theta \in [-M_a, M_a] \Rightarrow \theta' = \theta \frac{\pi}{2M_a} \\ &\theta \notin [-M_a, M_a] \Rightarrow \theta' = \pi \end{aligned} \tag{10}$$

$$\theta = \cos^{-1} \frac{v_i . v_j}{| v_i || v_j |} \tag{11}$$

M_a and M_p define the maximum angular and positional differences such that $d(P_i, P_j)$ may be one or less. Prototypes within this bound are considered similar enough to form a basis for a prediction, *i.e.*, if $d(P_i, P_j)$ is greater than 1 for all i then no suitably similar prototype exists. The metric compares the position of two prototypes, and the direction of their velocities. Two prototypes are closest if they describe a trajectory traveling in the same direction, in the same place. In practice, the values of 15cm and $\pi/4$ radians for M_p and M_a respectively were found to be appropriate. -A trajectory with exactly the same direction as the developing trajectory constitutes a match up to a displacement of 15cm, a trajectory with no displacement constitutes a match up to an angular discrepancy of $\pi/4$ radians, and within those thresholds there is some leeway between the two characteristics. The threshold values must be large enough to permit some generalisation of observed trajectories, but not so large that totally unrelated motions are considered suitable for prediction when extrapolation would be more appropriate.

The absolute velocity, and bending characteristics are not compared in the metric. Predictions are therefore general with respect to the path leading a trajectory to a certain position with a certain direction and velocity, so branching points are not problematic. Also the speed at which an observed trajectory was performed does not affect the way it can be generalised to new trajectories. This applies equally to the current trajectory and previously observed trajectories.

When seeking a prototype we might naïvely compare *all* recorded prototypes with P'_{j-1} to find the closest. If none exist within a distance of 1 we use P'_{j-1} itself to extrapolate as above. Needless to say however, it would be computationally over-burdensome to compare P'_{j-1} with *all* the recorded prototypes. To optimise this search procedure we defined a voxel array to store the prototypes. The array encompassed a cuboid enclosing the reachable space of the robot, partitioning it into a $50 \times 50 \times 50$ array of cuboid voxels indexed by three integer coordinates. The storage requirement of the empty array was 0.5MB. New prototypes were placed in a list attached to the voxel containing their positional component p_i. Given P_{j-1} we only needed to consider prototypes stored in voxels within a distance of M_p from p_{j-1} since prototypes in any other voxels would definitely exceed the maximum distance according to the metric. Besides limiting the total number of candidate prototypes, the voxel array also facilitated an optimal ordering for considering sets of prototypes. The voxels were considered in an expanding sphere about p_j. A list of integer-triple voxel index offsets was presorted and used to quickly identify voxels close to a given centre voxel ordered by minimum distance to the centre voxel. The list contained voxels up to a minimum distance of M_p. This ensures an optimal search of the voxel array since the search may terminate as soon as we encounter a voxel that is too far away to contain a prototype with a closer minimum distance than any already found. It also permits the search to be cut short if time is unavailable. In this case the search terminates optimally since the voxels most likely to contain a match are considered first. This facilitates the parameterisable time bound since the prototype search is by far the dominant time expense of the learning algorithm.

2.3.3 Creation and maintenance

Prototypes were continually created based on the stream of input position samples describing the observed trajectory. It was possible to create a new prototype for each new sample, which we placed in a cyclic buffer. For each new sample we extracted the average prototype of the buffer to reduce sampling noise. A buffer of 5 elements was sufficient. The averaged prototypes were shunted through a delay buffer, before being added to the voxel array. This prevented prototypes describing a current trajectory from being selected to predict its development (extrapolation) when other prototypes were available. The delay buffer contained 50 elements, and the learning algorithm was iterated at 10Hz so that new prototypes were delayed by 5 seconds.

Rather than recording every prototype we limited the total number stored by averaging certain prototypes. This ensures the voxel array does not become clogged up and slow, and reduces the memory requirement. Therefore before inserting a new prototype into the voxel array we first searched the array for a similar prototype. If none was found we added the new prototype, otherwise we blended it with the existing one. We therefore associated a count of the number of blends applied to each prototype to facilitate correct averaging with new prototypes. In fact we performed a non-linear averaging that capped the weight of the existing values, allowing the prototypes to tend towards newly evolved motion patterns within a limited number of demonstrations. Suppose P_a incorporates n blended prototypes, then a subsequent blending with P_b will yield:

$$P_a' = P_a \frac{D(n)-1}{D(n)} + P_b \frac{1}{D(n)} \tag{12}$$

$$D(n) = 1 + A_M - \frac{A_M}{1 + nA_G} \tag{13}$$

$A_M/(1 + nA_M)$ defines the maximum weight for the old values, and A_G determines how quickly it is reached. Values of 10 and 0.1 for A_M and A_G respectively were found to be suitable. This makes the averaging process linear as usual for small values but ensures the contribution of the new prototype is worth at least $1/11^{th}$.

We facilitated an upper bound on the storage requirements using a deletion indexing strategy for removing certain prototypes. An integer clock was maintained, and incremented every time a sample was processed. New prototypes were stamped with a deletion index set in the future. A list of the currently stored prototypes sorted by deletion index was maintained, and if the storage bounds were reached the first element of the list was removed and the corresponding prototype deleted. The list was stored as a *heap* (Cormen et al.) since this data structure permits fast $O(\log(\text{num elements}))$ insertion, deletion and repositioning. We manipulated the deletion indices to mirror the reinforcement aspect of human memory. A function $R(n)$ defined the period for which a prototype reinforced n times should be retained (n is equivalent to the blending count). Each time a prototype was blended with a new one we calculated the retention period, added the current clock and re-sorted the prototype index. $R(n)$ increases exponentially up to a maximum asymptote.

$$R(n) = D_M - \frac{D_M}{1 + D_G n^{D_P}} \tag{14}$$

D_M gives the maximum asymptote. D_G and D_P determine the rate of increase. Values of 20000, 0.05 and 2 were suitable for D_M,D_G and D_P respectively. The initial reinforcement thus extended a prototype's retention by 2 minutes, and subsequent reinforcements roughly doubled this period up to a maximum of about half an hour (the algorithm was iterated at 10Hz).

3. Results

The initial state and state after playing Sticky Hands with a human partner are shown in Fig. 7. Each prototype is plotted according to its position data. The two data sets are each viewed from two directions and the units (in this and subsequent figures) are millimeters. The *X, Y* & *Z* axes are positive in the robot's left, up and forward directions respectively. The point *(0,0,0)* corresponds to the robot's sacrum. The robot icons are intended to illustrate orientation only, and not scale. Each point represents a unique prototype stored in the motion predictor's memory, although as discussed each prototype may represent an amalgamation of several trajectory samples. The trajectory of the hand loosely corresponds to the spacing of prototypes but not exactly because sometimes new prototypes are blended with old prototypes according to the similarities between each's position and velocity vectors.

The initial state was loaded as a default. It was originally built by teaching the robot to perform an approximate circle 10cm in radius and centred in front of the left elbow joint (when the arm is relaxed) in the frontal plane about 30cm in front of the robot. The prototype positions were measured at the robot's left hand, which was used to play the game and was in contact with the human's right hand throughout the interaction. The changes in the trajectory mostly occur gradually as human and robot slowly and cooperatively develop cycling motions. Once learned, the robot can switch between any of its previously performed trajectories, and generalise them to interpret new trajectories.

The compliant positioning system, and its compatibility with motions planned by the prediction algorithm was assessed by comparing the Sticky Hands controller with a 'positionable hand' controller that simply maintains a fixed target for the hand in a compliant manner so that a person may reposition the hand.

Fig. 8 shows a force/position trace where the width of the line is linearly proportional to the magnitude of the force vector (measured in all 3 dimensions), and Table 1 shows corresponding statistics. Force measurements were averaged over a one minute period of interaction, but also presented are 'complied forces', averaging the force measurements over only the periods when the measured forces exceeded the compliance threshold. From these results it is clear that using the force transducer yielded significantly softer compliance in all cases. Likewise the 'positionable hand' task yielded slightly softer compliance because the robot did not attempt to blend its own trajectory goals with those imposed by the human.

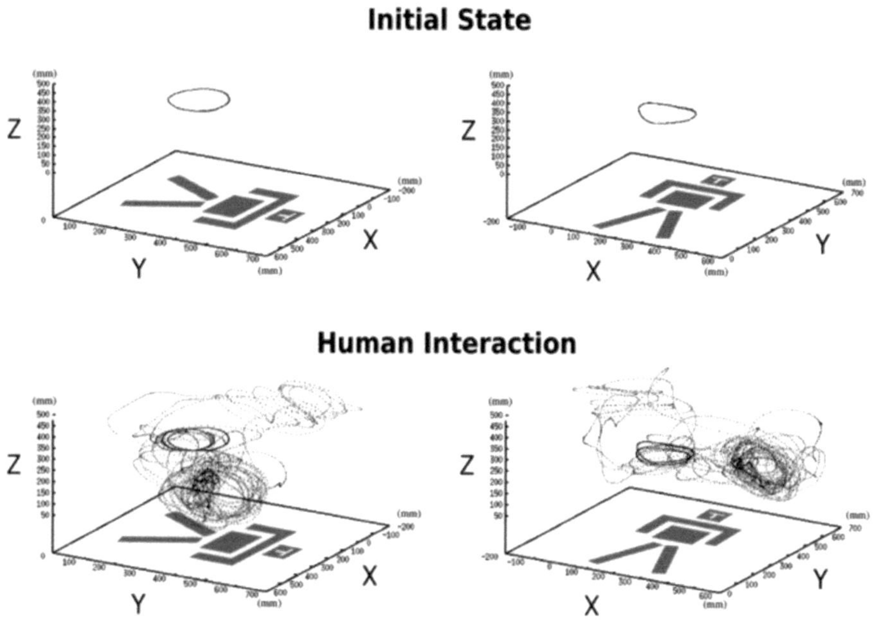

Fig. 7. Prototype state corresponding to a sample interaction.

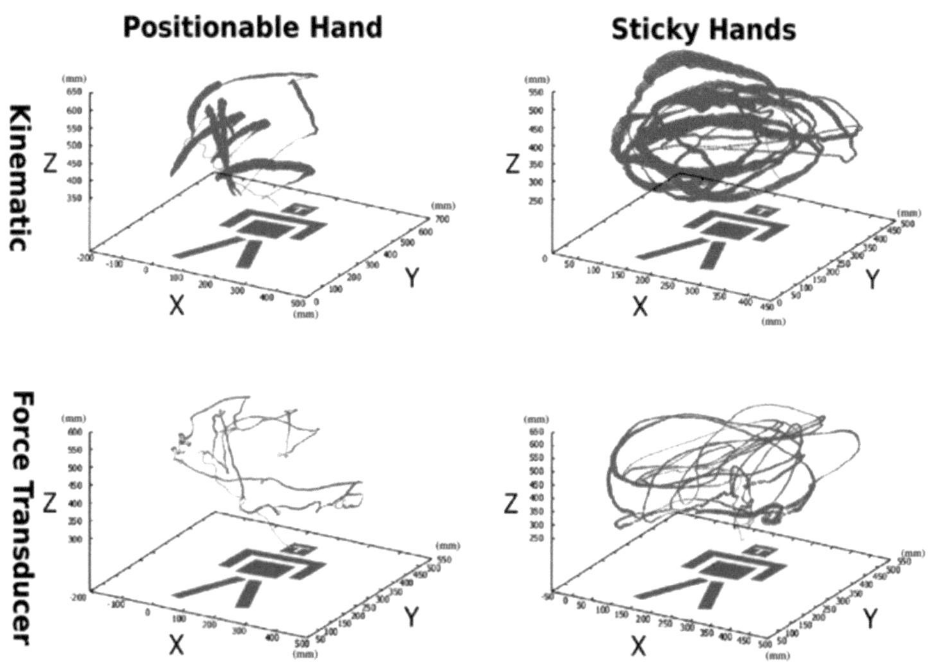

Fig. 8. Force measured during 'positionable hand' and Sticky Hands tasks.

Task	Contact force (N)		Complied forces (N)	
	Mean	Var.	Mean	Var.
Force Transducer Sticky Hands	4.50	4.83	5.72	4.49
Force Transducer 'Positionable Hand'	1.75	2.18	3.23	2.36
Kinematically Compliant Sticky Hands	11.86	10.73	13.15	10.73
Kinematically Compliant 'Positionable Hand'	8.90	10.38	12.93	11.40

Table 1 Forces experienced during 'positionable hand' and Sticky Hands tasks.

Examining a sequence of interaction between the robot and human reveals many of the learning system's properties. An example sequence during which the robot used the kinematic compliance technique is shown in Fig. 9. The motion is in a clockwise direction, defined by progress along the path in the a-b-c direction, and was the first motion in this elliptical pattern observed by the prediction system. The 'Compliant Adjustments' graph shows the path of the robot's hand, and is marked with thicker lines at points where the compliance threshold was exceeded. *i.e.*, points where the prediction algorithm was mistaken about the motion the human would perform. The 'Target Trajectory' graph shows in lighter ink the target sought by the robot's hand along with in darker ink the path of the robot's hand. The target is offset in the Z (forwards) direction in order to bring about a contact force against the human's hand. At point *(a)* there is a kink in the actual hand trajectory, a cusp in the target trajectory, and the beginning of a period during which the robot experiences a significant force from the human. This kink is caused by the prediction algorithm's expectation that the trajectory will follow previously observed patterns that have curved away in the opposite direction, the compliance maintaining robot controller adjusts the hand position to attempt to balance the contact force until the curvature of the developing trajectory is sufficient to extrapolate its shape and the target trajectory well estimates the path performed by the human. At point *(b)* however, the human compels the robot to perform an elliptical shape that does not extrapolate the curvature of the trajectory thus far. At this point the target trajectory overshoots the actual trajectory due to its extrapolation. Once again there is a period of significant force experienced against the robot's hand and the trajectory is modified by the compliance routine. At point *(c)* we observe that, based on the prototypes recorded during the previous ellipse, the prediction algorithm correctly anticipates a similar elliptical trajectory offset positionally and at a somewhat different angle.

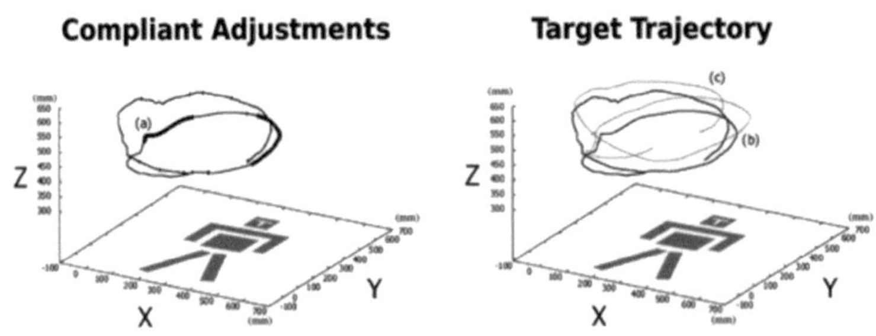

Fig. 9 Example interaction showing target trajectory and compliance activation

4. Discussion

We proposed the 'Sticky Hands' game as a novel interaction between human and robot. The game was implemented by combining a robot controller process and a learning algorithm with a novel internal representation. The learning algorithm handles branching trajectories implicitly without the need for segmentation analysis because the approach is not pattern based. It is possible to bound the response time and memory consumption of the learning algorithm arbitrarily within the capabilities of the host architecture. This may be achieved trivially by restricting the number of prototypes examined or stored. The ethos of our motion system may be contrasted with the work of Williamson (1996) who produced motion controllers based on positional primitives. A small number of postures were interpolated to produce target joint angles and hence joint torques according to proportional gains. Williamson's work advocated the concept of ``behaviours or skills as coarsely parameterised atoms by which more complex tasks can be successfully performed''. Corresponding approaches have also been proposed in the computer animation literature, such as the motion verbs and adverbs of Rose et al. (1998). Williamson's system is elegant, providing a neatly bounded workspace, but unfortunately it was not suitable for our needs due to the requirements of a continuous interaction incorporating more precise positioning of the robot's hand.

By implementing Sticky Hands, we were able to facilitate physically intimate interactions with the humanoid robot. This enables the robot to assume the role of playmate and partner assisting in a human's self-development. Only minimal sensor input was required for the low-level motor controller. Only torque and joint position sensors were required, and these may be expected as standard on most humanoid robots. With the addition of a hand mounted force transducer the force results were also obtained. Our work may be viewed as a novel communication mechanism that accords with the idea that an autonomous humanoid robot should accept command input and maintain behavioral goals at the same level as sensory input (Bergener et al. 1997). Regarding the issue of human instruction however, the system demonstrates that the blending of internal goals with sensed input can yield complex behaviors that demonstrate a degree of initiative. Other contrasting approaches (Scassellati 1999) have achieved robust behaviors that emphasize the utility of human instruction in the design of reinforcement functions or progress estimators.

The design ethos of the Sticky Hands system reflects a faith in the synergistic relationship between humanoid robotics and neuroscience. The project embodies the benefits of cross-fertilized research in several ways. With reference to the introduction, it may be seen that (i) neuroscientific and biological processes have informed and inspired the development of the system, *e.g.,* through the plastic memory component of the learning algorithm, and the control system's "intuitive" behaviour which blends experience with immediate sensory information as discussed further below; (ii) by implementing a system that incorporates motion based social cues, the relevance of such cues has been revealed in terms of human reactions to the robot. Also, by demonstrating that a dispersed representation of motion is sufficient to yield motion learning and generalization, the effectiveness of solutions that do not attempt to analyze nor segment observed motion has been confirmed; (iii) technology developed in order to implement Sticky Hands has revealed processes that could plausibly be used by the brain for solving motion tasks, *e.g.,* the effectiveness of the system for blending motion targets with external forces to yield a compromise between the motion modeled internally and external influences suggests that humans might be capable of performing learned motion patterns according to a consistent underlying model subject to forceful external influences that might significantly alter the final motion; (iv) the Sticky Hands system is in itself a valuable tool for research since it provides an engaging cooperative interaction between a human and a humanoid robot. The robot 's behaviour

may be modulated in various ways to investigate for example the effect of less compliant motion, different physical cues, or path planning according to one of various theories of human motion production.
The relationship between the engineering and computational aspect of Sticky Hands and the neuroscientific aspect is thus profound. This discussion is continued in the following sections which consider Sticky Hands in the context of relevant neuroscientific fields: human motion production, perception, and the attribution of characteristics such as naturalness and affect. The discussion is focused on interaction with humans, human motion, and lastly style and affect.

4.1 Interacting with humans

The Sticky Hand task requires two partners to coordinate their movements. This type of coordination is not unlike that required by an individual controlling an action using both their arms. However, for such bimanual coordination there are direct links between the two sides of the brain controlling each hand. Though surprisingly, even when these links are severed in a relatively rare surgical intervention known as callosotomy, well-learned bimanual processes appear to be remarkably unaffected (Franz, Waldie & Smith, 2000). This is consistent with what we see from experienced practitioners of Tai Chi who perform Sticky Hands: that experience with the task and sensory feedback are sufficient to provide graceful performance. It is a reasonable speculation that the crucial aspect of experience lays in the ability to predict which movements are likely to occur next, and possibly even what sensory experience would result from the actions possible from a given position.
A comparison of this high level description with the implementation that we used in the Sticky Hands task is revealing. The robot's experience is limited to the previous interaction between human and robot and sensory information is limited to either the kinematics of the arm and possibly also force information. Clearly the interaction was smoother when more sensory information was available and this is not entirely unexpected. However, the ability of the robot to perform the task competently with a very minimum of stored movements is impressive. One possibility worth considering is that this success might have been due to a fortunate matching between humans' expectations of how the game should start and the ellipse that the robot began with. This matching between human expectations and robot capabilities is a crucial question that is at the heart of many studies of human-robot interaction.
There are several levels of possible matching between robot and human in this Sticky Hands task. One of these, as just mentioned is that the basic expectations of the range of motion are matched. Another might be that the smoothness of the robot motion matches that of the human and that any geometric regularities of motion are matched. For instance it is known that speed and curvature are inversely proportional for drawing movements (Lacquaniti et al. 1983) and thus it might be interesting in further studies to examine the effect of this factor in more detail. A final factor in the relationship between human and robot is the possibility of social interactions. Our results here are anecdotal, but illustrative of the fact that secondary actions will likely be interpreted in a social context if one is available. One early test version of the interaction had the robot move its head from looking forward to looking towards its hand whenever the next prototype could not be found. From the standpoint of informing the current state of the program this was useful. However, there was one consequence of this head movement that likely was exacerbated by the fact that it was the more mischievous actions of the human partner that would confuse the robot. This lead the

robot head motion to fixate visually on its own hand, which by coincidence was where most human partners were also looking, leading to a form of mutual gaze between human and robot. This gestural interaction yielded variable reports from the human players as either a sign of confusion or disapproval by the robot.
This effect is illustrative of the larger significance of subtle cues embodied by human motion that may be replicated by humanoid robots. Such actions or characteristics of motion may have important consequences for the interpretation of the movements by humans. The breadth of knowledge regarding these factors further underlines their value. There is much research describing how humans produce and perceive movements and many techniques for producing convincing motion in the literature of computer animation. For example, there is a strong duality between dynamics based computer animation and robotics (Yamane & Nakamura 2000). Computer animation provides a rich source of techniques for generating (Witkin & Kass 1988; Cohen 1992; Ngo & Marks 1993; Li et al. 1994; Rose et al. 1996; Gleicher 1997) and manipulating (Hodgins & Pollard 1997) dynamically correct motion, simulating biomechanical properties of the human body (Komura & Shinagawa 1997) and adjusting motions to display affect or achieve new goals (Bruderlin & Williams 1995; Yamane & Nakamura 2000).

4.2 Human motion

Although the technical means for creating movements that appear natural and express affect, skill, *etc.* are fundamental, it is important to consider the production and visual perception of *human* movement. The study of human motor control for instance holds the potential to reveal techniques that improve the replication of human-like motion. A key factor is the *representation* of movement. Interactions between humans and humanoids may improve if both have similar representations of movement. For example, in the current scenario the goal is for the human and robot to achieve a smooth and graceful trajectory. There are various objective ways to express smoothness. It can be anticipated that if both the humanoid and human shared the same representation of smoothness then the two actors may converge more quickly to a graceful path. The visual perception of human movement likewise holds the potential to improve the quality of human-robot interactions. The aspects of movement that are crucial for interpreting the motion correctly may be isolated according to an analysis of the features of motion to which humans are sensitive. For example, movement may be regarded as a complicated spatiotemporal pattern, but the recognition of particular styles of movement might rely on a few isolated spatial or temporal characteristics of the movement. Knowledge of human motor control and the visual perception of human movement could thus beneficially influence the design of humanoid movements. Several results from human motor control and motor psychophysics inform our understanding of natural human movements. It is generally understood several factors contribute to the smoothness of human arm movements. These include the low-pass filter characteristics of the musculoskeletal system itself, and the planning of motion according to some criteria reflecting smoothness. The motivation for such criteria could include minimizing the wear and tear on the musculoskeletal system, minimizing the overall muscular effort, and maximizing the compliance of motions. Plausible criteria that have been suggested include the minimization of jerk, *i.e.*, the derivative of acceleration (Flash & Hogan 1985), minimizing the torque change (Uno et al. 1989), the motor-command change (Kawato 1992), or signal dependent error (Harris & Wolpert 1998). There are other consistent properties of human motion besides smoothness that have been observed. For example, that the endpoint trajectory of the hand behaves like a concatenation of piecewise planar segments (Soechting & Terzuolo 1987a; Soechting & Terzuolo 1987b). Also, the movement speed is related to its geometry in terms of curvature and torsion. Specifically, it has been reported that for planar segments velocity is inversely proportional to curvature raised to the 1/3rd power, and that for non-planar segments the velocity is inversely proportional to the 1/3rd power of curvature multiplied by 1/6th power of torsion (Lacquaniti et al. 1983; Viviani & Stucchi 1992; Pollick & Sapiro 1996; Pollick et al. 1997; Handzel & Flash, 1999). Extensive psychological experiments of the paths negotiated by human-humanoid dyads could inform which principles of human motor control are appropriate for describing human-humanoid cooperative behaviours.

4.3 Style and affect

Recent results examining the visual recognition of human movement are also of relevance with regard to the performance of motion embodying human-like *styles*. By considering the relationship between movement kinematics and style recognition, it has been revealed that recognition can be enhanced by exaggerating temporal (Hill & Pollick 2000), spatial (Pollick et al. 2001a), and spatiotemporal (Giese & Poggio 2000; Giese & Lappe 2002) characteristics of motion. The inference of style from human movement (Pollick et al. 2001b) further supports the notion that style may be specified at a kinematic level. The kinematics of motion may thus be used to constrain the design of humanoid motion.

However, the meaningful kinematic characteristics of motion may rely on dynamic properties in a way that can be exploited for control purposes. The brief literature review on human motor control and visual perception of human movement above provides a starting point for the design of interactive behaviours with humanoid robots. The points addressed focus on the motion of the robot and may be viewed as dealing with the problem in a bottom up fashion. In order to make progress in developing natural and affective motion it is necessary to determine whether or not a given motion embodies these characteristics effectively. However, it is possible that cognitive factors, such as expectancies and top down influences might dominate interactions between humans and humanoids, *e.g.*, the humanoid could produce a natural movement with affect but the motion could still be misinterpreted if there is an expectation that the robot would not move naturally or display affect.

5. Conclusion

Having described the Sticky Hands project: it's origin, hardware and software implementation, biological inspiration, empirical evaluation, theoretical considerations and implications, and having broadened the later issues with a comprehensive discussion, we now return to the enquiries set forth in the introduction.

The Sticky Hands project itself demonstrates a natural interaction which has been accomplished effectively -the fact that the objectives of the interaction are in some aspects open-ended creates leeway in the range of acceptable behaviours but also imposes complex high-level planning requirements. Again, while these may be regarded as peculiar to the Sticky Hands game they also reflect the breadth of problems that must be tackled for advanced interactions with humans. The system demonstrates through analysis of human motion, and cooperation how motion can be rendered naturally, gracefully and aesthetically. These characteristics are both key objectives in Sticky Hands interaction, and as we have indicated in the discussion also have broader implications for the interpretation, quality and effectiveness of interactions with humans in general for which the attribution of human qualities such as emotion engender an expectation of the natural social cues that improve the effectiveness of cooperative behaviour through implicit communication.

We have drawn considerable knowledge and inspiration from the fields of computer graphics, motion perception and human motion performance. The benefit that the latter two fields offer for humanoid robotics reveal an aspect a larger relationship between humanoid robotics and neuroscience. There is a synergistic relationship between the two fields that offers mutual inspiration, experimental validation, and the development of new experimental paradigms to both fields. We conclude that exploring the depth of this relationship is a fruitful direction for future research in humanoid robotics.

8. References

Adams, B.; Breazeal, C.; Brooks, R.A. & Scassellati, B. (2000). Humanoids Robots: A New Kind of Tool. *IEEE Intelligent Systems*, 25-31, July/August

Atkeson, C.G.; Hale, J.G.; Kawato, M.; Kotosaka, S.; Pollick, F.E.; Riley, M.; Schaal, S.; Shibata, T.; Tevatia, G.; Ude A. & Vijayakumar, S. (2000). Using humanoid robots to study human behavior. *IEEE Intelligent Systems*, 15, pp46-56

Bergener, T.; Bruckhoff, C.; Dahm, P.; Janben, H.; Joublin, F. & Menzner, R. (1997). Arnold: An Anthropomorphic Autonomous Robot for Human Environments. *Proc. Selbstorganisation von Adaptivem Verhalten (SOAVE 97)*, 23-24 Sept., Technische Universitt Ilmenau

Bruderlin, A. & Williams, L. (1995). Motion Signal Processing. *Proc. SIGGRAPH 95, Computer Graphics Proceedings*, Annual Conference Series, pp97-104

Cohen, M.F. (1992). Interactive Spacetime Control for Animation. *Proc. SIGGRAPH 92, Computer Graphics Proceedings*, Annual Conference Series, pp293-302

Coppin P.; Pell, R.; Wagner, M.D.; Hayes, J.R.; Li, J.; Hall, L. ; Fischer, K.D.; Hirschfield & Whittaker, W.L. (2000). EventScope: Amplifying Human Knowledge and Experience via Intelligent Robotic Systems and Information Interaction. *IEEE International Workshop on Robot-Human Interaction*, Osaka, Japan

Cormen, T.H.; Leiserson, C.E. & Rivest, R.L. Introduction To Algorithms. McGraw-Hill, ISBN 0-07-013143-0

Flash, T. & Hogan, N. (1985). The coordination of arm movements: An experimentally confirmed mathematical model. *Journal of Neuroscience*, 5, pp1688-1703

Giese, M.A. & Poggio, T. (2000). Morphable models for the analysis and synthesis of complex motion patterns. *International Journal of Computer Vision*, 38, 1, pp59-73

Giese, M.A. & Lappe, M. (2002). Perception of generalization fields for the recognition of biological motion. *Vision Research*, 42, pp1847-1858

Gleicher, M. (1997). Motion Editing with Spacetime Constraints. *Proc. 1997 Symposium on Interactive 3D Graphics*

Handzel, A. A. & Flash, T. (1999). Geometric methods in the study of human motor control. *Cognitive Studies* 6, 309-321

Harris, C.M. & Wolpert, D.M. (1998). Signal-dependent noise determines motor planning. *Nature* 394, pp780-784

Hikiji, H. (2000). Hand-Shaped Force Interface for Human-Cooperative Mobile Robot. *Proceedings of the 2000 IEICE General Conference*, A-15-22, pp300

Hill, H. & Pollick, F.E. (2000). Exaggerating temporal differences enhances recognition of individuals from point light displays. *Psychological Science*, 11, 3, pp223-228

Hodgins, J.K. & Pollard, N.S. (1997). Adapting Simulated Behaviors For New Characters. *Proc. SIGGRAPH 97, Computer Graphics Proceedings*, Annual Conference Series, pp153-162

Kawato, M. (1992). Optimization and learning in neural networks for formation and control of coordinated movement. In: *Attention and performance*, Meyer, D. and Kornblum, S. (Eds.), XIV, MIT Press, Cambridge, MA, pp821-849

Komura, T. & Shinagawa, Y. (1997). A Muscle-based Feed-forward controller for the Human Body. *Computer Graphics forum* 16(3), pp165-176

Lacquaniti, F.; Terzuolo, C.A. & Viviani, P. (1983). The law relating the kinematic and figural aspects of drawing movements. *Acta Psychologica*, 54, pp115-130

Li, Z.; Gortler, S.J. & Cohen, M.F. (1994). Hierarchical Spacetime Control. *Proc. SIGGRAPH 94, Computer Graphics Proceedings*, Annual Conference Series, pp35-42

Ngo, J.T. & Marks, J. (1993). Spacetime Constraints Revisited. *Proc. SIGGRAPH 93, Computer Graphics Proceedings*, Annual Conference Series, pp343-350

Pollick, F.E. & Sapiro, G. (1996). Constant affine velocity predicts the 1/3 power law of planar motion perception and generation. *Vision Research*, 37, pp347-353

Pollick, F.E.; Flash, T.; Giblin, P.J. & Sapiro, G. (1997). Three-dimensional movements at constant affine velocity. *Society for Neuroscience Abstracts*, 23, 2, pp2237

Pollick, F.E.; Fidopiastis, C.M. & Braden, V. (2001a). Recognizing the style of spatially exaggerated tennis serves. *Perception*, 30, pp323-338

Pollick, F.E.; Paterson, H.; Bruderlin, A. & Sanford, A.J. (2001b). Perceiving affect from arm movement. *Cognition*, 82, B51-B61

Rose, C.; Guenter, B.; Bodenheimer, B. & Cohen, M.F. (1996). Efficient Generation of Motion Transitions using Spacetime Constraints. *Proc. SIGGRAPH 96, Computer Graphics Proceedings*, Annual Conference Series, pp147-154

Rose, C.; Bodenheimer, B. & Cohen, M.F. (1998). Verbs and Adverbs: Multidimensional Motion Interpolation. *IEEE Computer Graphics & Applications*, 18(5)

Scassellati, B. (1999). Knowing What to Imitate and Knowing When You Succeed. *Proc. of AISB Symposium on Imitation in Animals and Artifacts*, Edinburgh, Scotland

Scassellati, B. (2000). Investigating models of social development using a humanoid robot. In: *Biorobotics*, Webb, B. and Consi, T. (Eds.), MIT Press, Cambridge, MA

Soechting, J.F. & Terzuolo, C.A. (1987a). Organization of arm movements. Motion is segmented. *Neuroscience*, 23, pp39-51

Soechting, J.F. & Terzuolo, C.A. (1987b). Organization of arm movements in three-dimensional space. Wrist motion is piecewise planar. *Neuroscience*, 23, pp53-61

Stokes, V.P.; Lanshammar, H. & Thorstensson, A. (1999). Dominant Pattern Extraction from 3-D Kinematic Data. *IEEE Transactions on Biomedical Engineering* 46(1)

Takeda H.; Kobayashi N.; Matsubara Y. & Nishida, T. (1997). Towards Ubiquitous Human-Robot Interaction. *Proc. of IJCAI Workshop on Intelligent Multimodal Systems*, Nagoya Congress Centre, Nagoya, Japan

Tevatia, G. & Schaal, S. (2000). Inverse kinematics for humanoid robots. *IEEE International Conference on Robotics and Automation*, San Francisco, CA

Uno, Y.; Kawato, M. & Suzuki, R. (1989). Formation and control of optimal trajectory in human multijoint arm movement. *Biological Cybernetics*, 61, pp89-101

Viviani, P. & Stucchi, N. (1992). Biological movements look uniform: Evidence of motor-perceptual interactions. *Journal of Experimental Psychology: Human Perception and Performance*, 18, pp602-623

Williamson, M.M. (1996). Postural primitives: Interactive Behavior for a Humanoid Robot Arm. *Proc. of SAB 96*, Cape Cod, MA, USA

Witkin, A. & Kass, M. (1988). Spacetime Constraints. *Proc. SIGGRAPH 88, Computer Graphics Proceedings*, Annual Conference Series, pp159-168

Yamane, K. & Nakamura, Y. (2000). Dynamics Filter: Towards Real-Time and Interactive Motion Generator for Human Figures. *Proc. WIRE 2000*, 27-34, Carnegie Mellon University, Pittsburgh, PA

17

Central Pattern Generators for Gait Generation in Bipedal Robots

Almir Heralić[1], Krister Wolff[2], Mattias Wahde[2]
[1]*University West, Trollhättan, Sweden*
[2]*Chalmers University of Technology, Göteborg, Sweden*

1. Introduction

An obvious problem confronting humanoid robotics is the generation of stable and efficient gaits. Whereas wheeled robots normally are statically balanced and remain upright regardless of the torques applied to the wheels, a bipedal robot must be actively balanced, particularly if it is to execute a human-like, dynamic gait. The success of gait generation methods based on classical control theory, such as the zero-moment point (ZMP) method (Takanishi et al., 1985), relies on the calculation of reference trajectories for the robot to follow. In the ZMP method, control torques are generated in order to keep the zero-moment point within the convex hull of the support area defined by the feet. When the robot is moving in a well-known environment, the ZMP method certainly works well. However, when the robot finds itself in a dynamically changing real-world environment, it will encounter unexpected situations that cannot be accounted for in advance. Hence, reference trajectories can rarely be specified under such circumstances. In order to address this problem, alternative, biologically inspired control methods have been proposed, which do not require the specification of reference trajectories. The aim of this chapter is to describe one such method, based on central pattern generators (CPGs), for control of bipedal robots.

Clearly, walking is a rhythmic phenomenon, and many biological organisms are indeed equipped with CPGs, i.e. neural circuits capable of producing oscillatory output given tonic (non-oscillating) activation (Grillner, 1996). There exists biological evidence for the presence of central pattern generators in both lower and higher animals. The lamprey, which is one of the earliest and simplest vertebrate animals, swims by propagating an undulation along its body. The wave-like motion is produced by an alternating activation of motor neurons on the left and right sides of the segments along the body. The lamprey has a brain stem and spinal cord with all basic vertebrate features, but with orders of magnitude fewer nerve cells of each type than higher vertebrates. Therefore, it has served as a prototype organism for the detailed analysis of the nervous system, including CPGs, in neurophysiological studies (Grillner, 1991; Grillner, 1995). In some early experiments by Brown (Brown, 1911, Brown, 1912), it was shown that cats with transected spinal cord and with cut dorsal roots still showed rhythmic alternating contractions in ankle flexors and extensors. This was the basis of the concept of a spinal locomotor center, which Brown termed the half-center model (Brown, 1914). Further biological support for the existence of a spinal CPG structure in vertebrates is presented in (Duysens & Van de Crommert, 1998).

However, there is only evidence by inference of the existence of human CPGs. The strongest evidence comes from studies of newborns, in which descending supraspinal control is not yet fully developed, see e.g. (Zehr & Duysens, 2004) and references therein. Furthermore, advances made in the rehabilitation of patients with spinal cord lesions support the notion of human CPGs: Treadmill training is considered by many to rely on the adequate afferent activation of CPGs (Duysens & Van de Crommert, 1998). In view of the results of the many extensive studies on the subject, it seems likely that primates in general, and humans in particular, would have a CPG-like structure.

In view of their ability to generate rhythmic output patterns, CPGs are well suited as the basis for bipedal locomotion. Moreover, CPGs exhibit certain properties of adaptation to the environment: Both the nervous system, composed of coupled neural oscillators, and the musculo-skeletal system have their own nonlinear oscillatory dynamics, and it has been demonstrated that, during locomotion, some recursive dynamics occurs between these two systems. This phenomenon, termed mutual entrainment, emerges spontaneously from the cooperation among the systems' components in a self-organized way (Taga et al., 1991). That is, natural periodic motion, set close to the natural (resonant) frequency of the mechanical body, is achieved by the entrainment of the CPGs to a mechanical resonance by sensory feedback. The feedback is non-essential for the rhythmic pattern generation itself, but rather modifies the oscillations in order to achieve adaptation to environmental changes.

In the remainder of this chapter, the use of CPGs in connection with bipedal robot control will be discussed, with particular emphasis on CPG network optimization aimed at achieving the concerted activity needed for bipedal locomotion. However, first, a brief introduction to various CPG models will be given.

2. Biological and analytical models for CPGs

2.1 Models from biology

From biological studies, three main types of neural circuits for generating rhythmic motor output have been proposed, namely the closed-loop model, the pacemaker model, and the half-center model.

The closed-loop model was originally proposed for the salamander (Kling & Székely, 1968). In some way it resembles the half-center model (see below), but the interneurons are organized in a closed loop of inhibitory connections. There are corresponding pools of motor neurons activated, or inhibited, in sequence, allowing for a finer differentiation in the activation of the flexors and extensors, respectively.

In the pacemaker model, rhythmic signals result as an intrinsic cell membrane property, involving complex interaction of ionic currents, of a group of pacemaker cells. The electrical impulses that control heart rate are generated by such cells. The pacemaker cells drive flexor motor neurons directly, and bring about concurrent inhibition of extensor motor neurons through inhibitory interneurons. These two models are further described in (Shephard, 1994).

The half-center model, mentioned above, was suggested by Brown (Brown, 1914) in order to account for the alternating activation of flexor and extensor muscles of the limbs of the cat during walking. Each pool of motor neurons for flexor or extensor muscles is activated by a corresponding half-center of interneurons, i.e. neurons that send signals only to neurons and not to other body parts (such as muscles). Another set of neurons provides a steady excitatory drive to these interneurons. Furthermore, inhibitory connections between each half-center of interneurons ensure that when one half-center is active, the other is being suppressed. It was hypothesized that, as activity in the first half-center progressed, a process of fatigue would build up in the inhibitory connections between the two half-centers, thereby switching activity from one half-center to the other (Brown, 1914). Since then, support for the half-center model has been found in experiments with cats (Duysens & Van de Crommert, 1998).

2.2 Computational CPG Models

In mathematical terms, CPGs are usually modeled as a network of identical systems of differential equations, which are characterized by the presence of attractors, i.e. bounded subsets of the phase space to which the dynamics becomes confined after a sufficiently long time (Ott, 1993). Usually, a periodic gait of a legged robot is a limit cycle attractor, since the robot periodically returns to (almost) the same configuration in phase space.

Several approaches for computational modeling of the characteristics of CPGs can be found in the literature: Drawing upon neurophysiological work on the lamprey spinal cord, Ekeberg and co-workers have studied CPG networks based on model neurons ranging from biophysically realistic neuronal models, describing the most important membrane currents and other mechanisms of importance (Ekeberg et al., 1991), to simple connectionist-type non-spiking neurons (Ekeberg, 1993). The use of the biophysical models makes it possible to compare the simulation results directly with corresponding experimental data. The advantage of using the simpler model, on the other hand, is the weak dependence of certain parameters that are hard to measure experimentally.

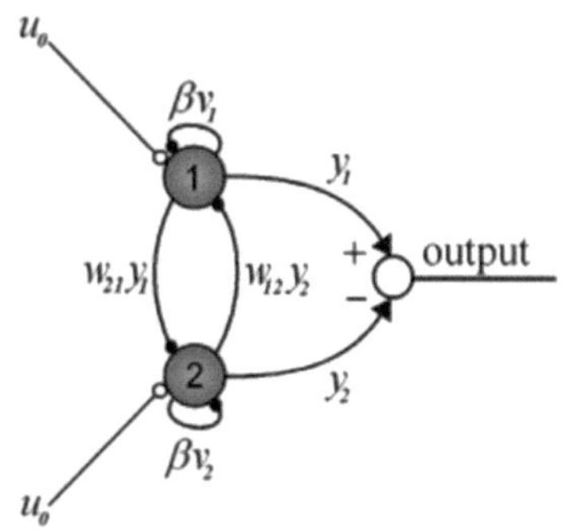

Fig. 1. The Matsuoka oscillator unit. The nodes (1) and (2) are referred to as neurons, or cells. Excitatory connections are indicated by open circles, and inhibitory connections are indicated by filled disks.

However, in this work the CPG model formulated in mathematical terms by Matsuoka (Matsuoka, 1987) has been used for the development of CPG networks for bipedal walking. The Matsuoka model is a mathematical description of the half-center model. In its simplest form, a Matsuoka CPG (or oscillator unit) consists of two neurons arranged in mutual inhibition, as depicted in Fig. 1. The neurons in the half-center model are described by the following differential equations (Taga, 1991):

$$\tau_u \dot{u}_i = -u_i - \beta v_i + \sum_{j=1}^{N} \omega_{ij} y_j + u_0 \tag{1}$$

$$\tau_v \dot{v}_i = -v_i + y_i \tag{2}$$

$$y_i = \max(0, u_i) \quad (3)$$

where u_i is the inner state of neuron i, v_i is an auxiliary variable measuring the degree of self-inhibition (modulated by the parameter β) of neuron i, τ_u and τ_v are time constants, u_0 is an external tonic (non-oscillating) input, w_{ij} are the weights connecting neuron j to neuron i, and, finally, y_i is the output of neuron i. Two such neurons arranged in a network of mutual inhibition (a half-center model), form an oscillator, in which the amplitude of the oscillation is proportional to the tonic input u_0. The frequency of the oscillator can be controlled by changing the values of the two time constants τ_u and τ_v. If an external oscillatory input is applied to the input of a Matsuoka oscillator, the CPG can lock onto its frequency. Then, when the external input is removed, the CPG smoothly returns to its original oscillation frequency. This property, referred to as entrainment, is highly relevant for the application of the Matsuoka oscillator in adaptive locomotion (Taga, 1991).

3. CPGs in bipedal robot control

Generating robust gaits for bipedal robots using artificial counterparts to biological CPGs is an active field of research. The first results in this field were obtained using simple 2D models, and somewhat later, simplified 3D models. The most recent results, however, cover the use of realistic 3D simulations often corresponding to real, physical robots (Righetti & Ijspeert, 2006). Several results have also been implemented using real robots, involving both 2D locomotion (Endo et al., 2004; Lewis et al., 2005) and full 3D locomotion (Ogino et al. 2004).

3.1 CPG-based control of simulated robots

In works by Taga and co-workers (Taga et al., 1991; Taga, 2000), a gait controller based on the half-center CPG model was investigated for a 2D simulation of a five-link bipedal robot. By creating global entrainment between the CPGs, the musculo-skeletal system, and the environment, robustness against physical perturbations as well as the ability to walk on different slopes were achieved (Taga et al., 1991). Moreover, the possibility to regulate the step length was realized and demonstrated in an obstacle avoidance task (Taga, 2000).
Reil and Husbands (Reil & Husbands, 2002) used genetic algorithms (GAs) in order to optimize fully connected recurrent neural networks (RNNs), which were used as CPGs to generate bipedal walking in 3D simulation. The GA was used for optimizing weights, time constants and biases in fixed-architecture RNNs. The bipedal model consisted of a pair of articulated legs connected with a link. Each leg had three degrees-of-freedom (DOFs). The resulting CPGs were capable of generating bipedal straight-line walking on a planar surface. Furthermore, by integrating the gait controller with an auditory input for sound localization, directional walking was achieved.
In a recent work by Righetti and Ijspeert, a system of coupled nonlinear oscillators was used as programmable CPGs in a bipedal locomotion task (Righetti & Ijspeert, 2006). The CPG parameters, such as intrinsic frequencies, amplitudes, and coupling weights, were adjusted to replicate a teaching signal corresponding to pre-existing walking trajectories. Once the teaching signal was removed, the trajectories remained embedded as the limit cycle of the dynamical system. The system was used to control 10 joints in a 25 DOF simulated HOAP-2 robot (the remaining joints were locked). It was demonstrated that, by varying the intrinsic frequencies and amplitudes of the CPGs, the gait of the robot could be modulated in terms

of walking speed and step length. By continuously decreasing the speed, the robot could be brought to a halt in a controlled way. It was also possible to generate backwards walking by simply inverting the sign of all frequency parameters in the CPG controller.

3.2 CPG-based control of real, physical robots

The examples given above involve simulated robots. However, some studies involving CPG-based control of real, physical robots have also been made. In work by Endo and co-workers (Endo et al., 2004) and Lewis and co-workers (Lewis et al., 2005), controllers based on the half-center CPG model were employed. The robots used were, however, somewhat simplified, having only 4 DOFs each. Further simplification included a supporting rod that was attached to the robots and to the floor, in order to prevent the robots from falling over sideways. A more complex robot was used in (Ogino et al., 2004) in a gait generation task. Here, reinforcement learning based on visual feedback was used to adjust the parameters of a CPG-based controller for a HOAP-1 robot, in order to achieve directional walking.

3.3 Optimization of CPG-based controllers

The examples given above confirm that CPGs are indeed suitable for generation of gaits and other types of repetitive motions. However, in most studies, the design of the CPG networks has commonly been carried out manually, in an intuitive manner, e.g. (Taga, 2000; Ogino et al., 2004). This is a time-consuming and difficult process, and it may also lead to suboptimal performance. By contrast, GAs are well suited for structural optimization, i.e. modification of the CPG network structure during optimization. However, even in cases where GAs have been applied, e.g. (Reil & Husbands, 2002), the approach has generally been restricted to parametric optimization in a network of fixed architecture. Furthermore, the use of predefined trajectories, as in (Righetti & Ijspeert, 2006), limits the approach only to those situations for which such trajectories exist. Nevertheless, the results presented in (Righetti & Ijspeert, 2006) reveal one of the advantages associated with a CPG-based approach, namely the ability of the corresponding controllers to smoothly change the gait pattern online, by simple parameter modification.

The work presented in the remainder of this chapter concerns simultaneous parametric and structural optimization of CPG networks in a gait generation task, using GAs as the optimization tool. Some early results from this study can also be found in (Wolff et al., 2006). When generating bipedal locomotion by artificial evolution from a starting point of essentially random CPG networks (or other controllers, for that matter), a great challenge concerns the definition of the fitness function. Using, for example, the distance covered in the initial forward direction as the fitness measure commonly results in controllers that simply throw the body of the robot forward, rather than walking. While actual walking would certainly result in higher fitness values, such solutions are very hard to find, given the easily accessible local optimum found by throwing the robot's body forward. Adding constraints on body posture as part of the fitness measure will in most cases only lead to other non-desirable gaits that display very limited similarity with human-like gaits. Thus, in the experiments presented below, rather than trying to evolve an upright, human-like bipedal gait by an ingenious definition of a fitness function, a supporting structure will be used for the purpose of helping the robot to balance as it starts to walk. Some different strategies for subsequently removing this support, while maintaining a dynamically stable gait, will then be investigated.

In the simulation experiments presented here, a fully three-dimensional bipedal robot with 14 DOFs, shown in the leftmost panel of Fig. 2, was used. The simulated robot weighs 7 kg and its height is 0.75 m. The distance between the ground and the hips is 0.45 m. The waist has 2 DOFs, each hip joint 3 DOFs, the knee joints 1 DOF each, and the ankle joints 2 DOFs each, as illustrated in the in Fig. 2 (second panel from the left).
The simulations were carried out using the EvoDyn simulation library (Pettersson, 2003), developed at Chalmers University of Technology. Implemented in object-oriented Pascal, EvoDyn runs both on Windows and Linux platforms and is capable of simulating tree-structured rigid-body systems. Its dynamics engine is based on a recursively formulated algorithm that scales linearly with the number of rigid bodies in the system (Featherstone, 1987). For numerical integration of the state derivates of the simulated system, a fourth order Runge-Kutta method is used. Visualization is achieved using the OpenGL library.

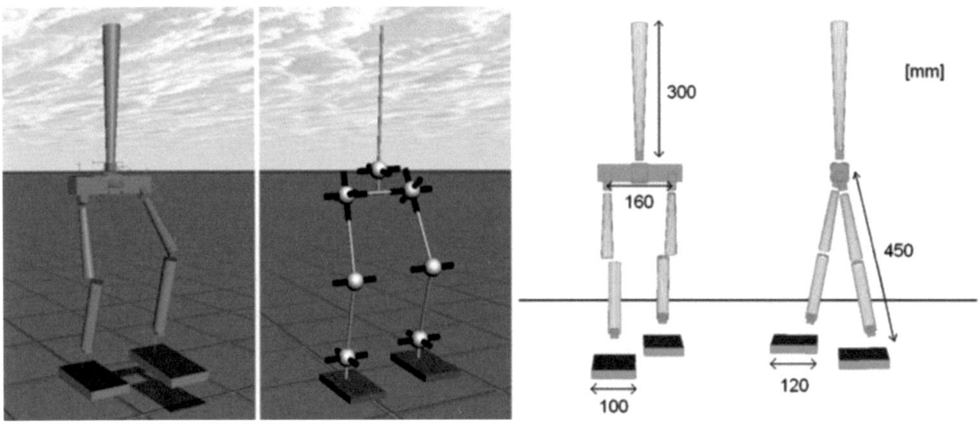

Fig. 2. The two panels to the left show the simulated robot and its kinematics structure with 14 DOFs, used in this work. The panels on the right show the dimensions of the robotic body.

In the simulation experiments, the output of the CPG network was either interpreted as motor torques, which were applied directly to the joint actuators, or as desired joint angles for the robot's joints to follow. In the latter case, a standard PD controller was implemented in order to generate the adequate torques to the joint actuators.

4. Optimization of CPG networks

This section presents the results from experiments conducted with the 14 DOF simulated robot introduced above. First, however, the setup of the experiments will be described in some detail, starting with a description of the fitness measure used in connection with the GA.
The method for CPG network generation presented here is intended for use in connection with fully three-dimensional bipedal robots, with many DOFs. For such robots, designing a CPG network by hand is a daunting task. Instead, in this method, a GA is used for carrying out structural and parametric optimization of the CPG network, with the aim of achieving bipedal walking, without specific reference trajectories. The approach poses many challenges, particularly since no *a priori* knowledge of the needed control signals is available. In fact, in the experiments presented below, the CPG networks have been evolved

starting from randomly generated initial populations (the members of which are referred to as *individuals*, following standard GA terminology) of such networks. Thus, for experiments of this kind, the first challenge is to choose a suitable fitness measure that will favor CPG networks capable of executing an upright bipedal gait.

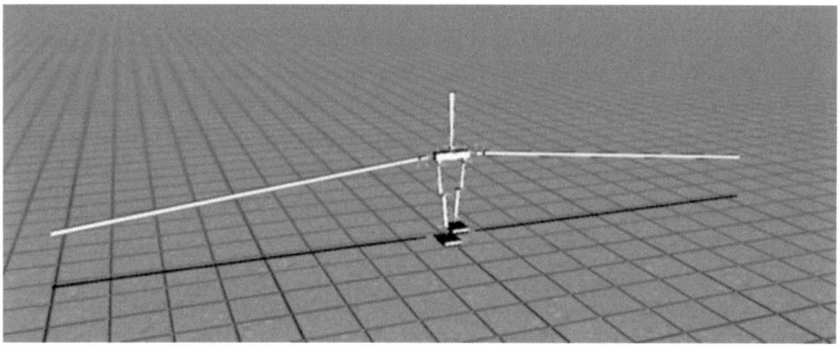

Fig. 3. The posture-support structure used in this work for helping the robot to balance as it starts to walk.

4.1 Fitness measure

How should one judge the motion patterns generated in the beginning of the evolutionary process, given the fact that, in all likelihood, none of the individuals are able to walk at that stage? The simplest way, perhaps, would be to judge the individuals by their ability to move forward, i.e. to use, as the fitness measure, simply the covered distance in the initial forward direction, starting each evaluation with the simulated robot in an upright position. By thus favoring individuals that can carry out some form of movement in the right direction, one may assume that, sooner or later, individuals will appear that are able to walk indefinitely. This, however, is not what typically happens. Instead, it is more common that the evolutionary process quickly discovers a local optimum in the fitness space, such that the robot simply throws its body forward, thus receiving a relatively high fitness compared to robots that simply fall down at the starting position. Once this local optimum has been found, the evolutionary process often gets stuck in some sort of crawling motion, or even a tripod-like gait in which the torso is used as a third leg. Of course, such mediocre results can, in principle, be avoided by adding constraints on body posture as part of the fitness measure. However, finding relevant constraints is a difficult task, especially in fully three-dimensional simulations. Usually, a modification of the fitness measure simply leads to the discovery of other unnatural (and thus undesirable) gaits.

In order to deal with this problem, an approach has been proposed that uses a supporting structure (shown in Fig. 3) added to the robot to help it maintain its balance as it starts to walk (Wolff et al., 2006). Introducing posture support enables those individuals that produce some sort of repetitive leg motion (but are initially unable to maintain balance) resulting in forward motion to gain high fitness. Another benefit of this approach is that the fitness measure can be left quite simple, e.g. the distance covered in the forward direction (even though some additional punishment must be introduced as well, as described in Section 4 below). On the other hand, the problem of when, and how, to remove the support must be considered instead. Several strategies for eliminating the support have therefore been investigated. Despite its appearance, it is modeled as a *massless* structure and therefore it does not affect the robot's dynamics, unless it starts to fall.

4.2 Structural and parametric optimization

The elementary CPG network structure, shown in Fig. 4, has been designed in a way that each joint is assigned a half-center oscillator (depicted in Fig. 1). Thus, in this case, the structural optimization part consists of deciding the presence, or absence, of the connections between the CPGs for different joints. Furthermore, motivated by the fact that the movements of the left and right parts of a bipedal robot are symmetrical during walking, symmetry constraints on the CPG network were imposed. Thus, the structure of the CPGs on the right side of the robot mirrors that of the left side, considerably reducing the size of the search space for the GA. For example, the connection weight between the left *hip* and the left *knee* is equal in value to the weight connecting the right *hip* to the right *knee*.

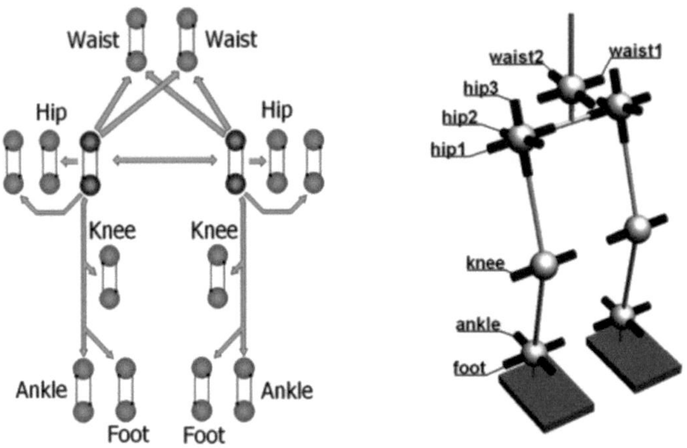

Fig. 4. Left panel: the structure of the CPG network used in the simulations, with the hip_1 joint CPG marked with a darker shade of gray. The connections between the individual oscillators are represented, in a slightly simplified way, by the arrows in the figure; see the text for a more detailed explanation. Right panel: The robot's kinematics structure with joint names indicated.

The CPG network is further constrained such that only the hip_1 CPG, responsible for rotation in the sagittal plane, can be connected to all the other ipsilateral[1] joint CPGs, the corresponding contralateral hip_1 CPG, and the *waist* CPGs. Note, however, that the hip_1 CPG on a given side, can only receive connections from the corresponding contralateral hip_1 CPG, see Fig. 4 for more details. The remaining joint oscillators do not transmit any signals at all. The introduction of these constraints was motivated by several preliminary experiments, the results of which indicated that the evolutionary process was considerably less efficient when a fully connected CPG network was used, yet without improving the final results.

For simplicity, in Fig. 4, the connections between CPG units are depicted with a single arrow. In fact, each such arrow represents a maximum of four unique connections between two joint oscillators, as indicated in Fig. 5. This figure depicts, as an example, the possible connections between the hip_1 CPG and the *knee* CPG. Thus, given the symmetry and connectivity constraints presented above, a total of 8x4 = 32 connections are to be determined by the GA. Note, that it is possible that the GA may disable all connections, w_1 –

[1] The term *ipsilateral* refers to the same side of the body, and is thus the opposite of the term *contralateral*.

w_4 between two CPG units. If this happens, one connection will be forced between the two CPGs. This is motivated by the fact that, without any interconnection, no synchronization between the two joints will be possible. In this setup, the forced connection will be added between the flexor neuron in the transmitting CPG, and the flexor neuron in the receiving CPG. There is, however, no particular reason for preferring this connection; Choosing any other connection would not make any fundamental difference.

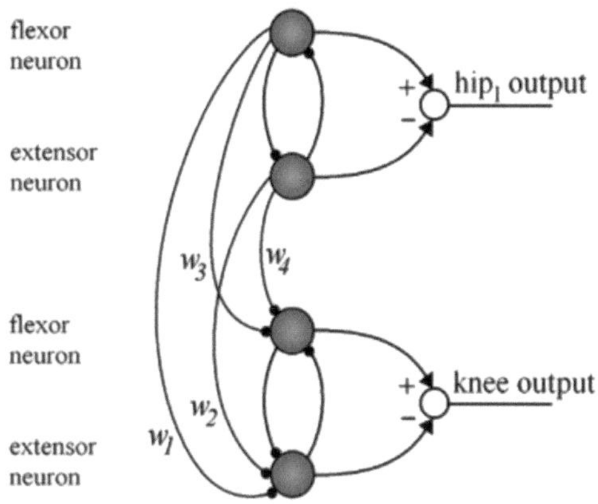

Fig. 5. All possible connections between the *hip*$_1$ CPG and the *knee* CPG.

For optimization of the CPG network, a standard genetic algorithm, (Holland, 1975), was chosen. In the GA, the CPG network of a given individual was represented by two chromosomes; one binary-valued chromosome determining the presence or absence of each of the 32 evolvable connections, and one real-valued chromosome determining the parameter values for those connections that are actually used in a given individual. In addition to these two chromosomes, a third (real-valued) chromosome was used for determining the sign and strength of the different feedback paths, as described in the next subsection.

In early experiments, the internal parameters (see Eqs. (1) and (2)) of the individual half-center CPGs were also evolved by the GA. However, with the supporting structure present, evolution always promoted parameters values that produced unnaturally large steps, so that the support could be exploited for reaching higher fitness values. In such cases, the robot became highly dependent on the supporting structure for balance, making it even more difficult to find an appropriate removal strategy for the support. For this reasons, the internal parameters of the individual half-center CPGs were set to fixed values, generating a motion frequency of the robot's legs approximating that of normal walking. The chosen parameters are shown in Fig. 6 along with the resulting output from the half-center oscillator. These parameters were applied to all CPGs, except for the *knee* joint CPGs and the *waist*$_1$ CPG. In analogy with human gait, the *knee* joint oscillator and the *waist*$_1$ oscillator were tuned to generate a rhythmic pattern with double frequency compared to the other CPGs. Thus, for these joints' CPGs, the τ_u and τ_v values were set to half of the values used for the other CPGs.

In each run, a population of 180 individuals was used. Selection of individuals (for reproduction), was carried out using tournament selection, with a tournament size of 8 individuals. In a given tournament, the individual with the highest fitness value was selected with a probability of 0.75, and a random individual (among the 8) with 0.25

probability. After each selection step, the mutation operator was applied, randomly changing a gene's value with probability $10/N$, where N denotes the total number of genes in the three chromosomes of the individual.

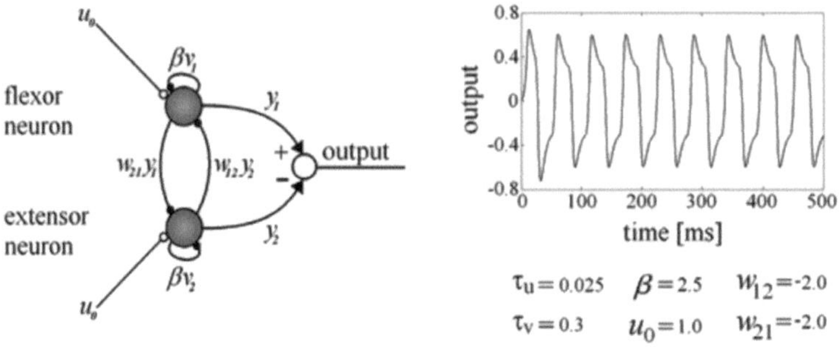

Fig. 6. Left panel: The half-center oscillator assigned to each joint. Right panel: The parameter values used, along with a typical output signal. Note that the *knee* and *waist*$_1$ joint oscillators used different values of the time constants (see the main text for details).

The fitness measure was taken as the distance walked by the robot in the initial forward direction, x, decreased by the sideways deviation y. The fitness function F, for a given individual i, can thus be expressed as:

$$F(i) = x - |y| \tag{4}$$

In addition to the fitness function, several indirect punishments were introduced during evaluation. For example, since the two-point supporting structure provided no support in the sagittal plane, the robot could still fall to the ground during evaluation, leading to the emergence of motion patterns such as crawling. Avoiding this was important, particularly in the beginning of the evolutionary process, since these individuals could gain better fitness than the very first individuals that walked with an upright body posture. Thus, even though the support was used, the resulting gaits could develop towards gait patterns other than upright walking. Thus, a rule was introduced such that if a robot's hips collided with the ground, the evaluation of that particular individual was terminated. Other punishments or modifications that were also introduced are described in the experiment section.

4.2 Feedback signals

In order to guide the evolutionary process towards an upright and stable bipedal gait, feedback was introduced by measuring the waist angle, thigh angle, and lower leg angle, all relative to the vertical axis. The introduction of feedback paths for the generation of bipedal locomotion has proven to be important in achieving adaptation to the environment and forming the overall gait pattern (Taga et al., 1991; Righetti & Ijspeert, 2006). In the experiments presented here, the structure of the feedback network was determined in advance. However, the actual type of the connection, i.e. whether it is inhibitory or excitatory, and the strength of the feedback paths were determined by the GA. Symmetry constraints were applied also in this case, meaning that the feedback structure of the right side of the robot mirrored that of the left side. Furthermore, motivated by biological findings indicating that tactile feedback from the foot is essential for human locomotion (Van Wezel et al., 1997), a touch sensor in each foot was introduced in the simulations. The signal from the touch sensor was used both as tactile feedback, i.e. to indicate foot-to-ground contact, but also to enable, or prohibit, other feedback signals to be connected to certain CPGs during a specific phase of the gait; see Fig. 7 for an example.

In this section, only some illustrative examples of the feedback network structure are given. For a detailed description of the feedback paths see (Wolff et al., 2006). The feedback signals to an individual CPG are introduced by adding an extra term to Eq. (1), which then becomes:

$$\tau_u \dot{u}_i = -u_i - \beta v_i + \sum_{j=1}^{N} \omega_{ij} y_j + u_0 + f \tag{5}$$

where f denotes the sum of all feedback signals connected to the particular CPG.

When designing the feedback network, the main idea was to trigger the different phases in a gait cycle. For example, for the left hip_1 CPG, the feedback structure was designed in the following way: The flexor and the extensor neuron receive two feedback signals, as shown in the left panel of Fig. 7: One signal from the right-foot touch sensor, and one signal measuring the inclination angles of the left and right upper leg, scaled by the strength of the feedback connections. The feedback signal from the touch sensor is intended to trigger the swing motion of the left leg, as soon as the right foot touches the ground. Feedback from the inclination angles of the upper legs is intended to promote repetitive leg motion in such a way that, when the left leg is moving forward, the right leg should move backwards, and vice versa. The feedback paths connected to the hip_1 CPG are depicted in the left panel of Fig. 7. Note, that the type of connections shown in the figure are chosen to easier demonstrate how the feedback signals were *intended* to affect the motion of the hip_1 joint. However, since the dynamical system of a single CPG unit becomes rather complex in an interaction with other CPGs, the robot, and the environment, it is very hard to predict if the chosen feedback configuration will work as intended. For this reason, both the sign and the strength of each connection were determined by the GA (but not the structure of the feedback paths). Fig. 7 shows an example in which the feedback connection to the flexor neuron of a certain CPG happened to be of opposite sign compared with the feedback connection to the extensor neuron of the same CPG.

In the right panel of Fig. 7, the feedback paths connected to the right hip_2 CPG are illustrated. In same way as above, the illustrated types of connections in the figure intend only to demonstrate how the structure of the feedback for this joint was planned to affect the motion of the hip_2 joint. It does not represent the best connection configuration (in terms of the sign and the strength) chosen by the GA. In the case of hip_2, only the feedback signal measuring the inclination angle of the right upper leg is connected to the oscillator. However, the feedback is only transmitted to the CPG unit if the touch sensor in the right foot is activated, i.e. when the right foot is on the ground. This configuration was chosen with the intention of ensuring that the hip_2 CPG was able to generate more torque in the stance phase, i.e. when the right leg is moving backwards, during which the entire body of the robot is resting on the right leg.

One should notice that by choosing an adequate feedback structure, the feedback signals may indirectly force synchronization between different joints, even if the connection weights between the joints' CPG units may not necessarily ensure this by themselves. In this work, a total of 20 parameters, determining the sign and the strength of the different feedback paths were evolved by the GA; see (Wolff et al., 2006) for details.

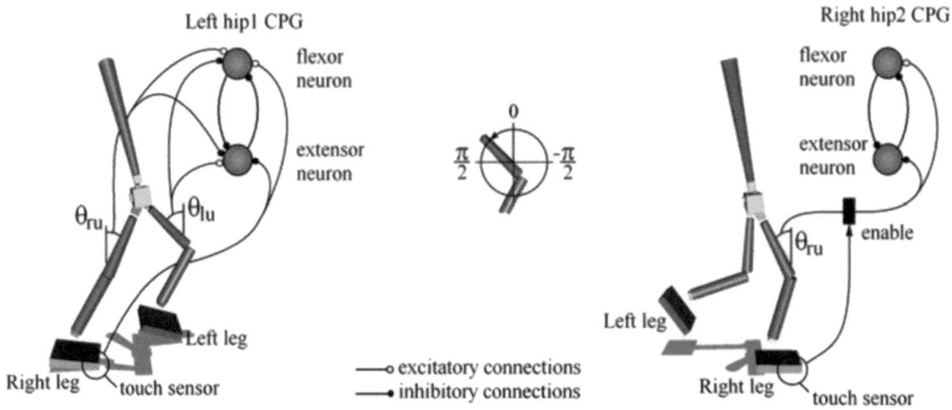

Fig. 7. The simulated robot with the hip_1 and hip_2 joint CPGs and the corresponding feedback, with specific connection types (i.e. signs) chosen to illustrate how the feedback was intended to affect the corresponding joint's motion (see the main text for details). The hip_1 flexor neuron rotates the leg in the counter-clockwise direction in the sagittal plane[2], while the hip_2 flexor neuron rotates the leg in the clockwise direction in the frontal plane (seen from behind the robot).

4.3 Experimental setup

In this subsection, the setup used in the simulation experiments will be discussed. Two approaches for generating dynamically balanced bipedal locomotion will be described, the difference being the method for removing the support structure.

In both approaches, a two-point supporting structure, as depicted in Fig. 3, was used. Initially, a four-point support was used, in an open-loop fashion without feedback signals. However, the GA exploited the support to such an extent that no useful results were obtained (Wolff et al., 2006). For example, the individuals obtained walked with unnaturally large steps and were unable to balance without constant use of the supporting structure. Thus, instead a two-point structure was introduced in order to minimize exploitation (Wolff et al., 2006), while still providing some support. The contact points of the support structure were placed 2 m from the robot and 0.25 m above the ground. This configuration was chosen to ensure small sideways leaning angle and at the same time allow the robot to bend its knees without the support touching the ground.

Method 1: Evolution in two steps

In this approach, the evolutionary process was divided into two steps. During the first step, the posture support was present and the hip_2, hip_3, and *ankle* joints were locked. In this step, the intention was to evolve a CPG network capable of producing a stable upright gait in the sagittal plane. Once a stable individual had been obtained, it was cloned, creating a new population consisting of copies of this individual. At this stage, the second step was initiated, in which the support was removed and the GA was assigned the task of finding a

[2] The **sagittal** plane is a vertical plane running from front to back through the body, while the **frontal** plane is a vertical plane running from side to side through the body, perpendicular to the sagittal plane

way to balance the robot in the frontal plane as well. Before the second step of evolution was started, the *hip*$_2$ and *ankle* joints were unlocked, and the corresponding genes were set randomly for each individual in the population. Since the *waist* joints also affect the frontal plane balance, the corresponding genes were randomly initiated as well. The remaining genes (which take identical values for all individuals) ensured sagittal plane balance and were therefore left unchanged in the second step of evolution. In this approach the *hip*$_3$ joint remained locked during the entire procedure. The fitness measure used in both steps was the distance walked by the robot in the initial forward direction, decreased by the sideways deviation, as formulated in Eq. (4).

Since the two-point supporting structure provides no support in the sagittal plane, the evolutionary procedure found several easily accessible solutions for maintaining sagittal balance. In some cases, the evolved individuals used the torso as a third leg, or the knees, for maintaining balance. Other examples include forward motion using somersaults or walking with unnaturally large steps. In order to prevent the individuals from developing other means of locomotion than upright, human-like gaits, several constraints had to be introduced. In Method 1, the following rules and modifications were added:

a) The contact points in the knee joints and the torso, used for detecting ground collisions, were removed.
b) If the robot's hip collided with the ground, the evaluation run of that individual was aborted.
c) If the robot's hips were located less than a 0.15 m above the ground, the supporting structure was immediately removed, and was not further used during the evaluation of the individual.

The removal of the robot's contact points in the torso and the knees eliminated the possibility of misusing these body parts for support. Ending a run as soon as the robot's hip collided with the ground, efficiently removed all crawling solutions. Finally, punishing the individuals for having the hips too close to the ground successfully removed those individuals trying to take unnaturally large steps (for improved sagittal plane balance). If the step length is large and the support is removed in this way, the robot will most likely be unable to maintain the frontal plane balance. Thus it will fall to the ground, terminating its evaluation. Using these rules, the evolutionary procedure was strongly biased towards individuals walking with an upright posture.

Initially, the posture support was intended to be present during the entire evaluation time of the first step, and then completely removed in the second step. However, it turned out that such an approach gave no obvious way of deciding at what point to interrupt the first step; This simply had to be judged by the experimenter in an *ad hoc* manner. Evolving for too long, for example, led to individuals that were indeed walking rapidly but, at the same time, over-exploiting the posture support, making the produced gait unsuitable for the second step, where the support was removed. For this reason, the time during which the support was present in the first step was changed from the entire evaluation period to only the first two seconds of the evaluation. This arrangement is motivated by the assumption that it is during the starting sequence, before entering the gait cycle, that the individuals are most vulnerable in terms of frontal plane balance. For this reason, in these experiments, the posture support was also present during the initial two seconds in the next step. The details regarding the experiments carried out using Method 1 are shown in Table 1. The results from the simulations are given in the next subsection.

Case	Support	t	DOFs	CPG genes	Feedback genes
1.1	2 sec, 2-point, 0.25, 2	40	8	40	14
1.2	2 sec, 2-point, 0.25, 2	40	12	32	10

Table 1. Details regarding the experiments made using Method 1. The first column indicates which step in Method 1 is being considered. In the column labeled *support*, the term *2 sec* indicates that the (two-point) support was only present during the first two seconds of the evaluation time. The numbers i, j in the same column denote the initial placement of the contact points in a given run, where i is the height above the ground [m], and j is the horizontal distance from the hip [m]. The evaluation time (third column) is denoted by t [s]. The fourth column (DOFs) gives the number of active joints, and the last two columns indicates the number of genes evolved by the GA.

Due to symmetry constraints, described in subsection 4.2, only those connections in the CPG network that correspond to one side of the robot, and the *waist* joints, must be evolved. Since there are four unique connections for each joint, a total of 8 genes per joint must be evolved, i.e. four genes determining the presence or absence of the connections, and four genes determining the parameters of the connections that are actually used. Thus, in the first step of Method 1, five unique joint configurations (hip_1, *knee*, *foot*, $waist_1$ and $waist_2$) must be evolved, leading to a total of 8x5=40 genes. Based on the feedback equations given in (Wolff et al., 2006) the number of genes needed to determine the characteristics of the feedback paths for these joints sums up to 14.
In the second step of Method 2, the genes corresponding to the hip_1, *knee* and the *foot* joints were not changed. Instead, the $waist_1$, $waist_2$, hip_2, and the *ankle* joints were evolved, bringing the total number of CPG genes to 4x8=32. The number of genes needed for the feedback network sums up to 10 since the feedback paths connected to hip_1, *knee* and the *foot* joints require four more parameters than the hip_2, and the *ankle* joints (Wolff et al., 2006).

Method 2: One-step evolution with 12 active DOFs

In Method 1, the supporting structure was only allowed in the beginning of an evaluation run, motivated by the assumption that it is during the starting sequence that the individuals are most vulnerable in terms of balance. Since it was found that this assumption turned out to be correct, the somewhat cautious two-step approach used in Method 1 is perhaps unnecessary. Therefore, in Method 2, the evolution of CPG networks was carried out in one single step during which all joints except the hip_3 joint were unlocked. Apart from this change, the additional rules, presented in connection with Method 1, were also valid here.

Case	Support	t	DOFs	CPG genes	Feedback genes	Type of CPG
2	2 sec, 2-point, 0.25, 2	40	12	56	18	torque
3	2 sec, 2-point, 0.25, 2	40	12	56	18	angle

Table 2. Details regarding the experiments made using Method 2. As in Method 1 the term *2 sec* in the column labeled *Support* indicates that the (two-point) support was only present during the first two seconds of the evaluation time. In the last column, the type of output produced by the CPG network is given.

In addition, both a torque-generating CPG network and a network for generating joint angles were tested. In the latter case, the appropriate torques were generated using a standard PD controller, implemented as

$$T_i = k_p(\theta_r - \theta_i) - k_d\dot{\theta}_i \ , \quad (6)$$

where T_i is the torque applied to joint i, k_p is the proportional constant, k_d is the derivative constant, θ_r is the desired angle, and θ_i is the current angle of the ith joint. In the experiment described in the 2nd row of Table 2, k_p and k_d were chosen as 20.0 and 4.0, respectively.
In both cases, the fitness measure was the distance walked by the robot in the initial forward direction, decreased by the sideways deviation, as formulated in Eq. (4).

4.4 Results

Method 1:
In Method 1, the generation of upright bipedal locomotion was carried out in a two-step evolutionary procedure. In the first step, during which the *hip*$_2$, *hip*$_3$ and *ankle* joints were locked, an individual walking with a speed of 0.58 m/s was evolved after 400 generations. The results for this individual are shown in Table 3, case 1.1. As Table 3 implies, the individual evolved here was fairly unstable and able to maintain balance only for one minute. Note that the stability in the frontal plane was only ensured by the *waist* joint, since the *hip*$_2$ and *ankle* joints were locked. Even though slower, but stable, individuals were found earlier in evolution, i.e. individuals that maintained balance during the testing time of 20 minutes, they were not suitable as a starting condition for the second step because of the way these individuals were walking: They were heavily leaning the torso forward, or backwards, keeping it motionless to create a sort of counterweight for balance. By contrast, the fastest individual maintained an active upright posture during the entire walking sequence.

Case	*Method*	*F*	*v*	*DOFs*	*Resulting gait*
1.1	1, step1	19.54	0.56	8	slow, stable for 60 sec.
1.2	1, step2	23.09	0.58	12	slow, stable for 12 min.
2	2, torque	35.56	0.90	12	fast, stable for 42 sec.
3	2, angle	52.46	1.31	12	fast, stable for >20 min.

Table 3. Results from the trials made using Method 1 and Method 2. In the column labeled *Method*, the method and the current step (or type of output) is given. *F* is the obtained fitness [m], and *v* denotes the average locomotion speed [m/s] of the robot during the evaluation period. DOFs is the number of active joints, and the last column gives a short description of the resulting gaits.

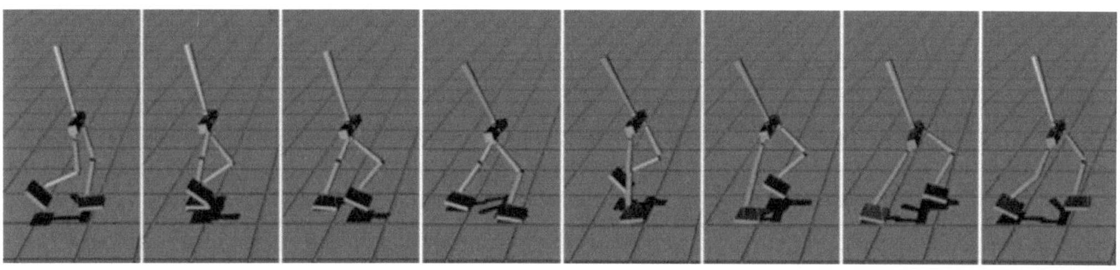

Fig. 8. The best evolved gait pattern, using Method 1, with 12 active DOFs. The details of the corresponding individual are shown on the 2nd row of Table 3.

In the second step, the *hip*$_2$ and *ankle* joints were unlocked and the best individual, found within 100 generations, was able to walk with a speed of 0.58 m/s. However, when the individual was tested for longer evaluation times, it fell after 12 minutes, see case 1.2 in Table 3. The resulting gait is depicted in Fig. 8.

Method 2, torque-generating CPGs:
While the CPG networks in Method 1 were evolved in two steps, with the hip_2, hip_3 and *ankle* joints locked at first, in Method 2, all joints, except the hip_3 joint, were evolved in one step. When Method 2 was tested with a torque-generating CPG network, the main difference compared to Method 1 was the difficulty to evolve stable individuals. While the average speed of the best individual was significantly improved, the balance capabilities were at the same time reduced, and it was only after 800 generations that the best individual was able to balance during the entire evaluation time of 40 seconds. However, this result was not satisfactory since the individual did not walk more than 42 seconds when the evaluation time was expanded; see Table 3, case 2 for details.
The resulting motion pattern resembled the one obtained in Method 1, shown in Fig. 8, with the exception of the foot joint, which was now more active during the lift-off phase. Nevertheless, the motion of the hip_2 and *ankle* joints appeared rather nervous, suggesting that more restraining feedback to these joints' CPGs is necessary, preferably from their own joint angles, something that was not included in the current feedback network.

Method 2, angle-generating CPGs:
When Method 2 was tested in connection with angle-generating CPGs, the results were significantly improved, compared to the previous results, both in terms of locomotion speed and stability of the best individual, but also in terms of evolution time. The best individual found with this approach reached a distance of 52.46 m, having an average locomotion speed of 1.31 m/s (see Table 3, case 3). The corresponding individual, in the case of torque-generating CPGs, reached a distance of 35.56 m walking at an average locomotion speed of 0.90 m/s. Furthermore, stable individuals capable of walking during the entire evaluation time of 40 seconds emerged quite early, around generation 25, compared to the case of torque-generating CPGs, where such results were obtained only after 800 generations. While the best individual using torque-generating CPGs was unable to maintain balance for more than 42 seconds, the best individual equipped with the angle-generating CPGs was able to walk without falling during a testing time of 20 minutes. The resulting gait is depicted in Fig. 9.

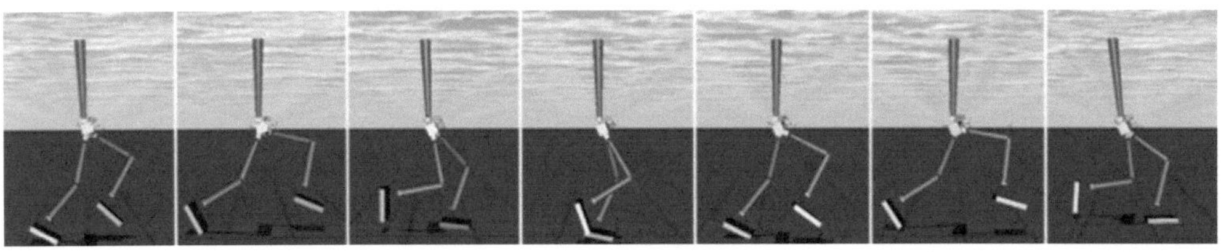

Fig. 9. The best evolved gait pattern obtained using Method 2, with 12 active DOFs and with angle-generating CPGs. The details of the corresponding individual are shown on 4th row of Table 3.

4.5 Discussion
The results of the study described above demonstrate the feasibility of using artificial evolution for designing CPG networks, as well as optimizing the parameters of such networks, in order to achieve efficient and stable bipedal walking for a simulated robot.

Both methods introduced here solved the problem of generating robust bipedal gaits. However, the first method showed some limitations in generating gaits with long-term stability. This drawback was a result of splitting up the evolutionary process into two parts. Since the hip_2 and *ankle* joints were locked during the first phase of Method 1, stability in the frontal plane was only ensured by the torso. As a result, the gaits had to be less dynamic in order for the robot to maintain balance in the frontal plane. Thus, these gaits were less suitable for generalization to full 3D, i.e. with all joints unlocked. Yet, the gaits evolved with this method were more stable than the solutions found in Method 2, when torque-generating CPGs were used in the latter method. The frontal plane balance seems to be the most critical point when trying to maintain an upright body posture. Thus, more feedback, based on the inclination of the body, should preferably be introduced to restrain the motion of some critical joints like the hip_2 and the *ankle*.
The best results were obtained with angle-generating CPGs. The major contribution to the improved results was the better motion control of the hip_2 and *ankle* joints, which were now easily restrained using the PD controller. However, a problem with this approach is that the individuals start to take larger and larger steps, as evolution progresses. In order to prevent the occurrence of such large steps, small perturbations should perhaps be introduced during the evaluation of the individuals, preferably in the frontal plane. This should force the robot to walk in a more cautious manner. The need for the supporting structure during the initial seconds indicates that the CPG network handling the gait cannot fully handle also the start-up of the walking cycle. Thus, an extra controller, based on CPGs or other approaches, should be used for the start-up sequence. It should then be tuned to enter the walking cycle and hand over control to a CPG network in a smooth way.

5. Current and future directions

Throughout the CPG experiments presented here, the connection paths of the feedback network were pre-defined. Since the feedback paths were specified in an *ad hoc* manner, there is certainly no guarantee that these paths are optimal. Therefore, a topic for further work would be to investigate whether the structure of the feedback network could be improved by applying an optimization method. Furthermore, the constraints that were added to the CPG networks, e.g. the restricted number of connections, could also be removed in order to evolve gait patterns from fully connected CPG networks instead.
In the case of bipedal robot control in a dynamically changing environment, more advanced control of posture and balance would be required. Thus, investigating how different kinds of feedback, such as information regarding foot joint angles and foot-to-ground contact, can influence the quality of the generated gaits is another relevant topic, particularly in view of the importance of tactile sensory feedback in the case of human bipedal walking (Ogihara & Yamazaki, 2001).
As shown in Fig. 6 above, the internal parameters of each CPG unit were pre-defined as well, in order to prevent the optimization procedure from promoting parameter values that generate unnaturally large steps, in which case the support could be exploited for reaching higher fitness values. However, in Section 4.4, it was shown that the support was only necessary during the first seconds of each evaluation run (see Case 3, Method 2). Thus, with the removal of the support after a few seconds of walking, the possibility of exploiting it was strongly reduced, making evolution of individual CPG parameters

possible. Results from preliminary experiments of that kind indicate similar performance of the evolved gaits (in terms of walking speed and stability) as that obtained in Case 3 of Method 2. However, an interesting difference in the evolution of the gaits was observed. In the experiments described in Section 4.4, the first stable gaits that emerged had basically the same walking pattern as the final gait evolved in the same run. The major contribution to the fitness increase was the increase in walking speed, rather than step length. However, in the approaches where the individual CPG parameters are also evolved, the first stable gaits that emerge have considerably smaller step length than the final gait evolved in that run.

Another topic that requires further development concerns gait transition from, say, a walking gait to a running gait. Two preliminary experiments have recently been carried out. In the first case, a stop-and-go routine was accomplished by manually changing the bias values and time constants appropriately. In the second case, a complete and smooth gait transition was realized by abruptly switching control from one CPG network to another. However, those maneuvers required quite some manual tuning of the CPG parameters in order to work fully. In a robot intended for real-world applications, the transitions must be carried out in a more automatic fashion.

Ultimately, the CPG-based controllers should, of course, be implemented in real, physical robots. Preliminary experiments with a small humanoid robot with 17 DOFs have recently been carried out (Wolff et al., 2007). In these experiments the robot's gait was optimized using an evolutionary approach, including structural modifications of the gait control program.

6. References

Brown, T. G. (1911). The intrinsic factors in the act of progression in the mammal, *Proceedings of the Royal Society of London Series B*, Vol. 84, pp. 308–319

Brown, T. G. (1912). The factors in rhythmic activity of the nervous system, *Proceedings of the Royal Society of London Series B*, Vol. 85, pp. 278–289

Brown, T. G. (1914). On the nature of the fundamental activity of the nervous centers: Together with an analysis of the conditioning of rhythmic activity in progression, and a theory of the evolution of function in the nervous system, *Journal of Physiology*, Vol. 48, pp. 18–46

Duysens, J. & Van de Crommert, H. W. A. A. (1998). Neural control of locomotion, Part 1: The central pattern generator from cats to humans, *Gait and Posture,* Vol. 7, No. 2, pp. 131-141

Ekeberg, Ö; Wallén, P.; Lansner, A.; Tråvén, H.; Brodin, L. & Grillner, S. (1991). A computer based model for realistic simulations of neural networks. I: The single neuron and synaptic interaction, Biological Cybernetics, Vol. 65, No. 2, pp. 81–90

Ekeberg, Ö. (1993). A combined neuronal and mechanical model of fish swimming, Biological Cybernetics, Vol. 69, No. 5-6, pp. 363–374

Endo, G.; Morimoto, J.; Nakanishi, J. & Cheng, G. (2004). An empirical exploration of a neural oscillator for biped locomotion control. In *Proceedings of the 2004 IEEE International Conference on Robotics and Automation (ICRA'04).*

Featherstone, R. (1987). *Robot Dynamics Algorithms*. Kluwer Academic Publishers.

Grillner, S. (1996). Neural networks for vertebrate locomotion, Scientific American, Vol. 274, pp. 64-69

Grillner, S.; Wallén, P.; Brodin, L. & Lansner, A. (1991). Neuronal network generating locomotor behavior in lamprey: Circuitry, transmitters, membrane properties, and simulation, *Annual Review of Neuroscience*, Vol. 14, pp. 169–199

Grillner, S.; Deliagina, T.; Ekeberg, Ö.; El Manira, A.; Hill, R.; Lansner, A.; Orlovsky, G. & Wallen, P. (1995). Neural networks that coordinate locomotion and body orientation in lamprey, *Trends in Neurosciences*, Vol. 18, No. 6, pp. 270–279

Holland, J. (1975). *Adaptation in Natural and Artificial Systems*. The University of Michigan Press, Ann Arbor, MI, USA.

Kling, U. & Székely, G. (1968). Simulation of rhythmic nervous activities. I. function of networks with cyclic inhibitions. *Kybernetik*, Vol. 5, No. 3, pp. 89–103

Lewis, M.; Tenore, F. & Eienne-Cummings, R. (2005). CPG design using inhibitory networks. In *Proc Int Conf on Robotics and Automation (ICRA'05)*, IEEE-RAS. Barcelona, Spain.

Matsuoka, K. (1987). Mechanisms of frequency and pattern control in the neural rhythm generators. Biological Cybernetics, Vol. 47, No. 2–3, pp. 345–353

Ogihara, N. & Yamazaki, N. (2001). Generation of human bipedal locomotion by a bio-mimetic neuro-musculo-skeletal model. *Biological Cybernetics*, Vol.84, No.1, pp. 1–11

Ogino, M.; Katoh, Y.; Aono, M.; Asada, M. & Hosoda, K. (2004). Reinforcement learning of humanoid rhythmic walking parameters based on visual information. *Advanced Robotics*, Vol. 18, No. 7, pp. 677-697

Ott, E. (1993). *Chaos in Dynamical Systems*, Cambridge University Press, Cambridge, UK.

Pettersson, J. (2003). EvoDyn: A simulation library for behavior-based robotics, Department of Machine and vehicle systems, Chalmers University of Technology, Göteborg, Technical Report

Reil, T. & Husbands, P. (2002). Evolution of central pattern generators for bipedal walking in a real-time physics environment. *IEEE Transactions in Evolutionary Computation*, Vol. 6, No. 2, pp. 159-168

Righetti, L. & Ijspeert, A. J. (2006). Programmable central pattern generators: an application to biped locomotion control. In *Proceedings of the 2006 IEEE International Conference on Robotics and Automation.*

Shepard, G. M. (1994). *Neurobiology*, chapter 20, pages 435-451. Oxford University Press, 3rd edition, New York.

Taga, G.; Yamaguchi, Y. & Shimizu, H. (1991). Self-organized control of bipedal locomotion by neural oscillators in unpredictable environment, *Biological Cybernetics,* Vol. 65, pp. 147–159

Taga, G. (2000). Nonlinear dynamics of the human motor control - real-time and anticipatory adaptation of locomotion and development of movements. In *Proc 1st Int Symp on Adaptive Motion of Animals and Machines (AMAM'00).*

Takanishi, A.; Ishida, M.; Yamazaki, Y. & Kato, I. (1985). The realization of dynamic walking by the biped walking robot WL-10RD, *Proceedings of the International Conference on Advanced Robotics (ICAR'85)*, pp. 459-466

Van Wezel, B. M. H.; Ottenhoff, F. A. M. & Duysens, J. (1997) Dynamic control of location-specific information in tactile cutaneous reflexes from the foot during human walking. *The Journal of Neuroscience*, Vol. 17, No. 10, pp. 3804-3814

Wolff, K.; Pettersson, J.; Heralić, A. & Wahde, M. (2006). Structural evolution of central pattern generators for bipedal walking in 3D simulation. In *Proc of the 2006 IEEE International Conference on Systems, Man, and Cybernetics*, pp. 227-234

Wolff, K.; Sandberg, D. & Wahde, M. (2007). Evolutionary optimization of a bipedal gait in a physical robot. Submitted to: *IEEE Transactions on Robotics.*

Zehr, E. P. & Duysens, J. (2004). Regulation of arm and leg movement during human locomotion., *Neuroscientist,* Vol. 10, No. 4, pp. 347-361

18

Copycat Hand - Robot Hand Generating Imitative Behaviour at High Speed and with High Accuracy

Kiyoshi Hoshino
University of Tsukuba
Japan

1. Introduction

In recent years, robots demonstrating an aping or imitating function have been proposed. Such functions can estimate the actions of a human being by using a non-contact method and reproduce the same actions. However, very few systems can imitate the behaviour of the hand or the fingers (Bernardin et al., 2005). On the other hand, reports in Neuroscience state that mirror neurons (Rizzolatti et al., 1996; Gallese et al., 1996), which participate in the actions imitated by chimpanzees, are activated only for actions of the hand and fingers, such as cracking peanut shells, placing a sheet of paper over an object, tearing the sheet into pieces, etc.. Moreover, since human beings perform intelligent actions using their hands, it is important to attempt to artificially imitate the actions of hands and fingers in order to understand the dexterity and intelligence of human hands from an engineering viewpoint. The object actions in the case where one "looks at an action performed by others and imitates it" include not only grasping or manipulating objects but also the actions involving "imitating shapes and postures of hands and fingers" of others such as sign language and dancing motions. The latter two types of actions are often more complicated and require dexterity as compared to the former actions.

In the action of imitating "the shape of hands and fingers" such as sign language, it is essential to estimate the shape of the hands and fingers. Furthermore, as compared to the imitation of the actions of the lower limbs, estimating the shape of the hands is significantly more important and difficult. In particular, human hands, which have a multi-joint structure, change their shapes in a complicated manner and often perform actions with self-occlusion in which the portion of one's own body renders other portions invisible. In the case of an artificial system, we can utilize a multi-camera system that records a human hand for imitative behaviours by surrounding it with a large number of cameras. However, all the animals that mimic the motions of others have only two eyes. To realize the reproduction of the actions of hands and fingers by imitating their behaviour, it is desirable to adopt a single-eye or double-eye system construction.

To roughly classify conventional hand posture estimation systems, the following two types of approaches can be used. The first approach is a 3D-model-based approach (Rehg & Kanade, 1994; Kameda & Minoh, 1996; Lu et al., 2003) that consists of extracting the local characteristics, or silhouette, in an image recorded using a camera and fitting a 3D hand model, which has been constructed in advance in a computer, to it. The second approach is a 2D-appearance-based approach (Athitos & Scarloff, 2002; Hoshino & Tanimoto, 2005) that consists of directly comparing the input image with the appearance of the image stored in a database. The former is capable of high-accuracy estimations of the shape, but it is weak against self-occlusion and also requires a long processing time. The latter can reduce the computation time; however, if 3D changes in the appearance of hands are not an issue, which also include the motions of the wrist and the forearm, a large-scale reference database is required, and it becomes difficult to control the robot hand by means of imitation. However, if the fundamental difficulty in the estimation of the hand posture lies in the complexity of the hand shape and self-occlusion, the high-accuracy estimation of the shape will become theoretically possible; this requires the preparation of an extensive database that includes hand images of all the possible appearances with complexity and self-occlusion. The feasibility of this approach depends on the search algorithm used for rapidly finding similar images from an extensive database.

Therefore, in this study, we aim at realizing a system for estimating the human hand shape capable of reproducing actions that are the same as those of human hands and fingers to a level reproducible using a robot hand at high speeds and with high accuracy. With regard to the processing speed, we aimed to achieve short-time processing of a level enabling the proportional derivative (PD) control of a robot. With regard to the estimation accuracy, we aimed at achieving a level of estimation error considered almost equivalent to that by the visual inspection of a human being, namely, limiting the estimation error in the joint angle to within several degrees. In addition to those two factors, dispersion in the estimation time due to the actions of human hands and fingers causes difficulties in the control of a robot. Therefore, we also aimed to achieve uniformity in the estimation time. In particular, in order to conduct high-speed searches of similar data, we constructed a large-scale database using simple techniques and divided it into an approximately uniform number of classes and data by adopting the multistage self-organizing map (SOM) process (Kohonen, 1988), including self-multiplication and self-extinction, and realized the high-speed and high-accuracy estimation of the shape of a finger within a uniform processing time. We finally integrated the hand posture estimation system (Hoshino & Tanimoto, 2005; 2006) with the humanoid robot hand (Hoshino & Kawabuchi, 2005; 2006) designed by our research group; our final system is referred to as the "copycat hand".

2. System construction

2.1 Construction of a large-scale database

First, we conducted measurements of the articular angle data (hereafter referred to as the "key angle data"). In the present study, we determined the articular angle at 22 points per hand in the form of a Euler angle by using a data glove (Cyberglove, Virtual Technologies). For the articular data in the shape of a hand that requires a high detecting accuracy, especially in the search for similar images, a large number of measurements were made. Furthermore, we generated CG images of the hand on the basis of the key articular angle data. The CG editing software Poser 5 (Curious Labs Incorporated) was used for the generation of images.

Second, from the two key angle data, we interpolated a plurality of the articular angle data in optional proportions. The interpolation of the articular data is linear. Moreover, we also generated the corresponding CG images of the hand on the basis of the interpolated data. This operation enables an experimenter equipped with the data glove to obtain CG images of the hand in various shapes with the desired fineness without having to measure all the shapes of the hand. Fig.1 shows a schematic chart of the interpolation of the articular angle data and the CG images of the hand. Furthermore, Fig.2 shows an example of the interpolated CG images of the hand. This figure represents an example of a case where the articular angle was measured at three different points in time for the actions of changing from 'rock' to 'scissors' in the rock-paper-scissors game, and the direct generation of CG and the generation of CG using interpolation were made from two adjoining data. In both these figures, the three images surrounded by a square represent the former, while the other images represent the latter.

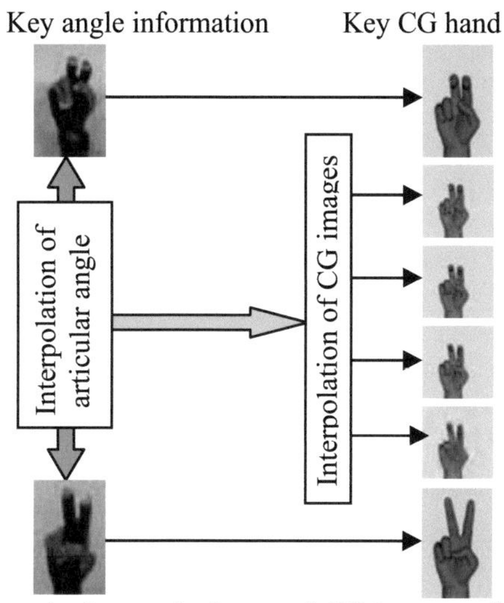

Fig. 1. Interpolation of the articular angle data and CG images of the hand.

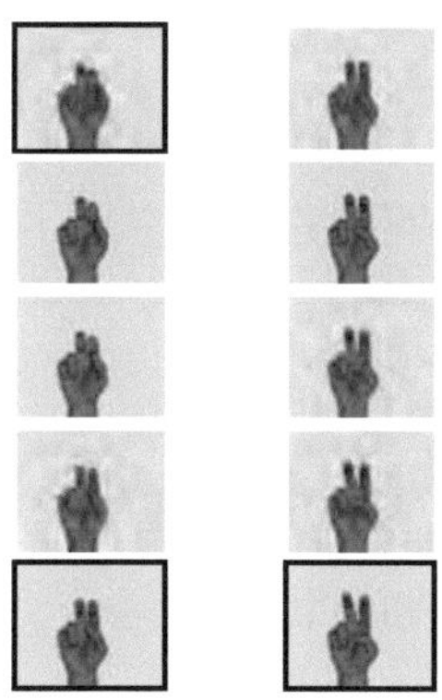

Fig. 2. Examples of the interpolated CG images of the hand.

Third, we added the data describing the differences among individuals. Because of the differences that exist among individuals (as shown in Fig.3), a wide variety of data is required for a database intended for searching similar images. For example, in the hand shape representing 'rock' in the rock-paper-scissors game, a significant difference among individuals is likely to appear in (1) the curvature of the coxa position of the four fingers other than the thumb and (2) the manner of protrusion of the thumb coxa.

Moreover, differences are likely to appear in (3) the manner of opening of the index and the middle finger and (4) the standing angle of the reference finger in the 'scissors' shape, and also in (5) the manner of opening and (6) the manner of warping, etc. of the thumb in the 'paper' shape. In order to express such differences among individuals in the form of the CG hand, we need to adjust the parameters of the length of the finger bone and the movable articular angle; therefore, we generated the CG images of hands having differences among individuals on the basis of the articular angle data obtained by the procedure described above. Fig.4 indicates an example of the additional generation of the CG hand in different shapes. In the figure, the X axis shows CG hands arranged in the order starting from those with larger projections of the thumb coxa, while the Y axis represents those with larger curvature formed by the coxa of the four fingers other than the thumb, respectively.

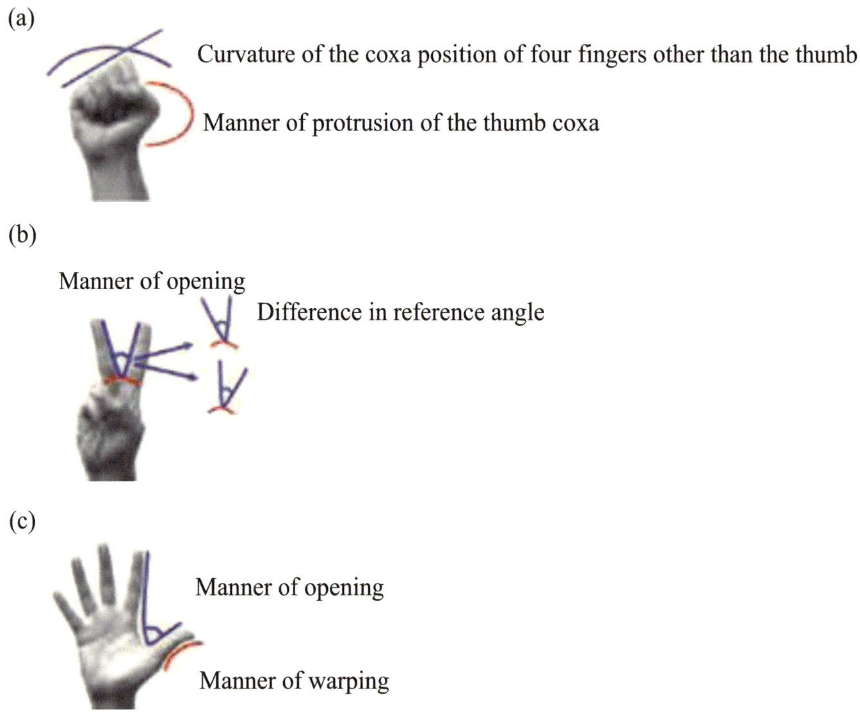

Fig. 3. Examples of the differences among individuals.

By performing the first to third steps mentioned above, we generated a total of 15,000 CG hand images using this system.

Then, the resolution was changed. Although the CG image generated this time had a resolution of 320 x 240 pixels, a substantial calculation time is required in order to estimate the posture and for applying various image processing techniques. In the present study, a reduced resolution of 64 x 64 was used. The pixel value after the resolution was changed is given by the following expression:

$$gr(i,j) = \frac{1}{r}\sum_{k}\sum_{l} go(i*320/64+k,\ j*320/64+l) \tag{1}$$

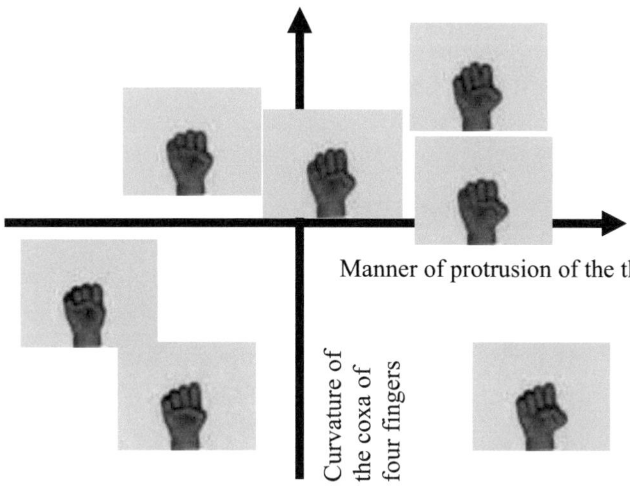

Fig. 4. Examples of the supplemented data of the differences among individuals.

Here, $gr(i,j)$ and $go(i,j)$ are the pixel values at row i and column j after and before altering the resolution, respectively. Here, the calculation has also been vertically conducted with 320 pixels in order to match the aspect ratio since the pixel resolution was altered to 64 x 64. Furthermore, k and l correspond to the row and column, respectively, within the respective regions before changing the resolution, and $r = k \times l$.

Finally, the contour was extracted. Differences exist in the environmental light, colour of human skin, etc. in the input images. The abovementioned factors were eliminated by extracting the contour in order to fix the width and the edge values, and the estimation errors were reduced by reducing the difference between the hand images in the database and in the input data.

2.2 Characterization

In the present study, we used the higher-order local autocorrelational function (Otsu & Kurita, 1998). The characteristics defined using the following expression were calculated with respect to the reference point and its vicinity:

$$x^N(a_1, a_2, \cdots, a_N) = \int f(r) f(r + a_1) \cdots f(r + a_N) dr \tag{2}$$

Here, x^N is the correlational function in the vicinity of the point r in dimension N. Since the pixels around the object point are important when a recorded image is generally used as the processing object, the factor N was limited up to the second order in the present study. When excluding the equivalent terms due to parallel translation, x^N is possibly expressed using 25 types of characteristic quantities, as shown in Fig.5. However, patterns M1 through M5 should be normalized since they have a smaller scale than the characteristic quantities of patterns M6 and thereafter. By further multiplying the pixel values of the reference point for patterns M2 through M5 and by multiplying the square of the pixel value of the reference point for pattern M1, a good agreement with the other characteristic quantities was obtained. In the present study, an image was divided into 64 sections in total – 8 x 8 each in the vertical and lateral directions - and the respective divided images were represented by 25 types of characteristic quantities using the higher-order local autocorrelational function.

Therefore, a single image is described using the characteristic quantities of 25 patterns x 64 divided sections. The image characteristics of the CG hand and the joint angle data were paired as a set for preparing the database.

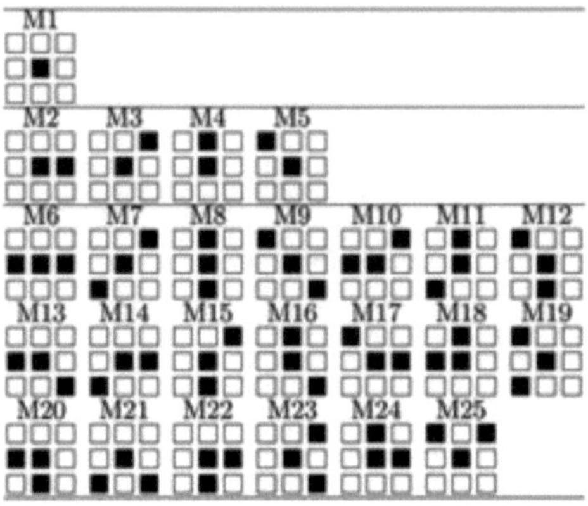

Fig. 5. Patterns of the higher-order local autocorrelational function.

2.3 Self-organization of the database

If the database prepared in the preceding sections is directly used for searching, it increases the search time together with a larger database. Hence, we intend to narrow the search space by clustering data with similar characteristics in the database. For example, sorting by using a dichotomizing search may be feasible for ordinary data; however, in the case where the characteristics range over multiple dimensions, a limitation is that the number of searches during a retrieval becomes the same as that in the total search. Therefore, we constructed a database using Kohonen's SOM (Kohonen, 1988).

Each database entry has a joint angle and a number of image characteristics; however, only the image characteristics are used in the search during estimation. There is a possibility that there exist data that have similar characteristics but significantly different joint angles; such data may be included in the same class if the classification is made on the basis of the characteristics during the self-organization of the database. On the other hand, there also exist data having significantly different characteristics, although the joint angles are similar. Therefore, we performed self-organization for both these types of data and conducted preliminary experiments; the obtained results are listed in Table 1. The mean value of the errors and the standard deviation are the values for the middle finger. The data for the other fingers are omitted from the table since they exhibited similar tendencies. Degree is used as the unit of the mean value of the errors and the standard deviation. As shown in the table, the case of self-organization on the basis of characteristics yielded better results. Consequently, we performed data clustering using self-organization on the basis of characteristics in the present study.

	processing time [ms]	mean error [degree]	standard deviation
joint angle	0.842	0.792	6.576
characteristics	0.764	0.373	5.565

Table 1. Performance of self-organization on the basis of joint angles and characteristics in the preliminary experiment.

First, we prepared classes having the representative angle, representative number of characteristics and neighbourhood class information as classes in the initial period. For the initial angles and the number of characteristics, random numbers in the range of 0 to 1 were used. With regard to the neighbourhood class information, we calculated the distance between classes in the angles by using the Euclidean distance and determined classes close to one another in this distance as neighbouring classes; this information was retained as the class number. Although the number of neighbouring classes depends on the scale of the database and the processing performance of the PC, we studied it heuristically in this experiment, and determined classes up to that close to the eighth as the neighbour classes. Next, we calculated the distance in the characteristics between the data and the classes and selected the closest class by using the data in a secondary database. This class will hereafter be referred to as the closest neighbour class. Moreover, the used date will be considered as those belonging to the closest neighbour class. The representative angle and representative number of characteristics of the closest neighbour class were renewed by using the expression below so that they may be placed closer to the data.

$$\begin{aligned} CA_{ij} &= CA_{ij} - \alpha(CA_{ij} - DA_{rj}) \\ CF_{ij} &= CF_{ij} - \alpha(CF_{ij} - DF_{rj}) \end{aligned} \tag{3}$$

where CA_{ij} denotes the representative angle j of class i; DA_{rj}, the angle j of data r; CF_{ij}, the representative number of characteristics j of class i; DF_{rj}, the representative number of characteristics j of data r; and a, the coefficient of learning.

In this experiment, a was heuristically determined as 0.0001. Next, a similar renewal was also made in the classes included in the neighbour class information of the closest neighbour class. However, their coefficient of learning was set to a value lower than that of the closest neighbour class. In the present study, it was heuristically selected as 0.01. This was applied to all the data in the primary database. In order to perform self-organization, the abovementioned operation was repeated until there was almost no change in the representative angle and the representative number of characteristics of the class.

Narrowing and acceleration of the search process can be realized to some extent, even if the database is used without self-organization. However, if such a database is used, dispersion is observed in the amount of data included in each class, thereby inducing dispersion in the processing time. Therefore, we intended to avoid the lack of uniformity in the processing time by introducing an algorithm for self-multiplication and self-extinction during self-organization. After selecting the class of adherence for all the data, we duplicate the classes that contain an amount of data exceeding 1.5 times the ideal amount. In addition, we deleted the classes containing an amount of data no more than one-half the ideal amount of data. Therefore, the amount of data belonging to each class was maintained within a certain range without significant dispersion, and the processing time was maintained within a certain limit, irrespective of the data in the class that was used for searching during the estimation. In case the algorithm for self-multiplication and self-extinction is introduced, a change is produced in the relationships among the classes, which remains unchanged in ordinary self-organization, making it necessary to redefine the relationships among the classes. Therefore, we newly prepared the neighbour class information by a method similar to that used during initialization in which we duplicated and deleted the classes.

Estimations made by using a database obtained in this manner can considerably increase the search speed as compared to the complete search of ordinary data. However, considering further increases in the database and acceleration of the searches, the database clustering was

performed not only in a single layer but also in multiple layers. Fig. 6 shows the schematic structure of the multiple layers. The class obtained with the aforementioned processing is defined as the second-layer class and is considered as data. A third-layer class is prepared by clustering the second-layer classes as the data. The third-layer class is prepared by following the same procedure as that used in the preparation of the second-layer class. Further, a fourth-layer class is prepared by clustering the third-layer classes. The lesser the amount of data in one class (or the number of classes in the lower layers), the higher the layer in which clustering can be performed. However, to absorb the dispersion of data, etc., it is preferable to prepare classes having an amount of data with a certain volume. Table 2 lists the results of the preliminary experiment in which clustering was performed by setting the amount of data in a class at 5, 10 and 20. Although the search time is reduced if the clustering is performed with a small amount of data, the estimation accuracy also reduces accordingly; therefore, we set an ideal amount of data as 10 in the present study as a trade-off between the two parameters.
The clustered database obtained using the abovementioned operation was termed as a tertiary database. This tertiary database will hereafter be simply referred to as the database. In this system, we finally constructed a database comprising 5, 10 and 10 classes in order from the upper layers, where each class has approximately 10 data items.

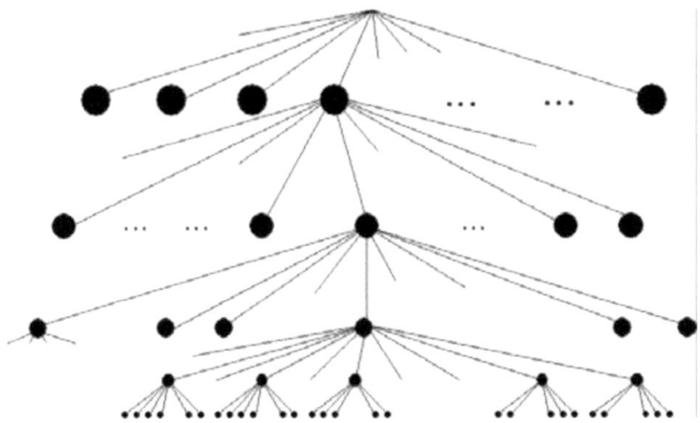

Fig. 6. Schematic structure of a database with multiple layers.

	processing time [ms]	mean error [degree]	standard deviation
5	0.656	-0.035	5.868
10	0.764	0.373	5.565
20	1.086	0.145	5.400

Table 2. Performance according to the amount of data in a class in the preliminary experiment.

2.4 Search of similar images

During estimation, sequential images were acquired using a high-speed camera. In a manner similar to the preparation of the database, image processing techniques were applied to these images to obtain their characteristic quantities. By comparing each quantity with that in the database by means of a processing technique described later, the joint angle information that formed a pair with the most similar image were defined as each result was estimated.
To estimate the similarity at the first search, the distance was calculated by using the characteristic quantity for all classes in the database. The calculation was performed by simply using the Euclidean distance that is derived using the expression below:

$$E_r = \sum_{i=1}^{25*n} (x_{ri} - x_{ti})^2 \tag{4}$$

Here, both x_{ri} and x_{ti} are characteristic quantities i with the higher-order local autocorrelational functions of the class r and at the time t, respectively. The class that minimizes E_r was selected as the most vicinal class at time t. With respect to the affiliated data of the most vicinal class and all the vicinal classes of the most vicinal class, the distances from the characteristic quantities obtained from the image were calculated using expression (4). At each instance, the angle of the data with the shortest distance was regarded as the estimated angle. From the second search, the distance was not calculated by using the characteristic quantity for all the classes in the database. Instead, only the vicinal classes of the most vicinal class and the affiliated data were selected as the candidates for the search according to the histories at t-1, as shown in Fig.7.

(a) at first search: all classes are candidates for the search.

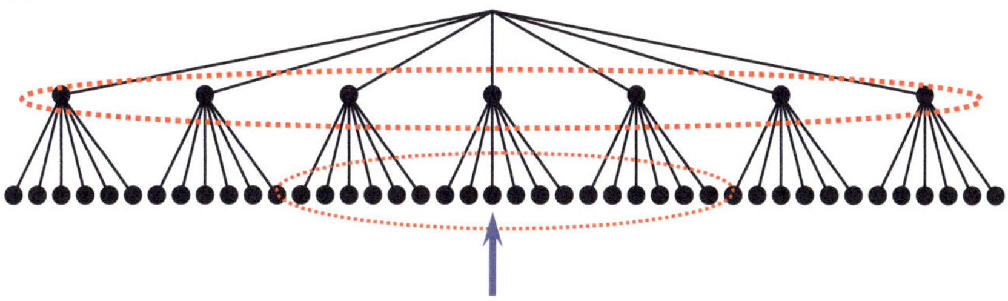

(b) from second search, the vicinal classes of the most vicinal class are candidates.

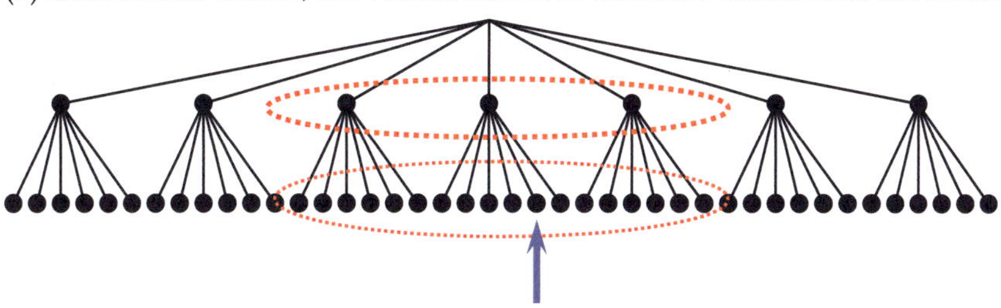

(c) if the result moves to and affiliates with another class,

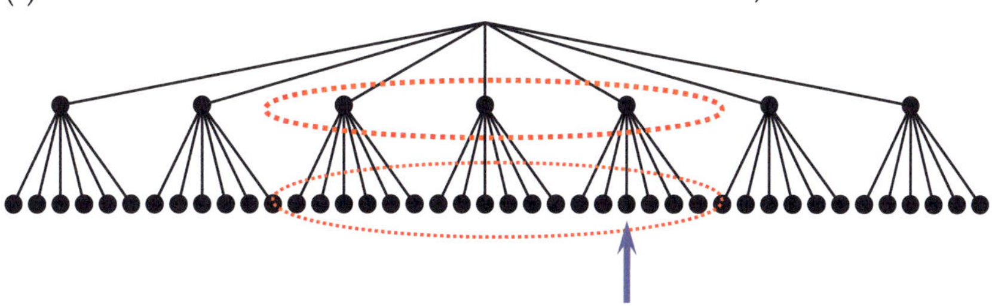

(d) then, the search space and candidate classes moves.

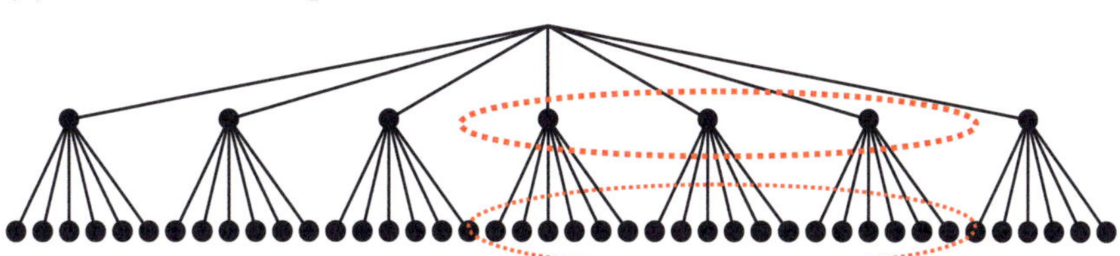

Fig. 7. Differences in the search spaces between the first search and the succeeding searches.

3. Experiment of posture estimation

3.1 Methods and procedures

In order to verify the effectiveness of this system, the actual images were subjected to experimental estimation. A subject held up a hand at a position approximately 1 m in front of the high-speed camera and moved the fingers freely provided the palm faced the camera. A slight motion of the hand was allowed in all the directions provided the hand was within the field angle of the camera. We employed a PC (CPU: Pentium 4, 2.8 GHz; main memory: 512 MB) and a monochromatic high-speed camera (ES-310/T manufactured by MEGAPLUS Inc.) in the experiments.

3.2 Results and discussions

Fig.8 shows the examples of the estimation. Each estimated result plotted using the wireframe model was superimposed on the actual image of a hand. It is evident that the finger angles have possibly been estimated with a high precision when the hand and fingers were continuously moved. It was verified that the estimation could be performed, provided the hand image did not blend into the background, even if the illuminating environment was changed.

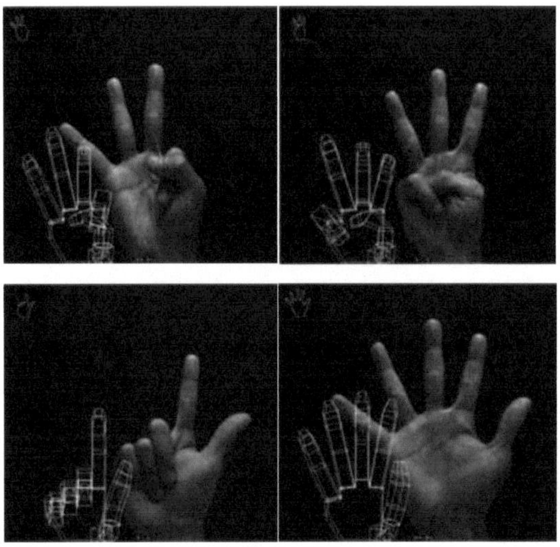

Fig. 8. Captured hand images and the results of the hand posture estimation.

For the purpose of a quantitative assessment of the system, the measured and estimated values have to be compared. However, in an ordinary environment using this system, it is impossible to acquire the measured values of the joint angle information from the human hand and fingers moving in front of the camera. Consequently, we performed the estimation experiment by wearing the data glove and a white glove above it. The results are shown in Fig.9, which reveals the angular data measured using the data glove and the estimated results. Fig.9(a) shows the interphalangeal (IP) joint of the thumb; Fig.9(b), the abduction between the middle and ring fingers; and Fig.9(c), the proximal interphalangeal (PIP) joint of the middle finger. The state where the joint is unfolded was set as 180 degrees. The system at this time operates at more than 150 fps and thus enables realtime estimation.

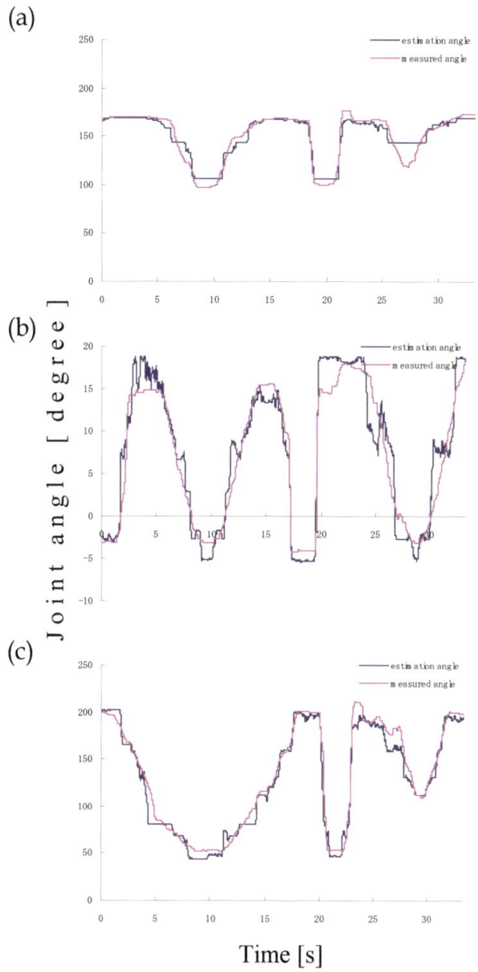

Fig. 9. Examples of the joint angle data measured using the data glove and the estimated results.

As evident from the figure, the standard deviation of the errors in the estimated angles was 4.51 degrees when we avoided the fluorescent light and used the angular data obtained by means of the data glove as the actual values; the results obtained did not have highly precise numerical values. We observed a trend of poor estimations, particularly for parts with little variation in the image (for example, the shape of the rock in the rock-paper-scissors game) against the angular variation. This may be expected, considering that a human is performing the figure estimation. In other words, we can hardly observe any difference visually for an angular difference of 10 degrees when each finger has a difference of 10 degrees. Therefore, the errors in this system, which conducts estimation on the basis of the camera image, may be considered as being within the allowable range. On the contrary, it can be observed from this figure that highly precise estimations are made in the region where visual differences are observed, namely, where the image changes significantly with the angular variations and where it is located in between the flexion and the extension.

Next, the comparative experiments were conducted. The difference between the previous experiment and these comparative experiments is that the hand position agrees with or closely resembles the database image since the object for estimation is set by selecting the CG hand image from the database. Consequently, we can determine the expected improvement in the estimating precision when the processing for positioning the input image is integrated into this system. The standard deviation of the errors when estimating the object was set to 2.86 degrees by selecting the CG image from the database, thus allowing very high-precision estimation. It is expected that the estimation error can be reduced to this extent in the future by integrating the processing for correcting the position into this system. Moreover, the processing time for the search, except for the image processing, is 0.69 ms per image. From the viewpoint of precision and processing speed, the effectiveness of the multi-step search using the self-organized database has been proved.

As mentioned above, the estimation error for unknown input images had a standard deviation of 4.51 degrees. Since this is an image processing system, small variations in the finger joints in the rock state of the rock-paper-scissors game will definitely exhibit a minimal difference in the appearance; these differences will numerically appear as a large error in the estimation. However, this error possibly contains calibration errors arising from the use of the data glove, as well as the errors caused by slight differences in the thickness, colour, or texture of the data glove covered with the white glove. Therefore, the output of the data glove or the actual value of the quantitative assessment requires calibration between the strain gauge output and the finger joint value whenever the glove is worn since the joint angle is calculated from a strain gauge worn on the glove. No such calibration standards exist, particularly for the state in which the finger is extended; therefore, the measured angle can be easily different from the indicated value. Even when the estimation is newly calibrated, it is possible that the state of calibration may be different in each experiment. On the other hand, it is not necessary to apply calibration to the second experiment that selects the CG hand image from the database. It is highly possible that this influences the standard deviation value of 4.51 degrees; therefore, it is possible to consider that the standard deviation of the errors lies between 4.51 and 2.86 degrees even if the system has not been subjected to corrective processing for the hand position.

The scheme of the present study allows you to add new data even without understanding the system. Another advantage is that the addition of new data does not require a long time since it is unnecessary to reorganize the database even when several new data items are added; this is because the database can sequentially self-organize itself by using the algorithm for self-multiplication and self-extinction of database classes. Furthermore, it is possible to search the neighbouring classes having angular similarities since each class possesses information about the vicinal classes in this system. This fact can also be regarded as the best fit for estimating the posture of a physical object that causes successive temporal angular variations, such as estimating the posture of the human hand. We attempted to carry out the hand posture estimation when the hand is rotated, although the number of trials was inadequate. Fig.10 shows an example of the result, which suggests that our system functions when a subject is in front of the camera and is rotating his/her hand. A subject can also swing the forearm, and our system can effectively estimate the shape of the fingers, as shown in Fig.11.

The image information and the joint angle information are paired in the database in our system. Once we output the results of the hand posture estimation to a robot hand, the robot can reproduce the same motions as those of the fingers of a human being and mimic them.

Fig.12 shows a dexterous robot hand (Hoshino & Kawabuchi, 2005) imitating the human hand motions without any sensors attached to it. We refer to this integrated system as the "copycat hand". This system can generate imitative behaviours of the hand because the hand posture estimation system performs calculations at high speeds and with high accuracy.

Fig. 10. An example of hand posture estimation using the rotating motion of the wrist.

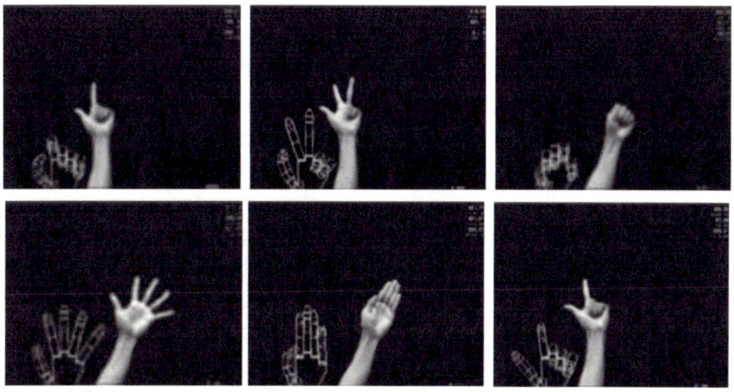

Fig. 11. Examples of the estimation when a subject is swinging his/her forearm.

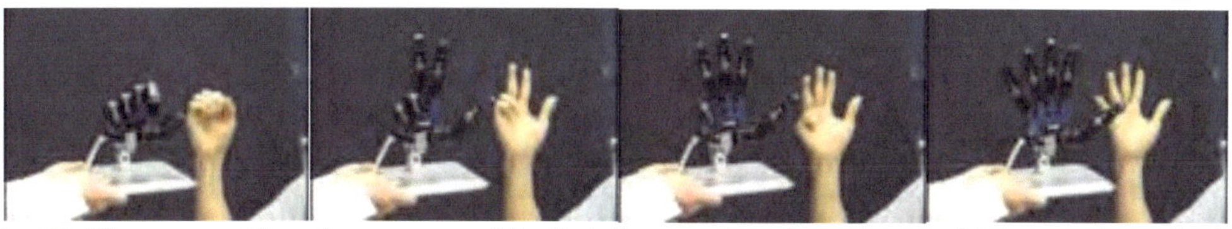

Fig. 12. The copycat hand can ape and imitate human hand motions at high speeds and with high accuracy.

4. Conclusion

To realize a robot hand capable of instantly imitating human actions, high speed, high accuracy and uniform processing time in the hand posture estimation are essential. Therefore, in the present study, we have developed a method that enables the searching of similar images at high speeds and with high accuracy and the search involves uniform processing time, even in the case where a large-scale database is used. This is achieved by (1) clustering databases having approximately uniform amounts of data using self-organization, including self-multiplication and self-extinction and (2) by collating the input images with the data in the database by means of the low-order image characteristics, while narrowing the search space in accordance with the past history.
In the preliminary construction of the database, we generated CG images of the hand by measuring the joint angles using a data glove and interpolating them; furthermore, we

extracted the contours, image characteristics and the characteristics that change only in the hand shape, irrespective of the environmental light or skin colour. The image was divided into several images and was converted into a number of characteristics by using the high-order local autocorrelation function; the image was then saved in the database in a state paired with the joint angle data obtained from a data glove. By clustering this database using self-organization depending on the number of characteristics and by the self-organization of classes in multiple stages, a multistage search was enabled using the representative numbers of classes in several layers. Moreover, by incorporating self-multiplication and self-extinction algorithms, we achieved a unification of the amount of data belonging to each class as well as the number of classes in the lower layers to avoid the dispersion of the search time in the classes.
The input image at the time of an actual estimation of the hand finger shape was subjected to various types of image processing techniques in the same manner as that at the time of construction of the database, and it was converted into a number of characteristics. The distance from the number of characteristics obtained from the picture was calculated by using a representative number of characteristics. Classes at close distances were selected as candidate classes for the estimated angle, and a similar distance calculation was also performed in the classes in each layer belonging to a candidate class for the estimated angle. Among the respective data belonging to the candidate classes for the estimated angle in the lowest class, the angle data of the data with the closest distance between the number of characteristics was considered as the estimation result. Furthermore, for the selection of a candidate class, we attempted to reduce the search space by using the previous estimation results and the neighbour information.
By estimating the sequential images of the finger shape by using this method, we successfully realized a process involving a joint angle estimation error within several degrees, a processing time of 150 - 160 fps, and an operating time without dispersion by using a PC having a CPU clock frequency of 2.8 GHz and a memory capacity of 512 MB. Since the image information and the joint angle information are paired in the database, the system could reproduce the same actions as those of the fingers of a human being by means of a robot without any time delay by outputting the estimation results to the robot hand.

5. Acknowledgement

This work is partly supported by Proposal-Oriented Research Promotion Program (PRESTO) of Japan Science and Technology Agency (JST) and Solution-Oriented Research for Science and Technology (SORST) project of JST.

6. References

Athitos, V. & Scarloff, S. (2002). An appearance-based framework for 3D hand shape classification and camera viewpoint estimation, *Proc. Automatic Face and Gesture Recognition*, pp.40-45

Bernardin, K.; Ogawara, K.; Ikeuchi, K. & Dillmann, R. (2005). A sensor fusion approach for recognizing continuous human grasping sequences using Hidden Markov Models, *IEEE Transactions on Robotics*, Vol.21, No.1, pp.47-57

Gallese, V.; Fadiga, L.; Fogassi, L. & Rizzolatti, G. (1996). Action recognition in the premotor cortex, *Brain*, Vol.119, pp.593-60

Hoshino, K. & Tanimoto, T. (2005). Real time search for similar hand images from database for robotic hand control, *IEICE Transactions on Fundamentals of Electronics, Communications and Computer Sciences*, Vol.E88-A, No.10, pp.2514-2520

Hoshino, K. & Tanimoto, T. (2006). Method for driving robot, *United Kingdom Patent Application No.0611135.5*, (PCT/JP2004/016968)

Hoshino, K. & Kawabuchi, I. (2005). Pinching at finger tips for humanoid robot hand, *Journal of Robotics and Mechatronics*, Vol.17, No.6, pp.655-663

Hoshino, K. & Kawabuchi, I. (2006). Hobot hand, *U.S.A. Patent Application No.10/599510*, (PCT/JP2005/6403)

Kameda, Y. & Minoh, M. (1996). A human motion estimation method using 3-successive video frames, *Proc. Virtual Systems and Multimedia*, pp.135-140

Kohonen, T. (1988). The neural phonetic typewriter, *IEEE computer*, Vol.21, No.3, pp.11-22

Lu, S.; Metaxas, D.; Samaras, D. & Oliensis, J. (2003). Using multiple cues for hand tracking and model refinement, *Proc. CVPR2003*, Vol.2, pp.443-450

Otsu, N. & Kurita, T. (1998). A new scheme for practical, flexible and intelligent vision systems, *Proc. IAPR. Workshop on Computer Vision*, pp.431-435

Rehg, J. M. & Kanade, T. (1994). Visual tracking of high DOF articulated structures: an application to human hand tracking, *Proc. European Conf. Computer Vision*, pp.35-46

Rizzolatti, G.; Fadiga, L.; Gallese, V. & Fogassi, L. (1996). Premotor cortex and the recognition of motor actions, *Cognitive Brain Research*, Vol.3, pp.131-141

Appendix: Composition of the humanoid robot hand

(Hoshino & Kawabuchi, 2005; 2006)

As compared to walking, the degree of freedom (DOF) assigned to manipulation functions and to fingers is extremely low. The functions of the hands are mostly limited to grasping and holding an object and pushing a lever up and down. The robot hand itself would tend to become larger and heavier and it would be almost impossible to design a slender and light-weight robot if currently available motors and reduction gears are used with a number of DOFs equivalent to that of the human hand. It is important to determine where and how to implement the minimum number of DOFs in a robot hand.

We have designed the first prototype of a dexterous robot hand. The length from the fingertip to the wrist is approximately 185 mm and the mass of the device is 430 g, which includes mechanical elements such as motors with encoders and reduction gears without electrical instrumentation such as motor control amplifiers, additive sensors, or cables for external connection.

Fig.13 shows two examples of generating movements involved in Japanese sign language. In the case of the numeral 2, the index finger and the middle finger should be stretched during abduction and pass through a clearance generated by the thumb. Generating the numeral 30 involves a difficulty. A ring is formed by the thumb and the fourth finger and the other three fingers are stretched while exhibiting abduction and then bent to a suitable angle. As for the two examples generated by this system, movements were carried out promptly while maintaining an appropriate accuracy in order to facilitate a reasonable judgment of the numerals created by using the sign language. The time duration of the movement is slightly over 1 s for the numeral 2 and approximately 2 s for the numeral 30.

An important function for the robot hand is picking up small, thin, or fragile items using only the fingertips. This capability is equally or even more important than the ability to securely grasp a heavy object. Therefore, we designed a second prototype focusing on the terminal joint of the fingers and the structure of the thumb.

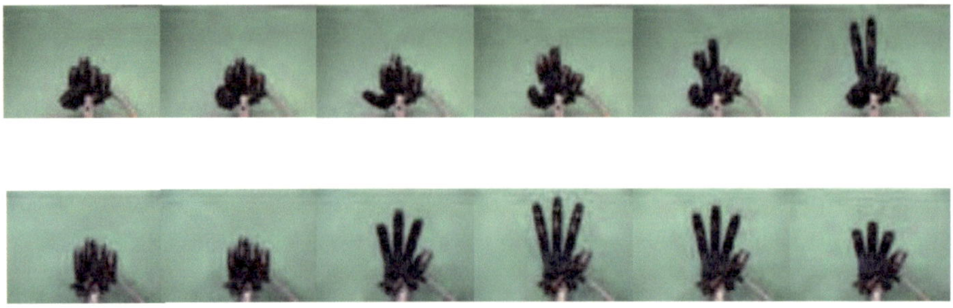

Fig. 13. Examples of the sign language movements.

As an independent DOF, we implemented a small motor at every fingertip joint, namely at distal interphalangeal (DIP) joints of four fingers and interphalangeal (IP) joint of the thumb. The mass of the motor is approximately 10 g with a gear. Although the maximum motor torque is very small (0.5 Nmm), the maximum fingertip force is 2 N because of the high-speed reduction ratio and the short distance between the joint and the fingertip, which provides sufficient force for picking up an object. Moreover, it has a wide movable range. Each fingertip joint can bend inward as well as outward, which, for instance, enables the robot hand to stably pick up a business card on a desk.

We also added a twisting mechanism to the thumb. When the tips of the thumb and fingers touch, the contact is at the fingertip and the thumb pads; however, this may not provide a sufficient contact with the other fingertip pads since the thumb cannot twist to make this contact. The human hand has soft skin and padding at the fingertips and the high control of motion and force at the fingertips enables stable pinching even if the finger pads are not in complete mutual contact. However, we expect that the fingertip force produced by the terminal joint drive at the tip of the two finger groups will act in opposite directions at the same point, implying that the two fingertips will oppose each other exactly at the pad.

Fig.14 shows the snapshots of the performance of the second type of robot hand, which repeated the series of movements and stably pinched the business card. The mass of the hand is approximately 500 g and the length from the fingertip to the wrist is approximately 185 mm, which are almost equivalent to those of the human hand.

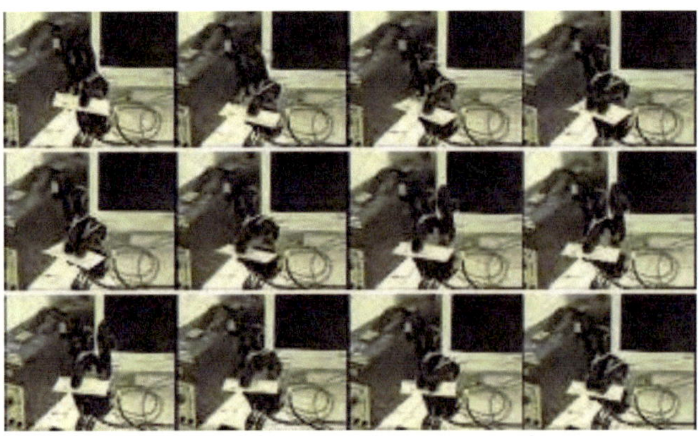

Fig. 14. Snapshots of the robot hand handling a business card using two or three fingers.

19

Energy-Efficient Walking for Biped Robot Using Self-Excited Mechanism and Optimal Trajectory Planning

Qingjiu Huang & Kyosuke Ono
Tokyo Institute of Technoligy, Japan

1. Introduction

Recently, a lot of researches which aim at realization of dynamic biped walking are being performed. There is Honda's ASIMO as a representative of researches of humanoid robots. ASIMO has joints of many degrees of freedom that are near to a human being, high environment adaptability and robustness, and can do various performances. However, it needs many sensors, complicated control, and walks with bending a knee joint to keep the position of a centre of gravity constant. Therefore, it walks unnaturally and consumes much energy.

On the other hand, McGeer performed the research of passive dynamic walking from the aspect of that it is natural motion in a gravitational field (McGeer, T., 1990). This robot which could go down incline only using potential energy was developed and realized the energy-efficient walking. However, it needs incline, and its applied range is small because it has no actuator. Therefore, the researches that aimed at energy-efficient biped walking on level ground have been performed. S.H.Collins exploited the robot which had actuators at only ankles (Collins, S. H. & Ruina A., 2005). M.Wisse exploited the robot which used pneumatic actuators (Wisse, M. & Frankenhuyzen, J. van, 2003). Ono exploited the self-excitation drive type robot which had an actuator only at hip joint (Ono, K. et al, 2001); (Ono, K. et al, 2004); (Kaneko, Y. & Ono K., 2006). And then, Osuka and Asano performed the level ground walking from a point of view to mechanical energy for joints which is the same with the energy consumed of passiveness walk (Osuka, K. et al, 2004); (Asano, F. et al, 2004). These biped robot's studies used the technique of the passive dynamic walking which used inertia and gravity positively by decreasing the number of actuators as much as possible. However, in order to adapt the unknown ground, the biped robot needs actuators to improve the environment adaptability and robustness. Here, Ono proposed the optimal trajectory planning method based on a function approximation method to realize an energy-efficient walking of the biped robot with actuators similar to a passive dynamic walking (Imadu, A. & Ono, K. 1998); (Ono, K. & Liu, R., 2001); (Ono, K. & Liu, R., 2002); (Peng, C. & Ono K., 2003). Furthermore, Huang and Hase verified the optimal trajectory planning method for energy-efficient biped walking by experiment, and proposed the inequality state constraint to obtain better solution which is desirable posture in the intermediate time of the walking period (Hase, T. & Huang, Q., 2005); (Huang, Q. & Hase, T., 2006); (Hase, T., et al., 2006).

In this chapter, we introduce the newest researches on energy-efficient walking of the biped robot for level ground form two viewpoints, one is semi-passive dynamic walking with only hip actuator using self-excited mechanism, another is active walking with actuators using optimal trajectory planning.
The chapter is organized as follows. In section 2, the self-excited walking of a four-link biped robot and the self-excitation control algorithm enables the robot to walk on level ground by numerical simulation and experiment will be introduced. In section 3, we aim at realizing an energy-efficient walking of the four-link biped robot with actuators similar to a passive dynamic walking. An optimal trajectory planning method based on a function approximation method applied to a biped walking robot will be shown. And then, we use the inequality state constraint in the intermediate time and restrict the range of joint angles. In this way, a better solution which is desirable posture in the intermediate time can be obtained. Furthermore, in section 4, with "Specific Cost", we show that the biped walking with the above two methods have more efficient energy than the other methods which use geared motors. Finally, the conclusions will be presented in section 5.

2. Self-Excited Walking for Biped Mechanism

In this section, we introduce a study on the self-excited walking of a four-link biped mechanism that proposed by Ono (Ono, K. et al, 2001). And then, we show that the self-excitation control enables the three-degree-of-freedom planar biped model to walk on level ground by numerical simulation. From the parameter study, it was found that stable walking locomotion is possible over a wide range of feedback gain and link parameter values and that the walking period is almost independent of the feedback gain. Various characteristics of the self-excited walking of a biped mechanism were examined in relation to leg length and length and mass ratios of the shank. Next, a biped mechanism was manufactured similar to the analytical model. After parameter modification the authors demonstrated that the biped robot can perform natural dynamic walking on a plane with a 0.8 degree inclination. The simulated results also agree with the experimental walking locomotion.

2.1 Analytical Model of Biped Walking Robot and Kinetic Process

2.1.1 Features of Biped Locomotion and Possibility of Its Self-Excitation

Fig.1 shows a biped mechanism to be treated in this study. Here we focus only on the biped locomotion in the sagittal plane. The biped mechanism does not have an upper body and consists of only two legs that are connected in a series at the hip joint through a motor. Each leg has a thigh and a shank connected at a passive knee joint that has a knee stopper. By the knee stopper, an angle of the knee rotation is restricted like the human knee. The legs have no feet, and the tip of the shank has a small roundness. The objective of this study is to make the biped mechanism perform its inherent natural walking locomotion on level ground not by gravitational force but by active energy through the hip motor.
The necessary conditions for the biped mechanism to be able to walk on level ground are as follows: (1) The inverted pendulum motion for the support leg must synchronize with the swing leg motion. (2) The swing leg should bend so that the tip does not touch the ground. (3) The dissipated energy of the mechanism through collisions at the knee and the ground, as well as friction at joints, should be supplied by the motor. (4) The knee of the support leg

should not be bent by the internal force of the knee stopper. (5) The synchronized motion between the inverted pendulum motion of the support leg and the two-DOF pendulum motion of the swing leg, as well as the balance of the input and output energy, should have stable characteristics against small deviations from the synchronized motion.

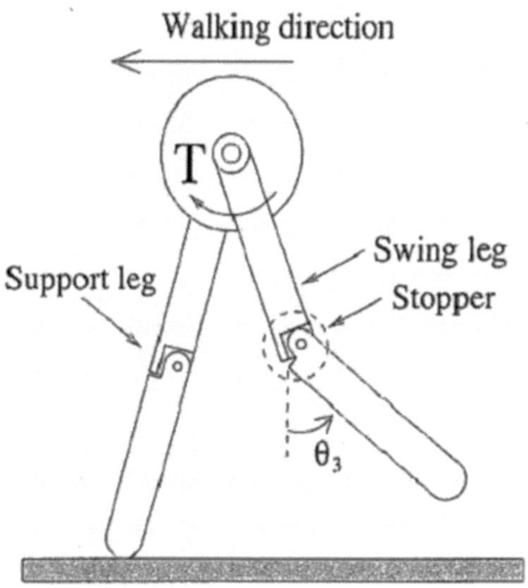

Fig.1. Three-degree-of freedom walking mechanism on a sagittal plane (Ono, K. et al, 2001)

First we pay attention to the swing leg and try to generate a swing leg motion that can satisfy the necessary conditions (2) and (3) by applying the self-excitation control to the swing leg motion. Ono and Okada (Ono, K. & Okada, T., 1994) have already investigated two kinds of self-excitation control of two-DOF vibration systems and showed that the Van der Pol-type self-excitation can evoke natural modes of the original passive system, while the asymmetrical stiffness matrix type can excite the anti-resonance mode that has a phase shift of about 90 degrees between input and output positions.

The two-DOF pendulum of a swing leg has the first-order mode with an in-phase at each joint and the second-order mode with an out-of-phase at each joint. Thus, it will be impossible to generate a walking gait by applying the Van der Pol-type self-excitation. In contrast, by means of the negative feedback from the shank joint angle θ_3 to the input torque T at the thing joint, the system's stiffness matrix becomes asymmetrical. Thus, the swing motion would change so that the shank motion delays at about 90 degrees from the thigh motion. Through this feedback, it is also expected that the kinetic energy of the swing leg increases and that the reaction torque (=T) will make the support leg rotate in the forward direction in a region where $\theta_3 > 0$. The self-excitation of the swing leg based on the asymmetrical matrix is explained in detail below.

2.1.2 Self-Excitation of the Swing Leg

Fig.2 depicts the two-DOF swing leg model whose first joint is stationary. To make Fig..2 compatible with Fig.3 (b), the upper and lower links are termed the second and third links, respectively. To generate a swing motion like a swing leg, only the second joint is driven by the torque T2, which is given by the negative position feedback of the form

$$T_2 = -k\theta_3 . \tag{1}$$

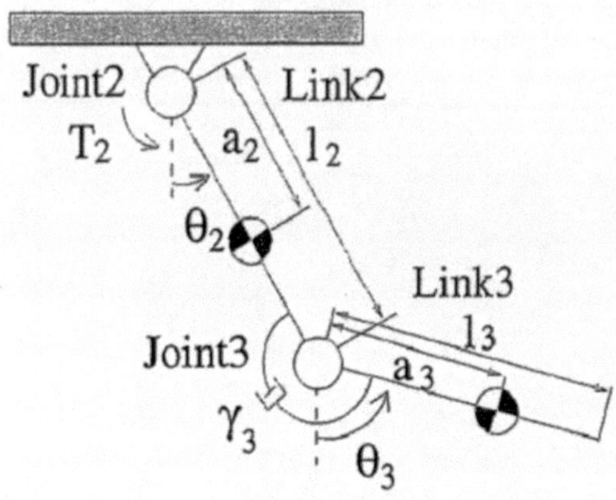

Fig.2. Analytical model of two-degree-of-freedom swing leg (Ono, K. et al, 2001)

From the fundamental study of the asymmetrical stiffness matrix-type self-excitation (Ono, K. & Okada, T., 1994), it is known that damping plays an important role in inducing the self-excited motion. Thus, the viscous rotary damper with coefficient γ_3 is applied to joint 3.

Taking Eq. (1) into account, the equation of motion for the two-DOF pendulum system is written as

$$\begin{bmatrix} I_2 + m_2a_2^2 + m_3l_2^2 & m_3a_3l_2 \cos(\theta_3 - \theta_2) \\ m_3a_3l_2 \cos(\theta_3 - \theta_2) & I_3 + m_3a_3^2 \end{bmatrix} \begin{bmatrix} \ddot{\theta}_2 \\ \ddot{\theta}_3 \end{bmatrix} + \begin{bmatrix} \gamma_3 & -m_3a_3l_2 \sin(\theta_3 - \theta_2)\dot{\theta}_3 - \gamma_3 \\ m_3a_3l_2 \sin(\theta_3 - \theta_2)\dot{\theta}_2 - \gamma_3 & \gamma_3 \end{bmatrix} \begin{bmatrix} \dot{\theta}_2 \\ \dot{\theta}_3 \end{bmatrix} + \begin{bmatrix} (m_2a_2 + m_3l_2) g \sin\theta_2/\theta_2 & k \\ 0 & m_3a_3 g \sin\theta_3/\theta_3 \end{bmatrix} \begin{bmatrix} \theta_2 \\ \theta_3 \end{bmatrix} = 0. \tag{2}$$

We note from Eq. (2) that the stiffness matrix of the system becomes asymmetrical because of the feedback gain k.

To examine the value of k to excite a swing leg motion autonomously and the natural frequency and mode at the threshold, Eq. (2) is linearized about $\theta_2 = \theta_3 = 0$. The linearized equation of motion becomes

$$\begin{bmatrix} I_2 + m_2a_2^2 + m_3l_2^2 & m_3a_3l_2 \\ m_3a_3l_2 & I_3 + m_3a_3^2 \end{bmatrix} \begin{bmatrix} \ddot{\theta}_2 \\ \ddot{\theta}_3 \end{bmatrix} + \begin{bmatrix} \gamma_3 & -\gamma_3 \\ -\gamma_3 & \gamma_3 \end{bmatrix} \begin{bmatrix} \dot{\theta}_2 \\ \dot{\theta}_3 \end{bmatrix} + \begin{bmatrix} (m_2a_2 + m_3l_2) g & k \\ 0 & m_3a_3g \end{bmatrix} \begin{bmatrix} \theta_2 \\ \theta_3 \end{bmatrix} = 0. \tag{3}$$

Putting $\theta_2 = \Theta_2 e^{\lambda t}$ and $\theta_3 = \Theta_3 e^{\lambda t}$ into Eq. (3), we have

$$\begin{bmatrix} a_{11}\lambda^2 + \gamma_3\lambda + k_{11} & a_{12}\lambda^2 - \gamma_3\lambda + k \\ a_{21}\lambda^2 - \gamma_3\lambda & a_{22}\lambda^2 + \gamma_3\lambda + k_{22} \end{bmatrix} \begin{bmatrix} \Theta_2 \\ \Theta_3 \end{bmatrix} = 0. \tag{4}$$

Where $a_{11} = I_2 + m_2 a_2^2 + m_3 a_2^2$, $a_{12} = a_{21} = m_3 a_3 l_2$, $a_{22} = I_3 + m_3 a_3^2$, $k_{11} = (m_2 a_2 + m_3 l_2) g$, $k_{22} = m_2 a_3 g$.

From the condition that the determinant in Eq. (4) is zero, we can obtain the characteristic equation of the linearized system of the form

$$A_0 \lambda^4 + A_1 \lambda^3 + A_2 \lambda^2 + A_3 \lambda + A_4 = 0. \tag{5}$$

As k increases, one of the eigenvalues becomes unstable. At the boundary, λ becomes an imaginary number. Thus, putting $\lambda = \omega i$ into Eq. (5), we can get the critical k value and the natural frequency at the inst ability threshold.

Here we assume that both links are a uniform bar whose length is 0.4 m ($= l_2 = l_3$), and the mass is 2 kg ($= m_2 = m_3$). For these link parameters, one of the natural modes becomes unstable when k > 1.43 or k < -3.14. The natural frequency of the unstable swing motion at k= 1.43 is 0.71 Hz, whereas that at k = -3.14 is 0.61 Hz. If the damping does not exist at joint 3, one of the natural modes becomes unstable when 30.4 > K > 6.3. Note that the damping coefficient γ_3, however small it might be, can significantly reduce the critical k value. Strangely enough, the magnitude of the damping coefficient γ_3 does not change the critical k value and the natural frequency at the threshold. However, the damping coefficient value slightly influences the natural mode of the unstable vibration. By putting the natural frequency and k value into Eq. (4) , we found a swing motion such that the lower link delays from the upper link are excited when k ≥ 1.43. The magnitude and phase of Θ_2/Θ_3 are 0.60 and 8.89 degrees, respectively, when γ_3 = 0.15 Nms/rad. As the unstable vibration grows in magnitude and approaches a limit cycle, the phase of the Θ_2/Θ_3 tends to be 90 degrees (Ono, K. & Okada, T., 1994).

2.2 Analytical Model of Biped Locomotion and Basic Equations

The next question is whether the biped mechanism, which combines the two-DOF swing leg discussed in Section 2.1 and the single-DOF support leg, can generate a biped locomotion that satisfies the conditions (1), (4), and (5). Since it is difficult to derive this solution analytically, we numerically show that the nonlinear biped system excited by the asymmetrical feedback exhibits a stable biped locomotion that satisfies the three conditions as a limit cycle of the system.

Fig.3 (a) shows the representative postures of a biped mechanism during half of the period of biped locomotion. From an aspect of the difference of the basic equation, one step process can be divided into two phases: (1) from the start of the swing leg motion to the collision at the knee (the first phase) and (2) from the knee collision to the touchdown of the straight swing leg (the second phase). We assume that the change of the support leg to the swing leg occurs instantly and that the end of the second phase is the beginning of the first phase. The self-excitation feedback of (1) is applied only during the first phase. We assume that the support leg is kept straight because of internal reaction torque at the knee for simplifying the forward dynamic simulation. The validity of this assumption will be examined later.

Under the assumption of a straight support leg, the analytical model during the first phase is represented as a three-DOF link system, as shown in Fig.3 (b). Viscous damping is applied to the knee joint of the swing leg. The equation of motion in this system during the first phase is written as

$$\begin{bmatrix} M_{11} & M_{12} & M_{13} \\ & M_{22} & M_{23} \\ sym & & M_{33} \end{bmatrix} \begin{bmatrix} \ddot{\theta}_1 \\ \ddot{\theta}_2 \\ \ddot{\theta}_3 \end{bmatrix} \tag{6}$$

Where the elements M_{ij}, C_{ij}, and K_i of the matrices are shown in Appendix A. T2 is feedback input torque given by Eq. (1).

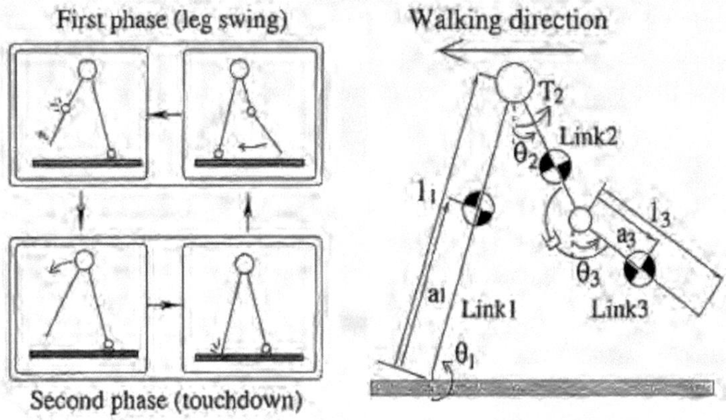

(a) Two phases of biped walking (b) Three-DOF model

Fig.3. Analytical model of three-degree-of-freedom walking mechanism (Ono, K. et al, 2001)

When the shank becomes straight in line with the thigh at the end of the first phase, it is assumed that the knee collision occurs plastically. From the assumption of conservation of momentum and angular momentum before and after the knee collision, angular velocities after the knee collision are calculated from the condition $\dot{\theta}_2^+ = \dot{\theta}_3^+$ and the equation of the form

$$\begin{bmatrix} \dot{\theta}_1^+ \\ \dot{\theta}_2^+ \\ \dot{\theta}_3^+ \end{bmatrix} = [M]^{-1} \begin{bmatrix} f_1\left(\theta_1, \dot{\theta}_1^-\right) \\ f_2\left(\theta_2, \dot{\theta}_2^-\right) - \tau \\ f_3\left(\theta_3, \dot{\theta}_3^-\right) + \tau \end{bmatrix}, \tag{7}$$

Where the elements of the matrix [M] are the same as M_{ij} in Eq. (6) . f_1, f_2, and f_3 are presented in Appendix B. τ is the moment impulse at the knee.

During the second phase, the biped system can be regarded as a two-DOF link system. Thus, the basic equation becomes

$$\begin{aligned} &\begin{bmatrix} I + m_1 a_1^2 + m_2 l_1^2 & m_2 a_2 l_1 \cos(\theta_3 - \theta_2) \\ m_2 a_2 l_1 \cos(\theta_2 - \theta_1) & I_2 + m_2 a_2^2 \end{bmatrix} \begin{bmatrix} \ddot{\theta}_2 \\ \ddot{\theta}_3 \end{bmatrix} \\ &+ \begin{bmatrix} 0 & -m_2 a_2 l_1 \sin(\theta_2 - \theta_1)\dot{\theta}_2 \\ m_2 a_2 l_1 \sin(\theta_2 - \theta_1)\dot{\theta}_1 & 0 \end{bmatrix} \begin{bmatrix} \dot{\theta}_2 \\ \dot{\theta}_3 \end{bmatrix} \\ &+ \begin{bmatrix} (m_1 a_1 + m_2 l_1) g \sin\theta_1/\theta_1 & 0 \\ 0 & m_2 a_2 g \sin\theta_2/\theta_2 \end{bmatrix} \begin{bmatrix} \theta_1 \\ \theta_2 \end{bmatrix} = 0. \end{aligned} \tag{8}$$

We assume that the collision of the swing leg with the ground occurs plastically (the translational velocity at the tip of the swing leg becomes zero) and that there is a sufficient friction between the tip and the ground surface to prevent slippage. The angular velocities of the links after the collision are derived from conservation laws of momentum and angular momentum. They are calculated from Eq. (7) by putting $\tau = 0$.
To investigate the self-excited biped locomotion, a set of basic equations of (6), (7), and (8) were numerically solved based on the fourth-order Runge-Kutta method. The standard values of the link parameters used in the calculation are shown in Table 1. The values of mass and moment of inertia were calculated assuming uniform mass distribution along the link. Since the magnitude of the damping coefficient scarcely influences the biped gait, $\gamma_3 = 0.15$ Nms/rad was used in the calculation.

		Link1	Link2	Link3
m_i	mass [kg]	4.0	2.0	2.0
I_i	length [kgm^2]	0.21	0.027	0.027
l_i	Moment of inertia [m]	0.80	0.40	0.40
a_i	Offset of mass center [m]	0.40	0.20	0.20

Table 1. Link Parameter Values Used for Simulation Study (Ono, K. et al, 2001).

2.3 Simulation Results and Characteristics of Self-Excited Walking

In this simulation, we assumed that the support leg is kept straight due to internal torque during the first and second phases. To check whether the support leg can be lept straight by the passive knee stopper, the internal torque at the knee of the two-DOF support leg model was inversely calculated by using the calculated stable biped locomotion. Fig.4 plots the angular displacement of $\theta_1 - \pi$, θ_2, and θ_3 and the internal torques at the knees of the support and swing legs, which are calculated from θ_1, θ_2, and θ_3 based on the inverse dynamic analysis. The torque curve at $\phi_{off} = \theta_1 - \theta_2 \le 0$. If the knee torque is negative, the reaction torque exists at the knee stpooer so that the support leg can be regarded as a single-DOF link. We observe that at the beginning of the first phase, the knee torque of the support leg becomes positive for a short time of less than 0.1 s.
Then, we calculated the knee torque of the support leg by assuming an offset angle of $\phi_{off} = \theta_2 - \theta_3 = 2$ degrees at the knee stopper. The calculated result is also shown in Fig.4. We note that the knee torque of the support leg shifts downward and changes to a negative value during the entire period. Moreover, it is seen from the negative knee torque of the swing leg that the swing leg can be kept straight during the secondphase.
In the numerical sumulation, steady biped locomotion can be obtained as a limit cycle for the feedback gain value in a range of 4.8 Nm/rad $\le k \le$ 7.5 Nm/rad. When k is smaller than 4.8 Nm/rad, the initial step length gradually decreases and the biped mechanism falls down forward. When k is larger than 7.5 Nm/rad, the step changes irregularly and the biped mechanism falls down backwards because the swing leg is raised too high. Fig.5 shows the walking period, average velocity, and efficiency as a function of feedback gain k. The efficiency is defined as the ratio of average walking velocity to average input power, where the average input power is given by

$$P = \frac{1}{t_4 - t_0} \int_{t=t_0}^{t=t_4} \left| \dot{\theta}_2 T_2 \right| dt \, . \tag{9}$$

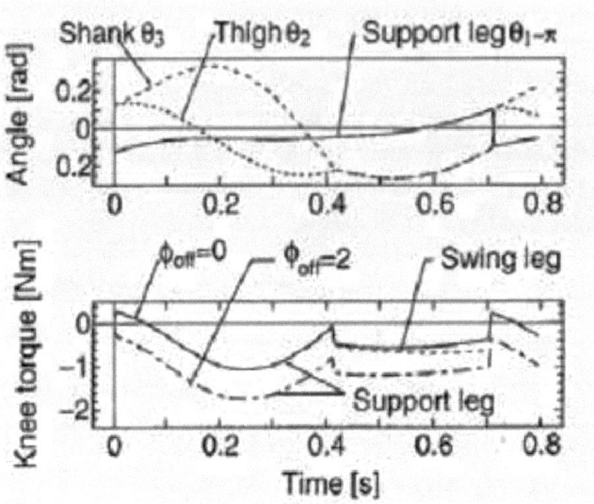

Fig.4. Motion of legs and internal torque at knee during one step (Ono, K. et al, 2001).

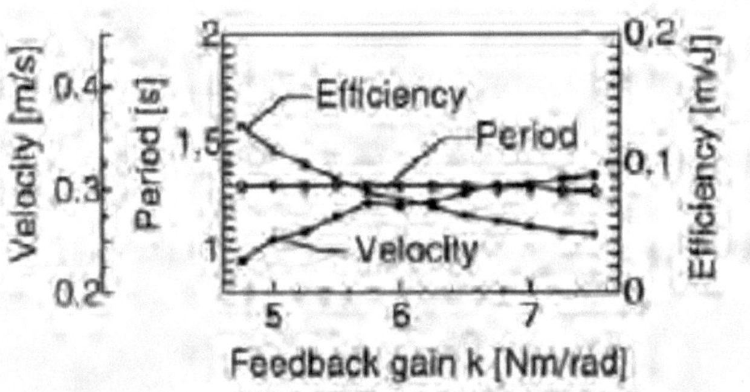

Fig.5. Effect of feedback gain on walking period, velocity, and efficiency (Ono, K. et al, 2001).

We note from this figure that the walking period shows an almost constant value of 1.3 s at any k value. This is because the self-excited biped locomotion is a natural motion inherent in the biped system. On the other hand, the average velocity increases gradually because of the increase in step length. However, the efficiency decreases with an increase in the feedback gain k because the average input power increases more rapidly with an increase in k. It is worthy to note that the average input power to generate biped locomotion with a velocity of 0.28 m/s is only about 4 W (=0.28/0.07) at k = 6.0 Nm/rad. If the additional movement of 0.6 m by feet during one period of 1.3 s is taken into account, the natural walking velocity becomes 2.7 km/h. the natural walking period of an adult with a 0.9 m leg height is 1.3 s (Yang 1994). Thus, it can be said that the calculated results of the walking period and velocity are comparable with the natural walking of an adult with a 0.8 leg height.

Validity of the self-excited walking stated above was confirmed experimentally (Ono, K., 2001). Extensions of the self-excited walking to a biped with feet and bent knee walking mode are described in (Ono, K., 2004) and (Kaneko, Y. & Ono K., 2006), respectively.

3. Energy-Efficient Trajectory Planning for Biped Walking Robot

In this section, we aim at realizing a energy-efficient walking of biped with actuators similar to a passive dynamic walking. We suggested optimal trajectory planning method based on a function approximation method, and applied for a biped walking robot analytically. We obtained the solutions which made the square integral value of input torque minimum. Until this point, we clarified the application of optimal trajectory planning method for biped walking robots, but they did not constrain the clearance between toe and ground at intermediate time of one period. As is generally known, it is necessary to give the clearance between toe and ground in the intermediate time of one step period. However, because there is the redundant degree of freedom by only constraining the clearance with equality state constraint, the solutions of undesired postures may be obtained. Therefore we suggest the optimal trajectory planning method with inequality state constraint that can restrict the joint angles to the desired range in a certain time. Concretely, we use inequality state constraint in the intermediate time and restrict the range of joint angles and may obtain a better solution which is desirable posture in the intermediate time.

With this method, we calculate optimal walking gait for biped with upper body mass. This model is attached the same masses to the support leg and swing leg in forward direction. And we examine the influence of installation position of the upper body mass.

And We perform a walking experiment by a prototype robot to inspect the effectiveness of this method. At first, in order to realize a stable walking, we attach a sensor at a sole to measure the time of grounding and compensate the error between reference trajectory and experimental trajectory. And, we install adjustment masses to revise parameters of the prototype robot. Furthermore, we use "Specific Cost" which shows energy consumption per unit mass and per unit movement velocity for evaluation of energy efficiency in section 4. By comparing the data of the other robots with this method, we show that our robot walks more efficiently than the other robots which use geared motors.

3.1 Analytical Model of Biped Walking Robot and Kinetic Process

3.1.1 Analytical Model and Kinetic Process

Fig.6 shows the three link biped model used in this research. Link 1 is the support, link 2 is the thigh, and link 3 is the shank. This model has upper body masses which is the same masses attached to the support leg and swing leg in forward direction. Table.3 shows the link parameters used in this analysis.

Table.2 shows symbols used in this paper (Hase, T.; Huang, Q. & Ono, K., 2006).

One step period is from Posture 1 to Posture 4 shown in Fig.7. State variables of Posture 4 are same as one of Posture 1. When it changes from Posture 3 to Posture 4, a collision between toe and ground happens, and link parameters exchange. In Posture 3, the swing leg collides with the ground, and becomes the support leg and the support leg leaves from the ground and comes back to Posture 4 (Posture 1). Leg exchange is performed by perfectly inelastic collision momentarily.

Symbols	Description
m_i	*i*-th link mass
I_i	*i* -th link moment of inertia
l_i	*i* -th link length
a_i	Length between *i*-th joint and *i*-th link COG
b_i	Length between (*i*+1)-th joint and *i*-th link COG
m_d	Upper body mass
d	Length between upper body mass and hip joint
θ_i	*i*-th link absolute angle
α_i	Offset angle between *i*-th link and *i*-th link COG from joint *i*
β_i	Offset angle between *i*-th link and *i*-th link COG from joint (*i*+1)
x_i	x-coordinate of *i*-th link COG
y_i	y-coordinate of *i*-th link COG
g	Acceleration of gravity
u_i	*i*-th joint input torque
J	Performance function
ζ_i	Dummy variable used for inequality constraint
δ_i	Clearance between toe and ground
$\boldsymbol{p}_i$	*i*-th coefficient vector of basis function
$\boldsymbol{h}_i$	*i*-th basis function
T	Period of one step
S	Length of ste

Table 2. Symbols used in this section (Hase T. & Huang, Q., 2005)

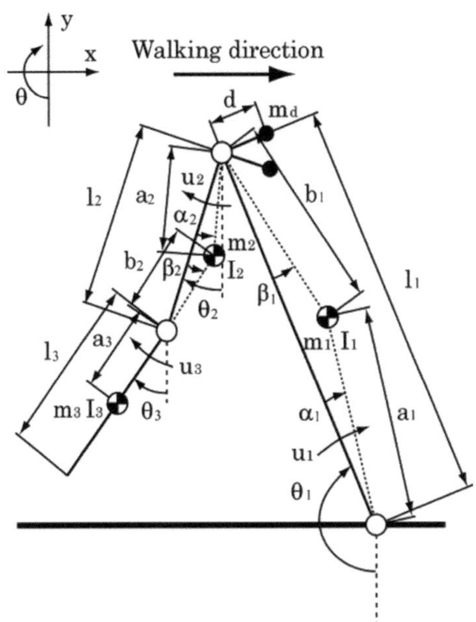

Fig.6. 3 link model with upper body mass (Hase T. & Huang, Q., 2005).

		Link1	Link2	Link3
m_i	[kg]	4.2	1.2	1.2
I_i	[kgm^2]	0.127	0.029	0.029
l_i	[m]	0.60	0.30	0.30
a_i	[m]	0.339	0.115	0.125

Table 3. Link parameters (Hase T. & Huang, Q., 2005).

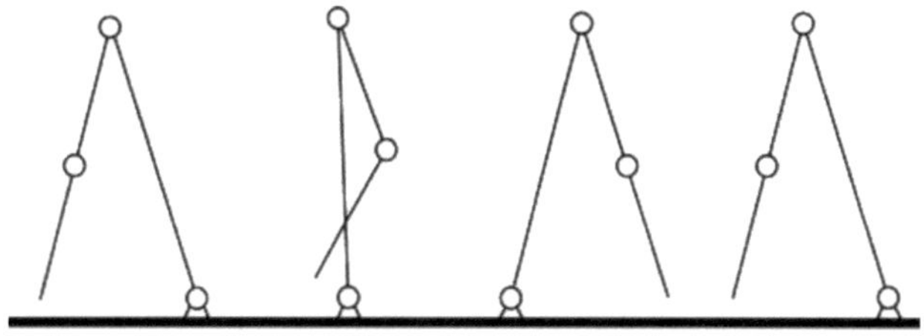

Fig.7. Process of one step walking (Hase T. & Huang, Q., 2005).

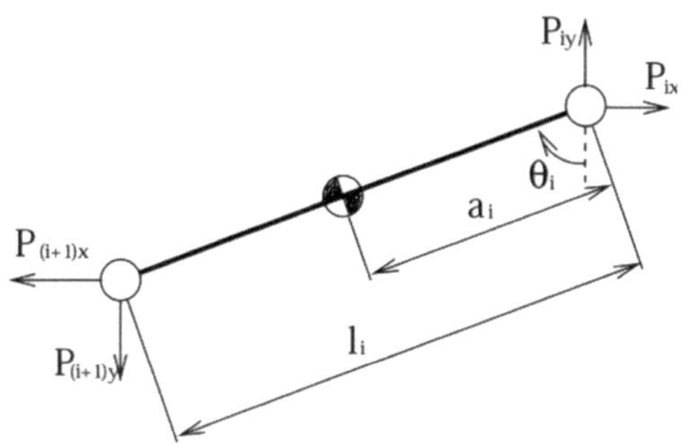

Fig.8. Impulse for *i*-th link (Hase T. & Huang, Q., 2005).

3.1.2 Dynamic Equations and Analysis of Collision between Toe and Ground

By the analytical model shown in Fig.6, dynamic equations are lead as follows. We assumed that there is no friction in each joint.

$$\boldsymbol{M}(\theta)\ddot{\theta} + \boldsymbol{C}(\theta)\dot{\theta}^2 + \boldsymbol{K}(\theta) = \boldsymbol{u} \tag{10}$$

When collision between toe and ground occurs, it is assumed that a momentum and angular momentum of each link are conserved. Leg exchange happens shown as Fig.7.

Fig.8 shows impulse for i-th link. "v_{ix}", "v_{iy}" is x, y ingredient of translation velocity of a center of gravity of each link. "+" indicates the state just after the swing leg collides with the ground, and "-" indicates the state just before that.

Using the conservation law of momentum and angular momentum, following equations work out.

$$\begin{cases} m_i(v_{ix}^+ - v_{ix}^-) = P_{ix} - P_{(i+1)x} \\ m_i(v_{iy}^+ - v_{iy}^-) = P_{iy} - P_{(i+1)y} \\ I_i(\theta_i^+ - \theta_i^-) = \boldsymbol{a}_i \times \boldsymbol{P}_i + (\boldsymbol{l}_i - \boldsymbol{a}_i) \times \boldsymbol{P}_{i+1} \end{cases} \tag{11}$$

When Eq.(10) are organized, angular velocity just before and after collision are as follows.

$$M(\theta)\dot{\theta}^{+} = H(\theta)\dot{\theta}^{-} \tag{12}$$

Matrices in Eq.(10), (12) are written in (Ono, K. & Liu, R., 2002).

3.2 Optimal Trajectory Planning Method with Inequality Constraint

3.2.1 Conventional Optimal Trajectory Planning Method (Ono, K. & Liu, R., 2002)

We apply the function approximation method to express the joint angle trajectory by Hermite polynomial function set. Then the optimal trajectory planning problem is converted into the parameter optimization problem of the coefficients of the basis function. The joint angle trajectory is expressed as follows:

$$\theta(\boldsymbol{p},t) = \boldsymbol{h}(t)^T \boldsymbol{p} \tag{13}$$

where $\boldsymbol{h}(t)$ is the Hermite basis function vector and $\boldsymbol{p}$ is the coefficient vector of the basis function. Here, the coefficient vector is the boundary state variable vector.

The total constraint conditions and the performance index which are function of $\boldsymbol{p}$ are expressed as follows:

$$\boldsymbol{C}(\boldsymbol{p}) = 0 \tag{14}$$

$$J = J(\boldsymbol{p}) \tag{15}$$

then, we obtain the solution which make Eq.(15) minimum while Eq.(14) is satisfied.

The approximated solution $\boldsymbol{p}_{k+1}$ in the (k+1)-th step of Eq.(14) is solved by Newton-Raphson iteration method in the following form:

$$\frac{\partial \boldsymbol{C}(\boldsymbol{p}_k)}{\partial \boldsymbol{p}} \Delta \boldsymbol{p} = -\boldsymbol{C}(\boldsymbol{p}_k) \qquad (\Delta \boldsymbol{p} = \boldsymbol{p}_{k+1} - \boldsymbol{p}_k) \tag{16}$$

In Eq.(16), the unknown variable number is far more than the equation number. Therefore, the number of the solutions which satisfy Eq.(14) is infinite. A particular solution is solved using the singular value decomposition (SVD).

Then, the general solutions $\Delta \boldsymbol{p}$ become as follows:

$$\Delta \boldsymbol{p} = \Delta \boldsymbol{p}_0 - \boldsymbol{B}\boldsymbol{q} \tag{17}$$

$\boldsymbol{p}_0$ is a particular solution. $\boldsymbol{B}$ is solved by using the SVD method, and $\boldsymbol{q}$ is the null-space vector to be determined from the J minimum condition as follows:

$$\frac{\partial J}{\partial \boldsymbol{q}} = \frac{\partial J}{\partial \boldsymbol{p}} \boldsymbol{B} = 0 \tag{18}$$

In order to solve $\boldsymbol{q}$ to satisfy Eq.(18), we use Newton-Raphson iteration method. The second order partial differentiation of J with respect to $\boldsymbol{q}$ is calculated as follows:

$$\frac{\partial^2 J}{\partial \boldsymbol{q}^2} = \boldsymbol{B}^T \frac{\partial^2 J}{\partial \boldsymbol{p}^2} \boldsymbol{B} \tag{19}$$

Then, we obtain $\boldsymbol{p}$ from $\boldsymbol{q}$ and obtain the optimal solution.

3.2.2 Introducing the Inequality Constraint

Conventionally, we used only equality constraint at the beginning, middle and ending time for a biped walking robot. But there is such a case that the joint angle should pass the desirable ranges without being restricted precisely. In that case, the better solution will be obtained when the joint angle is restricted with inequality constraint than equality constraint. When we use inequality constraint, we convert Eq.(14) into follow equations with dummy variables.

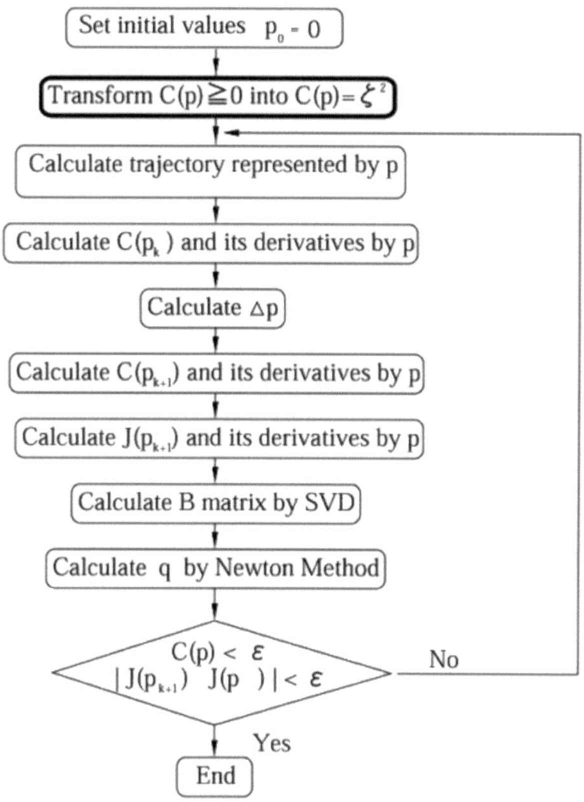

Fig.9. Flowchart of this algorism (Hase T. & Huang, Q., 2005).

$$\begin{aligned} C(p) \geq 0 \quad &\Rightarrow \quad C(p) = \varsigma^2 \\ C(p) \leq 0 \quad &\Rightarrow \quad C(p) = -\varsigma^2 \end{aligned} \tag{20}$$

Then, we use the dummy variables as optimization parameters.
For example, if you restrict the state variable of joint angle at boundary time t_i as follows:

$$\theta(t_i) \geq 0 \tag{21}$$

Then, the dummy variable introduced as follows:

$$\theta(t_i) = \varsigma^2 \tag{22}$$

The state variables at boundary time t_i are expressed as j-th optimization parameter p_j as follows:

$$\theta(t_i) = p_j \tag{23}$$

Here, Eq.(22) and Eq.(23) are organized as follows:

$$p_j = \varsigma^2 \tag{24}$$

Then, we convert p_j into ς^2 in a coefficient parameter of the performance index, basis function, and other constraints. Optimization calculation is performed with respect to ς and other parameters (not p_j). Fig.9 shows the flowchart of this algorithm.

3.3 Application for Biped Walking Robot

3.3.1 Performance Function

In order to obtain a natural walking solution with the lowest input torque, the performance index in this paper is defined as the sum of the integration of square input torque as follows:

$$J = \sum_{i=1}^{3} k_i \int_0^T u_i^2(t)dt \tag{25}$$

where k_i is the weighting factor corresponding to joint *i*, and u_i is the input torque at joint *i*. When joint *i* is passive joint, we relatively make k_i larger and bring the passive joint torque close to 0.

3.3.2 The Results of Using Equality Constraint at Beginning and Ending Time

At first, the period *T* of one step and the length of the step *S* are determined. The beginning and ending posture is that both legs become straight. This posture is determined when the length of the step is determined.

Fig.10 and Fig.11 show stick figure of trajectory, joint input torque, angular velocity, and clearance between toe and ground at *T*=0.60[s], *S*=0.15[m].

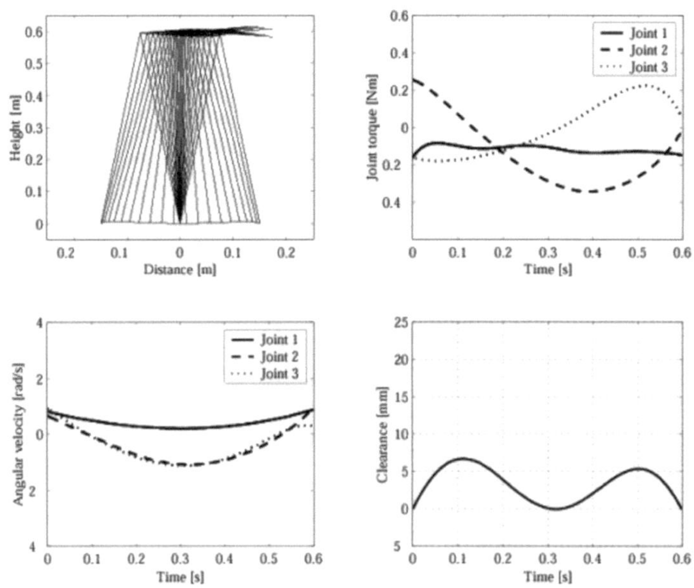

Fig.10. The result of using equality constraint at beginning and ending time (*S*=0.150[m] *T*=0.60[s]) (Hase T. & Huang, Q., 2005).

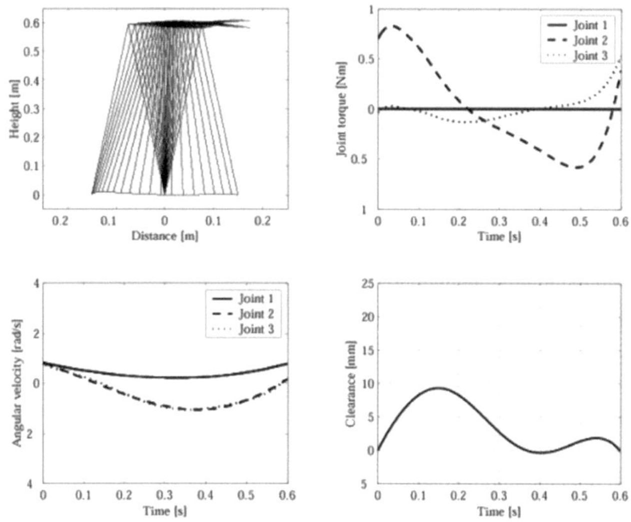

Fig.11. The result of using equality constraint at beginning and ending time (u_1 =0, S=0.150[m], T=0.60[s]) (Hase T. & Huang, Q., 2005).

3.3.3 The Results of Using Inequality Constraint at Intermediate Time

From the result of preceding section, there is the moment when toe comes in contact with ground. There is not utility in this result.

Then, we give the clearance (δ) between toe and ground by constraining the posture at intermediate time (0.5T). The constraint is expressed as follows:

$$\delta = -l_1 \cos\theta_1(0.5T) - l_2 \cos\theta_2(0.5T) - l_3 \cos\theta_3(0.5T) \tag{26}$$

This constraint has three variables and becomes redundant. It is desirable that the joint angles come to the following range so that it is natural walking.

$$\begin{aligned} \theta_2(0.5T) &\le 0 \\ \theta_3(0.5T) &\ge 0 \end{aligned} \tag{27}$$

These inequality constraints are converted into following equality constraints.

$$\begin{aligned} \theta_2(0.5T) &= -\zeta_1^2 \\ \theta_3(0.5T) &= \zeta_2^2 \end{aligned} \tag{28}$$

Figs.12 and 13 show the results of δ =20[mm]. These results show that clearance condition is satisfied.

3.3.4 A Change of the Evaluation Function for a Change of a Period

Figs.14 and 15 show the changes of performance function J for change of period T. Conventionally, we constrained the swing leg posture at intermediate time with equality constraint. These graphs compare this method (inequality) with a conventional method (equality), and the performance function J by this method is smaller than that of a conventional method. They have local minimum solutions between 0.6s from 0.5s.

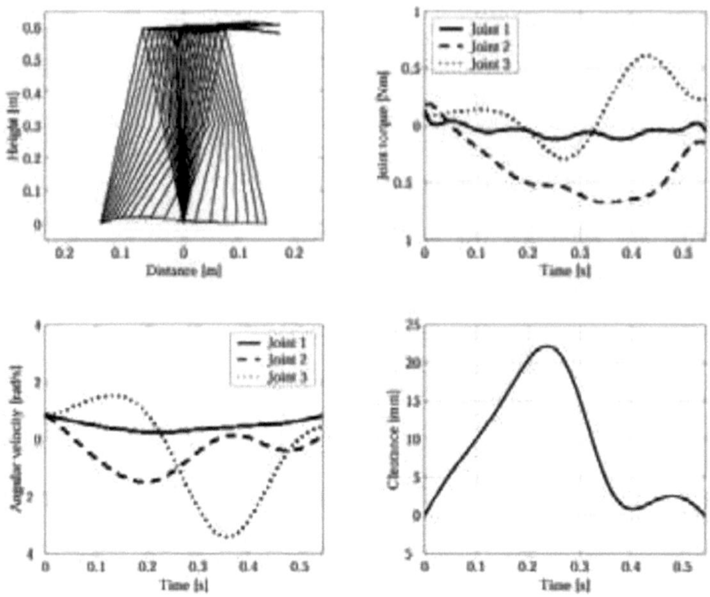

Fig.12. The result of using inequality constraint at intermediate time (δ =20[mm], S=0.150[m], T=0.55[s]) (Hase T. & Huang, Q., 2005).

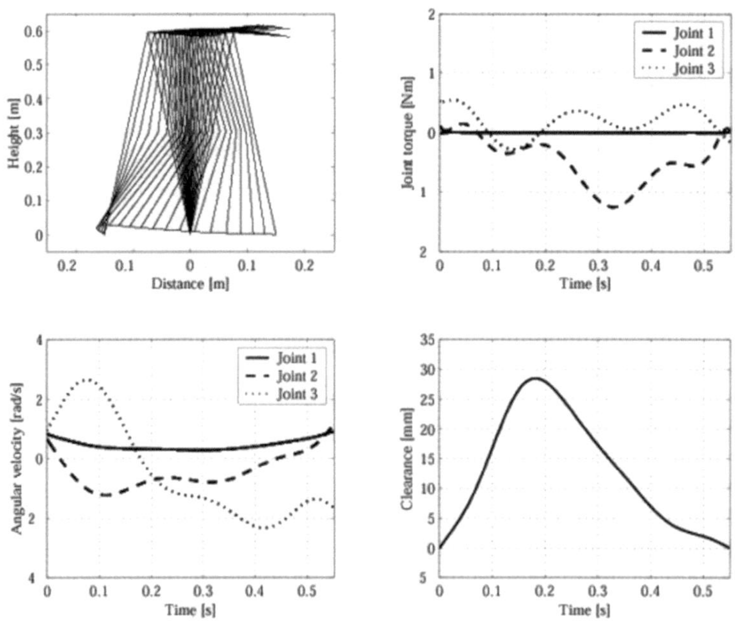

Fig.13. The result of using inequality constraint at intermediate time (u_1 =0, δ =20[mm], S=0.150[m], T=0.55[s]) (Hase T. & Huang, Q., 2005).

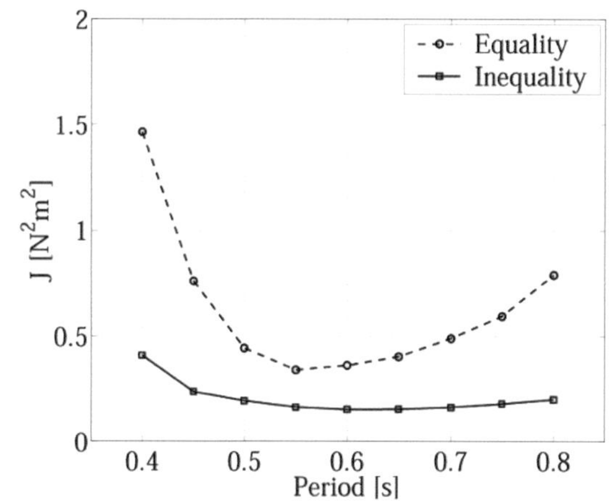

Fig.14. Change of performance function J for change of period T (δ=20[mm], S=0.15[m]) (Hase T. & Huang, Q., 2005).

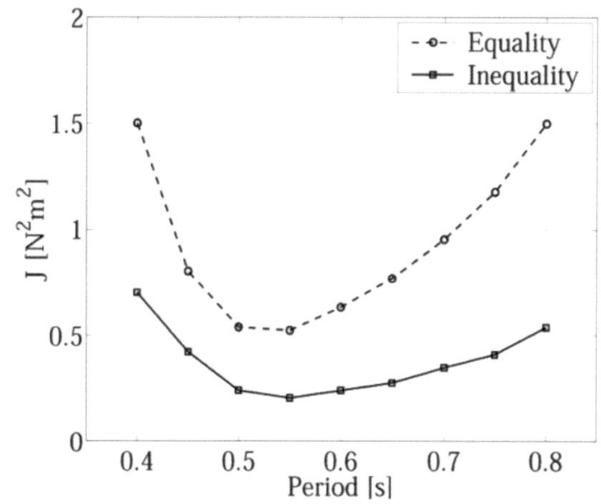

Fig.15. Change of performance function J for change of period T (u_1 =0, δ =20[mm], S=0.15[m]) (Hase T. & Huang, Q., 2005).

3.3.5 The Influence of Installation position of the upper body mass

Fig.16 shows influence of installation position of the upper body mass. When u_1 is 0, the input becomes larger as the installation position becomes longer. When u_0 is 0, there is an optimal installation position at 80mm.

3.4. Walking Experiment for Energy-Efficient Trajectory Planning

3.4.1 Hardware

Fig.17 shows the summary of the prototype robot. This robot has three legs to constraint the motion of roll direction. Hip joint and knee joint have actuators. These actuators are DC motors with gear reducer. They are controlled by motor drivers. Table.4 shows the parameters of the prototype robot.

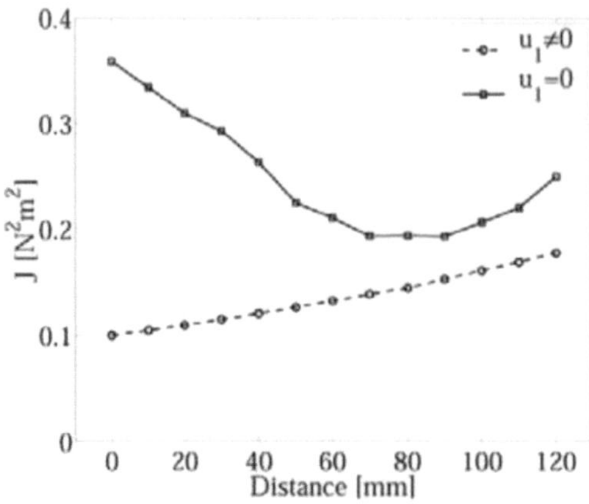

Fig.16. The influence of installation position of the upper body mass (δ =20[mm], T=0.55[s]) (Hase T. & Huang, Q., 2005).

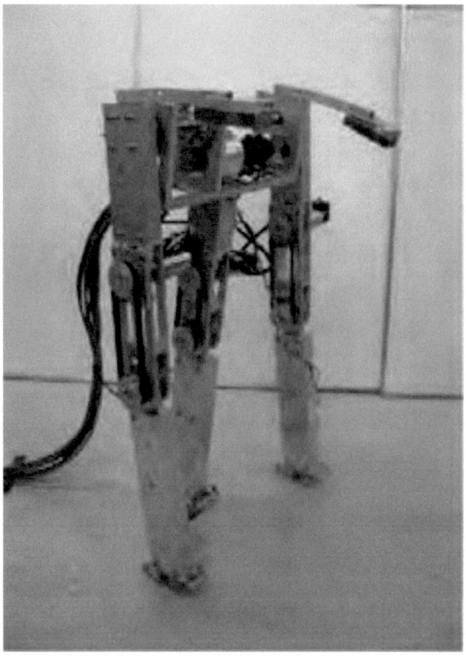
Fig.17. Prototype robot (Hase T. & Huang, Q., 2005).

	Inner		Outer	
	Thigh	Shank	Thigh	Shank
m [kg]	3.02	1.22	2.92	1.20
I [kgm^2]	0.027	0.015	0.031	0.017
l [m]	0.30	0.30	0.30	0.30
a [m]	0.10	0.12	0.13	0.13

Table 4. Link parameters (Hase T. & Huang, Q., 2005).

3.4.2 Software

We performed an optimization calculation in offline beforehand and save the data of relative joint angle trajectory. It is the best that robot controlled with feed forward control of optimal torque. But it is difficult to measure the parameters of experimental robot and collision between toe and ground precisely. Therefore, we used PD feedback control for each joint and let the experimental trajectory follow the optimal trajectory.

In the experiment, the timing of a landing does not agree with the analysis and the delay occurs for optimal trajectory. Then, we attached a touch sensor at a sole and the input of the next period starts when a sensor becomes on. The robot keeps the ending posture until the toe lands.

The analysis model has only three links and assumed the support leg to be one link. We attached stoppers to knee joints and fixed it by pushing it to stoppers with motors when the leg becomes support leg.

3.4.3 The Results of Walking Experiment

Figs.18 and 19 show the results of eight steps of walking experiment. Fig.18 is relative joint angle. It has delay, but almost follows the optimal trajectory. Because we used PD feed back control in order to cancel model simplifications (no backlash, friction at joints, perfect inelastic collision between toe and ground). Fig.19 shows the PD feedback input torque. When the legs collide with the ground, they vibrate and PD feedback input becomes larger.

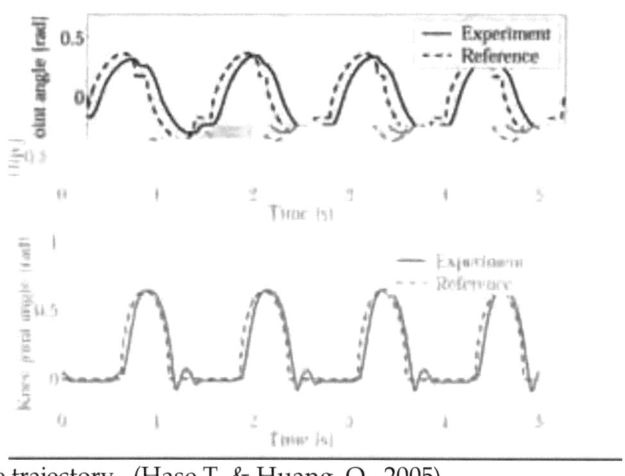

Fig.18. Joint angle trajectory (Hase T. & Huang, Q., 2005).

4. Energy Efficiency

Energy efficiency of walking can be evaluated with Specific Cost "C" (the energy consumed per unit mass and per transport velocity) as follows:

$$C = \frac{P}{mgv} \tag{29}$$

Here, P is the energy consumed per unit of time [W], m is mass[kg], v is transport velocity[m/s].

In the case of the self-excited mechanism, there is one type of Specific Cost C_{mt} that calculated by the simulation results in section 2.3. In this case, C_{mt} is obtained with the mechanical work energy (4W) and walking velocity (0.28m/s) for the 8kg weight of the robot model. Then, C_{mt} become about 0.18.

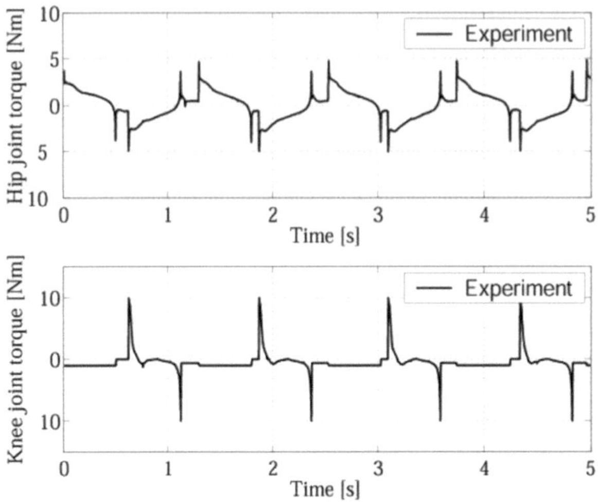

Fig.19. Joint input torque (Hase T. & Huang, Q., 2005).

And then, in the case of the optimal trajectory planning, there are two types of Specific Cost (C_{et}, C_{mt}). C_{et} is obtained with the energy that consumed at the motors (11.7W), microcomputer, and sensors (5W). C_{et} become 0.68. C_{mt} is obtained with the mechanical work energy (8.8W). Then, C_{mt} become 0.36. Table.5 shows C_{et} and C_{mt} of some types of biped walking robots (Collins, S. H. & Ruina A., 2005).

	C_{et}	C_{mt}	Actuator
Honda's Asimo	3.2	1.6	
Our Robot (Self-Excited Machanism)	-	0.18	Geared motor
Our Robot (Optimal trajiectry planning)	0.68	0.36	
T.U.Delft's Denise	5.3	0.08	Pneumatic McKibben muscle
MIT's Spring Flamingo	2.8	0.07	Series elastic actuator
Comell's Biped	0.2	0.055	Push-off actuator
McGeer's Dynamite	-	0.04	

Table 5. Specific cost of transport and mechanical energy efficiency of biped walking robots.

As our robots uses gears of large reduction ratio, the energy consumption by friction grows large, and Specific Cost grows larger than the other robots which used the mechanism of passive dynamic walking. However, it is very smaller than the other type robot using the gears of large reduction ratio.

5. Conclusion

In this chapter, we introduced the researches on energy-efficient walking of the biped robot for level ground form two viewpoints, one is semi-passive dynamic walking with only hip actuator using self-excited mechanism, another is active walking with actuators using optimal trajectory planning. We can make the conclusions as follow.

In the case of energy-efficient walking using the self-excited mechanism:

1. To develop an efficient control method of natural biped locomotion, the self-excitation control of asymmetry stiffness type was applied to the biped mechanism to generate natural walking.
2. In a two-DOF swing leg whose thigh and shank are connected by a passive knee joint, the input torque at the hip joint can generate a walking gait of the swing leg and can increase the swing amplitude and its kinetic energy.
3. From the simulation for a biped mechanism whose leg is as tall as 0.8 m (similar to a standard adult), the walking period is 1.3s and walking velocity 0.28m/s at an average input power of 4W. According to the Specific Cost data comparing with the other types of biped walking robots, we showed that the energy-efficient walking is possible using the self-excited mechanism for biped locomotion.

In the case of energy-efficient walking using the optimal trajectory planning:

1. We proposed optimal trajectory planning method based on a function approximation method, and applied for a biped walking robot analytically. And in order to avoid the solutions of undesired postures when constrain the clearance between toe and ground at intermediate time of one period, we proposed an optimal trajectory planning method with inequality constraint, and applied it to a biped walking robot with upper body mass.
2. We performed a walking experiment by a prototype robot to inspect the effectiveness of this method. In the walking experiment, we realized the stable walking. One step period was 0.6s, and walking velocity was 0.3m/s.
3. We showed the energy-efficient walking using the proposed optimal trajectory planning method for biped locomotion by comparing Specific Cost of other robots with geared motors.

8. References

McGeer, T. (1990). Passive Dynamic Walking, *International Journal of Robotics Research*, Vol.9, pp.62-82

Collins, S. H. & Ruina A. (2005). A Bipedal Walking robot with efficient and human-like gait, *Proceedings of IEEE International Conference of Robotics and Automation*, Barcelona, Spain, Biped Locomotion I, CD-ROM No.1935

Wisse, M. & Frankenhuyzen, J. van (2003). Design and construction of MIKE; a 2d autonomous biped based on passive dynamic walking, *Proceedings of Conference of Adaptive Motion of Animals and Machines*, Kyoto, Japan, Analysis & Control of Bipedal Locomotion, CD-ROM No.WeP-I-1

Ono, K. & Okada, T. (1994). Self-excited vibratory actuator (1st report: Analysis of two-degree-of-freedom self-excited systems), *Transactions of the JSME(C)*, Vol.60, No.577, pp.92-99 (in Japanese)

Ono, K.; Takahashi, R.; Imadu, A. & Shimada, T. (2001). Self-excited walking of a biped mechanism, *International Journal of Robotics Research*, Vol.20, No.12, pp.953-966

Ono, K.; Furuichi, T. & Takahashi, R. (2004). Self-Excited Walking with Feet, *International Journal of Robotics Research*, Vol.23, No.1, pp.55-68

Kaneko, Y. & Ono K. (2006). Study of Self-Excited Biped Mechanism with Bending Knee, *Transactions of the Japan Society of Mechanical Engineers*, Series C, Vol.72, No.714, pp.416-434 (in Japanese)

Osuka, K.; Sugimoto Y. & Sugie T. (2004). Stabilization of Semi-Passive Dynamic Walking based on Delayed Feedback Control, *Journal of the Robotics Society of Japan*, Vol.22, No.1, pp.130-139 (in Japanese)

Asano, F.; Luo, Z.-W. & Yamakita, M. (2004). Gait Generation and Control for Biped Robots Based on Passive Dynamic Walking, Journal of the Robotics Society of Japan, Vol.22, No.1, pp.130-139

Imadu, A. & Ono, K. (1998). Optimum Trajectory Planning Method for a System that Includes Passive Joints (1st Report, Proposal of a Function Approximation Method), *Transactions of the Japan Society of Mechanical Engineers*, Series C, Vol.64, No.618, pp.136-142 (in Japanese)

Ono, K. & Liu, R. (2001). An Optimal Walking Trajectory of Biped Mechanism (1st Report, Optimal Trajectory Planning Method and Gait Solutions Under Full-Actuated Condition), *Transactions of the Japan Society of Mechanical Engineers*, Series C, Vol.67, No.660, pp.207-214 (in Japanese)

Liu, R & Ono, K. (2001). An Optimal Trajectory of Biped Walking Mechanism (2nd Report, Effect of Under-Actuated Condition, No Knee Collision and Stride Length), *Transactions of the Japan Society of Mechanical Engineers*, Series C, Vol.67, No.661, pp.141-148 (in Japanese)

Ono, K. & Liu, R. (2002). Optimal Biped Walking Locomotion Solved by Trajectory Planning Method, Transactions of the ASME, *Journal of Dynamic Systems, Measurement and Control*, Vol.124, pp.554-565

Peng, C. & Ono K. (2003). Numerical Analysis of Energy-Efficient Walking Gait with Flexed Knee for a Four-DOF Planar Biped Model, *JSME International Journal*, Series C, Vol.46, No.4, pp.1346-1355

Hase T. & Huang, Q. (2005). Optimal Trajectory Planning Method for Biped Walking Robot based on Inequality State Constraint, *Proceedings. of 36th International Symposium on Robotics*, Biomechanical Robots, CD-ROM, WE413, Tokyo, Japan

Hase, T.; Huang, Q. & Ono, K. (2006). An Optimal Walking Trajectory of Biped Mechanism (3rd Report, Analysis of Upper Body Mass Model under Inequality State Constraint and Experimental Verification), *Transactions of the Japan Society of Mechanical Engineers*, Series C, Vol.72, No.721, pp.2845-2852 (in Japanese)

Huang, Q.. & Hase, T. (2006). Energy-Efficient Trajectory Planning for Biped walking Robot, *Proceedings. of the 2006 IEEE International Conference on Robotics and Biomimetics*, pp.648-653, Kunming, China

20

Geminoid: Teleoperated Android of an Existing Person

Shuichi Nishio*, Hiroshi Ishiguro*†, Norihiro Hagita*
** ATR Intelligent Robotics and Communication Laboratories*
† Department of Adaptive Machine Systems, Osaka University
Japan

1. Introduction

Why are people attracted to humanoid robots and androids? The answer is simple: because human beings are attuned to understand or interpret human expressions and behaviors, especially those that exist in their surroundings. As they grow, infants, who are supposedly born with the ability to discriminate various types of stimuli, gradually adapt and fine-tune their interpretations of detailed social clues from other voices, languages, facial expressions, or behaviors (Pascalis et al., 2002). Perhaps due to this functionality of nature and nurture, people have a strong tendency to anthropomorphize nearly everything they encounter. This is also true for computers or robots. In other words, when we see PCs or robots, some automatic process starts running inside us that tries to interpret them as human. The media equation theory (Reeves & Nass, 1996) first explicitly articulated this tendency within us. Since then, researchers have been pursuing the key element to make people feel more comfortable with computers or creating an easier and more intuitive interface to various information devices. This pursuit has also begun spreading in the field of robotics. Recently, researcher's interests in robotics are shifting from traditional studies on navigation and manipulation to human-robot interaction. A number of researches have investigated how people respond to robot behaviors and how robots should behave so that people can easily understand them (Fong et al., 2003; Breazeal, 2004; Kanda et al., 2004). Many insights from developmental or cognitive psychologies have been implemented and examined to see how they affect the human response or whether they help robots produce smooth and natural communication with humans.

However, human-robot interaction studies have been neglecting one issue: the "appearance versus behavior problem." We empirically know that appearance, one of the most significant elements in communication, is a crucial factor in the evaluation of interaction (See Figure 1). The interactive robots developed so far had very mechanical outcomes that do appear as "robots." Researchers tried to make such interactive robots "humanoid" by equipping them with heads, eyes, or hands so that their appearance more closely resembled human beings and to enable them to make such analogous human movements or gestures as staring, pointing, and so on. Functionality was considered the primary concern in improving communication with humans. In this manner, many studies have compared robots with different behaviors. Thus far, scant attention has been paid to robot appearances. Although there are many empirical discussions on such very simple static robots as dolls, the design of a robot's appearance, particularly to increase its human likeness, has always been the role of industrial designers; it has seldom been a field of study. This is a serious problem for developing and evaluating interactive robots. Recent neuroimaging studies show that certain brain activation does not occur when the observed actions are performed by non-human agents (Perani et al., 2001; Han et al., 2005). Appearance and behavior are tightly coupled, and concern is high that evaluation results might be affected by appearance.

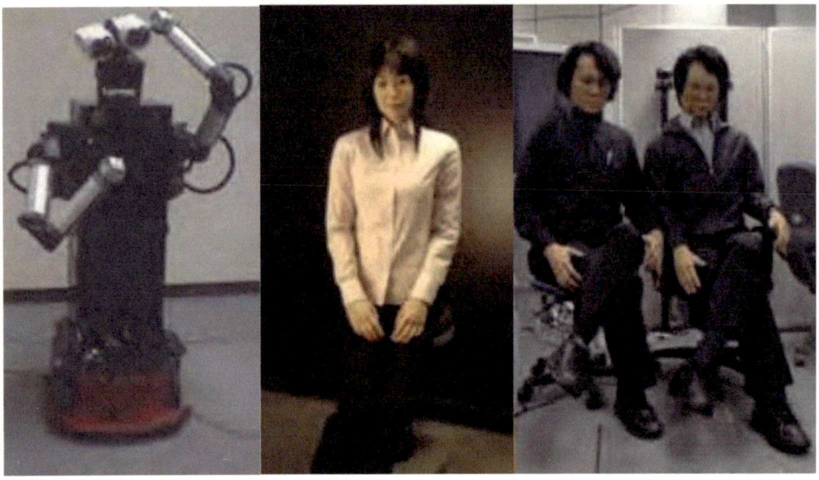

Fig. 1. Three categories of humanlike robots: humanoid robot Robovie II (left: developed by ATR Intelligent Robotics and Communication Laboratories), android Repliee Q2 (middle: developed by Osaka University and Kokoro corporation), geminoid HI-1 and its human source (right: developed by ATR Intelligent Robotics and Communication Laboratories).

In this chapter, we introduce *android science,* a cross-interdisciplinary research framework that combines two approaches, one in robotics for constructing very humanlike robots and androids, and another in cognitive science that uses androids to explore human nature. Here androids serve as a platform to directly exchange insights from the two domains. To proceed with this new framework, several androids have been developed so far, and many researches have been done. At that time, however, we encountered serious issues that sparked the development of a new category of robot called *geminoid.* Its concept and the development of the first prototype are described. Preliminary findings to date and future directions with geminoids are also discussed.

2. Android Science

Current robotics research uses various findings from the field of cognitive science, especially in the human-robot interaction area, trying to adopt findings from human-human interactions with robots to make robots that people can easily communicate with. At the same time, cognitive science researchers have also begun to utilize robots. As research fields extend to more complex, higher-level human functions such as seeking the neural basis of social skills (Blakemore, 2004), expectations will rise for robots to function as easily controlled apparatuses with communicative ability. However, the contribution from robotics to cognitive science has not been adequate because the appearance and behavior of current robots cannot be separately handled. Since traditional robots look quite mechanical and very different from human beings, the effect of their appearance may be too strong to ignore. As a result, researchers cannot clarify whether a specific finding reflects the robot's appearance, its movement, or a combination of both.

We expect to solve this problem using an android whose appearance and behavior closely resembles humans. The same thing is also an issue in robotics research, since it is difficult to clearly distinguish whether the cues pertain solely to robot behaviors. An objective, quantitative means of measuring the effect of appearance is required.

Androids are robots whose behavior and appearance are highly anthropomorphized. Developing androids requires contributions from both robotics and cognitive science. To realize a more humanlike android, knowledge from human sciences is also necessary. At the same time, cognitive science researchers can exploit androids for verifying hypotheses in understanding human nature. This new, bi-directional, cross-interdisciplinary research framework is called *android science* (Ishiguro, 2005). Under this framework, androids enable us to directly share knowledge between the development of androids in engineering and the understanding of humans in cognitive science (Figure 2).

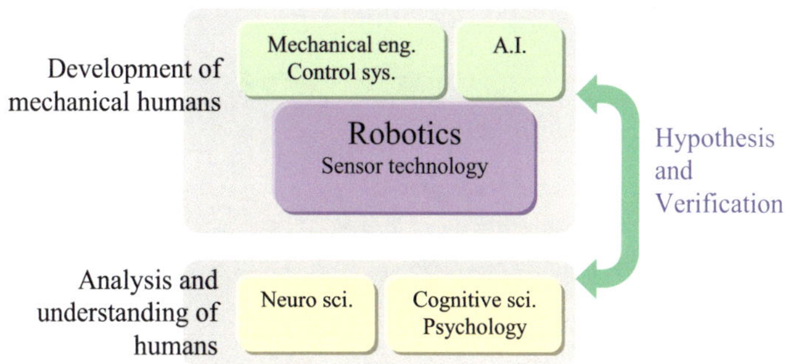

Fig. 2. Bi-directional feedback in *Android Science.*

The major robotics issue in constructing androids is the development of humanlike appearance, movements, and perception functions. On the other hand, one issue in cognitive science is "conscious and unconscious recognition." The goal of android science is to realize a humanlike robot and to find the essential factors for representing human likeness. How can we define human likeness? Further, how do we perceive human likeness? It is common knowledge that humans have conscious and unconscious recognition. When we observe objects, various modules are activated in our brain. Each of them matches the input sensory data with human models, and then they affect reactions. A typical example is that even if we recognize a robot as an android, we react to it as a human. This issue is fundamental both for engineering and scientific approaches. It will be an evaluation criterion in android development and will provide cues for understanding the human brain's mechanism of recognition.

So far, several androids have been developed. Repliee Q2, the latest android (Ishiguro, 2005), is shown in the middle of Figure 1. Forty-two pneumatic actuators are embedded in the android's upper torso, allowing it to move smoothly and quietly. Tactile sensors, which are also embedded under its skin, are connected to sensors in its environment, such as omnidirectional cameras, microphone arrays, and floor sensors. Using these sensory inputs,

the autonomous program installed in the android can make smooth, natural interactions with people near it.
Even though these androids enabled us to conduct a variety of cognitive experiments, they are still quite limited. The bottleneck in interaction with human is its lack of ability to perform long-term conversation. Unfortunately, since current AI technology for developing humanlike brains is limited, we cannot expect humanlike conversation with robots. When meeting humanoid robots, people usually expect humanlike conversation with them. However, the technology greatly lags behind this expectation. AI progress takes time, and such AI that can make humanlike conversation is our final goal in robotics. To arrive at this final goal, we need to use currently available technologies and understand deeply what a human is. Our solution for this problem is to integrate android and teleoperation technologies.

3. Geminoid

Fig. 3. Geminoid HI-1 (right).

We have developed *Geminoid,* a new category of robot, to overcome the bottleneck issue. We coined "geminoid" from the Latin "geminus," meaning "twin" or "double," and added "oides," which indicates "similarity" or being a twin. As the name suggests, a geminoid is a robot that will work as a duplicate of an existing person. It appears and behaves as a person and is connected to the person by a computer network. Geminoids extend the applicable field of android science. Androids are designed for studying human nature in general. With geminoids, we can study such personal aspects as presence or personality traits, tracing their origins and implemention into robots. Figure 3 shows the robotic part of HI-1, the first geminoid prototype. Geminoids have the following capabilities:

Appearance and behavior highly similar to an existing person
The appearance of a geminoid is based on an existing person and does not depend on the imagination of designers. Its movements can be made or evaluated simply by referring to the original person. The existence of a real person analogous to the robot enables easy comparison studies. Moreover, if a researcher is used as the original, we can expect that

individual to offer meaningful insights into the experiments, which are especially important at the very first stage of a new field of study when beginning from established research methodologies.

Teleoperation (remote control)
Since geminoids are equipped with teleoperation functionality, they are not only driven by an autonomous program. By introducing manual control, the limitations in current AI technologies can be avoided, enabling long-term, intelligent conversational human-robot interaction experiments. This feature also enables various studies on human characteristics by separating "body" and "mind." In geminoids, the operator (mind) can be easily exchanged, while the robot (body) remains the same. Also, the strength of connection, or what kind of information is transmitted between the body and mind, can be easily reconfigured. This is especially important when taking a top-down approach that adds/deletes elements from a person to discover the "critical" elements that comprise human characteristics. Before geminoids, this was impossible.

3.1 System overview

The current geminoid prototype, HI-1, consists of roughly three elements: a robot, a central controlling server (geminoid server), and a teleoperation interface (Figure 4).

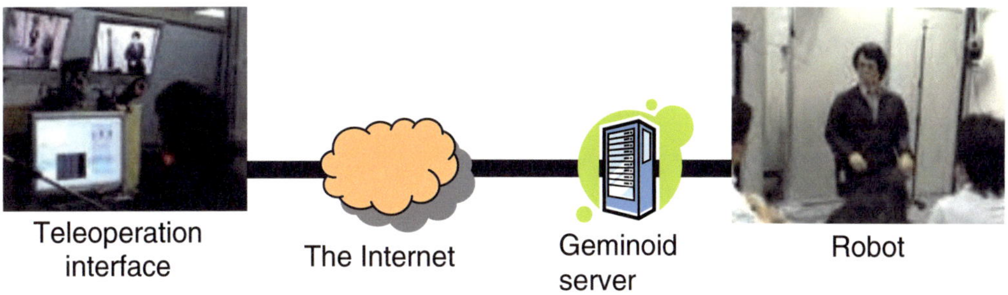

Fig. 4. Overview of geminoid system.

A robot that resembles a living person
The robotic element has essentially identical structure as previous androids (Ishiguro, 2005). However, efforts concentrated on making a robot that appears—not just to resemble a living person—to be a copy of the original person. Silicone skin was molded by a cast taken from the original person; shape adjustments and skin textures were painted manually based on MRI scans and photographs. Fifty pneumatic actuators drive the robot to generate smooth and quiet movements, which are important attributes when interacting with humans. The allocations of actuators were decided so that the resulting robot can effectively show the necessary movements for human interaction and simultaneously express the original person's personality traits. Among the 50 actuators, 13 are embedded in the face, 15 in the torso, and the remaining 22 move the arms and legs. The softness of the silicone skin and the compliant nature of the pneumatic actuators also provide safety while interacting with humans. Since this prototype was aimed for interaction experiments, it lacks the capability to walk around; it always remains seated. Figure 1 shows the resulting robot (right) alongside the original person, Dr. Ishiguro (author).

Teleoperation interface

Figure 5 shows the teleoperation interface prototype. Two monitors show the controlled robot and its surroundings, and microphones and a headphone are used to capture and transmit utterances. The captured sounds are encoded and transmitted to the geminoid server by IP links from the interface to the robot and vice versa. The operator's lip corner positions are measured by an infrared motion capturing system in real time, converted to motion commands, and sent to the geminoid server by the network. This enables the operator to implicitly generate suitable lip movement on the robot while speaking. However, compared to the large number of human facial muscles used for speech, the current robot only has a limited number of actuators on its face. Also, response speed is much slower, partially due to the nature of the pneumatic actuators. Thus, simple transmission and playback of the operator's lip movement would not result in sufficient, natural robot motion. To overcome this issue, measured lip movements are currently transformed into control commands using heuristics obtained through observation of the original person's actual lip movement.

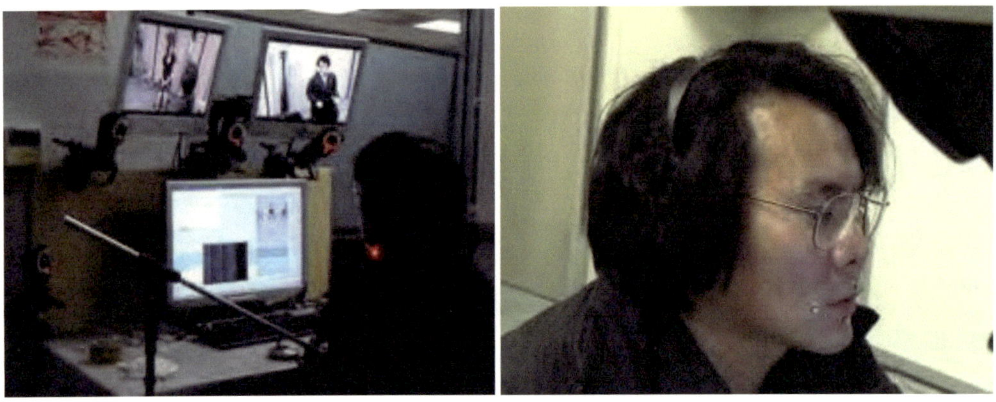

Fig. 5. Teleoperation interface.

The operator can also explicitly send commands for controlling robot behaviors using a simple GUI interface. Several selected movements, such as nodding, opposing, or staring in a certain direction can be specified by a single mouse click. This relatively simple interface was prepared because the robot has 50 degrees of freedom, which makes it one of the world's most complex robots, and is basically impossible to manipulate manually in real time. A simple, intuitive interface is necessary so that the operator can concentrate on interaction and not on robot manipulation. Despite its simplicity, by cooperating with the geminoid server, this interface enables the operator to generate natural humanlike motions in the robot.

Geminoid server

The geminoid server receives robot control commands and sound data from the remote controlling interface, adjusts and merges inputs, and sends and receives primitive controlling commands between the robot hardware. Figure 6 shows the data flow in the geminoid system. The geminoid server also maintains the state of human-robot interaction and generates *autonomous* or *unconscious* movements for the robot. As described above, as the robot's features become more humanlike, its behavior should also become suitably sophisticated to retain a "natural" look (Minato et al., 2006). One thing that can be seen in every human being, and that most robots lack, are the slight body movements caused by an autonomous system, such as breathing or blinking. To increase the robot's naturalness, the geminoid server emulates the human autonomous system and automatically generates these micro-movements, depending on the state of interaction each time. When the robot is "speaking," it shows different micro-movements than when "listening" to others. Such automatic robot motions, generated without operator's explicit orders, are merged and adjusted with *conscious* operation commands from the teleoperation interface (Figure 6). Alongside, the geminoid server gives the transmitted sounds specific delay, taking into account the transmission delay/jitter and the start-up delay of the pneumatic actuators. This adjustment serves synchronizing lip movements and speech, thus enhancing the naturalness of geminoid movement.

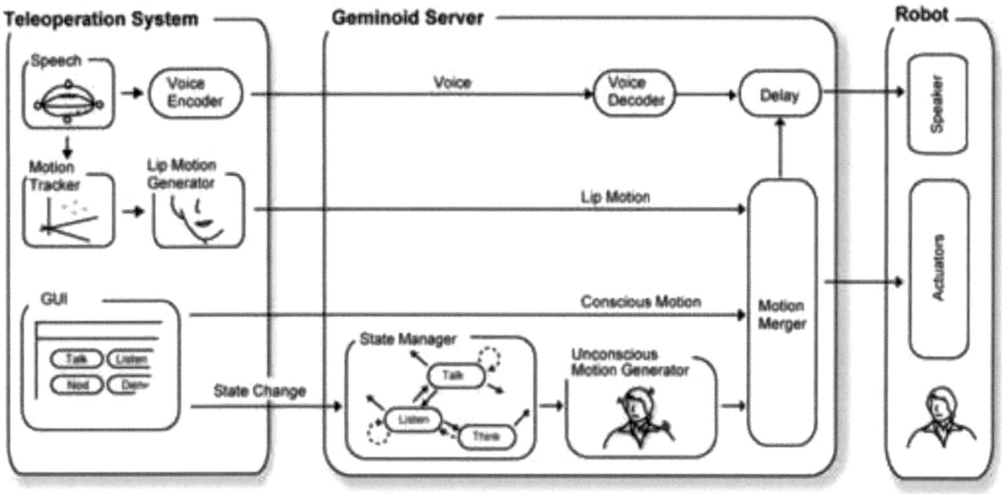

Fig. 6. Data flow in the geminoid system.

3.2 Experiences with the geminoid prototype

The first geminoid prototype, HI-1, was completed and press-released in July 2006. Since then, numerous operations have been held, including interactions with lab members and experiment subjects. Also, geminoid was demonstrated to a number of visitors and reporters. During these operations, we encountered several interesting phenomena. Here are some discourses made by the geminoid operator:

- When I (Dr. Ishiguro, the origin of the geminoid prototype) first saw HI-1 sitting still, it was like looking in a mirror. However, when it began moving, it looked like somebody else, and I couldn't recognize it as myself. This was strange, since we copied my movements to HI-1, and others who know me well say the robot accurately shows my characteristics. This means that we are not objectively recognizing our unconscious movements ourselves.
- While operating HI-1 with the operation interface, I find myself unconsciously adapting my movements to the geminoid movements. The current geminoid cannot move as freely as I can. I felt that, not just the geminoid but my own body is restricted to the movements that HI-1 can make.

- In less than 5 minutes both the visitors and I can quickly adapt to conversation through the geminoid. The visitors recognize and accept the geminoid as me while talking to each other.
- When a visitor pokes HI-1, especially around its face, I get a strong feeling of being poked myself. This is strange, as the system currently provides no tactile feedback. Just by watching the monitors and interacting with visitors, I get this feeling.

We also asked the visitors how they felt when interacting through the geminoid. Most said that when they saw HI-1 for the very first time, they thought that somebody (or Dr. Ishiguro, if familiar with him) was waiting there. After taking a closer look, they soon realized that HI-1 was a robot and began to have some weird and nervous feelings. But shortly after having a conversation through the geminoid, they found themselves concentrating on the interaction, and soon the strange feelings vanished. Most of the visitors were non-researchers unfamiliar with robots of any kind.

Does this mean that the geminoid has overcome the "uncanny valley"? Before talking through the geminoid, the initial response of the visitors seemingly resembles the reactions seen with previous androids: even though at the very first moment they could not recognize the androids as artificial, they nevertheless soon become nervous while being with the androids. Are intelligence or long-term interaction crucial factors in overcoming the valley and arriving at an area of natural humanness?

We certainly need objective means to measure how people feel about geminoids and other types of robots. In a previous android study, Minato et al. found that gaze fixation revealed criteria about the naturalness of robots (Minato et al., 2006). Recent studies have shown different human responses and reactions to natural or artificial stimuli of the same nature. Perani et al. showed that different brain regions are activated while watching human or computer graphic arms movements (Perani et al., 2001). Kilner et al. showed that body movement entrainment occurs when watching human motions, but not with robot motions (Kilner et al., 2003). By examining these findings with geminoids, we may be able to find some concrete measurements of human likeliness and approach the "appearance versus behavior" issue.

Perhaps HI-1 was recognized as a sort of communication device, similar to a telephone or a TV-phone. Recent studies have suggested a distinction in the brain process that discriminates between people appearing in videos and existing persons appearing live (Kuhl et al., 2003). While attending TV conferences or talking by cellular phones, however, we often experience the feeling that something is missing from a face-to-face meeting. What is missing here? Is there an objective means to measure and capture this element? Can we ever implement this on robots?

4. Summary and further issues

In developing the geminoid, our purpose is to study *Sonzai-Kan* , or human presence, by extending the framework of android science. The scientific aspect must answer questions about how humans recognize human existence/presence. The technological aspect must realize a teleoperated android that works on behalf of the person remotely accessing it. This will be one of the practical networked robots realized by integrating robots with the Internet.

The following are our current challenges:

Teleoperation technologies for complex humanlike robots

Methods must be studied to teleoperate the geminoid to convey existence/presence, which is much more complex than traditional teleoperation for mobile and industrial robots. We are studying a method to autonomously control an android by transferring motions of the operator measured by a motion capturing system. We are also developing methods to autonomously control eye-gaze and humanlike small and large movements.

Synchronization between speech utterances sent by the teleoperation system and body movements

The most important technology for the teleoperation system is synchronization between speech utterances and lip movements. We are investigating how to produce natural behaviors during speech utterances. This problem is extended to other modalities, such as head and arm movements. Further, we are studying the effects on non-verbal communication by investigating not only synchronization of speech and lip movements but also facial expressions, head, and even whole body movements.

Psychological test for human existence/presence

We are studying the effect of transmitting Sonzai-Kan from remote places, such as meeting participation instead of the person himself. Moreover, we are interested in studying existence/presence through cognitive and psychological experiments. For example, we are studying whether the android can represent the authority of the person himself by comparing the person and the android.

Application

Although being developed as research apparatus, the nature of geminoids can allow us to extend the use of robots in the real world. The teleoperated, semi-autonomous facility of geminoids allows them to be used as substitutes for clerks, for example, that can be controlled by human operators only when non-typical responses are required. Since in most cases an autonomous AI response will be sufficient, a few operators will be able to control hundreds of geminoids. Also because their appearance and behavior closely resembles humans, in the next age geminoids should be the ultimate interface device.

5. Acknowledgement

This work was supported in part by the Ministry of Internal Affairs and Communications of Japan.

6. References

Blakemore, S. J. & Frith, U. (2004). How does the brain deal with the social world? *Neuroreport*, 15, 119-128

Breazeal, C. (2004). Social Interactions in HRI: The Robot View, *IEEE Transactions on Man, Cybernetics and Systems: Part C*, 34, 181-186

Fong, T., Nourbakhsh, I., & Dautenhahn, K. (2003). A survey of socially interactive robots, *Robotics and Autonomous Systems*, 42, 143–166

Han, S., Jiang, Y., Humphreys, G. W., Zhou, T., and & Cai, P. (2005). Distinct neural substrates for the perception of real and virtual visual worlds, *NeuroImage*, 24, 928- 935

Ishiguro, H. (2005). Android Science: Toward a New Cross-Disciplinary Framework. *Proceedings of Toward Social Mechanisms of Android Science: A CogSci 2005 Workshop*, 1–6

Kanda, T., Ishiguro, H., Imai, M., & Ono, T. (2004). Development and Evaluation of Interactive Humanoid Robots, *Proceedings of the IEEE*, 1839-1850

Kilner, J. M., Paulignan, Y., & Blakemore, S. J. (2003). An interference effect of observed biological movement on action, *Current Biology*, 13, 522-525

Kuhl, P. K., Tsao. F. M., & Liu, H. M. (2003). Foreign-language experience in infancy: Effects of short-term exposure and social interaction on phonetic learning. *Proceedings of the National Academy of Sciences*, 100, 9096-9101

Minato, T., Shimada, M., Itakura, S., Lee, K., & Ishiguro, H. (2006). Evaluating the human likeness of an android by comparing gaze behaviors elicited by the android and a person, *Advanced Robotics*, 20, 1147-1163

Pascalis, O., Haan, M., and Nelson, C. A. (2002). Is Face Processing Species-Specific During the First Year of Life?, *Science*, 296, 1321-1323

Perani, D., Fazio, F., Borghese, N. A., Tettamanti, M., Ferrari, S., Decety, J., & Gilardi, M.C. (2001). Different brain correlates for watching real and virtual hand actions, *NeuroImage*, 14, 749-758

Reeves, B. & Nass, C. (1996). The Media Equation, CSLI/Cambridge University Press

21

Obtaining Humanoid Robot Controller Using Reinforcement Learning

Masayoshi Kanoh[1] and Hidenori Itoh[2]
[1]*Chukyo University*, [2]*Nagoya Institute of Technology*
Japan

1. Introduction

Demand for robots is shifting from their use in industrial applications to their use in domestic situations, where they "live" and interact with humans. Such robots require sophisticated body designs and interfaces to do this. Humanoid robots that have multi-degrees-of-freedom (MDOF) have been developed, and they are capable of working with humans using a body design similar to humans. However, it is very difficult to intricately control robots with human generated, preprogrammed, learned behavior. Learned behavior should be acquired by the robots themselves in a human-like way, not programmed manually. Humans learn actions by trial and error or by emulating someone else's actions. We therefore apply reinforcement learning for the control of humanoid robots because this process resembles a human's trial and error learning process.

Many existing methods of reinforcement learning for control tasks involve discrediting state space using BOXES (Michie & Chambers, 1968; Sutton & Barto, 1998) or CMAC (Albus, 1981) to approximate a value function that specifies what is advantageous in the long run. However, these methods are not effective for doing generalization and cause perceptual aliasing. Other methods use basis function networks for treating continuous state space and actions.

Networks with sigmoid functions have the problem of catastrophic interference. They are suitable for off-line learning, but are not adequate for on-line learning such as that needed for learning motion (Boyan & Moore, 1995; Schaal & Atkeson, 1996). On the contrary, networks with radial basis functions are suitable for on-line learning. However, learning using these functions requires a large number of units in the hidden layer, because they cannot ensure sufficient generalization. To avoid this problem, methods of incremental allocation of basis functions and adaptive state space formation were proposed (Morimoto & Doya, 1998; Samejima & Omori, 1998; Takahashi et al., 1996; Moore & Atkeson, 1995).

In this chapter, we propose a dynamic allocation method of basis functions called Allocation/Elimination Gaussian Softmax Basis Function Network (AE-GSBFN), that is used in reinforcement learning to treat continuous high-dimensional state spaces. AE-GSBFN is a kind of actor-critic method that uses basis functions and it has allocation and elimination processes. In this method, if a basis function is required for learning, it is allocated dynamically. On the other hand, if an allocated basis function becomes redundant, the function is eliminated. This method can treat continuous high-dimensional state spaces because the allocation and elimination processes reduce the number of basis functions required for evaluation of the state space.

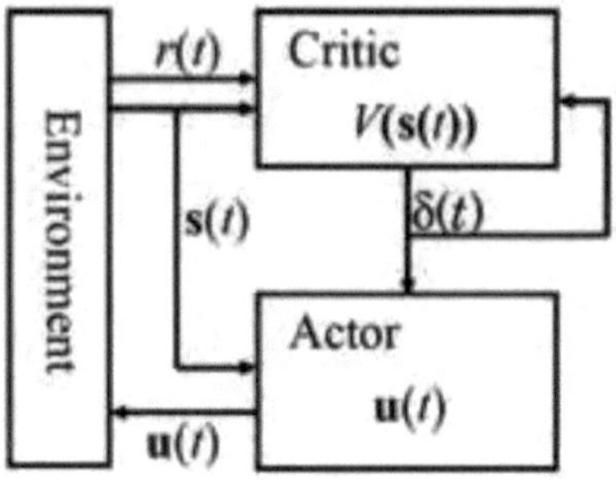

Fig. 1. Actor-critic architecture.

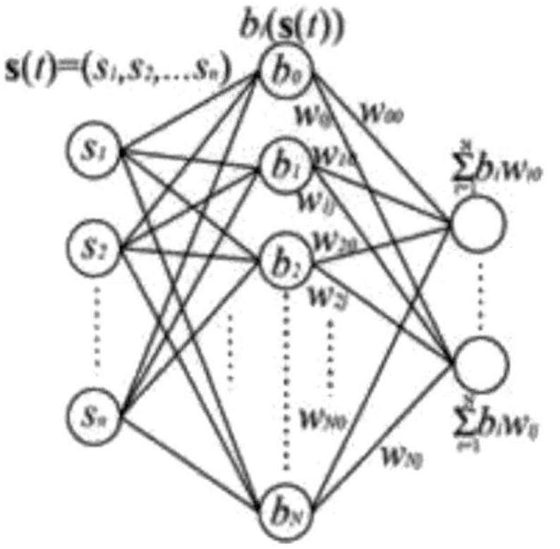

Fig. 2. Basis function network.

To confirm the effectiveness of our method, we used computer simulation to show how a humanoid robot learned two motions: a standing-up motion from a seated position on chair and a foot-stamping motion.

2. Actor-Critic Method

In this section, we describe an actor-critic method using basis functions, and we apply it to our method.
Actor-critic methods are temporal difference (TD) methods that have a separate memory structure to explicitly represent the policy independent of the value function (Sutton & Barto, 1998). Actor-critic methods are constructed by an actor and a critic, as depicted in Figure 1. The policy structure is known as the actor because it is used to select actions, and the estimated value function is known as the critic because it criticizes the actions made by the actor.
The actor and the critic each have a basis function network for learning of continuous state spaces. Basis function networks have a three-layer structure as shown in Figure 2, and basis functions are placed in middle-layer units. Repeating the following procedure, in an actor-critic method using basis function networks, the critic correctly estimates the value function $V(\mathbf{s})$, and then the actor acquires actions that maximize $V(\mathbf{s})$.

1. When state $\mathbf{s}(t)$ is observed in the environment, the actor calculates the j-th value $u_j(t)$ of the action $\boldsymbol{u}(t)$ as follows (Gullapalli, 1990):

$$u_j(t) = u_j^{\max} g\left(\sum_i^N \omega_{ij} b_i(\mathbf{s}(t)) + n_j(t)\right), \tag{1}$$

where $u_j^{\max}$ is a maximal control value, N is the number of basis functions, $b_i(\mathbf{s}(t))$ is a basis function, ω_{ij} is a weight, $n_j(t)$ is a noise function, and $g()$ is a logistic sigmoid activation function whose outputs lie in the range (−1, 1). The output value of actions is saturated into $u_j^{\max}$ by $g()$.

2. The critic receives the reward $r(t)$, and then observes the resulting next state $\mathbf{s}(t+1)$. The critic provides the TD-error $\delta(t)$ as follows:

$$\delta(t) = r(t) + \gamma V(\mathbf{s}(t+1)) - V(\mathbf{s}(t)), \tag{2}$$

where γ is a discount factor, and $V(\mathbf{s})$ is an estimated value function. Here, $V(\mathbf{s}(t))$ is calculated as follows:

$$V(\mathbf{s}(t)) = \sum_i^N v_i b_i(\mathbf{s}(t)), \tag{3}$$

where v_i is a weight.

3. The actor updates weight ω_{ij} using TD-error:

$$\omega_{ij} \leftarrow \omega_{ij} + \beta\delta(t) n_j(t) b_i(\mathbf{s}(t)), \tag{4}$$

where β is a learning rate.

4. The critic updates weight v_i:

$$v_i \leftarrow v_i + \alpha\delta(t) e_i, \tag{5}$$

where α is a learning rate, and e_i is an eligibility trace. Here, e_i is calculated as follows:

$$e_i \leftarrow \gamma\lambda e_i + b(\mathbf{s}(t)), \tag{6}$$

where λ is a trace-decay parameter.

5. Time is updated.

$$t \leftarrow t + \Delta t. \tag{7}$$

Note that Δt is 1 in general, but we used the description of Δt for the control interval of the humanoid robots.

3. Dynamic Allocation of Basis Functions

In this chapter, we propose a dynamic allocation method of basis functions. This method is an extended application of the Adaptive Gaussian Softmax Basis Function Network (A-GSBFN) (Morimoto & Doya, 1998, 1999). A-GSBFN only allocates basis functions, whereas our method both allocates and eliminates them. In this section, we first briefly describe A-GSBFN in Section 3.1; then we propose our method, Allocation/Elimination Gaussian Softmax Basis Function Network (AE-GSBFN), in Section 3.2.

3.1 A-GSBFN

Networks with sigmoid functions have the problem of catastrophic interference. They are suitable for off-line learning, but not adequate for on-line learning. In contrast, networks with radial basis functions (Figure 3) are suitable for on-line learning, but learning using these functions requires a large number of units, because they cannot ensure sufficient generalization. The Gaussian softmax functions (Figure 4) have the features of both sigmoid functions and radial basis functions. Networks with the Gaussian softmax functions can therefore assess state space locally and globally, and enable learning motions of humanoid robots.

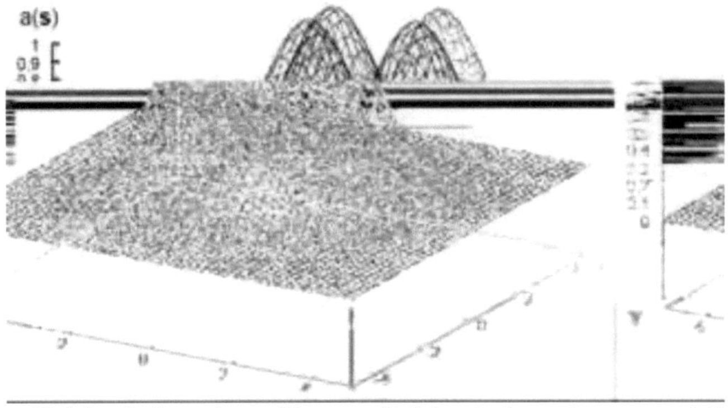

Fig. 3. Shape of radial basis functions. Four radial basis functions are visible here, but it is clear that the amount of generalization done is insufficient.

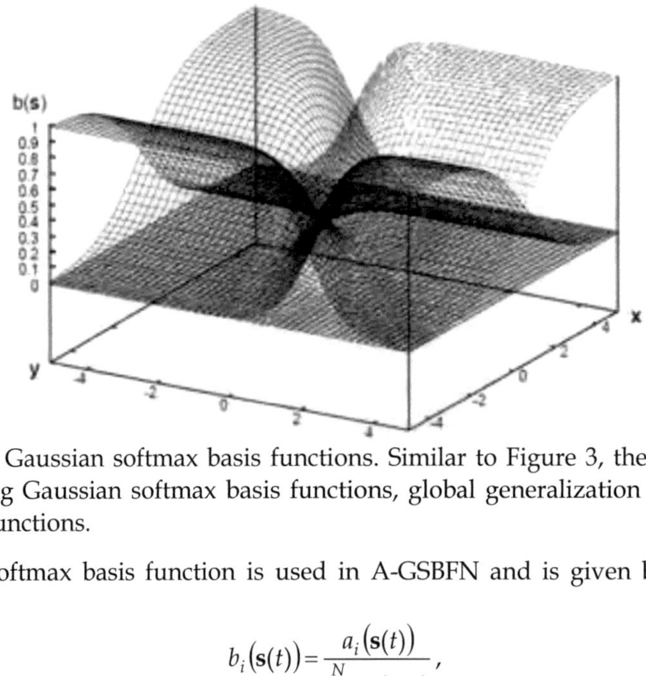

Fig. 4. Shape of Gaussian softmax basis functions. Similar to Figure 3, there are four basis functions. Using Gaussian softmax basis functions, global generalization is done, such as using sigmoid functions.

The Gaussian softmax basis function is used in A-GSBFN and is given by the following equation:

$$b_i(\mathbf{s}(t)) = \frac{a_i(\mathbf{s}(t))}{\sum_k^N a_k(\mathbf{s}(t))}, \tag{8}$$

where $a_i(\mathbf{s}(t))$ is a radial basis function, and N is the number of radial basis functions. Radial basis function $a_i(\mathbf{s}(t))$ in the i-th unit is calculated by the following equation:

$$a_i(\mathbf{s}(t)) = \exp\left(-\frac{1}{2}\left\|M(\mathbf{s}(t) - \mathbf{c}_i)\right\|^2\right), \tag{9}$$

where $\mathbf{c}_i$ is the center of the i-th basis function, and M is a matrix that determines the shape of the basis function.

In A-GSBFN, a new unit is allocated if the error is larger than threshold $\delta_{\max}$ and the activation of all existing units is smaller than threshold $a_{\min}$:

$$|h(t)| > \delta_{\max} \quad \text{and} \quad \max_i a_i(\mathbf{s}(t)) < a_{\min}, \tag{10}$$

where $h(t)$ is defined as $h(t) = \delta(t) n_j(t)$ at the actor, and $h(t) = \delta(t)$ at the critic. The new unit is initialized with $\mathbf{c}_i = \mathbf{s}$, and $\omega_i = 0$.

3.2 Allocation/Elimination GSBFN

To perform allocation and elimination of basis functions, we introduce three criteria into A-GSBFN: trace ε_i of activation of radial basis functions, additional control time η, and existing time τ_i of radial basis functions. The criteria ε_i and τ_i are prepared for all basis functions, and η is prepared for both actor and critic networks. A learning agent can gather further information on its own states by using these criteria.
We now define the condition of allocation of basis functions.

Definition 1 - Allocation
A new unit is allocated at $\mathbf{c}_i = \mathbf{s}(t)$ if the following condition is satisfied at the actor or critic networks:

$$|h(t)| > \delta_{\max} \quad \text{and} \quad \max_i a_i(\mathbf{s}(t)) < a_{\min} \quad \text{and} \quad \eta > T_{\text{add}}, \tag{11}$$

where T_{add} is a threshold.

■

Let us consider using condition (10) for allocation. This condition is only considered for allocation, but it is not considered as a process after a function is eliminated. Therefore, when a basis function is eliminated, another basis function is immediately allocated at the near state of the eliminated function. To prevent immediate allocation, we introduced additional control time η into the condition of allocation. The value of η monitors the length of time that has elapsed since a basis function was eliminated. Note that η is initialized at 0, when a basis function is eliminated.
We then define the condition of elimination using ε_i and τ_i.

Definition 2 - Elimination
The basis function $b_i(\mathbf{s}(t))$ is eliminated if the following condition is satisfied in the actor or critic networks.

$$\varepsilon_i > \varepsilon_{\max} \quad \text{and} \quad \tau_i > T_{\text{erase}}, \tag{12}$$

where $\varepsilon_{\max}$ and T_{erase} are thresholds.

□

The trace ε_i of the activation of radial basis functions is updated at each step in the following manner:

$$\varepsilon_i \leftarrow \kappa \varepsilon_i + a_i(\mathbf{s}(t)), \tag{13}$$

where κ is a discount rate. Using ε_i, the learning agent can sense states that it has recently taken. The value of ε_i takes a high value if the agent stays in almost the same state. This situation is assumed when the learning falls into a local minimum. Using the value of ε_i, we consider how to avoid the local minimum. Moreover, using τ_i, we consider how to inhibit a basis function from immediate elimination after it is allocated. We therefore defined the condition of elimination using ε_i and τ_i.

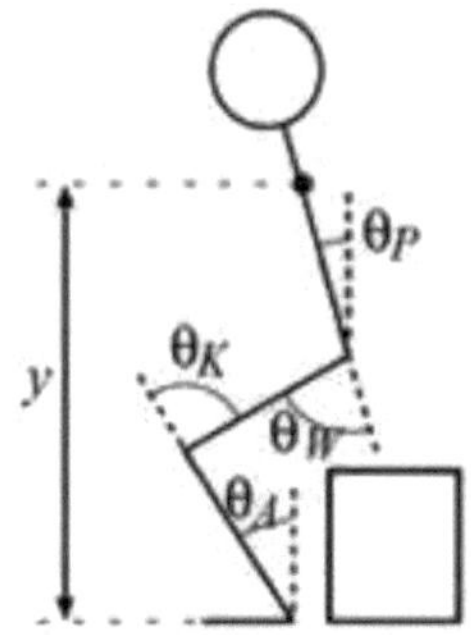

Fig. 5. Learning motion; standing up from a chair.

4. Experiments

4.1 Standing-up motion learning

In this section, as an example of learning of continuous high-dimensional state spaces, AE-GSBFN is applied to a humanoid robot learning to stand up from a chair (Figure 5). The learning was simulated using the virtual body of the humanoid robot HOAP-1 made by Fujitsu Automation Ltd. Figure 6 shows HOAP-1. The robot is 48 centimeters tall, weighs 6 kilograms, has 20 DOFs, and has 4 pressure sensors each on the soles of its feet. Additionally, angular rate and acceleration sensors are mounted in its chest. To simulate learning, we used the Open Dynamics Engine (Smith).

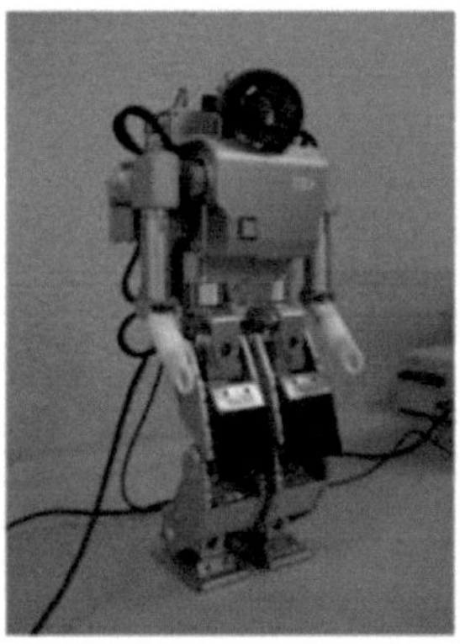

Fig. 6. HOAP-1 (Humanoid for Open Architecture Platform).

The robot is able to observe the following vector $\mathbf{s}(t)$ as its own state:

$$\mathbf{s}(t) = \left(\theta_W, \dot{\theta}_W, \theta_K, \dot{\theta}_K, \theta_A, \dot{\theta}_A, \theta_P, \dot{\theta}_P\right), \tag{14}$$

where θ_W, θ_K, and θ_A are waist, knee, and ankle angles respectively, and θ_P is the pitch of its body (see Figure 5). Action $\mathbf{u}(t)$ of the robot is determined as follows:

$$\mathbf{u}(t) = \left(\dot{\theta}_W, \dot{\theta}_K, \dot{\theta}_A\right), \tag{15}$$

One trial ended when the robot fell down or time exceeded t_{total} = 10 [s]. Rewards $r(t)$ were determined by height y [cm] of the robot's chest:

$$r(t) = \begin{cases} -20\left|\dfrac{l_{stand} - y}{l_{stand} - l_{down}}\right| & (\textit{during trial}) \\ -20\left|t_{total} - t\right| & (\textit{on failure}) \end{cases}, \tag{16}$$

where l_{stand} = 35 [cm] is the position of the robot's chest in an upright posture, and l_{down} = 20 [cm] is its center in a falling-down posture. We used $u_j^{max} = \pi/36$, $\gamma = 0.9$, $\beta = 0.1$, $\alpha = 0.02$, $\lambda = 0.6$, and $\Delta t = 0.01$ [s] for parameters in Section 2, M=(1.0, 0.57, 1.0, 0.57, 1.0, 0.57, 1.0, 0.57), $\delta_{max} = 0.5$, and $a_{min} = 0.4$ in Section 3.1, and $T_{add} = 1$ [s], $\kappa = 0.9$, $\varepsilon_{max} = 5.0$, and $T_{erase} = 3$ [s] in Section 3.2.

Figure 7 shows the learning results. First, the robot learned to fall down backward, as shown in i). Second, the robot intended to stand up from a chair, but fell forward, as shown in ii), because it could not yet fully control its balance. Finally, the robot stood up while maintaining its balance, as shown in iii). The number of basis functions in the 2922nd trial was 72 in both actor and critic networks. Figure 8 shows the experimental result with the humanoid robot HOAP-1. The result shows that HOAP-1 was able to stand up from a chair, as in the simulation.

We then compared the number of basis functions in AE-GSBFN with the number of basis functions in A-GSBFN. Figure 9 shows the number of basis functions of the actor, averaged over 20 repetitions. In these experiments, motion learning with both AE-GSBFN and A-GSBFN was successful, but the figure indicates that the number of basis functions required by AE-GSBFN was fewer than that by A-GSBFN. That is, high dimensional learning is possible using AE-GSBFN. Finally, we plotted the height of the robot's chest in successful experiments in Figures 10 and 11. In the figures, circles denote a successful stand-up. The results show that motion learning with both AE-GSBFN and A-GSBFN was successful.

4.2 Stamping motion learning

In Section 4.1, we described our experiment with learning of transitional motion. In this section, we describe our experiment with periodic motion learning. We use a stamping motion as a periodic motion (Figure 12). Periodic motions, such as locomotion, are difficult to learn only through reinforcement learning, so in many cases, a Central Pattern Generator (CPG), etc., is used in addition to reinforcement learning (e.g., Mori et al., 2004). In this experiment, we use inverse kinematics and AE-GSBFN to obtain a stamping motion.

Inverse kinematics calculates the amount $\dot{\theta}$ of change of each joint angle from the amount $\dot{P}$ of change of the coordinates of a link model:

$$\dot{\theta} = J^{-1}(\theta)\dot{P}, \tag{17}$$

where $J(\theta)$ is the Jacobian matrix. Generally, since the dimension of $\dot{\theta}$ differs from the dimension of $\dot{P}$, $J(\theta)$ does not become a regular matrix, and its inverse matrix cannot be calculated. Moreover, even if it could be calculated, a motion resolution by $J^{-1}(\theta)$ cannot be performed in the neighborhood of singular points, which are given by θ around $\det J(\theta) = 0$. To solve these problems, we used the following function (Nakamura & Hanafusa, 1984) in this section:

$$\dot{\theta} = J^T(JJ^T + k_s I)^{-1}\dot{P}, \tag{18}$$

i) 300th trial

ii) 1031st trial

iii) 1564th trial

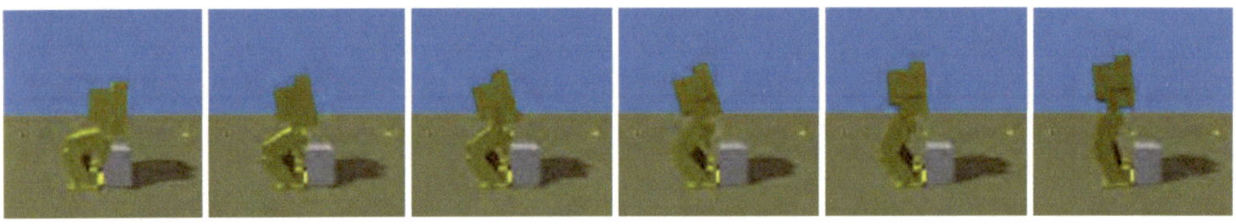

Fig. 7. Learning results.

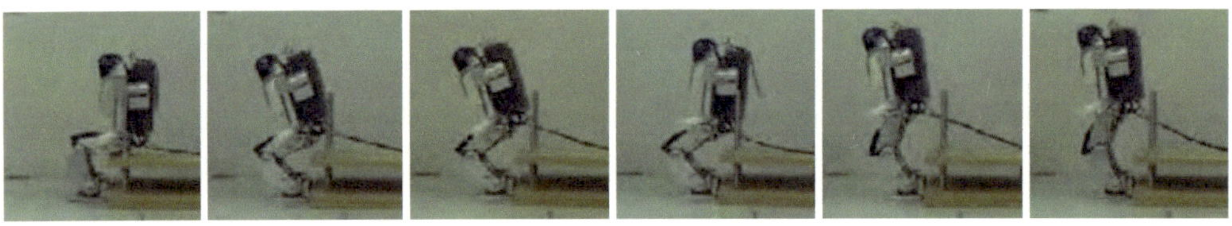

Fig. 8. Experimental result with HOAP-1.

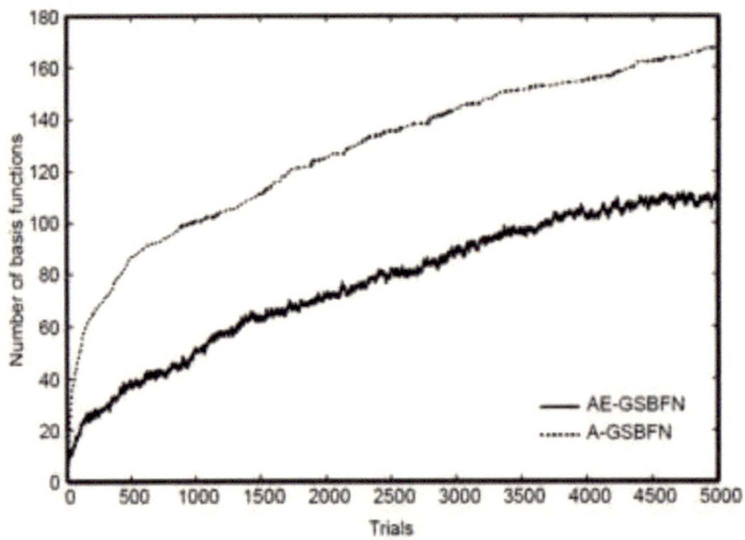

Fig. 9. Number of basis functions in the actor network (averaged over 20 repetitions).

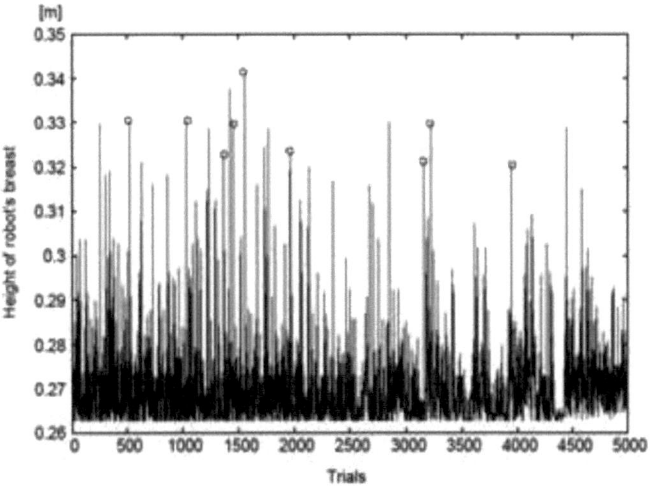

Fig. 10. Height of robot's chest with AE-GSBFN. Circles denote successful stand-ups.

where k_s is a scalar function with which it becomes a positive value near singular points and becomes 0 otherwise:

$$k_s = \begin{cases} k_0\left(1 - \dfrac{w}{w_0}\right)^2 & (w < w_0) \\ 0 & (otherwise) \end{cases}, \tag{19}$$

where k_0 is a positive parameter, w_0 is a threshold that divides around singular points from the others, and w is given by $w = \sqrt{\det J(\theta) J^T(\theta)}$.

In this experiment, the coordinate of the end of the legs is given by inverse kinematics (i.e., up-down motion of the legs is given), and motion of the horizontal direction of the waist is learned by AE-GSBFN. The coordinate value was acquired from the locomotion data of HOAP-1. Concretely, motion is generated by solving inverse kinematics from p_w to the idling leg, and from the supporting leg to p_w (Figure 12 (a)). The change of supporting and idling legs is also acquired from HOAP-1's locomotion data.

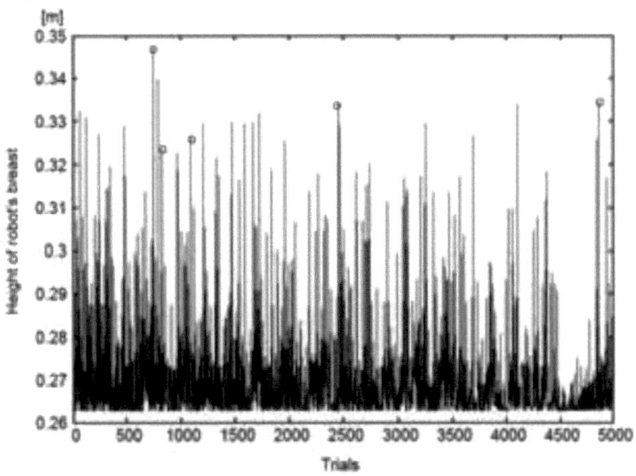

Fig. 11. Height of robot's chest with A-GSBFN. Circles denote successful stand-ups.

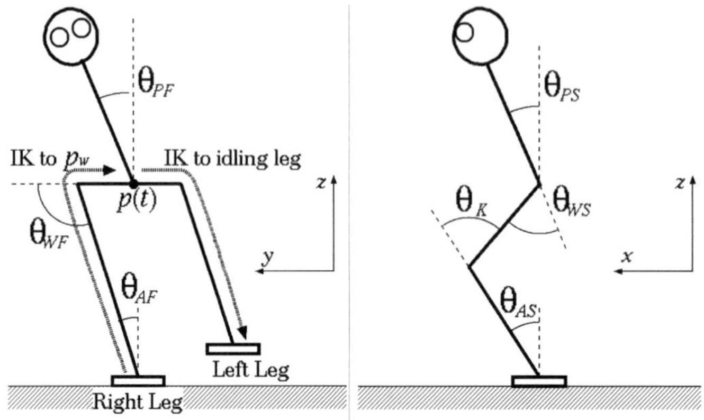

(a) Frontal plane (b) Sagittal plane

Fig. 12. Learning motion; stamping the foot.

The robot is able to observe the following vector $\mathbf{s}(t)$ as its own state:

$$\mathbf{s}(t) = (\theta_{WF}^{(R)}, \dot{\theta}_{WF}^{(R)}, \theta_{WF}^{(L)}, \dot{\theta}_{WF}^{(L)}, \theta_{AF}^{(R)}, \dot{\theta}_{AF}^{(R)}, \theta_{AF}^{(L)}, \dot{\theta}_{AF}^{(L)}, \theta_{PF}, \dot{\theta}_{PF}, \theta_{WS}^{(R)}, \dot{\theta}_{WS}^{(R)}, \theta_{WS}^{(L)}, \dot{\theta}_{WS}^{(L)}, \theta_{K}^{(R)}, \dot{\theta}_{K}^{(R)}, \theta_{K}^{(L)}, \dot{\theta}_{K}^{(L)}, \theta_{PS}, \dot{\theta}_{PS}), \quad (20)$$

where $\theta_{WF}^{(\cdot)}$ (right leg: $\theta_{WF}^{(R)}$, left leg: $\theta_{WF}^{(L)}$) and $\theta_{AF}^{(\cdot)}$ (right leg: $\theta_{AF}^{(R)}$, left leg: $\theta_{AF}^{(L)}$) are angles of the waist and ankle about the roll axis, respectively, and θ_{PF} is the pitch of its body about the roll axis. Also $\theta_{WS}^{(\cdot)}$ (right leg: $\theta_{WS}^{(R)}$, left leg: $\theta_{WS}^{(L)}$) and $\theta_{K}^{(\cdot)}$ (right leg: $\theta_{K}^{(R)}$, left leg: $\theta_{K}^{(L)}$) are angles of the waist and knee about the pitch axis, respectively, and θ_{PS} is the pitch of its body about the pitch axis. Note that the angle of the ankle of each leg about the pitch axis was controlled to be parallel to the ground.

Action $\mathbf{u}(t)$ of the robot is determined as follows:

$$\mathbf{u}(t) = \dot{p}(t), \quad (21)$$

where $\dot{p}(t)$ is the amount of change of $p(t)$ which is the position of the center of the robot's waist. Note that the value of $p(t)$ is a y-coordinate value, and does not include x- or z-coordinate values.

One trial terminated when the robot fell down or time exceeded t_{total} = 17.9 [s]. Rewards $r(t)$ were determined by the following equation:

$$r(t) = \begin{cases} -20|p(t) - p(0)| & \textit{(during trial)} \\ -20|t_{\text{total}} - t| & \textit{(on failure)} \end{cases}. \quad (22)$$

We can use the value of the difference between its supporting leg and $p(t)$ as rewards, but these rewards may represent the ideal position of $p(t)$ because of the use of inverse kinematics. Therefore, we used the above equation. Using the equation (22), the closer $p(t)$ is to $p(0)$, the more the rewards increases. Intuitively, it is unsuitable for rewards of stamping motion learning, but acquiring a stamping motion only brings more rewards, because an up-down motion of the leg is given forcibly by inverse kinematics, and it is necessary to change $p(t)$ quite a lot to make the robot stay upright without falling down.

We used $u_j^{\max} = 1.0\times 10^{-4}$, $\gamma = 0.9$, $\beta = 0.1$, $\alpha = 0.02$, $\lambda = 0.6$, and $\Delta t = 0.01$ [s] for parameters in Section 2, M=diag(2.0, 1.1, 2.0, 1.1, 2.0, 1.1, 2.0, 1.1, 2.0, 1.1, 2.0, 1.1, 2.0, 1.1, 2.0, 1.1, 2.0, 1.1, 2.0, 1.1), $\delta_{\max} = 0.5$, and $a_{\min} = 0.135$ in Section 3.1, and $T_{\mathrm{add}} = 1.0$ [s], $\kappa = 0.9$, $\varepsilon_{\max} = 5.0$, and $T_{\mathrm{erase}} = 2.0$ [s] in Section 3.2. We also used k_0 = 0.01 and w_0=0.003 for the parameters of inverse kinematics.

Figure 13 shows the learning results. The robot can control its balance by moving its waist right and left. Figure 14 plots the amount of time taken to fall down. You can see that the time increases as the learning progresses. Figure 15 shows the value of $p(t)$ in the 986th trial. It is clear that $p(t)$ changes periodically. These results indicate that a stamping motion was acquired, but the robot's idling leg does not rise perfectly when we look at the photos in Figure 13. We assume that the first reason for these results is that it is difficult to control the angle of ankle using inverse kinematics (since inverse kinematics cannot control $\theta_{AF}^{(R)}$ and $\theta_{AF}^{(L)}$ to be parallel to the ground). The second reason is that we only used y-coordinate values of the waist for learning, and the third is because we used equation (22) for rewards. To solve the second issue, we can use its z-coordinate value. Using equation (22), the third reason, a small periodic motion is obtained (Figure 16). To solve this problem, we should consider another reward function for this experiment. We will explore these areas in our future research.

Fig. 13. Simulation result (986th trial).

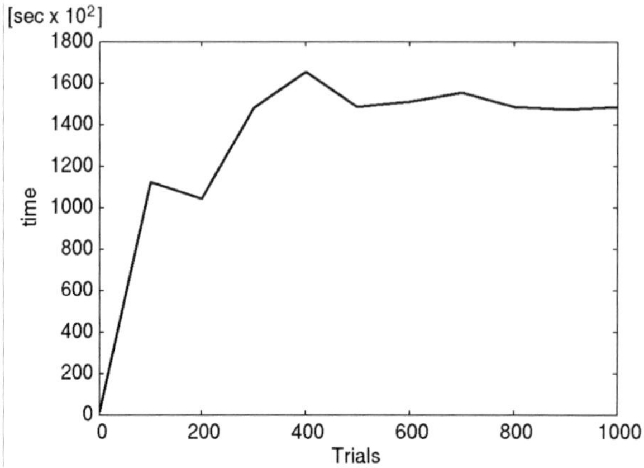

Fig. 14. Time during trial (averaged over 100 repetitions). If the value of a vertical axis is large, the stamping motion extends for a long time.

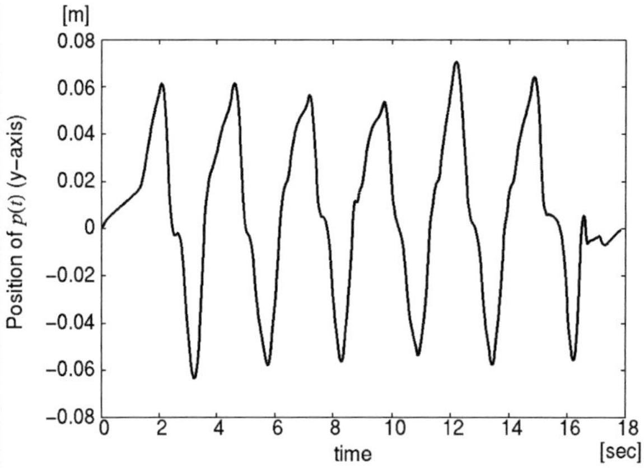

Fig. 15. Position of $p(t)$ in horizontal direction in 986th trial.

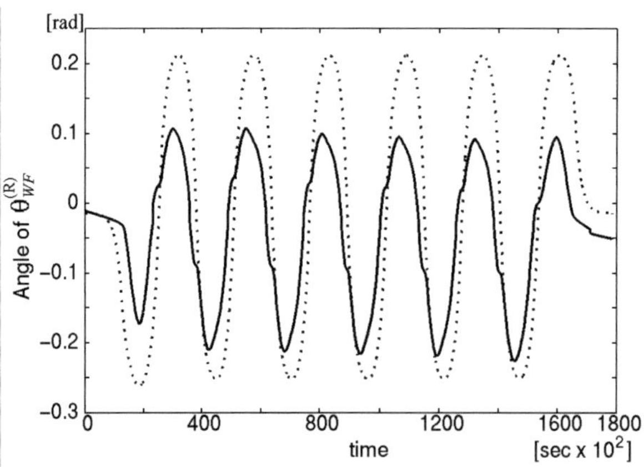

Fig. 16. Ideal angle and output angle of $\theta_{WF}^{(R)}$ with AE-GSBFN in 986th trial. The dotted line indicates an ideal motion and the solid line indicates the acquired motion with AE-GSBFN. It is clear that the acquired motion consists of small periodic motions compared with the deal motion.

5. Conclusion

In this chapter, we proposed a dynamic allocation method of basis functions, AE-GSBFN, in reinforcement learning. Through allocation and elimination processes, AE-GSBFN overcomes the curse of dimensionality and avoids a fall into local minima. To confirm the effectiveness of AE-GSBFN, we applied it to the motion control of a humanoid robot. We demonstrated that AE-GSBFN is capable of providing better performance than A-GSBFN, and we succeeded in enabling the learning of motion control of the robot.
The future objective of this study is to do some general comparisons of our method with other dynamic neural networks, for example, Fritzke's "Growing Neural Gas" (Fritzke, 1996) and Marsland's "Grow When Required Nets" (Marsland et al., 2002). An analysis of the necessity of hierarchical reinforcement learning methods proposed by Morimoto and Doya (Morimoto & Doya, 2000) in relation to the standing up simulation is also an important issue for the future study.

6. References

Albus, J. S. (1981). *Brains, Behavior, and Robotics*, Byte Books

Boyan, J. A. & Moore, A. W. (1995). Generalization in reinforcement learning: Safely approximating the value function, *Advances in Neural Information Processing Systems*, Vol. 7, 369–376

Fritzke, B. (1996). Growing self-organizing networks -- why?, *European Symposium on Artificial Neural Networks*, 61–72

Gullapalli, V. (1990). A stochastic reinforcement learning algorithm for learning real valued functions, *Neural Networks*, Vol. 3, 671–692

Marsland, S. ; Shapiro, J. & Nehmzow, U. (2002). A self-organizing network that grows when required, *Neural Networks*, Vol. 15, 1041–1058

Michie, D. & Chambers, R. A. (1968). BOXES: An Experiment in Adaptive Control, In: *Machine Intelligence* 2, E. Dale and D. Michie (Ed.), pp. 137-152, Edinburgh

Moore, A. W. & Atkeson, C. G. (1995). The parti-game algorithm for variable resolution reinforcement learning in multidimensional state space, *Machine Learning*, Vol. 21, 199–234

Mori, T.; Nakamura, Y., Sato, M., & Ishii, S. (2004). Reinforcement Learning for CPG-driven Biped Robot, *Nineteenth National Conference on Artificial Intelligence (AAAI2004)*, pp. 623-630

Morimoto, J. & Doya, K. (1998). Reinforcement learning of dynamic motor sequence: Learning to stand up, *IEEE/RSJ International Conference on Intelligent Robots and Systems*, pp. 1721–1726

Morimoto, J. & Doya, K. (1999). Learning dynamic motor sequence in high-dimensional state space by reinforcement learning -- learning to stand up -- , *IEICE Transactions on Information and Systems*, Vol. J82-D2, No. 11, 2118–2131

Morimoto, J. & Doya, K. (2000). Acquisition of stand-up behavior by a real robot using hierarchical reinforcement learning, *International Conference on Machine Learning*, pp. 623–630

Nakamura, K. & Hanafusa, H. (1984). Singularity Low-Sensitive Motion Resolution of Articulated Robot Arms, *Transactions of the Society of Instrument and Control Engineers*, Vol. 20, No. 5, pp. 453–459 (in Japanese)

Samejima, K. & Omori, T. (1998). Adaptive state space formation method for reinforcement learning, *International Conference on Neural Information Processing*, pp. 251-255

Schaal, S. & Atkeson, C. C. (1996). From isolation to cooperation: An alternative view of a system of experts, *Advances in Neural Information Processing System*, Vol. 8, 605–611

Smith, R. Open Dynamics Engine, http://opende.sourceforge.net/ode.html

Sutton, R. S. & Barto, A. G. (1998). *Reinforcement Learning: An Introduction*, MIT Press

Takahashi, Y. ; Asada, M. & Hosoda, K. (1996). Reasonable performance in less learning time by real robot based on incremental state space segmentation, *IEEE/RSJ International Conference on Intelligent Robots and Systems*, pp. 1518–1524

22

Reinforcement Learning Algorithms In Humanoid Robotics

Duško Katić, Miomir Vukobratović
Robotics Laboratory, Mihailo Pupin Institute
Belgrade, Serbia

1. Introduction

Many aspects of modern life involve the use of intelligent machines capable of operating under dynamic interaction with their environment. In view of this, the field of biped locomotion is of special interest when human-like robots are concerned. Humanoid robots as anthropomorphic walking machines have been in operation for more than twenty years. Currently, research on the design and the humanoid robots are one of the most exciting and challenging topics in the field of robotics. .The potential applications of this research area are very foremost in the middle and long term. Humanoid robots are expected to be servants and maintenance machines with the main task to assist human activities in our daily life and to replace humans in hazardous operations. It is as obvious as interesting that anthropomorphic biped robots are potentially capable to effectively move in all unstructured environments where humans do. There also raises strong anticipations that robots for the personal use will coexist with humans and provide supports such as the assistance for the housework, care of the aged and the physically handicapped. Consequently, humanoid robots have been treated as subjects of robotics researches such as a research tool for human science, an entertainment/mental-commit robot or an assistant/agent for humans in the human living environment.

Humanoid robot are autonomous systems capable of extracting information from their environments and using knowledge about the world and intelligence of their duties and proper governing capabilities. Intelligent humanoid robots should be autonomous to move safely in a meaningful and purposive manner, i.e. to accept high-level descriptions of tasks (specifying what the user wants to be done, rather than how to do it) and would execute them without further human intervention. Future humanoid robots are likely to have greater sensory capabilities, more intelligence for valid reasoning and decision making, higher levels of manual dexterity and adequate mobility as compared to humans. Naturally, the first approach to making humanoid robots more intelligent was the integration of sophisticated sensor systems as computer vision, tactile sensing, ultrasonic and sonar sensors, laser scanners and other smart sensors. However, today's sensor products are still very limited in interactivity and adaptability to changing environments. As the technology and algorithms for real-time 3D vision and tactile sensing improve, humanoid robots will be able to perform tasks that involve complex interaction with the environment (e.g. grasping and manipulating the objects). A major reason is that uncertainty and dynamic changes make the development of reliable artificial systems particularly challenging. On the other hand, to design robots and systems that best adapt to their environment, the necessary research includes investigations in the field of mechanical robot design (intelligent mechanics), environment perception systems and embedded intelligent control that ought to cope with the task complexity, multi-objective decision making, large volume of perception data and substantial amount of heuristic information. Also, in the case when the robot performs in an unknown environment, the knowledge may not be sufficient. Hence, the robot has to adapt to the environment and to be capable of acquiring new knowledge through the process of learning. The robot learning is essentially concerned with equipping robots with the capacity of improving their behavior over time, based on their incoming experiences.

Although there has been a large number of the control methods used to solve the problem of humanoid robot walking , it is difficult to detect a specific trend. Classical robotics and also the more recent wave of humanoid and service robots still rely heavily on teleoperation or fixed behavior-based control with very little autonomous ability to react to the environment. Among the key missing elements is the ability to create control systems that can deal with a large movement repertoire, variable speeds, constraints and most importantly, uncertainty in the real-world environment in a fast, reactive manner. There are several intelligent paradigms that are capable of solving intelligent control problems in humanoid robotics. Connectionist theory (NN - neural networks), fuzzy logic (FL), and theory of evolutionary computation (GA - genetic algorithms), are of great importance in the development of intelligent humanoid robot control algorithms (Katić & Vukobratovic, 2003a; Katić & Vukobratovic, 2003b; Katić & Vukobratović, 2005). Due to their strong learning and cognitive abilities and good tolerance of uncertainty and imprecision, intelligent techniques have found wide applications in the area of advanced control of humanoid robots. Also, of great importance in the development of efficient algorithms are the hybrid techniques based on the integration of particular techniques such as neuro-fuzzy networks, neuro-genetic algorithms and fuzzy-genetic algorithms.

One approach of departing from teleoperation and manual 'hard coding' of behaviors is by learning from experience and creating appropriate adaptive control systems. A rather general approach to learning control is the framework of *Reinforcement Learning*, described in this chapter. Reinforcement learning offers one of the most general framework to take traditional robotics towards true autonomy and versatility.

Reinforcement learning typically requires an unambiguous representation of states and actions and the existence of a scalar reward function. For a given state, the most traditional of these implementations would take an action, observe a reward, update the value function, and select as the new control output the action with the highest expected value in each state (for a greedy policy evaluation). Updating of value function and controls is repeated until convergence of the value function and/or the policy. This procedure is usually summarized under "value update - policy improvement" iterations. The reinforcement learning paradigm described above has been successfully implemented for many well-defined, low dimensional and discrete problems (Sutton & Barton, 1998; Bertsekas & Tsitsiklis, 1996) and has also yielded a variety of impressive applications in rather complex domains in the last 20 years. Reinforcement learning is well suited to training mobile robots, in particular teaching a robot a new behavior (e.g. avoid obstacles) from scalar feedback. Robotics is a very challenging domain for reinforcement learning, However, various pitfalls have been encountered when trying to scale up these methods to high dimension, continuous control problems, as typically faced in the domain of humanoid robotics.

2. Control Problem in Humanoid Robotics

In spite of a significant progress and accomplishments achieved in the design of a hardware platform of humanoid robot and synthesis of advanced intelligent control of humanoid robots, a lot of work has still to be done in order to improve actuators, sensors, materials, energy accumulators, hardware, and control software that can be utilized to realize user-friendly humanoid robots. We are still in an initial stage when the understanding of the motor control principles and sensory integration subjacent to human walking is concerned.

Having in mind the very high requirements to be met by humanoid robots, it is necessary to point out the need for increasing the number of degrees of freedom (DOFs) of their mechanical configuration and studying in detail some previously unconsidered phenomena pertaining to the stage of forming the corresponding dynamic models. Besides, one should emphasize the need for developing appropriate controller software that would be capable of meeting the most complex requirements of accurate trajectory tracking and maintaining dynamic balance during regular (stationary) gait in the presence of small perturbations, as well as preserving robot's posture in the case of large perturbations. Finally, one ought to point out that the problem of motion of humanoid robots is a very complex control task, especially when the real environment is taken into account, requiring as a minimum, its integration with the robot's dynamic model.

There are various sources of control problems and various tasks and criteria that must be solved and fulfilled in order to create valid walking and other functions of humanoid robots. Previous studies of biological nature, theoretical and computer simulation, have focussed on the structure and selection of control algorithms according to different criteria such as energy efficiency, energy distribution along the time cycle, stability, velocity, comfort, mobility, and environment impact. Nevertheless, in addition to these aspects, it is also necessary to consider some other issues: capability of mechanical implementation due to the physical limitations of joint actuators, coping with complex highly-nonlinear dynamics and uncertainties in the model-based approach, complex nature of periodic and rhythmic gait, inclusion of learning and adaptation capabilities, computation issues, etc.

The major problems associated with the analysis and control of bipedal systems are the high-order highly-coupled nonlinear dynamics and furthermore, the discrete changes in the dynamic phenomena due to the nature of the gait. Irrespective of the humanoid robot structure and complexity, the basic characteristic of all bipedal systems are: a) the DOF formed between the foot and the ground is unilateral and underactuated ; b) the gait repeatability (symmetry) and regular interchangeability of the number of legs that are simultaneously in contact with the ground. During the walk, two different situations arise in sequence: the statically stable double-support phase in which the mechanism is supported on both feet simultaneously, and statically unstable single-support phase when only one foot of the mechanism is in contact with the ground. Thus, the locomotion mechanism changes its structure during a single walking cycle from an open to a closed kinematic chain. Also, it is well known that through the process of running the robot can be most of the time in no-support phase. In this case, the control schemes that are successful for walking problem are not necessarily successful for the running problem. All the mentioned characteristics have to be taken into account in the synthesis of advanced control algorithms that accomplish stable, fast and reliable performance of humanoid robots.

The stability issues of humanoid robot walking are the crucial point in the process of control synthesis. In view of this humanoid walking robots can be classified in three different categories. First category represents static walkers, whose motion is very slow so that the system's stability is completely described by the normal projection of the Centre of Gravity, which only depends on the joint's position. Second category represents dynamic walkers, biped robots with feet and actuated ankles. Postural stability of dynamic walkers depends on joint's velocities and acceleration too. These walkers are potentially able to move in a static way provided they have large enough feet and the motion is slow. The third category represents purely dynamic walkers, robots without feet. In this case the support polygon during the single-support phase is reduced to a point, so that static walking is not possible. In the walk with dynamic balance, the projected centre of mass is allowed outside of the area inscribed by the feet, and the walker may essentially fall during parts of the walking gait. The control problems of dynamic walking are more complicated than in walking with static balance, but dynamic walking patterns provide higher walking speed and greater efficiency, along with more versatile walking structures.

The rotational equilibrium of the foot is the major factor of postural instability with legged robots. The question has motivated the definition of several dynamic-based criteria for the evaluation and control of balance in biped locomotion. The most common criterion are the centre of pressure (CoP), the zero-moment point (ZMP) concept, that has gained widest acceptance and played a crucial role in solving the biped robot stability and periodic walking pattern synthesis (Vukobratović and Juričić, 1969). The ZMP is defined as the point on the ground about which the sum of all the moments of the active forces equals zero. If the ZMP is within the convex hull of all contact points between the foot and the ground, the biped robot can walk.

For a legged robot walking on complex terrain, such as a ground consisting of soft and hard uneven parts, a statically stable walking manner is recommended. However, in the cases of soft terrain, up and down slopes or unknown environment, the walking machine may lose its stability because of the position planning errors and unbalanced foot forces. Hence, position control alone is not sufficient for practical walking, position/force control being thus necessary. Foot force control can overcome these problems, so that foot force control is one of the ways to improve the terrain adaptability of walking robots. For example, in the direction normal to the ground, foot force has to be controlled to ensure firm foot support and uniform foot force distribution among all supporting legs; foot force in the tangential direction has to be monitored to avoid slippage.

A practical biped needs to be more like a human - capable of switching between different known gaits on familiar terrain and learning new gaits when presented with unknown terrain. In this sense, it seems essential to combine force control techniques with more advanced algorithms such as adaptive and learning strategies. Inherent walking patterns must be acquired through the development and refinement by repeated learning and practice as one of important properties of intelligent control of humanoid robots. Learning enables the robot to adapt to the changing conditions and is critical to achieving autonomous behaviour of the robot.

Many studies have given weight to biped walking which is based only on stability of the robot: steady-state walking, high-speed dynamic walking, jumping, and so on. A humanoid robot is however, a kind of integrated machine: a two-arm and two-leg mechanism. Hence, we must not only focus on the locomotion function but also on arm's function with this kind of machines; manipulation and handling being major functions of robot's arms.

When the ground conditions and stability constraint are satisfied, it is desirable to select a walking pattern that requires small torque and velocity of the joint actuators. Humanoid robots are inevitably restricted to a limited amount of energy supply. It would therefore be advantageous to consider the minimum energy consumption, when cyclic movements like walking are involved. With this in mind, an important approach in research is to optimise simultaneously both the humanoid robot morphology and control, so that the walking behaviour is optimised instead of optimising walking behaviour for the given structure of humanoid robot. Optimum structures can be designed when the suitable components and locomotion for the robot are selected appropriately through evolution. It is well known that real-time generation of control algorithms based on highly-complex nonlinear model of humanoid robot commonly suffers from a large amount of computation. Hence, new time-efficient control methods need to be discovered to control humanoid robots in real time, to overcome the mentioned difficulty.

In summary, conventional control algorithms is based on a kinematics and dunamic modeling of the mechanism structure.(Vukobraović et al, 1990). This implies precise identification of intrinsinc parameters of biped's robot mechanism , requires a high precise measurement of humanoid state variables and needs for precise evaluation of interaction forces between foot and ground. Moreover, these methods require a lot of computation together with some problems related to mathematical tractability, optimisation, limited extendability and limited biological plausibility. The second approach based on intelligent control techniques have a potential to overcome the mentioned constraints. In this case, it is not necessary to know perfectly the parameters and characteristics of humanoid mechanism. Also, these methods take advantage from off-line and on-line learning capabilities. This last point is very important because generally the learning ability allows increasing the autonomy of the bioed robot.

3. Reinforcement Learning Framework and Reinforcement Learning Algoritgms in Humanoid Robotics

Recently, reinforcement learning has attracted attention as a learning method for studying movement planning and control (Sutton & Barton, 1998; Bertsekas & Tsitsiklis, 1996). Reinforcement learning is a kind of learning algorithm between supervised and unsupervised learning algorithms which is based on Markov decision process (MDP). Reinforcement learning concept is based on trial and error methodology and constant evaluation of performance in constant interaction with environment.

In many situations the success or failure of the controller is determined not only by one action but by a succession of actions. The learning algorithm must thus reward each action accordingly. This is referred to as the problem of delayed reward. There are two basic methods that are very successful in solving this problem, *TD learning* (Sutton & Barto, 1998) and *Q learning* (Watkins & Dayan, 1992). Both methods build a state space value function that determines how close each state is to success or failure. Whenever the controller outputs an action, the system moves from one state to an other. The controller parameters are then updated in the direction that increases the state value function.

For the solution of large-scale MDPs or continuous state and action spaces, it's impossible for reinforcement learning agent to go through all the states and actions. In order to realize the optimal approximation for value functions of continuous states and actions respectively, therefore, learning agent must have generalization ability. In other words, such an agent should be able to utilize finite learning experience to acquire and express a good knowledge of a large-scale space effectively. How to design a function approximator with abilities of high generalization and computation efficiency has become a key problem for the research

field of reinforcement learning. Using prior knowledge about the desired motion can greatly simplify controller synthesis. Imitation-based learning or learning from demonstration allow for policy search to focus only on the areas of the search space that pertain to the task at hand. Both, model-based and model-free approaches exist to find optimal policies when agents are allowed to act for unlimited time. For physical agents, such as humanoid robots acting in the real world, it is much more difficult to gain experience. Hence, the exhaustive exploration of highdimensional state and action spaces is not feasible. For a physical robot, it is essential to learn from few trials in order to have some time left for exploitation.
The robot learning is essentially concerned with equipping robots with the capacity of improving their behavior over time, based on their incoming experiences. For instance, it could be advantageous to learn dynamics models, kinematic models, impact models, for model-based control techniques. Imitation learning could be employed for the teaching of gaits patterns, and reinforcement learning could help tuning parameters of the control policies in order to improve the performance with respect to given cost functions.
Dynamic bipedal walking is difficult to learn for a number of reasons. First, biped robots typically have many degrees of freedom, which can cause a combinatorial explosion for learning systems that attempt to optimize performance in every possible configuration of the robot. Second, details of the robot dynamics such as uncertainties in the ground contact and nonlinear friction in the joints must be only experimentally validated. Since it is only practical to run a small number of learning trials on the real robot, the learning algorithms must perform well after obtaining a very limited amount of data. Finally, learning algorithms for dynamic walking must deal with dynamic discontinuities caused by collisions with the ground and with the problem of delayed reward -torques applied at one time may have an effect on the performance many steps into the future.
In area of humanoid robotics, there are several approaches of reinforcement learning (Benbrahim & Franklin, 1997; Chew & Pratt, 2002; Li et al. , 1992; Mori et al., 2004; Morinoto et al., 2004; Nagasaka et al., 1999; Nakamura et al., 2003; Salatian et al., 1997; Zhou & Meng, 2000) with additional demands and requirements because high dimensionality of the control problem.
In the paper (Benbrahim & Franklin, 1997), it is shown how reinforcement learning is used within a modular control architecture to enable a biped robot to walk. The controller structure consists of central (CPG) and peripheral controllers. The learning architecture succeeded in dealing with the problems of large numbers of inputs, knowledge integration and task definition. The central controller controls the robot in nominal situations, and the peripheral controllers intervene only when they consider that the central controller's action contradicts their individual control policies (Figure 1). The action is generated by computing the average of the outputs of all controllers that intervene including the central controller. Each peripheral controller's role is to correct the central controller's mistakes and issue an evaluation of the general behaviour. The central controller then uses the average of all evaluations to learn a control policy that accommodates the requirements of as many peripheral controllers as possible.
The central controller as well as some of the peripheral controllers in this study use adaptive CMAC neural networks. Because of modular nature, it is possible to use several neural networks with small numbers of inputs instead of one large neural network. This dramatically increases the learning speed and reduces the demand on memory and computing power. The architecture also allows easy incorporation of any knowledge by adding a peripheral controller that represents that knowledge The CPG uses reinforcement learning in order to learn an optimal policy. The CMAC weights are updated using the reinforcement signals received from the peripheral controllers. Reinforcement learning is

well suited for this kind of application. The system can try random actions and choose those that yield good reinforcement. The reinforcement learning algorithm uses the actor-critic configuration. It searches the action space using a Stochastic Real Valued (SRV) unit at the output. The reinforcement signal is generated using TD Learning. The CMAC neural networks used in the biped's learning are pre-trained using a biped walking robot simulator and predefined simple walking gates. The size of the search domain is determined by the standard deviation of the Gaussian unit. If the standard deviation is too small, the system will have a very small search domain. This decreases the learning speed and increases the system's vulnerability to the local minima problem. If the factor is too large, the system's performance will not reach its maximum because there will always be a randomness even if the system has learned an optimal solution. It is in general safer to use a large factor than a small one. Even though this learning algorithms and architecture have successfully solved the problem of dynamic biped walking, there are many improvements that can be added to increase learning speed, robustness, and versatility. The performance must also be improved by dynamically setting the PID gains to deal with each specific situation.

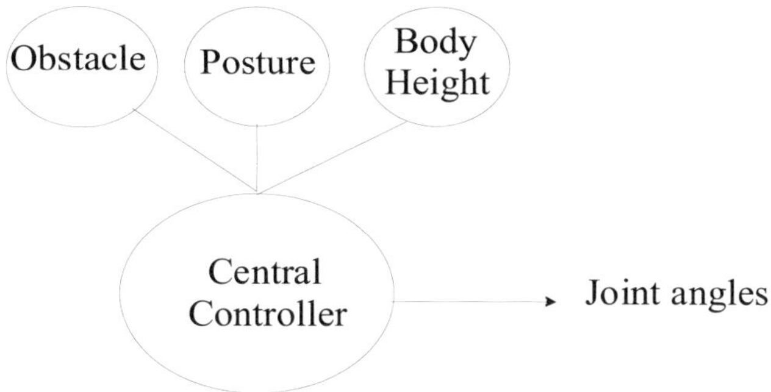

Fig.1. Controller Architecture Benbrahim & Franklin.

More recently, (Kun & Miller III, 1999) has developed a hierarchical controller that combines simple gait oscillators, classical feedback control techniques and neural network learning, and does not require detailed equations of the dynamics of walking. The emphasis is on the real-time control studies using an experimental ten-axis biped robot with foot force sensors. The neural network learning is achieved using CMAC controller, where CMAC neural networks were used essentially as context sensitive integral errors in the controller, and the control context being defined by the CMAC input vector. There are 3 different CMAC neural networks for humanoid posture control. The Front/Back Balance CMAC neural network was used to provide front/back balance during standing, swaying and walking. The training of this network is realized using data from foot sensors. The second CMAC neural network is used for Right/Left Balance, to predict the correct knee extension required to achieve sufficient lateral momentum for lifting the corresponding foot for the desired length of time. The training of this network is realized using temporal difference method based on the difference between the desired and real time of foot rising. The third CMAC network is used to learn kinematically consistent robot postures. In this case, training is also realized by data from foot sensors. The results indicated that the experimental biped was able to learn the closed-chain kinematics necessary to shift body weight side-to-side while maintaining good foot contact. Also, it was able to learn the quasi-static balance required to avoid falling forward or backward while shifting body weight side-to-side at different speeds. It was able to learn the dynamic balance in order to lift a foot off the floor for a desired length of time and different initial conditions. There were, however, many limitations (limited step length, slow walking, no adaptation for left-right balance, no possibility of walking on sloping surfaces). The new dynamically balanced scheme for handling variable-speed gait was proposed based on the preplanned but adaptive motion sequences in combination with closed-loop reactive control. This allows the algorithm to improve the walking performance over consecutive steps using adaptation, and to react to small errors and disturbances using reactive control. New sensors (piezoresistive accelerometers and two solid-state rate gyroscopes) are mounted on the new UNH biped (Fig. 2).

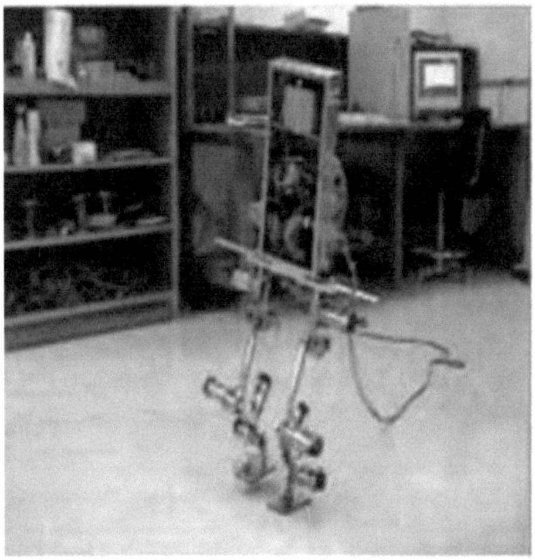

Fig. 2. The UNH biped walking.

Training of neural networks is realized through the process of temporal difference learning using information about ZMP from robot foot sensors. The CMAC neural networks were first trained during repetitive foot-lift motion similar to marching in place. Then, training was carried out during the attempts at walking for increased step length and gait speeds.

The experimental results indicate that the UNH biped robot can walk with forward velocities in the range of 21 - 72 cm/min, with sideways leaning speed in the range of 3.6 - 12.5 cm/s. The main characteristic of this controller is the synthesis of the control signal without dynamic model of biped. The proposed controller could be used as a basis for similar controllers of more complex humanoid robots in the future research. However, this controller is not of a general nature, because it is suitable only for the proposed structure of biped robot and must be adapted for the bipeds with different structures. More research efforts are needed to simplify the controller structure, to increase the gait speed, and to ensure stability of dynamic walking.

The policy gradient method is one of the reinforcement learning methods successfully applied to learn biped walking on actual robots (Benbrahim & Franklin, 1997; Tedrake et al., 2004). However, this method requires hours to learn a walking controller, and approach in (Tedrake et al., 2004) requires a mechanically stable robot.

There are direct (model-free) and model-based reinforcement learning (Doya, 2000; Sutton & Barto, 1998) . The direct approach to RL is to apply policy search directly without learning a model of the system. In principle, the model-based reinforcement learning is more data efficient than direct reinforcement learning. Also, it was concluded that model-based reinforcement learning finds better trajectories, plans and policies, and handles changing goals more efficiently. On the other hand, reported that a model-based approach to reinforcement learning is able to accomplish given tasks much faster than without using knowledge of the environment.

The problem of biped gait synthesis using the reinforcement learning with fuzzy evaluative feedback is considered in paper (Zhoy & Meng, 2000). As first, initial gait from fuzzy rules is generated using human intuitive balancing scheme. Simulation studies showed that the fuzzy gait synthesizer can only roughly track the desired trajectory. A disadvantage of the proposed method is the lack of practical training data. In this case there are no numerical feedback teaching signal, only evaluative feedback signal exists (failure or success), exactly when the biped robot falls (or almost falls) down. Hence, it is a typical reinforcement learning problem. The dynamic balance knowledge is accumulated through reinforcement learning constantly improving the gait during walking. Exactly, it is fuzzy reinforcement learning that uses fuzzy critical signal. For human biped walk, it is typical to use linguistic critical signals such as "near-fall-down", "almost-success", "slower", "faster", etc. In this case, the gait synthesizer with reinforcement learning is based on a modified GARIC (Generalized Approximate Reasoning for Intelligent Control) method. This architecture of gait synthesizer consists of three components: action selection network (ASN), action evaluation network (AEN), and stochastic action modifier (SAM) (Fig. 3) The ASM maps a state vector into a recommended action using fuzzy inference. The training of ASN is achieved as with standard neural networks using error signal of external reinforcement. The AEN maps a state vector and a failure signal into a scalar score which indicates the state goodness. It is also used to produce internal reinforcement. The SAM uses both recommended action and internal reinforcement to produce a desired gait for the biped. The reinforcement signal is generated based on the difference between desired ZMP and real ZMP in the x-y plane. In all cases, this control structure includes on-line adaptation of gait synthesizer and local PID regulators. The approach is verified using simulation experiments. In the simulation studies, only even terrain for biped walking is considered, hence the approach should be verified for irregular and sloped terrain.

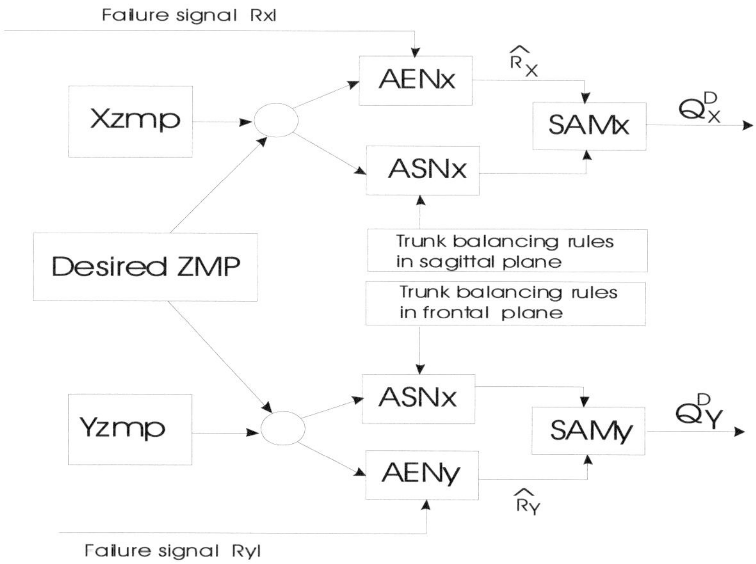

Fig. 3. The architecture of the reinforcement learning based gait synthesizer.

where $Xzmp, Yzmp$ are the ZMP coordinates; $\theta^d_{zmp}, \theta^d_{zmp}$ are the desired joint angles of the biped gait.
There are some research works that include the application of reinforcement learning control algorithms for passive or semi-passive biped walkers (Fig.4) (Tedrake et al., 2004; Morimoto et al., 2005; Schuitema et al., 2005).

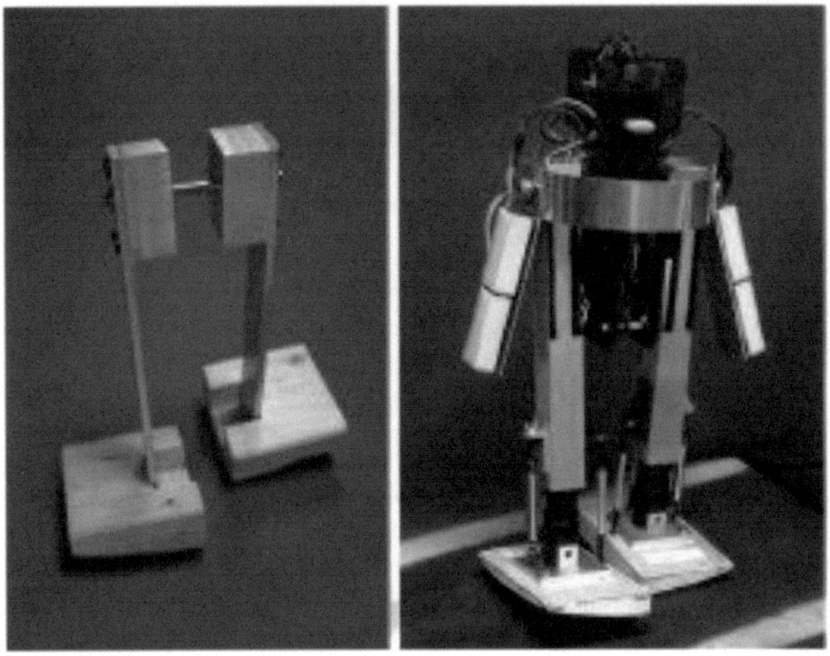

Fig. 4. Simple Passive Dynamic Walker.

In (Schuitema et al. , 2005), Reinforcement Learning passive walker is proposed as model free approach with fully driven optimization method. This approach is adaptive, in the sense that when the robot or its environment changes without notice, the controller can adapt until performance is again maximal. Optimization relatively easily towards several goals, such as: minimum cost of transport, largest average forward speed, or both. Statistical learning algorithm makes small changes in control parameters and uses correlations between control changes and system error changes. Stochasticity is added to deterministic feedback control policy. Gradient following algorithm based on temporal difference error was applied.
In paper (Morimoto et al., 2004) a model-based reinforcement learning algorithm for biped walking in which the robot learns to appropriately place the swing leg was proposed. This decision is based on a learned model of the Poincare map of the periodic walking pattern. The model maps from a state at the middle of a step and foot placement to a state at next middle of a step. Actor-Critic algorithms of reinforcement learning has a great potential in control of biped robots. For a example, a general algorithm for estimating the natural gradient, the Natural Actor-Critic algorithm, is introduced in paper (Peters et al., 2003). This algorithm converges to the nearest local minimum of the cost function with respect to the Fisher in formation metric under suitable conditions. It offers

a promising route for the development of reinforcement learning for truly high-dimensionally continuous state-action systems. In paper (Tedrake et al., 2004) a learning system which is able to quickly and reliably acquire a robust feedback control policy for 3D dynamic walking from a blank-slate using only trials implemented on physical robot. The robot begins walking within a minute and learning converges in approximately 20 minutes. This success can be attributed to the mechanics of our robot, which are modelled after a passive dynamic walker, and to a dramatic reduction in the dimensionality of the learning problem. The reduction of the dimensionality was realized by designing a robot with only 6 internal degrees of freedom and 4 actuators, by decomposing the control system in the frontal and sagittal planes, and by formulating the learning problem on the discrete return map dynamics. A stochastic policy gradient algorithm to this reduced problem was applied with decreasing the variance of the update using a state-based estimate of the expected cost. This optimized learning system works quickly enough that the robot is able to continually adapt to the terrain as it walks. The learning on robot is performed by a policy gradient reinforcement learning algorithm (Baxter & Bartlett, 2001; Kimura & Kobayashi, 1998; Sutton et al., 2000).
Some researxhers (Kamio & Iba, 2005) were efficiently applied hybrid version of reinforcement learning structures, integrating genetic programming and Q-Learning method on real humanoid robot.

4. Hybrid Reinforcement Learning Control Algorithms for Biped Walking

The new integrated hybrid dynamic control structure for the humanoid robots will be proposed, using the model of robot mechanism. Our approach consists in departing from complete conventional control techniques by using hybrid control strategy based on model-based approach and learning by experience and creating the appropriate adaptive control systems. Hence, the first part of control algorithm represents some kind of computed torque control method as basic dynamic control method, while the second part of algorithm is reinforcement learning architecture for dynamic compensation of ZMP (Zero-Moment-Point) error.
In the synthesis of reinforcement learning strycture, two algorithms will be shown, that are very successful in solving biped walking problem: adaptive heuristic approach (AHC) approach, and approach based on Q learning, To solve reinforcement learning problem, the most popular approach is temporal difference (TD) method (Sutton & Barto, 1998). Two TD-based reinforcement learning approaches have been proposed: The adaptive heuristic critic (AHC) (Barto et al., 1983) and Q-learning (Watkins & Dayan, 1992). In AHC, there are two separate networks: An action network and an evaluation network. Based on the AHC, In (Berenji & Khedkar, 1992), a generalized approximate reasoning-based intelligent control (GARIC) is proposed, in which a two-layer feedforward neural network is used as an action evaluation network and a fuzzy inference network is used as an action selection network. The GARIC provides generalization ability in the input space and extends the AHC algorithm to include the prior control knowledge of human operators. One drawback of these actor-critic architectures is that they usually suffer from the local minimum problem in network learning due to the use of gradient descent learning method.
Besides the aforementioned AHC algorithm-based learning architecture, more and more advances are being dedicated to learning schemes based on Q-learning. Q-learning collapses the two measures used by actor/critic algorithms in AHC into one measure referred to as

the Q-value. It may be considered as a compact version of the AHC, and is simpler in implementation. Some Q-learning based reinforcement learning structures have also been proposed (Glorennec & Jouffe, 1997; Jouffe, 1998; Berenji, 1996).. In (Berenji & Jouffe, 1997), a dynamic fuzzy Q-learning is proposed for fuzzy inference system design. In this method, the consequent parts of fuzzy rules are randomly generated and the best rule set is selected based on its corresponding Q-value. The problem in this approach is that if the optimal solution is not present in the randomly generated set, then the performance may be poor. In (Jouffe, 1998), fuzzy Q-learning is applied to select the consequent action values of a fuzzy inference system. For these methods, the consequent value is selected from a predefined value set which is kept unchanged during learning, and if an improper value set is assigned, then the algorithm may fail. In (Berenji, 1996), a GARIC-Q method is proposed. This method works at two levels, the local and the top levels. At the local level, a society of agents (fuzzy networks) is created, with each learning and operating based on GARIC. While at the top level, fuzzy Q-learning is used to select the best agent at each particular time. In contrast to the aforementioned fuzzy Q-learning methods, in GARIC-Q, the consequent parts of each fuzzy network are tunable and are based on AHC algorithm. Since the learning is based on gradient descent algorithm, it may be slow and may suffer the local optimum problem.

4.1 Model of the robot's mechanism

The mechanism possesses 38 DOFs. Taking into account dynamic coupling between particular parts (branches) of the mechanism chain, a relation that describes the overall dynamic model of the locomotion mechanism in a vector form:

$$P + J^T(q)F = H(q)\ddot{q} + h(q,\dot{q}) \tag{1}$$

where: $P \in R^{nx1}$ is the vector of driving torques at the humanoid robot joints; $F \in R^{nx1}$ is the vector of external forces and moments acting at the particular points of the mechanism; $H \in R^{nxn}$ is the square matrix that describes 'full' inertia matrix of the mechanism; $h \in R^{nx1}$ is the vector of gravitational, centrifugal and Coriolis moments acting at n mechanism joints; $J \in R^{nxn}$ is the corresponding Jacobian matrix of the system; $n = 38$, is the total number of DOFs; $q \in R^{nx1}$ is the vector of internal coordinates; $\dot{q} \in R^{nX1}$ is the vector of internal velocities.

4.2 Definition of control criteria

In the control synthesis for biped mechanism, it is necessary to satisfy certain natural principles. The control must to satisfy the following two most important criteria: (i) accuracy of tracking the desired trajectories of the mechanism joints (ii) maintenance of dynamic balance of the mechanism during the motion. Fulfillment of criterion (i) enables the realization of a desired mode of motion, walk repeatability and avoidance of potential obstacles. To satisfy criterion (ii) it means to have a dynamically balanced walk.

4.3. Gait phases and indicator of dynamic balance

The robot's bipedal gait consists of several phases that are periodically repeated. Hence, depending on whether the system is supported on one or both legs, two macro-phases can be distinguished: (i) single-support phase (SSP) and (ii) double-support phase (DSP). Double-support phase has two micro-phases: (i) weight acceptance phase (WAP) or heel strike, and (ii) weight support phase (WSP). Fig. 5 illustrates these gait phases, with the projections of the contours of the right (RF) and left (LF) robot foot on the ground surface, whereby the shaded areas represent the zones of direct contact with the ground surface.

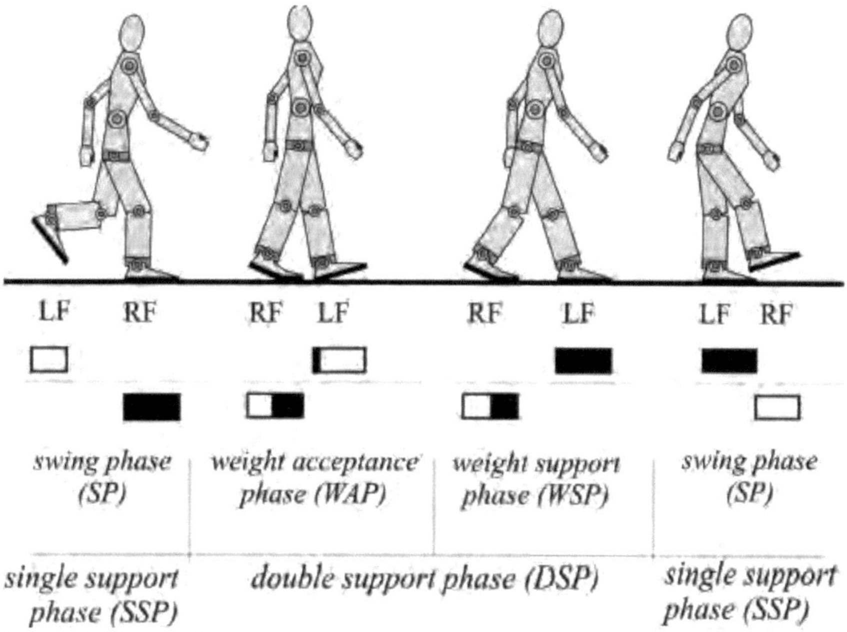

Fig. 5. Phases of biped gait.

The indicator of the degree of dynamic balance is the ZMP, i.e. its relative position with respect to the footprint of the supporting foot of the locomotion mechanism. The ZMP is defined (Vuobratović & Juričić, 1969) as the specific point under the robotic mechanism foot . at which the effect of all the forces acting on the mechanism chain can be replaced by a unique force and all the rotation moments about the x and y axes are equal zero. Figs 6a and 6b show details related to the determination of ZMP position and its motion in a dynamically balanced gait. The ZMP position is calculated based on measuring reaction forces $F_i . i = 1,..,4$ under the robot foot. Force sensors are usually placed on the foot sole in the arrangement shown in Fig. 6 a. Sensors' positions are defined by the geometric quantities l_1, l_2 and l_3 . If the point 0_{zmp} is assumed as the nominal ZMP position (Fig. 6a), then one can use the following equations to determine the relative ZMP position with respect to its nominal:

$$\Delta M_x^{(zmp)} = \tfrac{l_3}{2}\left[\left(F_2 + F_4\right) - \left(F_2^0 + F_4^0\right)\right] - \tfrac{l_3}{2}\left[\left(F_1 + F_3\right) - \left(F_1^0 + F_3^0\right)\right],$$

$$\Delta M_y^{(zmp)} = l_2\left[\left(F_3 + F_4\right) - \left(F_3^0 + F_4^0\right)\right] - l_1\left[\left(F_1 + F_2\right) - \left(F_1^0 + F_2^0\right)\right],$$

$$F_r^{(z)} = \sum_{i=1}^{4} F_i, \quad \Delta x^{(zmp)} = \frac{-\Delta M_y^{(zmp)}}{F_r^{(z)}}, \quad \Delta y^{(zmp)} = \frac{\Delta M_x^{(zmp)}}{F_r^{(z)}}$$

where F_i and $F_i^0 . i = 1,..,4$, are the measured and nominal values of the ground reaction force; $\Delta M_x^{(zmp)}$ and $\Delta M_y^{(zmp)}$ are deviations of the moments of ground reaction forces around the axes passed through the 0_{zmp}; $F_r^{(z)}$ is the resultant force of ground reaction in the vertical z-direction, while $\Delta x^{(zmp)}$ and $\Delta y^{(zmp)}$ are the displacements of ZMP position from its nominal 0_{zmp}. The deviations $\Delta x^{(zmp)}$ and $\Delta y^{(zmp)}$ of the ZMP position from its nominal position in x- and y-direction are calculated from the previous relation . The instantaneous position of ZMP is the best indicator of dynamic balance of the robot mechanism. In Fig. 6b are illustrated certain areas (Z_0, Z_1 and Z_2), the so-called safe zones of dynamic balance of the locomotion mechanism. The ZMP position inside these "safety areas" ensures a dynamically balanced gait of the mechanism whereas its position outside these zones indicates the state of loosing the balance of the overall mechanism, and the possibility of its overturning. The quality of robot balance control can be measured by the success of keeping the ZMP trajectory within the mechanism support polygon (Fig. 6b).

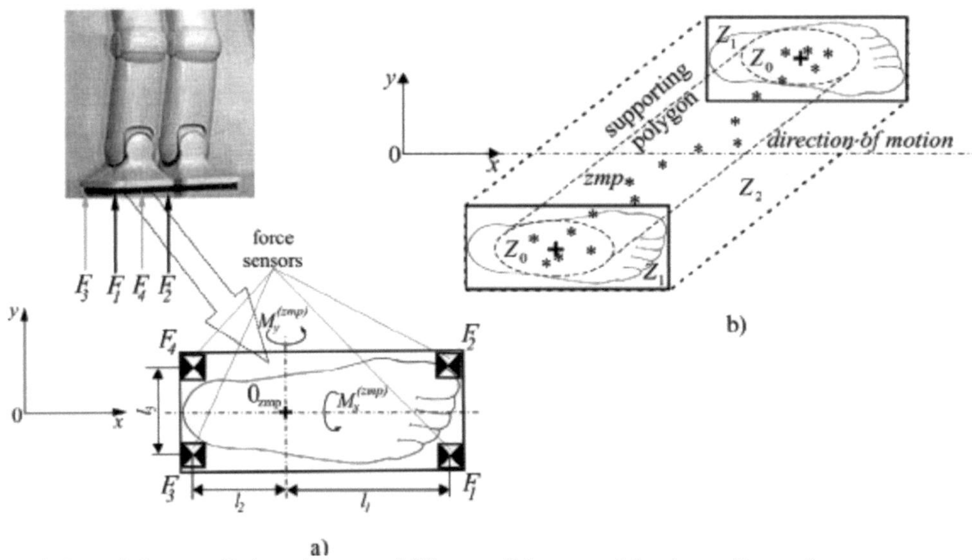

Fig. 6. Zero-Moment Point: a) Legs of "Toyota " humanoid robot ; General arrangement of force sensors in determining the ZMP position; b) Zones of possible positions of ZMP when the robot is in the state of dynamic balance.

4.4 Hybrid intelligent control algorithm with AHC reinforcement structure

Biped locomotion mechanism represents a nonlinear multivariable system with several inputs and several outputs. Having in mind the control criteria, it is necessary to control the following variables: positions and velocities of the robot joints and ZMP position . In accordance with the control task, we propose the application of the hybrid intelligent control algorithm based on the dynamic model of humanoid system . Here we assume the following: (i) the model (1) describes sufficiently well the behavior of the system; (ii) desired (nominal) trajectory of the mechanism performing a dynamically balanced gait is known. (iii) geometric and dynamic parameters of the mechanism and driving units are known and constant. These assumptions can be taken as conditionally valid, the rationale being as follows: As the system elements are rigid bodies of unchangeable geometrical shape, the parameters of the mechanism can be determined with a satisfactory accuracy.

Based on the above assumptions, in Fig. 7 a block-diagram of the intelligent controller for biped locomotion mechanism is proposed. It involves two feedback loops: (i) basic dynamic controller for trajectory tracking, (ii) intelligent reaction feedback at the ZMP based on AHC reinforcement learning structure. The synthesized dynamic controller was designed on the basis of the centralized model. The vector of driving moments $\hat{P}$ represents the sum of the driving moments $\hat{P}_1, \hat{P}_2$,. The torques $\hat{P}_1$ are determined so to ensure precise tracking of the robot's position and velocity in the space of joints coordinates. The driving torques $\hat{P}_2$ are calculated with the aim of correcting the current ZMP position with respect to its nominal. The vector $\hat{P}$ of driving torques represents the output control vector.

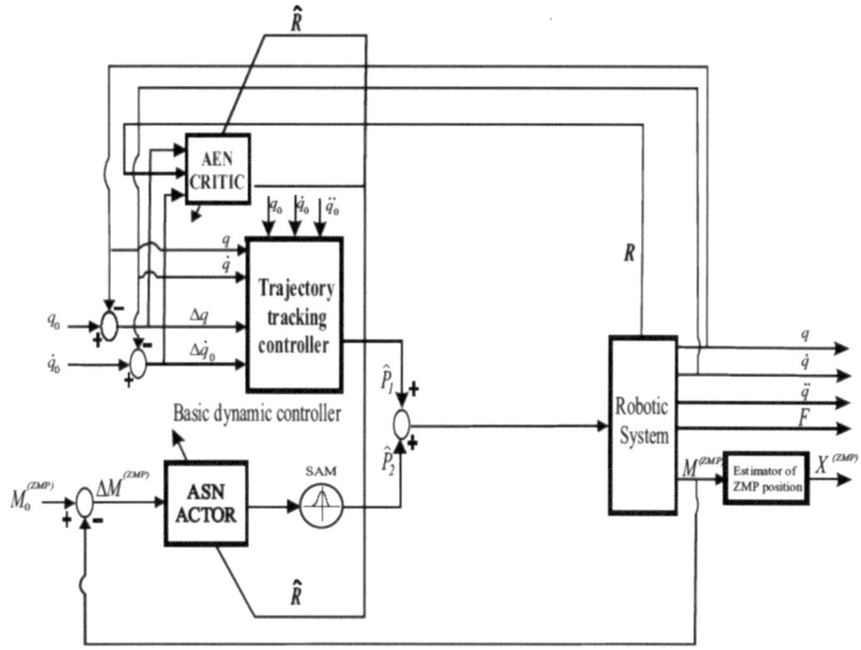

Fig. 7 Hybrid Controller based on Actor-Critic Method for trajectory tracking.

4.5 Basic Dynamic Controller

The proposed dynamic control law ha the following form:

$$\hat{P} = \hat{H}(q)[\ddot{q}_0 + K_v(\dot{q} - \dot{q}_0) + K_p(q - q_0)] + \hat{h}(q, \dot{q}) \tag{2}$$

where $\hat{H}, \hat{h}$ and $\hat{J}$ are the corresponding estimated values of the inertia matrix, vector of gravitational, centrifugal and Coriolis forces and moments and Jacobian matrix from the model (1). The matrices $K_p \in R^{nxn}$ and $K_v \in R^{nxn}$ are the corresponding matrices of position and velocity gains of the controller. The gain matrices K_p and K_v can be chosen in the diagonal form by which the system is decoupled into n independent subsystems. This control model is based on centralized dynamic model of biped mechanism.

4.6 Compensator of dynamic reactions based on reinforcement learning structure

In the sense of mechanics, locomotion mechanism represents an inverted multi link pendulum. In the presence of elasticity in the system and external environment factors, the mechanism's motion causes dynamic reactions at the robot supporting foot. Thus, the state of dynamic balance of the locomotion mechanism changes accordingly. For this reason it is essential to introduce dynamic reaction feedback at ZMP in the control synthesis. There are relationship between the deviations of ZMP positions $(\Delta x^{(zmp)}, \Delta y^{(zmp)})$ from its nominal position 0_{zmp} in the motion directions x and y and the corresponding dynamic reactions $M_x^{(zmp)}$ and $M_y^{(zmp)}$ acting about the mutually orthogonal axes that pass through the point 0_{zmp}. $M_x^{(zmp)} \in R^{1x1}$ and $M_x^{(zmp)} \in R^{1x1}$ represent the moments that tend to overturn the robotic mechanism. $M_0^{(zmp)} \in R^{2x1}$ and $M^{(zmp)} \in R^{2x1}$ are the vectors of nominal and measured values of the moments of dynamic reaction around the axes that pass through the ZMP (Fig. 6a). Nominal values of dynamic reactions, for the nominal robot trajectory, are determined off-line from the mechanism model and the relation for calculation of ZMP; $\Delta M^{(zmp)} \in R^{2x1}$ is the vector of deviation of the actual dynamic reactions from their nominal values; $P_{dr} \in R^{2x1}$ is the vector of control torques, ensuring the state of dynamic balance.

The control torques P_{dr} has to be displaced to the some joints of the mechanism chain. Since the vector of deviation of dynamic reactions $\Delta M^{(zmp)}$ has two components about the mutually orthogonal axes x and y, at least two different active joints have to be used to compensate for these dynamic reactions. Considering the model of locomotion mechanism, the compensation was carried out using the following mechanism joints: 9, 14, 18, 21 and 25 to compensate for the dynamic reactions about the x-axis, and 7, 13, 17, 20 and 24 to compensate for the moments about the y-axis. Thus, the joints of ankle, hip and waist were taken into consideration. Finally, the vector of compensation torques $\hat{P}_2$ was calculated on the basis of the vector of the moments P_{dr} in the case when compensation of ground dynamic reactions is performed using all six proposed joints, using the following relation

$$\hat{P}_2(9) = \hat{P}_2(14) = \hat{P}_2(18) = \hat{P}_2(21) = \hat{P}_2(25) = 1/5P_{dr} \quad (3)$$

$$\hat{P}_2(7) = \hat{P}_2(13) = \hat{P}_2(17) = \hat{P}_2(20) = \hat{P}_2(24) = 1/5P_{dr} \quad (4)$$

In nature, biological systems use simultaneously a large number of joints for correcting their balance. In this work, for the purpose of verifying the control algorithm, the choice was restricted to the mentioned ten joints: 7, 9, 13, 14, 17, 18, 20, 21, 24, and 25. Compensation of ground dynamic reactions is always carried out at the supporting leg when the locomotion mechanism is in the swing phase, whereas in the double-support phase it is necessary to engage the corresponding pairs of joints (ankle, hip, waist) of both legs.

On the basis of the above the fuzzy reinforcement control algorithm is defined with respect to the dynamic reaction of the support at ZMP.

4.7. Reinforcement Actor-Critic Learning Structure

This subsection describes the learning architecture that was developed to enable biped walking. A powerful learning architecture should be able to take advantage of any available

knowledge. The proposed reinforcement learning structure is based on Actor-Critic Methods (Sutton & Barto, 1998).
Actor-Critic methods are *temporal difference (TD)* methods, that have a separate memory structure to explicitly represent the control policy independent of the value function. In this case, control policy represents policy structure known as **Actor** with aim to select the best control actions. Exactly, the control policy in this case, represents the set of control algorithms with different control parameters. The input to control policy is state of the system, while the output is control action (signal). It searches the action space using a Stochastic Real Valued (SRV) unit at the output. The unit's action uses a Gaussian random number generator. The estimated value function represents a **Critic**, because it criticizes the control actions made by the actor. Typically, the critic is a state-value function which takes the form of TD error necessary for learning. TD error depends also from reward signal, obtained from environment as result of control action. The TD Error can be scalar or fuzzy signal that drives all learning in both actor and critic.
Practically, in proposed humanoid robot control design, it is synthesized the new modified version of GARIC reinforcement learning structure (Berenji & Khedkar, 1992) . The reinforcement control algorithm is defined with respect to the dynamic reaction of the support at ZMP, not with respect to the state of the system. In this case external reinforcement signal (reward) R is defined according to values of ZMP error.
Proposed learning structure consists from two networks: AEN(Action Evaluation Network) - CRITIC and ASN(Action Selection Network) - ACTOR. AEN network maps position and velocity tracking errors and external reinforcement signal R in scalar or fuzzy value which represent the quality of given control task. The output scalar value of AEN is important for calculation of internal reinforcement signal. $\hat{R}$ AEN constantly estimate internal reinforcement based on tracking errors and value of reward. AEN is standard 2-layer feedforward neural network (perceptron) with one hidden layer. The activation function in hidden layer is sigmoid, while in the output layer there are only one neuron with linear function. The input layer has a bias neuron. The output scalar value v is calculated based on product of set C of weighting factors and values of neurons in hidden later plus product of set A of weighting factors and input values and bias member. There are also one more set of weighting factors B between input layer and hidden layer. The number of neurons on hidden later is determined as 5. Exactly, the output v can be represented by the following equation:

$$v = \sum_i B_i \Delta M_i^{(zmp)} + \sum C_j f(\sum_{ji} A_i \Delta M_i^{(zmp)}) \qquad (5)$$

where f is sigmoid function.

The most important function of AEN is evaluation of TD error, exactly internal reinforcement. The internal reinforcement is defined as TD(0) error defined by the following equation:

$$\hat{R}(t+1) = 0, \quad begining \;\; state \qquad (6)$$

$$\hat{R}(t+1) = R(t) - v(t), \quad failure \;\; state \qquad (7)$$

$$\hat{R}(t+1) = R(t) + \gamma v(t+1) - v(t), \quad otherwise \qquad (8)$$

where γ is a discount coefficient between 0 and 1 (in this case γ is set to 0.9).

ASN (action selection network) maps the deviation of dynamic reactions $\Delta M^{(zmp)} \in R^{2x1}$ in recommended control torque. The structure of ASN is represented by The ANFIS - Sugeno-type adaptive neural fuzzy inference systems. There are five layers: input layer. antecedent part with fuzzification, rule layer, consequent layer, output layer wit defuzzification. This system is based on fuzzy rule base generated by expert knoOwledge with 25 rules. The partition of input variables (deviation of dynamic reactions) are defined by 5 linguistic variables: *NEGATIVE BIG, NEGATIVE SMALL, ZERO, POSITIVE SMALL* and *POSITIVE BIG*. The member functions is chosen as triangular forms.

SAM (Stochastic action modifier) uses the recommended control torque from ASN and internal reinforcement signal to produce final commanded control torque P_{dr} . It is defined by Gaussian random function where recommended control torque is mean, while standard deviation is defined by following equation:

$$\sigma(\hat{R}(t+1)) = 1 - \exp(-|\hat{R}(t+1)|) \tag{9}$$

Once the system has learned an optimal policy, the standard deviation of the Gaussian converges toward zero, thus eliminating the randomness of the output.

The learning process for AEN (tuning of three set of weighting factors A, B, C) is accomplished by step changes calculated by products of internal reinforcement, learning constant and appropriate input values from previous layers, i.e. according to following equations:

$$B_i(t+1) = B_i(t) + \beta\hat{R}(t+1)\Delta M_i^{(zmp)}(t) \tag{10}$$

$$C_j(t+1) = C_j(t) + \beta\hat{R}(t+1)f(\sum_{ji} A_{ij}(t)\Delta M_i^{(zmp)}(t)) \tag{11}$$

$$A_{ij})t+1) = A_{ij}(t) +$$
$$\beta\hat{R}(t+1)f(\sum_{ji} A_{ij}(t)\Delta M_i^{(zmp)})(t)(1-f(\sum_{ji} A_{ij}(t)\Delta M_i^{(zmp)})(t))sgn(C_j\Delta M_j^{(zmp)}(t)) \tag{12}$$

where β is learning constant. The learning process for ASN (tuning of antedecent and consequent layers of ANFIS) is accomplished by gradient step changes (back propagation algorithms) defined by scalar output values of AEN, internal reinforcement signal, learning constants and current recommended control torques.

In our research, the precondition part of ANFIS is online constructed by special clustering approach. The general grid type partition algorithms perform either with training data collected in advance or cluster number assigned a priori. In the reinforcement learning problems, the data are generated only when online learning is performed. For this reason, a new clustering algorithm based on Euclidean Distance measure, with the abilities of online learning and automatic generation of number of rules is used.

4.8 Hybrid intelligent control algorithm with Q reinforcement structure

From the perspective of ANFIS Q-learning, we propose a method, as combination of automatic precondition part construction and automatic determination of the consequent parts of a ANFIS system. In application, this method enables us to deal with continuous state and action spaces. It helps to solve *the curse of dimensionality* encountered in high-dimensional continuous state space and provides smooth control actions. Q-learning is a widely-used reinforcement learning method for an agent to acquire optimal policy. In this learning, an agent tries an action, $a(t)$, at a particular state, $x(t)$, and then evaluates its consequences in terms of the immediate reward $R(t)$.To estimate the discounted cumulative reinforcement for taking actions from given states, an evaluation function, the Q-function, is used. The Q-function is a mapping from state-action pairs to predict return and its output for state x and action a is denoted by the Q-value, $Q(x,a)$. Based on this Q-value, at time t, the agent selects an action $a(t)$. The action is applied to the environment, causing a state transition from $x(t)$ to $x(t+1)$, and a reward $R(t)$ is received. Then, the Q-function is learned through *incremental dynamic programming*. The Q-value of each state/action pair is updated by

$$\begin{aligned} Q(x(t),a(t)) &= Q(x(t),a(t)) + \alpha(R(t) \\ &\quad + \gamma\, Q^{*}(x(t+1)) \; - Q(x(t),a(t))) \\ Q^{*}(x(t+1)) &= \max_{b\in A(x(t+1))} \mathrm{Q}(\mathrm{x}(\mathrm{t}+1),\mathrm{b}) \end{aligned} \tag{13}$$

where $A(x(t+1))$ is the set of possible actions in state ; α is the learning rate; γ is the discount rate.

Based on the above facts, in Fig. 8 a block-diagram of the intelligent controller for biped locomotion mechanism is proposed. It involves two feedback loops: (i) basic dynamic controller for trajectory tracking, (ii) intelligent reaction feedback at the ZMP based on Q-reinforcement learning structure

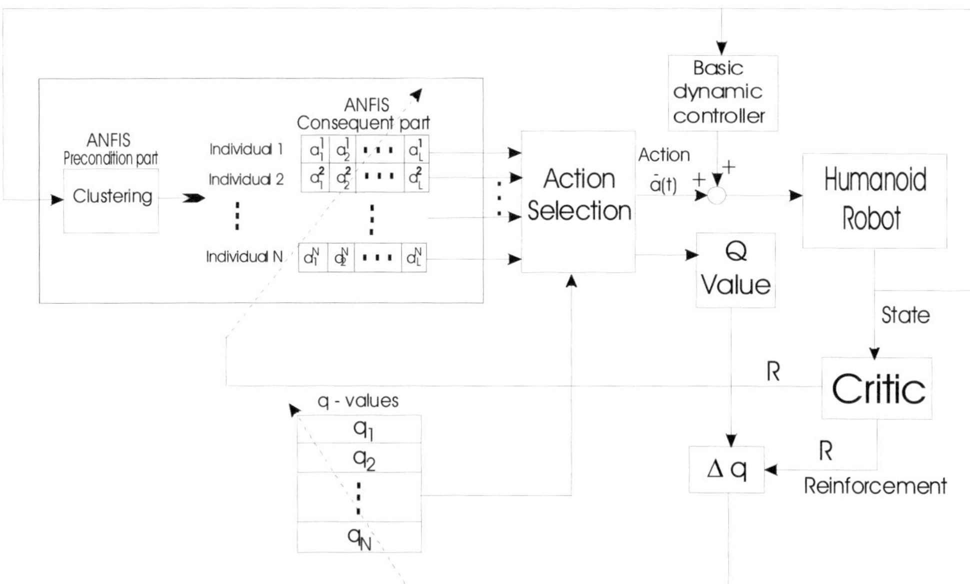

Fig. 8. Hybrid Controller based on Q-Learning Method for trajectory tracking.

4.9. Reinforcement Q-Learning Structure

The precondition part of the ANFIS system is constructed automatically by the clustering algorithm . Then, the consequent part of this newly generated rule is designed. In this methods, a population of candidate consequent parts is generated. Each individual in the population represents the consequent part of a fuzzy system. Since we want to solve reinforcement learning problems, a mechanism to evaluate the performance of each individual is required. To achieve this goal, each individual has a corresponding Q-value. The objective of the Q-value is to evaluate the action recommended by the individual. A higher Q-value means a higher reward that will be achieved. Based on the accompanying Q-value of each individual, at each time step, one of the individuals is selected. With the selected individual (consequent part), the fuzzy system evaluates an action and a corresponding system Q-value. This action is then applied to the humanoid robot as part of hybrid control algorithm with a reinforcement returned. Based on this reward, the Q-value of each individual is updated based on temporal difference algorithm. The parameters of consequent part of ANFIS is also updated based on back propagation algorithm and value of reinforcement The previous process is repeatedly executed until success.
Each rule in the fuzzy system is presented in the following form:

$$\textbf{Rule:} \text{ If } x_1(t) \text{ is } \mathrm{A}_{i1} \text{ And} \ldots \mathrm{x}_n(\mathrm{t}) \text{ is } \mathrm{A}_{in} \text{ Then a(t) is } \mathrm{a}_i(\mathrm{t}) \tag{14}$$

Where $x(t)$ is the input value, $\overline{a}$ (t) is the output action value, A is a fuzzy set and a(t) is a recommended action is a fuzzy singleton. If we use a Gaussian membership function as fuzzy set , then for given an input data $x = (x_1, x_2, ..., x_n)$, the firing strength $\Phi_i(x)$ of rule i is calculated by

$$\Phi_i(x) = \exp\left\{ -\sum_{j=1}^{n} \left(\frac{x_j - m_{ij}}{\sigma_{ij}}\right)^2 \right\} \tag{15}$$

where m_{ij} and σ_{ij} denote the mean and width of the fuzzy set.

Suppose a fuzzy system consists of L rules. By weighted average defuzzification method, the output of the system is calculated by

$$\overline{a} = \frac{\sum_{i=1}^{L} \Phi_i(x) a_i}{\sum_{i=1}^{L} \Phi_i(x)} \tag{16}$$

A population of recommended actions , involving individuals is created. Each individual in the population represents the consequent values, $a_1,..,a_L$ of a fuzzy system. The Q-value used to predict the performance of individual i is denoted as q_i. An individual with a higher Q-value means a higher discounted cumulative reinforcement will be obtained by this individual. At each time step, one over these N individuals is selected as the consequent part of a fuzzy system based on their corresponding Q-values. This fuzzy system with competing consequences may be written as

If (**Precondition Part**) Then (**Consequence**) is

$$\text{Individual 1 } (a_1^1,\ldots,a_L^1 \quad \text{with } q_i$$

$$\text{Individual 2 } (a_1^2,\ldots,a_L^2 \quad \text{with } q_2$$

....

$$\text{Individual N } (a_1^N,\ldots,a_L^N \quad \text{with } q_N$$

To accomplish the selection task, we should find the individual i^* whose Q-value is the largest, i.e. We call this a greedy individual, and the corresponding actions for rules are called greedy actions. The greedy individual is selected with a large probability 1-ε. Otherwise, the previously selected individual is adopted again. Suppose at time , the individual $\hat{i}$ is selected, , i.e., actions $a_1^{\hat{i}}(t),\ldots,a_L^{\hat{i}}(t)$ are selected for rules 1, ..., L, respectively. Then, the final output action of the fuzzy system is

$$\bar{a}(t) = \frac{\sum_{i=1}^{L} \Phi_i(x(t)) a_i^{\hat{i}}(t)}{\sum_{i=1}^{L} \Phi_i(x(t))} \tag{17}$$

The Q-value of this final output action should be a weighted average of the Q-values corresponding to the actions $a_1^{\hat{i}}(t),\ldots,a_L^{\hat{i}}(t)$ .i.e.,

$$Q(x(t),\bar{a}(t)) = \frac{\sum_{i=1}^{L} \Phi_i(x(t)) q_i(t)}{\sum_{i=1}^{L} \Phi_i(x(t))} = \mathrm{q_i(t)} \tag{18}$$

From this equation, we see that the Q-value of the system output is simply equal to $q_i(t)$, the Q-value of the selected individual i . This means q_i that simultaneously reveals both the performance of the individual and the corresponding system output action. In contrast to traditional Q-learning, where the Q-values are usually stored in a look-up table, and can deal only with discrete state/action pairs, here both the input state and the output action are continuous. This can avoid the impractical memory requirement for large state-action spaces. The aforementioned selecting, acting, and updating process is repeatedly executed until the end of a trial.

Every time after the fuzzy system applies an action $\bar{a}(t)$ to the environment and a reinforcement $R(t)$, learning of the Q-values is performed. Then, we should update $q_i(t)$ based on the immediate reward $R(t)$ and the estimated rewards from subsequent states. Based on the Q-learning rule, we can update $q_{\hat{i}}$ as

$$q_{\hat{i}}(t) = q_{\hat{i}}(t) + \alpha(R(t) + \gamma Q^*(x(t+1)) - q_{\hat{i}}(t))$$
$$Q^*(x(t+1)) = \max_{i=1,..,N} Q(x(t+1), \bar{a}^i)$$
$$= \max_{i=1,..,N} q_i(t) = q_{i^*}(t) \quad (19)$$

That is

$$q_{\hat{i}}(t) = q_{\hat{i}}(t) + \alpha(R(t) + \gamma q_{i^*}(x(t+1)) - q_{\hat{i}}(t)) = q_{\hat{i}}(t) + \alpha \Delta q_{\hat{i}}(t) \quad (20)$$

Where $\Delta q_i(t)$ is regarded as the temporal error.

To speed up the learning, the eligibility trace is combined with Q-learning . The eligibility trace for individual i at time t is denoted as $e_i(t)$. On each time step, the eligibility traces for all individuals are decayed by λ, and eligibility trace for the selected individual $\hat{i}$ on the current step increased by 1, that as

$$e_i(t) = \lambda e_i(t+1), \qquad \text{if } i \neq \hat{i}$$
$$= \lambda e_i(t-1) + 1. \qquad \text{if } i = \hat{i}$$

λ is a trace-decay parameter. The value $e_i(t)$ can be regarded as an accumulating trace for each individual i since it accumulates whenever an individual is selected, then decays gradually when the individual is not selected. It indicates the degree to which each individual is eligible for undergoing learning changes. With eligibility trace, (20) is changed to

$$q_{\hat{i}}(t) = q_{\hat{i}}(t) + \alpha \Delta q_{\hat{i}}(t) e_i(t) \quad (21)$$

for all i=1,…,N. Upon receiving a reinforcement signal, the Q-values of all individuals are updated by (21).

4.10 Fuzzy Reinforcement Signal

The detailed and precise training data for learning is often hard to obtain or may not be available in the process of biped control synthesis. Furthermore, a more challenging aspect of this problem is that the only available feedback signal (a failure or success signal) is obtained only when a failure (or near failure) occurs, that is, the biped robot falls down (or almost falls down). Since no exact teaching information is available, this is a typical reinforcement learning problem and the failure signal serves as the reinforcement signal.

For reinforcement learning problems, most of the existing learning methods for neural networks or fuzzy-neuro networks focus their attention on numerical evaluative information. But for human biped walking, we usually use linguistic critical signal, such as "near fall down", "almost success", "slower", "faster" and etc., to evaluate the walking gait. In this case, using fuzzy evaluation feedback is much closer to the learning environment in the real world. Therefore, there is a need to explore possibilities of the reinforcement learning with fuzzy evaluative feedback, as it was investigated in paper (Zhou & Meng, 2000). Fuzzy reinforcement learning generalizes reinforcement learning to fuzzy environment where only the fuzzy reward function is available.

The most important part of algorithm represent the choice of reward function - external reinforcement. It is possible to use scalar critic signal (Katić & Vukobratovič, 2007), but as one of solution, the reinforcement signal was considered as a fuzzy number R(t). We also assume that R(t) is the fuzzy signal available at time step t and caused by the input and action chosen at time step t-1 or even affected by earlier inputs and actions. For more effective learning, a error signal that gives more detail balancing information should be given, instead of a simple "go -no go" scalar feedback signal. As an example in this paper, the following fuzzy rules can be used to evaluate the biped balancing according to following table.

$\Delta x^{(zmp)}$ / $\Delta y^{(zmp)}$	SMALL	MEDIUM	HUGE
SMALL	EXCELLENT	GOOD	BAD
MEDIUM	GOOD	GOOD	BAD
HUGE	BAD	BAD	BAD

Fuzzy rules for external reinforcement

The linguistic variables for ZMP deviations $\Delta x^{(zmp)}$ and $\Delta y^{(zmp)}$ and for external reinforcement *R* are defined using membership functions that are defined in Fig.9.

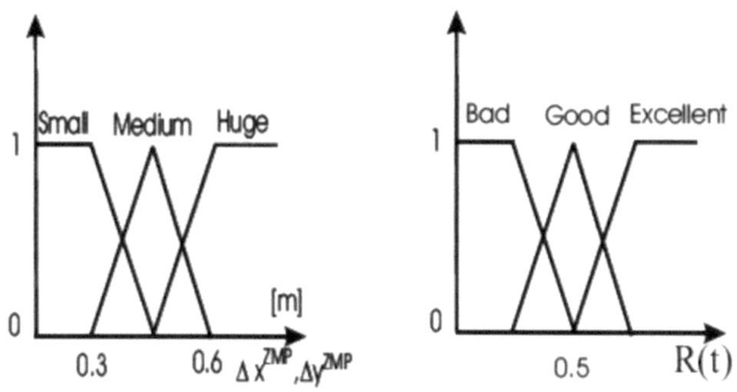

Fig. 9. The Membership functions for ZMP deviations and external reinforcement.

5. Experimental and Simulation Studies

With the aim of identifying a valid model of biped locomotion system of anthropomorphic structure, the corresponding experiments were carried out in a caption motion studio (Rodić et al., 2006). For this purpose, a middle-aged (43 years) male subject, 190 [cm] tall, weighing 84.0728 [kg], of normal physical constitution and functionality, played the role of an experimental anthropomorphic system whose model was to be identified . The subject's geometrical parameters (the lengths of the links, the distances between the neighboring joints and the particular significant points on the body) were determined by direct measurements or photometricaly. The other kinematic parameters, as well as dynamic ones, were identified on the basis of the biometric tables, recommendations and empirical relations (Zatsiorsky et al., 1990). A summary of the geometric and dynamic parameters identified on the considered experimental bio-mechanical system is given in Tables 1 and 2. The selected subject, whose parameters were identified , performed a number of motion tests (walking, staircase climbing, jumping), whereby the measurements were made under the appropriate laboratory conditions. Characteristic laboratory details are shown in Fig. 10. The VICON caption motion studio equipment was used with the corresponding software package for processing measurement data. To detect current positions of body links use was made of the special markers placed at the characteristic points of the body/limbs (Figs. 10a and 10b). Continual monitoring of the position markers during the motion was performed using six VICON high-accuracy infra-red cameras with the recording frequency of 200 [Hz] (Fig. 10c). Reactive forces of the foot impact/contact with the ground were measured on the force platform (Fig. 10d) with a recording frequency of 1.0 [GHz]. To mimic a rigid foot-ground contact, a 5 [mm] thick wooden plate was added to each foot (Fig. 10b).

Link	Length [m]	Mass [kg]	CM Position	
			Sagittal	Longitudinal
Head	0.2722	5.8347	0.0000	0.1361
Trunk	0.7667	36.5380	0.0144	0.3216
Thorax	0.2500	13.4180	0.0100	0.1167
Abdomen	0.3278	13.7291	0.0150	0.2223
Pelvis	0.1889	9.3909	0.0200	0.0345
Arm	0.3444	2.2784	0.0000	-0.1988
Forearm	0.3222	1.3620	0.0000	-0.1474
Hand	0.2111	0.5128	0.0000	-0.0779
Thigh	0.5556	11.9047	0.0000	-0.2275
Shank	0.4389	3.6404	0.0000	-0.1957
Foot	0.2800	1.1518	0.0420	-0.0684

Table 1. The anthropometric data used in modeling of human body (kinematic parameters and mass of links).

Link	Radii of gyration [m]	Moments of inertia
	Sagitt Trans. Longit.	Ix Iy Iz
Head	0.0825 0.0856 0.0711	0.0397 0.0428 0.0295
Trunk	0.2852 0.2660 0.1464	2.9720 2.5859 0.7835
Thorax	0.1790 0.1135 0.1647	0.4299 0.1729 0.3642
Abdomen	0.1580 0.1255 0.1534	0.3427 0.2164 0.3231
Pelvis	0.1162 0.1041 0.1109	0.1267 0.1017 0.1155
Arm	0.0982 0.0927 0.0544	0.0220 0.0196 0.0067
Forearm	0.0889 0.0854 0.0390	0.0108 0.0099 0.0021
Hand	0.1326 0.1083 0.0847	0.0090 0.0060 0.0037
Thigh	0.1828 0.1828 0.0828	0.3977 0.3977 0.0816
Shank	0.1119 0.1093 0.0452	0.0456 0.0435 0.0074
Foot	0.0720 0.0686 0.0347	0.0060 0.0054 0.0014

Table 2. The anthropometric data used in modeling of human body (dynamic parameters - inertia tensor and radii of gyration).

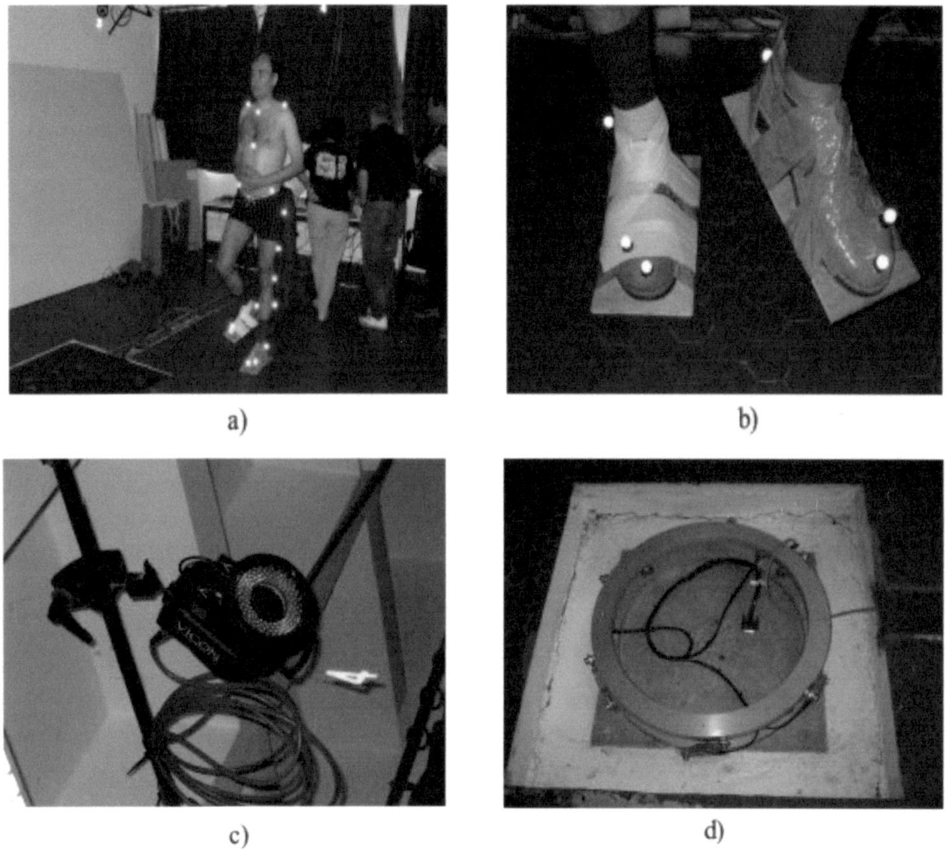

Fig. 10. Experimental capture motion studio in the Laboratory of Biomechanics (Univ. of La Reunion, CURAPS, Le Tampon, France): a) Measurements of human motion using fluorescent markers attached to human body; b) Wooden plates as feet-sole used in locomotion experiments; c) Vicon infra-red camera used to capture the human motion; d) 6-DOFs force sensing platform -sensor distribution at the back side of the plate.

A moderately fast walk (v =1.25 [m/s]) was considered as a typical example of task which encompasses all the elements of the phenomenon of walking. Having in mind the experimental measurements on the biological system and, based on them further theoretical considerations, we assumed that it is possible to design a bipedal locomotion mechanism (humanoid robot) of a similar anthropomorphic structure and with defined (geometric and dynamic) parameters. In this sense, we have started from the assumption that the system parameters presented in Tables 1 and 2 were determined with relatively high accuracy and that they reflect faithfully characteristics of the considered system. Bearing in mind mechanical complexity of the structure of the human body, with its numerous DOFs, we adopted the corresponding kinematic structure (scheme) of the biped locomotion mechanism (Fig. 11) to be used in the simulation examples. We believe that the mechanism (humanoid) of the complexity shown in Fig. 11 would be capable of reproducing with a relatively high accuracy any anthropomorphic motion -rectilinear and curvilinear walk, running, climbing/descending the staircase, jumping, etc. The adopted structure has three active mechanical DOFs at each of the joints -the hip, waist, shoulders and neck; two at the ankle and wrist, and one at the knee, elbow and toe. The fact is that not all available mechanical DOFs are needed in different anthropomorphic movements. In the example considered in this work we defined the nominal motion of the joints of the legs and of the trunk. At the same time, the joints of the arms, neck and toes remained immobilized. On the basis of the measured values of positions (coordinates) of special markers in the course of motion (Figs. 10a, 10b) it was possible to identify angular trajectories of the particular joints of the bipedal locomotion system. These joints trajectories represent the nominal, i.e. the reference trajectories of the system i. The graphs of these identified/reference trajectories are shown in Figs. 12 and 13. The nominal trajectories of the system's joints were differentiated with respect to time, with a sampling period of $\Delta t = 0.001$ [ms]. In this way, the corresponding vectors of angular joint velocities and angular joint accelerations of the system illustrated in Fig. 11 were determined. Animation of the biped gait of the considered locomotion system, for the given joint trajectories (Figs. 12 and 13), is presented in Fig. 14 through several characteristic positions. The motion simulation shown in Fig. 14 was determined using kinematic model of the system. The biped starts from the state of rest and then makes four half-steps stepping with the right foot once on the platform for force measurement.Simulation of the kinematic and dynamic models was performed using Matlab/Simulink R13 and Robotics toolbox for Matlab/Simulink. Mechanism feet track their own trajectories (Figs. 12 and 13) by passing from the state of contact with the ground (having zero position) to free motion state.

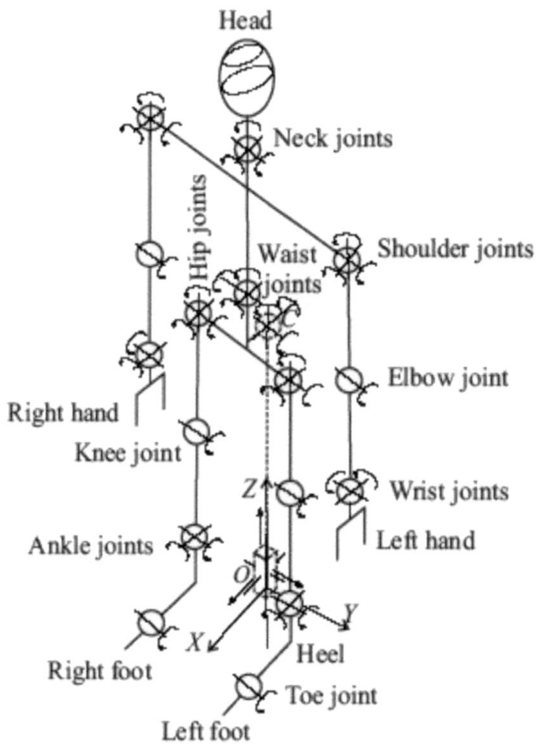

Fig. 11. Kinematic scheme of a 38-DOFs biped locomotion system used in simulation as the kinematic model of the human body referred to in the experiments.

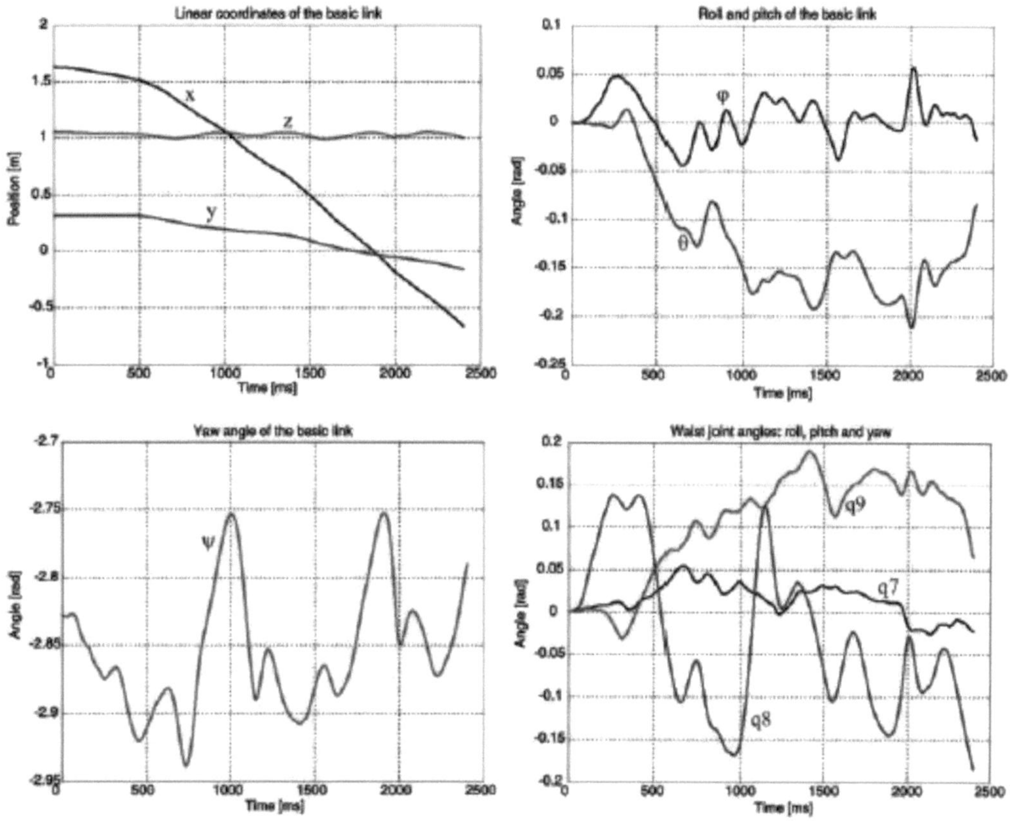

Fig. 12. Nominal trajectories of the basic link: x-longitudinal, y-lateral, z-vertical, φ-roll, θ-pitch, ψ-yaw; Nominal waist joint angles: q7-roll, q8-yaw, q9-pitch.

Some special simulation experiments were performed in order to validate the proposed reinforcement learning control approach. Initial (starting) conditions of the simulation examples (initial deviations of joints' angles) were imposed. Simulation results were analyzed on the time interval 0.1[s]. In the simulation example, two control algorithms were analyzed: (i) basic dynamic controller described by computed torque method (without learning) and (ii) hybrid reinforcement learning control algorithm. (with learning). The results obtained by applying the controllers (i) (without learning) and (ii) (with learning) are shown on Figs. 15 and Fig.16. It is evident , that better results were achieved with using reinforcement learning control structure.
The corresponding position and velocity tracking erros in the case of application reinforcement learning structure structure are presented on Figs. 17 and 18. The tracking errirs converge to zero values in the given time interval. It means that the controller ensures good tracking of the desired trajectory. Also, the application of reinforcement learning structure ensures a dynamic balance of the locomotion mechanism..
In Fig. 19 value of internal reinforcement through process of walking is presented. It is clear that task of walking within desired ZMP tracking error limits is achieved in a good fashion.

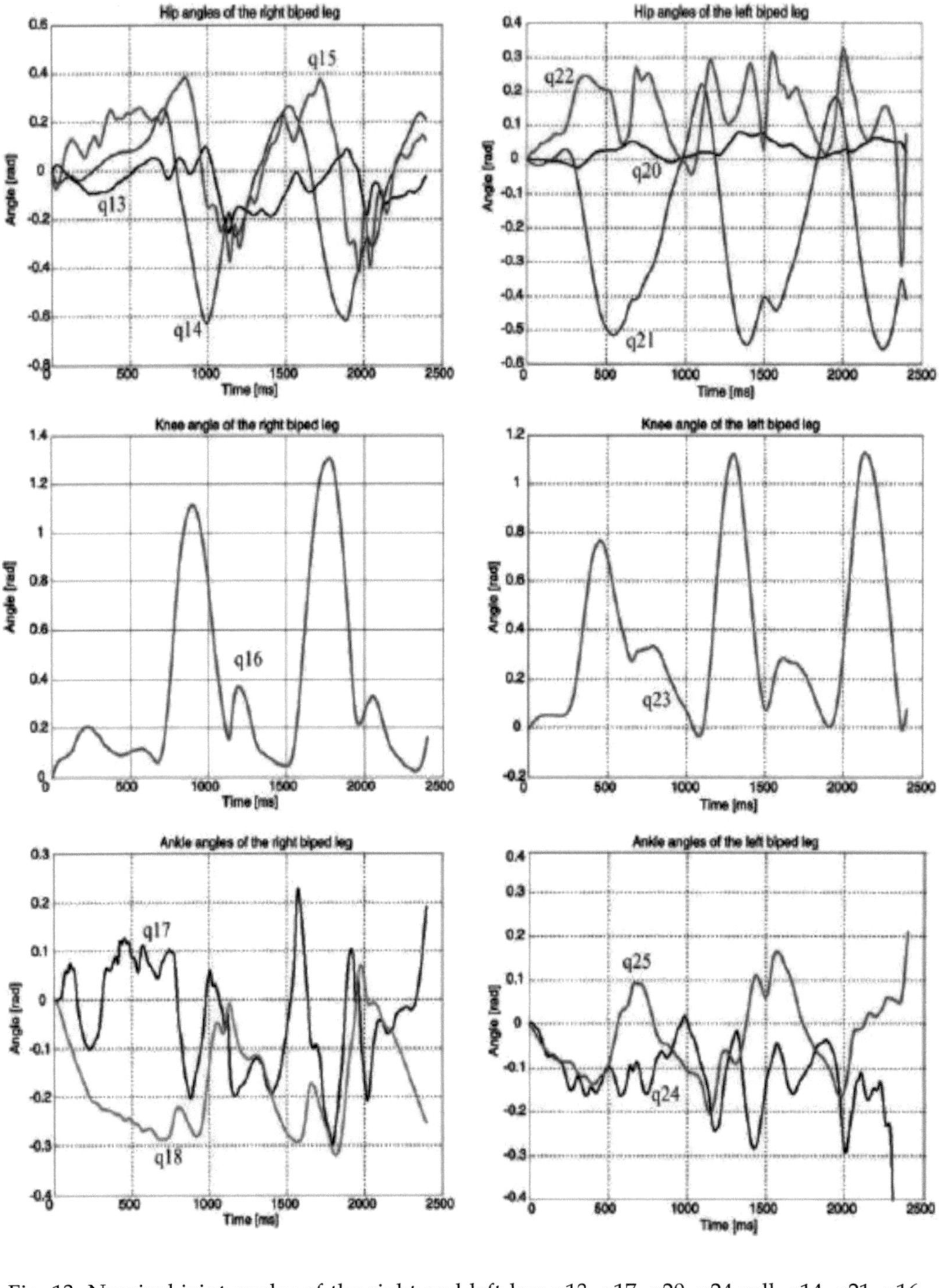

Fig. 13. Nominal joint angles of the right and left leg: q13, q17, q20, q24-roll, q14, q21, q16, q18, q23, q25-pitch, q15, q22-yaw.

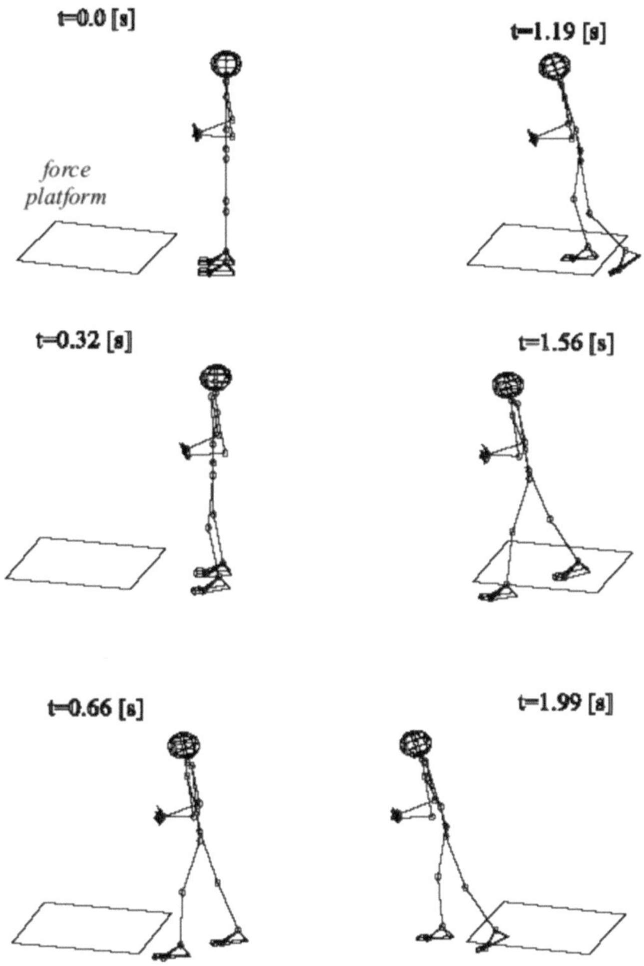

Fig.14. Model-based animation of biped locomotion in several characteristic instants for the experimentally determined joint trajectories.

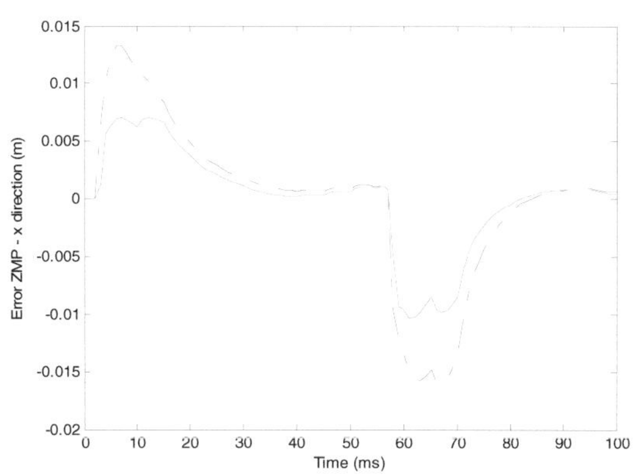

Fig. 15. Error of ZMP in x-direction (- with learning - . - without learning).

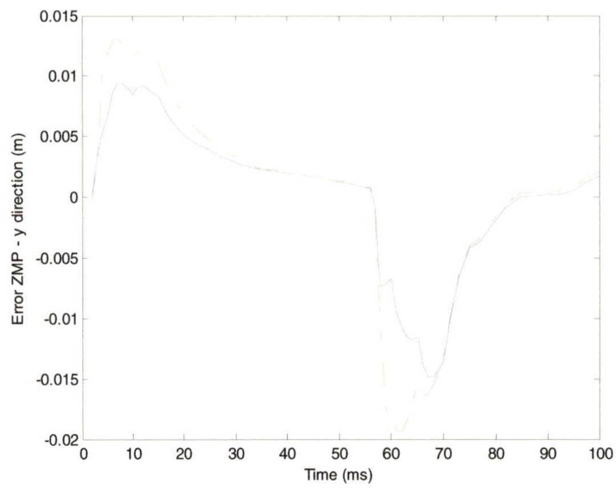

Fig. 16. Error of ZMP in y-direction (- with learning - . - without learning).

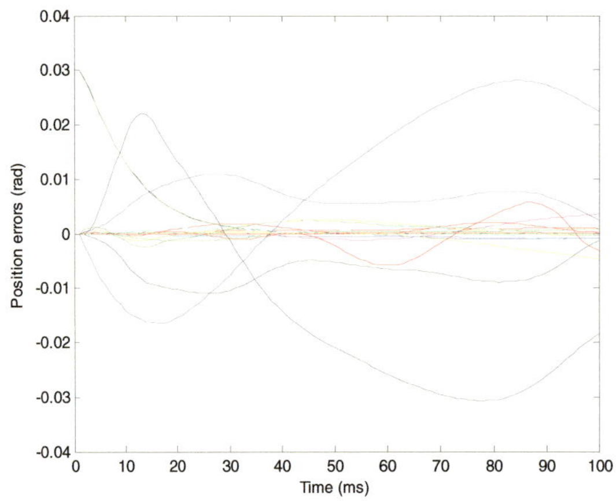

Fig. 17. Position tracking errors.

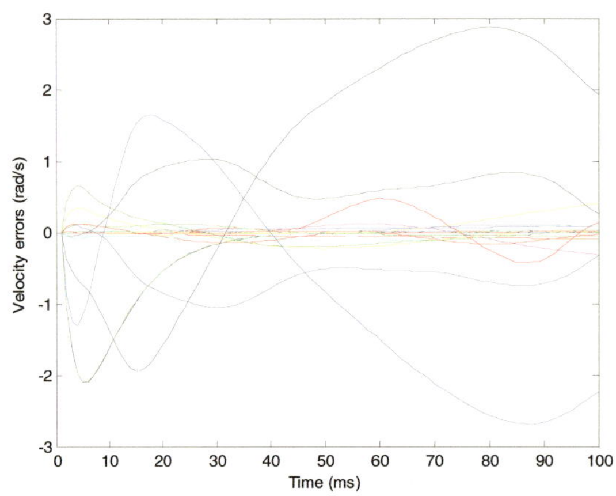

Fig. 18. Velocity tracking errors.

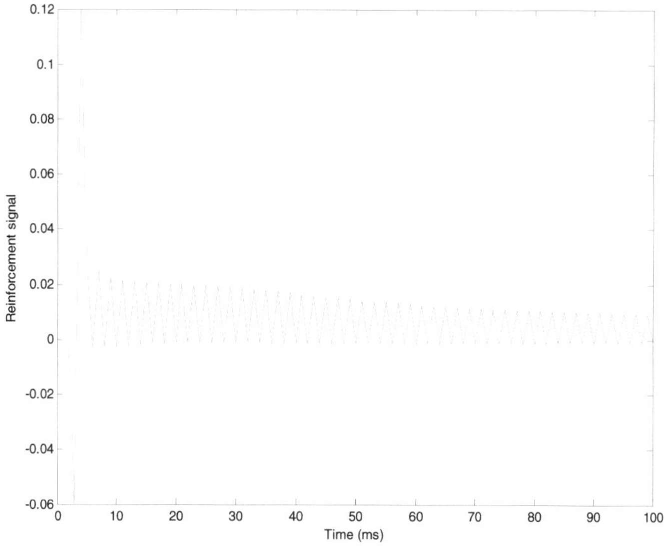

Fig. 19. Reinforcement through process of walking.

6. Conclusions

This study considers a optimal solutions for application of reinforcement learning in humanoid robotics Humanoid Robotics is a very challenging domain for reinforcement learning, Reinforcement learning control algorithms represents general framework to take traditional robotics towards true autonomy and versatility. The reinforcement learning paradigm described above has been successfully implemented for some special type of humanoid robots in the last 10 years. Reinforcement learning is well suited to training biped walk in particular teaching a robot a new behavior from scalar or fuzzy feedback. The general goal in synthesis of reinforcement learning control algorithms is the development of methods which scale into the dimensionality of humanoid robots and can generate actions for biped with many degrees of freedom. In this study, control of walking of active and passive dynamic walkers by using of reinforcement learning was amalyzed.

Various straightforward and hybrid intelligent control algorithms based RL for active and passive biped locomotion is presented. The proposed RL algorithms use the learning elements that consists of various types of neural networks, fuzzy logic nets or fuzzy-neuro networks with focus on fast convergence properties and small number of learning trials.

Special part of study represents synthesis of hybrid intelligent controllers for biped walking. The hybrid aspect is connected with application of model-based and model free approaches as well as with combination of different paradigms of computational intelligence. These algorithms includes combination of a dynamic controller based on dynamic model and special compensators based on reinforcement structures. Two different reinforcement learning structures were proposed based on actor-critic approach and Q-learning. The algorithms is based on fuzzy evaluative feedback that are obtained from human intuitive balancing knowledge. The reinforcement learning with fuzzy evaluation feedback is much closer to the human biped walking evaluation than the original one with scalar feedback.

The proposed hybrid intelligent control scheme fulfills the preset control criteria. Its application ensures the desired precision of robot's motion, maintaining dynamic balance of the locomotion mechanism during a motion

The developed hybrid intelligent dynamic controller can be potentially applied in combination with robotic vision, to control biped locomotion mechanisms in the course of fast walking, running, and even in the phases of jumping, as it possesses both the conventional position-velocity feedback and dynamic reaction feedback. Performance of the control system was analyzed in a number of simulation experiments in the presence of different types of external and internal perturbations acting on the system. In this paper, we only consider the flat terrain for biped walking. Because the real terrain is usually very complex, more studies need to be conducted on the proposed gait synthesis method for irregular and sloped terrain.

Dynamic bipedal walking is difficult to learn because combinatorial explosion in order to optimize performance in every possible configuration of the robot., uncertainties of the robot dynamics that must be only experimentally validated, and because coping with dynamic discontinuities caused by collisions with the ground and with the problem of delayed reward -torques applied at one time may have an effect on the performance many steps into the future. Hence, for a physical robot, it is essential to learn from few trials in order to have some time left for exploitation. It is thus necessary to speed the learning up by using different methods (hierarchical learning, subtask decomposition, imitation,…).

7. Acknowledgments

The work described in this conducted was conducted within the national research project "Dynamic and Control of High-Performance Humanoid Robots: Theory and Application". and was funded by the Ministry of Science and Environmental Protection of the Republic of Serbia. The authors thank to Dr. Ing. Aleksandar Rodić for generation of experimental data and realization of humanoid robot modeling and trajectory generation software.

8. References

Barto, A.G., Sutton, R.S & Anderson, C.W., (1983), Neuron like adaptive elements that can solve difficult learning control problems, *IEEE Transactions on Systems, Man,and Cybernetics*, SMC-13, 5, September 1983, 834–846.

Baxter, J. and Bartlett, P., (2001). Infinite-horizon policy-gradient estimation , *Journal of Artificial Intelligence Research*, 319-350.

Benbrahim, H. & Franklin, J.A. (1997), Biped Dynamic Walking using Reinforcement Learning, *Robotics and Autonomous Systems*, 22, 283-302.

Berenji,H.R. & Khedkar,P., (1992), Learning and Tuning Fuzzy Logic controllers through Reinforcements, *IEEE Transactions on Neural Networks*, 724-740

Berenji, H.R., (1996), Fuzzy Q-learning for generalization of reinforcement," in *Proc. IEEE Int. Conf. Fuzzy Systems*, 2208–2214.

Bertsekas,D.P. and Tsitsiklis,J.N. (1996), *Neuro-Dynamic Programming*, Athena Scientific, Belmont, USA.

Chew, C. & Pratt, G.A., (2002), Dynamic bipedal walking assisted by learning, *Robotica*, 477-491

Doya, K., (2000), Reinforcement Learning in Continuous Time and Space, *Neural Computation*, 219-245.

Glorennec, P.Y. & ouffe, L. (1997), Fuzzy Q - Learning, in *Proceedings of FUZZ-IEEE'97*, Barcelona, July 1997.

Jouffe, L., (1998), Fuzzy inference system learning by reinforcement methods, *IEEE Transactions on Systems, Man, Cybernetics:Part C, Appl. Rev.*, 28, 3, August 1998, 338–355.

Kamio, S., & Iba, H., (2005), Adaptation Technique for Integrating Genetic Programming and Reinforcement Learning fir Real Robot, *IEEE Transactions on Evolutionary Computation*, 9, 3, June 2005, 318 - 333.

Katić, D. & Vukobratović, M., (2003a), Survay of Intelligent Control Techniques for Humanoid Robots, *Journal of Intelligent and Robotic Systems*, 37, 2003, 117 -141.

Katić, D. & Vukobratović, M. (2003b), *Intelligent Control of Robotic Systems*, Kluwer Academic Publishers, Dordrecht, Netherlands

Katić, D. & Vukobratović, M.: (2005), Survey of Intelligent Control Algorithms For Humanoid Robots", in *Proceedings of the 16th IFAC World Congress*, Prague, Czech Republic, July 2005.

Katić, D. & Vukobratović, M., (2007), Hybrid Dynamic Control Algorithm for Humanoid Robots Based on Reinforcement Learning, accepted for publication in *Journal of Intelligent and Robotic Systems* .

Kimura, H. & Kobayashi, S. (1998), An analysis of actor/critic algorithms using eligibility traces: Reinforcement learning with imperfect value functions ,in *Proceedings of the International Conference on Machine Learning (ICML '98)*, 278-286.

Kun , A.L. and Miller,III, W.T., (1999), Control of Variable -Speed Gaits for a Biped Robot, *IEEE Robotics and Automation Magazine*, 6, September 1999, 19-29.

Li, Q., Takanishi, A. & Kato, I., (1992), Learning control of compensative trunk motion for biped walking robot based on ZMP, in *Proceedings of the 1992 IEEE/RSJ Intl.Conference on Intelligent Robot and Systems*, 597–603.

Mori, T., Nakamura, Y., Sato, M. and Ishii, S., (2004), Reinforcement learning for a cpg-driven biped robot, in *Proceedings of the Nineteenth National Conference on Artificial Intelligence (AAAI)*, 623–630.

Morimoto, J., Cheng, G., Atkeson, C.G. & Zeglin,G., (2004) A Simple Reinforcement Learning Algorithm For Biped Walking, in *Proceedings of the 2004 IEEE International Conference on Robotics & Automation*, New Orleans, USA.

Nagasaka, K., Inoe, H. and Inaba, M., (1999), Dynamic walking pattern generation for a humanoid robot based on optimal gradient method, in *Proceedings of the International Conference on Systems, Man. and Cybernetics*, 908-913.

Nakamura, Y., Sato, M. & Ishii, S., (2003), Reinforcement learning for biped robot, *in Proceedings of the 2nd International Symposium on Adaptive Motion of Animals and Machines.*

Peters, J., Vijayakumar, S., and Schaal, S., (2003), Reinforcement Learning for Humanoid Robots, in *Proceedings of the Third IEEE-RAS International Conference on Humanoid Robots*, Karlsruhe & Munich.

Rodić, A., Vukobratović, M., Addi, K. & Dalleau, G. (2006), Contribution to the Modelling of Non-Smooth, Multi-Point Contact Dynamcs of Biped Locomotion - Theory and Experiments, submitted to journal *Robotica.*

Salatian, A.W., Yi, K.Y. and Zheng, Y.F., (1997) , Reinforcement Learning for a Biped Robot to Climb Sloping Surfaces, *Journal of Robotic Systems*, 283-296.

Schuitema, E., Hobbelen, D..G.E, Jonker, P.P., Wisse, M. & Karssen, J.G.D., (2005), Using a controller based on reinforcement learning for a passive dynamic walking robot, *IEEE International conference on Humanoid Robots 2005*, Tsukuba, Japan.

Sutton, R.S. and Barto,A.G. (1998) *Reinforcement Learning: An Introduction*, The MIT Press, Cambridge, USA.

Sutton, R.S., McAllester, D., & Singh, S., (2000), Policy Gradient Methods for Reinforcement learning with Function Approximation, in *Advances in Neural Information Processing Systems*, 12, MIT Press, Cambrdige, USA, 1057-1063.

Tedrake, R.,Zhang, T.W. & Seung, H.S., (2004), Stochastic policy gradient reinforcement learning on a simple 3d biped, in *Proceedings of the 2004 IEEE/RSJ International Conference on Intelligent Robots and Systems.*

Vukobratović, M, & Juričić, D. (1969), Contribution to the Synthesis of Biped Gait, *IEEE Transactions on Biomedical Engineering*, BME-16, 1, 1-6.

Vukobratović, M., Borovac, B., Surla, D., & Stokić, D., (1990), *Biped Locomotion - Dynamics, Stability, Control and Application*, Springer Verlag, Berlin, Germany.

Watkins, C.J.C.H. & Dayan, P., (1992), Q Learning, *Machine Learning*,, 279-292.

Zatsiorsky, V., Seluyanov, V. & Chugunova, L. (1990) ,Methods of Determining Mass - Inertial Characteristics of Human Body Segments, *Contemporary Problems of Biomechanics*, 272-291, CRC Press.

Zhou, C. & Meng, D. (2000), Reinforcement Learning with Fuzzy Evaluative Feedback for a Biped Robot. *In:Proceedings of the IEEE International Conference on Robotics and Automation*, San Francisko, USA, 3829-3834.

23

A Designing of Humanoid Robot Hands in Endo skeleton and Exoskeleton Styles

Ichiro Kawabuchi
KAWABUCHI Mechanical Engineering Laboratory, Inc.
Japan

1. Introduction

For a serious scientific interest or rather an amusing desire to be the creator like Pygmalion, human being has kept fascination to create something replicates ourselves as shown in lifelike statues and imaginative descriptions in fairy tales, long time from the ancient days. At the present day, eventually, they are coming out as humanoid robots and their brilliant futures are forecasted as follows. 1) Humanoid robot will take over boring recurrent jobs and dangerous tasks where some everyday tools and environments designed and optimised for human usage should be exploited without significant modifications. 2) Efforts of developing humanoid robot systems and components will lead some excellent inventions of engineering, product and service. 3) Humanoid robot will be a research tool by itself for simulation, implementation and examination of the human algorithm of motions, behaviours and cognitions with corporeality.
At the same time, I cannot help having some doubts about the future of the humanoid robot as extension of present development style. Our biological constitution is evolved properly to be made of bio-materials and actuated by muscles, and present humanoid robots, on the contrary, are bounded to be designed within conventional mechanical and electric elements prepared for industrial use such as electric motors, devices, metal and plastic parts. Such elements are vastly different in characteristics from the biological ones and are low in some significant properties: power/weight ratio, minuteness, flexibility, robustness with self-repairing, energy and economic efficiency and so on. So the "eternal differences" in function and appearance will remain between the human and the humanoid robots.
I guess there are chiefly two considerable ways for developing a preferable humanoid robot body. One is to promote in advance a development of artificial muscle that is exactly similar to the biological one. It may be obvious that an ideal humanoid robot body will be constructed by a scheme of putting a skeletal model core and attaching the perfect artificial muscles on it (Weghel et al., 2004, e.g.). Another is to establish some practical and realistic designing paradigms within extensional technology, in consideration of the limited performance of mechanical and electric elements. In this way, it will be the key point that making full use of flexible ideas of trimming and restructuring functions on a target. For example, that is to make up an alternative function by integrating some simple methods, when the target function is too complex to be a unitary package. Since it seems to take long time until the complete artificial muscles will come out, I regard the latter way is a good prospect to the near future rather than just a compromise.

In searching the appropriate designing paradigms for humanoid robots, it may be just the stage of digging and gathering many diverse and unique studies, styles and philosophies of the designing. After examining their practicability, reasonability and inevitability through the eyes of many persons over the long time, it will work out the optimised designing paradigms. I believe the most outstanding ingenuities in mechanical design are brought out in robot hands, as they are most complex mechanical systems on a humanoid robot body. Fortunately, I have had some experiences of designing them in each endoskeleton and exoskeleton styles. To contribute on such evolutive process as a mechanical engineer, I bring up here some of my ideas of designing robot hands.

2. Humanoid Robot Hand in Endoskeleton Style

2.1 Basic Design Conditions

There are two opposite orientations in planning specifications of a robot hand. One is emphasizing similarity in motion and function to the human hand. Another is emphasizing similarity in size and weight. A robot hand following the former orientation tends to become rather large against the human hand. Although attempts to make it small and light have been done by putting the actuators away into a forearm segment and actuating the fingers via links or wires (Jacobsen, 1984, e.g.), the whole system is still large and heavy as a load at an end of arm. A robot hand following the latter orientation tends to lack functions as shown in general prosthetic hands. Most research robot hands, except some modern prosthetic ones (Harada hand, e.g.), are designed following the former orientation to enlarge their universality on study scenes. And it is rational estimation that the properties aimed in the latter orientation will be resolved spontaneously as the industrial progress leads general mechanical and electric elements high-performance. However, I have taken the latter orientation, because I believe it should be also constitutive approach to an ideal robot hand, moreover it is challenging for me to contrive novel designs of complex mechanisms. Then I have set the basic design conditions as shown on Table 1.

1. Outline should be similar to the human hand with 5 fingers.
2. Every motor should be embedded within the outline, and connections with outside controller and power source are limited to some slim and limber electric cables.
3. Compactness and lightness should be equal to the average naked human hand. Concretely, mass is at most 500g, longitudinal length from fingertip to prospective center of wrist is at longest 200mm, and diameter of fingers is at most 20mm.
4. As far as the third condition is satisfied, number of motors and range of finger movements should be enriched for realizing some practical functions like holding bulk object, pinching thin card, and the most sign language. As key finger motions, all fingers move widely from opening to clenching, and the thumb and each of other fingers contact opposing on both fingertips.
5. As far as the fourth condition is satisfied, power of motors and robustness of mechanisms should be augmented. The desirable output force at each fingertip is over 5N under a straight posture.

Table 1. The basic design conditions.

2.2 Kinematical Arrangement of Joints

Fig. 1 shows the kinematical arrangement of joints in my robot hand. I identified each joint with two numbers and describe it as symbol $J_{n,m}$, in which the first suffix means finger number and the second one means joint number on each finger. The reason I avoid the popular joint nomenclature with the anatomical term like MP (metacarpal phalangeal joint) and DIP (distal interphalangeal joint) is that some of original human joints are complex one with multi degree of freedom (DOF) and hardly be substituted by a unitary joint mechanism.

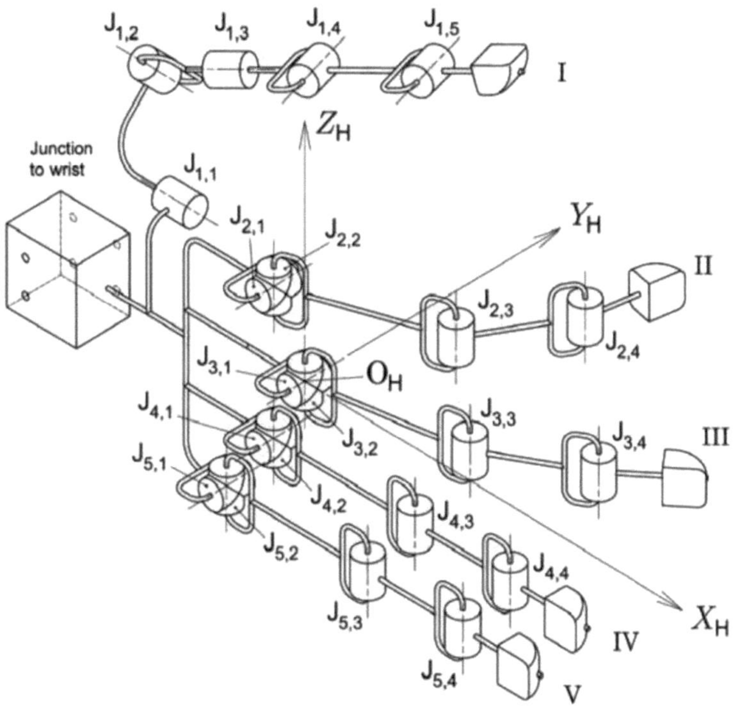

Fig. 1. The kinematical arrangement of joints.

The joint arrangement on the four fingers except the thumb would be common to most humanoid robot hand. At the same time, difference of thumb structures between several competitive robot hands (Fig. 2) shows clearly the diverse design philosophies following individual purposes and available technologies. The every thumb structure on Fig. 2 has a complex joint that consists of a pair of revolute joints intersecting their axes at one point. The complex joint is an influential technology and the Shadow hand, above all, has two ones to realize the most similar action and appearance to those of human. On the other hand, I avoided such joint with estimating its complexity will require large space that violates the basic design conditions when it would be designed by my present technical capabilities.

The order of each direction of joint axis and fingertip is also an influential characteristic for range and variety of the thumb motion. Especially in the NASA hand, to compensate disadvantages by low of DOF, the order is decided advisedly with considering a reasonable emulation of human thumb motion (Lovchik & Diftler, 1999). In my robot hand, the axis direction of joint $J_{1,3}$ is set for twisting the thumb tip that will conduce a stable pinching mentioned after. The axis directions of two joints $J_{1,1}$ and $J_{1,2}$ are decided mainly because of convenience to embed reducers with wide movable range.

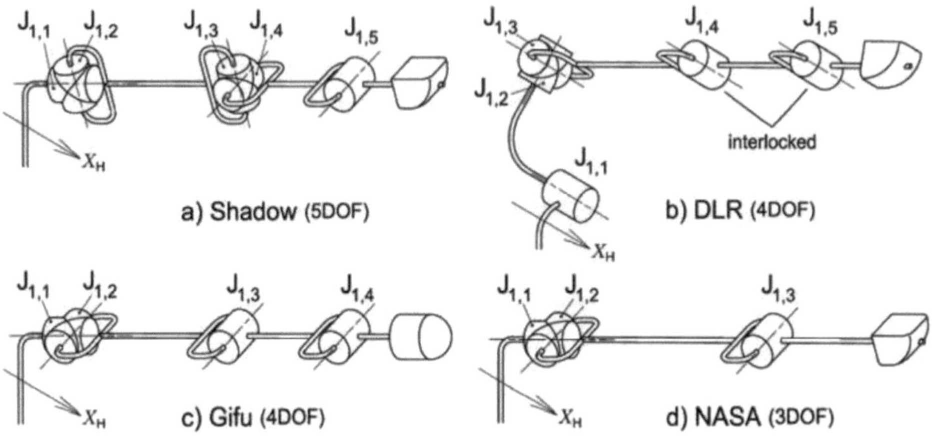

Fig. 2. Thumb structures of competitive robot hands.

2.3 Compression of Independent DOF

An "actuator assembly" that consists of a motor, an encoder and a reducer occupies considerable space in robot hand rigidly. Therefore, to plan a compact robot hand, it is a reasonable way to compress the number of equipped assemblies. This means furthermore compression of the independent DOF on the robot hand, so I pursued this idea boldly with keeping a reasonable emulation to the human hand motion.

In the four fingers except the thumb, most robot hands are designed with the two distal joints $J_{n,3}$ and $J_{n,4}$ coupled-driven by one motor, since those joints in the human hand often move together. I furthered this idea of interlocking and made all three joints $J_{n,2}$, $J_{n,3}$ and $J_{n,4}$ interlocked. In the same way, the relative opening-closing motion among the four fingers at the three root joints $J_{2,1}$, $J_{4,1}$ and $J_{5,1}$ is also interlocked. Moreover the joint $J_{3,1}$ is eliminated since the human middle finger rarely moves at the joint. In the thumb, the two distal joints $J_{1,4}$ and $J_{1,5}$ are interlocked and the joint $J_{1,3}$ is fixed in a certain angle.

Based on the consideration above, the essential independent DOF was extracted into 8. The member joints for each DOF are listed on Table 2(a) and named the "essential set". Although the essential set can realize almost all required finger motions like making a loop with the thumb and the little finger, it lacks sophisticated dexterity for some delicate handlings like pinching a thin card. At last, I have added some DOF on the essential set as listed on the Table 2(b) and it is named the "latest set".

In the following two sections I explain mechanisms corresponding to the essential set, and in the later sections I introduce more complex mechanisms corresponding to the latest set.

	Member		Member
1	$J_{1,1}$	5	$J_{2,2}+J_{2,3}+J_{2,4}$
2	$J_{1,2}$	6	$J_{3,2}+J_{3,3}+J_{3,4}$
3	$J_{1,4}+J_{1,5}$	7	$J_{4,2}+J_{4,3}+J_{4,4}$
4	$J_{2,1}+J_{4,1}+J_{5,1}$	8	$J_{5,2}+J_{5,3}+J_{5,4}$

(a) The essential set (8DOF)

	Member		Member		Member
1	$J_{1,1}$	6	$J_{2,1}+J_{4,1}+J_{5,1}$	11	$J_{3,4}$
2	$J_{1,2}$	7	$J_{2,2}$ (differential)	12	$J_{4,2}+J_{4,3}$
3	$J_{1,3}$	8	$J_{2,2}+J_{2,3}$	13	$J_{4,4}$
4	$J_{1,4}$	9	$J_{2,4}$	14	$J_{5,2}+J_{5,3}$
5	$J_{1,5}$	10	$J_{3,2}+J_{3,3}$	15	$J_{5,4}$

(b) The latest set (15DOF)

Table 2. The independent DOF set and member joints.

2.4 Global Finger Flexion Mechanism

As mentioned in the section 2.2 a complex joint in which two rotational axes intersect was avoided in the thumb mechanism, however, the other four fingers retain such complex joint at the root. The difficulties in developing such complex joint include how electric wirings are passed through the joint. The delicate wirings should not be overstretched due to joint rotation, so the wirings should be passed just or near the intersection point of two axes. I introduced a scheme that no part of an actuator assembly is placed around the intersection but enough empty space only for the wirings, at the same time the rotational power for the two axes is provided via wire or link mechanism from an off-site motors.

Fig. 3 shows the internal structure of a basic finger, all the four fingers have the same mechanism. A small DC motor with a built-in encoder (Faulhaber, model 1516SR or 1319SR) is embedded in the largest finger segment lengthwise, like filling the segment volume. A reducer that consists of a crown gear train and two-stage planetary gear train is built in the joint $J_{n,3}$ and drives it. For optimum integration: a combination of larger and smaller gears is effective in uniting higher reduction ratio and efficiency, axes of all gear trains are arranged coaxial to that of joint $J_{n,3}$, so that volume of the joint part admits possibly largest diameter of gears without making the finger segment longer. In the actual reducer, the crown gear which diameter is almost equal to that of finger segment engages with the motor pinion and derives 10 times torque efficiently. In a general gear train with axes mutually orthogonal, the rigidity is low and the play is generated easily; so it is reasonable putting the crown gear train at the first stage viewed from the motor, where the transmission torque is low and the influence of play is negligible. Moreover the planetary gear train is constructed as a unique mixture of planetary-type and star-type for most compactness, so this reducer gets high reduction ratio 1/350 and the maximum joint torque is at least 0.5Nm.

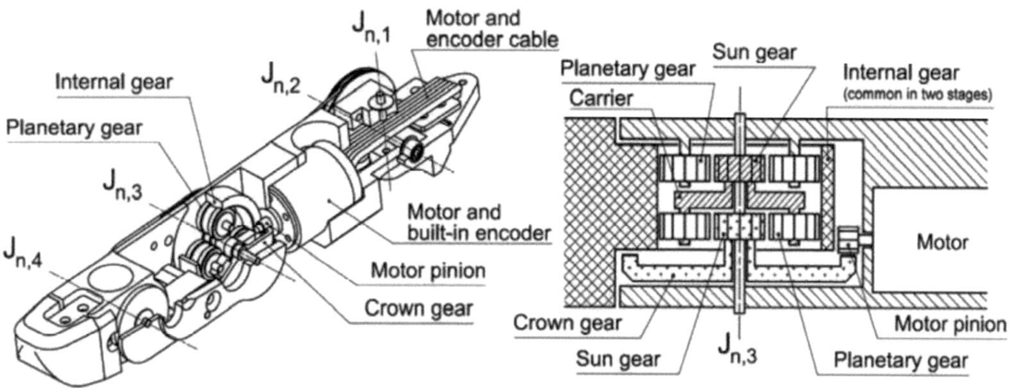

Fig. 3. The internal structure of a basic finger.

It is very important feature of a robot hand that it has smooth back-drivability at each joint for stable control of contacting force on a fingertip. I think it is more acceptable a slight play at gears as a consequence of high back-drivability than a large frictional resistance of gears as a consequence of elimination of the play, thanks to the general robot hand is demanded to perform relatively small force and low absolute accuracy in finger motions. So the actual fingers of my robot hand are manufactured with slight play to get most back-drivability; each finger can be moved passively by small force on the fingertip at most about 0.3N.

The interlocking mechanism among the three joints $J_{n,2}$, $J_{n,3}$ and $J_{n,4}$ consists of wire-pulley mechanisms with pulleys carved on the sidewall (Fig. 4(a)). Since they are thin enough the finger segments afford some internal space for motor, sensor and electric component. Although identical transmission can be constructed using a link mechanism, it tends to become larger due to restriction on facilitating smooth transmission in large rotational angle near 90degree. Considering a reasonable emulation of the human motion, the transmission ratios are set 7/10 from $J_{n,3}$ to $J_{n,2}$, and 5/7 from $J_{n,3}$ to $J_{n,4}$ respectively (same (b)).

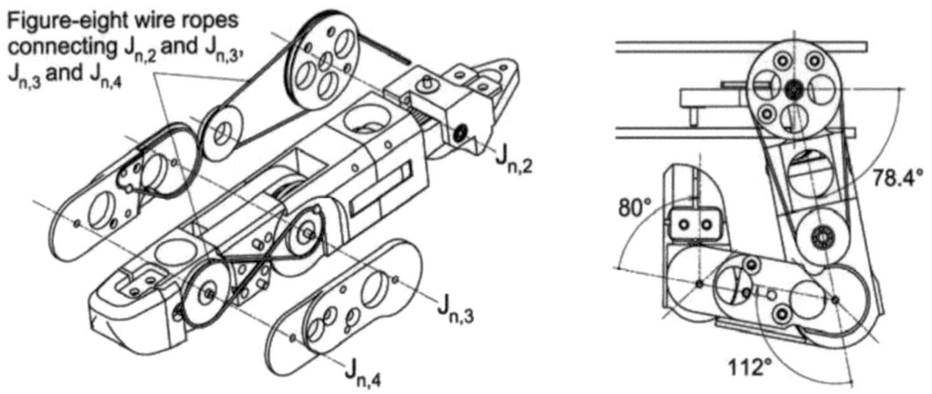

(a) Wire-pulley mechanisms (b) Maximum flexion range

Fig. 4. Interlocking mechanism of the global finger flexion.

2.5 Abduction-Adduction Mechanism

Since the relative opening-closing motion among the four fingers except the thumb is called abduction-adduction in the anatomical term, the interlocking mechanism among the three joints $J_{2,1}$, $J_{4,1}$ and $J_{5,1}$ was named after it. Introduction of this mechanism has mainly two reasons as follows. The first is cutting back on the actuator assembly. The second is a practical idea that the ring and little fingers do not need such motion independently, because they on the human hand rarely take part in a dexterous pinching function. So it can be said that the target of this mechanism is only generating a relative opening-closing motion between the index finger and the middle finger.

Rotational angle of those joints is small, so this mechanism is constructed as a link mechanism (Fig. 5), and in quest of efficiency and compactness it is designed as below. The joint $J_{4,1}$ is selected as driven by a motor directly, so that the power transmission to other joints becomes the shortest route with high efficiency and low adverse influences like play. To obtain a large gear ratio with possibly least stages of gear train, a large sector gear that passes over the palm longitudinal length is adopted and fixed on the joint $J_{4,1}$ (Fig. 6(a)), and the entire reduction ratio is implemented as 1/400. Thanks to the drive motor is located on the edge of the palm, over half of the space in the palm can be free for electric components. Even when each finger is bent to the maximum flexion as shown in the Fig. 4(b), the opening-closing motion can act independently thanks to the clearance between fingertip and palm; the robot hand can perform V-sign actions, e.g. freely. An electric printed circuit board (PCB) that contains a microcomputer for local processing, motor amplifiers, motor current sensors and an interface unit communicating with an outside controller through a serial signal line, is embedded like filling almost all the space in the palm (Fig. 6(b)).

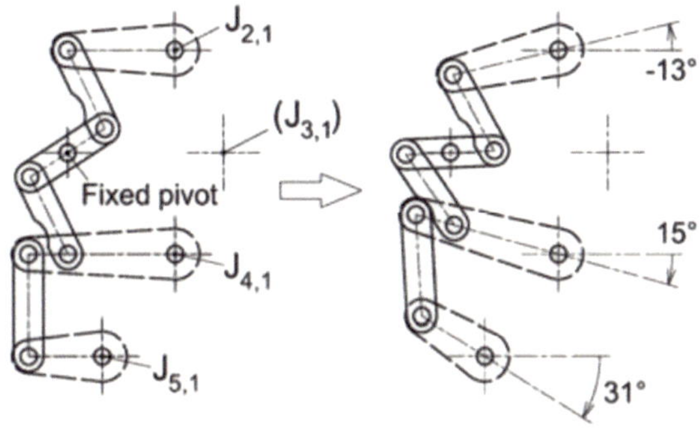

Fig. 5. Connecting linkage for the abduction-adduction motion.

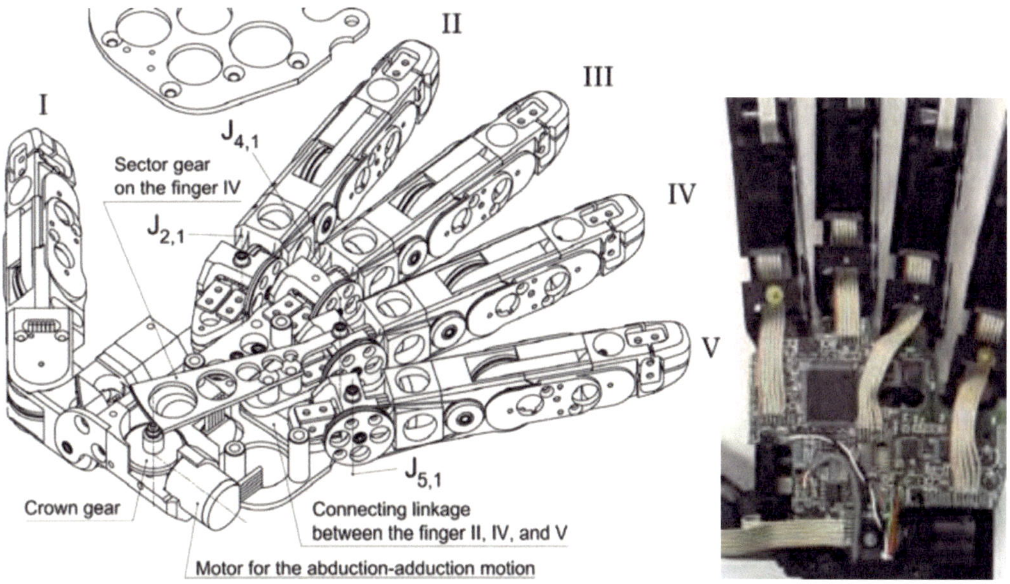

(a) Location of the reduction gears and the linkage

(b) Embedded PCB

Fig. 6. Mechanism and electric equipment in the palm.

2.6 Introduction of Independent Fingertip Motion

Stably and softly pinching thin paper and handling small piece in complicated shape are the vital functions for a humanoid robot hand, when it could be recognised dexterous enough. To realize theses delicate functions, the key feature is a fine force and motion control on the fingertips. The human finger has excellent feature of controlling them finely in any direction, and furthermore it can generate large force in case of necessity. On the contrary, usual small actuator assembly that consists of motor and geared power train never has such large "dynamic range" in force control. In case of my robot finger, the main actuator assembly for the global finger flexion should be designed to generate possibly larger force for moving itself in sufficient speed against its mass and inertia, and moreover grasping a heavy object. Then it is hard for the actuator to be expected to generate fine force on the fingertip.

After considering above, I focused attention on the joint $J_{n,4}$ ($J_{1,5}$ in case of the thumb), as it could be free from generating the large force above; that means it could concentrate to generate a minute and fine force suitable for delicate pinching. Then I introduced a special independent DOF at the joint $J_{n,4}$ ($J_{1,5}$) so that it can control force and motion on the fingertip independently and finely. It is surely a little strange assignment of DOF while general robot hand has interlocking between the two distal joints $J_{n,3}$ and $J_{n,4}$ ($J_{1,4}$ and $J_{1,5}$), however, I think that it maintains the basic design conditions as general harmony as a humanoid robot hand; the human finger also has latent ability to move its terminal joint independently.

In addition, I propose a new strategy of the handling that combines a powerful grasping and a delicate pinching in consideration of that the small additional actuator at the joint $J_{n,4}$ ($J_{1,5}$) cannot sustain large contacting force at the fingertip.

In power grasping (Fig. 7(a)), powerful grasping forces generated by the main actuator are transmitted to an object through the surface of finger segments except the fingertip. At the same time, the delicate force generated by the joint $J_{n,4}$ ($J_{1,5}$) lets the fingertip compliant to the object, and enlarge the effect of slip-free grasping. The movable range of the joint $J_{n,4}$ ($J_{1,5}$) is broadened enough into the backside, so that the function of fingertip can work well in wide range of its posture. The shape of surface around the terminal joint is formed to be suitable for the direct contacting against the object.

In delicate pinching (Fig. 7(b)), the contacting forces on the object are generated only by the joint $J_{n,4}$ ($J_{1,5}$). When larger pinching forces that the joint $J_{n,4}$ ($J_{1,5}$) cannot sustain are necessary, rotating the joint to the dead end of movable range where a mechanical stopper sustains the large force generated by the main actuator; strong pinching is executed.

This strategy can be categorized as a general idea for a two-stage mechanism that compiles separated rough and fine control methods for enlarging the total dynamic range.

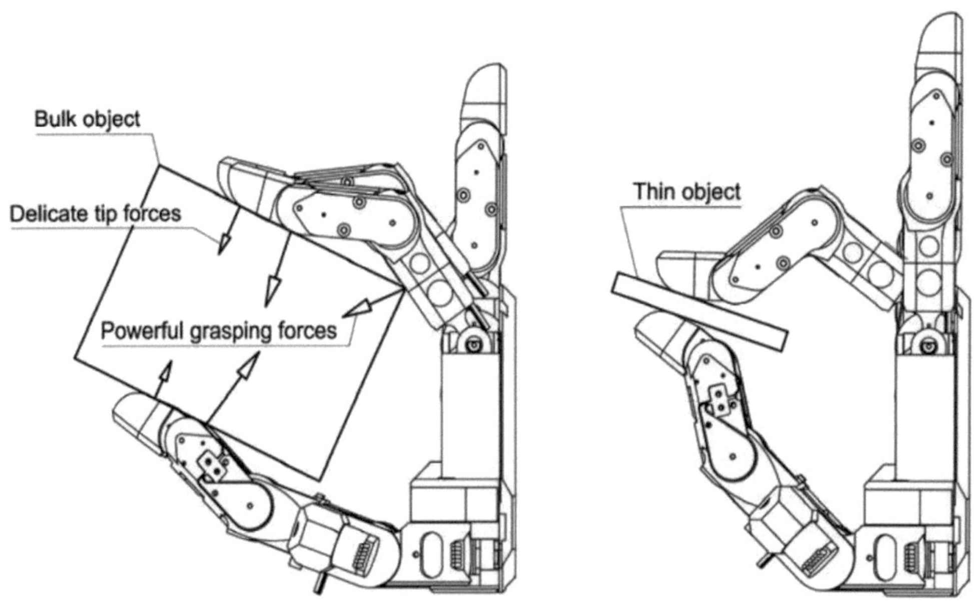

(a) Power grasping (b) Delicate pinching

Fig. 7. New strategy of object handling with independent fingertip motion.

2.7 Independent Terminal Joint Mechanism

The mechanical feature necessary for the new handling strategy above is a smooth compliance of fingertip to an object, in concrete terms a following motion to keep a soft contacting force. Although a similar feature could be produced with adding a spring on the way of power transmission to the fingertip, the spring occupies not negligible space in the small finger segment. Another way introducing a force sensor on the fingertip is also hard in acquiring appropriate sensor and tuning up a force feedback controller robust. So the most practical way to realize the compliance is exploiting a simple open-loop torque control method of the motor only by observing its current.

In this case, it requires a high efficient reducer with low frictional resistance and play of gears. I adopted a gear train that uses only high precision spur gears with expectation of their high efficiency. The motor length is restricted to the width of the finger segment, because the motor must be laid widthwise to be parallel to the joint axis. I searched on the web a possibly higher performance motor that fits into the limited length, and found a small DC motor (Sanyo Precision, model KS-2717) in production for small electric appliances. Since the output torque of the motor is very small 0.5Nmm in maximum, the reduction ratio is needed to be possibly higher. To realize high gear ratio avoiding deterioration of efficiency and inflation of volume, a two-stage thin gear train is built in against a sidewall with combining possibly largest and smallest gears for maximizing the gear ratio (Fig. 8); the gear module is 0.16 and the number of smallest gear's teeth is only nine. Consequently, the gear ratio is 3/250, and the maximum fingertip force is almost 2N.

A thin potentiometer is introduced to measure the angular displacement and put against another sidewall. By this alignment of the reducer and the potentiometer on the both sidewalls, the finger segment retains enough empty space inside.

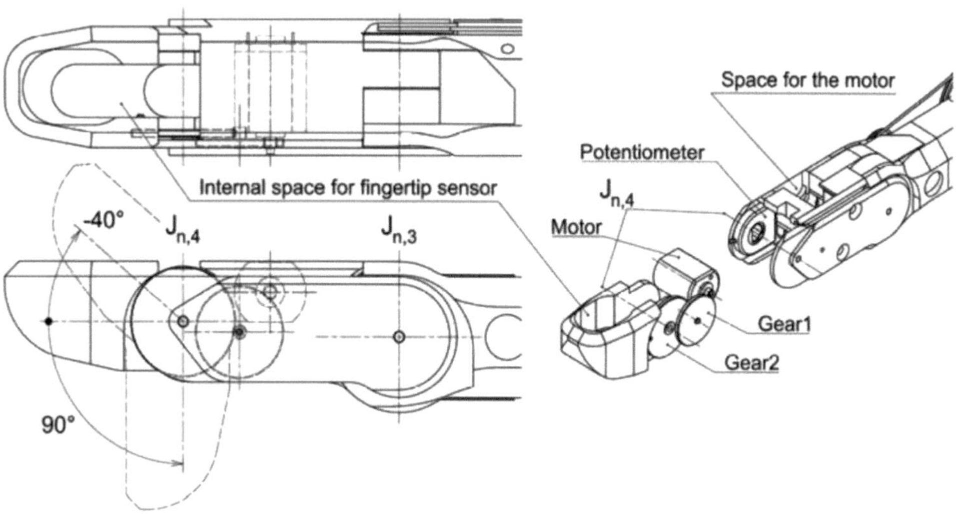

Fig. 8. Actuator assembly for the independent terminal joint.

To confirm the smooth compliance of a fingertip, a primary experiment was conducted. Concretely I examined active spring effect of the joint $J_{n,4}$ generated as follows. The joint $J_{n,4}$ was always position-controlled to keep its angular displacement $\theta_{n,4}$ at a certain target. The original target was set 0, that meant the terminal segment and middle segment were in line.

When the motor current of the joint $J_{n,4}$ exceeded a certain threshold, the target was shifted gradually to a temporary target by adding/subtracting a small variation proportional to the difference between the threshold and the present motor currents, so that the excess of motor current got reduced. When the motor current fell below the threshold and besides the temporary target was different from the original one, the temporary target was restored gradually to the original one by adding/subtracting a small variation proportional to the difference between the latest temporary target and the original one. In consequence of repeating this control in high frequency, the active spring effect was created stably.

In the experiment, the external pressure for examining the compliance was provided by the other finger part. Concretely, the finger part between the joints $J_{n,2}$ to $J_{n,4}$ was moved repeating a slight sine wave motion by a position-control of the main motor, and kept pressing the fingertip against a fixed object. Contacting force between the fingertip and the object was measured using a film-like force sensor (Nitta, FlexiForce) placed on the object; as a matter of course the observed force was not used in the motor control.

Fig. 9 shows the result of experiment with transitions of the contacting force and the angular displacement $\theta_{n,4}$. The threshold of motor current was set in two values as examples that provide the limits of contacting force as about 0.9N and 1.4N respectively. Effectiveness of the smooth compliance of fingertip is confirmed by analysing each transition as follows. When the contacting force just exceeded the limit, the joint $J_{n,4}$ started to rotated to release the excess, so that the contacting force was stuck at the limit. The joint $J_{n,4}$ started to return to the original position when the contacting force fell below the limit. These movements are just the characteristics of desirable compliance. And stable constancy of the maximum contacting force at the limit reveals high efficiency of force transmission through the reducer, and prove adequacy of the estimating way of contacting force from the motor current.

This smooth compliance cannot be produced by a simple spring mechanism because it cannot change the output force freely, then this practical function with minimum additional parts is one of the distinctive features on my robot hand.

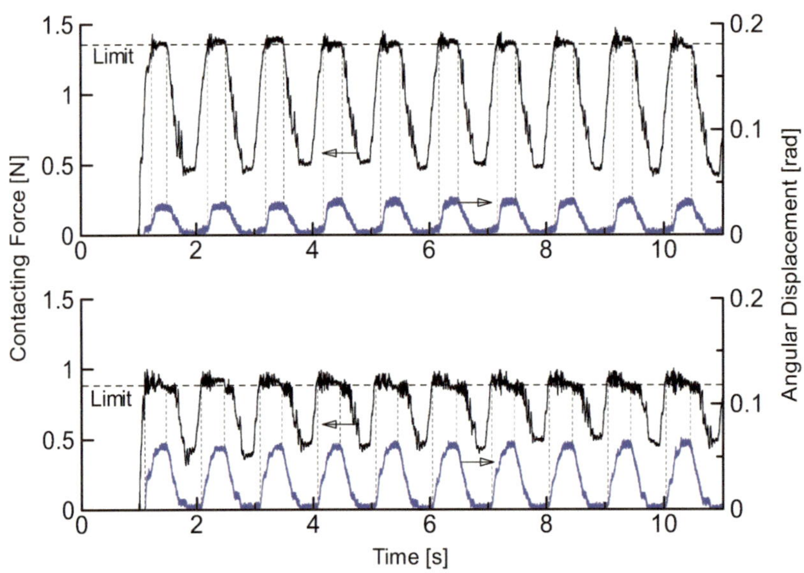

Fig.9. Experimental result of the contacting force and the angular displacement of a fingertip.

2.8 Thumb Twisting Mechanism

To fully maximize the capability of the smooth compliance generated by the independent terminal joint mechanism, the direction of fingertip force should be oriented rightly in the normal line of contacting surface on an object. This manner furthermore increases the stability of pinching in pressing the surface by entire flat pad on the fingertip. In order to realize this manner concretely, either or both of cooperating fingers in a pinching need to twist its fingertip along its longitudinal axis in a considerable angular displacement. Then I gave the thumb the distinctive joint $J_{1,3}$ for this twisting function.

Fig. 10(a) shows 5 segments in the thumb; each segment is identified with the number starting from the root, and described as symbol $S_{1,n}$. In the primitive robot hand with the essential 8DOF, the two segments $S_{1,2}$ and $S_{1,3}$ are integrated by fixing the twist angle. Avoiding deterioration of the original mechanical harmony, the actuator assembly for the joint $J_{1,3}$ was designed additionally. To minimize the joint $J_{1,3}$ volume, cylindrical case of the motor for the joint $J_{1,4}$ that is already embedded over segments $S_{1,2}$ and $S_{1,3}$ is exploited as the axis of $J_{1,3}$. Unlike the terminal joint mechanism, this twist function requires relatively large motor for moving large mass of the finger structure, so a bulge is added on the segment $S_{1,2}$ to mount a motor same as that for the global finger flexion (Fig. 10(b)). Although this bulge is a little unbecoming, I accept it as a reasonable way for preventing stretch of the thumb length. The movable range of twisting is 45degree.

By the way, the two joints $J_{1,1}$ and $J_{1,2}$ have the same actuator assembly for the global finger flexion shown in the Fig. 3, and each has adequate movable range of over 120degree and 60degree respectively.

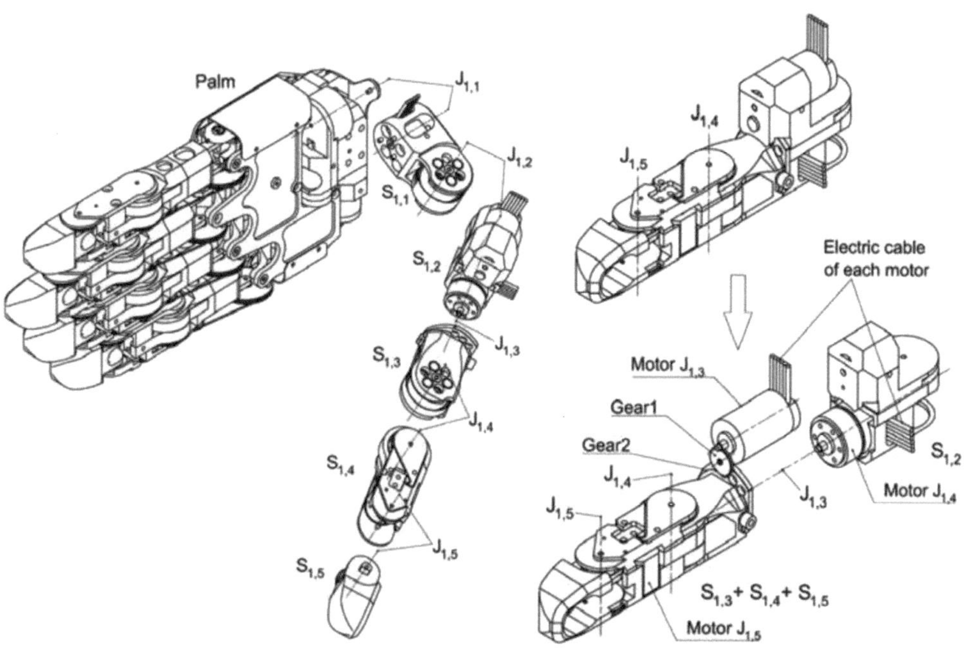

(a) Segments in the thumb (b) Actuator assembly

Fig. 10. Thumb mechanism with twisting function.

2.9 Independent Root Joint Mechanism

In the four fingers except the thumb, since the both joints $J_{n,2}$ and $J_{n,3}$ need no little power in the global finger flexion, the idea of interlocking these two joints and actuating them by one relatively large motor has adequate rationality, as far as the finger has no more capacity to accept two motors for actuating them independently. However, in some cases, the independent motion of each joint is required to realize some slight motion like adjusting the contacting place of a fingertip on an object. In order to demonstrate my technical capability to realize such complex requirement additionally, an actuator assembly was introduced at the joint $J_{2,2}$ particularly.

As a matter of course there is no capacity to accept a large motor, the additional motor is selected as the same small one driving the terminal joint. As the global finger flexion should be generated by the existing mechanism, the additional small actuator assembly should be designed to generate a differential motion as being overlapped on the global finger flexion. Well, the pulley on the joint $J_{2,2}$ is existing as a basement of the global finger flexion and its shape is round and coaxial to the axis of joint $J_{2,2}$, so it is convenient for realizing the differential motion by rotating the pulley around the axis.

Fig. 11 shows the actuator assembly to rotate the pulley. To sustain the large torque around the joint $J_{2,2}$ for the global finger flexion, it needs possibly larger reduction ratio. Therefore a worm gear train, that generally has large gear ratio, is introduced, so that the entire reduction ratio gets 1/1000. Although a worm gear train has no back-drivability, it is also an advantage in this case because that gear train can support any large torque in case of necessity. The movable range of the pulley is +15 to -15degree that makes useful adjusting motion at the fingertip in 10mm order.

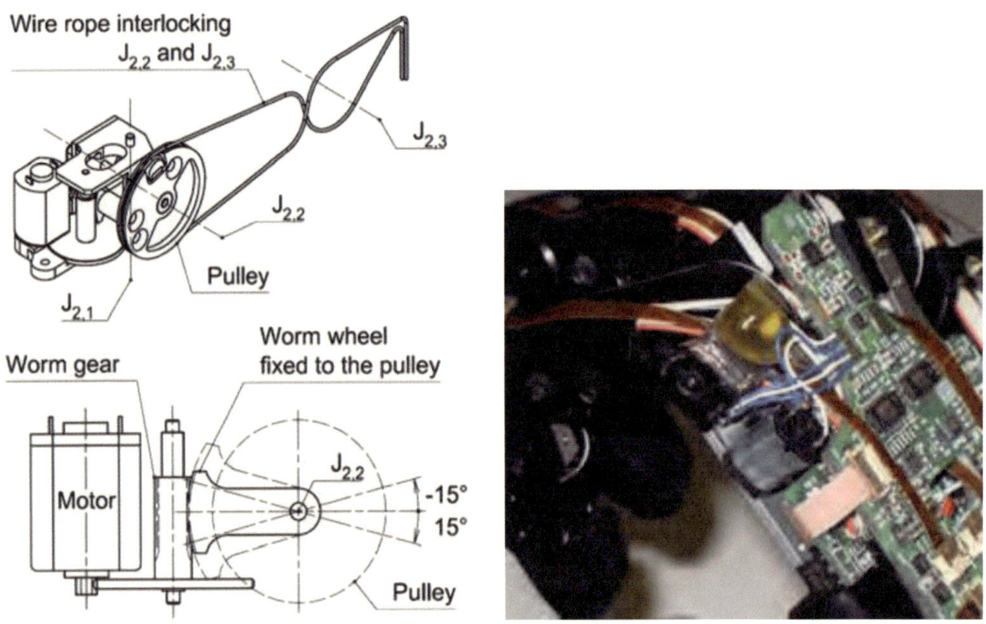

(a) Worm gear mechanism to drive the pulley (b) Actual embedded situation

Fig. 11. Differential mechanism for the independent root joint motion.

2.10 Smart Wiring for Bypassing Reducer

The quality of a robot system is evaluated from many kinds of dimension including neatness of the electric wiring, since its weight and volume can bring recognizable deterioration in the performance of high-speed motion and indisputably deteriorate the appearance. The lack of space for containing the wiring is the most common cause of this problem because expelling the wiring outside makes its weight and volume to increase. In my robot hand, as mentioned in the section 2.4, the discussion about the designing root joint structure of each finger was started by consideration of this problem. And more problem is outstanding around the joint filled with the large reducer of ratio 1/350 meaning $J_{1,1}$, $J_{1,2}$, $J_{1,4}$, $J_{2,3}$, $J_{3,3}$, $J_{4,3}$ and $J_{5,3}$. Recognizing the significance of this problem, a unique and practical design of wiring is introduced.

The role of the wiring is electric connection between the motor and sensor for the terminal joint and the main PCB in the palm, and a thin flexible PCB with 3.5mm width makes it. When the wiring is led as going around the reducer's circular outline, the change of shortest path length due to the finger flexion is remarkable, and then the method to retract and extract the corresponding length of wiring becomes the practical problem. My robot hand, fortunately, has enough margin space in the finger segments, and it can be formed an empty space where the wiring can adapt to the change of path length with changing the curving line by itself as shown in Fig. 12.

By the way, this wiring style cannot be adopted on the two thumb root joints $J_{1,1}$ and $J_{1,2}$ because of lack of the internal space, and then the wirings through these joints are forced to go outside in a wide circle unbecomingly. This problem will be solved in the next development step waiting for an investment opportunity.

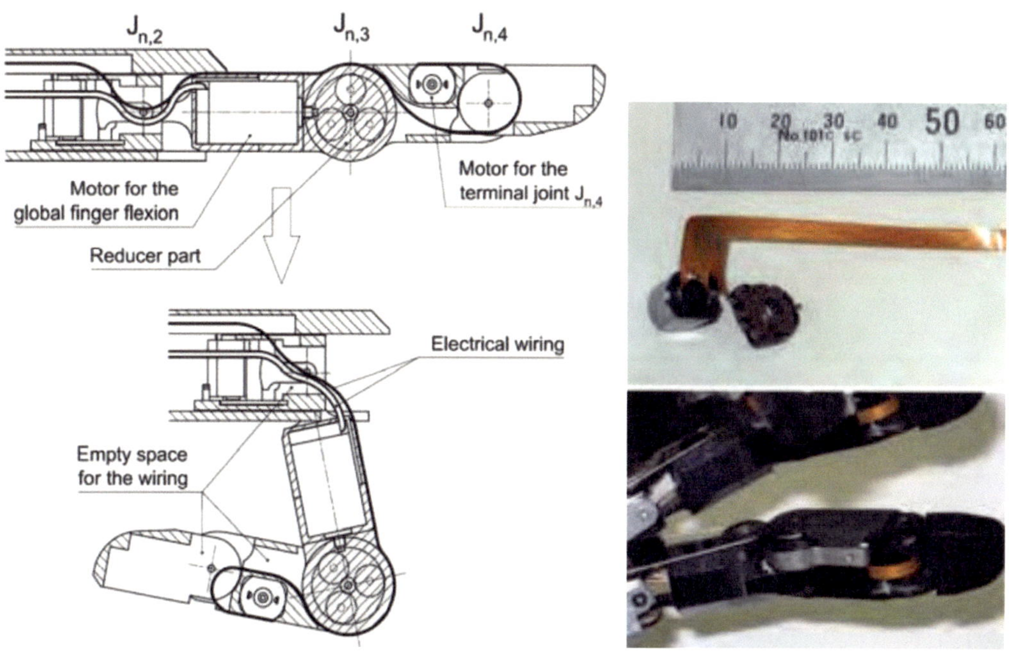

(a) Change of wiring path due to the finger flexion (b) Flexible PCB

Fig. 12. Design of the wiring around the joint that contains the large reducer.

2.11 Overall view of the Humanoid Robot Hand

As a conclusion of all previous considerations the latest model of my robot hand is built up as shown in Fig. 13; it has 15DOF as defined on the Table 2(b) while it satisfies the basic design conditions on the Table 1. The total mass including the internal electric equipment except the long cable connecting outside controllers is just 500g. The connections to outside systems are only $\phi 2.4$ signal cable and $\phi 4.5$ power cable. Some dimensions of details like the length of each finger segment are referred to my hand.

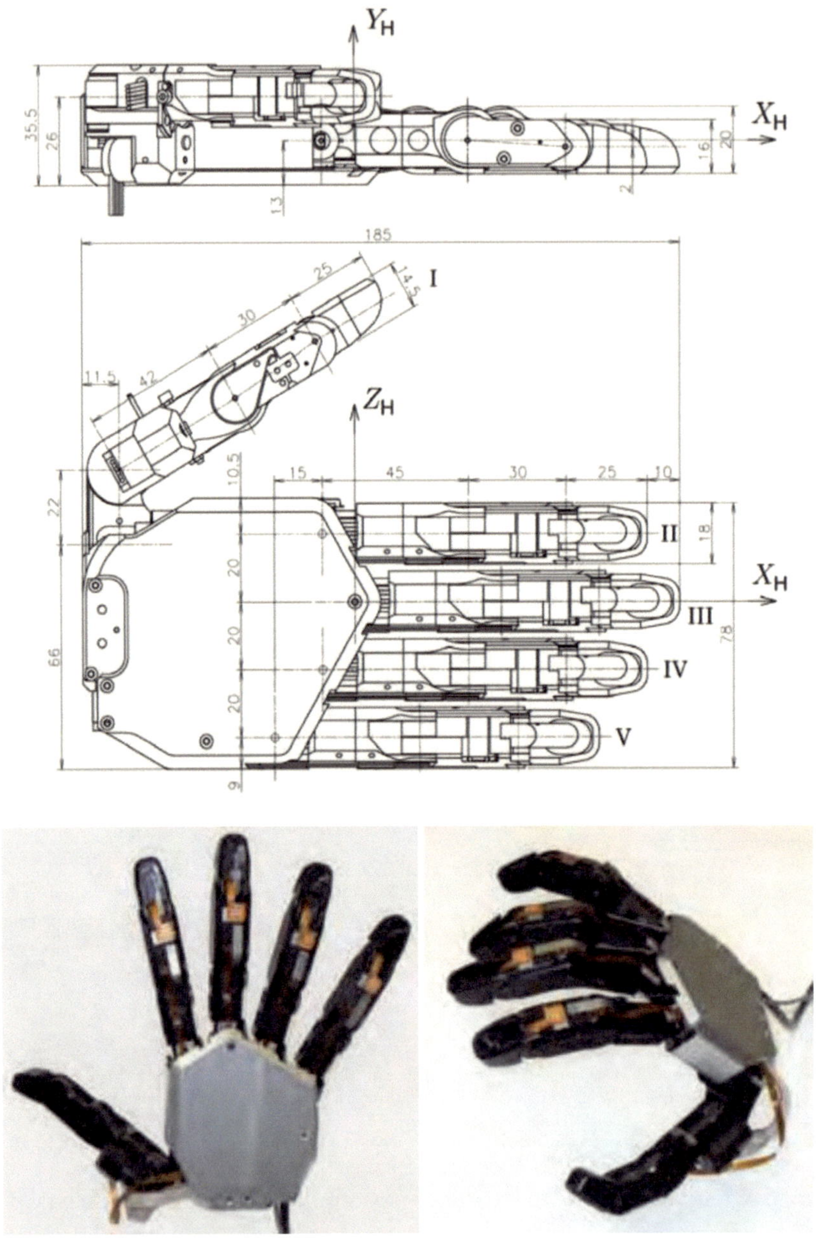

Fig. 13 Overall profile of the latest model.

To confirm dexterity of the robot hand, some experiments of representative and practical handling motions were conducted; this paper displays two handling types: pinching a business card and holding a pen (Fig. 14). The key evaluation items in these experiments were the two distinctive functions: the smooth compliance on a fingertip and the twisting of the thumb. All the fingertip forces were generated by the simple open-loop torque control method explained in the section 2.7 without force sensors.
By the way, the smart wiring style explained in the section 2.10 is installed only to the latest model, and the robot hand used in the experiments did not have it unfortunately.

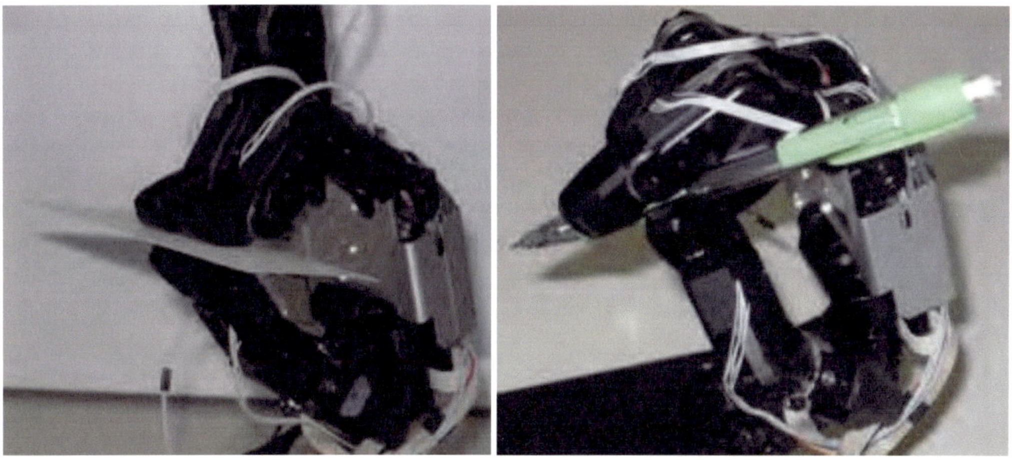

(a) Pinching a business card (b) Holding a pen

Fig. 14 The representative and practical handling motions.

In the experiment of pinching a business card, the robot hand performed switching several times two couples of pinching fingers: the thumb and the index finder/the thumb and the middle finger (Fig. 15). In the junction phase when all the three fingers contacted on the card, the thumb slid its fingertip under the card from a position opposing a fingertip to another position opposing another fingertip. In the experiment of holding a pen, the robot hand moved the pen nib up and down and sled the thumb fingertip along the side of the pen (Fig. 16). In both experiments, the objects: card and pen were held stably, and these achievements prove the contacting force appropriate in both strength and direction could be generated at each fingertip.

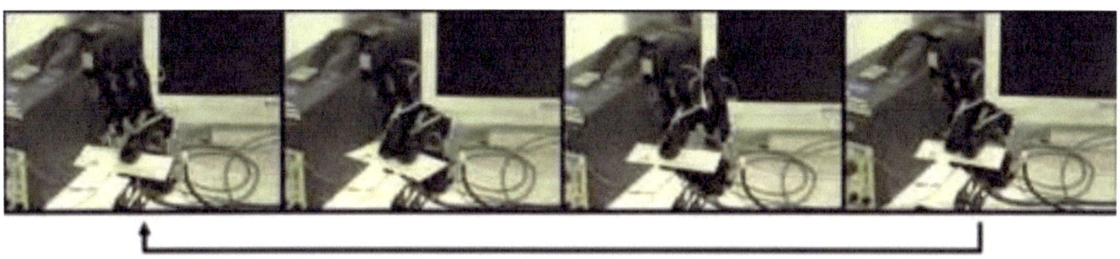

Fig. 15 Cyclical steps in the experiment of pinching a business card.

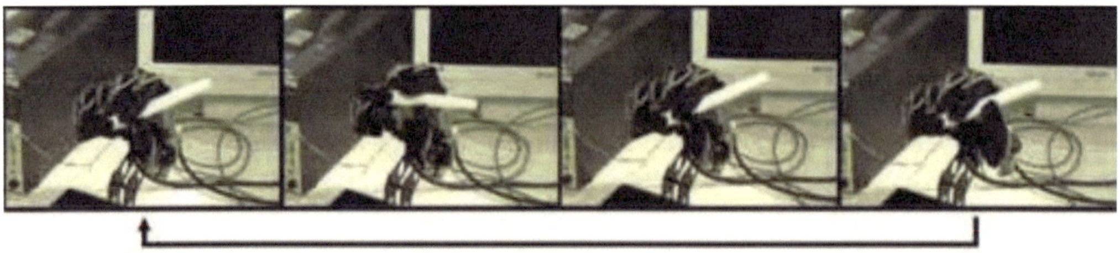

Fig. 16 Cyclical steps in the experiment of holding a pen.

At the SIGGRAPH 2006, I got an opportunity to join into a participating party of the "Hoshino K. laboratory in the university of Tsukuba" which introduced my humanoid robot hand for the first time. The robot hand was demonstrated on a humanoid robot arm that is actuated by pneumatic power, and has 7DOF wide movable range, slender structure and dimensions like an endoskeleton of a human arm (Fig. 17). While its power is low and the practical payload at the wrist joint is about 1kg, it could move the robot hand smoothly.
The conclusive advantage of the robot hand is that many complex functions are condensed in the humanlike size, weight and appearance, and realize the sophisticated dexterity. As the robot hand has rich suitability for delicate robot arms, after more sophistication, it will be developed to a good prosthetic hand in the near future.

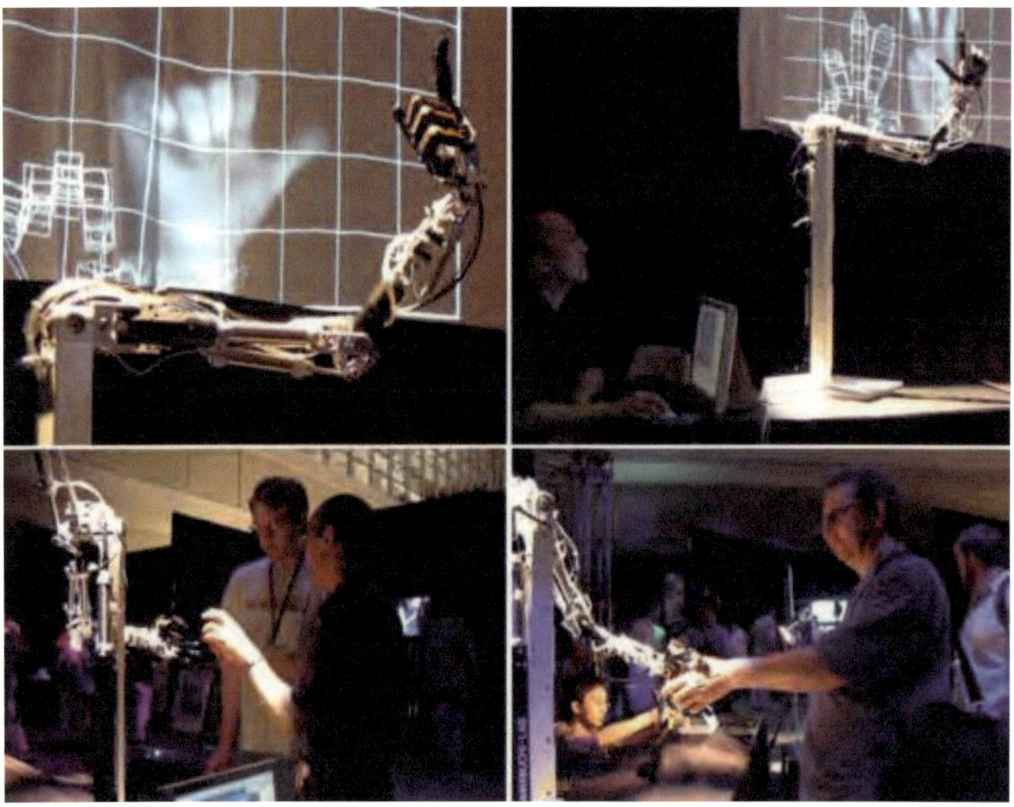

Fig. 17 Demonstration in the international exhibition SIGGRAPH 2006 in Boston.

3. Master Hand in Exoskeleton Style

3.1 Introduction of Circuitous Joint

As a dream-inspiring usage, the dexterous humanoid robot hand will be employed into a super "magic hand" with which an operator can manipulate objects freely from far away and get feedback of handling force and tactile sensations. Such intuitive master-slave control method of a humanoid robot with feedback of multi-modal perceptions is widely known as the Telexistence/Telepresence, however, developments of adequate master controllers for them have been rare in comparison with slave humanoid robots. I guess one of major reasons is a difficult restriction in mechanical design that any mechanism cannot interfere operator's body. To solve this problem an idea of exoskeleton is brought up by association of a suit of armour that can follow wide movable range of human body with covering it.

The most popular and practical master hand in exoskeleton style is the CyberGrasp, and most conventional master hands in exoskeleton style have the similar structure to it. They are designed to be lighter and slenderer with less material, so they have no core structure and cannot sustain their form as a hand without parasitism on operator's hand. This means they gives some constriction feeling to the operator and the slight force sensation in the feedback is masked. Then I have tried to design an ideal exoskeleton that fulfils every of lightness, slenderness and self-sustainability in its form.

In designing such exoskeleton, the main theme is focused on joint mechanisms. The most practical joint is a revolute one that consists of an axis and bearings, and general ways to place it corresponding to an operator's joint are in parallel on backside or in coaxial beside. However, the former tends to deteriorate the movable range of operator's joint (Fig. 18(a)) and the latter cannot find an existing space between operator's fingers. Therefore I propose a novel joint mechanism named "circuitous joint" that has a virtual axis coincided with the axis of operator's skeleton while the all mechanism exists on backside of operator's finger. Technically this virtual axis is the instantaneous center of relative rotation of two segments. Fig. 18(b) shows the principle of the circuitous joint that realizes the virtual axis by stretching displacement s of two segments in proportion to the joint angular displacement θ.

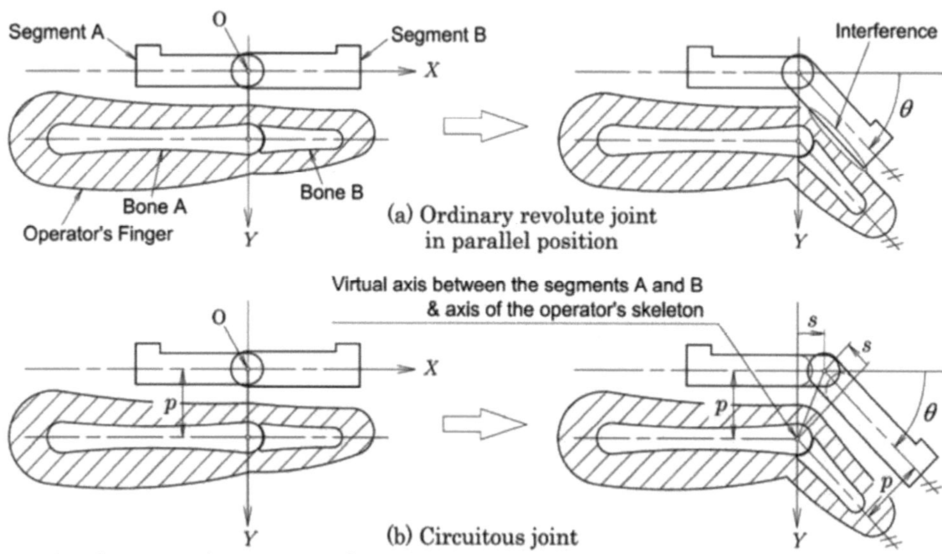

Fig. 18 Behaviour of two types of revolute joint in following operator's finger.

3.2 Fundamental Mechanism of the Circuitous Joint

In order to realize the principle of the circuitous joint mentioned above, rack and gearwheel mechanism was adopted in consideration of high rigidity of structure, certainty of motion, and facility of manufacturing. Fig. 19 shows the fundamental mechanism prepared for a principle study. A gearwheel is rotated on a rack by relative rotation of two segments, and shifting of its axis provides stretching of a segment that has the rack (Fig. 20). Since the two segments should make same stretching displacement together, two sets of the mechanism are combined in opposite direction. The gearwheel is formed to be sector gear by removing unnecessary part. We may note, in passing, this mechanism is an "over-constrained" mechanism, so it can keep its behaviour even without the actual axis.

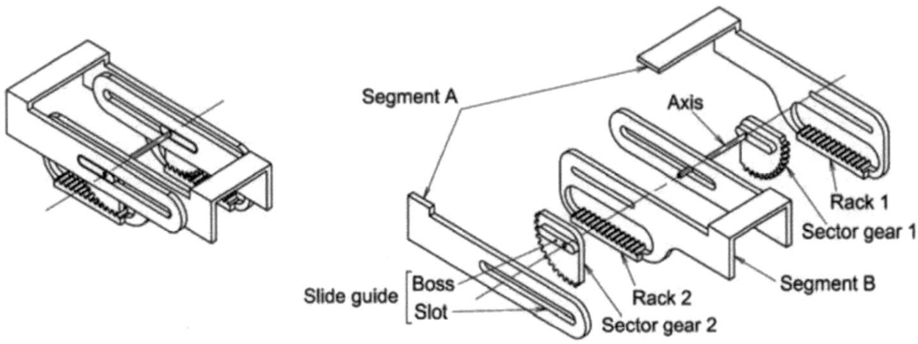

Fig. 19 The fundamental mechanism as a unit of the circuitous joint.

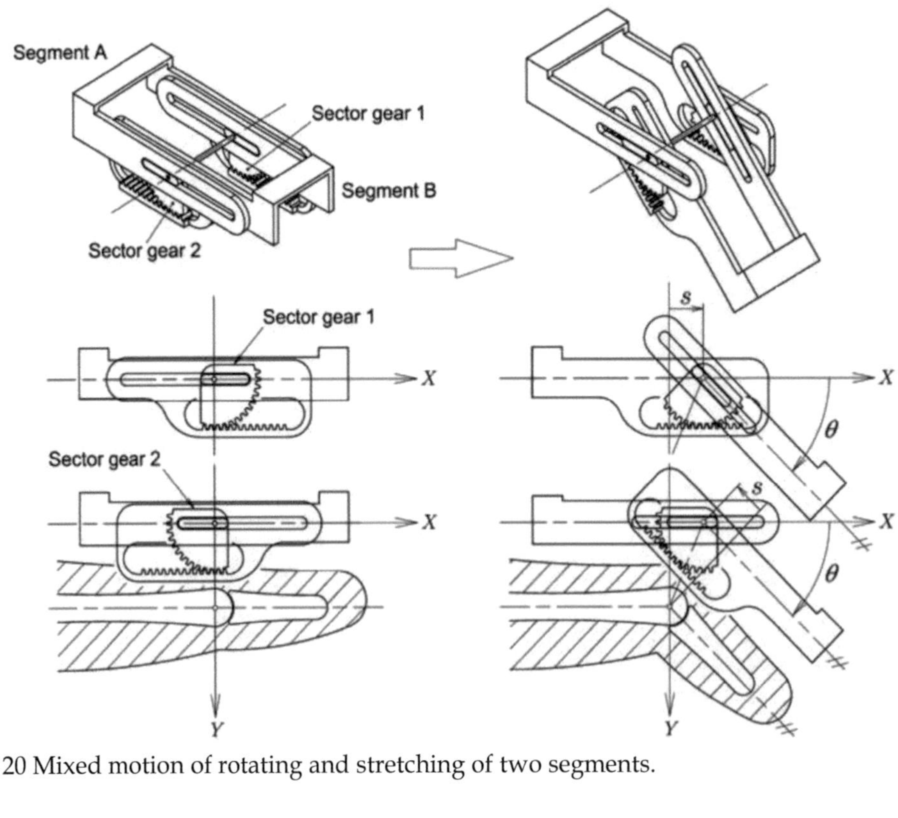

Fig. 20 Mixed motion of rotating and stretching of two segments.

3.3 Kinematical Design of the Optimal Circuitous Joint

To make the virtual axis coincide exactly to the axis of operator's skeleton, the relationship between the angular displacement θ and the stretching displacement s must be non-linear. This means the rectilinear rack and the circular gearwheels should not be adopted, however, they can get practical use with optimal specifications calculated as follows.

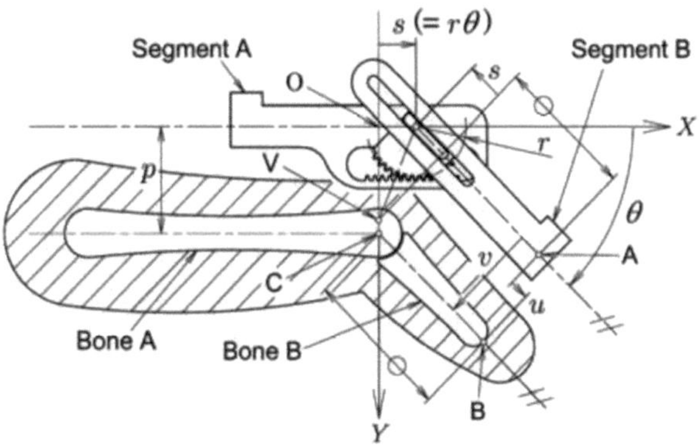

Fig. 21 Kinematical symbols in the circuitous joint.

Fig. 21 shows definition of kinematical symbols of parts and parameters; for example, point V is the virtual axis. The specifications that provide the shape of rack and sector gear are only the pitch circle radius r of the sector gear and the standard offset p between the center-lines of the Segment A and the Bone A. Since the standard offset p is decided 10mm due to convenience of practical design of mechanism, only the radius r is an object to be optimised. The point V moves on the Y-axis by change of θ and its behaviour is divided into three types according to the size of r (Fig. 22). Considering its nearest trajectory to the point C, the preferable range of r is presumed as $0.5p \leq r \leq (2/\pi)p$.

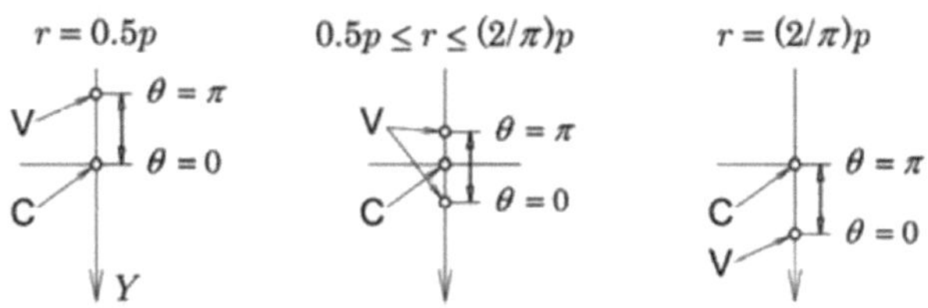

Fig. 22 Motion of the virtual axis V on the Y-axis by change of θ.

The evaluation item for the optimisation was set a deviation d defined by next formula that means deformation of kinematical relationship between two datum points A and B as shown in the Fig. 21, and the optimal radius r should minimise it.

$$d = \sqrt{u^2 + (p - v)^2} \quad \{\text{where } u = -r\theta\cos\theta + p\sin\theta - r\theta,\ v = p\cos\theta + r\theta\sin\theta\} \qquad (1)$$

Fig. 23 shows curves of the deviation d vs. θ in several settings of the radius r. The radius r is set within the presumed range. To generalise the optimisation each parameter is dealt as dimensionless number by dividing with the offset p. Screening many curves and seeking a curve which peak of d during a movable range of θ is minimum among them, the optimal r

is found as the value that makes the sought curve. For example, when the movable range is $0 \leq \theta \leq \pi/2$ the optimal radius r is $0.593p$ and the peak deviation d is $0.095p$, and when the movable range is $0 \leq \theta \leq \pi/3$ the optimal radius r is $0.537p$ and the peak deviation d is $0.029p$. As the offset p is set 10mm, the peak of d is below acceptable 1mm; therefore, the mechanism with rectilinear rack and circular gearwheels has practicability enough.

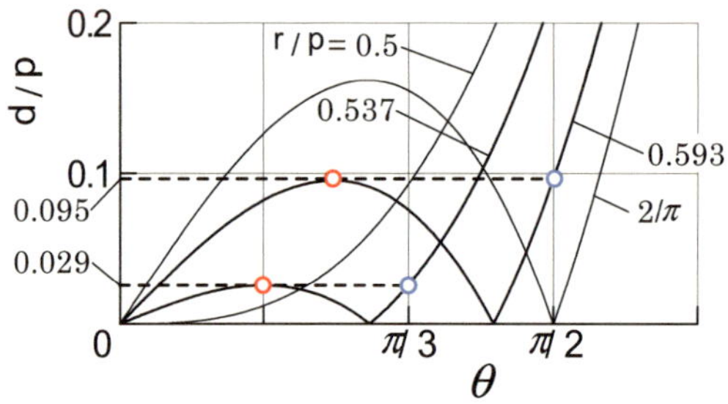

Fig. 23 Variation of curves of the deviation d.

3.4 Driving Method of the Circuitous Joint

To design the joint mechanism light and slender, a method to drive it from away via a wire rope is introduced. The wire rope is set along two segments veering by a pulley on the sector gear's axis, and one end is fixed on a segment and another end is retracted/extracted by a winding drum set at a stationary root (Fig. 24(a)). Since the wire rope can generate only pulling force that rotates the joint in straightening direction, a spring is added to generate pushing force that rotates it in bending direction (same (b)). This driving method has further conveniences to be applied to a tandem connection model (same (c)). A wire rope to a distal joint from the root can be extended easily through other joints. Its tensile force shares accessorily a part of driving force of other joints they are nearer to the root and need stronger driving force. Moreover, a coupled-driving method of plural joints can be realized only by winding their wire ropes together with one drum. The rate of each rotation can be assigned separately by independent radii on the drum.

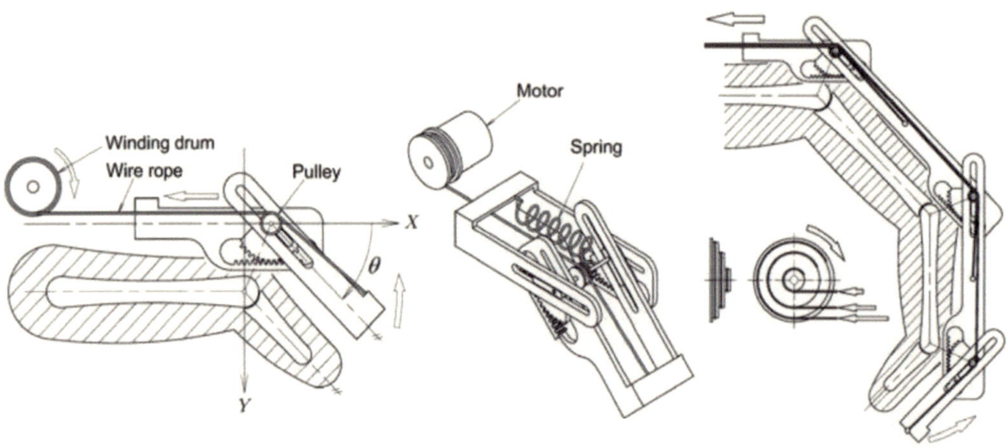

(a) Path of the wire rope (b) Pushing spring (c) Tandem connection

Fig.24 Driving method of the circuitous joint.

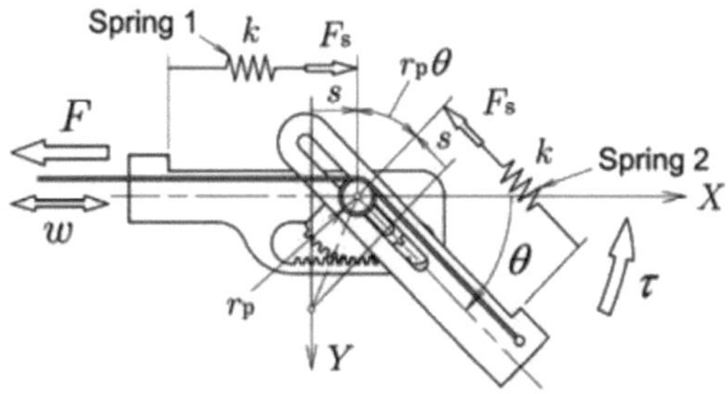

r_p : Radius of the pulley (constant)
k : Spring constant of the compression spring (constant)
Fs : Spring force generated by the spring (intermediate variable)
Fs' : Spring force generated by the spring when $\theta = 0$ (constant)
w : Retracting/extracting displacement of the wire rope (input variable)
F : Pulling force of the wire rope (input variable)
θ : Joint angular displacement (output variable)
τ : Joint torque (output variable)

Fig. 25 Statical symbols in the circuitous joint.

The definition of statical symbols is shown in Fig. 25, and the formulas for inverse statics calculating the input (manipulated) variables: w and F, from the output (controlled) variables: θ and τ are derived as follows.

$$w = (2r + r_P)\theta \tag{2}$$

$$F = \frac{1}{2r + r_P}\tau - \frac{2k \cdot r^2}{2(2r + r_P)}\theta + \frac{2Fs' \cdot r}{2r + r_P} \tag{3}$$

As these formulas show simple and linear relationship between the input and output valuables, this driving method promises further advantage that the algorithm of controlling both position and force is fairly simple. When the spring effect is negligible, as the second and third terms on the right side of formula (3) are eliminated, we would be able to control the output torque τ by using only the motor torque as the controlled variable.

3.5 Master Finger Mechanism (MAF)

Fig. 26 shows the practical master finger mechanism (MAF hereafter) corresponding to a middle finger of my hand and my humanoid robot hand, and proves the mechanism can follow them in wide movable range from opening to clenching. MAF is constructed with three discrete joint units, so that they are connected adapting to various pitch of operator's finger joints (Fig. 27). To make MAF narrow and short enough, each unit is designed possibly thin and aligned with partly overlapping. In this instance, all joints are coupled-driven by one relatively large motor (Faulhaber, model 1724SR).
As shown in Fig. 28, the actual rack is placed in opposite side viewed from the axis in comparison with the previous illustrations. The reason is to dissolve the interference between the mechanism and operator's finger that has came up in the previous arrangements. Inverse gear is added to correct the stretching direction of each segment and carried on a slider to keep the position at midpoint of the rack and the sector gear.

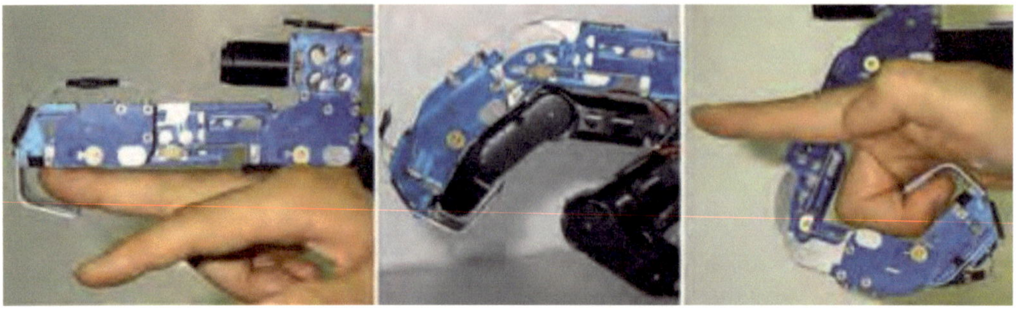

Fig. 26 Master finger mechanism (MAF) following various finger flexions.

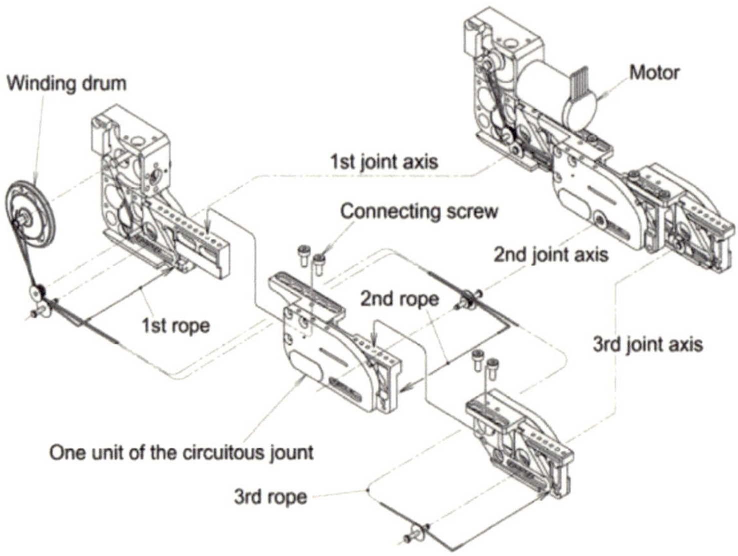

Fig. 27 Adjustable tandem connection of three joint units.

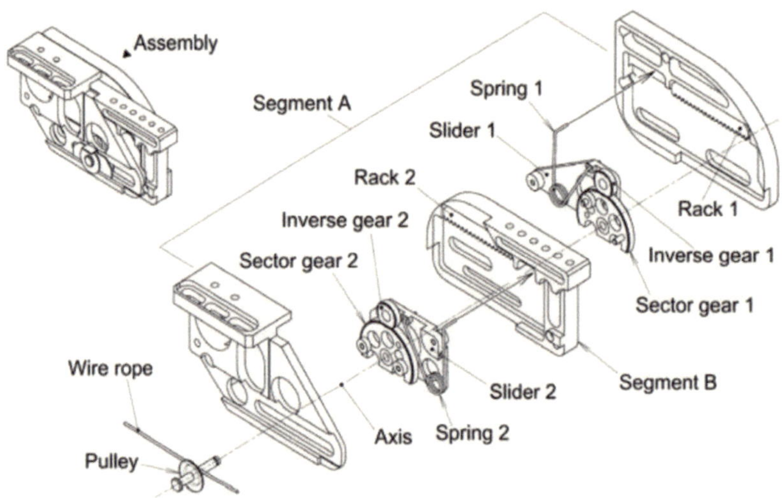

Fig. 28 Internal mechanism of the joint unit.

3.6 Master-Slave Control with Encounter-Type Force Feedback

As an ideal scheme of force display to the operator, the "encounter-type" has been proposed (McNeely, 1993, Tachi et al., 1994); that means a small object held up by a robot arm is approached and pressed to a part of operator's body where tactile sensation is necessary as occasion demands. Its chief advantages are making the operator to discriminate clearly the two phases "non-contact" and "contact", and free from constriction feeling during the non-contact phase. As it is suitable for the feature of my desired master hand, MAF introduced a function of non-contact following to the operator's finger.

Since the present MAF has only 1DOF, the target motion of operator's finger is reduced to same 1DOF, and a gap between both fingertips of MAF and the operator is set as the controlled variable during the non-contact phase. Concretely, a sensor at fingertip of MAF measures the gap, and MAF is position-controlled to keep the gap at the desired value 2mm. Fig. 29 shows the fingertip assembly that contains a micro optical displacement sensor (Sanyo Electric, SPI-315-34), technically that detects motion of a swing reflector moved by the operator's nail in slight force, and the gap is presumed from the motion.

During the contact phase, on the other hand, MAF should generate a desired contacting force against the operator's fingertips at the contact tip of the fingertip assembly. So a film-like force sensor (Nitta, FlexiForce) on the contact tip measures the contacting force, and MAF is force-controlled by changing the motor torque of winding the rope in proportion to the difference between the measured and desired contacting forces.

An experimental master-slave system between MAF and a slave humanoid robot finger (SLF hereafter) was constructed as follows. SLF is always position-controlled to realize the same motion of MAF. The two phases of contact/on-contact on controlling MAF are switched according to detecting existence/non-existence of the contacting force on SLF. A film-like force sensor on the surface of SLF's fingertip measures the contacting force, and the desired contacting force that MAF should generate is given as equal to that of SLF.

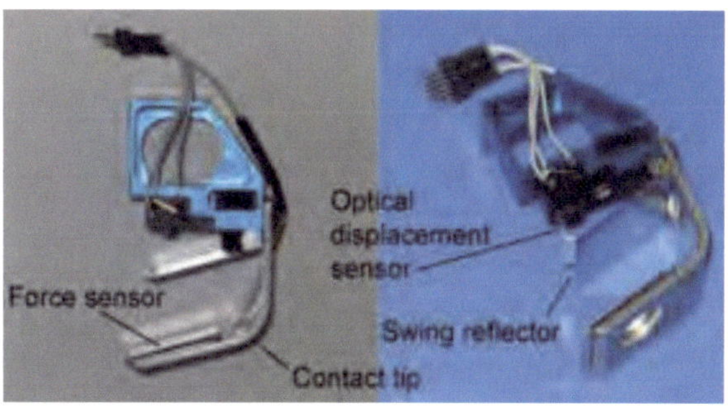

Fig. 29 Fingertip assembly for the master finger mechanism (MAF).

In order to confirm practicability of the master-slave system, an experiment was conducted. Fig. 30 shows the coupled motion of MAF and SLF in the non-contact phase; MAF was following the operator's finger with keeping a small gap at the fingertips. MAF and SLF could follow the operator's finger exactly as high as a less drastic speed. Since MAF had only 1DOF, SLF was prepared as the 1DOF mechanism interlocked all three joints. Moreover, the operator should also make his/her finger motion interlocking the three joints roughly similar to the behaviour of MAF. Though, I could forget an uncomfortable feeling by the fixed behaviour after familiarization, and enjoyed this experience.

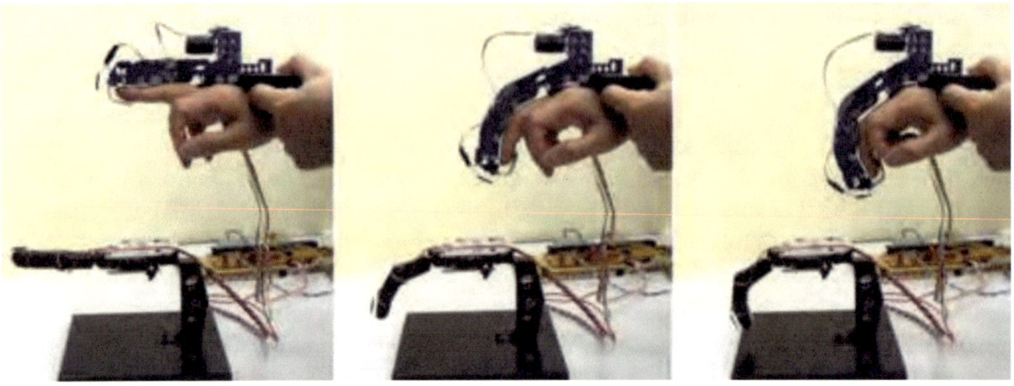

Fig. 30 Circumstance of the experimental master-slave control.

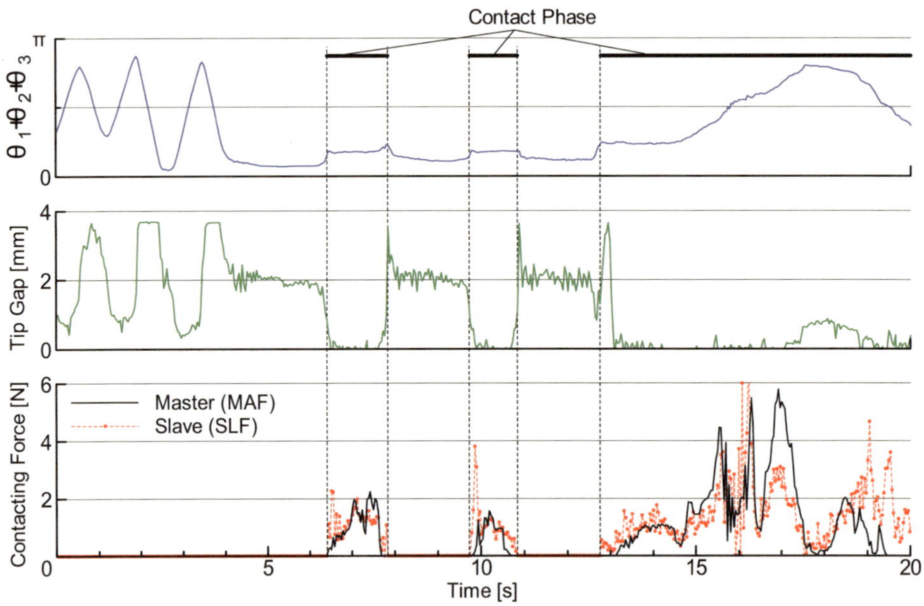

Fig. 31 Experimental result of transferring the contacting force.

Fig. 31 shows an experimental result; $\theta_1+\theta_2+\theta_3$ means the sum of three joint angular displacements on MAF. The two vital features are shown: prompt switching of contact/non-contact phases, and transferring the contacting force from SLF to MAF. The contacting force at the fingertip of SLF was given by an assistant pushing on it; for example, the two contact phases at the time 7s and 10s were caused by assistant's tapping. While the algorithm switching the phases was a primitive bang-bang control, an oscillation iterating contact/non-contact did not occur. I guess the season: since the gap between the fingertips is kept small during the non-contact phase, the impact at the encounter that will lead the oscillation is not so serious, moreover the human fingertip has effective damping to absorb it.

As shown by the curves after the time was 13s, the operator's finger could be moved by the assistant's force; the master-slave (bilateral) control with force feedback was verified.

In conclusion of this experiment, MSF has enough performance as a principal part of the master hand for the Telexistence/Telepresence.

3.7 Overall view of the Master Hand

Since it comes to the end of width of this paper, I describe briefly the overall view of the master hand. By the way, the nomenclature of each joint is same as shown in the Fig. 1.

I gave four fingers to the master hand (Fig. 32); the little finger was omitted due to its little worth in general activities. The three finger mechanisms are same as shown in the Fig. 26, and the second and fourth finger have the abduction-adduction motion with active joints at $J_{2,1}$ and $J_{4,1}$. The each joint is position-controlled to follow lateral motion of the operator's finger detected at fingertip with similar sensor mechanism as shown in the Fig. 29; however, the additional sensor put beside the fingertip is omitted in the Fig. 32.

In the thumb mechanism, the distal three segments are constructed with two circuitous joints at $J_{1,4}$ and $J_{1.5}$. At the same time, elated ingenuity is exercised to design the joint mechanism corresponding to the carpo-metacarpal (CM) joint of operator; to make the two joint axes $J_{1,1}$ and $J_{1.2}$ intersected in an empty space for containing the CM joint, a slider mechanism is introduced where a motor-driven carriage runs on a sector rail in a wide circle. While the two joint axes $J_{1,3}$ and $J_{1.4}$ for the MP joint are not intersected, the order of each direction of joint axis and fingertip is identical to that of the Shadow hand (Fig. 2).

In the non-contact phase, the thumb mechanism is position-controlled to follow the operator's thumb opposing on both fingertips; each independent DOF has individual sensor similar to the previous one. As the mechanism does not touch the operator's thumb, slight deviation of the controlling is negligible. In the contact phase, only the joints $J_{1,4}$ and $J_{1.5}$ are switched its control mode to the force-control. More sophisticated control algorithm for this thumb mechanism is under study in the "Tachi S. laboratory of the university of Tokyo" where I started developing this master hand as a researcher in 2001.

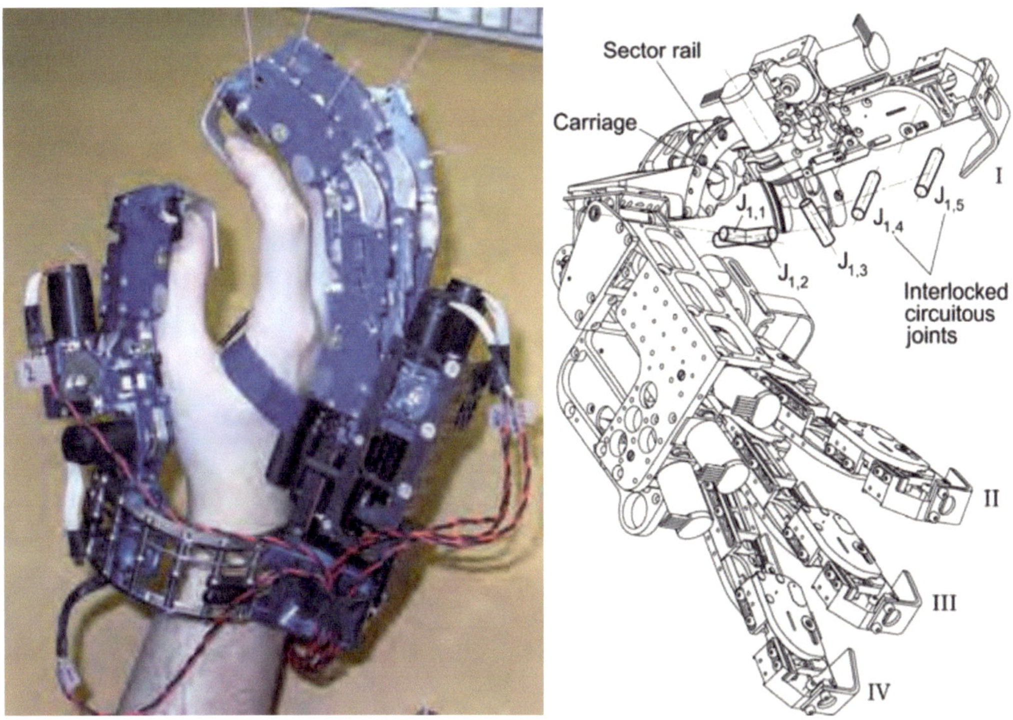

Fig. 32 Whole picture of the master hand.

4. Conclusion

To contribute on the evaluative process of searching the appropriate designing paradigms as a mechanical engineer, I bring up in this paper some of my ideas about the robot hand design concretely. While the designs of my robot hands may seem to be filled with eccentric, vagarious and serendipitous ideas for some people, I believe they are practical outcomes of flexible ingenuity in mechanical designing, so that they can take on pre-programmed but robust actuating roles for helping the programmable but limited actuators, and realize higher total balance in mechatronics. At the same time, for examining their practicability, reasonability and inevitability through the eyes of many persons, it will need to establish a standard definition and evaluation items in kinematics, dynamics, control algorithms and so on, that can subsume almost all humanoid robots. Concretely, a standard formats would be prepared to sort and identify any robot system by filling it. The Fig. 1 and 2 show my small trial of comprehensive comparison under a standard definition in the robot hand kinematics. And I hope the worldwide collaboration, so that it will promote developments of many sophisticated mechanical and electric elements that are easy to be used by many engineers like me who want any help to concentrate on his/her special fields.

5. Reference

CyberGrasp: Immersion Co.
http://www.immersion.com/3d/products/cyber_grasp.php

DLR Hand: German Aerospace Center
http://www.dlr.de/rm/en/desktopdefault.aspx/tabid-398/

Gifu Hand: Dainichi Co., Ltd.
http://www.kk-dainichi.co.jp/e/gifuhand.html

Harada Hand: Harada Electric Industry Inc.
http://www.h-e-i.co.jp/Products/e_m_g/ph_sh_2_004.html

Jacobsen, S.C. et al. (1984). The UTAH/M.I.T Dextrous Hand: Works in Progress, *Int. J. of Robotics Research,* Vol.3, No.4 (1984), pp.21-50

Lovchik, C.S. & Diftler, M.A.(1999). The Robonaut Hand: A Dexterous Robot Hand For Space, *Proc. of IEEE Int. Conf. on Robots & Automation,* Detroit, MI, May 1999

McNeely, W.A. (1993). Robotic Graphics: A New Approach to Force Feedback for Virtual Reality, *Proc. of IEEE Virtual Reality Annual Int. Symp.,* pp.336–341, Seattle, Sep 1993

NASA Hans: National Aeronautics and Space Administration
http://robonaut.jsc.nasa.gov/hands.htm

Shadow Hand: Shadow Robot Company Ltd.
http://www.shadowrobot.com/hand/

Tachi, S. et al. (1994). A Construction Method of Virtual Haptic Space, *Proc. of the 4th Int. Conf. on Artificial Reality and Telexistence (ICAT '94),* pp. 131-138, Tokyo, Jul 1994

Teleistence: Tachi, S. et al., Tele-existence (I): Design and evaluation of a visual display with sensation of presence, *Proc. of the RoManSy '84,* pp. 245-254, Udine, Italy, Jun 1984

Telepresence: Minsky, M., TELEPRESENCE, *OMNI,* pp. 44-52, Jun 1980

Weghel, M.V. et al. (2004). The ACT Hand : Design of the Skeletal Structure, *Proc. of IEEE Int. Conf. on Robots & Automation,* pp. 3375-3379, New Orleans, LA, Apr 2004

24

Assessment of the Impressions of Robot Bodily Expressions using Electroencephalogram Measurement of Brain Activity

A. Khiat[1], M. Toyota[2], Y. Matsumoto & T. Ogasawara
Nara Institute of Science and Technology - NAIST
JAPAN

1. Introduction

Recently, robotics research has focused on issues surrounding the interaction modalities with robots, how these robots should look like and how their behavior should adapt while interacting with humans. It is believed that in the near future robots will be more prevalent around us. Thus it is important to understand accurately our reactions and dispositions toward robots in different circumstances (Nomura et al., 2006). Moreover, the robot's correct production and perception of social cues is also important. Humans have developed advanced skills in interpreting the intentions and the bodily expressions of other human beings. If similar skills can be acquired by robots, it would allow them to generate behaviors that are familiar to us and thus increase their chances of being accepted as partners in our daily lives.

The expressiveness of a gesture is of great importance during an interaction process. We are often required to give special attention to these signs in order to keep track of the interaction. Humans have learned to adapt their behavior and to react to positive and negative bodily expressions (Bartenieff & Lewis, 1980). Although there has been remarkable work on the design issues of sociable robots (Breazeal, 2002) and affective autonomous machines (Norman et al., 2003), there has not been much work on investigating the real impact of robot bodily expressions on the human user in the context of human-robot interaction. Knowing the effect of a generated gesture, a robot can select more accurately the most appropriate action to take in a given situation. Besides, computer-animated characters have been used to evaluate human perception of the significance of gestures. However, animated characters and embodied ones should be treated differently since the latter are tangible entities (Shinozawa et al., 2005).

In this article we report a study on the relation between bodily expressions and their impacts on the observer. We also attempt to understand the effect that expressions have on the observer's brain activity. Its sensitivity to bodily expressions can be used during an interaction task since the brain is the source of every cognitive and emotional effort.

---

[1] Corresponding author: **Abdelaziz Khiat**, *Robotics laboratory, Graduate School of Information Science, Nara Institute of Science and Technology, Keihanna Science City, 630-0192 JAPAN.* Email: khiat@ieee.org
[2] Masataka Toyota is currently with Canon Corporation.

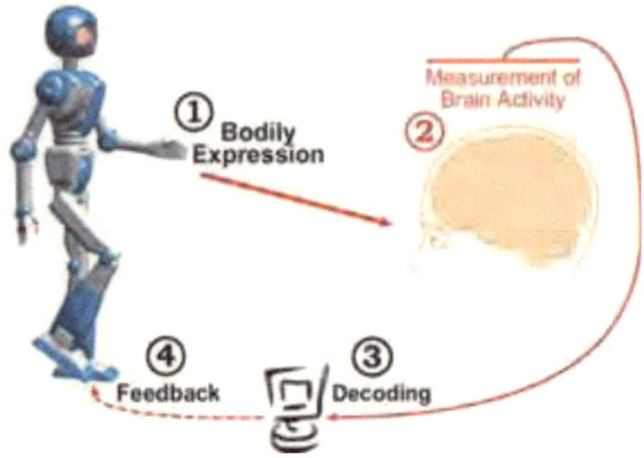

Fig. 1. Considered scenario for robot bodily expressions and its perceived impression.

In this work, we have conducted an experimental study where several users were asked to observe different robot bodily expressions while their brain activity was recorded. The results suggest the existence of a relation between the type of bodily expressions and the change in the level of low-alpha channel of brain activity. This result helped in the selection of features that were used to recognize the type of bodily expression an observer is watching at a certain time. The recognition rate was of about 80% for both cases of robot bodily expressions and of human bodily expressions. Potential applications include customized interface adaptation to the user, interface evaluation, or simple user monitoring.

2. Bodily expressions and their impressions

The considered scenario for this study is depicted in Fig. 1. First, we have a robot that is executing a series of movements. It transmits to the observer a meaningful expression which is called bodily expression ①. Second, we have a human observer that perceives the expression and interprets it using his/her a priori knowledge ②. Then, the observer gets an impression, which means that bodily expression affects him/her to a certain level, depending on its strength, his/her awareness or attention and his/her state of mind or mentality ③. It is important to emphasize the difference between how the observer perceives and interprets a bodily expression, and what impact this expression evokes in the observer. It is expected that the two are related, but there is no information about the nature of this relation or how it evolves and changes over time. One of the goals of this work is to clarify and explain certain aspects of this relation to open the possibility of generating an adaptive robot behavior based on this information.

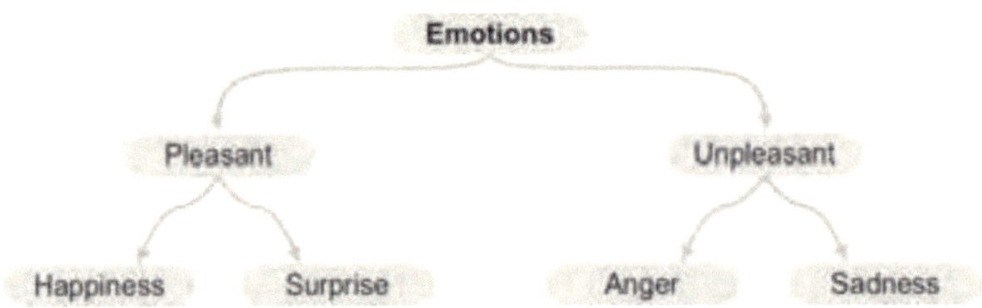

Fig. 2. The subset of Shaver's classification of emotions used in the categorization of Bodily Expressions.

2.1 Classification of bodily expressions

There is a need to classify bodily expressions generated by a robot in order to investigate their effects on the user. For this reason, salient differences among motions should be implemented. During an interaction process, humans go through different affective states, depending on several conditions such as degree of engagement, degree of awareness, and degree of interest among others. It is thus possible to classify every action taking place during an interaction process into the emotional effects that it would have on the observer. We adopted a simplified version of Whissel's wheel of activation-evaluation space described in (Whissel, 1989). We used the fact that we have two primary states for emotions: positive and negative ones, also known as pleasant and unpleasant emotions. The considered emotions are the following: happiness, surprise, sadness, and anger. In order to categorize these emotions we used a subset of Shaver's classification (see Figure 2), where happiness and surprise represent pleasant emotions while sadness and anger represent unpleasant emotions (Shaver et al., 1987). Bodily expressions were classified using one of the specified four emotions as pleasant or unpleasant.

2.2 Generation of robot bodily expressions

The humanoid robot ASKA (Ido et al., 2002) used in this study is shown in Figure 3. The body has a mobile platform and two arms and is based on the commercial robot TMSUK-4[3]. The head is a replica of the Infanoid robot (Kozima, 2002). This humanoid robot with its mobile platform has the advantage of being able to generate relatively fast motions compared to the currently available biped humanoid robots.

Since the pioneering work of (Johansson, 1973) on visual perception of biological motion, it has been known that humans can perceive a lot of information from body movements including the emotional state of the performer (Allison et al., 2000; Pollick et al., 2001). Recently, there is a growing interest in mathematically modeling emotion-based motion generation for real-world agents such as robots (Lim et al., 2004) and for virtual agents such as animated characters (Amaya et al., 1996). To be able to generate bodily expressions that reflect the selected emotions we rely on Laban features of movements (Bartenieff & Lewis, 1980). It has been shown by (Tanaka et al., 2001) that the qualitative Laban features of Effort and Shape correlate with the four basic emotions we have selected in section 2.1.

Based on the mathematical description of Laban features, shown in the Appendix, it is relatively easy to classify bodily expressions that reflect a certain emotion. Although there is no unique solution to this problem, the goal is to be able to generate a representative bodily expression for each one of the selected emotions.

The generated bodily expressions (BE) which reflect one of the basic emotions of happiness, surprise, sadness, anger or none are the following:

- **BE1:** The robot raises both arms and turns its body to the left, then to the right, twice. The goal is to show an expression of happiness.
- **BE2:** The robot raises its right hand and moves it in an arc toward the right side, then goes back to its initial position and lowers its right arm, the goal is to show an expression of no particular emotion.
- **BE3:** The robot raises both arms and its head, then moves backward for some distance, the goal is to show an expression of amazement or surprise.

---

[3] *TMSUK-4* is a trademark of tmsuk Co. Ltd, Kitakyushu.

- **BE4:** The robot lowers both arms and its head, then moves backward at low speed for some distance, the goal is to show an expression of sadness.
- **BE5:** The robot raises both arms gradually while advancing before stopping, then it lowers and raises its arms progressively for two cycles; the goal is to show an expression of happiness.
- **BE6:** The robot advances quickly, then goes back and raises its right arm while turning its head a bit to the right. It then lowers its arm and returns its head to the original position; the goal is to show an expression of anger.

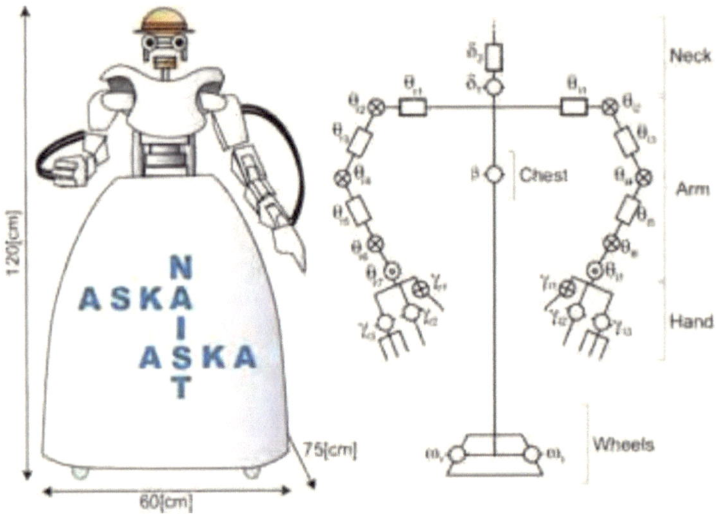

Fig. 3. Overview of the receptionist robot ASKA and its joints-link model.

The duration of each of these BEs was about 14[sec]. Their appropriateness and their expressiveness was tested experimentally using questionnaires (see section 3.1).

2.3 Assessment of impression and expressiveness of bodily expressions

There are mainly two types of methods to assess the effects of a particular action on a human. The classic self-reporting approach is widely used, while the assessment from measured physiological information is still an open research problem. The first type of methods gives subjective evaluation results; whereas the second type of methods is deemed to be more objective but suffers from inaccuracies. For our case, in order to assess expressiveness we adopted a self-reporting approach and asked the subjects to answer questionnaires. However, in order to assess impression the subjects answered questionnaires and their brain activity was also recorded.

Summarizing the subject's answers to questionnaires was used in order to assess expressiveness. Every subject had to select from: expression of happiness, expression of surprise, expression of sadness, expression of anger, or no meaningful expression. The subject also had to specify the degree of the expression in a scale of five: **1** for impertinent, **2** for slight, **3** for medium, **4** for strong and **5** for very strong. This selection of the degree of expression is a redundancy that was meant to confirm the subject's choice and assess the degree of confidence in his/her answer. These answers were then categorized into pleasant or unpleasant expressions using the subset of Shaver's classification shown in Figure 2.

As for impression assessment, spectral analysis method of electroencephalogram (EEG) data was used. A short EEG segment can be considered as a stationary process, which can be characterized by an autoregressive (AR) model. Let us denote $s(n)$ as a sample of EEG data of N points. We calculate $r_f(n)$ and $r_b(n)$, respectively the forward and backward prediction errors, as follows:

$$r_f(n) = \sum_{k=0}^{p} a(k) s(n+p-k) \tag{1}$$

$$r_b(n) = \sum_{k=0}^{p} a(k) s(n+k) \tag{2}$$

where $a(k)$ is the AR parameters and p is the order of the model. The order p is based on the "goodness of fit" criterion. We use the relative error variance (REV) criterion (Schlögl et al., 2000), defined as:

$$REV(p) = {MSE(p)} / {MSY} \tag{3}$$

$MSE(p)$ is the mean square error or variance of the error process of order p, and MSY is the variance of the total power of the signal sample. The optimal p is the one that minimizes $REV(p)$. In our case we take $p = 14$.

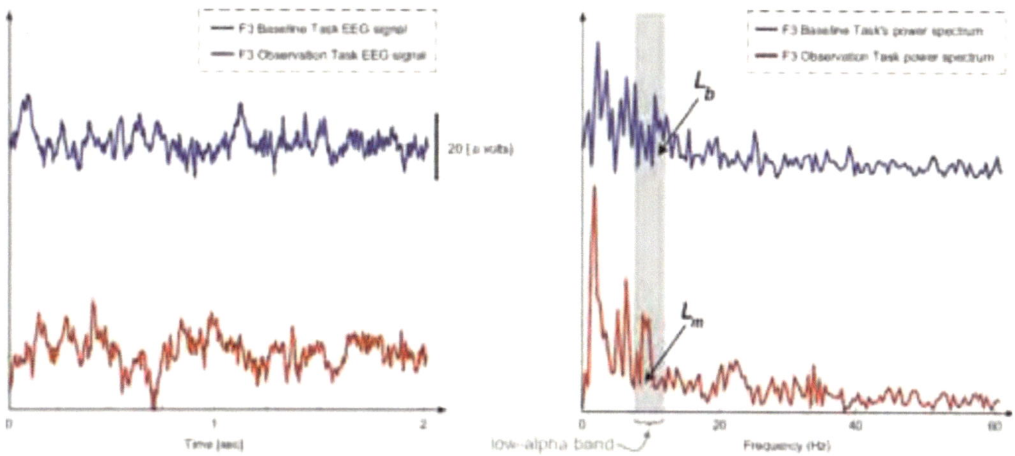

Fig. 4. An example illustrating the calculation of the power of low-alpha band for a 2[sec] data segment taken from electrode placement F3. The graph to the left shows the raw EEG signal for the baseline period and the observation period. The graph to the right shows the power spectra of the EEG signals, where low-alpha frequency band is highlighted.

We apply (1) and (2) to calculate an approximate estimation of the power spectrum $PS(f)$ of the signal s as follows:

$$PS(f) = \frac{V_p T}{\left| 1 + \sum_{k=0}^{p} a(k) e^{-j2\pi fkT} \right|^2} \tag{4}$$

$$V_p = \frac{1}{2} \sum_{n=0}^{N+p-1} \left([r_f(n)]^2 + [r_b(n)]^2 \right), \tag{5}$$

where V_p is the averaged sum of the forward and backward prediction error energies and T is the sampling period.

Research in cognitive neuroscience has shown that the power of low-alpha frequency band is the most reactive band to social cues such as movements (Allison et al., 2000; Cochin et al., 1998). We suppose that this frequency band reacts in a similar way to robot bodily expressions (Khiat et al., 2006). The next step in assessing the impression is to observe the amount of change in the power of low-alpha frequency band compared to the whole spectrum. The power L of a band between frequencies a and b is defined by:

$$L(a,b) = \frac{\int_a^b PS(f)df}{\int_0^\infty PS(f)df} \tag{6}$$

Using **(6)**, we calculate the power of low-alpha band frequency L_b for the data taken during the baseline period and L_m for the data taken during the period of the execution of a bodily expression. An example illustrating this calculation is shown in Fig. , where raw 2 seconds EEG signals collected during the baseline period and the observation period is shown to the left. The power spectrum of these signals is shown to the right, and the low-alpha frequency band is highlighted. A comparison between L_b and L_m would indicate the effect of a particular bodily expression on the user. This is used as the main evaluation criterion for impression.

3 Experimental study

3.1 Expressiveness of robot bodily expressions

The goal of this experiment is to evaluate the expressiveness of every generated robot bodily expression. Since this quality is highly subjective, the self-reporting approach is used.

Subjects. Seventeen (17) participants (two females and fifteen males aged between 20 and 50 years old) volunteered to take part in this experiment. They were either students or faculty members at the Graduate School of Information Science. They were all familiar with robots and had previous experiences of dealing with robots similar to the one used in the experiment.

Procedure. Every subject was shown a total of six bodily expressions, which were described in section 2.2. The execution of each of the bodily expressions by the humanoid robot ASKA lasted 14 seconds. After observing each robot bodily expression, enough time was given to the subject to answer two questions about the expressiveness of that robot bodily expression, and one more question about their impression after the observation. These answers were then summarized as explained in section 2.3 to assess their expressiveness.

BEs	*Pleasant*	*Unpleasant*	*Neither*
BE1	**100**%	0%	0%
BE2	6%	35%	**59**%
BE3	**94**%	6%	0%
BE4	0%	**94**%	0%
BE5	**65**%	12%	23%
BE6	0%	**82**%	18%

Table 1. Users' evaluations of the expression of each generated robot bodily expression (BE).

Results. Table 1 shows the results about bodily expressions after categorization into pleasant expressions, unpleasant expressions, or neither, clearly indicating the expressiveness as evaluated by the observers. The result about impressions is presented in Table 2 after categorizing the answers into pleased or unpleased.

These results demonstrate the existence of a strong correlation between the expressiveness of the robot bodily expressions as seen by the subjects and the target expression when these bodily expressions were generated (see section 2.2). BE1, which was created to express happiness, was classified as having a 100% pleasant expression. BE2, which was created to express a neutral emotion, was classified by 59% as neither pleasant nor unpleasant, and by 35% as unpleasant, suggesting that neutral bodily expressions can have a negative connotation. BE3, which was created to express surprise, was classified by 94% as a pleasant expression. BE4, which was generated to express sadness, was classified by 94% as being an unpleasant expression. Similarly, BE6 which was created to express anger was also classified by 82% as an unpleasant expression. The special case of BE5 was classified to a great extent as a pleasant expression by up to 65%. However, 23% said it did not express anything in particular and 12% claimed it was unpleasant.

BEs	*Pleased*	*Unpleased*	*Neither*
BE1	**65**%	35%	0%
BE2	30%	**70**%	0%
BE3	**68**%	32%	0%
BE4	19%	**81**%	0%
BE5	**100**%	0%	0%
BE6	47%	**53**%	0%

Table 2. Users' evaluations of their impressions after observing each robot bodily expression (BE).

The expressiveness of the generated BEs is confirmed to be in accordance with the target expressions for which they were created. BEs generated to express happiness and surprise expressions were classified as pleasant, and the BEs generated to express sadness and anger expressions were classified as unpleasant. Among the generated BEs we could choose one that is representative of each category in order to use it in the evaluation of its impressions on the observer.

3.2 Impressions of robot bodily expressions

The goal of this experiment is to evaluate the impression on the observer of the generated bodily expressions using a hybrid approach that combines the results of self-reporting and the analysis of brain activity.

Subjects. Seven (7) participants (one female and six males, 23~43 years old) volunteered to take part in this experiment. They were all students or faculty members at the Graduate School of Information Science, and only two of them had the experience of using electroencephalography to measure brain activity. Before starting the experiment each participant was fitted with electrodes and allowed to spend more than 20 minutes reading books of interest to familiarize and condition them to the electrodes' presence.

Procedure. During each session, 12 EEG channels (using sintered Ag/AgCl electrodes) were recorded by the 5200 Series DigitalAmp System[4]. The recording was performed from 10 placements, namely: Fp1, Fp2, F3, F4, T3, T4, P3, P4, O1, and O2 according to the international 10-20 standard (see Fig. 5). The placement Fz was used as the ground, and the signal from the left ear placement A1 was used as the reference signal. The contact impedance between all electrodes and the skull was kept below 5[kΩ]. The subjects were shown a total of six motions lasting 14 seconds each by the humanoid robot ASKA while their brain activity was recorded with 16-bit quantization at a sampling frequency of 200[Hz].

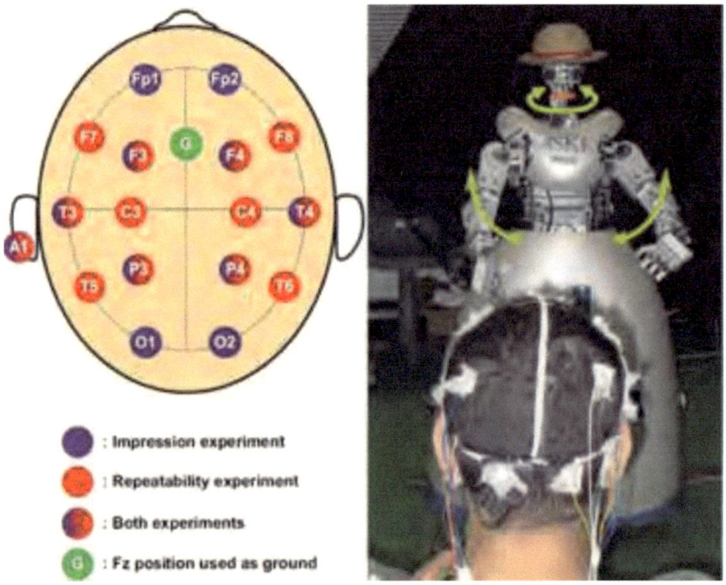

Fig. 5. The experimental setup where brain activity was measured according to the 10-20 international standard (Jasper, 1958).

The subjects were asked to relax as much as possible and think of nothing in particular when recording the baseline period, which lasted for 14[sec]. They were also told that they would be asked about the robot's movements and that they had to watch carefully when the robot was moving. This was important because we needed to make sure that the subjects attended to the task. After the observation of each bodily expression, the subjects described

[4] The 5200 Series DigitalAmp System is a trademark of NF Corporation, Yokohama.

their impression in their own words. Having no constraints to express themselves, the subjects gave more details about their impressions. These answers were used in categorizing the impressions into pleased or unpleased based on Shaver's original classification of emotions (Shaver et al., 1987).

Results. Table 2 shows the self-reporting result about the subjects' impressions after observing every robot bodily expression. There is a strong correlation between these results and the expression results, reported previously in section 3.1, with a coincidence level of 71%. For example, BE4 impression was considered to be unpleasant by up to 81% and its expressiveness was considered unpleasant by 94%. This is also the case for BE1 where its impression of being pleasant is 65%, and it expression of being pleasant is 100%. The same could be said for BE3, with a pleasant impression of 68% and a pleasant expression of 94%.
The case of BE6 is different from the previous ones. While its expression was considered unpleasant by 82%, its impression shows the small rate of 53% for being unpleasant and 47% for being pleasant. It is still inclined to the unpleasant side. However, its pleasant effect cannot be explained knowing that this bodily expression was created to express anger. The last case of BE2 shows a big difference between its 59% neutral expression and its 70% unpleasant impression.

Subject	Category	Electrodes									
		Fp1	Fp2	F3	F4	T3	T4	P3	P4	O1	O2
1	Pleasant	–	–	–	+	–	–	–	–	+	–
	Unpleasant	–	+	–	+	+	–	–	–	+	–
2	Pleasant	–	–	+	–	–	–	–	–	–	–
	Unpleasant	–	–	–	–	–	–	+	+	–	–
3	Pleasant	–	–	–	–	+	+	–	+	–	–
	Unpleasant	–	–	–	–	–	–	–	–	–	–
4	Pleasant	–	–	–	–	–	+	+	+	–	–
	Unpleasant	–	–	–	–	–	–	+	+	+	–
5	Pleasant	–	–	–	–	–	+	–	–	–	–
	Unpleasant	–	–	–	–	+	+	+	+	–	–
6	Pleasant	+	+	–	–	+	+	–	–	–	–
	Unpleasant	–	–	–	–	+	+	–	–	–	–
7	Pleasant	–	–	+	–	+	+	–	–	–	+
	Unpleasant	–	–	+	–	–	–	–	+	–	–

Table 3. Significant change in low-alpha power according to observed motion categories at every electrode and for each subject. (+: significant change p<.05; –: no significant change).

This suggests that bodily expressions with a neutral expression can be perceived negatively and can generate an unpleasant impression. The analysis of EEG data using the method described in section 2.3 allowed the calculation of the power L_m of low-alpha frequency band in each electrode channel and for each bodily expression. It also allowed the calculation of the power L_b of the same frequency band for the baseline period. Comparing L_m and L_b revealed the effect of observing a bodily expression in the change in the power of low-alpha frequency band for each electrode channel. Table 3 summarizes the results of this change in power, where only statistically significant change is indicated with the symbol +.

It can be seen that significant effect is mostly present at locations T3 and T4, then at P3 and P4, and finally at F3 and F4. Knowing that these positions are located above the superior temporal sulcus (STS) and above some specific parts of the prefrontal cortex (PFC) confirms previous research findings about the activation of STS in the perception of social cues (Cochin et al., 1998; Allison et al., 2000), and the activation of the mirror neurons located in the PFC during learning and imitation tasks (Rizolatti & Craighero, 2004). Some reaction can also be seen at other locations, for instance O1 and Fp2 for subject 1, O2 for subject 7, Fp1 and Fp2 for subject 6. The reaction at locations Fp1 and Fp2 are thought to be the result of blinking activity during the recording process, since these electrode positions are the closest to the eyes. It is important to assert that no preprocessing was done to avoid data with eye blinking artifacts. This approach was adopted because the goal is to conduct this investigation in natural conditions, where blinking activity is possible and should be considered. The reaction at locations O1 and O2 could be explained by the fact that during the vision process the visual cortex gets activated and this activation is usually captured at locations O1 and O2.

Nevertheless, the reactive locations were not always the same among different observers, suggesting high individual differences. A generalization cannot be derived at this point about the reaction of brain locations according to the category of the bodily expression that is being observed. However, the presence of a reaction is confirmed and another approach is necessary to achieve a more comprehensive result. On the other hand, there is a need to assess the repeatability of similar reactions from the same observer when he/she is shown the same bodily expression.

3.3 Repeatability of reaction in brain activity

The goal of this experiment is to confirm that the results obtained in the impression experiment (see section 3.2) are consistent over time for the same person. In other words, to make sure that brain reaction does happen all the time and at the same set of electrodes if a subject observes the same bodily expression several times.

Subject. One (1) student (male, 32 years old) volunteered to take part in this experiment. Similar to the previous experiment, the subject was fitted with an electro-cap and was given about 30 minutes to familiarize and get used to the presence of the cap.

Procedure. The subject participated in ten recording sessions. In each session, he was shown two bodily expressions, one for each category of bodily expressions, executed by the humanoid robot ASKA. Showing only representative bodily expressions is sufficient since the goal is to confirm the repeatability of brain reaction. Each bodily expression lasted for 14[sec], and the baseline period was recorded during the 14[sec] before the execution of each bodily expression. BE1 was chosen as a representative of pleasant bodily expressions, and BE4 was chosen as a representative of unpleasant bodily expressions. On one hand, BE1 was chosen because its expressiveness evaluation as pleasant (100%) was the highest among all the bodily expressions. Its impression evaluation (65%) was high enough to make sure it will have the desired effect on the observer, even though its impression was evaluated as the lowest among all the pleasant bodily expressions. In this case, the advantage was given to the expressiveness evaluation over the impression evaluation. On the other hand, BE4 was chosen because, similarly to BE1, its expressiveness evaluation as unpleasant (94%) was the highest among all the bodily expressions. Its impression evaluation (81%) was also the highest among all the bodily expressions, making it the perfect candidate to represent unpleasant bodily expressions. The recording was performed from 12 placements of an electro-cap[5], namely: F3, F4, C3, C4, P3, P4, F7, F8, T3, T4, T5, and T6 (see Fig. 5), using the Polymate AP1132[6] EEG recording device. The sampling frequency was set to 200[Hz] and the impedance was kept below 5[kΩ]. As a result of the experiment in section 3.2, the placements Fp1, Fp2, O1, and O2 were omitted, since they were shown to be not of a big importance. On the other hand, new placements were introduced, namely: F7, F8, C3, C4, T5, and T6, in order to get a more detailed coverage of the prefrontal and temporal cortices.

During the recording of the baseline period the subject was asked to relax as much as possible and to think of nothing in particular. To confirm the attendance to the task, the subject was told that he would be asked about the robot's movements and that he had to observe carefully. After the observation of each bodily expression, he was asked to explain the difference between the recently observed bodily expression and the one just before.

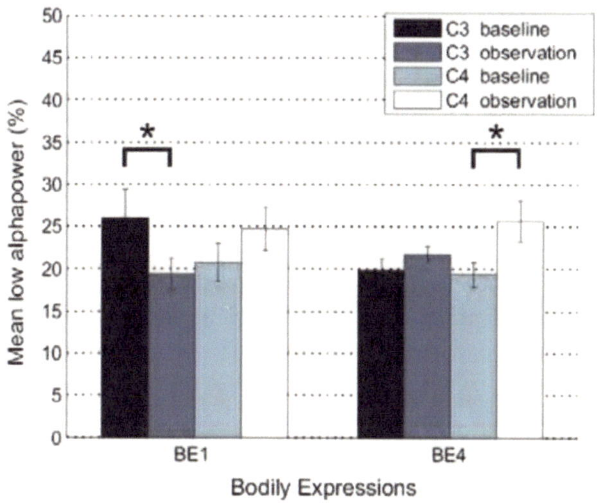

Fig.6. Mean alpha power calculated for the ten trials and for electrodes C3 and C4 (✶: $p< .05$).

Results. Analysis of the collected data, using the method described in section 2.3, resulted in identifying the electrode channels of placements C3 and C4 as the most reacting for this subject. Figure 6 shows the mean power of low-alpha frequency band calculated from the 10 trials for the electrode placements C3 and C4: where C3 reacted significantly to the pleasant bodily expression and C4 reacted significantly to the unpleasant bodily expression.

Fig. shows the overall result for the two bodily expressions by averaging the power change for all the electrodes over the 10 trials. The difference in means is significant between BE1 and BE4. Since BE1 is representative of pleasant bodily expressions and BE4 is representative of unpleasant ones, this result suggests an overall significant decrease in the power of low-alpha frequency band for pleasant motions, and a significant increase in power of low-alpha frequency band for unpleasant ones. This

---

[5] Electro-Cap is a trademark of Electro-Cap International Inc., USA.

[6] Polymate AP1132 was designed by Digitex Lab. Co. Ltd, and is commercialized by TEAC Corporation, Tokyo.

confirms that the change in low-alpha power happens every time the observer watches a bodily expression, and that this change is inversely proportional to the category of the observed bodily expression.

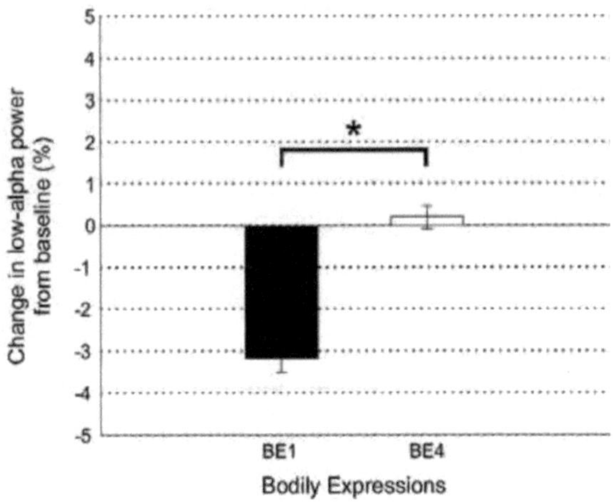

Fig. 7. Change in alpha power from the baseline using all electrodes, for each of the considered unpleasant and pleasant bodily expressions (✱: p < :05).

3.4 Discussion

The results presented in Table 1 confirmed the appropriateness of the expressiveness of the generated bodily expressions used in the experimental study. They show that the unpleasant bodily expressions were classified as unpleasant, and the pleasant bodily expressions were also classified as pleasant. During every experiment, the order of which the bodily expressions were shown to the observer was random so as not to allow the prediction of the nature of the next bodily expression. Although the subjects were not told anything about the bodily expressions, their answers agreed with the hypothesis. This implies that people tend to see bodily expressions in similar ways, which facilitates their interpretations and use in interactions. There exist a shared basic knowledge that allows humans the proper interpretation of similar expressions, although this knowledge is highly affected by the environment factors of culture and local customs. The bodily expressions were treated by the observers as if they were performed by a human even though it was the robot ASKA which actually performed them. It would be interesting to compare the interpretation of the same bodily expressions executed by both humans and robots to evaluate the existence of interpretation differences.

On the other hand, Table 2 correlates to a high extent with the results of Table 1. Here we can infer that observing a pleasant bodily expression will result in a pleasant impression on the observer and vice versa. This means that the observer is affected by what he sees even though the performer is just a robot. This effect on the observer is shown to be present in his/her brain activity with the results of section 3.2. Although, a generalization could not be concluded from the obtained results, the presence of a reaction in brain activity was proven.

It is important to acknowledge that the most reactive electrode positions were F3, F4, T3, T4, P3 and P4, which are located above the STS and PFC. This supports the claims that STS and mirror neurons get activated during the perception of social cues and the

observation of movements (Allison et al., 2000; Cochin et al., 1998), and that this can be used effectively in the implementation of Brain-Machine Interfaces (Nicolelis, 2001; Wessberg et al., 2000).
Finally, it was necessary to confirm the repeatability or the reproducibility of the same reaction in similar conditions. The results showed that the power level of low-alpha frequency band over all brain activity was inversely proportional to the category of the observed bodily expression. Particularly the most significant reaction was present at electrode positions C3 and C4 for the considered subject. These positions are close to the premotor and motor cortices. Due to the low spatial resolution of EEG, it is difficult to assert precisely which part of the brain is responsible for these reactions. However, current research findings confirm that the STS has an important role in the interpretation of social cues (Allison et al., 2000), and that mirror neurons are important during learning and imitation tasks (Rizolatti & Craighero, 2004).

4. Recognition of the impressions using Self-Organizing Maps

Self-Organizing Maps or Kohonen Networks are well suited to represent and generalize input data with an underlying structure that is not easily grasped (Kohonen, 1982). The unsupervised learning process tries to give a representation to the high-dimensional data rather than only a classification. Both the metric relation and the probability density of the data is approximated during this process, allowing the classification of newly measured data. Once the learning process has terminated, a labeling of the learned map can be done by using a relatively small set of labeled data items. The resulting map could be used to monitor the topographic patterns related to specific events as in (Joutsiniemi et al., 1995), or it could be used in the recognition of newly observed data. The similarity between a d-dimensional feature vector $X = (x_1, \ldots, x_d)$ and a prototype vector $M = (m_1, \ldots, m_d)$ in the learned map is calculated using the weighted Euclidean distance defined as:

$$D^2(X, M) = \sum_{j=1}^{d} w_j (x_j - m_j)^2 \tag{7}$$

where w_j is the weighting factor which can be used to give preference to certain features over others. This proved helpful and important in the semi-assisted learning of the data structure that was necessary for our EEG data. The usual approach in using SOM starts by preprocessing the selected data. Then, a feature extraction method is specified and used. After that, the map is calculated using competitive learning. Finally, the resulting map is labeled and used for recognition. It is important to note that in practical applications the selection and preprocessing of data is of extreme importance, because unsupervised methods only illustrate the structure in the data set, and the structure is highly influenced by the features chosen to represent data items (Kaski, 1997). In the following we will show how we used SOM in the recognition of bodily expressions executed by a humanoid robot and of similar bodily expressions executed by a human.

4.1 Recognition of the impressions of robot bodily expressions

4.1.1 Data acquisition

In the recognition of the impressions of robot bodily expressions we use the same data as in section 3.2. The data consists of all the signals collected at a sampling rate of 200[Hz] from the 10 EEG placements for twice 14 seconds and for each one of the seven subjects.

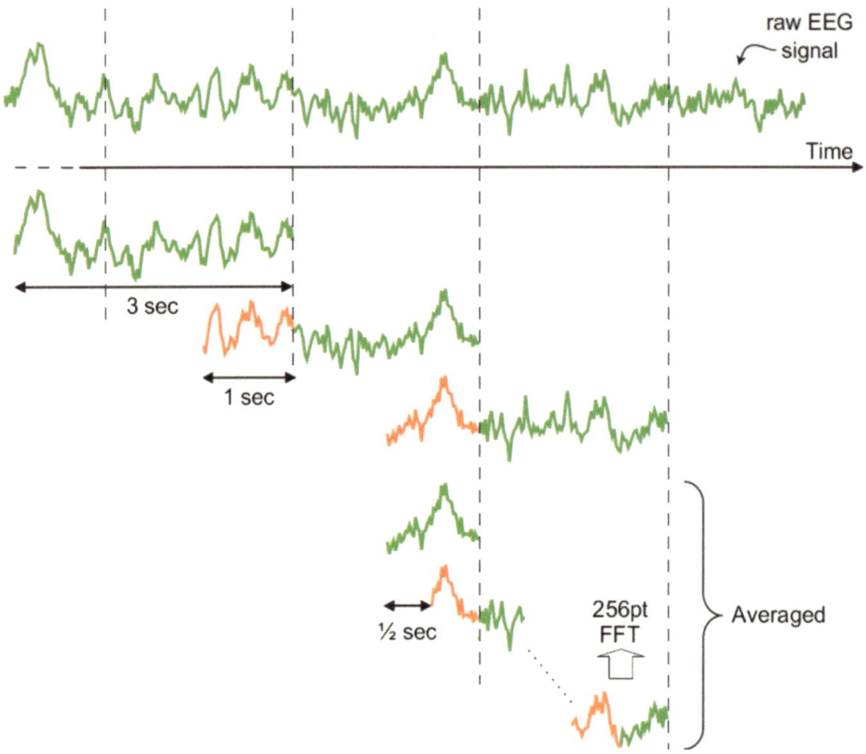

Fig. 8. Preprocessing EEG data for features extractions is done by calculating a moving average of overlapping windows of predefined length (3[sec]).

4.1.2 Data preprocessing

To prepare the collected data for the training task, we calculate the moving averaged power spectra of 10×2×7=140 signal sources. As shown in Fig. , we apply a 3[sec] (600-point) Hanning window on the signal with 1[sec] (200-point) overlap. The windowed 3[sec] epochs are further subdivided into several 1[sec] (200-point) sub-windows using the Hanning window again with 1/2[sec] (100-point) overlap, each extended to 256 points by zero padding for a 256-point fast Fourier transform (FFT). A moving median filter is then applied to average and minimize the presence of artifacts in all the sub-windows. The resulting moving averaged power spectrum is then reduced to six (6) features by integrating the spectral values weighted by six raised-sine shaped windows with an area normalized to unity. Consequently, the feature components do not need to be normalized. In a similarly way to (Joutsiniemi et al., 1995), the weighting windows are overlapping to ensure a smooth change of the features in accordance with the change in power spectrum. These windows cover the following frequency bands:

- Delta : 00~04[Hz]
- Theta : 04~08[Hz]
- Low-Alpha : 08~11[Hz]
- High-Alpha: 11~13[Hz]
- Beta : 13~30[Hz]
- Gamma : 30~50[Hz]

Notice that the alpha frequency band is divided into low and high. This will allow us to give different importance coefficients to each frequency band, according to each one's contribution in recognizing the effect of bodily expressions. The resulting time series of EEG power spectrum features consists of a vector of 10×6=60 features every 2[sec] (400-point) time intervals.

4.1.3 Map training and recognition results

The 2D map to learn using the collected features is arranged as a 2D lattice, with each location containing a 60-dimensional prototype vector. During the learning process or the self-organization, the importance coefficients w_j, used in the similarity metric **(7)**, were taken as 0.5, 0.5, 1.0, 0.9, 0.5, and 0.3 for the features delta, theta, low-alpha, high-alpha, beta, and gamma, respectively. Higher importance was given to the alpha bands with an emphasis on the low-alpha band, as a result of its sensitivity to the category of bodily expression being observed by a subject (see section 3.2). On the other hand, the higher frequency gamma band was given the lowest importance coefficient, since it was not proven to react significantly to social cues (Cochin et al., 1998).

After the training of the map, an approximation of the probability density of the input data is reached generating clusters which can be identified as associated to one of the following experimental conditions: observing unpleasant bodily expression, observing pleasant bodily expression, or baseline condition. 80% of the data was used for the training and the remaining 20% was used for the evaluation. The resulting recognition rate was of 65.2%, divided as 62.8% for data associated to unpleasant bodily expressions, 59.3% for data associated with pleasant bodily expressions, and 73.5% for data associated to baseline. Clearly, the rate of 65.2% is not satisfactory since this is a low rate to rely on when making a decision. However, when individual data was used separately in the learning process the recognition rate jumped high to the 79.5%. The previously realized low rate is explained by the existence of conflicting data items collected from different subjects. In order to understand the effect on the recognition rate of using data from different sources (subjects), all possible combinations of data sources were used to learn several maps and the recognition rates calculated. The change in recognition rates is shown in Fig. . It can be noticed that the addition of a new source of data decreased the recognition rate by about 5% for the first two additions. However, there is certain stability in the rate when the number of data sources was 3, 4 and 5. But again the rate decreased by a lower factor when adding more sources. This decline is explained by the existence of individual differences in the reaction to bodily expressions and probably even the interpretation. This outcome was also observed in the previous experiment about impressions (see section 3.2). To cope with this problem it is recommended to train several SOMs with a small number of data sources and use the totality in the recognition process.

4.2 Recognition of the impressions of human bodily expressions

There is little knowledge of the difference in the effect left on a person when observing humans and when observing robots. Most of the literature reports either cases (Ito & Tani, 2004; Allison et al., 2000; Schaal, 1999) and there is, to our knowledge, no previous work that tried to draw on the parallels between the two cases. Therefore, it is worthwhile to investigate similarities and differences in the recognition of the category of the impressions left on the observer for both cases.
In order to collect brain activity data when subjects are observing human bodily expressions, we conducted an experiment similar the one for the expressiveness of robot bodily expressions (see section 3.2). The goal is to evaluate the impression on the observer of the bodily expressions described in section 2.2.

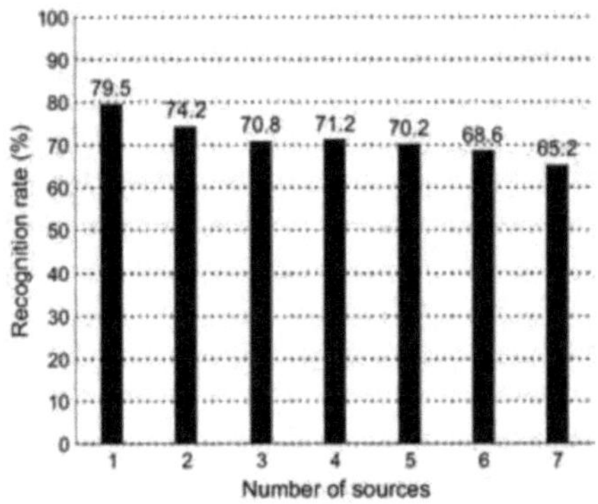

Fig. 9. Recognition rates when using data from different sources (subjects), when observing robot's bodily expressions.

4.2.1 Data acquisition

Subjects. Three (3) participants (males aged between 21 and 23 years old) volunteered to take part in this experiment. They were students at the graduate school of information science. They were all familiar with the experiment since they did participate in the previous one about the impressions of robot bodily expressions. Their brain activity was collected with an EEG measurement device and they were familiarized with the presence of the electrodes by letting them spend about 20 minutes reading books or surfing the Internet.

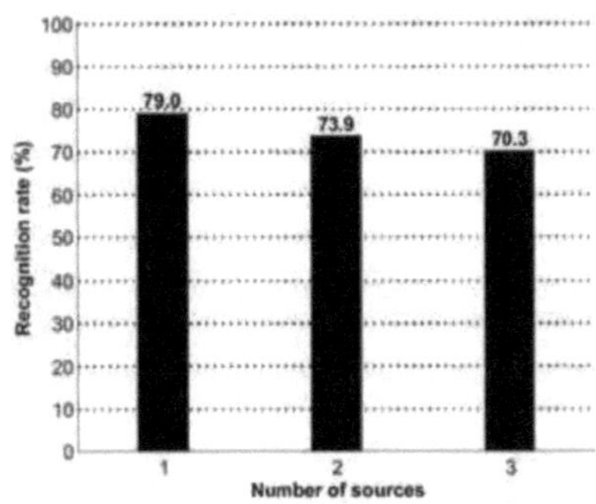

Fig. 10. Recognition rates when using data from different sources (subjects), when observing a human performer.

Procedure. The human bodily expressions to be shown to the subjects were prepared beforehand. A volunteer in a black tight suit performed these bodily expressions. He had a black cover on his head to make sure that no facial expressions get to the observer. These bodily expressions were a reproduction of the bodily expressions described in section 2.2 and were recorded on HDTV[7] video tapes. During the experiment, the bodily expressions were projected on a display big enough to ensure that the projected images of human performer is as close as possible to real size. Similar to the experiment in section 3.2, the recording was obtained from 12 electrode locations, namely: F3, F4, F7, F8, C3, C4, P3, P4, T3, T4, T5, T6, using the Polymate AP1132 EEG recording device. The sampling frequency was 200[Hz] and the impedance was kept below 5[kΩ]. The subjects were shown a total of six bodily expressions lasting 10[sec] each. During the recording of the baseline period, the subjects were asked to relax as much as possible and to think of nothing in particular. To help them achieve this state of mind, they were shown for 10[sec] the empty space of the room where the recording of the bodily expressions was performed. In addition, to confirm the subjects attended to the task they were told that they would be asked about the bodily expressions and that they had to observe carefully. After each observation, they were asked to explain the difference between the bodily expressions they just observed and the previous ones.

Performer of BEs	Human		Robot	
Data sources	Individual	Combined (3)	Individual	Combined (7)
Recognition rate (%)	79.0	70.3	79.5	65.2

Table 4. Comparison of the recognition rates for the cases of human bodily expressions and robot bodily expressions.

4.2.2 Recognition results

The preprocessing of the data was the same as in the previous case (section 4.1.2). Moreover, 80% of the data was used for training and the remaining 20% was used for evaluation. The resulting recognition rate was 70.3%, divided into 68.0% for data associated to the observation of unpleasant bodily expressions, 67.5% for data associated to the observation of pleasant bodily expressions, and 80.4% for data associated to the baseline. This rate is much better than the 65.2% rate achieved with the data of robot bodily expressions. However, the number of sources in this case is only 3, while it was 7 for the robot case. Thus, it is more accurate to compare this result to the 70.8% rate achieved when using only 3 data sources for the robot case, as shown in Fig. .

Using data from different sources showed degradation of the recognition rate, similar to the result of the robot, see Fig.. The addition of one source resulted in a decrease of 5.1%, then a decrease of 3.6% after adding a third source. This result supports the fact that individual differences remain present even in the case of observing bodily expressions performed by a human.

4.3 Discussion

Regardless of the performer, whether a robot or a human, the recognition rates of the category of the observed bodily expression was about 80% when using subject's data

[7] HDTV stands for High Definition television a.k.a High-Vision which allows the recording of a high resolution video stream.

individually. However, this rate decreased significantly when additional data from other subjects was used in the training process. To cope with this problem it would be interesting to train one SOM for each data source, and then combine the resulting SOMs into a bigger structure for the recognition task. Adopting this approach could result in keeping a high recognition rate while taking into consideration all the data that was collected so as not to loose the generality of the solution.

It is interesting to note that the average difference between the recognition rates for robot and human cases is relatively small as summarized in Table 4. This proves that SOMs are suitable to the generalization of the effect in the input EEG signals regardless of whether this effect is generated by a human or a robot. Even if differences appear clearly when analyzing raw EEG data, the SOMs succeed in eliminating these differences and keep only the important information. SOMs also succeed in separating the noise and artifacts from signals reflecting brain activity. This is another powerful characteristic that helps in the online processing of EEG signals for applications related to Brain-Machine Interfaces (BMI).

5 Conclusions and future work

In this paper, we investigated the relation between bodily expressions and their impressions on the observer. We started by generating six bodily expressions, then we classified them to belong to two categories according to their expressiveness, namely: pleasant and unpleasant. Their expressiveness was confirmed statistically by a self-reporting experiment where a number of volunteers answered questionnaires about the bodily expressions. Afterwards, we conducted an experiment to assess the impressions on the observers while watching the considered bodily expressions by collecting the observer's brain activity using electroencephalogram (EEG). The method adopted for spectral analysis revealed a correlation between the power level of low-alpha frequency band (8~11[Hz]) and the category of the observer bodily expression. The reproducibility or repeatability of this band's reaction was confirmed with a third experiment where a subject observed repeatedly candidate bodily expressions for each category. These results have opened the opportunity to utilize the change in the power level of low-alpha frequency band to examine the capacity of a humanoid robot in activating the social perception system in a human observer. A challenging problem that rises from this result is about the degree to which such reaction appears when observing robots with different human-like physical and behavioral characteristics. The understanding of which robot properties are necessary or sufficient to activate the social perception system in an observer is of particular interest.

Another important direction was to define a method which can assess and recognize the impression category from the observer's brain activity. For this reason, we presented a computational method to use for the recognition of the impression of bodily expressions.

The self-organizing maps (SOM), which allows at the simultaneous reduction of the amount of data and its projection to a lower dimensional space, was used for the recognition. It was shown that SOMs achieve relatively high recognition rates considering that the data we used is not filtered for noise elimination. The existence of high individual differences in the considered data was handled successfully with the SOM, because of its ability to separate signals resulting from different processes.

Future research directions should focus on improving the recognition rate to over 90% and to try to recognize a refined classification of bodily expressions. A link with human motion styles (Hsu et al., 2005) would be interesting to provide more details about the bodily expressions. There is an extensive work on generating emotional motions (Amaya et al. 1996; Lim et al. 2004) that could be incorporated in this research. Reaching this goal would enable us to build an adaptive interface that makes the robot judge the effect of its own gestures on the users and allows it to change its behavior accordingly.

6. Acknowledgements

This work was partly supported by the Ministry of Education, Culture, Sports, Science and Technology (MEXT) and the NAIST 21st Century COE program "Ubiquitous Networked Media Computing". The authors would like to thank Mr. Albert Causo for his comments and suggestions to improve the manuscript.

7. References

T. Allison, A. Puce & G. McCarthy (2000). Social perception from visual cues: Role of the STS region. Trends in Cognitive Sciences, 4(7):251-291.

K. Amaya, A. Bruderlin & T. Calvert (1996). Emotion from motion. In Graphics Interface '96, pages 222-229.

I. Bartenieff & D. Lewis (1980). Body Movement: Coping with the Environment. Gordon and Breach Science Publishers, USA.

C. Breazeal (2002). Designing Sociable Robots. The MIT Press, USA.

S. Cochin, C. Barthlemy, B. Lejeune, S. Roux & J. Martineau (1998). Perception of motion and qEEG activity in human adults. Electroencephalography and Clinical Neurophysiology, 107:287-295.

E. Hsu, K. Pulli, & J. Popovic (2005);. Style translation for human motion. ACM Transactions on Graphics, 24(3):1082-1089.

J. Ido, K. Takemura, Y. Matsumoto & T. Ogasawara (2002). Robotic receptionist ASKA: a research platform for human-robot interaction. In IEEE Workshop of Robot and Human Interactive Communication, pages 306-311.

M. Ito & J. Tani (2004). On-line imitative interaction with a humanoid robot using a dynamic neutral network model of a mirror system. Adaptive Behavior, 12(2):93-115.

H. H. Jasper (1958). The ten-twenty electrode system of the international federation. Electroencephalography and Clinical Neurophysiology, 10:371-375.

G. Johansson (1973). Visual perception of biological motion and a model for its analysis. Perception and Psychophysics, 14(2):201-211.

S.-L. Joutsiniemi, S. Kaski & T. A. Larsen (1995). Self-organizing map in recognition of topographic patterns of EEG spectra. IEEE Transactions on Biomedical Engineering, 42(11):1062-1068.

S. Kaski (1997). Data Exploration using Self-Organizing Maps. PhD thesis, Helsinki University of Technology.

A. Khiat, M. Toyota, Y. Matsumoto & T. Ogasawara (2006). Brain activity in the evaluation of the impression of robot bodily expressions. In IEEE/RSJ International Conference on Intelligent Robots and Systems, pages 5504-5508.

T. Kohonen (1982). Self-organized formation of topologically correct feature maps. Biological Cybernitics, 43:59-69.

H. Kozima (2002). Infanoid: A babybot that explores the social environment, pages 157-164. Socially Intelligent Agents: Creating Relationships with Computers and Robots. Kluwer Academic Publishers, The Netherlands.

H. Lim, A. Ishii & A. Takanishi (2004). Emotion-base biped walking. Robotica, 22:577-586.

M. L. Nicolelis (2001). Actions from thoughts. Nature, 409(18):403-407.

T. Nomura, T. Kanda & T. Suzuki (2006). Experimental investigation into influence of negative attitudes toward robots on human-robot interaction. AI & Society, 20(2):138-150.

D. A. Norman, A. Ortony, and D. M. Russell (2003). Affect and machine design: Lessons for the development of autonomous machines. IBM Systems Journal, 42(1):38-44.

F. E. Pollick, H. M. Paterson, A. Bruderlin & A. J. Sanford (2001). Perceiving affect from arm movement. Cognition, 82:B51-B61.

G. Rizolatti & L. Craighero (2004). The mirror-neuron system. Annual Reviews of Neuroscience, 27:169-192.

A. Schlögl, S. J. Roberts & G. Pfurtscheller (2000). A criterion for adaptive autoregressive models. In IEEE International Conference of Engineering in Medicine & Biology Society, pages 1581-1582.

S. Schaal (1999). Is imitation learning the route to humanoid robot? Trends in Cognitive Science, 3(6):233-242.

P. Shaver, J. Schwartz, D. Kirson & C. O'Connor (1987). Emotion knowledge: Further exploration of a prototype approach. Journal of Personality and Social Psychology, 52(6):1061-1086.

K. Shinozawa, F. Naya & K. Kogure (2005). Differences in effect of robot and screen agent recommendations on human decision-making. International Journal of Human Computer Study, 62(2):267-279.

T. Tanaka, T. Mori & T. Sato (2001). Quantitative analysis of impression of robot bodily expression based on laban movement theory. Journal of Robotics Society of Japan, 19(2):252-259. *(in Japanese)*

J. Wessberg, C. R. Stambaugh, J. D. Kralik, P. D. Beck, M. Laubach, J. K. Chapin, J. Kim, S. J. Biggs, M. A. Srinivasan & M. L. Nicolelis (2000). Real-time prediction of hand trajectory by ensembles of cortical neurons in primates. Nature, 408:361-365.

C. M. Whissel (1989). The Dictionary of Affect in Language, volume 4 of Emotion: Theory, research and experience. The measurement of emotions. Academic Press, USA.

Appendix: Mathematical expression of Laban features of the robot ASKA

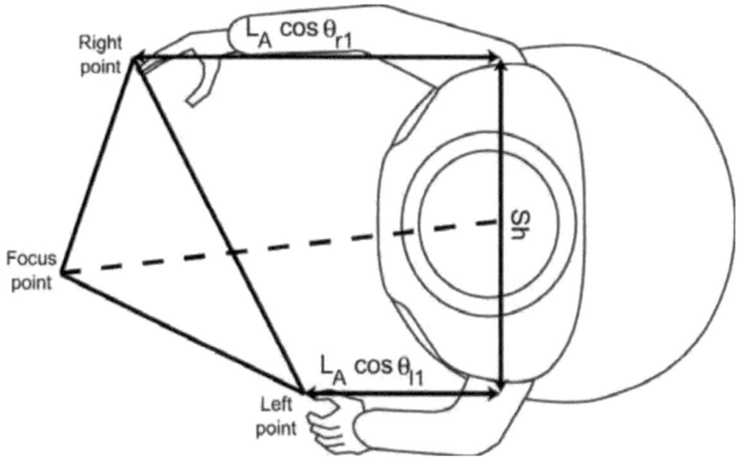

Fig. 11. Diagram of table plane superposed on a top view of the robot ASKA.

The mathematical definition of Laban features (Shape and Effort) using the robot's kinematic and dynamic information is given such that larger values describe fighting movement forms and smaller values describe indulging ones (Tanaka et al., 2001). Bartenieff and Lewis stated in (Bartenieff & Lewis, 1980) that the Shape feature describes the geometrical aspect of the movement using three parameters: table plane, door plane, and wheel plane. They also reported that the Effort feature describes the dynamic aspect of the movement using three parameters: weight, space, and time. The robot's link information which will be used in the features definitions is given in Fig. . In order to simplify the mathematical description, a limited number of joint parameters were considered in this definition, namely: the left arm θ_{l1}, the right arm θ_{r1}, the neck δ_1, the face δ_2, the left wheel ω_l, and the right wheel ω_r. The remaining parameters were fixed to default values during movement execution.

Using the diagram shown in Fig. , the table parameter of feature Shape represents the spread of silhouette as seen from above. It is defined as the scaled reciprocal of the summation of mutual distances between the tips of the left and the right hands along with a focus point, as shown in **(8)**.

$$Shape_{table} = {}^{s}\!\big/\!\left(T_{LF} + T_{RF} + T_{LR}\right) \tag{8}$$

where,

$$T_{LF} = \sqrt{\left(L_F \cos\delta_1 \cos\delta_2 - L_A \cos\theta_{l1}\right)^2 + \left({}^{Sh}\!\big/\!_2 - L_F \cos\delta_1 \sin\delta_2\right)^2},$$

$$T_{RF} = \sqrt{\left(L_F \cos\delta_1 \cos\delta_2 - L_A \cos\theta_{r1}\right)^2 + \left({}^{Sh}\!\big/\!_2 - L_F \cos\delta_1 \sin\delta_2\right)^2},$$

$$T_{LR} = \sqrt{Sh^2 + \left(L_A \cos\theta_{l1} - L_A \cos\theta_{r1}\right)^2},$$

The point of focus is set at the fixed distance $L_F = 44$ [cm] in the gaze line of the robot's head. $Sh = 33$ [cm] is the distance between the shoulders; $L_A = 44$ [cm] is the arm's length during movement execution and s is a scaling factor. The door parameter of feature Shape represents the spread of the silhouette as seen from the font. It is defined as the weighted sum of the elevation angles of both arms and the head as shown in **(9)**. The sine is used to reflect how upward or downward is each joint angle. The weights d_l, d_r, d_n were fixed empirically to 1.

$$Shape_{door} = d_l \sin\theta_{l1} + d_r \sin\theta_{r1} + d_n \sin\delta_{n1} \tag{9}$$

The wheel parameter of feature Shape represents the lengthwise shift of the silhouette in the sagittal plane. It is defined as the weighted sum of the velocities of the robot and the velocities of the arm extremities as shown in **(10)**. Weights were set empirically to -8 for w_t , to -1 for w_l and w_r.

$$Shape_{wheel} = w_t v_{tr} + w_l L_A \frac{d}{dt}\cos\theta_{l1} + w_r L_A \frac{d}{dt}\cos\theta_{r1} \tag{10}$$

The weight parameter of feature Effort represents the strength of the movement. It is defined in **(11)** as the weighed sum of the energies exhibited during movement per unit time at each part of the body. Weights were adjusted with respect of to the saliency of body parts. Relatively large weights $e_{rt} = e_{tr} = 5$ were given to the movement of the trunk and smaller weights were given to the movements of the arms $e_l = e_r = 2$ and the neck $e_n = 1$.

$$Effort_{weight} = e_l \dot{\theta}_{l1}^2 + e_r \dot{\theta}_{r1}^2 + e_n \dot{\delta}_{n1}^2 + e_{tr} v_{tr}^2 + e_{rt} v_{rt}^2 \tag{11}$$

where $v_{tr} = \dot{\omega}_l + \dot{\omega}_r$ is the translation velocity and $v_{rt} = \dot{\omega}_l - \dot{\omega}_r$ is the rotation velocity.
The space parameter of feature Effort represents the degree of conformity in the movement. It is defined in **(12)** as the weighed sum of the directional differences between elevation angles of the arms and the neck as well as the body orientation. Weights are also defined empirically by giving advantage to the arms' bilateral symmetry $s_{lr} = -5$ and body orientation $s_{rt} = -5$ over the other combinations $s_{ln} = s_{rn} = -1$.

$$Effort_{space} = s_{rt}|\omega_{rt}| + s_{lr}|\theta_{l1} - \theta_{r1}| + s_{ln}|\theta_{l1} - \delta_{n1}| + s_{rn}|\theta_{r1} - \delta_{n1}| \tag{12}$$

The time parameter of feature Effort represents the briskness in the movement execution and covers the entire span from sudden to sustained movements. It is defined in **(13)** as the ratio indicating the number of generated commands per time unit.

$$Effort_{time} = \frac{number\ of\ generated\ commands}{time\ span} \tag{13}$$

25

Advanced Humanoid Robot Based on the Evolutionary Inductive Self-organizing Network

Dongwon Kim, and Gwi-Tae Park
Department of Electrical Engineering, Korea University,
1, 5-ka, Anam-dong, Seongbuk-ku, Seoul 136-701,
Korea.

1. Introduction

The bipedal structure is one of the most versatile ones for the employment of walking robots. The biped robot has almost the same mechanisms as a human and is suitable for moving in an environment which contains stairs, obstacle etc, where a human lives. However, the dynamics involved are highly nonlinear, complex and unstable. So it is difficult to generate human-like walking motion. To realize human-shaped and human-like walking robots, we call this as *humanoid robot,* many researches on the biped walking robots have been reported [1-4]. In contrast to industrial robot manipulators, the interaction between the walking robots and the ground is complex. The concept of the zero moment point (ZMP) [2] is known to give good results in order to control this interaction. As an important criterion for the stability of the walk, the trajectory of the ZMP beneath the robot foot during the walk is investigated [1-7]. Through the ZMP, we can synthesize the walking patterns of humanoid robot and demonstrate walking motion with real robots. Thus ZMP is indispensable to ensure dynamic stability for a biped robot. The ZMP represents the point at which the ground reaction force is applied. The location of the ZMP can be obtained computationally using a model of the robot. But it is possible that there is a large error between the actual ZMP and the calculated one, due to the deviations of the physical parameters between the mathematical model and the real machine. Thus, actual ZMP should be measured to realize stable walking with a control method that makes use of it.

In this chapter, actual ZMP data throughout the whole walking phase are obtained from the practical humanoid robot. And evolutionary inductive self-organizing network [8-9] is applied. So we obtained natural walking motions on the flat floor, some slopes, and uneven floor.

2. Evolutionary Inductive Self-organizing Network

In this Section we will depict the evolutionary inductive self-organizing network (EISON) to be applied to the practical humanoid robot. Firstly the algorithm and its structure are shown and evaluation to show the usefulness of the method will be followed.

2.1 Algorithm and structure

The EISON has an architecture similar to feed-forward neural networks whose neurons are replaced by polynomial nodes. The output of the each node in EISON structure is obtained using several types of high-order polynomial such as linear, quadratic, and modified quadratic of input variables. These polynomials are called as partial descriptions (PDs). The PDs in each layer can be designed by evolutionary algorithm. The framework of the design procedure of the EISON [8-9] comes as a sequence of the following steps.

[Step 1] *Determine input candidates of a system to be targeted.*
[Step 2] *Form training and testing data.*
[Step 3] *Design partial descriptions and structure evolutionally.*
[Step 4] *Check the stopping criterion.*
[Step 5] *Determine new input variables for the next layer.*

In the following, a more in-depth discussion on the design procedures, step 1~step 5, is provided.

Step 1: *Determine input candidates of a system to be targeted*
We define the input variables such as $x_{1i}, x_{2i}, \cdots x_{Ni}$ related to output variables y_i, where N and i are the number of entire input variables and input-output data set, respectively.

Step 2: *Form training and testing data.*
The input - output data set is separated into training (n_{tr}) data set and testing (n_{te}) data set. Obviously we have $n = n_{tr} + n_{te}$. The training data set is used to construct a EISON model. And the testing data set is used to evaluate the constructed EISON model.

Step 3: *Design partial descriptions(PD) and structure evolutionally.*
When we design the EISON, the most important consideration is the representation strategy, that is, how to encode the key factors of the PDs, order of the polynomial, the number of input variables, and the optimum input variables, into the chromosome. We employ a binary coding for the available design specifications. We code the order and the inputs of each node in the EISON as a finite-length string. Our chromosomes are made of three sub-chromosomes. The first one is consisted of 2 bits for the order of polynomial (PD), the second one is consisted of 3 bits for the number of inputs of PD, and the last one is consisted of N bits which are equal to the number of entire input candidates in the current layer. These input candidates are the node outputs of the previous layer. The representation of binary chromosomes is illustrated in Fig. 1.

Bits in the 1st sub-chromosome	Order of polynomial(PD)
00	Type 1 - Linear
01 10	Type 2 - Quadratic
11	Type 3 - Modified quadratic

Table 1. Relationship between bits in the 1st sub-chromosome and order of PD.

The 1st sub-chromosome is made of 2 bits. It represents several types of order of PD. The relationship between bits in the 1st sub-chromosome and the order of PD is shown in Table 1. Thus, each node can exploit a different order of the polynomial.

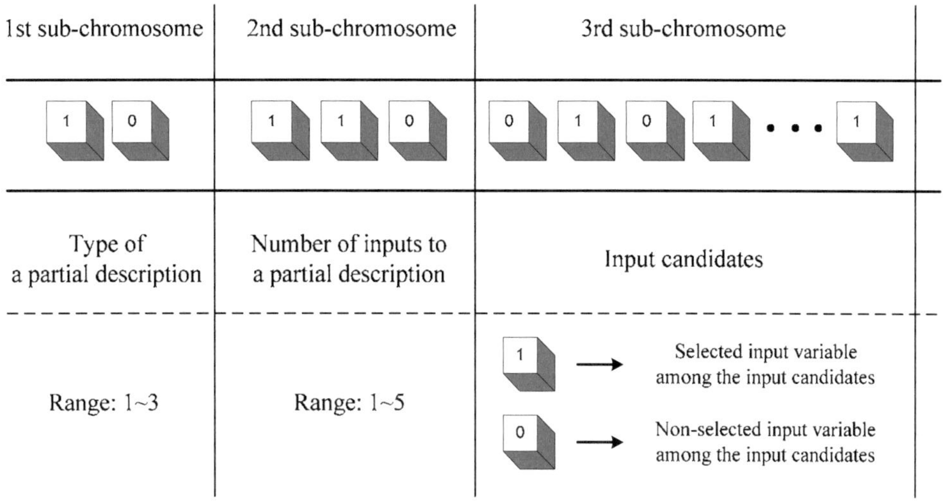

Fig. 1. Structure of binary chromosome for a PD

The 3rd sub-chromosome has N bits, which are concatenated a bit of 0's and 1's coding. The input candidate is represented by a 1 bit if it is chosen as input variable to the PD and by a 0 bit it is not chosen. This way solves the problem of which input variables to be chosen.

If many input candidates are chosen for model design, the modeling is computationally complex, and normally requires a lot of time to achieve good results. In addition, it causes improper results and poor generalization ability. Good approximation performance does not necessarily guarantee good generalization capability. To overcome this drawback, we introduce the 2nd sub-chromosome into the chromosome. The 2nd sub-chromosome is consisted of 3 bits and represents the number of input variables to be selected. The number based on the 2nd sub-chromosome is shown in the Table 2.

Bits in the 2nd sub-chromosome	Number of inputs to PD
000	1
001	2
010	2
011	3
100	3
101	4
110	4
111	5

Table 2. Relationship between bits in the 2nd sub-chromosome and number of inputs to PD.

Input variables for each node are selected among entire input candidates as many as the number represented in the 2nd sub-chromosome. Designer must determine the maximum number in consideration of the characteristic of system, design specification, and some prior knowledge of model. With this method we can solve the problems such as the conflict between overfitting and generalization and the requirement of a lot of computing time.

The relationship between chromosome and information on PD is shown in Fig. 2. The PD corresponding to the chromosome in Fig. 2 is described briefly as Fig. 3.
Fig. 2 shows an example of PD. The various pieces of required information are obtained its chromosome. The 1st sub-chromosome shows that the polynomial order is Type 2 (quadratic form). The 2nd sub-chromosome shows two input variables to this node. The 3rd sub-chromosome tells that x_1 and x_6 are selected as input variables. The node with PD corresponding to Fig. 2 is shown in Fig. 3. Thus, the output of this PD, $\hat{y}$ can be expressed as (1).

$$\hat{y} = f(x_1, x_6) = c_0 + c_1 x_1 + c_2 x_6 + c_3 x_1^2 + c_4 x_6^2 + c_5 x_1 x_6 \tag{1}$$

where coefficients c_0, c_1, …, c_5 are evaluated using the training data set by means of the standard least square estimation (LSE). Therefore, the polynomial function of PD is formed automatically according to the information of sub-chromosomes.

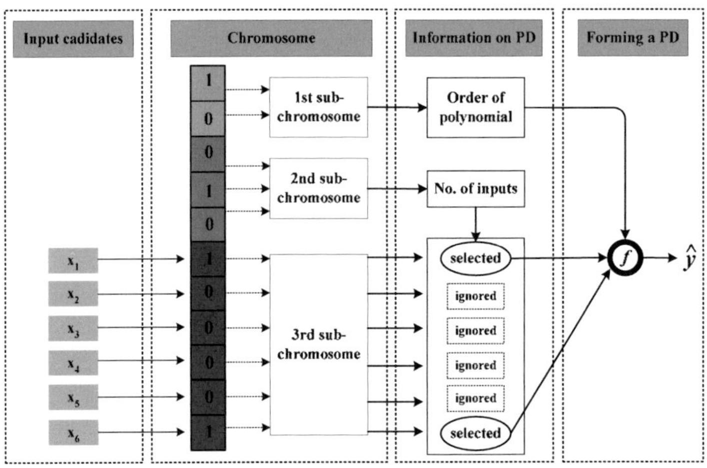

Fig. 2. Example of PD whose various pieces of required information are obtained from its chromosome.

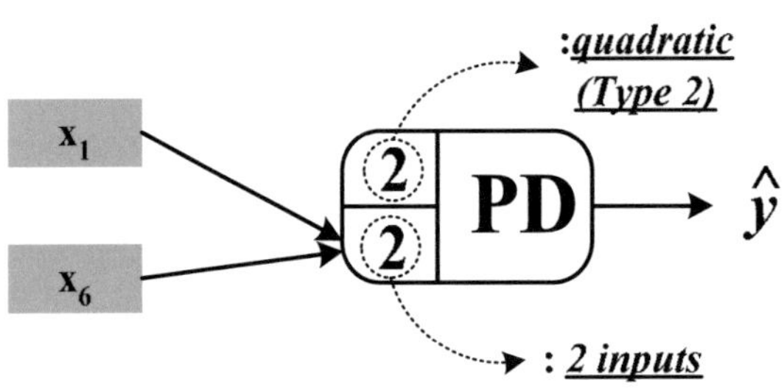

Fig. 3. Node with PD corresponding to chromosome in Fig. 2.

Step 4: *Check the stopping criterion.*
The EISON algorithm terminates when the 3rd layer is reached.

Step 5: *Determine new input variables for the next layer.*
If the stopping criterion is not satisfied, the next layer is constructed by repeating step 3 through step 4.

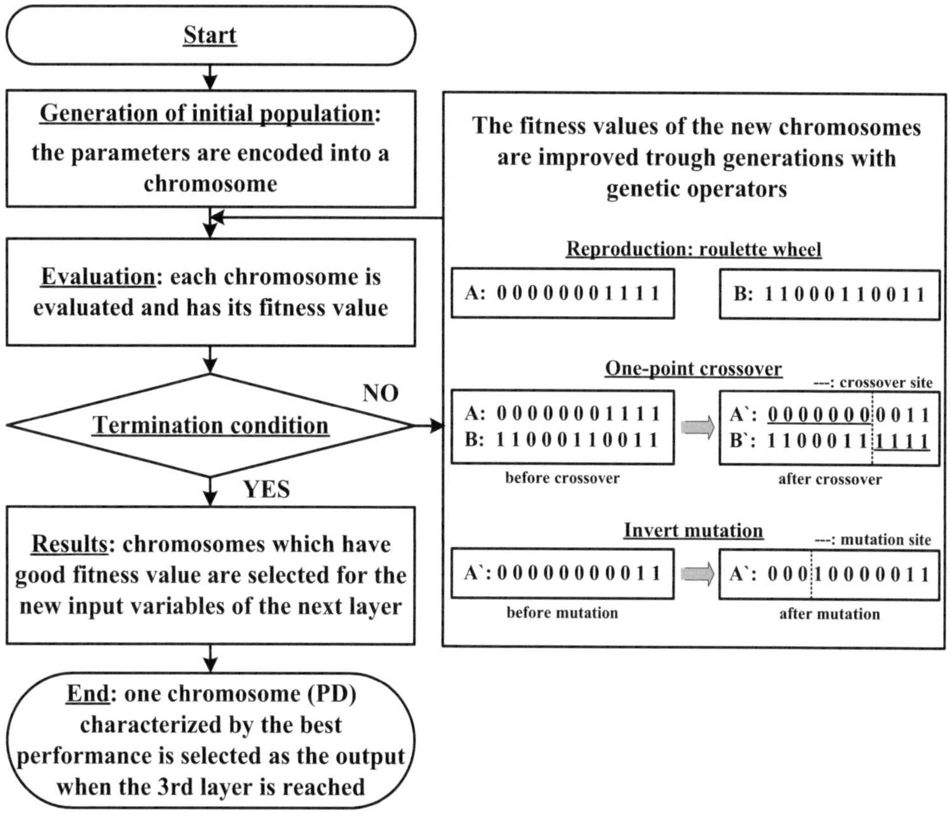

Fig. 4. Block diagram of the design procedure of EISON.

The overall design procedure of EISON is shown in Fig. 4. At the beginning of the process, the initial populations comprise a set of chromosomes that are scattered all over the search space. The populations are all randomly initialized. Thus, the use of heuristic knowledge is minimized. The assignment of the fitness in evolutionary algorithm serves as guidance to lead the search toward the optimal solution. Fitness function with several specific cases for modeling will be explained later. After each of the chromosomes is evaluated and associated with a fitness, the current population undergoes the reproduction process to create the next generation of population. The roulette-wheel selection scheme is used to determine the members of the new generation of population. After the new group of population is built, the mating pool is formed and the crossover is carried out. The crossover proceeds in three steps. First, two newly reproduced strings are selected from the mating pool produced by reproduction. Second, a position (one point) along the two strings is selected uniformly at random. The third step is to exchange all characters following the crossing site. We use one-point crossover operator with a crossover probability of P_c (0.85). This is then followed by the mutation operation. The mutation is the occasional alteration of a value at a particular bit position (we flip the states of a bit from 0 to 1 or vice versa). The mutation serves as an insurance policy which would recover the loss of a particular piece of information (any simple bit). The mutation rate used is fixed at 0.05 (P_m). Generally, after these three operations, the overall fitness of the population improves. Each of the population generated then goes through a series of evaluation, reproduction, crossover, and mutation, and the procedure is repeated until a termination condition is reached. After the evolution process, the final generation of population consists of highly fit bits that provide optimal solutions. After the termination condition is satisfied, one chromosome (PD) with the best performance in the final generation of population is selected as the output PD. All remaining other chromosomes are discarded and all the nodes that do not have influence on this output PD in the previous layers are also removed. By doing this, the EISON model is obtained.

2.2 Fitness function for EISON

The important thing to be considered for the evolutionary algorithm is the determination of the fitness function. The genotype representation encodes the problem into a string while the fitness function measures the performance of the model. It is quite important for evolving systems to find a good fitness measurement. To construct models with significant approximation and generalization ability, we introduce the error function such as

$$E = \theta \times PI + (1-\theta) \times EPI \tag{2}$$

where $\theta \in [0,1]$ is a weighting factor for PI and EPI, which denote the values of the performance index for the training data and testing data, respectively, as expressed in (5). Then the fitness value is determined as follows:

$$F = \frac{1}{1+E} \tag{3}$$

Maximizing F is identical to minimizing E. The choice of θ establishes a certain tradeoff between the approximation and generalization ability of the EISON.

2.3 Evaluation of the EISON

We show the performance of our EISON for well known nonlinear system to see the applicability. In addition, we demonstrate how the proposed EISON model can be employed to identify the highly nonlinear function. The performance of this model will be compared with that of earlier works. The function to be identified is a three-input nonlinear function given by (4)

$$y = (1 + x_1^{0.5} + x_2^{-1} + x_3^{-1.5})^2 \tag{4}$$

which is widely used by Takagi and Hayashi [10], Sugeno and Kang[11], and Kondo[12] to test their modeling approaches.

40 pairs of the input-output data sets are obtained from (4) [14]. The data is divided into training data set (Nos. 1-20) and testing data set (Nos. 21-40). To compare the performance, the same performance index, average percentage error (APE) adopted in [10-14] is used.

$$APE = \frac{1}{m}\sum_{i=1}^{m}\frac{|y_i - \hat{y}_i|}{y_i} \times 100 \qquad (\%) \tag{5}$$

where m is the number of data pairs and y_i and $\hat{y}_i$ are the i-th actual output and model output, respectively.

The design parameters of EISON in each layer are shown in Table 3. The simulation results of the EISON are summarized in Table 4. The overall lowest values of the performance index, PI=0.188 EPI=1.087, are obtained at the third layer when the weighting factor (θ) is 0.25.

Parameters	1st layer	2nd layer	3rd layer
Maximum generations	40	60	80
Population size:(w)	20:(15)	60:(50)	80
String length	8	20	55
Crossover rate (P_c)	0.85		
Mutation rate (P_m)	0.05		
Weighting factor: θ	0.1~0.9		
Type (order)	1~3		

Table 3. Design parameters of EISON for modeling.

w: the number of chosen nodes whose outputs are used as inputs to the next layer

Weighting factor	1st layer		2nd layer		3rd layer	
	PI	EPI	PI	EPI	PI	EPI
0.1	5.7845	6.8199	2.3895	3.3400	2.2837	3.1418
0.25	5.7845	6.8199	0.8535	3.1356	0.1881	1.0879
0.5	5.7845	6.8199	1.6324	5.5291	1.2268	3.5526
0.75	5.7845	6.8199	1.9092	4.0896	0.5634	2.2097
0.9	5.7845	6.8199	2.5083	5.1444	0.0002	4.8804

Table 4. Values of performance index of the proposed EISON model.

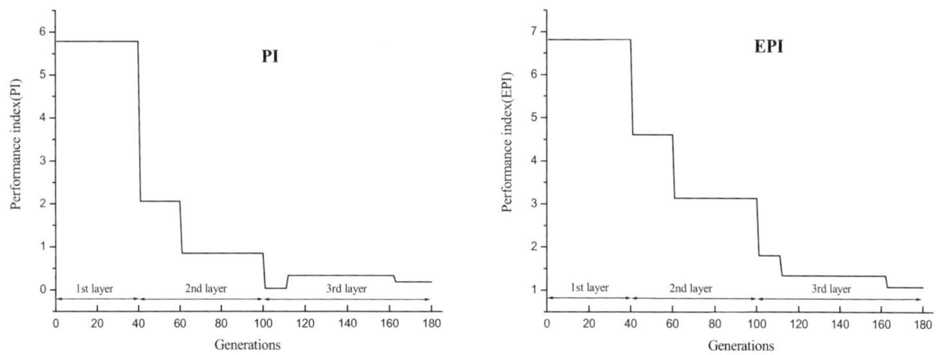

(a) training result (b) testing result

Fig. 5. Trend of performance index values with respect to generations through layers (θ =0.25).

Fig. 5 illustrates the trend of the performance index values produced in successive generations of the evolutionary algorithm when the weighting factor θ is 0.25. Fig. 6 shows the values of error function and fitness function in successive evolutionary algorithm generations when the θ is 0.25. Fig. 7 depicts the proposed EISON model with 3 layers when the θ is 0.25. The structure of EISON is very simple and has a good performance.

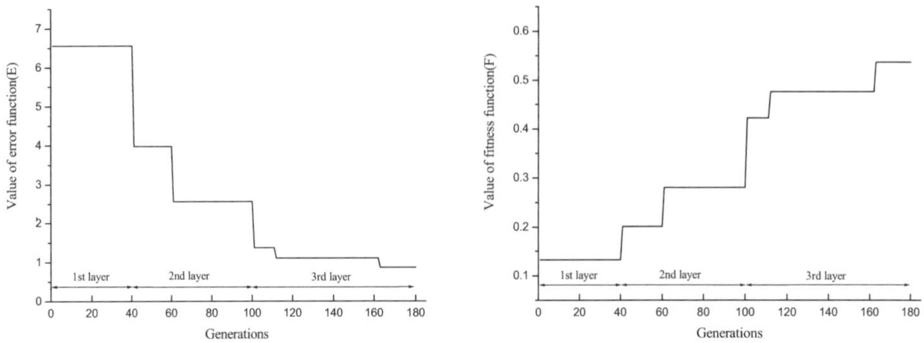

(a) error function (E) (b) fitness function (F)

Fig. 6. Values of the error function and fitness function with respect to the successive generations (θ =0.25).

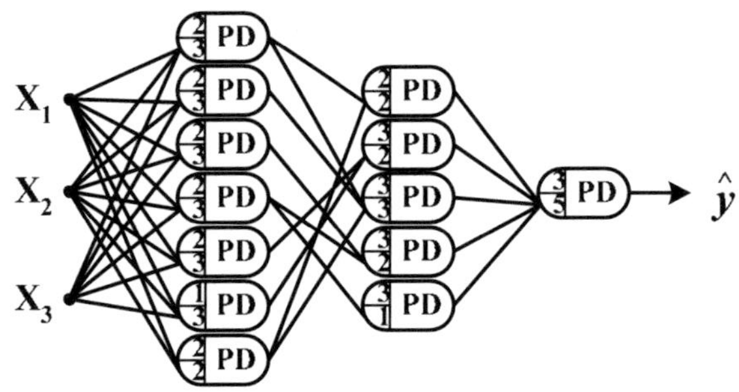

Fig. 7. Structure of the EISON model with 3 layers (θ =0.25).

Fig. 8 shows the identification performance of the proposed EISON and its errors when the θ is 0.25. The output of the EISON follows the actual output very well.

Table 5 shows the performance of the proposed EISON model and other models studied in the literature. The experimental results clearly show that the proposed model outperforms the existing models both in terms of better approximation capabilities (PI) as well as superb generalization abilities (EPI).

Model		APE	
		PI (%)	EPI (%)
GMDH model[12]		4.7	5.7
Fuzzy model [11]	Model 1	1.5	2.1
	Model 2	0.59	3.4
FNN [14]	Type 1	0.84	1.22
	Type 2	0.73	1.28
	Type 3	0.63	1.25
GD-FNN [13]		2.11	1.54
EISON		0.188	1.087

Table 5. Performance comparison of various identification models.

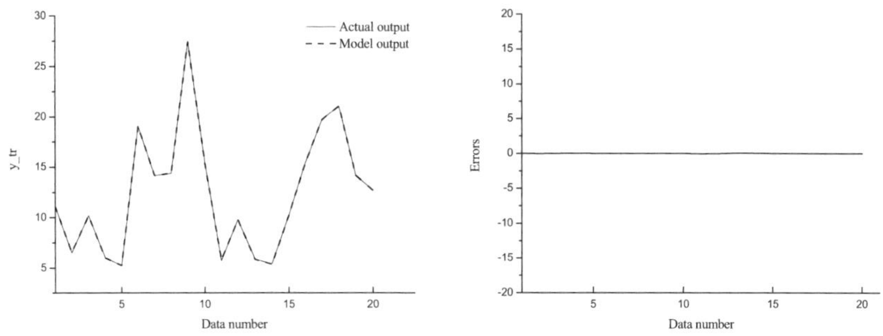

(a) actual versus model output of training data set (b) errors of (a)

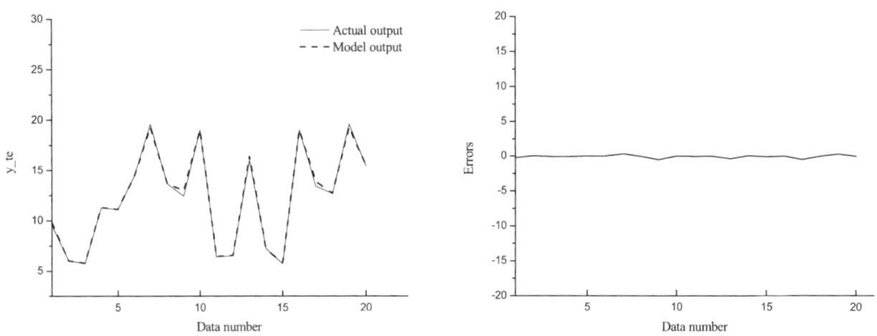

(c) actual versus model output of testing data set (d) errors of (c)

Fig. 8. Identification performance of EISON model with 3 layers and its errors

3. Practical Biped Humanoid Robot

3.1 Design

Biped humanoid robot designed and implemented is shown in Fig. 9. The specification of our biped humanoid robot is depicted in Table 6. The robot has 19 joints and the height and the weight are about 445mm and 3000g including vision camera. For the reduction of the weight, the body is made of aluminum materials. Each joint is driven by the RC servomotor that consists of a DC motor, gear, and simple controller. Each of the RC servomotors is mounted in the link structure. This structure is strong against falling down of the robot and it looks smart and more similar to a human.

Size	Height : 445mm
Weight	3kg
CPU	TMS320LF2407 DSP
Actuator (RC Servo motors)	HSR-5995TG (Torque : 30kg·cm at 7.4V)
Degree of freedom	19 DOF (Leg+Arm+Waist) = 2*6 + 3*2+1)
Power source	Battery
Actuator	: AA Size Ni-poly (7.4V, 1700mAh)
Control board	: AAA size Ni- poly (7.4V, 700mAh)

Table 6. Specification of our humanoid robot

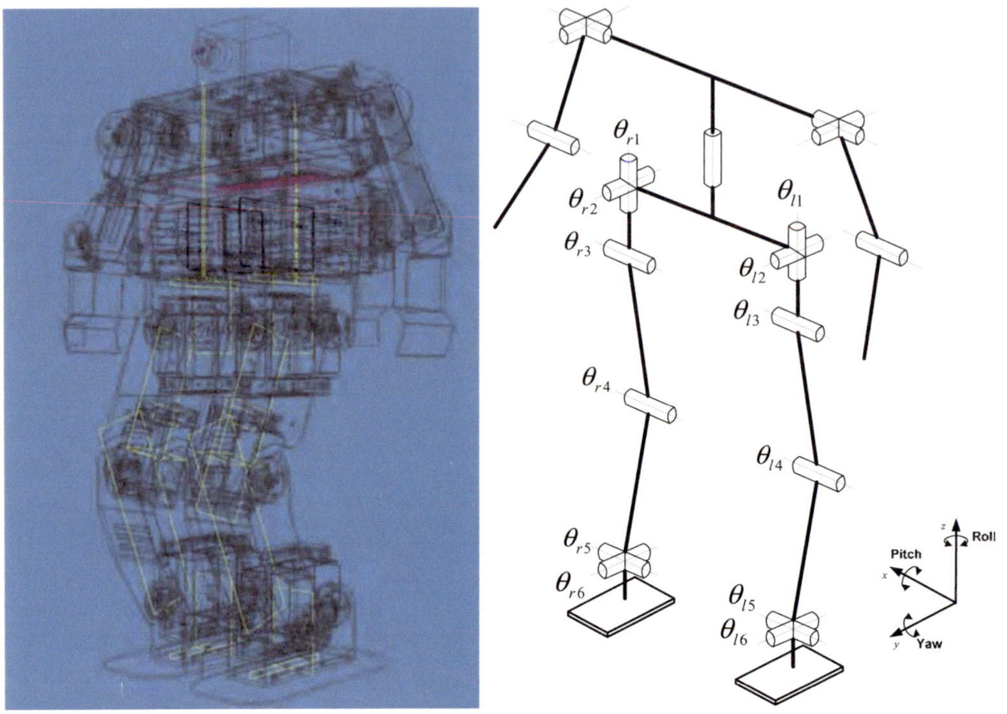

Fig. 9. Designed and implemented humanoid robot.

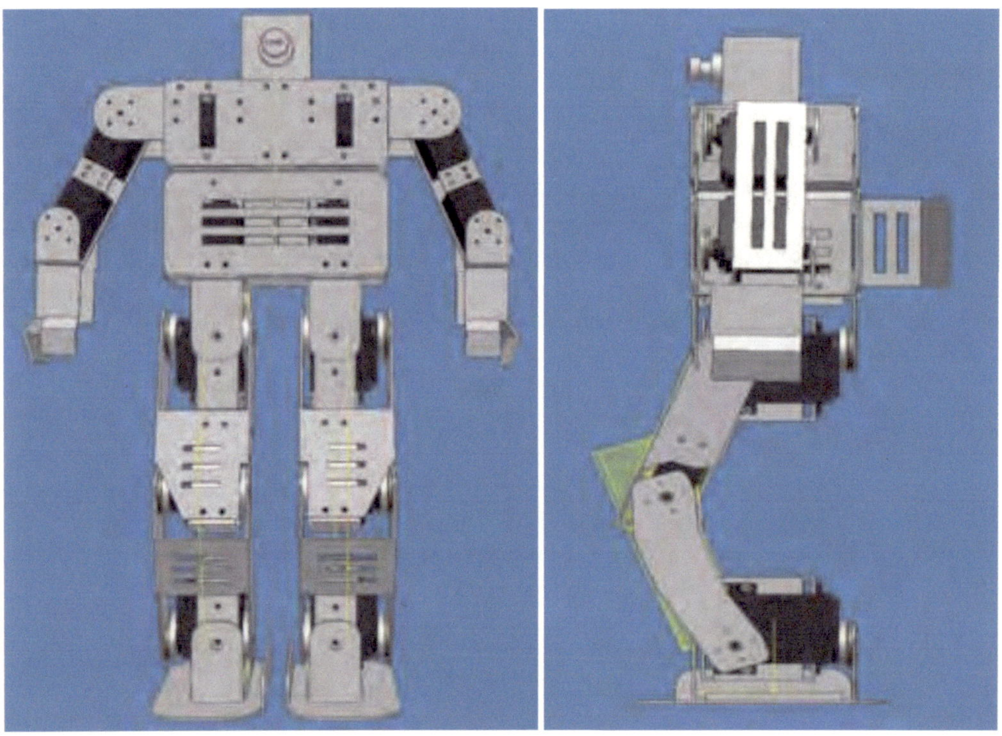

Fig. 10. 3D humanoid robot design and its practical figures

Front and side view of 3D robot and its practical figures are shown in Fig. 10. Block diagram of the biped walking robot and its electric diagram of control board and actuators are also shown in Figs. 11 and 12, respectively. Implemented control board and its electric wiring diagram schematic is presented in Fig. 13.

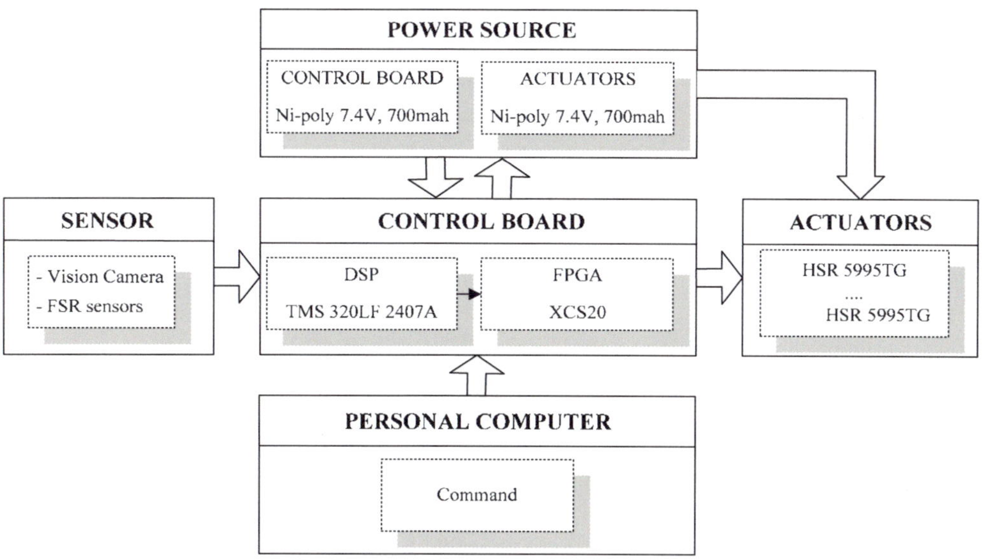

Fig. 11. Block diagram of the humanoid robot

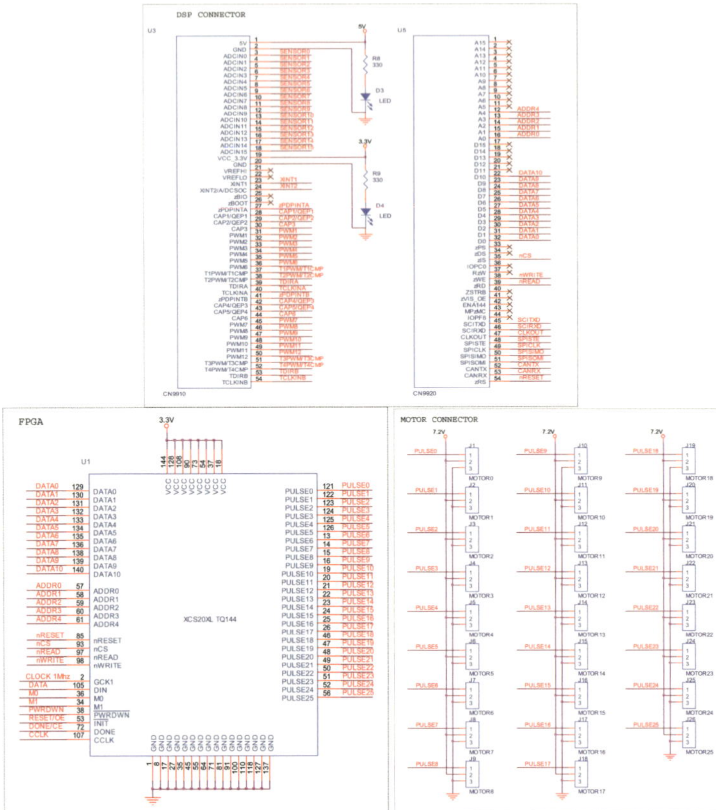

Fig. 12. Electric diagram of control board and actuators

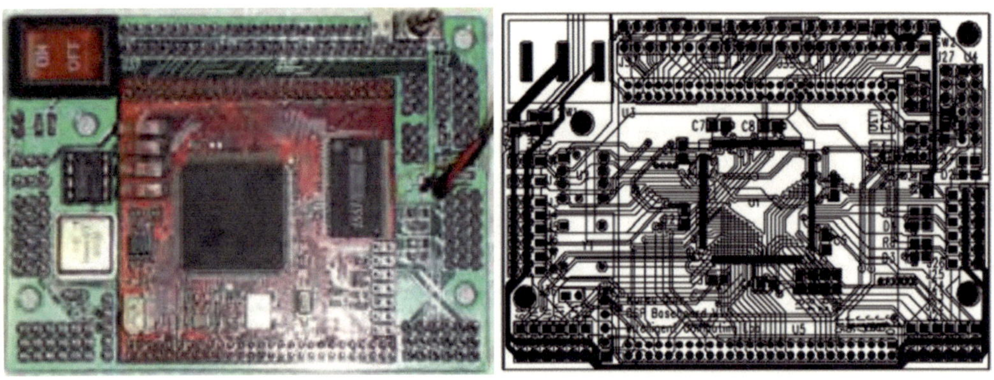

Fig. 13. Implemented control board and its electric wiring diagram schematic

3.2 Zero moment point measurement system

As an important criterion for the stability of the walk, the trajectory of the zero moment point (ZMP) beneath the robot foot during the walk is investigated. Through the ZMP, we can synthesize the walking patterns of biped walking robot and demonstrate walking motion with real robots. Thus ZMP is indispensable to ensure dynamic stability for a biped robot.

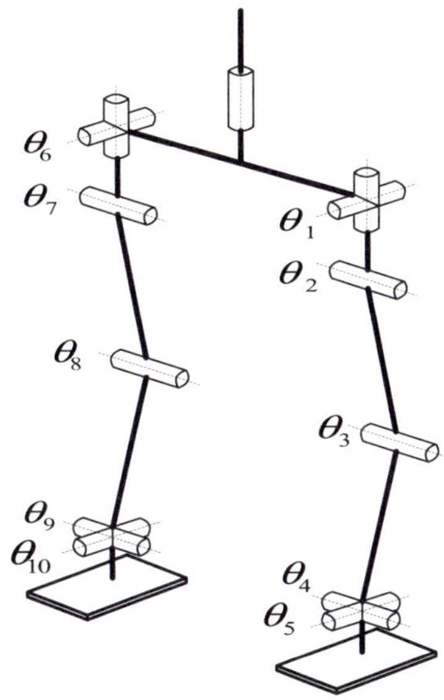

Fig. 14. Representation of joint angle of the 10 degree of freedoms.

The places of joints in walking are shown in Fig. 14. The measured ZMP trajectory data to be considered here are obtained from 10 degree of freedoms (DOFs) as shown in Fig. 14. Two DOFs are assigned to hips and ankles and one DOF to the knee on both sides. From these joint angles, cyclic walking pattern has been realized. This biped walking robot can walk continuously without falling down. The zero moment point (ZMP) trajectory in the robot foot support area is a significant criterion for the stability of the walk. In many studies, ZMP coordinates are computed using a model of the robot and information from the joint encoders. But we employed more direct approach which is to use measurement data from sensors mounted at the robot feet.

The type of force sensor used in our experiments is FlexiForce sensor A201 which is shown in Fig. 15. They are attached to the four corners of the sole plate. Sensor signals are digitized by an ADC board, with a sampling time of 10 ms. Measurements are carried out in real time.

Fig. 15. Employed force sensors under the robot feet.

The foot pressure is obtained by summing the force signals. By using the force sensor data, it is easy to calculate the actual ZMP data. Feet support phase ZMPs in the local foot coordinate frame are computed by equation 6

$$P = \frac{\sum_{i=1}^{8} f_i r_i}{\sum_{i=1}^{8} f_i} \tag{6}$$

where f_i is each force applied to the right and left foot sensors and r_i is sensor position which is vector.

4. Walking Pattern Analysis of the Humanoid Robot

The walking motions of the biped humanoid robot are shown in Figs. 16-18. These figures show series of snapshots in the front views of the biped humanoid robot walking on a flat floor, some slopes, and uneven floor, respectively. Fig. 16 gives the series of front views of this humanoid robot walking on a flat floor. In Fig. 17 depict the series of front views of this humanoid robot going up on an ascent. Fig. 18 shows another type of walking of biped humanoid robot, which is walking motion on an uneven floor.

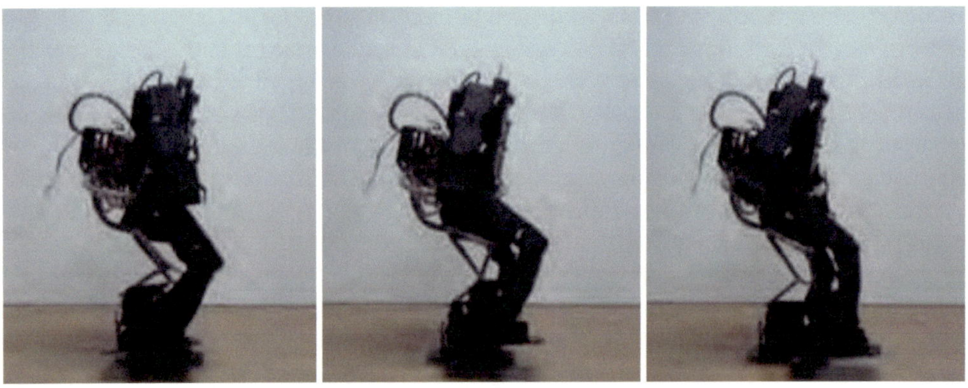

Fig. 16. Side view of the biped humanoid robot on a flat floor

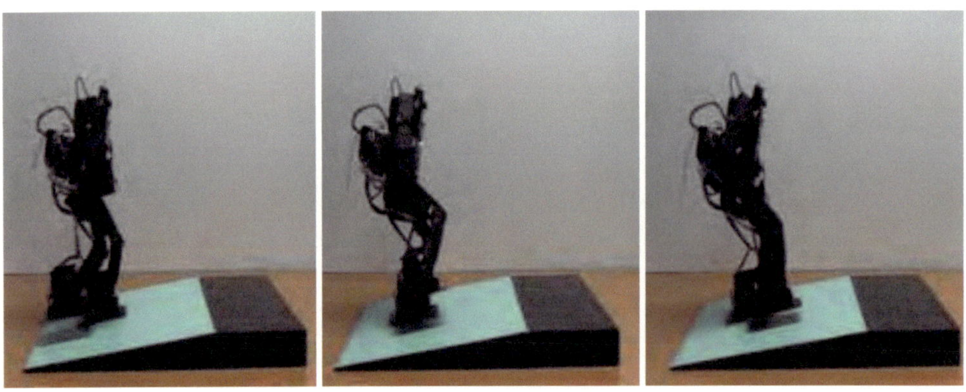

Fig. 17. Side view of the biped humanoid robot on an ascent.

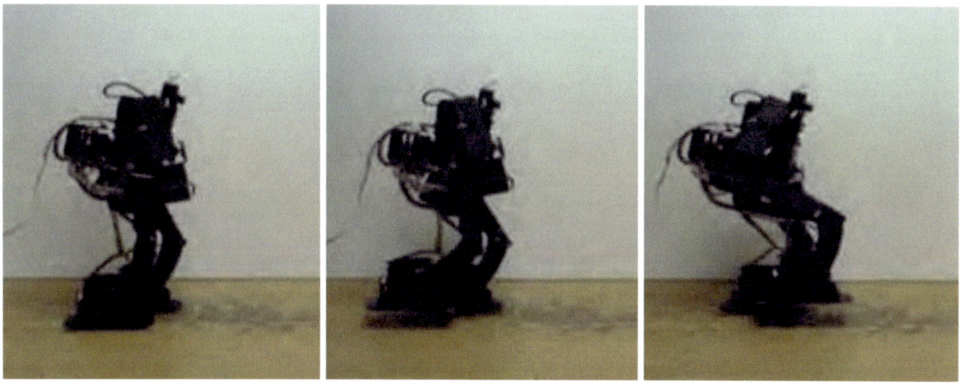

Fig. 18. Side view of the biped humanoid robot on an uneven floor.

Experiments using EISON was conducted and the results are summarized in tables and figures below. The design parameters of evolutionary inductive self-organizing network in each layer are shown in Table 7. The results of the EISON for the walking humanoid robot on the flat floor are summarized in Table 8. The overall lowest values of the performance indicies, 6.865 and 10.377, are obtained at the third layer when the weighting factor (θ) is 1. In addition, the generated ZMP positions and corresponding ZMP trajectory are shown in Fig. 19. Table 9 depicts the condition and results for the actual ZMP positions of our humanoid walking robot on an ascent floor. We can also see the corresponding ZMP trajectories in Fig. 20, respectively.

Parameters	1st layer	2nd layer	3rd layer
Maximum generations	40	60	80
Population size:(w)	40:(30)	100:(80)	160
String length	13	35	85
Crossover rate (P_c)	0.85		
Mutation rate (P_m)	0.05		
Weighting factor: θ	1		
Type (order)	1~3		

Table 7. Design parameters of evolutionary inductive self-organizing network.

w: the number of chosen nodes whose outputs are used as inputs to the next layer

Walking condition slope (deg.)	Layer	MSE (mm)	
		x-coordinate	y-coordinate
0°	1	9.676	18.566
	2	7.641	13.397
	3	6.865	10.377

Table 8. Condition and the corresponding MSE are included for actual ZMP position in four step motion of our humanoid robot.

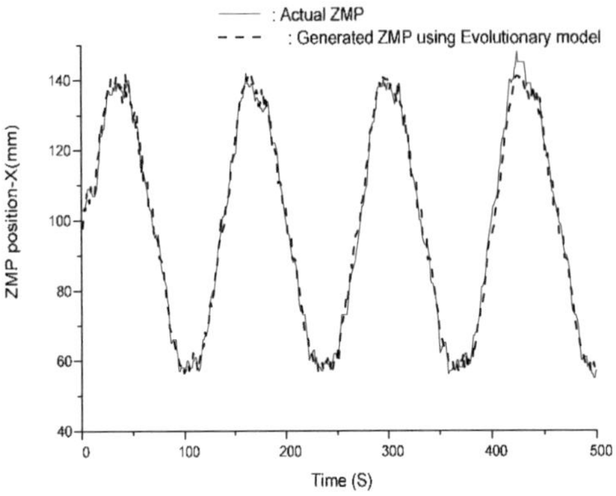

(a) x-coordinate

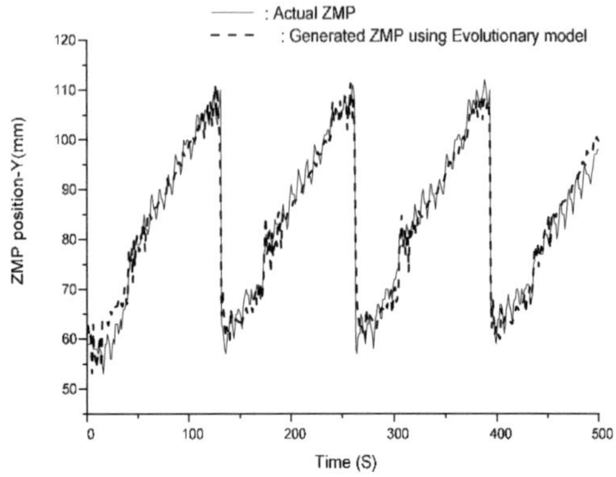

(b) y-coordinate

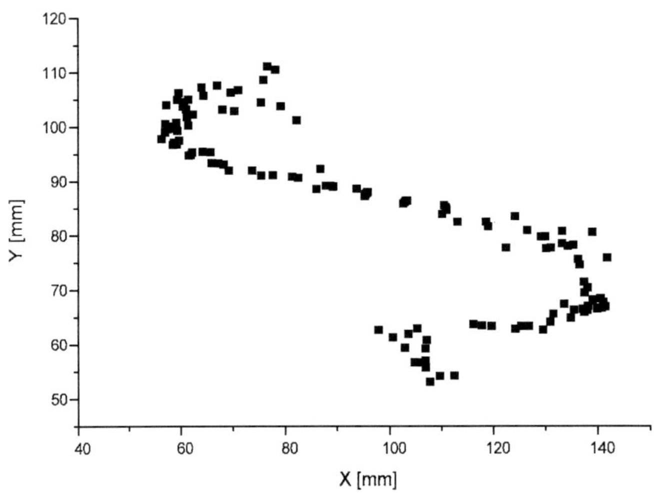

(c) Generated ZMP trajectories

Fig. 19. Generated ZMP positions and corresponding ZMP trajectories (0°).

5. Concluding remarks

This chapter deals with advanced humanoid robot based on the evolutionary inductive self-organizing network. Humanoid robot is the most versatile ones and has almost the same mechanisms as a human and is suitable for moving in an human's environment. But the dynamics involved are highly nonlinear, complex and unstable. So it is difficult to generate human-like walking motion. In this chapter, we have employed zero moment point as an important criterion for the balance of the walking robots. In addition, evolutionary inductive self-organizing network is also utilized to establish empirical relationships between the humanoid walking robots and the ground and to explain empirical laws by incorporating them into the humanoid robot. From obtained natural walking motions of the humanoid robot, EISON can be effectively used to the walking robot and we can see the synergy effect humanoid robot and evolutionary inductive self-organizing network.

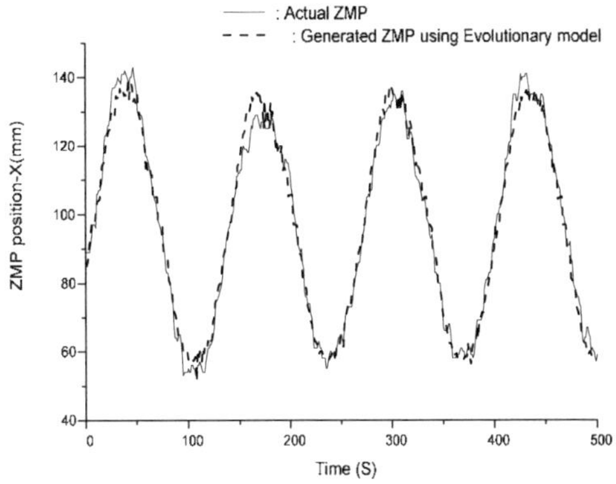

(a) x-coordinate

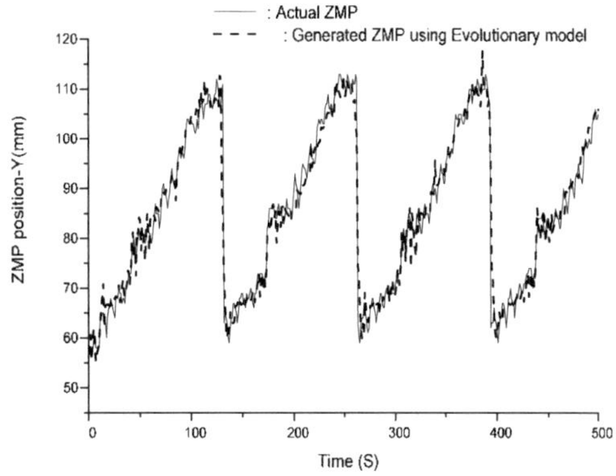

(b) y-coordinate

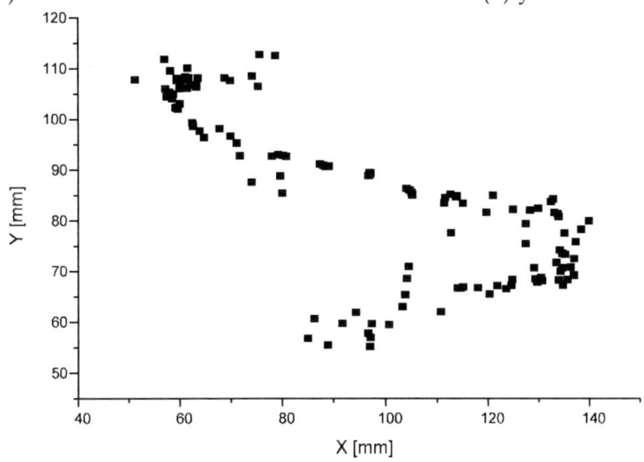

(c) Generated ZMP trajectories

Fig. 20. Generated ZMP positions and corresponding ZMP trajectories (10°).

6. Acknowledgements

The authors thank the financial support of the Korea Science & Engineering Foundation. This work was supported by grant No. R01-2005-000-11044-0 from the Basic Research Program of the Korea Science & Engineering Foundation.

7. References

[1] Erbatur, K.; Okazaki, A.; Obiya, K.; Takahashi, T. & Kawamura, A. (2002) A study on the zero moment point measurement for biped walking robots, *7th International Workshop on Advanced Motion Control*, pp. 431-436

[2] Vukobratovic, M.; Brovac, B.; Surla, D.; Stokic, D. (1990). *Biped Locomotion*, Springer Verlag

[3] Takanishi, A.; Ishida, M.; Yamazaki, Y.; Kato,I. (1985). The realization of dynamic walking robot WL-10RD, *Proc. Int. Conf. on Advanced Robotics*, pp. 459-466

[4] Hirai, K.; Hirose, M.; Haikawa, Y.; Takenaka, T. (1998). The development of Honda humanoid robot, *Proc. IEEE Int. Conf. on Robotics and Automation*, pp. 1321-1326

[5] Park, J. H.; Rhee, Y. K. (1998). ZMP Trajectory Generation for Reduced Trunk Motions of Biped Robots. *Proc. IEEE/RSJ Int. Conf. Intelligent Robots and Systems, IROS '98*, pp. 90-95

[6] Park, J. H.; Cho, H. C. (2000). An On-line Trajectory Modifier for the Base Link of Biped Robots to Enhance Locomotion stability, *Proc. IEEE Int. Conf. on Robotics and Automation*, pp. 3353-3358

[7] Kim, D.; Kim, N.H.; Seo, S.J.; Park, G.T. (2005). Fuzzy Modeling of Zero Moment Point Trajectory for a Biped Walking Robot, *Lect. Notes Artif. Int.*, vol. 3214, pp. 716-722 (**BEST PAPER AWARDED PAPER**)

[8] Kim, Dongwon; Park, Gwi-Tae. (2006). *Evolution of Inductive Self-organizing Networks*, Studies in Computational Intelligence Series: Volume 2: Engineering Evolutionary Intelligent Systems, Springer

[9] Kim, D. W.; and Park, G. T. (2003). A Novel Design of Self-Organizing Approximator Technique: An Evolutionary Approach, *IEEE Intl. Conf. Syst., Man Cybern. 2003, pp. 4643-4648* **(BEST STUDENT PAPER COMPETITION FINALIST AWARDED PAPER)**

[10] Takagi, H.; Hayashi, I. (1991). NN-driven fuzzy reasoning, *Int. J. Approx. Reasoning*, Vol. 5, No. 3, pp. 191-212

[11] Sugeno, M.; Kang, G. T. (1988). Structure identification of fuzzy model, *Fuzzy Sets Syst.*, Vol. 28, pp. 15-33

[12] Kondo, T. (1986). Revised GMDH algorithm estimating degree of the complete polynomial, *Tran. Soc. Instrum. Control Eng.*, Vol. 22, No. 9, pp. 928-934 (in Japanese)

[13] Wu, S.; Er, M.J.; Gao, Y. (2001). A Fast Approach for Automatic Generation of Fuzzy Rules by Generalized Dynamic Fuzzy Neural Networks, *IEEE Trans. Fuzzy Syst.*, Vol. 9, No. 4, pp. 578-594

[14] Horikawa, S. I.; Furuhashi, T.; Uchikawa, Y. (1992). On Fuzzy modeling Using Fuzzy Neural Networks with the Back-Propagation Algorithm, *IEEE Trans. Neural Netw.*, Vol. 3, No. 5, pp. 801-806

26

Balance-Keeping Control of Upright Standing in Biped Human Beings and its Application for Stability Assessment

Yifa Jiang Hidenori Kimura
Biological Control Systems Laboratory, RIKEN (The Institute of Physical and Chemical Research) Nagoya, 463-0003, Japan

1. Abstract

One of the most important tasks in biped robot is the balance-keeping control. A question arisen as how do our human beings make the balance-keeping possible in upright standing as human beings are the only biped-walking primates, which takes several million years of evolution period to achieve this ability. Studies on humans' balance-keeping mechanism are not only the work of physiologists but also a task of robot engineers since bio-mimetic approach is a shortcut for developing humanoid robot. This chapter will introduce some research progresses on balance-keeping control in upright standing. We will introduce the physical characteristics of human body at first, modeling the physical system of body, establishing a balance-keeping control model, and at last applying the balance-keeping ability assessment for falls risk prediction. We wish those studies make contributions to robotics.

2. Introduction

Scientist's interest and fascination in balance control and movement has a long history. Leonardo da Vinci emphasized the importance of mechanics in the understanding of movement and balance, wrote that "Mechanical science is the noblest and above all the most useful, seeing that by means of it all animated bodies which have movement perform all their actions[1]". In 1685, by using a balance board, Italian mathematician Johannes Alphonsus Borelli, in his "De motu animalium", determined the position of the center of gravity as being 57% of the height of the body taken from the feet in the erect position. Those early studies on human body mechanics determining the inertial properties of different body segments serve an important and necessary role in modern biomechanics.
Not like static upright stance in biped robot, upright standing in human is a high-skill task needed a precise control. In 1862 Vierordtand later Mosso (1884) demonstrated that normal standing was not a static posture but rather a continuous movement. Kelso and Hellebrandt in 1937 introduced a balance platform to obtain graphic recording of the involuntary postural sway of man in the upright stance and to locate the centre of weight with respect to the feet as a function of time. Using a force analysis platform and an accelerometer,

Whitney[2] (1958) concluded that the movement of the center of foot pressure must exaggerate the accompanying movement of the center of gravity of the body mass. Based on those studies, postural sway is regarded as presenting at the hip level suggests that the trunk rotates relative to the limbs during standing.
In other hand, clinical significance of postural sway was first observed by Babinski in 1899. He noted that a patient with the disorder termed "asynerhie cerebelleuse" could not extent his trunk from a standing position without falling backwards. He concluded that the ankle must perform a compensatory movement to prevent the subject from falling backwards. Thus, Babinski was one of the first to recognize the presence of an active postural control of muscular tone and its importance in balance control and in voluntary movements.
Modern theory on postural control was established on the work of Magnus[3] and De Kleijn[4]; they proposed that each animal species have a reference posture or stance, which is genetically determined. According to this view, postural control and its adaptation to the environment is based on the background postural tone and on the postural reflexes or reactions, which are considered to originate from inputs from the visual[5], vestibular[6] and proprioceptive[7] system. The gravity vector is considered to serve as a reference frame. The center nervous system needs to accomplish two tasks related to the force of gravity. First, it must activate extensor muscles to act against the gravity vector, creating postural tone. Second, it must stabilize the body's centre of gravity with respect to the ground.
Many studies[8-10] on balance-keeping control have been published in recent two decades. A lot of theoretical control models have been proposed for elucidating the body sway phenomenon during upright standing. Among them, Inverted pendulum is the most popular model for body sway analysis. However, those studies are seldom considered individual's physical conditions, and one-link inverted pendulum is dominated. However, a practicable control model for humans' balance-keeping ability assessment is still unavailable. For this reason, we proposed a PID model of upright standing, and a reality-like body sway was simulated. This chapter presents some recent research results in our laboratory in balance-keeping control and emphasizes its application for fall risk prediction in fall-prone subjects.

3. Upright Standing and Body Sway

Force-plate is a popular device for postural sway measurement[11]. A force-plate usually installs three or four force sensors for ground reaction force measurement. The center of ground reaction pressure (COP) can be calculated automatically by a computer program. However, COP is not equal to body sway because body's inertia is an important factor which influences the deviation of COP[12]. We developed an optical measuring system, which can directly record the trunk sway.
The body sway-measuring system was designed to record multi-channel video signal simultaneously (Fig.1). This device included three high-resolution CCD video cameras and one personal computer that were used for video signal recording and image processing. Three markers (white ball, 3.0cm in diameter) were used for imaging recognizing with one being put on subjects' back and two being put on legs 10cm above about the knee joint, and on the floor, a force-plate was installed for COP deviation assessment. During the measuring, subjects were told to stand on the force-plate and glance at a marker put on the front wall in the same level of their eyes.

Fig.1. The scheme shows of the body sway-measuring device. The device consists of three high-resolution CCD cameras (Cam1, Cam2 and Cam3) and a whiteboard. A white ball with diameter of 1.0cm was put on the whiteboard in subject's eye level. During the measurement, subjects were asked to fix their eyes on the ball. Postural sway analyzing was executed on one desktop PC, and an EMG recording PC was also installed too.

In this study, body sway in lateral direction was recorded and the body was regarded as two-link inversed pendulum. The motion of the first link was on ankle and the second link was on lumbosacral. The angular sway of the first link was measured as the averaged value of the two legs and the angular sway of the second link was calculated indirectly as shown in Fig.2, where point O is the ankle joint and point A is the lumbosacral joint. P1 represents the marker of legs and P2 represents the marker of subject's back. There have

$$x_1 = l_1 \sin\theta_1 \quad (1)$$

$$x_2 = L_1 \sin\theta_1 + l_2 \sin(\theta_1 + \theta_2). \quad (2)$$

Because the angular deviation scope is relatively small (usually < 2.0 degree), then we set $\sin\theta_1 = \theta_1, \sin(\theta_1 + \theta_2) = \theta_1 + \theta_2$ approximately, and θ_2 can be calculated as equation (3),

$$\theta_2 = \left(\frac{x_2}{L_1 + l_2} - \frac{x_1}{l_1}\right)\left(1 + \frac{L_1}{l_2}\right). \quad (3)$$

Here, we defined the height of lumbosacral as L_1, which is the distance from floor to 5th lumbar spine. The entire calculations can carried out online. Subjects kept a static upright stance on the force-plate with eyes open or eyes closed. For minimizing psychological influence, subjects were asked to numerate on mind while doing standing task.

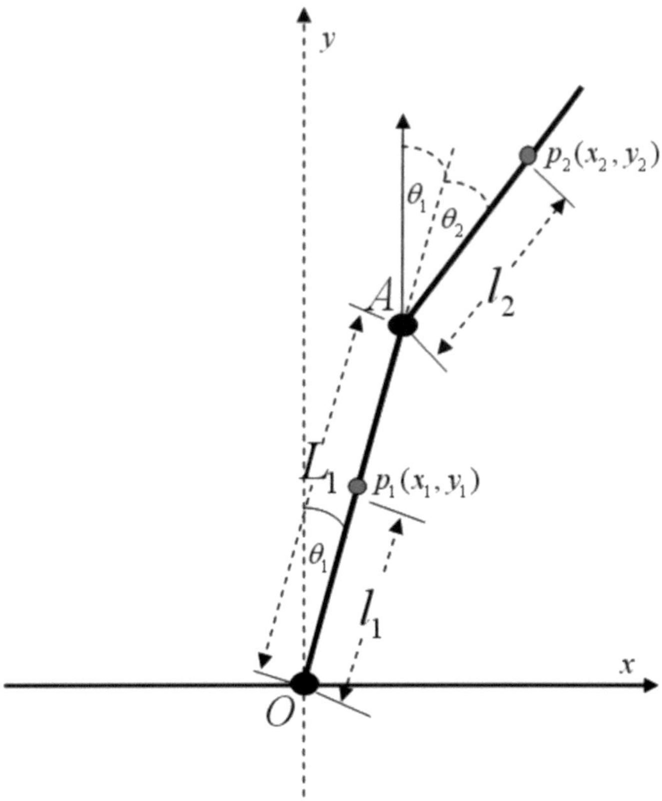

Fig.2. Relation of the sway angles during static upright stance. Points P1 and P2 are the positions of the markers, their coordinates can be calculated by the CCD video camera measuring system. Points O and A represent are the ankle joint and lumbosacral joint respectively. θ_1 is the angular sway of the ankle joint, and θ_2 is the angular sway of the lumbosacral joint. The postural sway is considered only in coronal direction.

A study from eight healthy young subjects[13] shown that the body sways scope of ankle and lumbosacral were 0.94±0.36degree (eye-open), 1.35±0.52degree (eye-closed) and 0.99±0.41degree (eye-open), 1.27±0.72degree (eye-closed), respectively. No significant difference existed between the ankle and lumbosacral. The ankle and lumbosacral sway approximately in the same degree during the static upright stance. Further more, Fourier transform of the sway time series showed that the phase differences between ankle and lumbosacral were approximately equal to π, i.e. i,e, ankle sways in opposite direction to the lumbosacral. The results help for establishing a structural inverted pendulum model.
We also analyzed the relationship between trunk sway and deviation of COP[12]. By setting $y(t)$ as the deviation of COP, $u(t)$ as the trunk sway, an approximate equation can be acquired, $y(t) \approx k\ddot{u}(t) + hu(t)$, where, k, h are constants. The result proven that COP deviation is different with trunk sway, and body's inertia effect does added on the effect of COP deviation as Whitney pointed out[5].

4. Musculoskeletal System of Human Body

From the viewpoint of structural specificity of the human body, pelvis and ankle are two major structures that play a pivotal role in balance control, the pelvic strategy and ankle strategy as hereby defined.

Human pelvis is composed of four irregular bones: two hip bones laterally and in front the sacrum and coccyx behind. Sacrum articulated with vertebral column formed lumbosacral joint, and also make joint with two femurs formed hip joints (Fig.3a). Muscles associated with lumbosacral are mainly ascribed to two pairs: posas major (PM) and glutaeus medius (GM), and in addition, also the erectors. In fact, many muscles surround hip joint are involved in the joint movement, and because the synergic effect[14, 15] of muscles' activity, it is difficult to deal each individual muscle separately. For model's simplicity, two pairs of muscles are selected for representing the total muscles' effects in the lateral direction movement (Fig.3a), i.e. PM and GM. It is regarded that the movement of lumbosacral is controlled by PMs and GM. This assumption can well explain the relationship between GM contraction and COP deviation in coronal direction.

The characteristic structures of femur are its head, greater trochanter and lesser trochanter. The greater trochanter is a large, irregular, quadrilateral eminence, situated at the junction of the neck with the upper part of the body, serves as the insertion of the tendon of the GM. The Lesser Trochanter is a conical eminence, projects from the lower and back part of the base of the neck, gives insertion to the tendon of the PM. The shape of femur looks like a letter of "Y" (Fig.3b). Based on the structural characteristics of pelvis, lumbosacral, hip and femurs an upright body model was constructed (Fig.3c). In this model, pelvis is expressed as a triangle connecting vertebral column and femurs with lumbosacral joint and hip joint respectively and driven by two symmetrical pairs of actuators, the PM and GM, form a closed multi-link system. In order to make the dynamics analysis concisely, the distance between two feet was supposed equal to the distance between two hip joints. Thus, the aim of central nerve system is to keep the angles θ_1 and θ_2 to be zero, i.e. to keep the body upright (Fig.3c).

From the structural model of body the position of vertical projection of center-of-mass (COM) defined as V_{cop} which can be calculated as equation (4).

$$V_{com} = \frac{\sum_i m_i g V_{m_i}}{\sum_i m_i g} \qquad (4)$$

When $|V_{com}| > d$, upright standing is impossible and falls may happen. In other words, upright stance is possible only when

$$|V_{com}| \leq d. \qquad (5)$$

Because the angular sway scope in ankle is approximately equal to the sway scope in lumbosacral, and their phase difference is π, the relationship of sway angles between ankle and lumbosacral can be considered as (Fig.3c)

$$\theta_1 = -\theta_2. \qquad (6)$$

Equation (6) also indicates that the trunk of body is always keeping perpendicular to the horizon. Based on this fact the structure model of body can be simplified as be shown in Fig.4.

In this model, PM and GM are the actuators keeping the inverted pendulum to be balanced. By intuition, right GM and left PM should activate simultaneously, which is an agonist against left GM and right PM. Its dynamics can be deduced by Lagrangian equation as equation (7).

$$\tau = [2m_1 l_g^{\,2} + (m_2 + m_3) l_1^{\,2}]\ddot{\theta} - [2m_1 l_g + (m_2 + m_3) l_1] g Sin\theta \qquad (7)$$

(τ : The torque generated from activities of GM and PM)

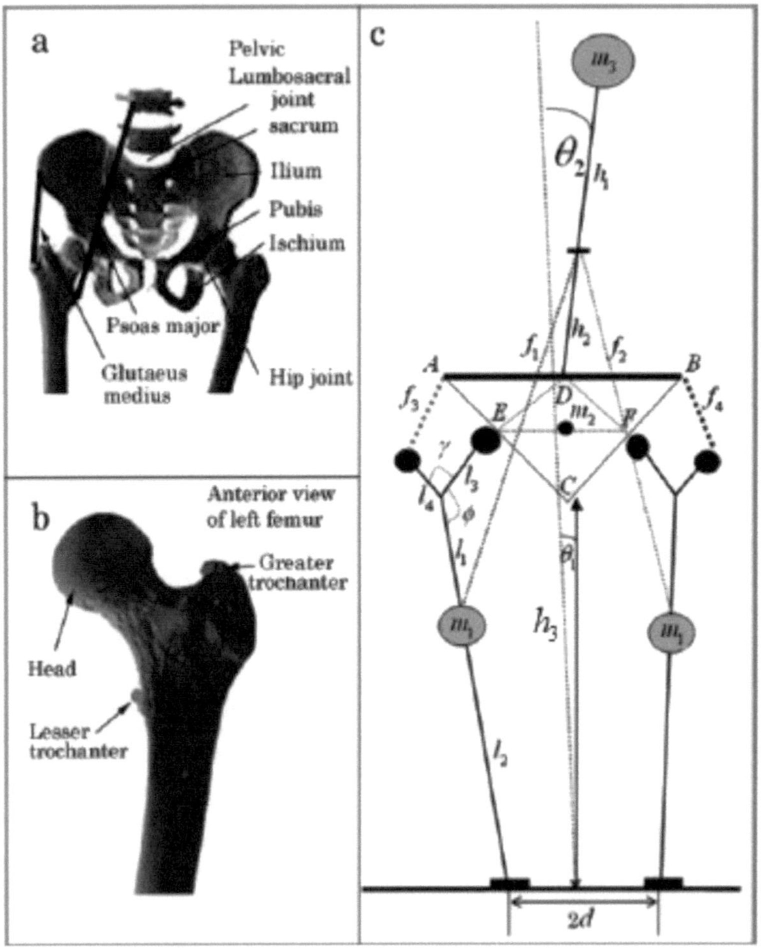

Fig.3. The pelvic anatomical structure and its responding structure model during static upright stance. The pelvic is fused by four bones, the sacrum, ilium, pubis and ischium, which linked with 5th lumbar spine and femur through lumbosacral and hip joint (a). The femur includes a head that is articulated with ilium and a greater trochanter and lesser trochanter where the GM and PM terminated (b). The structural model of pelvic is schemed (c). f_1, f_2 are the forces produced by PM, and f_3, f_4 are produced by GM. m_1, m_2 and m_3 represented the mass of lower leg, pelvic and upper body respectively. θ_1 is the sway angle of leg and θ_2 is the sway angle of upper trunk.

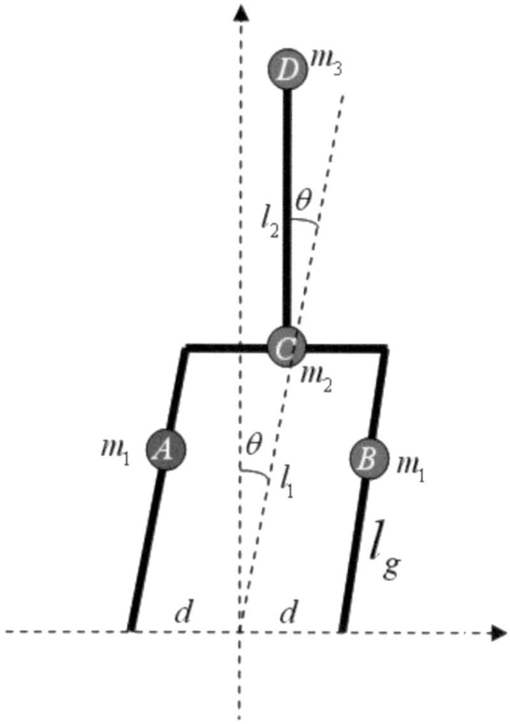

Fig.4. A simplified pelvic structural model during static upright stance. A, B are the masses of leg, C is the mass of pelvic and D is the mass of upper trunk. Because the lumbosacral always sways in inverse direction of the ankle joint with the same value of θ, the upper trunk is kept perpendicular to the horizon.

5. The Roles of Foot in Balance Control

Feet are special structure that evolved fit for standing and locomotion. The anatomy structure of human feet is quite complex which make up of a doze of muscles and near thirty bones. When subject stand on one foot, brisk EMG activities can be recorded on the surface of muscles around the ankle joint[16], by contrast, nearly no EMG activities can be recorded while standing on two feet. Using a force sheet to measure COP deviation, we recorded the surface EMG activities of m. tibialis anterior (TA), m. peroneus longus (PL) and Caput mediale m. gastrocnemii (MG) when asked subjects standing on one foot. The relationship between COP deviation and muscular activities was identified[13] (Fig.5). Based on the experimental data we established a model of muscular control of the COP deviation (Fig.6). Briefly, the ground reaction under the foot can be concentrated to three points; one is situated on the heel the other two are situated on the root of thumb and little figure respectively, which making up a triangle. The ground reaction forces acting on the vertex of the triangle are controlled by TA, PL and MG respectively as shown in the figure 6, which keeping the COP located in the scope of the triangle to make the inverted body balanced.

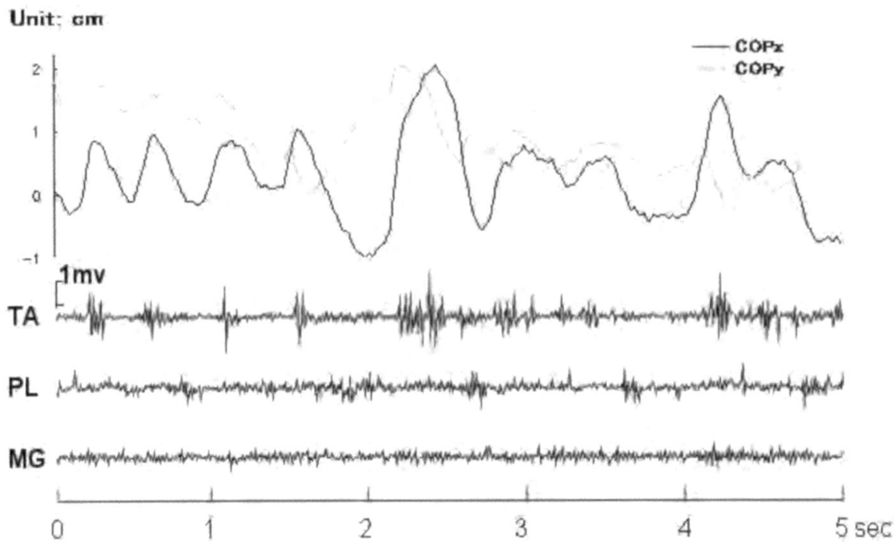

Fig.5. Showing COP deviation and surface EMG recordings from TA, PN and TS when standing on one foot. The COPx (coronal section) is closely related to TA and PN contraction.

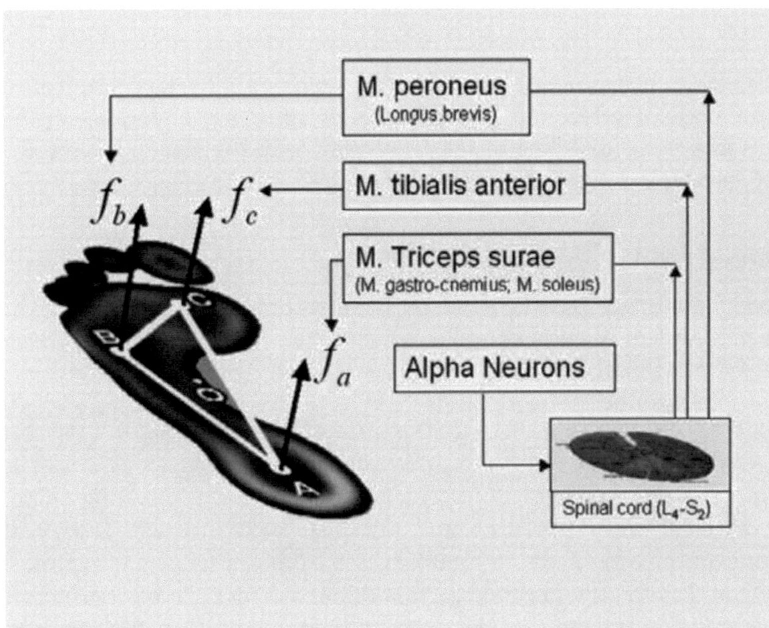

Fig.6. The model of COP control when standing on one foot. The scheme shows how the force under the three points is controlled by muscular contraction. Supposing the forces act on the three points as f_a, f_b and f_c, then the muscles bear the responsibility should be PN, TA, and TC respectively. The three muscles are innerved by neurons located at the L4-S2 level.

Nervous commands come from cerebral cortex reached to alpha motor neurons of the spinal anterior column to control the muscles around ankle joint to keep the body balanced. This kind of balance-keeping control is ankle strategy of balance control. In contract, nervous commands come from cerebrum to control the muscles around pelvis joints to keep the body balanced, hereby defined as pelvis strategy. It seems that the balance-keeping strategy is transformed from ankle to pelvis when the standing posture changes from one foot to two feet of upright standing.

Generally, the contact areas of the feet and ground play a key role in the static upright standing, with the more contact areas the feet have, the more stable of the posture[17] Will be. The contact surface is a polygon. Their vertexes are the contact points receive ground reaction force, and the area of the polygon was defined as effective contact area (Fig.7). What important is the physical stability of the inverted body, which determined by the height of the COM and the effective contact area of polygon. By defining the effective contact areas as s, the body mass as m and the height of COM as h, then $C = mgh/\sqrt{s}$ is an constrained value of the upright standing ability. In human beings, when standing on one foot, s value is approximately equal to 3000 cm^2, thus the C value reaches up to 10,000 Newton, which is far larger than the biped robot in the world in current time.

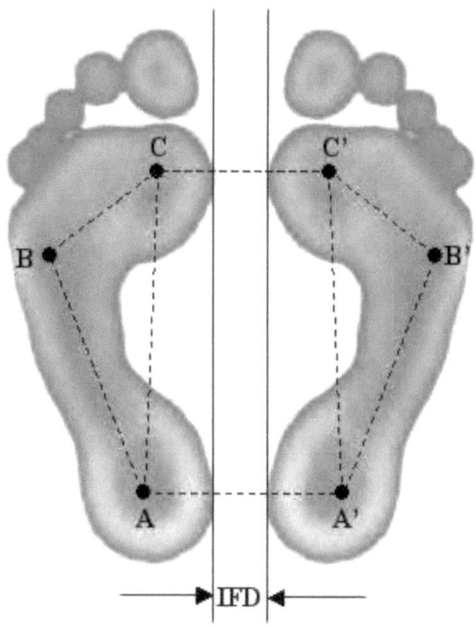

Fig.7. Scheme shows the effective contact area of support when standing in a static upright stance. The area is changed with different inter-foot-distance (IFD).

6. PID Model of Balance Control

As upright standing often modeled as an inverted pendulum, a feedback balance control is speculated to stabilize the pendulum system[18, 19]. Similar to control systems of any artifacts, balance-keeping control of human body is composed of sensors, actuators and a controller.

Physiologically, at least three sensory organs are contributed to balance-keeping, i.e. the vestibular organ[5], eyes[6] and proprioceptors[7] of muscles. Those sensors detect the gravity vector or body kinetic state and send information to the central nervous system (CNS). Actuators are muscles which affiliated to various body parts to produce torques. As mentioned ahead two pairs of muscle in pelvis, GM and PM are especially important in balance-keeping control (Figure 3a). The CNS works as a controller which controls muscles of whole body to produce torque needed for balance-keeping. Up now, the mechanism of body balance control is not clear clarified.
The body sway can be viewed as an output of body control process of humans' brain, which provides an interesting clue for investigating inner balance-keeping control mechanism. Using a classical control theoretical approach, a multi-link model of human body in coronal section as shown in Figure 3 has been constructed. The PID controller is modeled for the body's balance-keeping controller in upright standing. The dynamics of upright standing body is expressed as equation (7), the controller is

$$e(t) = r(t) - \theta(t) \qquad (r(t) = 0) \tag{8}$$

$$\tau(t) = K_P e(t - t_d) + K_D \frac{de(t - t_d)}{dt} + K_I \int_0^t e(t - t_d)dt, \tag{9}$$

where t_d is the time lag, K_P, K_D and K_I are unknown gains of the PID control. By combining the Eq. (7) and the Eq. (9), and set $\sin\theta(t) = \theta(t)$ yields

$$[2m_1 l_g^2 + (m_2 + m_3) l_1^2]\ddot{\theta}(t) - [2m_1 l_g + (m_2 + m_3) l_1]g\theta(t)$$
$$= K_P e(t - t_d) + K_D \frac{de(t - t_d)}{dt} + K_I \int_0^t e(t - t_d)dt. \tag{10}$$

The overall system is depicted in Figure 8.

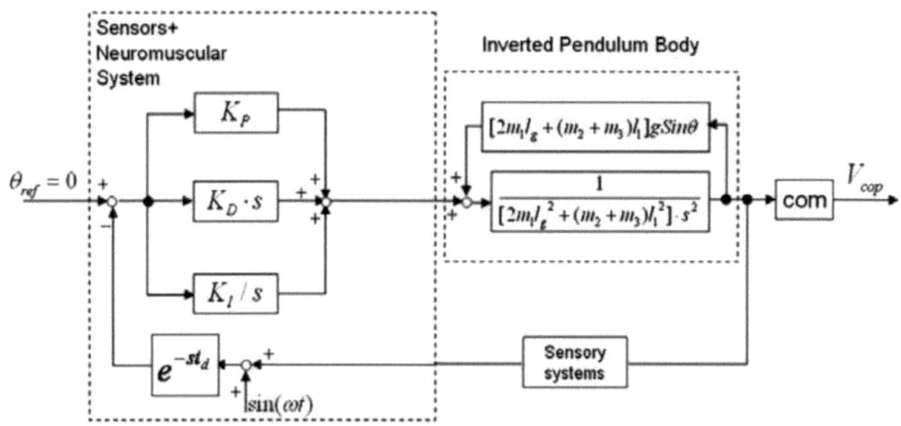

Fig.8. A block diagram of humans' static upright stance control. Because the aim of standing is to keep the body upright, the reference value of postural sway angle is set to zero. The central nervous system detects the error signal and sends output signal to muscles to keep the body upright, a state of equilibrium.

One task is to estimate its parameters of the PID model according to the measurement of body sway. This is a typical closed-loop identification problem to estimate controller parameters instead of plant parameters. Here, we take a modified lest square method for PID parameters estimation since a time lag parameter t_d is included in Eq. (10). The parameters of K_P, K_D, K_I and t_d can be identified based on the observation of θ. We use a linear regression scheme by rewriting the Eq. (10) as

$$y(t)= K_P u_1(t-t_d)+K_D u_2(t-t_d)+K_I u_3(t-t_d), \qquad (11)$$

$$(y=[2m_1 l_g^2+(m_2+m_3)l_1^2]\ddot{\theta}(t)-[2m_1 l_g+(m_2+m_3)l_1]g\theta(t),$$

$$u_1(t-t_d)=e(t-t_d),$$

$$u_2(t-t_d)=\frac{de(t-t_d)}{dt},$$

$$u_3(t-t_d)=\int_0^t e(t-t_d)dt).$$

Here, the $y(t)$, $u_1(t-t_d)$, $u_2(t-t_d)$ and $u_3(t-t_d)$ are measurable. Rewriting Eq. (11) as:

$$\mathbf{y}=\Omega\mathbf{k}+\mathbf{e}, \qquad (12)$$

here, $\mathbf{y},\mathbf{e}$ are $m\times 1$ vectors, $\mathbf{k}=[K_P\ K_D\ K_I]^t$ and Ω is a $m\times 3$ matrix ($m\geq 3$), then the parameters can be calculated as:

$$\begin{bmatrix} K_P \\ K_D \\ K_I \end{bmatrix} = A^{-1}\begin{bmatrix} (y(t),u_1(t-t_d)) \\ (y(t),u_2(t-t_d)) \\ (y(t),u_3(t-t_d)) \end{bmatrix}, \qquad (13)$$

where, $A=\Omega^t\Omega$. (,) is inner product. Lets $S=\mathbf{e}^T\mathbf{e}=\psi(t_d)$, t_d can be estimated because it satisfy

$$\frac{\partial S}{\partial t_d}=0. \qquad (14)$$

While the parameters are identified, selecting an initial function such as $\theta(t)=0.01\times\sin(t)$, $(-t_d<t<0)$, the body sway can be simulated. Figure 9a shows one of the simulated body sway with parameters estimate from the experimental data of a 37 years old healthy male. When a white noise is added to the input in this simulation, the time series of the simulated body sway are similar with the actual recorded in its amplitude and frequency (Figure 9b). This proved that the plant/controller model captures an essential mechanism of body sway and balance-keeping control.

As considering the sensor noise is input as shown in Figure 8 of $\sin(\omega t)$, the total transfer function from sensor noise to output (Fig. 10) becomes

$$T(s)=\frac{G_1G_2e^{-st_d}}{G_1G_2e^{-st_d}+1}, \qquad (15)$$

$$G_1(s)=\frac{K_Ds^2+K_Ps+K_I}{s}, \qquad (16)$$

$$G_2(s)=\frac{1}{Is^2+G}, \qquad (17)$$

$$(I=2m_1l_g{}^2+(m_2+m_3)l_1{}^2; G=-[2m_1l_g+(m_2+m_3)l_1]g; Sin\theta=\theta).$$

Its frequency response can be calculated, the gain $|T(j\omega)|$ becomes a function of ω and t_d as show in Figure 11. The peaks of gain appeared in both eyes open and eyes closed are consistent with the experimental recorded. The results suggest that the PID model is reasonable and useful for balance-keeping control analysis.

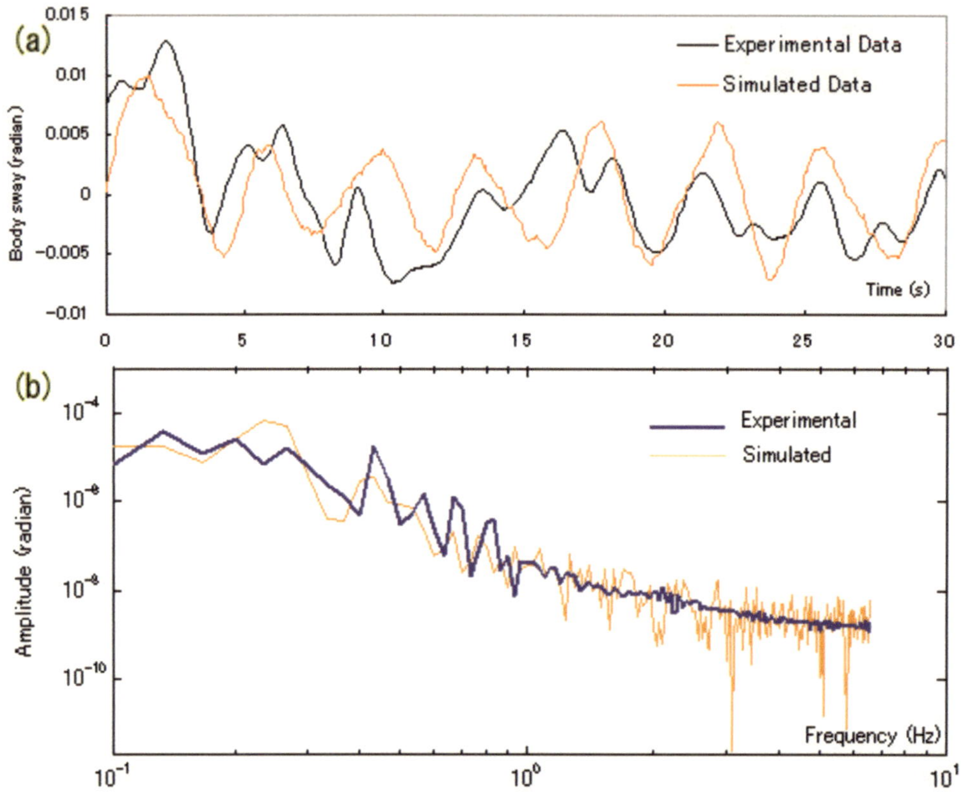

Fig. 9. Simulated and the Experimental recorded body sway in eyes open, time record (a) and frequency response (b). The subject is a 37-year old male. The body height is 1.65m, and body weight is 65Kg. K_P, K_D, K_I and t_d are estimated from the experimental data as 519.0Nm/rad., 72.3Nms/rad., 3.0 Nm/rad.s and 0.11s respectively. The frequency expression of simulated is similar to the experimental recorded.

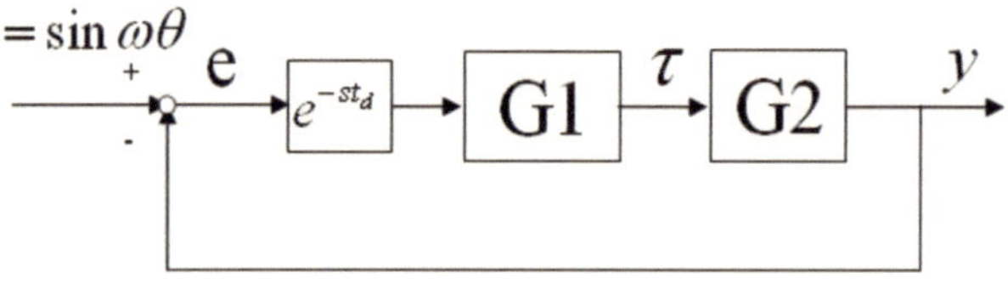

Fig. 10. While considering the sensory noise of $u = \sin \omega t$ as the input, the PID model can be simplified as three processes, i.e. a time-lag process, controller of G1 and dynamics of G2.

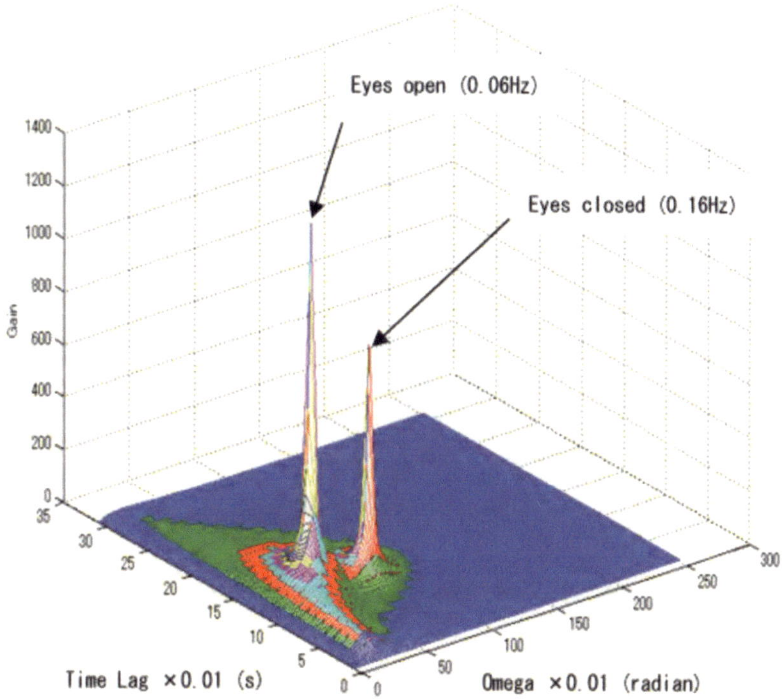

Fig. 11. The frequency response of the balance keeping control system simulated from the same subject. The gain is a function of ω and t_d. The peaks of gain in eyes open (0.06Hz) and eyes closed (0.16Hz) are different, which is roughly consistent with the experimental data as a peak of 0.02Hz in eyes open and a 0.12Hz in eyes closed.

7. Balance-Keeping Ability and fall prediction

Human beings' balance-keeping ability is developed in different age[20]. It is known that in age of 20 to 21 years, individual's balance-keeping ability get to his/her peak, and then decreased with aging[21]. One of main causing factors leads to falls happening in old subjects is the deterioration of the balance-keeping ability[22]. Because of this fact balance-keeping ability assessment becomes the first step for falls-happen prediction. Up to now, several versions of individual's balance-keeping procedures have been published[23], however satisfied and practicable method which can be used for falls-happen prediction is still unavailable. One reason of these failures attributed to the influence of individual's physical status on the balance-keeping ability assessment.

Previous works have shown that mean velocity of COP ($\dot{v}$ COP) is modified by a lot of factors including sex[24], aging, body weight[25], height[26], and also to subjects' grasping power[27]. These factors make the interpretation of the results uneasy when inter-individual comparisons are considered. An effort to adjust the results of COP examination by different age groups has been conducted and its results has been using for ordinary clinical practice[21]. However, subjects' physical characteristics, an important factor which affect the result of COP examination, are still not under specific consideration, even though a lot of models have been proposed[28, 29]. We introduced an index of averaged angular velocity (TSS) for stability estimation[30] and it was proven that TSS is a useful index. Based on a study on 68 subjects, we found that increased in body weight and height tended to decrease the body sway but aging increased body sway. The reciprocal of TSS, defined as trunk sway time (TST), kept a close correspondence with the $\dot{v}$ COP[30]. To reduce the influence of physical characteristics, especially the height on the measuring results, an empirical mathematic model had been introduced as

$$TST_s = k \times \frac{\sqrt{m} \times l^a}{\theta + B}, \qquad (18)$$

where m is body weight, l is height and k, θ are constants, B defined as $(\frac{y}{21} - 1)^2$, y is age (year). This model fit the recorded data quite well.
Another interesting result acquired from a study on aging effect on the PID controller. Parameters identification showed that k_D is deceased with aging[31] (Fig.12). The result suggests a promising method for individual's balance-keeping ability assessment in the future.

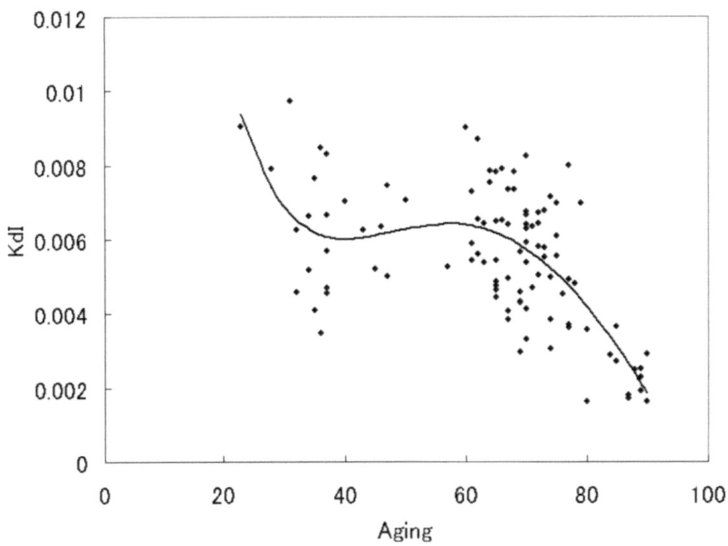

Fig.12. K_D values take from 102 healthy elderly subjects (averaged 64.1±16.3 years old). The K_D is normalized as $KdI = K_D / mgh$ (m: body weight, g: gravity acceleration, h: body height). The result shows the K_D decreased with aging just like the equilibrium ability.

A consequences of falls may be serious at any age, but to the elderly they have significant beyond that in younger people. These consequences can be direct physical effects such as injury or death, or less direct effects of increased dependency and impaired self-care, leading to demands on carers, or profound psychological effects. It is known that postural stability is related to the incidence of falls in elderly. Individuals with poor postural stability show high fall incidence. When defining the measured trunk-sway-time as TST_r, the value of $R_f = TST_r / TST_s$ represents individual's postural stability. From a study of 51 older health subjects[32], the relationship between individual's R_f value and the incidence of falls was investigated. Regressive analyses showed that the value of R_f and the incidence of falls have a reverse relationship, when $R_f \leq 1.12$ the falls incidence was significantly higher than $R_f > 1.12$. The results suggest that falls are predictable by R_f assessment, and this procedure appeared a useful means for fall-prevention especially in elder people.

8. Vision influences on Balance Control

Keeping upright stance is a basic task for other complex motions such as locomotion and running. The mechanisms of visual information benefit for motor control are still unknown. We investigated the visual influence on balance-keeping control in upright standing[33]. Ten healthy subjects (male 4, female 6, aged 37.7±7.21 years) take part in the study. Four kinds of visual stimulation patterns are designed (3 for central visual field stimulation, one is eyes closed, Fig.13).

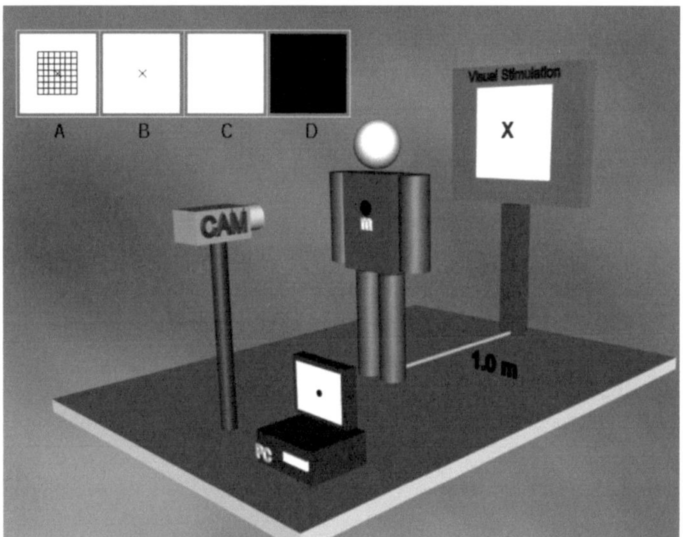

Fig.13. A shows a simplified body biomechanics structural model during static upright stance. Ⓐ, Ⓑ are the masses of leg, Ⓒ is the mass of pelvic and Ⓓ is the mass of upper trunk. Because the lumbosacral always sways in inverse direction of the ankle joint with the same value of θ, the upper trunk is kept perpendicular to the horizon. B is the block diagram of humans' static upright stance control. Because the aim of static upright standing is to keep the body upright, the reference value of postural sway angle is set to zero. The central nervous system detects the error signal and sends output signal to muscles to keep the body upright, a state of equilibrium. C shows a simplified block diagram of upright stance control. Here, the sense noise ($\sin(\omega t)$) is looked as an input, and the output is y,

the deviation of the body's center-of-mass. G1 is the transfer function of PID controller, and G2 is the transfer function of inverted pendulum model.

The results showed that the gain of the PID parameter K_D is decreased significantly in eyes closed (131.5±37.6Nms/rad in eyes open and 90.4±26.0Nms/rad in eyes closed , p<0.001, Fig.14), however, K_P and K_I are unchanged. Simulation results also proved that when decreasing the gain of K_D locus of simulated is more like the measured spontaneous body sway in eyes closed. The results suggested environmental visual cue is important for balance-keeping control, and this effect is pattern-dependent. Of cause, angular velocity is also increased when eyes are closed (Fig.15).

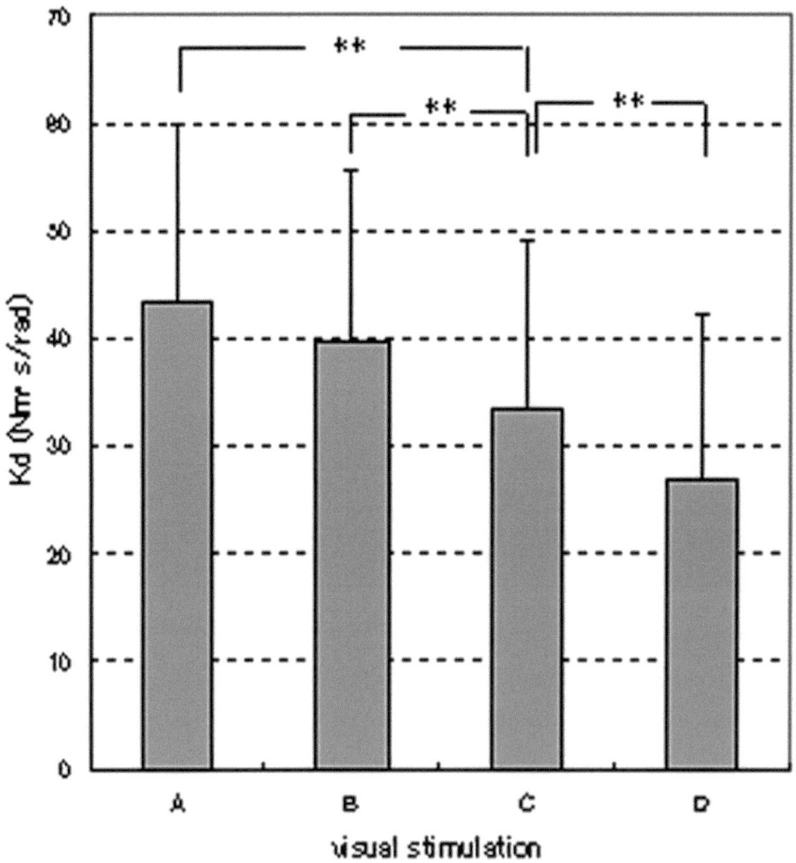

Fig.14. Averaged K_D values of the 10-subject decreased from A to D. No significant differences was found between A and B, however, others showed significant difference. **: $p < 0.01$.

There are two hypothesizes have been proposed about the mechanism of the visual effect on balance-keeping. One regards that information coming from proprioception of extra-ocular muscles is important for balance-keeping control[34]. The other theory is "retinal slip" insisted that images slip on the retinal is used as a cue for balance-keeping control[35]. Our present studies agreed with the retinal slip hypothesis.

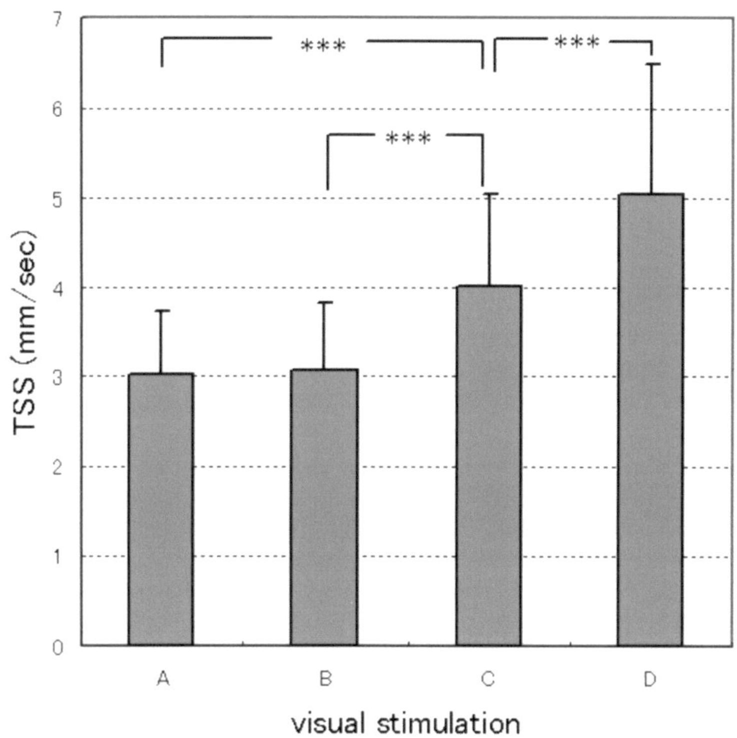

Fig.15. Averaged values of the TSS in different visual stimulations are increased from A to D. However, no significant difference between A and B. ***: $p < 0.001$.

9. Summary

Upright standing is a simple and basic posture of human beings. However, the relatively large mass of the upper body and its elevated position in relation to the area of support during standing accentuate the importance of an accurate control of trunk movements for the maintenance of equilibrium. The kinematics and control strategy of the central nervous system have been studied in recent decade, which brought a PID control algorithm to model the balance-keeping in upright standing and had successfully interpreted the phenomenon of spontaneously body sway. Modeling the human body as two-link inverted pendulum system, we successfully identified parameters of individual's PID parameters and make this model analyzable and practicable. The simulation results, both of the body sway and the spectral response, are quite consistent with experimental data. This proved that the PID model is a reasonable and a useful method as well as by measuring the averaged angular velocity. Both of the two methods help for falls prediction, and become a promising method for falls prevention.

Many authors have argued that complex architectures including feedforward/feedback is necessary for the maintenance of upright stance, however, our studies together with some other recent studies have shown that a model based primarily on a simple feedback mechanism with 120-ms to 150-ms time delay can account for postural control during a broad variety of disturbance[36]. Also, one interesting result is that k_D is a key parameter related to individual balance keeping ability. Since k_D is not just influenced by visual cue but also sensitive to aging. It seems that human balance keeping ability is mainly determined by the gains regulation of k_D, and still there have much works to be done in the future.

10. References

1) William, M. and Lissner, H.R. Boimechanics of human motion. W.B. Saunder Company, Philadephia, 1977.

2) Whitney, R.J. The strength of the lifting action in man. Ergonomics 1, 101-128, 1958.

3) Magnus, R. Korperstellung. Berlin: Sprinter, 1924.

4) De Kleijn, A., Experimental physiology of the labyrinth. J Laryngol. & Otol., 38, 646-663, 1923.

5) Redfern, M.S., Yardley, L. and Bronstein, A.M.. Visual influence on balance, J Anxiety Disord., 15, 81-94, 2001.

6) Fitzpatrick, R.C. and Day, B.L.. Probing the human vestibular system with galvanic stimulation. J Appl Physiol., 96, 2301-2306, 2004.

7) Maurer, C. and Peterka, R. J.. Multisensory control of human upright stance. Exp Brain Res., 171, 231-250, 2005.

8) Geurts, A.C. Haart, M. van Nes I.J. and Duysens, J.. A review of standing balance recovery from stroke. Gait Posture, 22, 267-281, 2004.

9) Morosso, P.G. and Schieppati, M.. Can muscle stiffness alone stabilize upright standing? J Neurophysiol. 82, 1622-1226, 1999.

10) Bloem, B.R., Van Dijk, J.G., and Beckley, D.J., et al. Altered postural reflexes in Parkinson's disease: a reverse hypothesis. Med Hypotheses 39, 243-247, 1992.

11) Ji, Z., Findley, T., Chaudhry H. and Bukiet, B.. Computational method to evaluate ankle postural stiffness with ground reaction forces. J Rehabil Res Dev. 41, 207-214, 2004.

12) Jiang, Y., Nagasaki, S. and Kimura, H.. The Relation Between Trunk Sway and the Motion of Centre of Pressure During Quiet Stance. *Jpn. J. Phys. Fit. Sport Med.*, 52, 533-542, 2003.

13) Jiang, Y., Nagasaki, S., Matsuura Y. and Zhou, J.. Dynamic studies on human body sway by using a simple model with special concerns on the pelvic and muscle Roles. *Asian J. Control*, 8, 297-306, 2006.

14) Massion, J., Postural Control Systems in Developmental Perspective. *Neurosci. Biobehav. Rev.*, 22, 465-472, 1998.

15) Krishnamoorthy, V., Latash, M.L., Scholz, J.P. and Zatsiorsky, V.M.. Muscle Synergies During Shifts of the Center of Pressure by Standing Persons. *Exp. Brain Res.*, 152, 281-292, 2003.

16) Jiang, Y., Nagasaki, S. and Matsuura Y. et al.. The role of ankle joint in the control of postural stability during upright standing on one-foot. *Jpn. J Educ. Health. Sci.,49,* 277-284, 2004.
17) Cote K. P., Brunet M. E., Gansneder B.M. and Shultz S.J.. Effects of Pronated and Supinated Foot Postures on Static and Dynamic Postural Stability. J Athl Train. 40, 41-46, 2005.
18) Masani K., Vette A.H., Popovic M.R.. Controlling balance during quiet standing: proportional and derivative controller generates preceding motor command to body sway position observed in experiments. Gait Posture. 23, 164-172, 2006.
19) Mergner T., Maurer C. and Peterka R.J.. A multisensory posture control model of human upright stance. Prog. Brain Res. 142, 189-201, 2003.
20) Dickin D.C., Brown L.A. and Doan J.B.. Age-dependent differences in the time course of postural control during sensory perturbations. Aging Clin. Exp. Res. 18, 94-99, 2006.
21) Imaoka K., Murase H. and Fukuhara M., Collection of data for healthy subjects in stabilometry. Equilibrium Res. Suppl., 12, 1-84, 1997.
22) Chaudhry H., Findley T., Quigley K.S. et al.. Postural stability index is a more valid measure of stability than equilibrium score. J Rehabil. Res. Dev. 42, 547-556, 2005.
23) Tinetti M.E., Williams T. F. and Mayewski R., Fall risk index for elderly patients based on number of chronic disabilities. Am. J. Med., 80, 429-434, 1986.
24) Paper S.A. and Soames R. W.. The influence of stationary auditory fields on postural sway behaviour in man. Eur. J. Appl. Physiol. Occup. Physiol., 63, 363-367, 1991.
25) Horstmann G. A and Dietz V.. A basic posture control mechanism: the stabilization of the centre of gravity. Electroencephalogr. Clin. Neurophysiol., 76, 165-176, 1990.
26) Berger W., Trippel M., Discher M. and Dietz V., Influence of subjects' height on the stabilization of posture. Acta Otolaryngol., 112, 22-30, 1992.
27) Kimura M., Okuno T., Okayama Y. and Tanaka Y.. Characteristics of the ability of the elderly to maintain posture. Rep. Res. Cent. Phys. Ed., 26, 103-114, 1998.
28) Giese M.A., Dijkstra T.M., Schoner G. and Gielen C.C.. Identification of the nonlinear state-space dynamics of the action-perception cycle for visually induced postural sway. Biol. Cybern. 74, 427-437, 1996.
29) Jeka J., Oie K., Schoner G., Dijkstra T. and Henson E.. Position and velocith coupling of postural sway to somatosensory drive. J Neurophysiol. 79, 1661-1674, 1988.
30) Jiang, Y., Nagasaki, S., Matsuura Y. et al.. Postural sway depends on aging and physique during upright standing in normals. *Jpn. J Educ. Health. Sci.,48,* 233-238, 2002.
31) Kimura Hidenori, Yifa Jiang, A PID model of human balance keeping. *IEEE Control System Magazine,* 26(6), 18-23, 2006.
32) Nagasaki, Jiang, Y., S., Yoshinori F. et al.. Falls risk prediction in old women: evaluated by trunk sway tests in static upright stance. *Jpn. J Educ. Health. Sci.,48,* 353-358, 2003.

33) Qin S., Nagasaki S., Jiang Y., et al. Body sway control and visual influence during quiet upright standing. Jpn. J Physic. Fitness & Sports Medicine, 55, 469-476, 2006.

34) Baron J.B., Ushio N. and Tangapregassom M.J.. Orthostatic postural activity disorders recorded by statokinesimeter in post-concussional syndromes: oculomoter aspect.Clin. Otolaryngol. Allied Sci., 4,169-174. 1979.

35) Glasauer S., Schneider E., Jahn K., Strupp M., and Brandt T. How the eyes move the body. Neurology. 65, 1291-1293, 2005.

36) Maurer C. and Peterka R.J.. A new interpretation of spontaneous sway measures based on a simple model of human postural control. J Neurophysiol., 93, 189-200,2005.

27

Experiments on Embodied Cognition: A Bio-Inspired Approach for Robust Biped Locomotion

Frank Kirchner, Sebastian Bartsch and José DeGea
German Research Center for Artificial Intelligence,
And University of Bremen,
Robotics Lab, Bremen
Germany

1. Introduction

Recently, the psychological point of view that grants the body a more significant role in cognition has also gained attention in artificial intelligence. Proponents of this approach would claim that instead of a 'mind that works on abstract problems' we have to deal with and understand 'a body that needs a mind to make it function' (Wilson, 2002). These ideas differ quite radically from the traditional approach that describes a cognitive process as an abstract information processing task where the real physical connections to the outside world are of only sub-critical importance, sometimes discarded as mere 'informational encapsulated plug-ins' (Fodor, 1983). Thus most theories in cognitive psychology have tried to describe the process of human thinking in terms of propositional knowledge. At the same time, artificial intelligence research has been dominated by methods of abstract symbolic processing, even if researchers often used robotic systems to implement them (Nilsson, 1984).

Ignoring sensor-motor influences on cognitive ability is in sharp contrast to research by William James (James, 1890) and others (see (Prinz, 1987) for a review) that describe theories of cognition based on motor acts, or a theory of cognitive function emerging from seminal research on sensor-motor abilities by Jean Piaget (Wilson, 2002) and the theory of affordances by (Gibson, 1977). In the 1980s the linguist Lakoff and the philosopher Johnson (Lakoff & Johnson, 1980) put forward the idea of abstract concepts based on metaphors for bodily, physical concepts; around the same time, Brooks (Brooks, 1986) made a major impact on artificial intelligence research by his concepts of behavior based robotics and interaction with the environment without internal representation instead of the sense-reason-act cycle. This approach has gained wide attention ever since and there appears to be a growing sense of commitment to the idea that cognitive ability in a system (natural or artificial) has to be studied in the context of its relation to a 'kinematically competent' physical body.

Among the most competent (in a multi functional sense) physical bodies around are certainly humans, so the study of humanoid robots appears to be a promising field for

understanding the mechanisms and processes involved in generating intelligence in technical systems.
In the following we will give an overview of the field of humanoid robot research.

2. State of the Art Humanoids

A review on humanoid robot systems, cannot be made without bearing in mind that many of the current developments concentrate on one or the other feature of human performance. Some of them are good at manipulating objects with anthropomorphic arms but move over a wheeled platform. Some others walk on two legs but lack of a torso and arms. Some combine those two features but lack a human appearance or communication abilities. Some other developments concentrate on human-like communication skills, like speech recognition, gestures and the generation of facial expressions that denote sadness, happiness, fear or any other state that a human is able to recognise. All these aspects are crucial for the final goal of attaining a robot that is perceived as humanoid. A robot that transmits its feelings, ideas or thoughts, that behaves like a human when performing a task or a movement and with which a human feels safe and confident to collaborate with are key points for the social acceptance of a humanoid robots.
A description of the state of the art might start chronologically since the development of the first complete humanoid, the Wabot-1 from the Waseda University in 1970 but we choose to list the developments on the humanoid field beginning with the systems that incorporate the more human-like features and are considered the most advanced systems to continue describing systems that work on single or a combination of several human-like aspects, all of them of major interest and importance.
Before describing the most important developments on the field, it is worth mentioning a few aspects about observable facts depending on the origin of the robot: namely, Asia, Europe or USA. There are differences on the complexity of the systems but also on the different approaches that are followed or the motivations that lead the development of the robot. Japan (and Korea in a minor extent) are seen as world leaders in the humanoid robot research. They have the most complex robots with the most similar human resemblance. They believe in a complete immersion of the robots in a future society, where robots do not differentiate easily from humans. The more remarkable points of their developments are the hardware (the mechanics), the physical appearance of the robot and the fact that the industry is leading the research on these robots, expecting a huge market in a near future. USA entered the humanoid era because of the needs posed by the claims of the 'modern' Artificial Intelligence: the need for a human-like body as a prerequisite for a robot to achieve human-like intelligence. It is the interaction with the environment and the gathered experience what is thought to be the basis for the appearance of intelligent behaviors. Europe, on the other side, is basically concerned with giving a real application to the development of humanoid robots and that is on the service robotics area, for rehabilitation and/or personal care of the elderly. The most advanced systems are heading towards that goal. At the same time, inspiration from biological systems is a very common term to describe the approaches used in those robots. Several European projects work on sensorimotor coordination, cognitive architectures and learning approaches that have their roots in the cooperation with scientists in the neuroscience, biology and psychology areas.
The ASIMO robot from Honda (Hirai, 1998) (cf. figure 1) is without a doubt the most advanced humanoid robot nowadays. Honda employed vast human and economical

resources to achieve a complete human-like looking robot, pushing forward the research in many areas. The current research model is 130cm tall, weights 54 kg and is able to run at 6km/h (December 2005). The research began in 1986, achieving a first ASIMO prototype in the year 2000. Nowadays, ASIMO is the only robot that is able to autonomously walk around and climb stairs and slopes. Furthermore, it is able to understand some human gestures and interact with people using its speech recognition system and some pre-programmed messages. ASIMO can also push a cart, keeping a fixed distance to it while moving and still maintaining the capability to change direction or speed of movement, walking hand-in-hand with a person or carrying a tray.

Fig. 1. Honda ASIMO (left picture), Sony QRIO (middle picture), and ROBONAUT (right picture).

The HRP-2P (Kaneko, 2003) robot specified by AIST (National Institute of Advanced Industrial Science and Technology, Japan) and whose hardware was manufactured by the Kawada Industries (a company that also worked with Honda and the University of Tokyo in the development of the ASIMO and the H6-H7 robots) is one of the most advanced humanoids nowadays. It differs from ASIMO on the fact that it is a research prototype whose software is open to any roboticist. Moreover, it was designed to walk on uneven terrains and recover from falling positions, features not yet possible for ASIMO. It weights 58kg and is 154cm tall.

Probably the third robot in importance in Japan is the H7 (Kagami, 2001) from the University of Tokyo. However, there is not much information available apart from videos showing its capabilities walking on a flat terrain. As above mentioned, Kawada Inc. was responsible for the hardware development. It weights 55kg, is 147cm tall and has 30 DoF.

Sony entered the humanoid world in 1997 with the SDR-1X series, achieving the SDR-4X version in 2003, named QRIO (Ischida, 2003) (cf. Fig. 1) as was intended to be commercially available. In 2006 Sony announced the decision to stop the further development of the robot. QRIO is comparable to ASIMO in its walking capabilities although since it was designed as an entertainment robot, its size is substantially smaller than ASIMO: its weights 7kg and is 58cm tall. Its main features include the ability to adapt its walking to the most difficult situations: from walking on irregular or tiled terrains to react to shocks and possible falling conditions. But since its origins as entertainment robot, the most remarkable features are those that enhanced its interaction capabilities with people: the robot is able to recognise faces, use memory to remember a previously seen person or his/her words, detect the person who is speaking and incorporates a vocabulary of more than 20,000 words that enables the robot to maintain simple dialogues with humans.

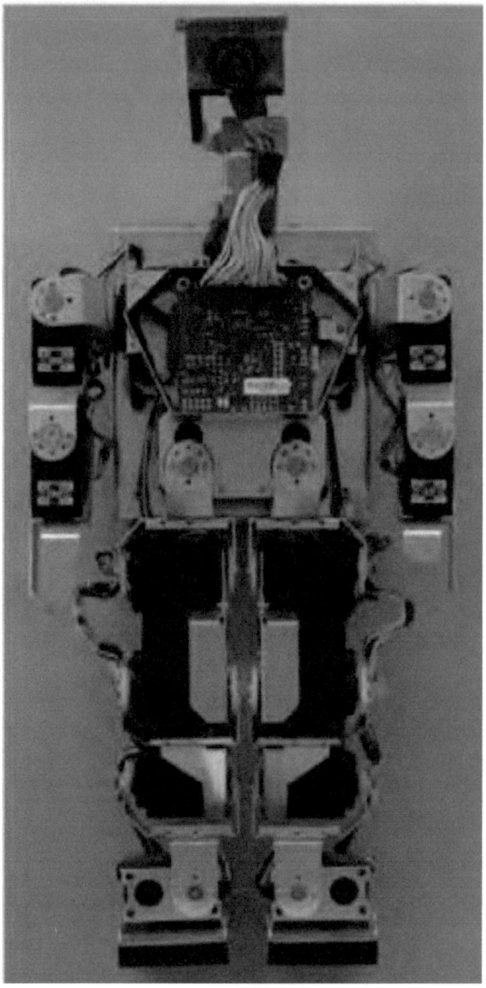

Fig. 2. The Robot BIN-HUR based on Kondo's KHR-1.

Hubo (Il Woo, 2005) is the most well-known humanoid robot in South Korea and one of the world's most advanced. It is the latest development of the series of KHR robots (KHR-1, KHR-2 and KHR-3 - Hubo). It is 125cm tall and weights 56kg, having 41 DoF. Apart from improving in this latest version its walking abilities, Hubo is now also able to talk to someone by using a speech recognition system.

Fujitsu also entered the humanoid area in 2003 with the HOAP-1 (Murase, 2001). Its major claim with it was its learning capabilities and the use of neural networks to control the locomotion implementing a Central Pattern Generator (CPG), proven to be one of the responsible neural circuitry for the locomotion on vertebrates. These artificial neural oscillators are used to create rhythmic motions to generate the appropriate gait. The major advantage is claimed to be its adaptation to the environment and new terrain configurations and the minimum computational effort to control the locomotion. No need for modelling kinematics, dynamics or generating stable trajectories using complex criteria are required. It was intended to be used in research labs and universities as an educational tool where to test different algorithms and for that reason provides an open source software and weights only 6kg and is 48cm tall. It can walk up to 2km/h and is sold at about 50,000€. In 2004, HOAP-2 received the Technical Innovation Award from the Robotics Society of Japan.

Toyota also presented a series of partner robots (2005), one of them walking in two legs, finding its application in the elderly care and rehabilitation. As a curious feature, Toyota included artificial lips with human finesse what, together with their hands, enables them to play trumpets in a similar way a human does.

WABIAN-RIII and WENDY (Ogura, 2004) are the latest developments from the Waseda University, as already mentioned, the pioneers in the humanoid field with the first full-scale humanoid robot, a project that began in 1970 and finished three years later with Wabot-1. WABIAN-RII continues the research in dynamical walking plus load carrying and the addition of emotional gesture while performing tasks. Likewise, WENDY incorporates emotional gestures to the manipulation task that is being carried out. WABIAN-RII weights 130kg and is 188cm tall while WENDY is 150cm tall and weights 170kg.

Johnnie (Löffler, 2000) is probably the most well-known and advanced humanoid robot in Europe. It was developed at the University of Munich with the aim of realising a human-like walking, in this case based on the well-established Zero Moment Point (ZMP) approach introduced by Honda in the ASIMO robot, but with the aid of a vision system. It is able to walk on different terrains and climb over some stairs. It is 180cm tall and weights 45kg.

Robonaut (Ambrose, 2001) (cf. figure 1) is a humanoid robot developed by the NASA with the aim of replacing a human astronaut in EVA tasks (outside the vehicle). The main feature of the robot is a human dexterous manipulation capability that enables it to perform the same tasks an astronaut would perform and with the same dexterity. The robot is not autonomous but tele-operated from inside the vehicle. Since legs have no utility in space, the robot is composed of two arms and a torso that is attached to a mechanical link enabling the positioning of the robot in any required position/orientation. Because of the bulky suits the astronauts have to wear to protect against radiations, their manipulation capabilities are greatly reduced and the handles, tools and interfaces they use are designed to be handled with their special gloves. A robot, even though needing some protection against radiation, would not required such a bulky suit thus recovering to a certain extent a human dexterity. Moreover, risks for astronauts are avoided on these missions outside the spatial vehicle. It has the size of a human torso and arm, with 54 DoF in total: 14 for each hand, 7 for each arm and the link to the vehicle, 2 in the neck and 3 on the waist.

In the field of human-robot interaction, the robot Cog (Brooks, 1998), from the MIT AI Lab, is the best example. Cog is composed of a torso, two arms and a head. The main focus of this project is to create a platform in order to prove the ideas exposed by Rodney Brooks claiming that human-like intelligence appearance requires a human-like body that interacts with the world in the same way a human does. Besides, for a robot to gain experience in interacting with people it needs to interact with them in the same way people do. One underlying hope in Brooks theory is that: Having a human-like shape, humans will more easily perceive the robot as one of them and will interact with it in the same way as with other people, providing the robot with valuable information for its interaction learning process. These ideas have been taken to the next level with Kismet (Breazeal, 1998), which is also a robot from the same Lab. In this case, a robot composed of a human-like head. The research focus is on natural communication with humans using, among others, facial expressions, body postures, gestures, gaze directions and voice. The robot interacts in an expressive face-to-face way with another person, showing its emotional state and learning from humans as kind of a parent-child relation.

In the same direction, the Intelligent Robotics Lab of the University of Osaka (Japan) developed in 2005 the most human-looking robot so far. It is a female robot named Repliee Q1Expo (Ishiguru, 2005). Its skin is made of silicon what gives it a more natural appearance and integrates a vast number of sensors and actuators to be able to interact with people in a very natural and friendly way. Even the chest is moved rhythmically to create the illusion that the robot is breathing.

However, as promising as some of these developments seem to be at first glance, one has to carefully evaluate what exactly can be learned for the field of 'embodied cognition' from the study of more or less isolated features of human behavior, whether that be in the field of complex locomotion, manipulation or interaction. In her paper (Wilson, 2002) identifies six viewpoints for the new so-called 'embodied cognition' approach:

1) Cognition is situated: All Cognitive activity takes part in the context of a real world environment.
2) Cognition is time pressured: How does cognition work under the pressures of real time interaction with the environment
3) Off-loading of cognitive work to the environment: Limits of our information processing capabilities demand for off-loading.
4) The environment is part of the cognitive system: because of dense and continuous information flow between the mind and the environment it is not meaningful to study just the mind.
5) Cognition is for action: Cognitive mechanisms (perception/memory, etc.) must be understood in their ultimate contribution to situation-appropriate behavior.
6) Off-line Cognition is body-based: Even when uncoupled from the environment, the activity of the mind is grounded in mechanisms that evolved for interaction with the environment.

We have cited all six viewpoints here, as they represent an interesting perspective on the state of the art in embodied cognition. In the experimental work presented here we focus our attention on viewpoint 2 that appears to have a crucial instantiation especially in humanoid robots that have to find a way to effectively and efficiently counteract the effects of gravity while walking. In fact looking at viewpoint 3 and 4 counteracting appears to be wrong from the beginning instead having gravity work for the system appears to be a better way of achieving robust locomotion in a technical two legged system.

3. Robust Biped Locomotion Control

In this section we describe an experiment with a humanoid robot that achieves robust locomotion in the absence of a kinematical model. For stable goal directed behavior it relies solely on two simple biological mechanisms that are integrated in an architecture for low level locomotion control.

3.1. The Hardware

The robot (see figure 2) is based on the Kondo KHR-1 construction kit and has 18 DOFs, 5 per leg, 3 per arm, and 2 as pan tilt unit for the head. The system is 40 cm high (30 cm shoulder-height) and has a total weight of 1.5 Kg. Its mechanics are mainly part of the Kondo KHR-1 construction kit. As control unit we use a custom-made microcontroller board.
An ADXL202 tilt sensor was integrated in the upper body to provide information about the pitch of the robot. Pressure sensors in the feet indicate if they have ground contact. For wireless communication we use a Bluetooth module. The head consist of a CMUCam2 which is used for color tracking affixed on a pan tilt unit.
The microcontroller board is composed of an MPC 565 PowerPC microcontroller mounted on a custom-designed mainboard. The MPC 565 is running at 40 MHz, has 2 MB flash memory and 8 MB RAM. Amongst others it is equipped with three time processing units (TPUs) each with 16 channels, two analog digital converter modules (ADCs) each with 64 channels, and two RS-232 interfaces.
The mainboard provides 32 Servo plug-in positions which are connected to two of the TPUs to generate pulse width modulated (PWM) signals for the activation of the Servos. Furthermore, the plug-in positions are connected to the ADCs to feed back the servo's current and the actual position on the basis of its potentiometer value.
Two more plug-in positions each with 8 pins are linked with the ADCs to connect additional analog sensors, e.g. tilt sensors or pressure sensors.
The CMUCam2 is an intelligent camera module which was developed at the Carnegie Mellon University. It has the integrated possibility to track colors and to communicate over a RS-232 connection. In our case we will use it to track the color of a ball, which will be provided as sensory information for a higher level behavior whose intention it is to follow the ball

3.2 The control architecture

Our architecture is based on two approaches to robust and flexible real world locomotion in biological systems, which seem to be contradictory at first sight. These are the Central Pattern Generator (CPG) model and the pure reflex driven approach.
A CPG is able to produce a rhythmic motor pattern even in the complete absence of sensory feedback. The general model of a CPG has been identified in nearly every species even though the concrete instantiations vary among the species to reflect the individual kinematical characteristics in the animals.
The idea therefore seems to be very promising as a concept to realize locomotion in kinematically complex robotic systems, see figure 3. As it resembles the divide and conquer strategies that are reflected in nearly all solutions to complex control problems.
Another model for the support of robust locomotion is also provided by evolution in the animal kingdom. This is the concept of reflex based control (Delcomyn, 1980). A reflex can be viewed as a closed loop control system with fixed input/output characteristics. In some animals, like the locust, this concept is said to actually perform all of the locomotion control and no further levels of control, like the CPG, are involved (Cruse, 1978).

Whether or not complex motion control can be achieved only via reflex systems is subject to further discussion, however, the concept of a set of fixed wired reactions to sensory stimuli is of high interest to roboticists who aim to gain stability in the systems locomotion.

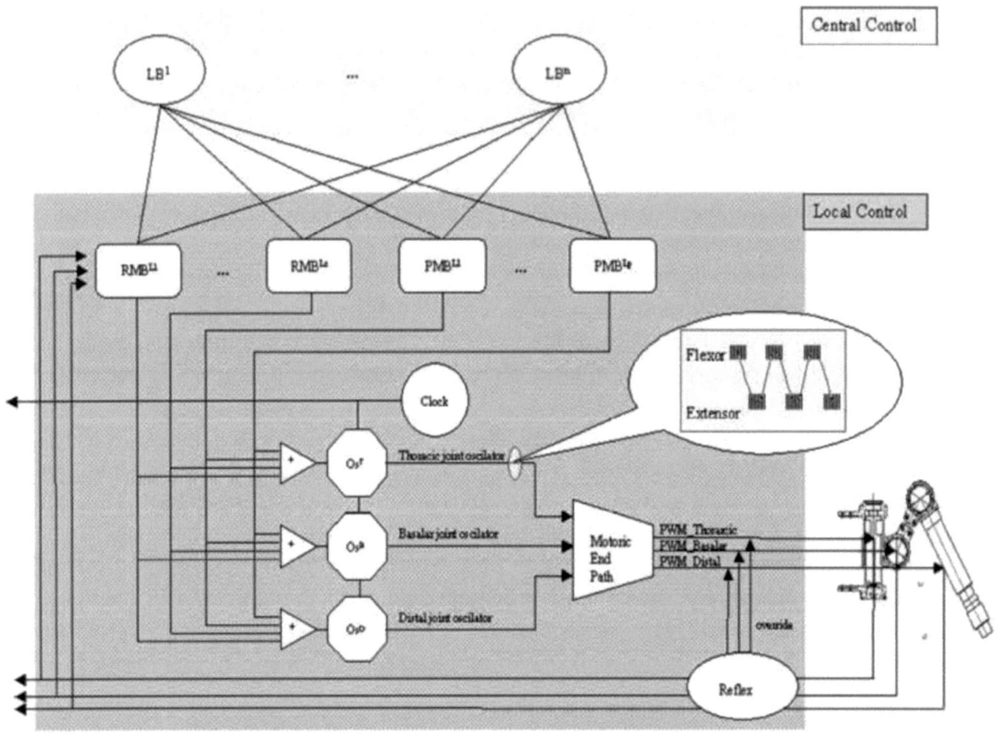

Fig. 3. The low level control architecture. On the global level (light gray area) we have implemented Locomotion Behaviors (LB's), typically (Forward, Backward and Lateral locomotion). These global behaviors are connected to all local leg controllers and activate the local single leg motion behaviors. The local level (dark gray area) implements Rhytmic Motion Behaviors (RMB's) and Postural Motion Behaviors (PMB's). These behaviors simultaneously influence the amplitude and frequency parameters of -in this case- three oscillating networks (OST, OSB and OSD). The oscillators are connected to a common clock which is used for local and global synchronization purposes. The oscillators output is a rhythmic, alternating flexor and extensor, stimulation signal (see callout box), which is translated into PWM signals via the motoric end path. Inline with the output of the motoric end path are a set of pertubartion specific reflexes, which override the signals on the end path with precompiled activation signals if the sensor information from the physical joints meets a set of defined criteria.

The design of the control architecture described here was thus driven by these two concepts. The CPG approach appeared to be interesting to generate rhythmic walking patterns which can be implemented computationally efficient, while the reflex driven approach seemed to provide a simple way to stabilize these walking patterns by providing: 1) a set of fixed situation-reactions rules to external disturbances and 2) as a way to bias leg coordination among multiple independent legs (Cruse, 1978). Figure 3 outlines the general idea.

This approach features the idea of continuous rhythmic locomotion as well as postural activity which is generated by spinal central pattern generators in vertebrate systems (Kirchner, 2002), (Spenneberg, 2005).

For our technical implementation, these activities are solely defined by 3 parameters: amplitude, frequency, and offset of the rhythmic movement. Please note the possibility to set amplitude and frequency to zero, just modifying the offset parameter, which would result in linear, directly controlled joint movements. In those cases where amplitude and frequency have non-zero values, the activation patterns will result in a rhythmic movement of the joint around the offset (or baseline) with given frequency and amplitude. To produce complex locomotion patterns, like forward, left, right, or backward movements, all joints of the robot have to be activated simultaneously, while some (legs, shoulder, and hip) actually produce rhythmic activities, others, (like neck, elbow, etc.) will have their amplitude and frequency values set to zero maintaining a position at the offset value. One important aspect of central pattern generators is their nature as feedback control loops, here the so-called proprioceptive information is fed back into the controller and modifies its activity.

4. Implementation Issues

Our implementation of the low-level control architecture, shown in figure 5, consists of a combination of drivers and behaviors, which are connected thru special functions (merge functions). Our concept for locomotion is a combination of balance control, see figure 4 and posture behaviors, which should keep the robot balanced while walking or during external interferences. Central Pattern Generator (CPG) behaviors are used to produce rhythmic motions for walking. The speed at which the rhythmic motions are performed is defined by a global clock.

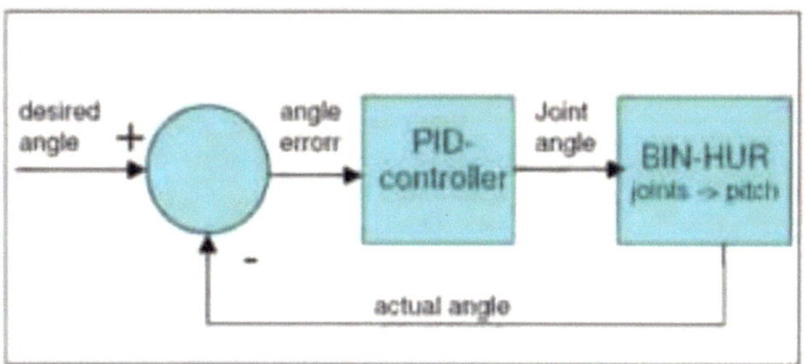

Fig. 4. The control cycle for balance control.

A higher level behavior 'walking' has the task to implement directional walking and another high level behavior 'ball-following' will use the sensory information of the CMUCam2 to follow the intention to track a ball by giving instructions to the walking behavior.

5. Experimental Results

To demonstrate a possible result for the activation pattern of a joint while overlaying different CPGs and modifying the posture, we first let the robot walk hanging in the air without any balancing behavior in order to get even curves. We could not let the robot walk on the ground without balance behaviors because then the system braces and topples down.

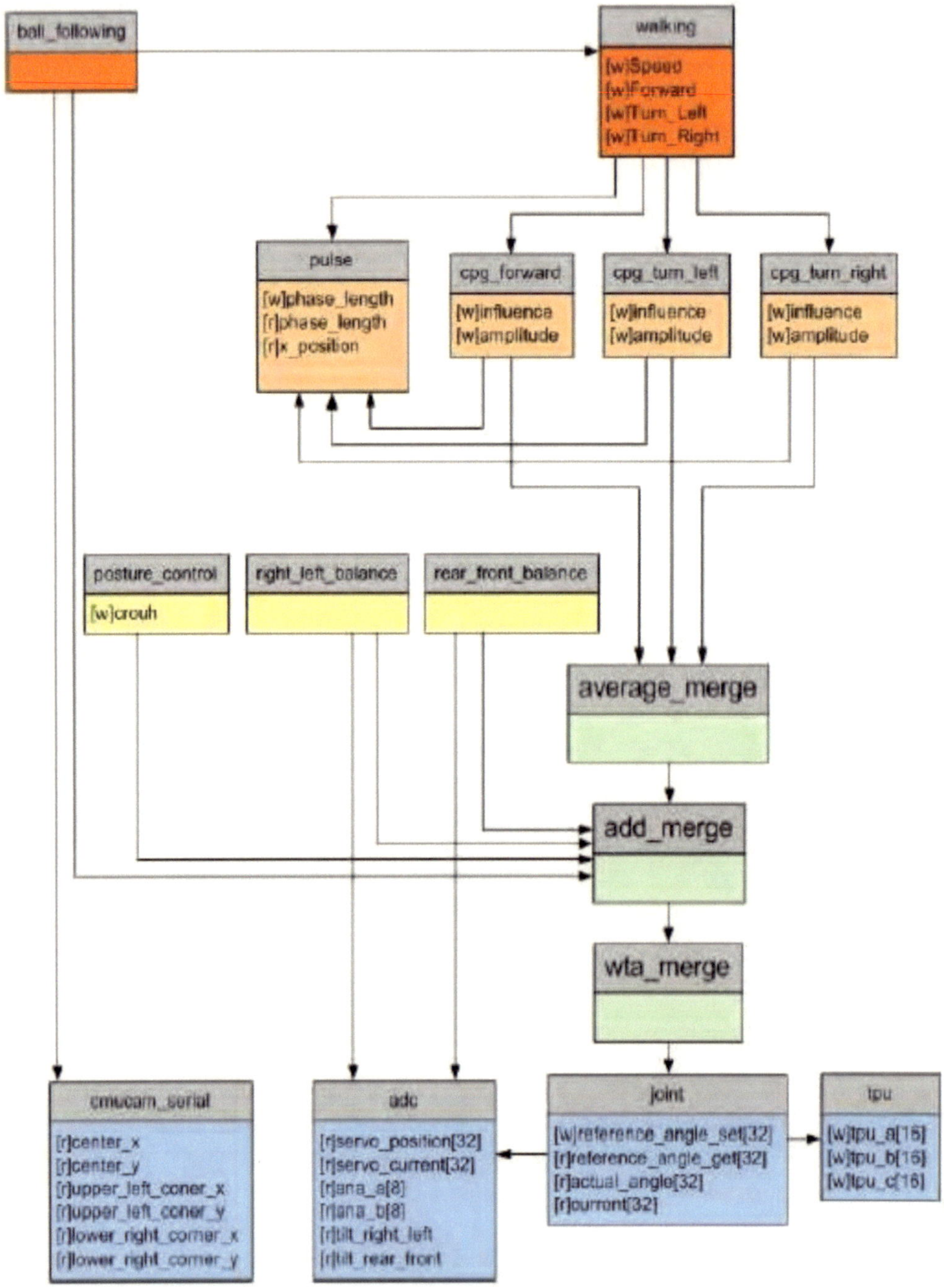

Fig. 5. Implementation of the low-level control concept on the humanoid robot.

Figure 6 shows the desired angles for both hip forward joints at a pulse of 2000 milliseconds. The right and left legs curve are shifted by half the period because one leg is in the swing phase while the other one is in the stance phase.
The rhythm from 0 ms to 15000 ms is only generated by the forward CPG, however, the offset value is set to 10 degrees from 5000 ms to 10000 ms and combined via add merge, thus resulting in a more ducked posture while walking.
From 15000 ms to 21000 ms the forward and turn left CPGs are active and mixed together with an average merge which has the effect that the robot takes a moderate left curve.
In the time segment from 21000 ms to 27000 ms just the CPG for turning left is active and after that there is only a basic posture value.

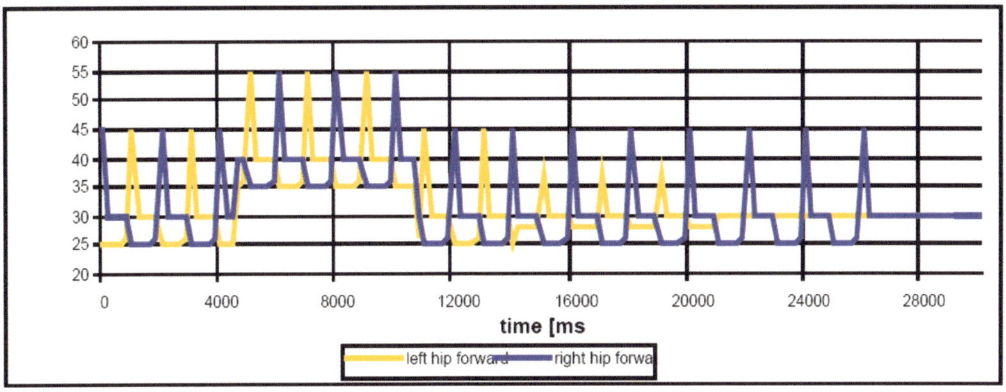

Fig. 6. Reference angle for right and left hip forward joint while walking hanging in the air without balance behaviors.

5.1. Walking with balance

When the robot walks on the ground with active balance behaviors, the desired joint angles of the balance behaviors are added to the values of the posture behavior and the active CPGs. In this experiment, we let the robot also walk with a pulse of 2000 milliseconds.
First we let it walk on the ground without active balance behavior to show the desired and real angles of the leg joints with resistance of the gravity and the robot's weight. During this run, shown in figure 9, we prevented the robot from toppling down by hand. From 0 ms to 6000 ms the robot walked forward, and then took a moderate left curve till 12000 ms, and after that it turns left on the spot.
In the second trial, shown in figure 10, we activated the balance behavior. The robot walked forward from 0 ms to 4000 ms, followed by a moderate left curve till 12000 ms, and then turned left on the spot.
As you can see in figure 10, the activation of the balance behavior results in noisier curves than just walking forward without balance behaviors like in figure 9 but it stabilizes the system and prevents it from bracing.
The balance behavior which is designed as a PID-Controller takes the tilt value shown in figure 7 as input for the controller and writes the controller's output values multiplied with a specific factor for each joint to the servos. Negative sensor values represent a right or rather rear leaning, and positive values a left and accordingly front leaning.

As you can see, the output of the PID-Controller shown in figure 8 is less noisy than the tilt sensor values and seems to be more rhythmic. The pattern is repeated every 2000 milliseconds which shows that frequency of the interferences and the retaliatory action depend on the pulse.

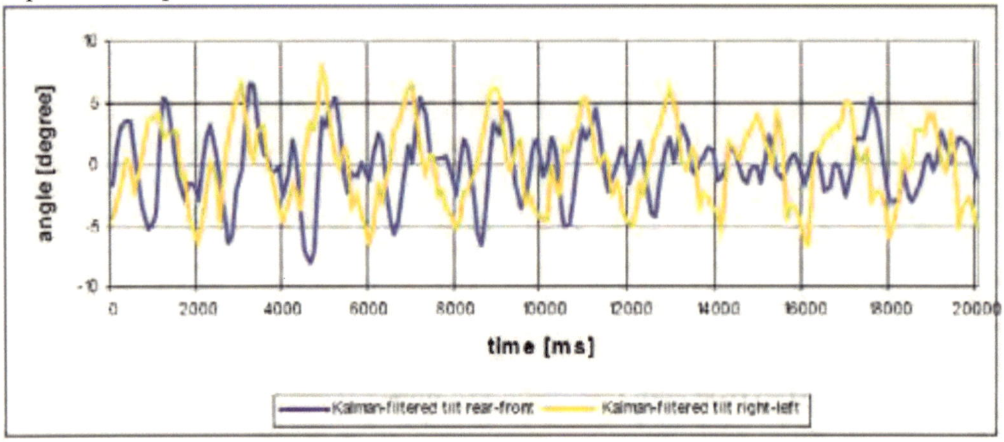

Fig. 7. Values for rear-front and right-left tilt, while walking on the ground with active balance behavior.

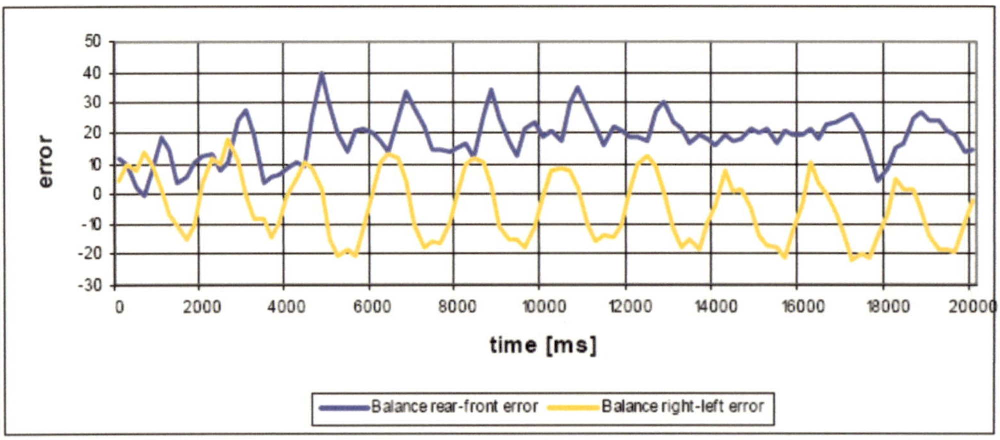

Fig. 8. Calculated error from the balance behavior's PID-Controller while walking on the ground with active balance behavior.

5.2. Reaction on a lateral hit

To test the static stability we used the following experimental setup. A ball with a weight of 250 grams was fixed as a pendulum over the robot. The band it is attached with has a length of 15 cm. Then we let the ball fall from a height of 15 cm 5 times and hit the robot at the right shoulder. If we do not activate the balance behaviors the robot cannot absorb the hit just by his stable standing resulting from the posture behavior and topples down. With active balance behaviors the robot tries to react against the hit because of the difference between the desired and the actual leaning and is able to stay and adjust his balance.

Figure 12 shows the desired and real angle of the left arm and leg joints resulting from the reaction of the balance behavior as an average of the 5 recurrences.
Figure 13 shows the perception of the hit by the tilt sensor whose values are used as input for the balance behavior's PID-Controller, shown in figure 14.

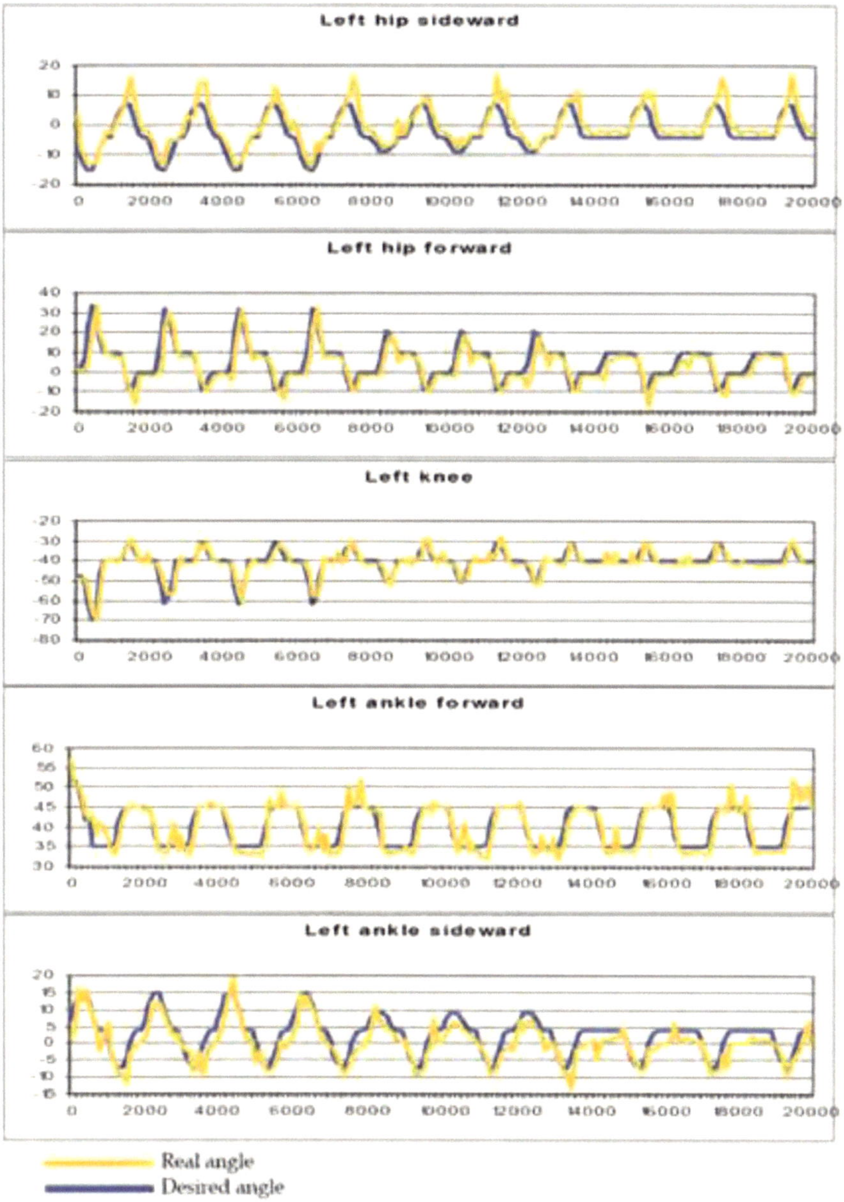

Fig. 9. Desired and real angle (degree) from left leg joints while walking on the ground without active balance behavior over 20000 ms. The robot was prevented from toppling down by hand.

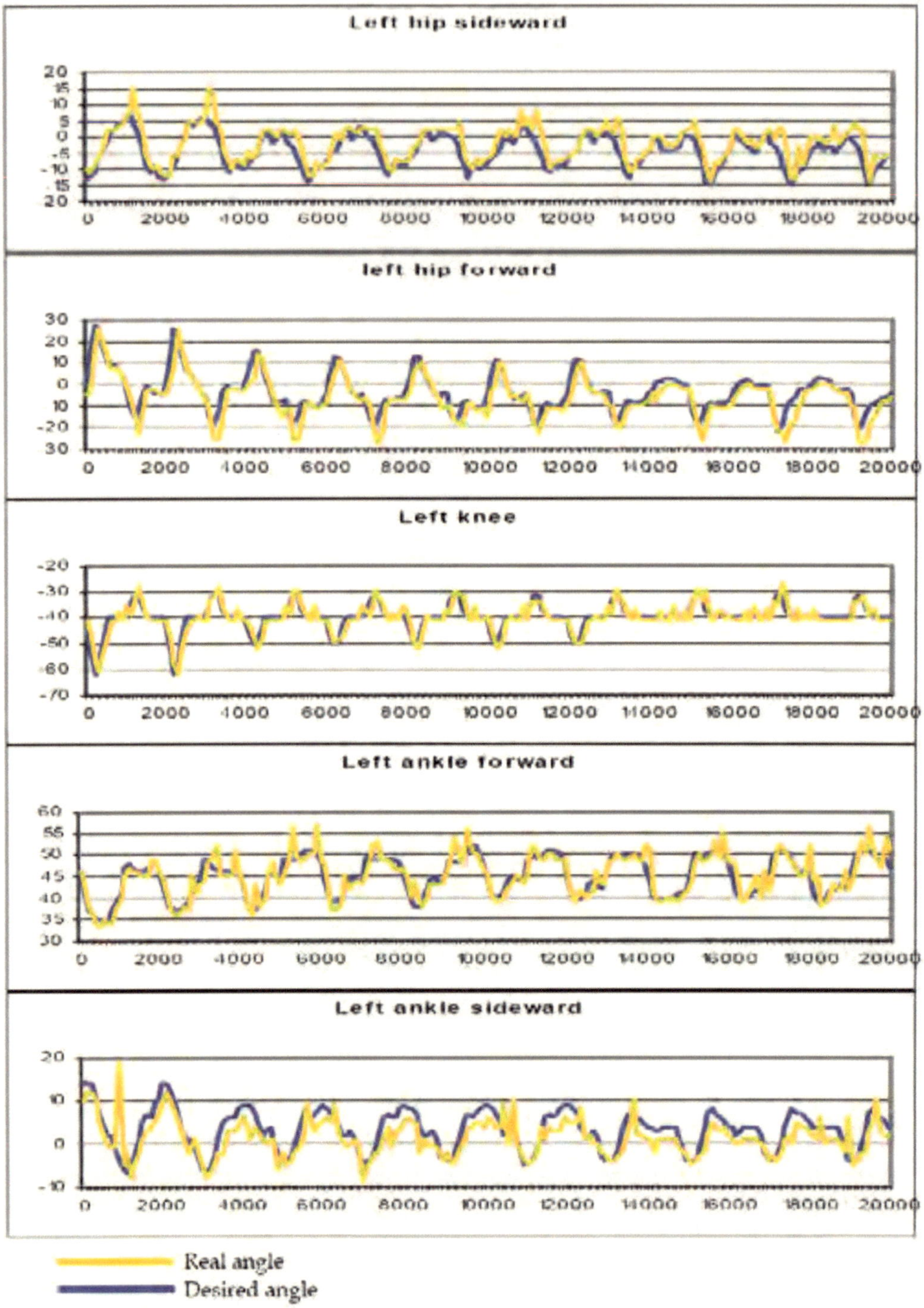

Fig. 10. Desired and real angle (degree) from left leg joints while walking on the ground with active balance behavior over 20000 ms.

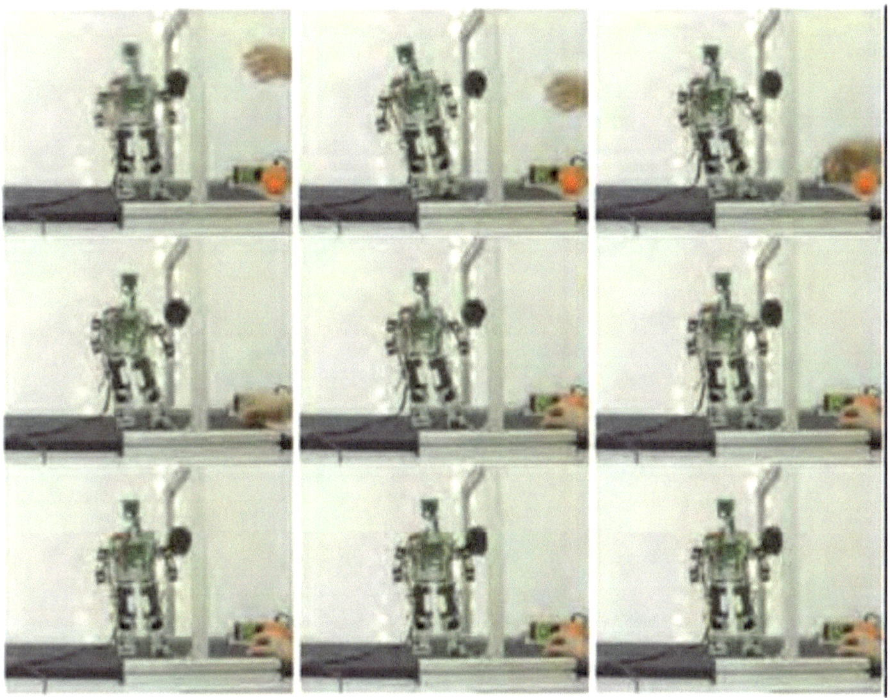

Fig. 11. Snapshots of the reaction on a lateral hit in 200ms intervals.

6. Discussions and future work

The CPG based approach for combining rhythmic movements in two-legged robotic systems does work and produces less calculating costs than inverse kinematics (Tevatia, 2000), it results in smoother movements than using simple look up tables, and is easier to realize than neural networks (Shan, 2002). However, direct, goal directed movements are difficult to implement and still require kinematic models of the system.

Recently, we are working on a biologically inspired hybrid learning architecture, see figure 15 for embodied cognition supporting recognition and representation on the basis of sensorimotor coordination. The notion of hybrid architectures is straightforward in the literature, born from the understanding that neither reactive nor deliberative systems provide a sufficient basis for truly cognitive agents. It is therefore natural to postulate systems that contain both types of systems. An important question in these architectures is where to draw the line between the reactive and the deliberative component, and what their relationship should be. Regarding this important question, the architecture we are working on meaningfully combines a reactive layer and a higher level deliberative layer. The reactive layer will be responsible for attention control, object categorization, and reflex triggering.

The locomotion approach we have developed for the humanoid robot will be integrated in the reactive layer and will be exploited in approaching and manipulating objects. The deliberative layer will provide a means for the robot to learn and adapt to new environments. The ball following behavior that we want to implement, for example, can be implemented in the deliberative layer, where behaviors running in the reactive layer can be modulated and combined to achieve the required behavior

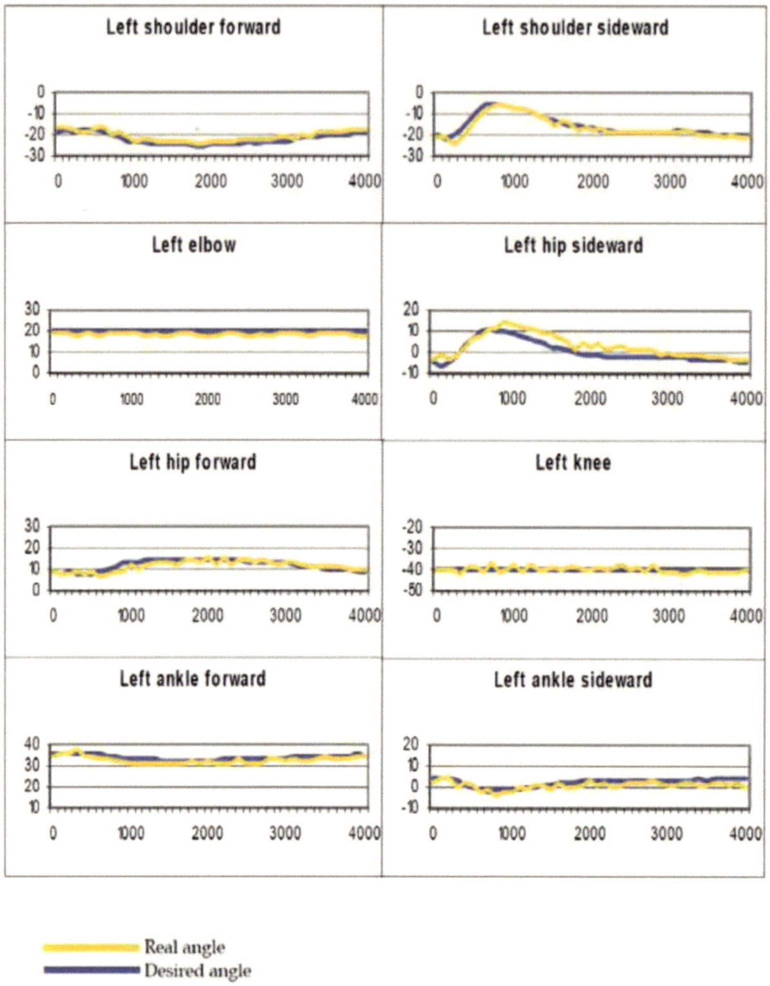

Fig. 12. Desired and real angle (degree) from left arm and leg joints over 4000 ms as an average over 5 recurrences.

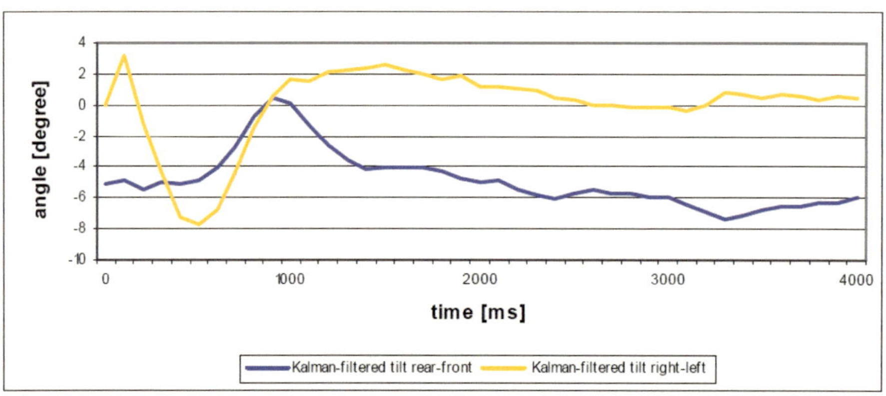

Fig. 13 Tilt values for rear-front and right-left pitch as an average over 5 recurrences.

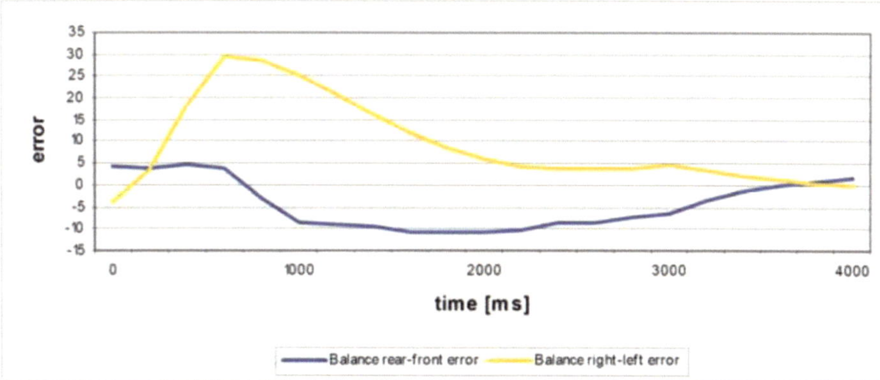

Fig. 14. Calculated error from the balance behavior's PID-controller as an average over 5 recurrences.

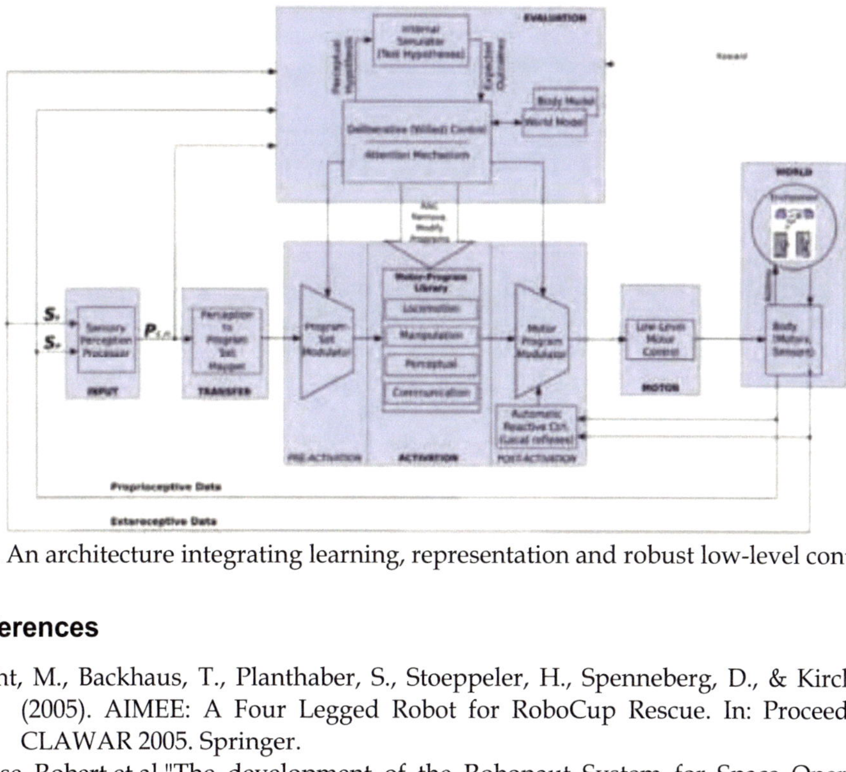

Fig. 15. An architecture integrating learning, representation and robust low-level control.

8. References

Albrecht, M., Backhaus, T., Planthaber, S., Stoeppeler, H., Spenneberg, D., & Kirchner, F. (2005). AIMEE: A Four Legged Robot for RoboCup Rescue. In: Proceedings of CLAWAR 2005. Springer.

Ambrose, Robert et al "The development of the Robonaut System for Space Operations", ICAR 2001, Invited Session on Space Robotics, Budapest, August 20, 2001.

Breazeal, C. and Velasquez, J. "Toward Teaching a Robot 'Infant' using Emotive Communication Acts" In Proceedings of 1998 Simulation of Adaptive Behavior, workshop on Socially Situated Intelligence, Zurich Switzerland. 25-40.

Brooks, R. et al "The Cog Project: Building a Humanoid Robot" In Computation for Metaphors, Analogy and Agents, Vol. 1562 of Springer Lecture Notes in Artificial Intelligence, Springer-Verlag, 1998.

Brooks, R. A. (1986). A robust layered control system for a mobile robot. IEEE Journal of Robotics and Automation, 2

Fodor, J.A. (1983). The modularity of mind, Cambridge, MIT Press

Hirai, K. et al "The development of Honda Humanoid Robot" In Proc. of the 1998 IEEE Int. Conf. on Robotics & Automation, p.1321-1326 (1998).

Ishida, T. "A small biped entertainment robot SDR-4X II" In Proceedings of the 2003 IEEE Internation Symposium on Computational Intelligence in Robotics and Automation, pp. 1046-1051, vol.3, 2003

II-Woo Park "Mechanical Design of Humanoid Robot platfrom KHR-3(KAIST Humanoid Robot-3: HUBO)", in Humanoids 2005, Japan

Ishiguro, H. and T. Minato, "Development of androids for studying on human-robot interaction" In Proceedings of 36th International Symposium on Robotics, TH3H1, Dec. 2005.

James, W. (1890). The Principles of Psychology. Henry Holt, New York.

Kirchner, F., Spenneberg, D., & Linnemann, R. (2002). A biologically inspired approach towards robust real world locomotion in an 8-legged robot. In: J. Ayers, J. Davis, & A. Rudolph (Eds.), Neurotechnology for Biomimetic Robots. MIT-Press, MA, USA.

Kaneko, K. et al"Humanoid Robot HRP-2" In Proceedings of the 2004 IEEE International Conference on Robotics& Automation, 2004

Kagami, S. et al"Design and Implementation of Software Research Platform for Humanoid Robotics: H6" In Proc. of International Conference on Robotics and Automation (ICRA'01), pp. 2431--2436 , 2001

Lakoff, G., & Johnson, M. (1980). Metaphors we Live by, University of Chicago Press.

Löffler, K. Gienger, M. "Control of a Biped Jogging Robot" In Proceedings of the 6th International Workshop on Advanced Motion Control, Nagoya, Japan, 307-323, 2000.

Murase, Y. et al "Design of a Compact Humanoid Robot as a Platform" In 19th Conf. of Robotics Society of Japan, p.789-790 (2001)

Nilsson, N.J. (1984). Shakey the robot, SRI Technical Note, no. 323, SRI, Menlo Park, CA, USA.

Ogura, Y. et al"Development of a Human-like Walking Robot Having Two 7-DOF Legs and a 2-DOF Waist" In Proceedings of the 2004 IEEE International Conference on Robotics and Automation, pp134-139, 2004

Prinz, W. (1987). Ideo-motor action. In Heuer, H., & Sanders, A.F. (Eds.) Perspectives on perception and action, p. 47-76, Lawrence Erlbaum Associates, Hillsdale.

Shan, J. Fumio Nagashima (2002). Neural Locomotion Controller Design and Implementation for Humanoid Robot HOAP-1. In: Proceedings of The 20th Annual Conference of the Robotics Society of Japan 2002.

Spenneberg, D. (2005). A Hybrid Locomotion Control Approach. In: Proceedings of the CLAWAR 2005 Conference.

Spenneberg, D., Albrecht, M., & Backhaus, T. (2005). M.O.N.S.T.E.R.: A new Behavior Based Microkernel for Mobile Robots. In: Proceedings of the ECMR 2005.

Spenneberg, D., Hilljegerdes, J., Strack, A., Zschenker, H., Albrecht, M., Backhaus, T., & Kirchner, F. (2005). ARAMIES: A Four-Legged Climbing and Walking Robot. In: Proceedings of 8th International Symposium iSAIRAS.

Tevatia, G. Schaal, S (2000). Inverse kinematics for humanoid robots. In: Proceedings of IEEE International Conference on Robotics and Automation (ICRA 2000)

Toyota Partner Robot. http://www.toyota.co.jp/en/special/robot.

Wilson, M. (2002). Six views of embodied cognition, University of California, Santa Cruz, In Psychonomic Bulletin & Review, 9, p. 625-636.

28

A Human Body Model for Articulated 3D Pose Tracking

Steffen Knoop, Stefan Vacek, Rüdiger Dillmann
University of Karlsruhe (TH)
Karlsruhe, Germany

1. Introduction

Within the last decade, robotic research has turned more and more towards flexible assistance and service applications. Especially when cooperating with untrained persons at small distances in the same workspace, it is essential for the robot to have a deep understanding and a reliable hypothesis of the intentions, activities and movements of the human interaction partner.

With growing computational capacities and new emerging sensor technologies, methods for tracking of articulated motion have become a hot topic of research. Tracking of the human body pose (often also referred to as Human Motion Capture) without invasive measurement techniques like attaching markers or accelerometers and gyroscopes demands (1) for algorithms that maximally exploit sensor data to resolve ambiguities that compulsorily arise in tracking of a high-degree-of-freedom system, and (2) for strong models of the tracked body that constrain the search space enough to enable fast and online tracking.

This chapter proposes a 3D model for tracking of the human body, along with an iterative tracking approach. The body model is composed of rigid geometric limb models, and joint models based on an elastic band approach. The joint model allows for different joint types with different numbers of degrees of freedom. Stiffness and adhesion can be controlled via joint parameters.

Effectiveness and efficiency of these models are demonstrated by applying them within an Iterative Closest Point (ICP) approach for tracking of the human body pose. Used sensors include a Time-of-Flight camera (depth camera), a mono colour camera as well as a laser range finder. Model and sensor information are integrated within the same tracking step for optimal pose estimation, and the resulting fusion process is explained, along with the used sensor model. The presented tracking system runs online at 20-25 frames per second on a standard PC.

We first describe related work and approaches, which partially form the basis for the presented models and methods. Then, a brief introduction into the ICP is given. The model for body limbs and joints is explained in detail, followed by a description of the full tracking algorithm. Experiments, examples and different evaluations are given. The chapter closes with a discussion of the achieved results and a conclusion.

2. Related work

Tracking of human body motion is a highly active field in current research. Depending on the target application, many different sensors and models have been used. This includes invasive sensors like magnetic field trackers (Ehrenmann et al., 2003; Calinon & Billard, 2005) that are fixed to the human body. Within the context of human robot interaction in every-day life, this approach is not feasible; non-invasive tracking approaches must be applied. Most of these are based on vision systems, or on multi-sensor fusion (Fritsch et al., 2003). Systems which rely on distributed sensors (Deutscher et al., 2000) are not practicable in the given domain; the tracking system must be able to rely only on sensors mounted on the robot.

Several surveys exist on the area of tracking humans (Aggarwal & Cai, 1999; Gavrila, 1999; Moeslund & Granum, 2001; Wang et al., 2003). Possible applications range from the mentioned human-robot interaction to surveillance and security domains. Hence, there is a big variety of methods ranging from simple 2d approaches such as skin colour segmentation (Fritsch et al., 2002) or background subtraction techniques (Bobick & Davis, 2001) up to complex reconstructions of the human body pose. (Ramanan & Forsyth, 2003) shows how to learn the appearance of a human using texture and colour.

Sidenbladh (Sidenbladh, 2001) used a particle filter to estimate the 3d pose in monocular images. Each particle represents a specific configuration of the pose which is projected into the image and compared with the extracted features. (Cheung et al., 2003) use a *shape-from-silhouette* approach to estimate the human's pose.

A similar particle filtering approach is used in (Azad et al., 2004). The whole body is tracked based on edge detection, with only one camera. The input video stream is captured with 60Hz, which implies only small changes of the configuration between two consecutive frames. As it is a 2d approach, ambiguities of the 3d posture can hardly be resolved.

An ICP-based approached for pose estimation is shown in (Demirdjian & Darrell, 2002). The authors use cylinders to model each body part. In (Demirdjian, 2003) the same authors show how they model joint constraints for their tracking process. However, it the effect of the ICP is partially removed when the constraints are enforced. Nevertheless, parts of the work described in this chapter are based on the work of Demirdjian.

3. Sensors and framework

For tracking of the human in a human-robot interaction context, only the sensors onboard the robot can be used. In our setup, we use several different sensors as input for the tracking algorithm, which fuses all available information to obtain an optimal estimation of the current pose of the human.

3.1 Sensors

3D point clouds are acquired by a Time-of-Flight camera. This depth camera called Swissranger (CSEM, 2006) has a resolution of 160 x 124 pixels and a depth range of 0.5 to 7.5 meters. Fig. 1 shows the depth image of an example scene. Alternatively, point clouds from reconstruction of stereo images can be used.

A standard single FireWire camera is used to obtain colour images. These are processed by a standard skin-colour based algorithm to track head and hands in the image. All image regions which are candidates for skin regions are provided to the tracking algorithm.

A SICK laser range finder which is mounted for navigation purposes delivers 3D points in its measurement plane. These points are also provided to the tracking.

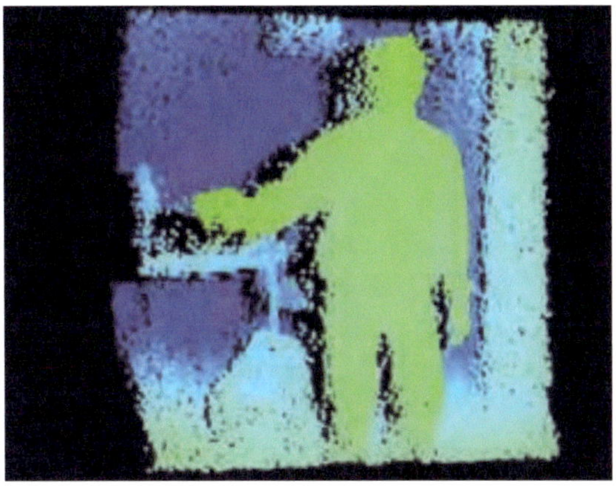

Fig. 1 Depth image, retrieved with the Time-of-Flight sensor. The point cloud is visualized in 3D, and depth is additionally encoded in colours. Green is equivalent to near, and blue to distant measurements.

3.2 Iterative Closest Point algorithm

This section gives a short introduction to the Iterative Closest Point (ICP) algorithm. The goal of the ICP is to match two indexed sets of the same points which are given in different coordinate systems and calculate the translation $\vec{t}$ and rotation R that transform the first coordinate system into the second.

For tracking, the first set corresponds to the data points of the sensor and the second set corresponds to points on the surface of a rigid body. Following (Besl & McKay, 1992), the first set is denoted $P=\{\vec{p}_i\}$, the second one $X=\{\vec{x}_i\}$. Both sets have the same size with $N_x = N_p = N$ and each point $\vec{p}_i$ corresponds to point $\vec{x}_i$.

With six degrees of freedom, at least three points are necessary to compute translation and rotation. Because sensor data is always corrupted with noise, no exact solution exists; instead, the problem is solved as a minimization problem for the sum of squared distances:

$$f(R,\vec{t}) = \frac{1}{N}\sum_{i=1}^{N}\left\| R(\vec{x}_i) + \vec{t} - \vec{p}_i \right\|^2 \tag{1}$$

With $\vec{\mu}_p$ and $\vec{\mu}_x$ as the mean value of P and N respectively, and with setting $\vec{p}_i' = \vec{p}_i - \vec{\mu}_p$, $\vec{x}_i' = \vec{x}_i - \vec{\mu}_x$ and $\vec{t}' = \vec{t} + R(\vec{\mu}_x) - \vec{\mu}_p$, equation (1) can be written as

$$f(R,\vec{t}) = \frac{1}{N}\sum_{i=1}^{N}\left\| R(\vec{x}'_i) - \vec{p}'_i + \vec{t}' \right\|^2 \tag{2}$$

This leads to

$$f(R,\vec{t}) = \frac{1}{N}\left(\sum_{i=1}^{N}\left\| R(\vec{x}'_i) - \vec{p}'_i \right\|^2 - 2\vec{t}\sum_{i=1}^{N}\left(R(\vec{x}_i') - \vec{p}_i' \right) + N\left\| \vec{t}' \right\|^2 \right) \tag{3}$$

In equation (3), the first part is independent from $\vec{t}\,'$, the second part reveals to zero. Therefore the function becomes minimal if $\vec{t}\,'=0$. Transformation yields

$$\vec{t} = \vec{\mu}_p - R(\vec{\mu}_x) \tag{4}$$

Having the optimal translation (giving $\vec{t}\,'=0$), equation (2) becomes:

$$f(R,\vec{t}) = \frac{1}{N}\sum_{i=1}^{N}\left\| R(\vec{x}\,'_i) - \vec{p}\,'_i \right\|^2 \tag{5}$$

Considering $\|R(\vec{x}_i\,')\| = \|\vec{x}_i\,'\|$, the equation can be rewritten as

$$f(R,\vec{t}) = \frac{1}{N}\left(\sum_{i=1}^{N}\left\|\vec{x}\,'_i\right\|^2 - 2\sum_{i=1}^{N} R(\vec{x}_i\,')\vec{p}_i\,' + \sum_{i=1}^{N}\left\|\vec{p}\,'_i\right\|^2 \right) \tag{6}$$

Maximizing

$$\sum_{i=1}^{N} R(\vec{x}\,'_i)\cdot\vec{p}\,'_i \tag{7}$$

Gives the optimal rotation. See (Horn, 1987) for details.
The described method can be used to compute the translation and rotation for two point sets P and N, where each $\vec{p}_i$ corresponds to $\vec{x}_i$. These point sets and their association has to be set up from the measurements and the surface model of the tracked body. The first set can be directly derived from the measured 3D point cloud. N is constructed by calculating for each data point $\vec{p}_i$ the *closest point on the model surface*, giving $\vec{x}_i$. This can be done with geometrical relations. Having these two indexed point sets, the optimal rotation and translation can be computed and applied to the model.
Association of model surface points to measurements can only be estimated. This estimation usually results in a non-optimal transformation, and the algorithm is iteratively applied until a minimum is reached. In summary, the *Iterative Closest Point* (ICP) algorithm works as follows:

1. For the given mode position and the given measurement M, calculate the closest points on the model giving CP_0.
2. Calculate the sum of squared distances between data points and model points, giving $d_0(M, CP_0)$.
3. Estimate rotation and translation and apply to the model.
4. Calculate the new set of closest points with the new model position, giving CP_i.
5. Calculate the sum of squared distances between data points and model points giving $d_i(M, CP_i)$.
6. If $d_{i-1}(M, CP_{i-1}) - d_i(M, CP_i) < \varepsilon$, stop iteration. Otherwise go to step 3.

Note that the computation of the closest points on the model is the most time consuming step in the ICP loop, since it includes geometrical calculations for each data point in the point cloud.
The ICP as mentioned above can only cope with one rigid body as model. For extension of the method to articulated models, the tracked body has to be modelled with a set of rigid bodies, which are connected to one articulated model. Two problems have to be solved in

this context: (i) data points must be associated with body parts, before they can be used for tracking, because each body part is transformed separately by the ICP. (ii) A mechanism must be established which introduces the joint constraints of the model in the tracking process, to avoid drift of the limbs due to separate tracking of each part.
The first issue is solved by checking each measurement point with all body part models, and assigning it to the geometrically closest body part. In addition, a threshold is used to discard measurements which have too high distance to the model.
To avoid that the body parts drift apart, we introduce a novel joint model. The complete body model together with the joint model will be explained in the next section.

4. Proposed body model

For the tracking system a 3d body model is used. Each body part is represented with a *degenerated cylinder*. The top and the bottom of each cylinder are described by an ellipse. The ellipses are not rotated to each other and the planes are parallel. In total such a body is described by five parameters: major and minor axis of each ellipse, plus the length of the cylinder.
The overall body model is built in a tree-like hierarchy starting with the torso as root body part. Each child is described with a degenerated cylinder and the corresponding transformation from its parent. Up to now the body model consists of ten body parts (torso, head, two for each arm and two for each leg) which is depicted in Fig. 2. It should be mentioned that this body model is not necessarily restricted to humans, and also other bodies can be modelled easily.

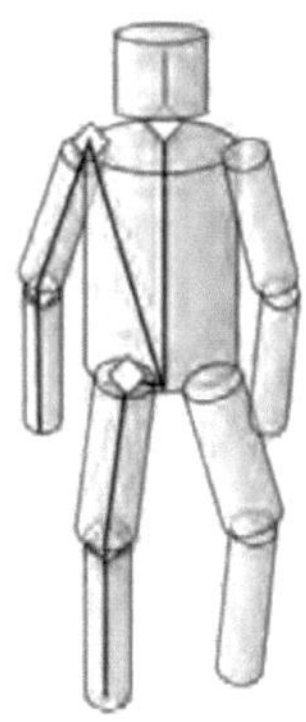

Fig. 2. Hierarchical body model, consisting of 10 cylinders and 9 joints. The torso describes the root limb.

If the fusion algorithm also incorporates data from feature trackers (like some vision based algorithms, or magnetic field trackers that are fixed on the human body), it is required to identify certain feature points on the human body. This is done following the H|Anim Specification (H|Anim, 2003).

4.1 Joint constraint model

The joint model we propose is based on the concept of introducing elastic bands into the body model. These elastic bands represent the joint constraints. For the ICP algorithm, the elastic bands can be modelled easily as artificial correspondences and will thus be considered automatically in each computation step.

For each junction of model parts, a set of elastic bands is defined (see Fig. 3). These relations set up corresponding points on both model parts. The corresponding points can then be used within the model fitting process to adjust the model configuration according to any sensor data input and to the defined constraints.

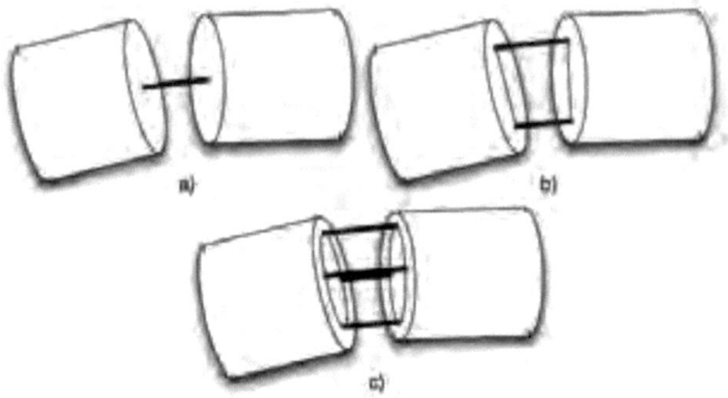

Fig. 3. Different joint types. Universal joint with 3 degrees of freedom (a), hinge joint with one full and 2 restricted degrees of freedom (b), and (c) elliptic joint with 3 restricted degrees of freedom.

4.2 Joint Constraint Types

With this approach, different types of joints can be modelled. Looking at a model for the human body, different kinds of joints with varying degrees of freedom are required:

- Universal Joints have 3 full degrees of freedom. This joint type can be found e.g. in the shoulder. The upper arm can rotate up/down, forward/backward and around its main axis. Universal joints are modelled by a point-to-point correspondence (one elastic band) between both body parts with one point on each, see Fig. 3 a).
- Hinge Joints have one real degree of freedom, the others being almost fixed. This can be found e.g. in the human knee or elbow (only 1 DoF), or in the hip (1 real DoF, the other two existing, but highly restricted in motion). Hinge joints are modelled by a set of correspondences which are distributed along a straight line on both body parts. The same restriction can be achieved with correspondences only at each end of the line (two elastic bands), see Fig. 3 b).
- Elliptic Joints have all degrees of freedom highly restricted. An example on the human body is the neck (or the wrist): Motion is possible in all 3 degrees of freedom: Left/right, forward/backward, and turning. Each direction is very limited in range. Elliptic joints are modelled by a set of correspondences distributed along an ellipse on both body parts. This restriction can be achieved with correspondences on each end of the main axes of the ellipse (four elastic bands), see Fig. 3 c).

Universal and hinge joints are special cases of the elliptic joint. For the hinge joint, one major axis of the ellipse is set to zero, resulting in a straight line. Setting both axes to zero produces a universal joint, because all correspondences are reduced to one point-to-point relation.

Following these definitions, each joint is modelled with a set of parameters describing the type of joint and its behaviour. This parameter set consists of the major axes of the ellipse, its position and orientation on both body parts, and the weight of the given correspondence. These parameters and the resulting behaviours are now described in detail.

Major axes: The model type (universal, hinge, elliptic) and the valid range of each degree of freedom control the choice for the major axes sizes and ratio. Universal joints are modelled with both ellipse axes set to zero. For hinge and elliptic joints, the axis direction defines the rotation axis, and the axis length defines the stiffness of the other two rotational degrees of freedom.

In Fig. 3 b), rotational flexibility around the z-axis (perpendicular to the image plane) and around the symmetric axis of the cylinders is very limited due to the modelled joint.

Position and orientation: Position and orientation of the point-to-point, hinge or elliptic joint model with respect to both body parts define the connection between both parts.

Weight: When the joint model is used within a bigger tracking framework the elastic bands can be used as correspondences which are included as tracking constraints. The use of measured correspondences together with artificial ones puts up the need for correct weighting strategies between input and model constraints.

To incorporate this, each joint model can be weighted with respect to measured input. This parameter is then used within the model-fitting algorithm to balance between measured input and joint constraints. The weight parameter is defined in relation to the number of 'natural' correspondences to keep the ratio between measurements and constraints.

To increase the weight of a joint model tightens the coupling between both model parts, by decreasing the coupling becomes looser. For hinge and elliptic joints, higher weight also increases stiffness of a kinematic chain. This makes sense especially for joints like the human neck or wrist which shows a very tight coupling and very limited angular range.

The proposed joint model provides a "soft" way to restrict the degrees of freedom for the model parts. It additionally provides means to control the "degree of restrictiveness" for each DoF in a joint. Applying elastic strips is tantamount to introducing a set of forces which hold the model parts together. The connected points and magnitude define the joint behaviour.

The soft joint model can e.g. be applied to the joint between human pelvis and thigh: While in forward/backward direction the movement is almost unrestricted, there is a high restriction for the left/right movement. But still a small movement is possible, and a 1-DoF model would not be sufficient.

Nevertheless, it is possible to model plain bending joints with a hinge joint model with one large axis.

4.3 Joint Model within the ICP

One of the main advantages of this joint model is that it can be very easily integrated in tracking algorithms. The joint model is added as a second data source, which adds correspondences between model and real world.

Introducing the joint model correspondences in the ICP framework (see sec. 3.2) is done by transforming the elastic band constraints into artificial input points according to the following rules:

- For each correspondence with the weight W, W artificial point pairs are generated.
- The artificial point pairs have to be added to the correspondence list after computation of the Closest Point Relations.
- Because each body part is processed separately, each joint model has to be added twice, once to each associated body part.

- The generated point pairs each represent one point on the model and the associated artificial data point. So each pair has to be added to one body part as Model - Data and to the other Data - Model relation to retrieve the desired forces (from the elastic bands) on both model parts. These forces then try to establish the modelled connections.
- The artificial correspondences are recalculated in each ICP step.
- The chosen weight of each joint depends on the desired stiffness of the model. To always achieve the same stiffness during tracking, the ratio between measured and artificial point relations has to be constant. This means that the number of generated artificial points for one body part in each step depends on the quantity of measurements for this part of the model. The generated relations are linearly scaled with the number of measurements.
- From our experience, the ratio r between measured and artificial points should be chosen as approximately $0.4 < r < 0.7$. This gives enough cohesion within the model without implying to hard and static relationships.

It is important to note that the introduction of multiple identical correspondences within the ICP does not increase computation time with the order of point counts (like a set of different measured points would). The only additional effort consists of one multiplication of the resulting point matrix (4x4) with the scalar weight W.

5. Data fusion for tracking

The goal of the tracking system is to track the posture of a human body in 3d by matching the internal 3d body model with the current input sensor data. Thus, the tracking system offers three interfaces: sensor data stream (input), parameter configuration (input), and current posture estimation (output). All sensor data formats that can be exploited are described in sec. 5.1. The configuration values we have identified will be described in sec. 5.2 along with the processing steps.

The current posture estimation output is given with respect to the hierarchical body model defined in sec. 4. In each time step, the whole body model is provided. This allows for changes not only in the body pose (joint angle space), but also for changes in the model itself (configuration and parameters of the body model). This may concern scaling of the model for different persons with varying body heights, or even addition and deletion of body parts in case of changing tracking targets or other effects. This can be useful e.g. if the tracked person is holding and handling a big object, which then can be added easily to the tracked configuration.

The fusion algorithm, which is implemented in a tracking system called *VooDoo*, is depicted in Fig. 4. The next section describes possible input data, while sec. 5.2 depicts the processing steps within the tracking loop.

5.1 Input data

The proposed tracking algorithm is able to include, process and fuse different kinds of sensor data (see also Fig. 4):

- *Free 3d points* from ToF-sensors or from pure stereo depth images. The system has to decide whether to use these points as measurements of the tracked model. For a point that is not discarded, the corresponding point on the model surface is computed.

- *3d points on the human body* that are e.g. generated by a stereo vision system that tracks a person in image space and generates the corresponding 3d points by stereo reconstruction.
- *3d points assigned to a single body part* may also be generated by a stereo vision system tracking special body parts like the face or the hands.
- *3d point-to-point relations* are 3d points that can be assigned to a given point on the tracked human body. Thus, tracking of special features or points (e.g. with markers, or magnetic field trackers attached to the human body) can be integrated.
- *2d point-to-line relations* can e.g. be derived from a 2d image space based tracker. The pixel in the image plane together with the focal point define a ray in 3d, which corresponds to the point on the human body that has been detected in the image.

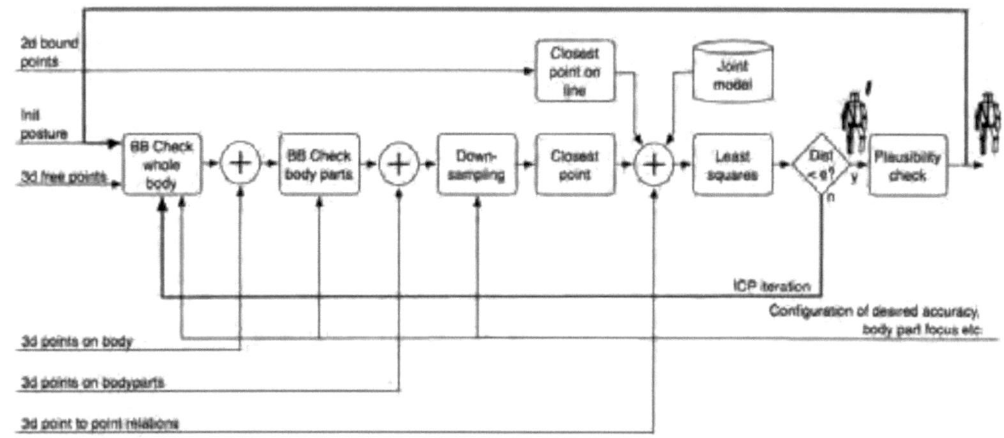

Fig. 4. The complete tracking algorithm, containing all input data streams. "BB" denotes Bounding Box.

This data can originate from any sensor that gives data in the described format. Obviously, all input data has to be transformed into the tracker coordinate system before it is used within the system. In our setup, we use the 3d point clouds from laser range finder and depth camera as free 3d points. Extracted features (hands, head) from the colour image are projected onto 2d point-to-line relations.

5.2 Processing

For the ICP matching algorithm, a list of corresponding point pairs has to be set up for each limb. Therefore, all "free" 3d points have to be analyzed in order to decide whether they correspond to points on the tracked model. Otherwise, they are discarded. Additionally, all given correspondences from other tracking procedures and the background knowledge on joint constraints have to be added to the correspondences list. Then, the optimal resulting model configuration has to be computed. These steps are performed iteratively until an optimum of the configuration is reached.

Before the input data of one time step is processed, it is possible to adjust internal model parameters. This can be e.g. the model scale factor, or particular cylinder sizes. Even limbs can be added to or removed from the model.

The processing steps are now described in detail.

Prefiltering free 3d points: The whole point cloud of free 3d points from used depth sensors is processed in order to remove all points that are not contained within the bounding box of the body model (see Fig. 4, step *BB Check whole body*). This is done on the assumption that the body configuration changes only locally between two time frames. A parameter defines an additional enlargement of the bounding box prior to this filtering step. The resulting point list is concatenated with any sensor data input that has already assigned its measured 3d points with the tracked. It results in a list of 3d points which are close to the body model and thus are candidates for measurements of the tracked body.
Assigning points to limb models: The point list is now processed in order to assign measured points to dedicated limb models based on the bounding box of each limb model (see Fig. 4, step *BB Check body parts*). Again, the bounding boxes can be enlarged by a parameter to take the maximum possible displacement into account. Points that do not fall in any bounding box are again removed. Several behaviours can be selected for points that belong to more than one bounding box (overlap): These points are either shared between limb models, exclusively assigned to one limb or shared only in case of adjacent limbs. This last method avoids collisions between limbs that are not directly connected. The resulting point list can be joined with any sensor data input that has already assigned its measured 3d points with dedicated limbs of the tracked body. The resulting point list contains candidates for measurements of each limb.
Point Number reduction: The resulting point list can be downsampled before the calculation of the closest points to reduce the overall number of points (see Fig. 4, step *Downsampling*). This step is controlled by three parameters: the sampling factor, and minimum and maximum number of points per limb. Thus, it is possible to reduce the number of points for limbs with many measurements, but maintain all points for limbs which have been measured with only a few points.
Closest point computation: The closest point calculation is the most time-consuming step in the whole loop. For each remaining data point, the corresponding model point on the assigned limb model has to be computed for the ICP matching step (see Fig. 4, step *Closest Point*). This involves several geometric operations. Depending on the resulting distance between data and model point, all points within a given maximum distance are kept and the correspondence pair is stored in the output list. All other points are deleted.
3d point-to-point relations from input data can now be added to the resulting list, which holds now corresponding point pairs between data set and model.
Addition of 2d measurements: Each 2d measure (e.g. tracked features in 2d image plane of a camera) of a feature on the human body defines a ray in 3d which contains the tracked feature. This fact is used to add the 2d tracking information to the 3d point correspondences (see Fig. 4, step *Closest point on line*): For each reference point on the body model, the closest point on the straight line is computed and added to the list.
Joint model integration: The joint model for each junction is added as artificial point correspondences for each limb, depending on the limb type (see Fig. 4, step *Joint model*). According to sec. 4.1, the correspondences can be interpreted as elastic bands which apply dedicated forces to the limbs to maintain the model constraints. Thus, artificial correspondences will keep up the joint constraints in the fitting step.
Model fitting: When the complete list of corresponding point pairs has been set up, the optimal transformation between model and data point set can be computed according to sec. 3.2 (Fig. 4, step *Least squares*). The transformation is computed separately for each limb.

When all transformations have been computed, they can be applied to the model. The quality measure defined in sec. 3.2 is used for the fitting. All are repeated until the quality measure is below a given threshold or a maximum number of steps has been performed.

6. Sensor model

Each used data source has its own stochastic parameters which have to be taken into account. The described approach offers a very simple method for this: each input date is weighted with a measure that describes its accuracy. The ICP algorithm then incorporates these weights in the model-fitting step. Thus it is possible to weight a 2d face tracker much higher than a single 3d point from a Time-of-Flight camera, or to weight 3d points from a Time-of-Flight-camera slightly higher than points from the stereo reconstruction due to the measuring principle and the sensor accuracy.
It is important to note that an increased weight for a single point does not affect the time needed for the computation. This is very important and is due to the fact that in the presented approach, each measurement is projected into model space. This is different to e.g. particle filtering approaches, where each particle is projected into each sensor's measurement space to compute the likelihood. In consequence, adding a sensor source to the tracking framework increases computation time only with the number of different measurements from the sensor.
An example configuration can be seen in Fig. 5. The model consists of two cylinders, connected by a linear joint. The measurements contain a 3d point cloud, and a 3d measurement of one end point. This configuration can e.g. result from a stereo depth image of a human arm and a colour based hand tracker.

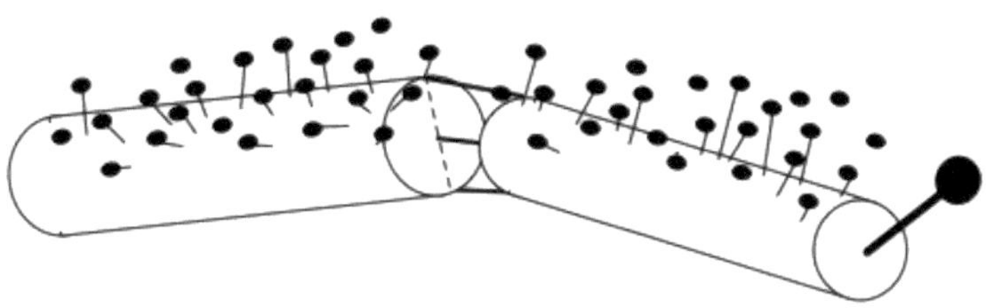

Fig. 5. Different weights for measurements from different sensors, projected into 3D model space. The depicted point sizes correspond to the sensor data weight, the lines indicate the closest-point relations. These pictures motivated the system name *VooDoo*.

7. Experiments, evaluation and results

The described tracking procedure has been implemented and tested with a Time-of-Flight camera and a stereo camera. The tracking runs online at a frame-rate of appr. 20-25Hz on a Pentium4 with 3.2GHz with a model of a human body, consisting of 10 cylinders with 9 joints. For the experiments, the same data sequences have been processed using different input sensor configurations to test the fusion.
Fig. 6 shows example images from a sequence of 15 seconds containing a "bow" and a "wave" movement. The first row shows the scene image, which has been also used for segmentation of face and hands. The second and third row contain the tracking result

with 3d data only (row 2) and 2d data only (row 3), where the 3d data has been acquired with the Time-of-Flight camera and the 2d data is derived from skin colour segmentation in one image of the stereo camera. The rays in 3d defined by the skin colour features can be seen here. Row 4 shows the tracking result with both inputs used.
For the shown results, the following weights for the input data have been used: 3d data points w = 1.0, face tracker w = 30.0, hand tracker w = 20.0.

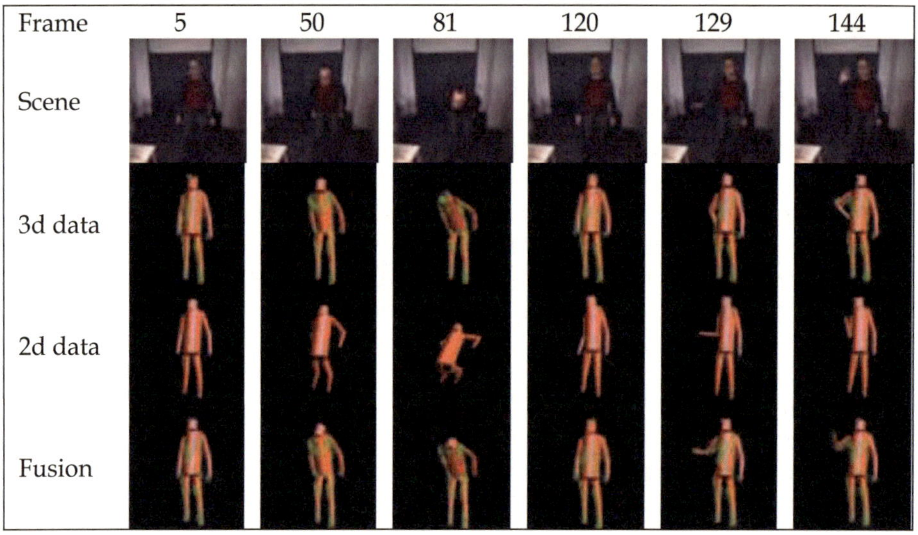

Fig. 6. Experiments with different sensor inputs, taken from a sequence containing a "bow" and a "wave" movement. The frame number is displayed on the top. The used 2d and 3d correspondences have been added to the resulting model images.

Different conclusions can be drawn from the results:

- Huge movements are easily detected by the 3d data based tracking: The ``bow'' movement is tracked quite well. On the other hand, fast movements with the extremities may cause failures when only 3d data is used, as with the ``wave'' movement.
- Tracking only with a 2d feature tracker works quite well for the tracked body parts. Nevertheless, the body configuration cannot be determined only from 2d features (see frame 81). To do this, a lot more background information on the human body would be needed.
- Fusion of both input sensors in 3d shows very good results: Huge body movements as well as fine and fast movements of the extremities can be recognized, and the algorithm is able to reliably track the body configuration.

As already stated in sec. 6, the computational effort and thus the frame-rate depends on the true number of different measurements, independent of the particular weights. To evaluate the computational performance and frame-rate of the presented method, several analyses have been carried out. The model corresponds again to the human body model used for the analysis in Fig. 6.

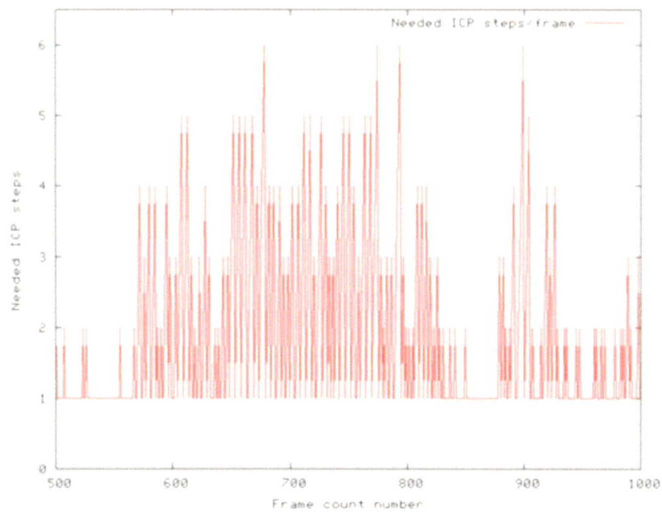

Fig. 7. Number of ICP steps required for a typical tracking sequence.

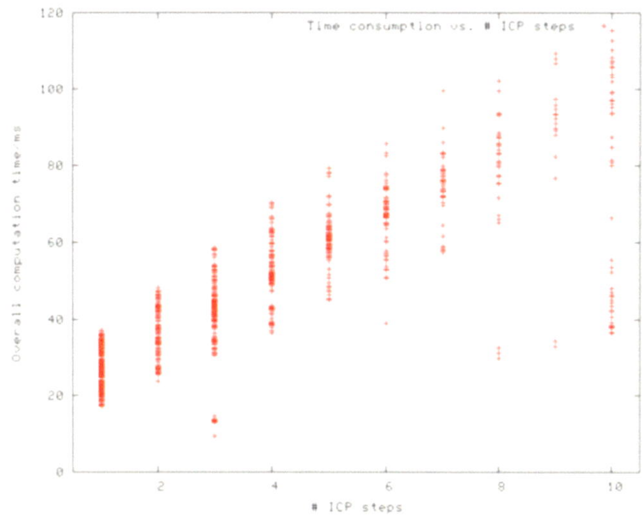

Fig. 8. Time consumption per ICP step vs. number of ICP steps.

The computational effort for one frame depends first of all on the number of ICP steps needed. The number of iterations again depends on the body displacement between two consecutive frames. Fig. 7 shows the number of required ICP steps during a typical tracking sequence for a human body model. During phases without large movements, one iteration is enough to approximate the body pose (frame 500 to 570). Extensive movements are compensated by more ICP iteration steps per frame (650 to 800).
The required time per frame obviously increases with the number of needed ICP steps. This relation is shown in Fig. 8. A maximum number of 6 ICP steps has turned out to be a good trade-off between time consumption per frame and tracking accuracy. This leads to a frame period of 20 - 70ms, which corresponds to a frame-rate of 14.2 to 50Hz. The maximum frame-rate in our framework is only constrained by the camera frame-rate, which is 30Hz.

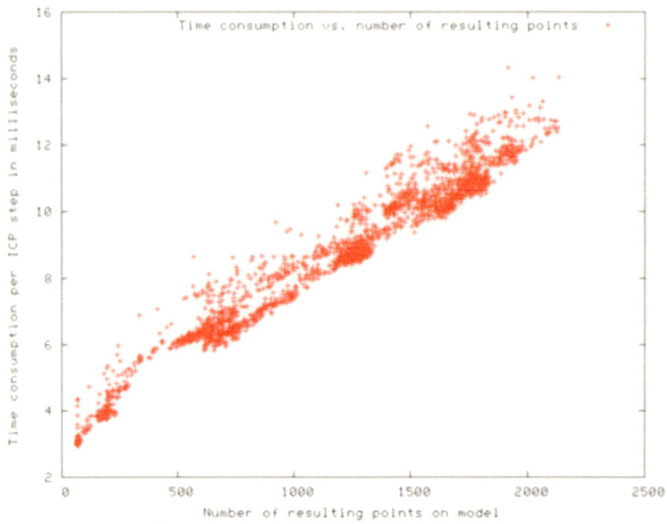

Fig. 9. Time consumption per frame vs. number of body measurements.

The relation between the number of body measurements and the computational effort for one ICP step is depicted in Fig. 9. For each measurement of the target, several computations have to be carried out. This leads to the dependency in Fig. 9. As expected, the time scales linearly with the number of measurements.

These results show that the presented tracking approach is able to incorporate several thousand measurements with reasonable computational effort. One disadvantage of the depicted iterative process is the negative dependency between target displacement and computational effort: The faster the target moves, the longer the tracking needs for one frame, which again leads to larger displacements due to the low frame-rate.

To overcome this, one has to find a good trade-off between accuracy and frame-rate. This compromise depends on the tracking target characteristics, as well as on the application which utilizes the Human Motion Capture data. It is also possible to switch between different behaviours, taking into account the requirements the applications which depend on the Motion Capture data: in case the data is used for physical interaction (e.g. handing over objects), the required accuracy is high, along with usually low dynamics. On the other hand, if the target is only to observe a human in the robot's environment, the required accuracy is low, but the person moves with high velocity.

8. Discussion and conclusion

This paper has proposed a geometric human body model, a joint model and a way for fusion of different input cues for tracking of an articulated body. The proposed algorithm is able to process 3d as well as 2d input data from different sensors like ToF-cameras, stereo or monocular images. It is based on a 3d body model which consists of a set of degenerated cylinders, which are connected by an elastic bands joint model. The proposed approach runs in real-time. It has been demonstrated with a human body model for pose tracking.

The main novelty and contribution of the presented approach lies in the articulated body model based on elastic bands with soft stiffness constraints, and in the notion of point correspondences as a general measurement and model format. Different joint behaviours can be modelled easily by distributing the elastic bands along two axes in the joint. The joint constraints are incorporated in the ICP as artificial measurements, so measurements and model knowledge are processed identically. The model can also be refined by adding cylindrical primitives for hands, fingers and feet. This is reasonable if the accuracy and resolution of the available sensors are high enough to resolve e.g. the hand posture, which is not the case in our approach due to the large distance between human and robot and the low measurement resolution.

The idea of introducing artificial correspondences into the fitting step can even be exploited further. Current works include further restriction of the joints in angular space by adding angular limits to certain degrees of freedom, which are maintained valid by artificial point correspondences. These will be generated and weighted depending on the current body configuration.

Our implementation of the described tracking framework has been released under the GPL license, and is available online at wwwiaim.ira.uka.de/users/knoop/VooDoo/doc/html/, along with sample sequences of raw sensor data and resulting model sequences.

9. References

Aggarwal, J. K.; Cai, Q. (1999) *Human motion analysis: A review,* Computer Vision and Image Understanding: CVIU, vol. 73, no. 3, pp. 428–440.

Azad, P.; Ude, A.; Dillmann, R.; Cheng, G. (2004) *A full body human motion capture system using particle filtering and on-the-fly edge detection,* in Proceedings of the IEEE-RAS/RSJ International Conference on Humanoid Robots. Santa Monica, USA.

Besl, P. J.; McKay, N. D. (1992) *A method for registration of 3-d shapes,* IEEE Transactions on pattern analysis and machine intelligence, vol. 14, no. 2, pp. 239–256, February.

Bobick, A. F.; Davis, J. W. (2001) *The recognition of human movement using temporal templates,* IEEE Transactions on Pattern Analysis and Machine Intelligence, no. 3, pp. 257–267.

Calinon, S.; Billard, A. (2005), *Recognition and reproduction of gestures using a probabilistic framework combining PCA, ICA and HMM,* in Proceedings of the International Conference on Machine Learning (ICML), Bonn, Germany

Cheung, G. K. M.; Baker, S.; Kanade, T. (2003) *Shape-from-silhouette of articulated objects and its use for human body kinematics estimation and motion capture,* in Computer Vision and Pattern Recognition.

CSEM (2006) Swissranger website. http://www.swissranger.ch

Demirdjian, D.; Darrell, T. (2002) *3-d articulated pose tracking to untethered deictic references,* in Multimodel Interfaces, pp. 267–272.

Demirdjian, D. (2003) *Enforcing constraints for human body tracking,* in Conference on Computer Vision and Pattern Recognition, Workshop Vol. 9, Madison, Wisconsin, USA, pp. 102–109.

Deutscher, J.; Blake, A.; Reid, I. (2000), *Articulated body motion capture by annealed particle filtering,* in Computer Vision and Pattern Recognition (CVPR), Hilton Head, USA, pp. 2126–2133.

Ehrenmann, M.; Zöllner, R.; Rogalla, O.; Vacek, S.; Dillmann, R. (2003). *Observation in programming by demonstration: Training and execution environment,* in Proceedings of Third IEEE International Conference on Humanoid Robots, Karlsruhe and Munich, Germany.

Fritsch, J.; Lang, S.; Kleinehagenbrock, M.; Fink, G. A.; Sagerer, G. (2002) *Improving adaptive skin color segmentation by incorporating results from face detection*, in Proc. IEEE Int. Workshop on Robot and Human Interactive Communication (ROMAN). Berlin, Germany

Fritsch, J.; Kleinehagenbrock, M.; Lang, S.; Plötz, T.; Fink, G.A.; Sagerer, G. (2003), *Multi-modal anchoring for human-robot-interaction*, Robotics and Autonomous Systems, Special issue on Anchoring Symbols to Sensor Data in Single and Multiple Robot Systems, vol. 43, no. 2–3, pp. 133–147.

Gavrila, D. M. (1999) *The visual analysis of human movement: A survey*, Computer Vision and Image Understanding, vol. 73, no. 1, pp. 82–98.

H|Anim (2003), *Information technology – Computer graphics and image processing – Humanoid animation (H-Anim), Annex B*, ISO/IEC FCD 19774, Humanoid Animation Working Group, Specification.

Horn, B. K. P. (1987) *Closed-form solution of absolute orientation using unit quaternions*, Optical Society of America Journal A, vol. 4, pp. 629–642, Apr. 1987.

Knoop, S.; Vacek, S. & Dillmann, R. (2005). *Modelling Joint Constraints for an Articulated 3D Human Body Model with Artificial Correspondences in ICP*, Proceedings of the International Conference on Humanoid Robots (Humanoids 2005), Tsukuba, Japan, December 2005, IEEE-RAS

Knoop, S.; Vacek, S. & Dillmann, R. (2006). *Sensor Fusion for 3D Human Body Tracking with an Articulated 3D Body Model*. Proceedings of the IEEE International Conference on Robotics and Automation (ICRA), Orlando, Florida, May 2006

Knoop, S.; Vacek, S. & Dillmann, R. (2006). *Sensor fusion for model based 3D tracking*. Proceedings of the IEEE International Conference on Multisensor Fusion and Integration for Intelligent Systems (MFI), Heidelberg, Germany, September 2006

Moeslund, T. B.; Granum, E. (2001) *A survey of computer vision-based human motion capture*, Computer Vision and Image Understanding, vol. 81, no. 3, pp. 231-268.

Ramanan, D.; Forsyth, D. A. (2003) *Finding and tracking people from the bottom up*, in Computer Vision and Pattern Recognition, vol. 2, 18-20 June, pp. II–467–II–474.

Sidenbladh, H. (2001) *Probabilistic tracking and reconstruction of 3d human motion in monocular video sequences*, Ph.D. dissertation, KTH, Stockholm, Sweden.

Wang, L.; Hu, W.; Tan, T. (2004) *Recent developments in human motion analysis*, Pattern Recognition, vol. 36, no. 3, pp. 585–601, 2003.and Electronics Engineers.

29

Drum Beating and a Martial Art Bojutsu Performed by a Humanoid Robot

Atsushi Konno, Takaaki Matsumoto, Yu Ishida, Daisuke Sato & Masaru Uchiyama
Tohoku University
Japan

1. Introduction

Over the past few decades a considerable number of studies have been made on impact dynamics. Zheng and Hemami discussed a mathematical model of a robot that collides with an environment (Zheng & Hemami, 1985). When a robot arm fixed on the ground collides with a hard environment, the transition from the free space to constrained space may bring instability in the control system. Therefore, the impact between robots and environments has been the subject of controversy. Asada and Ogawa analyzed the dynamics of a robot arm interacting with an environment using the inverse inertia matrices (Asada & Ogawa, 1987). In the early 90's, the optimum approach velocity for force-controlled contact has been enthusiastically studied (Nagata et al., 1990, Kitagaki & Uchiyama, 1992). Volpe and Khosla proposed an impact control scheme for stable hard-on-hard contact of a robot arm with an environment (Volpe & Khosla, 1993). Mills and Lokhorst proposed a discontinuous control approach for the tasks that require robot arms to make a transition from non-contact motion to contact motion, and from contact motion to non-contact motion (Mills & Lokhorst, 1993). Walker proposed measures named the dynamic impact measure and the generalized impact measure to evaluate the effects of impact on robot arms (Walker, 1994). Mandal and Payandeh discussed a unified control strategy capable of achieving a stable contact against both hard and soft environment (Mandal & Payandeh, 1995). Tarn et al. proposed a sensor-referenced control method using positive acceleration feedback and switching control strategy for robot impact control (Tarn et al., 1996). Space robots does not have fixed bases, therefore, an impact with other free-floating objects may bring the space robots a catastrophe. In order to minimize the impulsive reaction force or attitude disturbance at the base of a space robot, strategies for colliding using reaction null-space have been proposed (Yoshida & Nenchev, 1995, Nenchev & Yoshida, 1998).
Most of the researches have been made to overcome the problems introduced by impacts between robots and environments. Some researchers have tried to use the advantages of impacts. When a robot applies a force statically on an environment, the magnitude of force is limited by the maximum torque of the actuators. In order to exert a large force on the environment beyond the limitation, applying impulsive force has been studied by a few researchers. Uchiyama performed a nail task by a 3-DOF robotic manipulator (Uchiyama, 1975). Takase et al. developed a two-arm robotic manipulator named Robot Carpenter, and performed sawing a wooden plate and nailing (Takase, 1990). Izumi and Hitaka proposed to use a flexible link manipulator for nailing task, because the flexible link has an advantage in absorbing an impact (Izumi & Kitaka, 1993).

However, those works mentioned above were done using robotic manipulators fixed on the ground except for space robots, and thus, there was no need to take care about loosing a balance. Humanoid robots are expected to work on human's behalf. If a humanoid robot can do heavy works utilizing an impulsive force as well as a human does, the humanoid robot will be widely used in various application fields such as constructions, civil works, and rescue activities.
The first attempt on an impact motion by a humanoid robot was reported in (Hwang et al., 2003). Matsumoto et al. performed a Karate-chop using a small humanoid robot and broke wooden plates (Matsumoto et al., 2004). In order for a legged robot to effectively exert a large force to an environment without loosing a balance, working posture is important. Tagawa et al. proposed a firm standing of a quadruped for mobile manipulation (Tagawa et al., 2003). Konno et al. discussed an appropriate working posture of a humanoid robot (Konno et al., 2005).
This chapter addresses an impact motion performed by a humanoid robot HRP-2. A drum beating is taken as a case study, because it is a typical task that requires large impulsive forces. The drum beating motion is carefully designed to synchronize with music. The drum beating and a Japanese martial art Bojutsu were performed by a humanoid robot HRP-2 in the Prototype Robot Exhibition at Aichi Exposition 2005.

2. Why and Where Is an Impulsive Force Needed?

In order to show the advantages of using an impulsive force, a task of pushing a wall is taken as an example in this section. A model of a humanoid robot HRP-1 (the HONDA humanoid robot P3) is used in a simulation.
Fig. 1 shows the snapshots in a simulation in which the humanoid robot HRP-1 quasi-statically pushes a wall, while Fig. 2 shows the snapshots in a simulation in which the HRP-1 dynamically pushes a wall moving a body forward. In the simulation illustrated in Fig. 1, the body is fixed so that the projection of the centre of gravity (COG) comes on the middle of the fore foot and rear foot, while in the simulation illustrated in Fig. 2, the body is moved so that the projection of COG moves from the centre of rear foot to the centre of fore foot.
The results of the simulations are plotted in Fig. 3. Fig. 3 (a) shows the forces generated at the wrist (equal and opposite forces are generated on the wall) when the humanoid robot exerts a quasi-static force on a wall, while (b) shows the forces at the wrist when the humanoid robot dynamically exerts a force.

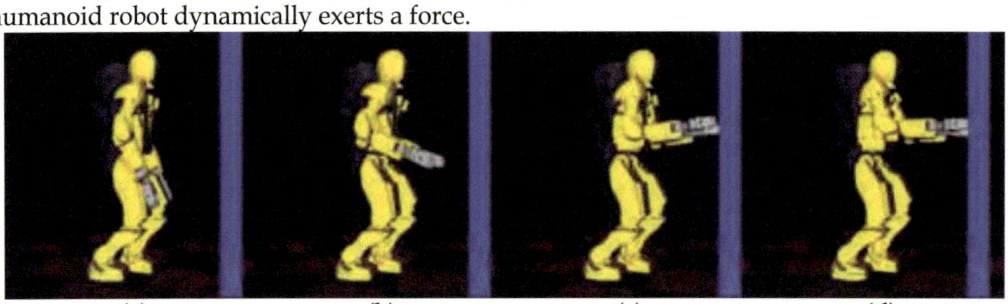

(a) (b) (c) (d)

Fig. 1. A humanoid robot quasi-statically pushes a wall. The body is fixed so that the projection of the center of gravity (COG) comes on the middle of the fore foot and rear foot. (a) at 0.0 [s], (b) at 2.0 [2], (c) at 4.0 [s], and (d) at 6.0 [s].

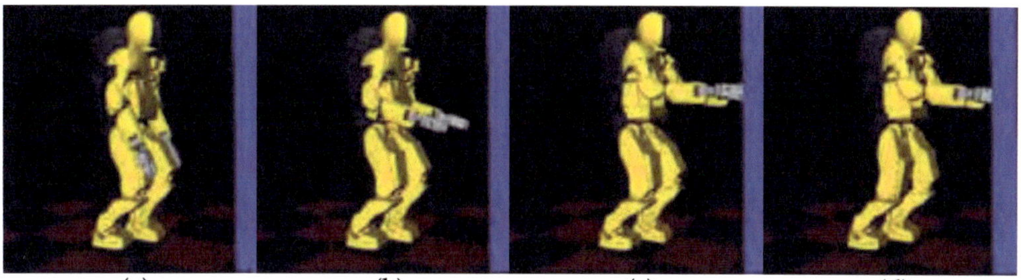

(a) (b) (c) (d)

Fig. 2. A humanoid robot pushes a wall moving the body to apply an impulsive force. In order to accumulate momentum, the body is moved so that the projection of COG moves from the center of rear foot to the center of fore foot. (a) at 0.0 [s], (b) at 2.0[2], (c) at 4.0 [s], and (d) at 6.0 [s].

As seen in Fig. 3, when the humanoid robot dynamically exerts a force on a wall, approximately 1.5 times larger force is generated compared with the case when the humanoid robot quasi-statically exerts a force.

There is a strong demand for the formulation of the impact dynamics of a humanoid robot to solve the following problems:

- *Working postures*: An optimum working posture at the impact tasks must be analyzed in order to minimize the angular momentum caused by an impulsive force. The angular momentum is more crucial than the translational momentum, because a humanoid robot easily falls down by a large angular momentum.
- *Impact motion synthesis*: Appropriate impact motions of a humanoid robot must be synthesized based on multibody dynamics, to exert a large force on an environment.
- *Stability analysis*: Exerting a large force on an environment, a humanoid robot must keep the balance. Therefore, stability analysis for the impact tasks is inevitable.
- *Shock absorbing control*: In order to minimize the bad effect caused by the discontinuous velocity, shock absorbing control algorithms must be studied.
- *Enrichment of applications*: Applications of the impact tasks must be developed to clearly show the advantages of using the impulsive force.

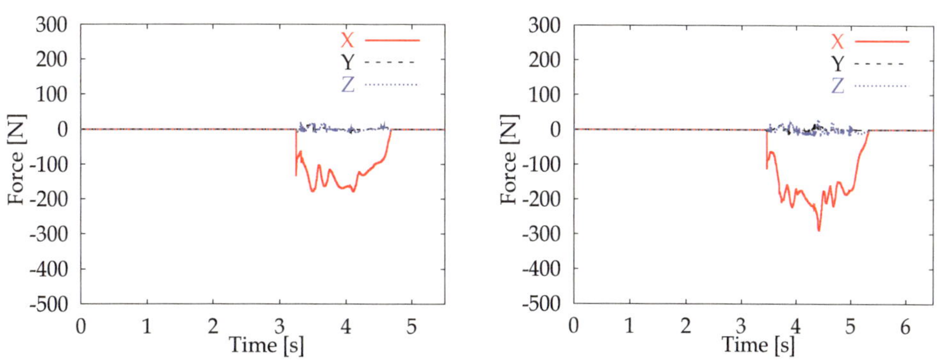

(a) (b)

Fig. 3. Force generated at the wrist. (a) When the humanoid robot exerts a quasi-static force on a wall. (b) When the humanoid robot exerts an impulsive force on a wall.

3. A Humanoid Robot HRP-2 and Control System Software

3.1 Specifications of the HRP-2

A humanoid robot HRP-2 was developed in the Humanoid Robotics Project (1998–2002) being supported by the Ministry of Economy, Trade and Industry (METI) through New Energy and Industrial Technology Development Organization (NEDO). The total robotic system was designed and integrated by Kawada Industries, Inc. and Humanoid Research Group of the National Institute of Advanced Industrial Science and Technology (AIST).

The height and weight of the HRP-2 are respectively 154 cm and 58 kg including batteries. The HRP-2 has 30 degrees of freedom (DOF). Please see the official web page of the HRP-2 (http://www.kawada.co.jp/global/ams/hrp_2.html) for more details.

In order to perform the drum beating and Bojutsu, small modifications are applied to the HRP-2. The arrangement of the wrist DOF is modified from the original, i.e. the last DOF at the wrist is pronated 90 $^{\circ}$. Furthermore, gloves are developed and attached to the hands to grip firmly the sticks.

3.2 Control system software

The control system software of the HRP-2 is supplied and supported by General Robotics Inc. The control system software provides a controller that can be used with the CORBA servers of OpenHRP (Hirukawa et al., 2003). As shown in Fig. 4, the controller is composed of many plugin softwares. The control system software also includes the I/O access library to access the lower level functions of the robot and a VRML simulator model of the HRP-2 and various utilities.

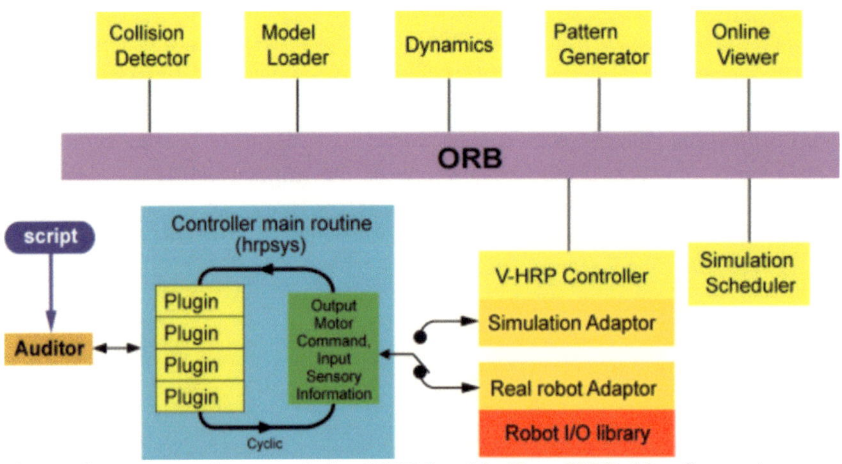

Fig. 4. Control system software of the HRP-2 with OpenHRP (the figure is quoted from http://www.generalrobotix.com/product/openhrp/products_en.htm).

Foundational plugins such as *Kalman Filter, Sequential Playback, Walk Stabilizer, Pattern Generator, Dynamics, Logger,* and *ZMPSensor* are also included in the control system software, however, users can develop own functions as a plugin to enrich the humanoid robot motions. Please see the official web page http://www.generalrobotix.com/product/openhrp/products_en.htm for more details of the control software.

4. Drum Beating

4.1 Primitive poses and motions

In order to generate drum beating motions of the humanoid robot HRP-2, the motion is decomposed into four primitive poses or motions: (a) initial pose, (b) swing, (c) impact, and (d) withdrawing, as shown in Fig. 5. Among the four primitive motions, impact and withdrawing are important to exert an impulsive force.

As presented in Fig. 6, three different swing patterns, (a) small swing, (b) middle swing and (c) big swing, are generated sharing the poses for the impact and withdrawing.

For these swing patterns, three different initial poses are given and the poses to pass through in swing motion are designed. Cubic spline is used to interpolate the given poses.

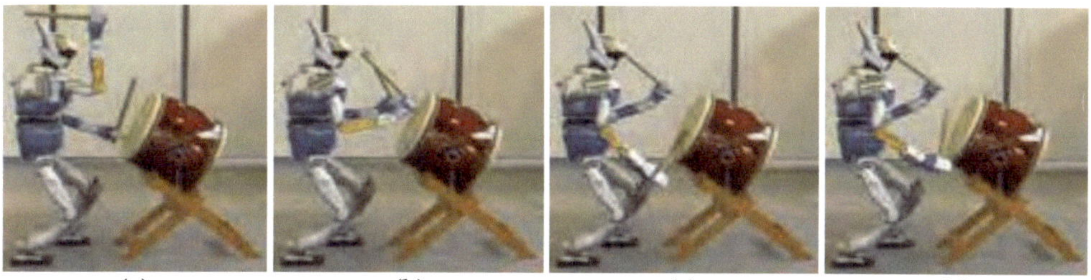

(a) (b) (c) (d)

Fig. 5. Four primitive poses or motions in a drum beating. (a) Initial pose. (b) Swing. (c) Impact. (d) Withdrawing.

4.2 Synchronization with music

The swing motion must be synchronized with music in the drum beating. For the synchronization, a beat timing script is prepared for each tune. An example of the script is listed as follows:

```
0.500   RS
1.270   LM
1.270   RM
0.635   LS
        ...
0.500   END
```

The numbers listed in the first column indicate the interval (s) to the next beating. The symbols listed in the second column indicate the way of beating. The first character 'R' or 'L' indicates the arm to move (Right or Left), while the second character 'S', 'M', 'B', or 'E' indicates the kinds of swing (Small swing, Middle swing, Big swing, or Edge beating, see Fig. 6).

For example, the third line of the script "1.270 RM" indicates "beat the drum after 1.270 s using the middle swing of the right arm." The period between the impact and the previous pose is fixed to 0.1 s to achieve the maximum speed at the impact. As shown in Fig. 6 (b), seven intermediate poses are designed for the middle swing between the initial pose and the impact, therefore, if the duration is specified to 1.270 s, each period ΔT_M between the poses is calculated as follows:

$$\Delta T_M = \frac{\text{duration} - 0.1}{\text{number of poses}} = \frac{1.270 - 0.1}{7}. \quad (1)$$

The duration time varies depending upon a tune.
There are two restrictions in the script: (i) the first beating must be RS (small swing of right arm), (ii) right arm and left arm must be alternating to beat.

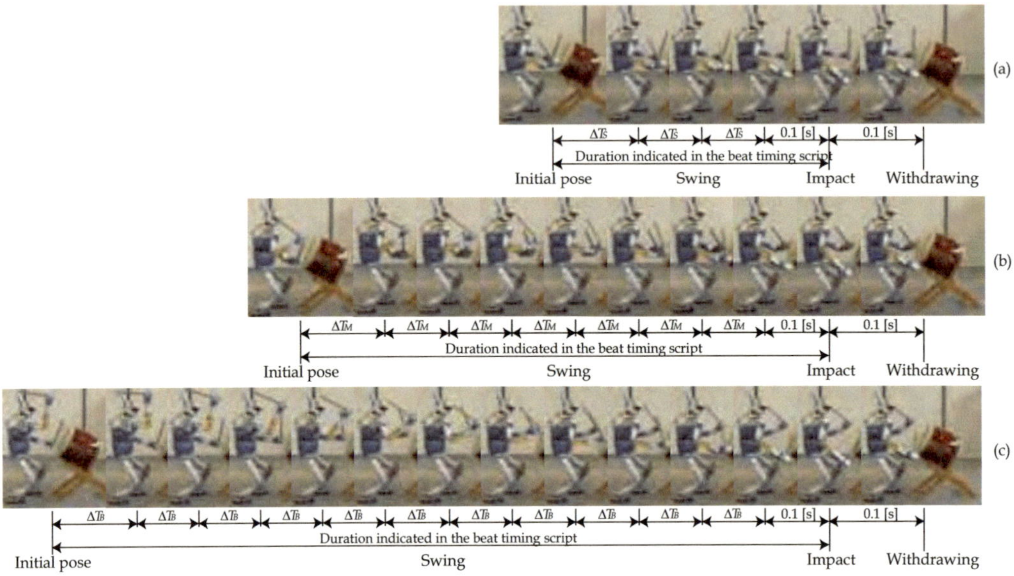

Fig. 6. Three swing patterns. The periods between impact and the previous pose, and between withdrawing and impact are fixed to 0.1 [s]. Other periods denoted by ΔT_S, ΔT_M, ΔT_B, are computed from the duration indicated in the beat timing script. (a) Small swing. (b) Middle swing. (c) Big swing.

4.3 Control software

Fig. 7 presents the flow of the control system. The components marked with red boundary boxes are developed in this work.
Firstly, wav files of the three tunes are prepared: (i) *ware wa umi no ko* (I am a son of the sea), (ii) *Tokyo ondo* (Tokyo dance song), and (iii) *mura matsuri* (village festival). They are very old and traditional tunes, and thus, copyright free. As soon as the Speak Server receives a queue from the robot control system, the server starts playing the tune. The queue is used to synchronize the tune with the drum beating motion.
Secondly, the timings of beating are scheduled by hand. In order to strictly count the timing, a time keeping software is newly developed. The time keeping software counts the rhythm of a tune. The timings of the beating are described in a script file as mentioned in Section 2.
Thirdly, a plugin software is developed as a shared object to generate drum beating motions interpreting the beat timing script.
Fourthly, interpolating the given poses presented in Fig. 6 using cubic spline, trajectories of all joints are produced online. The produced trajectories are given to the humanoid robot through a plugin *SeqPlay*.

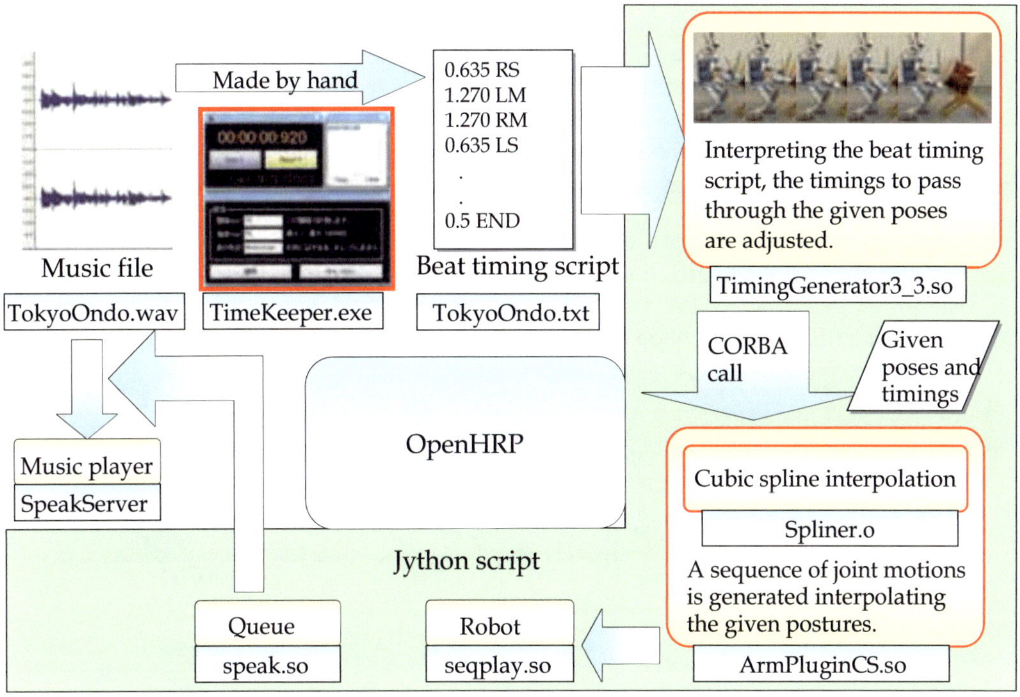

Fig. 7. A software diagram. The components marked with red boundary boxes are developed in this work.

4.3 Resultant joint trajectories

The reference and resultant joint trajectories of the elbow and wrist joints of the right arm are plotted in Fig. 8. The error in the impact time was approximately 30 [ms], which was not significant in the synchronization with music.

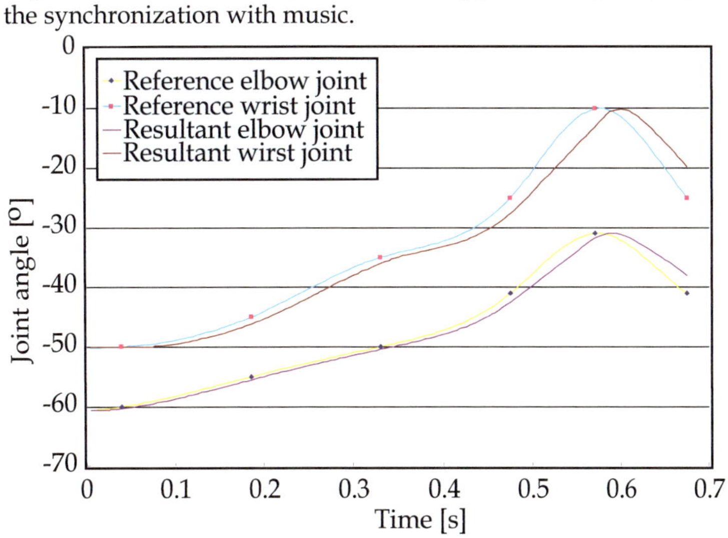

Fig. 8. A software diagram. The components marked with red boundary boxes are developed in this work.

As can be seen in Fig. 7, during the last 0.1 [s] before the impact (approximately from 0.5 to 0.6 [s]), gradients of the joint trajectories are steep compared with other periods. Since the period between the impact and the previous pose is set to 0.1 [s], maximum joint speed is almost achieved.

5. A Japanese Martial Art Bojutsu

In martial arts, impulsive forces are frequently used to fight with an antagonist. A Japanese martial art Bojutsu was also demonstrated by the humanoid robot HRP-2 in Aichi Exposition, although an impact was not performed in the demonstration. Some dynamic motions used in the demonstration are presented in Fig. 9.

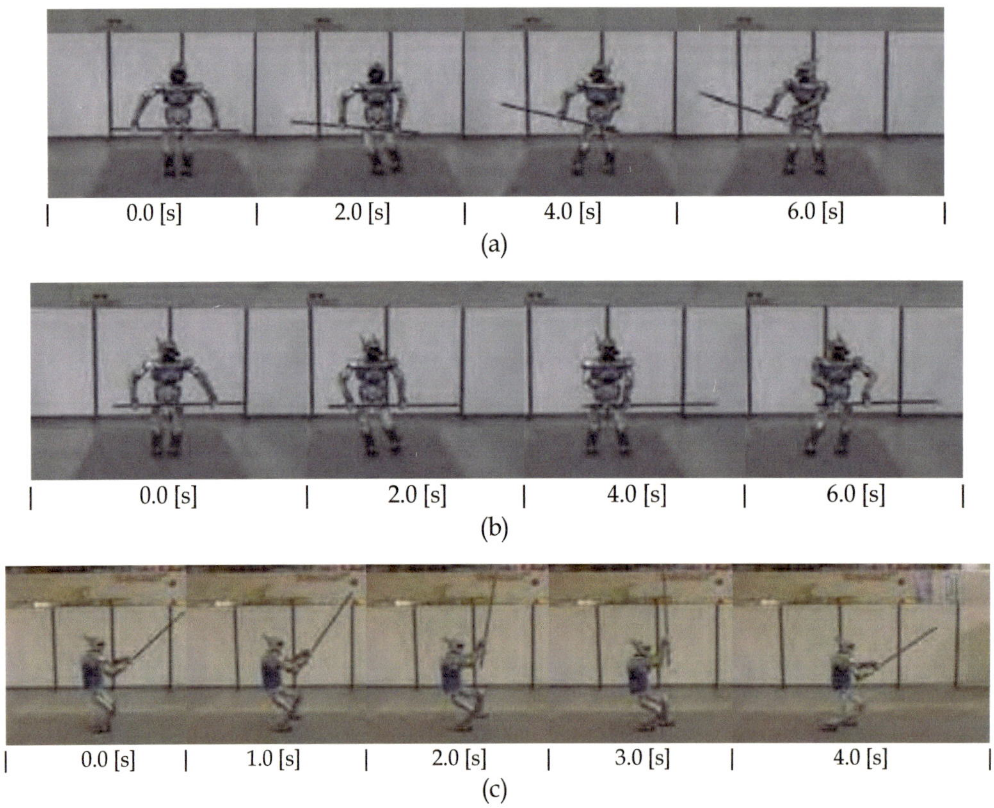

Fig. 9. The Japanese martial art Bojutsu motion patterns. (a) Thrusting a staff weapon rightward. (b) Thrusting a staff weapon leftward. (c) Banging down a staff weapon.

6. Demonstration at Aichi Exposition

The Prototype Robot Exhibition was held for 11 days from June 9 to 19, at the Morizo and Kiccoro Exhibition Center, a convention venue in the Aichi Expo site. The Prototype Robot Exhibition was organized by the Japan Association for the 2005 World Exposition and the New Energy and Industrial Technology Development Organization (NEDO). 63 prototypes performed demonstrations during the period.

The drum beating and Bojutsu demonstration was performed twice a day in the Prototype Robot Exhibition (Fig. 10).

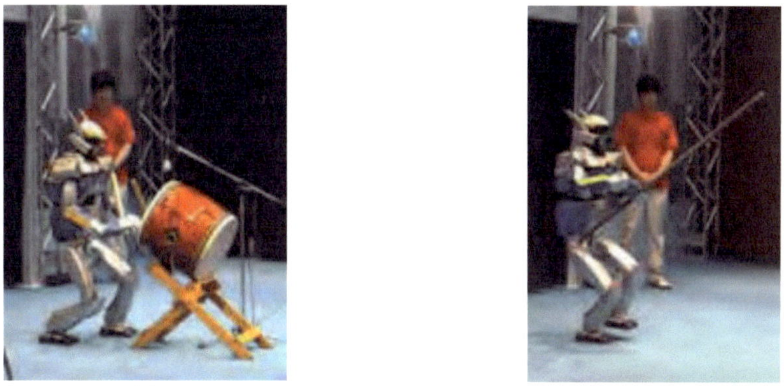

(a) (b)

Fig. 10. Demonstrations at Aichi Exposition 2005. (a) Drum beating performance. (b) A Japanese martial art Bojutsu performance.

7. Conclusion

This chapter proposed to utilize an impulsive force for humanoid robots to exerts a large force beyond the torque limitations of actuators. The problems of the impact tasks to be solved in the future work were brought up in Section 2.

A drum beating is taken as a case study, because it is a typical task that requires large impulsive forces. The details of the drum beating and a Japanese martial art Bojutsu performed by a humanoid robot HRP-2 in the Aichi Exposition were presented in this paper.

8. Acknowledgement

Authors would like to express special thanks to the staffs of Kawada Industries, Inc. and General Robotics Inc. for their kind and sincere support in this project. Authors also would like to express thanks to all the staffs who are related to the Prototype Robot Exhibition.

9. References

Asada, H. & and Ogawa, K. (1987). ON THE DYNAMIC ANALYSIS OF A MANIPULATOR AND ITS END EFFECTOR INTERACTING WITH THE ENVIRONMENT, *Proceedings of the IEEE Int. Conf. on Robotics and Automation*, pp. 751-756

Hirukawa, H.; Kanehiro, F. & Kajita, S. (2003). Open HRP: Open Architecture Humanoid Robotics Platform, *Robotics Research, STAR 6*, Springer-Verlag, Jarvis, R. A. & Zelinsky, A. Eds., pp. 99-112

Hwang, Y.; Konno, A. & Uchiyama, M. (2003). Whole Body Cooperative Tasks and Static Stability Evaluations for a Humanoid Robot, *Proceedings of IEEE/RSJ Int. Conf. on Intelligent Robots and Systems*, pp. 1901-1906

Izumi, T. & Kitaka, Y. (1993). Control of a Hitting Velocity and Direction for a Hammering Robot Using a Flexible Link, *Journal of the RSJ*, Vol. 11, No. 3, pp. 436-443, (In Japanese).

Kitagaki, K. & Uchiyama, M. (1992). OPTIMAL APPROACH VELOCITY OF END-EFFECTOR TO THE ENVIRONMENT, *Proceedings of the IEEE Int. Conf. on Robotics and Automation*, pp. 1928–1934

Konno, A.; Hwang, Y, Tamada, S. & Uchiyama, M. (2005). Working Postures for Humanoid Robots to Generate Large Manipulation Force, *Proceedings of IEEE/RSJ Int. Conf. on Intelligent Robots and Systems*, pp. 1788-1793

Mandal, N. & Payandeh, S. (1995). Control Strategies for Robotic Contact Tasks: An Experimental Study, *Journal of Robotic Systems*, Vol. 12, No. 1, pp. 67–92

Matsumoto, T.; Konno, A., Gou, L. & Uchiyama, M. (2006). A Humanoid Robot that Breaks Wooden Boards Applying Impulsive Force, *Proceedings of 2005 IEEE/RSJ Int. Conf. on Intelligent Robots and Systems*, pp. 5919–5924

Mills, J. K. & Lokhorst, D. M. (1993). Control of Robotic Manipulators During General Task Execution: A Discontinuous Control Approach, *The Int. Journal of Robotics Research*, Vol. 12, No. 2, pp. 146–163

Nagata, K.; Ogasawara, T. & Omata, T. (1990). Optimum Velocity Vector of Articulated Robot for Soft Bumping, *Journal of the SICE*, Vol. 26, No. 4, pp. 435–442, (In Japanese)

Nenchev, D. N. & Yoshida, K. (1998). Impact Analysis and Post Impact Motion Control Issues of a Free-Floating Space Robot Contacting a Tumbling Object, *Proceedings of IEEE Int. Conf. on Robotics and Automation*, pp. 913–919

Tagawa, T.; Aiyama, Y. & Osumi, H. (2003). Firm Standing of Legged Mobile Manipulator, *Proceedings of IEEE Int. Conf. on Robotics and Automation*, pp. 2031–2036

Takase, K. (1990). Task Execution by Robot Hand, *Journal of the SICE*, Vol. 29, No. 3, pp. 213–219, (In Japanese).

Tarn, T.-J.; Wu, Y., Xi, N. & Isidori, A. (1996). Force Regulation and Contact Transition Control, *IEEE Control Systems*, Vol. 16, No. 1, pp. 32–40

Uchiyama, M. (1975). A Control Algorithm Constitution Method for Artificial Arm and Dynamic Control Modes, *Biomechanism 3*, University of Tokyo Press, pp. 172–181, (In Japanese)

Volpe, R. & Khosla, P. (1993). A Theoretical and Experimental Investigation of Impact Control for Manipulators, *The Int. Journal of Robotics Research*, Vol. 12, No. 4, pp. 351-365

Walker, I. D. (1994). Impact Configurations and Measures for Kinematically Redundant and Multiple Armed Robot System, *IEEE Transactions on Robotics and Automation*, Vol. 10, No. 5, pp. 670–683

Yoshida K. & Nenchev, D. N. (1995). Space Robot Impact Analysis and Satellite-Base Impulse Minimization Using Reaction Null-Space, *Proceedings of IEEE Int. Conf. on Robotics and Automation*, pp. 1271–1277

Zheng, Y.-F. & Hemami, H. (1985). Mathematical Modeling of a Robot Collision with its Environment, *Journal of Robotic Systems*, Vol. 2, No. 3, pp. 289–307

30

On Foveated Gaze Control and Combined Gaze and Locomotion Planning

Kolja Kühnlenz, Georgios Lidoris, Dirk Wollherr, and Martin Buss
Institute of Automatic Control Engineering, Technische Universität München
D-80290 München, Germany

1. Introduction

This chapter presents recent research results of our laboratory in the area of vision and locomotion coordination with an emphasis on foveated multi-camera vision. A novel active vision planning concept is presented which coordinates the individual devices of a foveated multi-camera system. Gaze direction control is combined with trajectory planning based on information theoretic criteria to provide vision-based autonomous exploring robots with accurate models of their environment.

With the help of velocity and yaw angle sensors, mobile robots can update the internal knowledge about their current position and orientation from a previous time step; this process is commonly referred to as dead-reckoning. Due to measurement errors and slippage these estimations are erroneous and position accuracy degrades over time causing a drift of the estimated robot pose. To overcome the drift problem it is common to take absolute measurements evaluating visual information, which are fused dynamically with the odometry data by applying Kalman-filter or other techniques, e.g. (Dissanayake et al., 2001). The use of active vision systems for navigation is state-of-the-art providing a situation-related selective allocation of vision sensor resources, e.g. (Davison & Murray, 2002; Seara et al., 2003; Vidal-Calleja et al., 2006). Active vision systems comprising only one type of vision sensor face a trade-off between field of view and measurement accuracy due to limitations of sensor size and resolution, and of computational resources. In order to overcome this drawback the combined use of several vision devices with different fields of view and measurement accuracies is known which is called foveated, multi-resolution, or multi-focal vision, e.g. cf. (Dickmanns, 2003; Kühnlenz et al., 2006; Ude et al., 2006). Thereby, the individual vision devices can be independently controlled according to the current situation and task requirements. The use of foveated active vision for humanoid robot navigation is considered novel.

Active vision is also frequently utilized in the context of robotic exploration. Yet, gaze control and locomotion planning are generally decoupled in state-of-the-art approaches to simultaneous localization and mapping (SLAM). An integrated locomotion planning and gaze direction control concept maximizing the collected amount of information is presented in the second part of this chapter. This strategy results in more accurate autonomously acquired environment representations and robot position estimates compared to state-of-the-art approaches.

The chapter is organized as follows: In Section 2 vision-based localization and mapping in the context of humanoid robots is surveyed; Section 3 is concerned with foveated multi-camera coordination; novel concepts of gaze control and path planning coordination are presented in Section 4; evaluation studies comparing the novel concepts to conventional planning approaches and vision systems are presented in Section 5; conclusions are given in Section 6.

2. Vision-Based Localization and Mapping for Humanoid Robots

Most state-of-the-art humanoid robots are equipped with vision systems. The benefits of using these vision systems for providing absolute measurements of the robot pose in the environment are obvious: pose information on landmarks is provided and no additional devices as, e.g., laser scanners are necessary. Being equipped with internal sensors - angular sensors in the joints and widely used gyros and accelerometers in the trunk - humanoid robots are basically capable of dead-reckoning, i.e. the ability to update position and orientation known from previous measurements. Thus, common simultaneous localization and mapping techniques are applicable which are covered by common literature, e.g. (Sabe et al., 2004; Ozawa et al., 2005; Thomson & Kagami, 2005; Stasse et el., 2006).

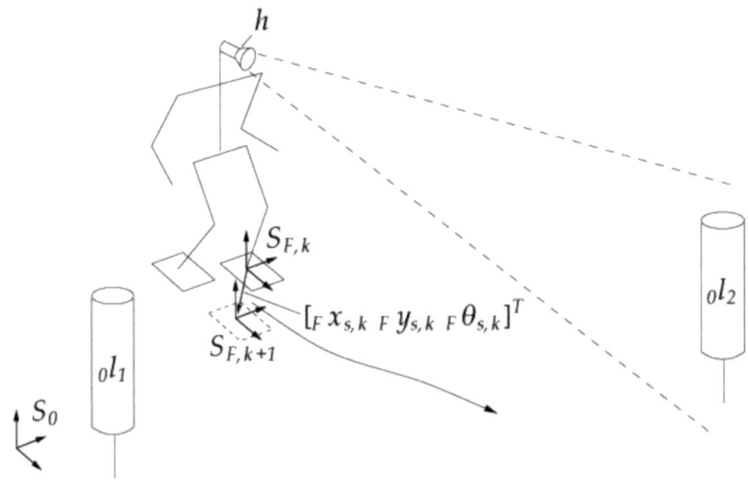

Fig. 1. Humanoid robot navigation scenario.

A fundamental aspect in simultaneous localization and mapping for humanoid walking is the formulation of a state-space model accounting for the footstep sequences of the robot. In vision-based SLAM, the system state, i.e. the robot pose and environment point positions, are predicted based on the dead-reckoning model of the mobile robot. Common Kalman-filter techniques are applied in order to obtain more accurate estimations accounting for uncertainties in the robot locomotion. Whenever visual measurements of environmental points are taken, updates of the robot state are computed. Changing ground contact situations of the feet, however, result in different kinematic chains from a world reference frame to measured environment points. This discontinuous movement of the humanoid robot requires an adaptation of the filter formulation. In earlier works we proposed a hybrid formulation of the state-space model in order to capture this locomotion principle (Seara et al., 2003). Thereby, the robot reference frame is placed in the foot currently in contact with the ground and is switched whenever the supporting foot changes. The dead-reckoning model is expressed by

$$x_{k+1} = x_k(1-\gamma_k) + f_s(x_k, u_k, d_{x,k})\gamma_k \quad (1)$$

where state-vector x contains the robot foot pose and the landmark positions, d represents system noise capturing dead-reckoning uncertainties, and $\gamma \in \{0; 1\}$ is a binary variable indicating a change of the supporting foot when $\gamma=1$. The commanded step u is expressed by

$$u_k = \left[{}_F x_{s,k} \quad {}_F y_{s,k} \quad {}_F \theta_{s,k}\right]^T, \quad (2)$$

including the commanded step position $[x_s\ y_s]^T$ and orientation θ_s with respect to the current supporting foot frame S_F. Figure 1 schematically shows a typical SLAM situation of a humanoid robot with the reference frame currently placed in the left foot.

In vision-based SLAM field of view restrictions of the vision device strongly limit the number of landmarks to be observed simultaneously. Yet, a larger field of view can only be realized accepting a lower measurement accuracy of the vision device mainly due to limitations of sensor size and resolution. Therefore, we propose the use of several vision devices which provide different fields of view and accuracies and a novel gaze control concept for coordinating the individual vision devices in order to provide both, large field of view and high measurement accuracy, simultaneously. These foveated active vision concepts for robot navigation are discussed in the following section.

3. Foveated Multi-Camera Coordination

3.1 Active Vision in SLAM

In order to gather an optimal situation-dependent amount of information the control of the vision system pose is common. To date, there are only few works in the area of active vision-based SLAM, e.g. (Davison & Murray, 2002; Se et el., 2002; Vidal-Calleja et el., 2006) which are based on measures representing the information gathered with respect to the SLAM task. All these approaches are greedy strategies only evaluating the current situation without considering future planning steps. In order to obtain an optimal gaze direction considering also some future planning steps, we proposed a gaze direction planning strategy with limited time horizon (Lidoris et al., 2006). Furthermore, in earlier works (Seara et al., 2003) we introduced a gaze control strategy considering concurrent tasks, localization, and obstacle avoidance for humanoid robots in order to account for navigation in physical environments.

3.2 Foveated Active Vision

Vision systems comprising only one type of vision sensors face a tradeoff between measurement accuracy and field of view due to limitations of sensor size and computational resources for image processing. Accuracy and field of view are mainly determined by the focal-length of the lens or mirror optics, respectively. Within the context of robot navigation this tradeoff implies a compromise between localization accuracy and keeping a large part of the scene in view.

With an active vision system this tradeoff could be compensated providing that a sufficiently accurate map of relevant landmarks or structures of interest to be observed is known a priori. Then the highest available focal-length and, thus, the highest measurement accuracy could be chosen. If additionally very fast gaze shifts can be realized, the narrow field of view would be acceptable as visual attention can be directed dynamically towards the most relevant structure in the current situation. Yet, in a variety of scenarios this approach is unsuitable or even unrealizable. In at least partially unknown environments and in exploration scenarios a sufficient map is not available and thus has to be created online. However, due to the strongly limited field of view the detection of new objects of potential interest is hardly possible. Another aspect are potentially relevant or even dangerous objects or activities in the local surroundings of the robot which cannot be detected.

In order to overcome the common drawback of trading field of view versus measurement accuracy, the combination of wide-angle and telephoto vision devices has been suggested. Such systems provide at the same time both, an observation of a large part of the environment and a selective examination with high accuracy. In common literature these systems are referred to as foveated, multi-resolution or multi-focal systems. The individual vision devices may be fixed with respect to each other or may be independently motion controllable in one or more degrees of freedom. Most common embodiments of foveated systems are used in state-of-the-art humanoid robots comprising two different cameras combined in each eye which are aligned in parallel, e.g. (Brooks et al., 1999; Ude et al., 2006; Vijayakumar et al., 2004). Systems for ground vehicles, e.g. (Apostoloff & Zelinsky, 2002; Maurer et al., 1996; Dickmanns, 2003) are another prominent class. An upcoming area are surveillance systems which strongly benefit from the combination of large scene overview and selective observation with high accuracy, e.g. (Bodor et al., 2004; Davis & Chen, 2003; Elder et al., 2004; Jankovic & Naish, 2005; Horaud et al., 2006). An embodiment with independent motion control of three vision devices with a total of 6 degrees-of-freedom (DoF) is the camera head of the humanoid robot LOLA developed at our laboratory which is shown in Figure 2 providing more flexibility and, due to directly driven gimbals, faster camera motions than other known systems, cf. e.g. (Kühnlenz et al., 2006).

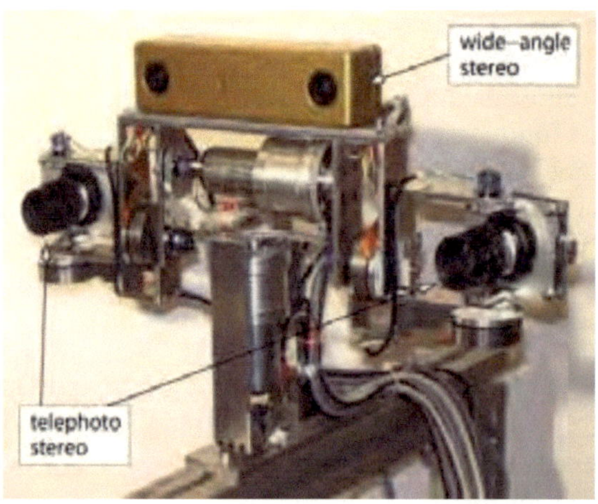

Fig. 2. Multi-focal vision system of humanoid LOLA (Kühnlenz et el. 2006).

Most known methods for active vision control in the field of foveated vision are concerned with decision-based mechanisms to coordinate the view direction of a telephoto vision device based on evaluations of visual data of a wide-angle device. For a survey on state-of-the-art methods cf. (Kühnlenz, 2006). A first approach towards a coordination of foveated multi-camera view direction planning for humanoid walking has been investigated in our laboratory which is presented in the following sections.

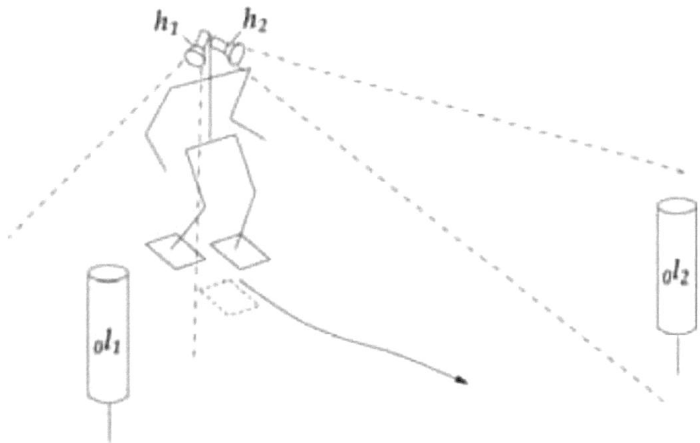

Fig. 3. Humanoid robot navigation scenario with multi-camera vision.

3.3 Considerations for Camera Coordination

In the area of foveated vision a large body of literature exists covering mechanisms to assess peripheral visual data in order to derive control commands to direct foveal attention towards regions of potential interest. The most prominent computational approaches in the biologically inspired field are computational neuroscience models of top-down modulated bottom-up attention weighting particular visual features of the environment, e.g. (Koch & Ullmann, 1984; Itti & Koch, 2001). In the technical field a larger variety of different methods is known. Common approaches solve optimization problems, assess the visual information content, or evaluate the environment towards particular visual features, e.g. (Bodor et al., 2004; Darrell, 1997; Pellkofer & Dickmanns, 2000; Scasselati, 1998; Shibata et al., 2001). To date, only few works have been presented on foveated and multi-camera attention considering locomotion tasks. Prominent examples are the works of (Pellkofer & Dickmanns, 2000) in the field of visual guidance of autonomous ground vehicles and gaze control concepts for the humanoid *LOLA* conducted in our laboratory (Kühnlenz, 2006), where optimal view directions are determined by maximizing the information gain.

In earlier works we proposed a task-related information measure as quality measure termed *incertitude* (Seara et al., 2003) which has been taken as the basis for the coordination of the two stereo-camera devices of *LOLA* with different characteristics. The mission of the humanoid robot is a locomotion task to walk along a certain path or to explore the world. A primary condition for view direction planning, thus, has to consider the quality of locomotion task accomplishment in order to determine an optimal view direction for the next time step. The concept of incertitude captures this task-dependence by evaluating the predicted certainty of the estimated robot foot pose. Therefore, the average of the main axes lengths of the foot pose covariance matrix confidence ellipsoid is computed

$$v_0 = \frac{1}{2}\sum_i \sqrt{e_i}\,, \qquad (3)$$

where counter i covers the considered components of the foot pose and e_i are the eigenvalues of the predicted foot pose covariance matrix P_{uu} which is a submatrix of the predicted covariance matrix of a possible target state as estimated by the Kalman-filter, e.g. cf. (Dissanayake et el., 2001)

$$P^i_{k|k} = \begin{pmatrix} P^i_{uu} & P^i_{um} \\ P^i_{mu} & P^i_{mm} \end{pmatrix}, \quad (4)$$

where $\boldsymbol{P}^i_{uu}$ is the error covariance matrix of the robot state estimate, $\boldsymbol{P}^i_{mm}$ is the map covariance matrix of the landmark state estimates and $\boldsymbol{P}^i_{um}$ is a cross-covariance matrix between robot and landmark states. Low values of the defined measure (3), thus, indicate a high certainty of the robot pose estimation and, therefore, good task performance for the locomotion task. Additional measures to assess the performance of secondary tasks have been proposed which also may have an indirect impact on the performance of the primary (locomotion) task, e.g. field of view limitations, presence of activities, etc., (Kühnlenz, 2006). These measures are all extensions to the central gaze control concept and, therefore, out of scope of this chapter.

Given such measures to assess the task performance of the humanoid robot the next task is to derive appropriate view directions for the individual vision devices in the following time step in order to achieve a particular desired task performance. This gaze control concept is topic of the following section.

3.4 Multi-Camera View Direction Planning

Common approaches to optimal view direction planning for mobile systems are based on a maximization of the information gain, e.g. (Davison, 1998; Pellkofer & Dickmanns, 2000; Seara et al., 2003), in order to determine either a selected gaze shift or a sequence of gaze behaviors. Particularly, in the field of foveated and multi-camera vision also visibility conditions are considered, e.g. (Pellkofer & Dickmanns, 2000; Kühnlenz, 2006).

The basic principle of multi-camera coordination in this chapter is an information maximization over a set of possible view directions of independent vision devices. The assumed task of the robot is to follow a path as closely as possible. As a consequence the estimation error of the robot pose within the environment during its motion has to be minimal in order to complete the mission optimally. The presumed objective for view direction planning is to gather the largest possible amount of information with respect to the task to be accomplished. An information gain corresponds to a reduction of uncertainty. In order to maximize the information gain the robot pose error has to be minimized by selecting appropriate view directions of the individual cameras of the foveated multi-camera vision system. Following this, an optimal configuration of view directions for the locomotion task in the next time step satisfies the condition of minimizing the robot pose estimation error. In terms of the task-related information measure defined in the previous section this gaze control strategy can be expressed by

$$\hat{\Omega}^* = \arg\min_{\hat{\Omega}} \hat{\nu}_0, \quad (5)$$

where $\Omega=[pan_1\ tilt_1\ \ldots\ pan_n\ tilt_n]^T$ is a configuration of pan- and tilt-angles of all vision devices, ν_0 is the incertitude information measure defined in the previous section, and $(.)^*$ denotes the optimal value. This method constitutes an extension to our earlier works on gaze control for humanoid robots (Seara et al., 2003) generalizing them to multi-camera vision systems. In Section 6, a comparative evaluation of this strategy is presented assuming a humanoid robot navigation scenario with sparsely distributed point landmarks.

The presented gaze control strategy considers a preplanned path of the humanoid robot which is not altered as the robot moves. The following section is concerned with combined planning of gaze direction and locomotion path in order to provide the mobile robot with capabilities of exploring unknown environments.

4. Combined Gaze Direction and Path Planning

In the previous section a foveated approach to active vision has been presented which optimally controls the devices such that the robot pose error is minimized. This section is concerned with a novel approach which combines locomotion planning and gaze direction control concepts in order to improve autonomous robotic exploration.

4.1 Locomotion Planning for Exploration and SLAM

Robotic exploration is largely understood as investigating an unknown environment such that the area is covered by the robot sensors and a representation is generated allowing the robot to perform its tasks with a certain amount of confidence. Early approaches focused on generating motion commands which minimize the time needed to cover the whole terrain. This was achieved by extracting frontiers between known and unknown areas (Yamauchi, 1998; Koenig & Tovey, 2001) and visiting the nearest unknown area. Such approaches only distinguish between previously visited and unknown terrain without taking into account the amount of information gathered after each action. To incorporate the uncertainty about the state of the environment, (Moorehead et al., 2001) try to minimize the uncertainty of the robot about grid cells, by using entropy as a criterion. Further, (Grabowski et al., 2003) present an approach in which the robot is forced to observe obstacles from different viewpoints so that sharper boundaries between objects and free-space are acquired. However, the techniques mentioned above assume the location of the robot as known.

Recently, some techniques have been proposed which actively control the motion of the robot while simultaneously creating a map of the environment and localizing the robot in it. In (Feder et al., 1999) information gain is introduced as a measure of utility for locally deciding on exploration actions. Formal information measures, as discussed in the introduction of this section, can be used to quantify uncertainty and therefore evaluate the effect of future control actions on the quality of the robot state estimate. Therefore, (Bourgault et al., 2002) introduced a utility function which trades off the cost of exploring new terrain with the utility of selecting future positions that reduce uncertainty. In (Stachniss et al., 2005) a similar decision theoretic framework is used in combination with a particle filter based SLAM solution for robotic exploration. Further, (Bryson & Sukkarieh, 2005) present simulated results which demonstrate the effect of different actions to information gain for unmanned aerial vehicles performing vision-based SLAM.

All the approaches mentioned above, perform a greedy optimization based on information theoretic criteria for the trajectory generation only over the next time step. However, some planning approaches have been introduced which demonstrate improved performance. Such a planning approach that introduces a new measure of map quality is described in (Sim & Little, 2006). However, some initial state estimate of all the landmarks is assumed and sensors are assumed to have an unlimited field of view. Another multi-step planning algorithm for SLAM is described in (Huang et al., 2005).

4.2 Combined Gaze Direction and Path Planning

Most conventional approaches in autonomous exploration and active vision either control the motion of the robot or the active sensors. In this chapter, we adapt the control inputs of the robot and the sensory system simultaneously so that the best state estimates possible are acquired and as much new terrain as possible is explored.

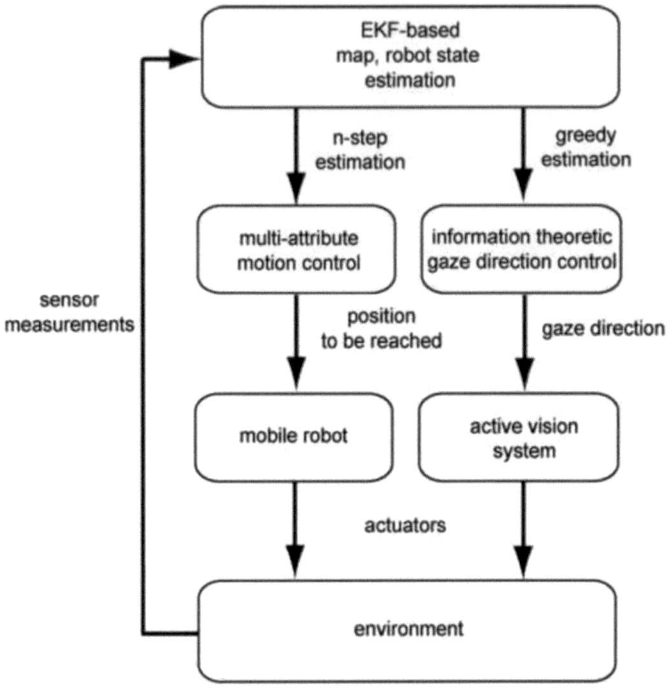

Fig. 4. Proposed motion and gaze direction control scheme

Figure 4 illustrates the proposed motion and gaze direction control scheme. The robot and its active vision system are controlled by two modules which use a common model of the environment. For trajectory planning, a multi-step prediction algorithm is introduced in order to evaluate all possible positions that can be reached by the robot over a finite given time horizon. This estimation forms a multi-attribute function which is used to decide where the robot should move next. A trade-off is made between localization, map accuracy, and proactivity of exploration. For the gaze direction control a greedy information-based optimization is used to choose those view directions that minimize position and map uncertainties. The robot depends on noisy data gained from the visual sensors and at the same time its actions affect the quality of the collected data and its environment.

For vision-guided robots one definition for optimally using sensory resources is selecting the next gaze direction such that measurements are obtained which are most informative about the state of the environment and the robot. This raises the question how information gain can be measured. A common measure of uncertainty is entropy. Which has been introduced by (Shannon, 1948). Entropy for a multivariate Gaussian distribution *p(x)*, with covariance $\boldsymbol{P}$ is defined as

$$H(p(x)) = \frac{1}{2}\log((2\pi)^n |P|) \tag{6}$$

Since the determinant of a matrix is a measure for its volume, the entropy measures the compactness and, thus, the informativeness of a distribution. In order to measure the utility of a gaze direction which will result in an observation *z*, we will use the mutual information gain $I[x,z]$. The gain of information between any two distributions can be computed as the change in entropy. In our case this change is the difference between the entropies of the state estimates before and after making an observation which are both multivariate Gaussians with covariances $P_{k+1|k}$ and $P_{k+1|k+1}$. Therefore, the information gain evaluates to

$$I[x,z] = H(x) - H(x|z) = \frac{1}{2}\log(|P_{k+1|k}|) - \frac{1}{2}\log(|P_{k+1|k+1}|) \tag{7}$$

This information gain can be calculated as a function of the state covariance matrix. From (7) it is obvious that information gain $I[x,z]$ becomes maximal, if the determinant of $P_{k+1|k+1}$ is minimized. Starting from the current state estimate the covariances of the states that can be observed next by the vision sensors are predicted. The equations for the prediction step of the classical SLAM algorithm based on an extended Kalman-filter (Dissanayake et al., 2001) are used. After all covariances are predicted the most informative state can be calculated by minimizing $|P_{k+1|k+1}|$. The new optimal gaze direction Ω of the active vision system corresponding to this state is then computed.

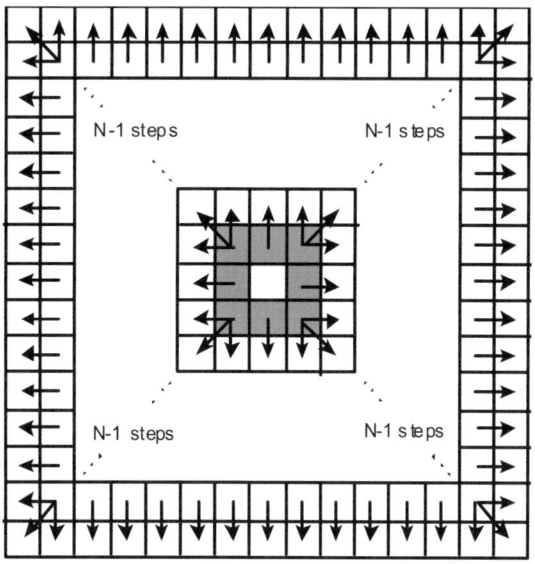

Fig. 5. Region covered while planning over a horizon of N steps. Highlighted grid cells show which cells are taken into account for gaze direction control.

The first step for choosing the next destination for the robot is to estimate the states and covariances of all possible positions that can be reached over its planning horizon. A discretized grid environment is used, where each grid represents a position that can be reached by the robot over future time steps. Therefore, the size of the grid cells depends on the physical properties of the robot. Based on this discretized environment the states and their covariances are computed. While the robot moves, observations are made and used to update the state estimation. This way, all available information is being used. More specifically, based on an initial state estimate and covariance matrix, we calculate all possible robot states and their covariances after N time steps and choose to move to the one that is most informative, namely the one that minimizes relative entropy as described in the previous section. A mathematical description of the algorithm used to produce the multi-step predictions, can be found in (Lidoris et al., 2007). The estimation procedure evolves in a

square-like manner, as shown in Figure 5. Starting from the currently estimated state the first eight neighboring states and covariances are computed. At each step, the estimated state and covariances of the neighboring states are used to infer the next ones until step *N*. By always using the nearest neighbor in the estimation process, the estimation error is kept minimum. Over each time step *k*, *8k* new states are calculated. The control signal, $u_{j_{i,k}}$ required in order to drive the robot from a state *j* to a state *i*, is chosen as indicated by the arrows in Figure 5.
Using the predicted covariance matrix (4) and the concept of relative entropy mentioned previously, each possible future position of the robot can be evaluated to choose the appropriate target position for the robot. The destination that maximizes the function

$$V_i = \frac{1}{2}\log(\frac{P_{uu}^i}{P_{uu}^0}) - \gamma\frac{1}{2}\log(\frac{P_{mm}^i}{P_{mm}^0}) \tag{8}$$

is chosen as a target for the robot. The first part of the function is a measure of the position uncertainty the robot will encounter in the future position and the second part is a measure of the map quality. The constant γ can be used to adjust the behavior of the robotic explorer. Setting γ to values smaller than one, results in a conservative exploration policy, since the robot will stays near to well-localized features giving more attention to localization. Large values of γ increase the proactivity of the explorer in the sense that it moves to unknown areas neglecting the lower localization accuracy. After selecting the target position which maximizes (8), the robot moves making observations which are used to update the estimated state and covariance. Each time a new state estimate is available, a recalculation of the gaze direction is made. This way we use all new information that becomes available during robot motion. Replanning takes place after *N* time steps when the target position is reached.

5. Comparative Evaluation Studies

In Sections 3 and 4 novel concepts of foveated active vision and combined gaze and locomotion coordination have been presented. This section is concerned with evaluation studies in order to assess the performance of the proposed approaches in comparison to state-of-the-art planning methods and vision systems.

5.1 Foveated Active Vision

In Section 3 a task-related information measure for the humanoid robot locomotion task and a multi-camera view direction planning strategy have been introduced. This section is concerned with an evaluation study comparing the performance of the novel approach to a conventional single stereo-camera strategy.
Considered is a typical locomotion task of a humanoid robot with the robot moving along a planned path. It has visual and odometrical capabilities such that it is able to localize itself and other objects within the environment. The robot is equipped with a foveated multi-camera vision system consisting of two stereo-camera devices with independently controllable pan- and tilt-angles, different focal-lengths, and different fields of view. The robot's mission is to follow the desired path. Therefore, it has to localize itself continually evaluating odometry data and visual information. Given a particular environmental situation, i.e. configuration of observable objects and robot pose, the objective is to dynamically select appropriate view directions for both vision devices. Figure 3 exemplarily shows a situation in the considered

navigation scenario where a humanoid robot fixates two landmarks with two vision devices of its foveated multi-camera vision system in order to localize itself in the world.
In order to demonstrate the benefits of foveated multi-camera view direction planning the proposed gaze control approach is now evaluated in a structured humanoid robot navigation scenario. Several vision system configurations are evaluated by comparison of the achieved navigation performances. The basic scenario is shown in Figure 6. Four landmarks are distributed within a rectangular environment. The mission of the robot is to follow the planned path in x-direction. In order to complete the mission successfully the robot has to localize itself within the environment evaluating available visual information on the positions of the identified landmarks. The robot pose is estimated dynamically using the Kalman-filter approach described in Section 2. In order to maximize the information gain optimal view directions of the individual vision devices are selected dynamically based on the proposed approach in Section 3.4. The positions of the landmarks are not known a priori nor are the number of landmarks. Configurations of the vision system in the considered scenario to be compared are: a) conventional single stereo-camera, focal-lengths 20mm, aperture angles 30°, stereo-base 25cm; b) foveated stereo-camera with two cameras per eye aligned in parallel, focal-lengths 2mm and 40mm, respectively, aperture angles 60° and 10°, respectively, stereo-bases 25cm; c) two independent stereo-cameras, focal-lengths 2mm and 40mm, respectively, aperture angles 60° and 10°, respectively, stereo-bases 25cm. All cameras are ideal, based on the pinhole camera model neglecting lens distortion and quantization effects. Gaussian vision sensor noise with a standard deviation of 1 pixel is considered. Dead-reckoning errors are taken from experiments with the humanoid robot *JOHNNIE*.

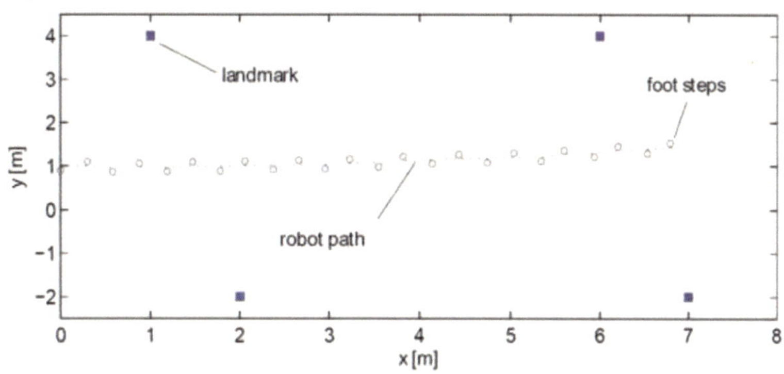

Fig. 6. Top-view of humanoid robot navigation scenario.

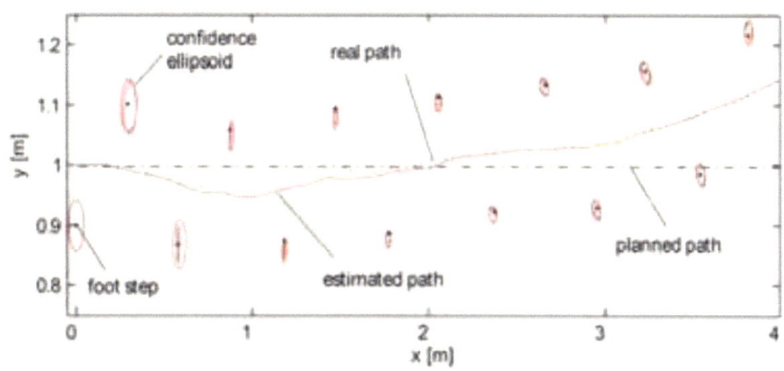

Fig. 7. Real, estimated, and planned paths and footsteps.

The navigation performance is rated assessing the localization accuracy. Therefore, the covariance matrix of the robot position is evaluated computing the areas of the 90%-confidence ellipses of the footstep position uncertainties of the humanoid robot. Figure 7 contains a cut-out of Figure 6 showing the planned and real paths, the path estimated by the Kalman-filter, the foot step positions, and their covariance ellipses. It is noted that due to dead-reckoning errors the real path deviates increasingly from the planned path as locomotion control is open loop. The estimated path follows the real path well.
Figure 8 and Figure 9 show the resulting view directions for each step of the robot for the individual vision systems. The propagations of the areas of the confidence ellipses are shown in Figure 10. Table 1 shows a comparison of the means of the confidence ellipse areas and the average number of landmarks visible for the vision systems for all scenarios. These results are discussed in the following.

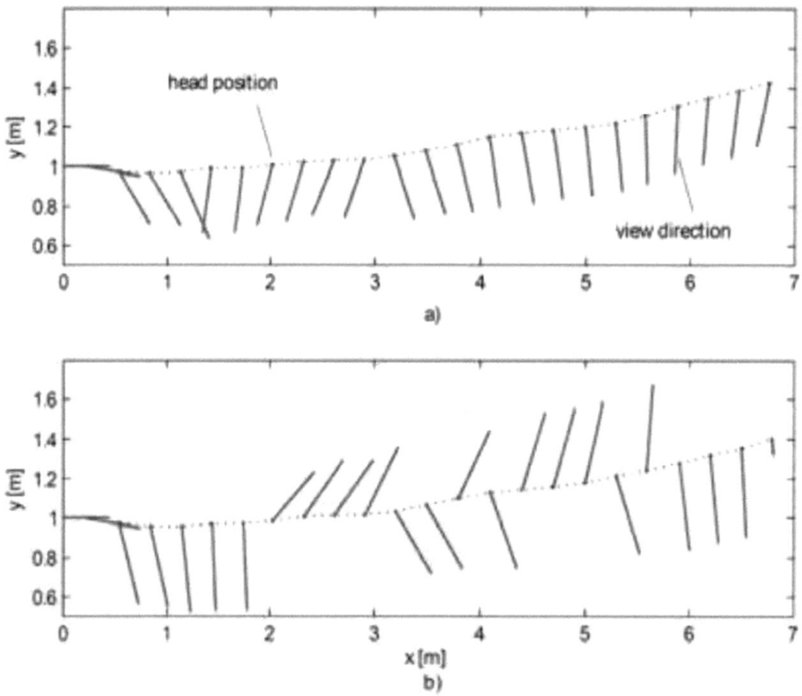

Fig. 8. Pan-angles of the a) conventional stereo-camera and b) of the foveated stereo-camera with two cameras per eye.

Mono-focal vision systems, i.e. systems comprising only one sensor type, suffer from a trade-off of accuracy versus field of view. In robot localization not only measurement accuracy, but also the number of visible landmarks is an important factor in order to determine the current robot position. Depending on the distribution of landmarks and the current situation it may be better to observe more landmarks with lower accuracy in one situation and fewer landmarks with higher accuracy in another. Thus, mono-focal systems are always a compromise working well only for a very limited class of environmental conditions and situations, however, failing in others. This problem is reflected by the results shown in Figure 8a and Table 1. The upper two of the sparsely distributed landmarks are not detected at the initial position and, thus, the planner only considers the lower landmarks. Most of the time only one landmark is visible and used for localization.

The foveated vision systems provide much more flexibility. Due to the large field of view of the wide-angle device more landmarks are detected in the initial position to be considered by the planner. So, at each step two or more landmarks are used for localization whereas at least one landmark is focused with telephoto cameras resulting in a significantly higher certainty of the estimated footstep positions of the humanoid robot as shown in Figure 10 and Table 1.

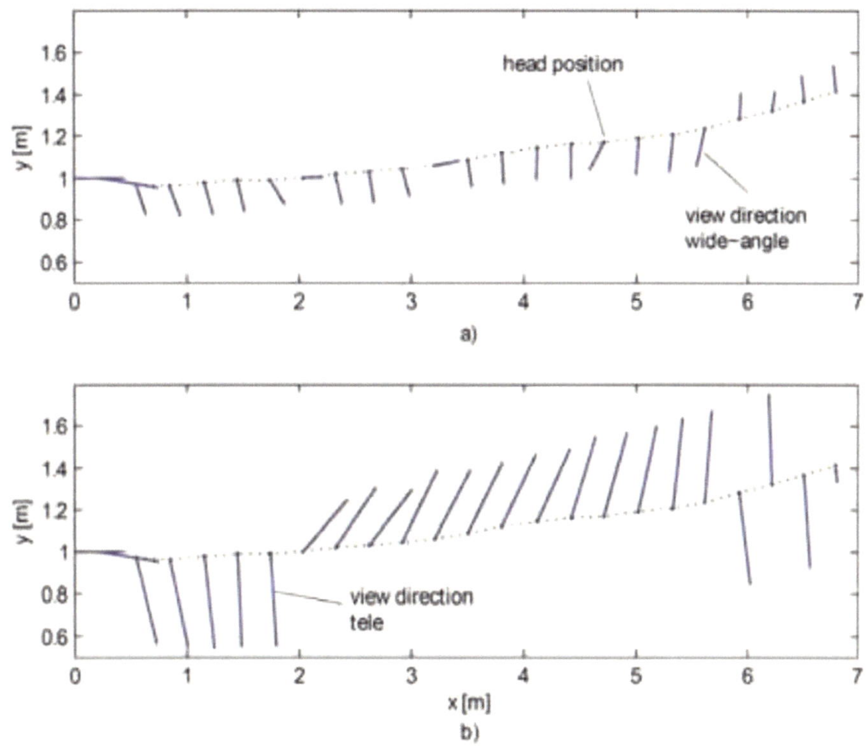

Fig. 9. Pan-angles of foveated vision system with two independent stereo cameras; a) telephoto and b) wide-angle stereo-camera.

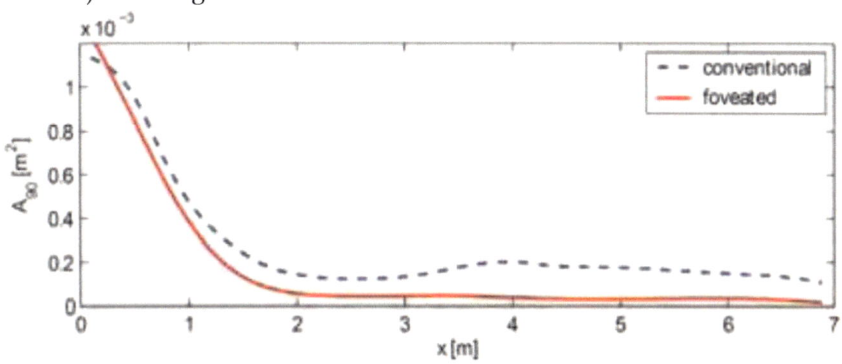

Fig. 10. Comparison of the areas A_{90} of the 90%-confidence ellipses of the footstep position covariance matrix using a conventional and a foveated vision system.

vision system	A_{90} [$10^{-4}m^2$]	average number of visible landmarks
a) conventional	2.7	1.1
b) foveated	1.7	2.3
c) foveated	1.6	2.4

Table 1. Mean of the areas A_{90} of the 90%-confidence ellipses of the footstep covariance matrix and average number of visible landmarks.

5.2 Combined Gaze and Trajectory Planning

In this section the performance of the proposed combined gaze direction and motion planning scheme described Section 4 is evaluated. The simulated environment consists of an area of size 20x20 meters with randomly allocated features. The active simulated head which is mounted on top of the robot is assumed to have a field of view of 60° and a maximum viewing range of 6 meters. No previous knowledge of the environment is assumed. Since it is out of the scope of this paper to deal with the correspondence problem of SLAM or feature selection, feature association is considered known and all observed features are used. A harsh odometry error of 10% is chosen, since are interested to see how our algorithm performs with very inaccurate robot models. A sensor model with a variance proportional to the distance for bearing and range measurements is used having the same high noise level. The active head can be moved with high angular velocities, so that saccadic movements are simulated. Finally, γ in (8) is chosen heuristically such that the robot balances between keeping good pose estimates and exploring the environment. The simulations time is set to 60s.

We first ran a simulation assuming a passive sensor, directed always straight ahead of the robot and a greedy policy for motion control which considers only the eight neighboring states. Only seven features were observed and map and localization uncertainty were very high as shown in Figure 11.

Next we conducted simulation with the proposed gaze direction control and again a greedy policy, followed by a simulation with a three-step planning horizon. In Figure 11 the absolute position error is depicted, for all three cases. It is evident that error reduces significantly as the planning horizon grows and gaze direction control is used.

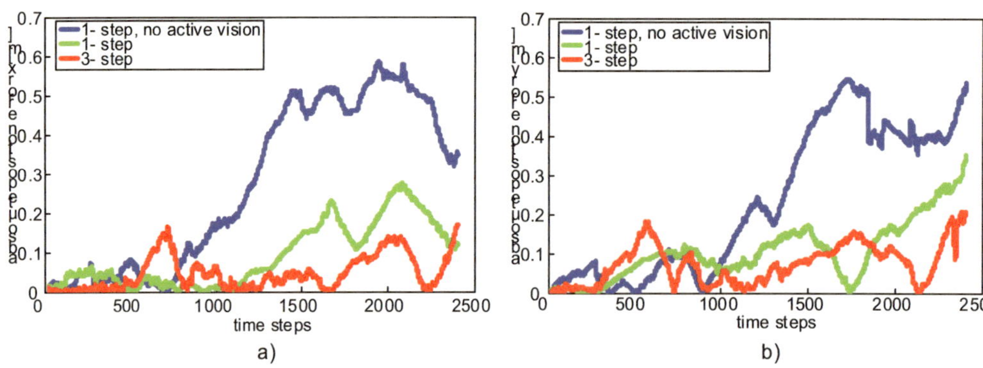

Fig. 11. Absolute position error as a function of time for one-step planning horizon without gaze direction control and one-step and three-step planning horizons with gaze direction control.

The map accuracy is illustrated in Figure 12 by the error ellipsoids for each observed feature for the final map. It is obvious that map accuracy grows as the planning horizon becomes larger. Also more features are observed if gaze direction control is used. From the final map acquired in the case of a three-step planning horizon with gaze direction control, it becomes clear that the proposed approach balances well by observing a large number of features and also building an accurate map.

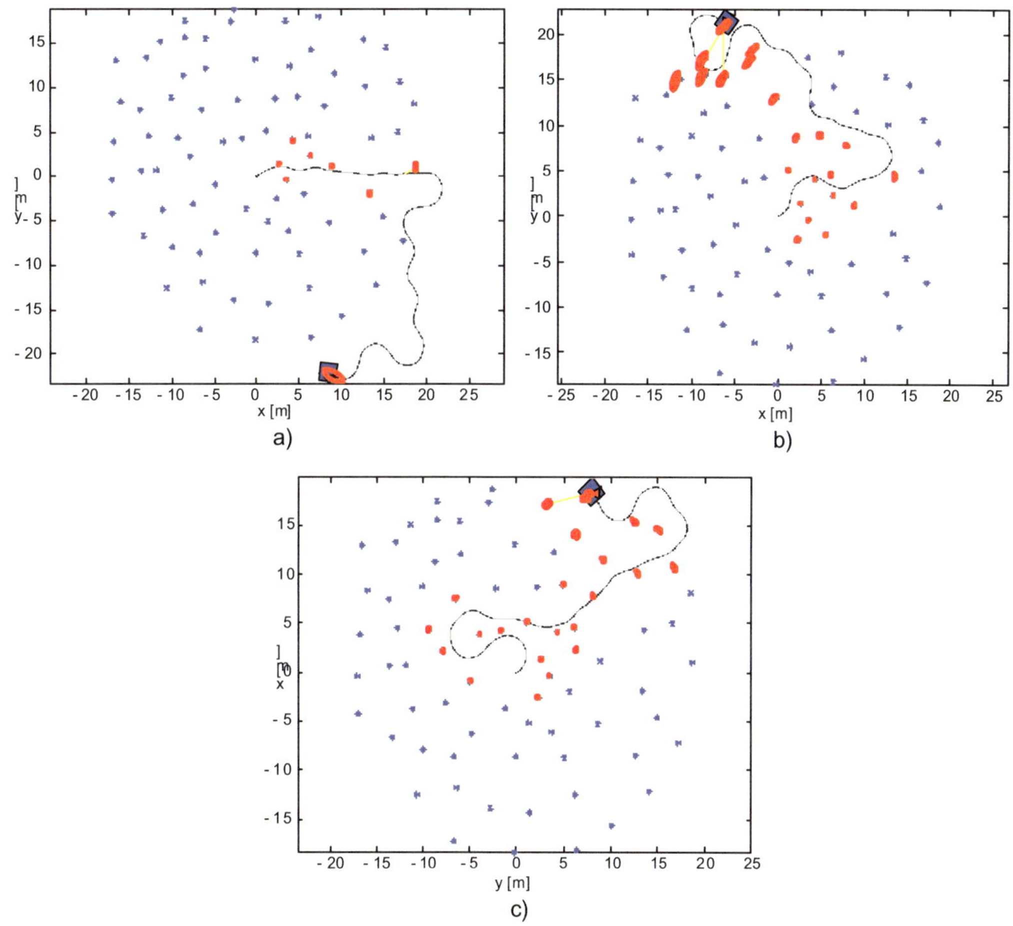

Fig. 12. Map accuracy illustrated by the error ellipsoids of each observed feature for the final map; a) one-step planning horizon without gaze direction control, b) one-step, and c) three-step planning horizons with gaze control. The estimated robot trajectory is illustrated by the black lines, while the red triangle on-top of the robot represents the active head and its gaze direction.

Figure 13 shows the reduction of the entropy over time. Each time a new feature is observed entropy reduces. For that reason it is step-formed. The greedy approaches need more time to reduce entropy and the larger the planning horizon is, the more entropy is reduced. Furthermore, when the planning horizon is small, more time is needed to observe the same number of features. Without gaze direction control entropy is not satisfactorily reduced. This results from the fact, that the gaze direction control module chooses to direct the sensor system mostly towards already observed and more certain features when the environment is known. Therefore, localization error and feature position uncertainties are kept at a minimum.

The results show that the proposed approach combining gaze direction control and motion planning based on information theoretic concepts for the exploration task, gives superior results in comparison to greedy approaches and approaches that neglect active sensor control.

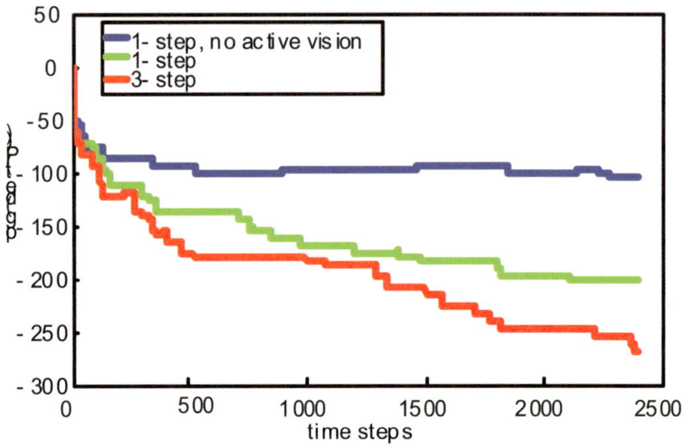

Fig. 13. Entropy measure $\log(|P_{k+1|k+1}|)$ over time for a one-step planning horizon without gaze direction control, one-step and three-step planning horizons with gaze direction control.

6. Conclusions

This chapter presents a foveated active vision planning concept for robot navigation in order to overcome drawbacks of conventional active vision when trading field of view versus measurement accuracy. This is the first approach of task-related control of foveated active vision in the context of humanoid robots as well as localization and mapping. In a typical robot navigation scenario the benefits of foveated active vision have been demonstrated: an improved localization accuracy combined with an extended visible field compared to conventional active vision. As a generic information maximization principle has been used the gaze control strategy is generalizable to other scenarios depending on the definition of the task-related information measures, thereby, allocating vision sensor resources optimally.

The second part of this chapter has presented novel concepts of coordinating gaze direction and locomotion planning. Through this planning procedure, all possible trajectories and viewing angles over a specific time horizon are anticipated. This results in significantly improved performance compared to known approaches, which neglect the limitations of the sensors. It is demonstrated that a very accurate environmental model is autonomously produced even if very inaccurate robot and sensor models are used. Such a scheme is ideal for vision-based humanoid robots.

The proposed approaches assume time-invariant scenarios. Path planning and gaze control methods for dynamically changing environments are not yet covered and subject to ongoing research.

7. Acknowledgments

The authors like to gratefully thank Dr. Javier Seara and Klaus Strobl-Diestro for reference simulation code for performance comparison. This work has been supported in part by the German Research Foundation (DFG) grant BU-1043/5-1 and the DFG excellence initiative research cluster *Cognition for Technical Systems - CoTeSys*, see also www.cotesys.org.

8. References

Apostoloff, N.; Zelinsky, A. (2002). Vision in and out of vehicles: Integrated driver and road scene monitoring, Proceedings of the 8th International Symposium on Experimental Robotics (ISER), 2002, Sant Angelo d'Iscia, Italy

Beis, J. F. & Lowe, D. G. (1997). Shape Indexing Using Approximate Nearest-Neighbour Search in High-Dimensional Spaces, Proceedings of the 1997 IEEE Conference on Computer Vision and Pattern Recognition (CVPR), 1997

Bodor, R.; Morlok, R. & Papanikolopoulos, N. (2004). Dual-camera system for multi-level activity recognition, Proceedings of the 2004 IEEE/RSJ International Conference on Intelligent Robots and Systems (IROS), 2004, pp.643-648, Sendai, Japan

Bourgault, F.; Makarenko, A. A.; Williams, S. B.; Grocholsky, B. & Durrant-Whyte, H. (2002). Information Based Adaptive Robotic Exploration, Proceedings of the IEEE/RSJ International Conference on Intelligent Robots and Systems (IROS), Lausanne, Switzerland

Bryson, M. & Sukkarieh, S. (2005). An Information-Theoretic Approach to Autonomous Navigation and Guidance of an Unihabited Aerial Vehicle in Unknown Environments, Proceedings of the IEEE/RSJ International Conference on Intelligent Robots and Systems (IROS), Edmonton, Canada

Brooks, R. A.; Breazeal, C.; Marjanovic, M.; Scasselati, B. & Williamson, M. M. (1999). The Cog Project: Building a Humanoid Robot, In: Computation for Methaphors, Analogy, and Agents, C. Nehaniv, (Ed.), pp. 52-87, Springer, Heidelberg, Berlin, Germany

Darrell, T. (1997). Reinforcement learning of active recognition behaviors, Interval Research Technical Report 1997-045,http://www.interval.com/papers/1997-045, 1997

Davis, J. & Chen, X. (2003). Foveated observation of shape and motion. Proceedings of the 2003 IEEE International Conference on Robotics and Automation (ICRA), pp. 1001-1005, 2003, Taipei, Taiwan

Davison, A. J. & Murray, D. W. (2002). Simultaneous Localization and Map-Building Using Active Vision, In IEEE Transactions on Pattern Analysis and Machine Intelligence, Vol. 24, No.7, pp. 865-880, 2002

Davison, A. J. (2003). Real-time simultaneous localization and Mapping with a single camera, Proceedings of the International Conference on Computer Vision, 2003, pp. 1403-1410, Nice

Dickmanns, E. D. (2003). An advanced vision system for ground vehicles. Proceedings of the International Workshop on In-Vehicle Cognitive Computer Vision Systems (IVC2VS), 2003, Graz, Austria

Dissanayake, G.; Newman, P.; Clark, S.; Durrant-Whyte, H. F. & Csorba, M. (2001). A Solution to the Simultaneous Localization and Map Building Problem, IEEE Transactions on Robotics and Automation, 2001, pp. 229-241

Doucet, A.; de Freitas, N.; Murphy, K. & Russell, S. (2000). Rao-blackwellised particle filtering for dynamic bayesian networks, Proceedings of the 16th Conference on Uncertainty in Artificial Intelligence, 2000

Elder, J. H.; Dornaika, F.; Hou, B. & Goldstein, R. (2004). Attentive wide-field sensing for visual telepresence and surveillance, In: Neurobiology of Attention, L. Itti, G. Rees & J. Tsotsos, (Eds.), 2004, Academic Press, Elsevier

Eustice, R.; Pizarro, O. & Singh, H. (2004). Visually augmented navigation in an unstructured environment using a delayed state history, Proceedings of the IEEE International Conference on Robotics and Automation, 2004, pp. 25-32, New Orleans, USA

Eustice, R.; Walter, M. & Leonard, J. (2005). Sparse extended information filters: insights into sparsification, Proceedings of the IEEE/RSJ International Conference On Intelligent Robots and Systems, 2005, pp. 641-648, Sendai, Japan

Feder, H. J. S.; Leonard J. J. & Smith, C. M. (1999). Adaptive Mobile Robot Navigation and Mapping, International Journal of Robotics and Research, Vol.18, No.7, pp. 650-668, 1999

Seara, J. F.; Strobl, K. H.; Martin E. & Schmidt, G. (2003). Task-oriented and situation-dependent gaze control for vision guided autonomous walking, Proceedings of the IEEE-RAS International Conference on Humanoid Robots (Humanoids), 2003, München and Karlsruhe, Germany

Grabowski, R.; Khosla, P. & Choset, H. (2003). Autonomous exploration via regions of interest, Proceedings of the IEEE/RSJ International Conference on Intelligent Robots and Systems (IROS), 2003, Las Vegas, USA

Gutmann, J.-S. & Konolige, K. (1999). Incremental Mapping of Large Cyclic Environments, Proceedings of International Symposium on Computational Intelligence in Robotics and Automation (CIRA), 1999, Monterey, CA, USA

Horaud, R.; Knossow, D. & Michaelis, M. (2006). Camera cooperation for achieving visual attention, Machine Vision and Applications, Vol. 15, No. 6, 2006, pp. 331-342

Huang, S.; Kwok, N. M.; Dissanayake, G.; Ha, Q. P. & Fang, G. (2005). Multi-Step Look-Ahead Trajectory Planning in SLAM: Possibility and Necessity, Proceedings of the International Conference on Robotics and Automation (ICRA), 2005, pp. 1103-1108, Barcelona, Spain

Itti, L. & Koch, C. (2001). Computational modeling of visual attention, Nature Reviews Neuroscience, Vol. 2, No. 3, 2001

Jankovic, N. D.; Naish, M. D. (2005). Developing a modular spherical vision system, Proceedings of the 2005 IEEE International Conference on Robotics and Automation (ICRA), pp. 1246-1251, 2005, Barcelona, Spain

Jung, I. & Lacroix, S. (2003). High resolution terrain mapping using low attitude aerial stereo imagery, Proccedings of the IEEE International Conference on Computer Vision, 2003, pp. 946-951

Karlsson, N.; Goncalves, L.; Munich M. E. & Pirjanian, P. (2005). The vSLAM Algorithm for Robust Localization and Mapping, Proceedings of the IEEE International Conference on Robotics and Automation (ICRA), 2005, Barcelona, Spain

Kim J. & Sukkarieh S. (2003). Airborne Simultaneous Localisation and Map Building, Proceedings of the IEEE International Conference on Robotics and Automation (ICRA), 2003, Taipei, Taiwan

Koch, C. & Ullmann, S. (1984). Selecting one among the many: A simple network implementing shifts in visual attention. MIT AI Memo No. 770, 1984

Koenig, S. & Tovey, C. (2003). Improved analysis of greedy mapping, Proceedings of the IEEE/RSJ International Conference on Intelligent Robots and Systems (IROS), 2003, Las Vegas, USA

Kühnlenz, K.; Bachmayer, M. & Buss, M. (2006). A multi-focal high-performance vision system, Proceedings of the 2006 IEEE International Conference on Robotics and Automation (ICRA), pp. 150-155, 2006, Orlando, FL, USA

Kühnlenz, K. (2006). Aspects of multi-focal vision, Ph.D. Thesis, Institute of Automatic Control Engineering, Technische Universität München, 2006, Munich, Germany

Kwok, N. M. & Dissanayake, G. (2004). An efficient multiple hypothesis filter for bearing-only SLAM, Proccedings of the IEEE/RSJ International Conference on Intelligent Robots and Systems, 2004, pp. 736-741

Leonard, J. J.; Feder H. J. S. (2000). A Computationally Efficient Method for Large-Scale Concurrent Mapping and Localization, Proceedings of the 9th International Symposium on Robotics Research, 2000, pp. 169-176, Salt Lake City, Utah

Leonard, J. J.; Rikoski, R. J.; Newman, P. M. & Bosse M. (2002) The International Journal of Robotics Research, Vol. 21, No. 10-11, 2002, pp. 943-975

Lidoris, G.; Kühnlenz, K.; Wollherr, D. & Buss, M. (2006). Information-Based Gaze Direction Planning Algorithm for SLAM, Proceedings of IEEE-RAS International Conference on Humanoid Robots (HUMANOIDS), 2006, Genova, Italy

Lidoris, G.; Kühnlenz, K.; Wollherr, D. & Buss, M. (2007). Combined trajectory planning and gaze direction control for robotic exploration, Proceedings of the International Conference of Robotics and Automation (ICRA), 2007, Rome, Italy

Lowe, D. G. (2004). Distinctive image features from scale-invariant keypoints, International Journal of Computer Vision, Vol. 60, No. 2, 2004, pp. 91-110

Lu, F. & Milios, E. (1997). Globally Consistent Range Scan Alignment for Environment Mapping, Autonomous Robots, Vol. 4, 1997, pp. 333-349

Maurer, M.; Behringer, R.; Furst, S.; Thomanek, F. & Dickmanns, E. D. (1996). A compact vision system for road vehicle guidance, Proceedings of the 13th International Conference on Pattern Recognition (ICPR), 1996

Montemerlo, M.; Thrun, S.; Koller, D. & Wegbreit, B. (2002). FastSLAM: A Factored Solution to Simultaneous Localization and Mapping, Proceedings of the National Conference on Artificial Intelligence (AAAI), Edmonton, Canada

Moorehead, S. J.; Simmons, R. & Whittaker, W. L. (2001). Autonomous exploration using multiple sources of information, Proceedings of IEEE International Conference on Robotics and Automation (ICRA), 2001, Seoul, Korea

Ozawa, R.; Takaoka, Y.; Kida, K.; Nishiwaki, J.; Chestnutt, J.; Kuffner, J.; Kagami, S.; Mizoguchi, H. & Inoue, H. (2005). Using visual odometry to create 3D maps for online footstep planning, Proceedings of IEEE International Conference on Systems, Man and Cybernetics (SMC), 2005, pp. 2643-2648, Hawaii, USA

Pellkofer, M. & Dickmanns, E. D. (2000). EMS-Vision: Gaze control in Autonomous vehicles, Proceedings of the IEEE Intelligent Vehicles Symposium, pp. 296-301, 2000, Dearborn, MI, USA

Sabe, K.; Fukuchi, M.; Gutmann, J. S.; Ohashi, T; Kawamoto, K. & Yoshigahara, T. (2004). Obstacle Avoidance and path planning for humanoid robots using stereo vision,

Proceedings of IEEE International Conference on Robotics and Automation (ICRA), 2004, pp. 592-597

Scassellati, B. (1998). Eye finding via face detection for a foveated, active vision system, Proceedings of the 15th National Conference on Artificial Intelligence (AAAI), pp. 969-976, 1998, Madison, WI, USA

Se, S.; Lowe D. & Little J. (2002). Mobile Robot Localization and Mapping with Uncertainty using Scale-Invariant Visual Landmarks, Intl. Journal of Robotics Research, Vol. 21, 2002, pp 735-758

Se, S.; Lowe, D. & Little, J. (2005). Vision-based Global Localization and Mapping for Mobile Robots, IEEE Transactions on Robotics, Vol. 21, No. 3, 2005, pp. 364-375.

Shannon, C. E. (1948). A mathematical theory of communication, Bell System Technical Journal, vol. 27, 1948, pp. 379-423 and 623-656

Shibata, T.; Vijayakumar, S.; Conradt, J. & Schaal, S. (2001). Humanoid oculomotor control based on concepts of computational neuroscience, Proceedings of the IEEE-RAS International Conference on Humanoid Robots (Humanoids), 2001, Tokyo, Japan

Sim, R. & Little, J. J. (2006). Autonomous vision-based exploration and mapping using hybrid maps and Rao-Blackwellised particle filters, Proceedings of the IEEE/RSJ Conference on Intelligent Robots and Systems (IROS), 2006, Beijing, China

Smith, R. C.; Self, M.; Cheeseman, P. (1990). Estimating Uncertain Spatial Relationships in Robotics, In Autonomous Robot Vehicles, I. J. Cox and G. T. Wilfong, eds., 1990, pp. 167-193, Springer-Verlag

Stachniss, C.; Grisetti, G. & Burgard, W. (2005). Information Gain-based Exploration Using Rao-Blackwellized Particle Filters, Proceedings of Robotics: Science and Systems (RSS), 2005, pp. 65-72, Cambridge, MA, USA

Stasse, O.; Davison, A. J.; Sellaouti, R. & Yokoi, K. (2006) Real-time 3D SLAM for humanoid robot considering pattern generator information, Proceedings of the IEEE/RSJ Conference on Intelligent Robots and Systems (IROS), 2006, Beijing, China

Thomson S. & Kagami, S. (2005). Humanoid robot localization using stereo vision, Proceedings of IEEE-RAS Intl. Conf. on Humanoid Robots (HUMANOIDS), 2005, pp. 19-25, Tsukuba, Japan

Thrun, S.; Liu, Y.; Koller, D.; Ng, A. Y.; Ghahramani, Z. & Durrant-Whyte, H. F. (2004). Simultaneous localization and mapping with sparse extended information filters, International Journal of Robotics Research, Vol. 23, No. 7-8, 2004

Ude, A.; Gaskett, C. & Cheng, G. (2006). Foveated Vision Systems with Two Cameras Per Eye, Proceedings of the 2006 IEEE International Conference on Robotics and Automation (ICRA), pp., 2006, Orlando, FL, USA

Yamauchi, B. (1998). Frontier-based exploration using multiple robots, Proceedings of International Conference on Autonomous Agents, pp. 47-53, Minneapolis, USA

Vidal-Calleja, T.; Davison, A. J.; Andrade-Cetto, J. & Murray, D.W. (2006). Active Control for Single Camera SLAM, Proceedings of the International Conference of Robotics and Automation (ICRA), Orlando, USA

Vijayakumar, S.; Inoue, M. & Souza, A. D. (2004). Maveric - Oculomotor experimental vision head, http://homepages.inf.ed.ac.uk/svijayak/projects/maveric/index.html, 2004

Vertical Jump: Biomechanical Analysis and Simulation Study

Jan Babič and Jadran Lenarčič
"Jožef Stefan" Institute
Slovenia

1. Introduction

Vertical jumping is a complex task requiring quick and harmonized coordination of jumper's body segments, first for the push-off, then for the flight and lastly for the landing. The prime criterion for vertical jump efficiency is the height of the jump that depends on the speed of the jumper's center of gravity (COG) in the moment when the feet detach from the ground. Besides maintaining the balance, the task of the muscles during the push-off phase of the jump is to accelerate the body's COG up in the vertical direction to the extended body position. During the push-off phase of the jump, the jumper's center of gravity must be above the supporting polygon that is formed by the feet (Babič et al., 2001). In contrast to the humans, today's humanoid robots are mostly unable to perform fast movements such as the vertical jump. They can mostly perform only slow and statically stable movements that do not imitate the human motion. Besides, these slow and statically stable movements are energy inefficient. With the understanding of the anatomy and the biomechanics of the human body, one can find out that, beside the shape, majority of today's humanoid robots and human bodies do not have a lot of common properties. To achieve a better imitation of the human motion and ability to perform fast movements such as the vertical jump or running, other properties and particularities, beside the shape of the body, should be considered in the design of the humanoid robot.

Lower extremities of today's humanoid robots are mostly serial mechanisms with simple rotational joints that are driven directly or indirectly by electrical servo drives. Such design of humanoid robot mechanism allows only rotational motion in joints to occur. This means that translations of the robot's center of gravity are solely a result of the transformation of rotations in joints into translations of the robot center of gravity. Especially in ballistic movements such as fast running or jumping where the robot center of gravity is to be accelerated from low or zero velocity to a velocity as high as possible, this transformation is handicapped. The transfer of the angular motion of the lower extremity segments to the desired translational motion of the robot center of gravity is less effective the more the joints are extended. When the joint is fully extended, the effect of this joint on the translational motion of the robot center of gravity in a certain direction equals zero. Besides, the motion of the segments should decelerate to zero prior to the full extension to prevent a possible damaging hyperextension. Where relatively large segments which may contain considerable amounts of rotational energy are involved, high power is necessary to decelerate the angular motion.

Servo drives that drive the joints of the humanoid robot represent a model of the group of monoarticular muscles that are passing a single joint and thus cause a rotational motion of the joint that are passing. Beside the monoarticular muscles that are passing only one joint, human lower extremity consists of another group of muscles that are passing two joints. These muscles are so called biarticular muscles. Their functions during human movement have been studied extensively by many researchers. One of such functions is the transportation of mechanical energy from proximal to distal joints. It is believed that this transportation causes an effective transformation of rotational motion of body segments into translation of the body centre of gravity (Schenau, 1989). Gastrocnemius muscle is a biarticular muscle that is passing the knee and the ankle joints and acts as a knee flexor and ankle extensor (see Fig. 1).

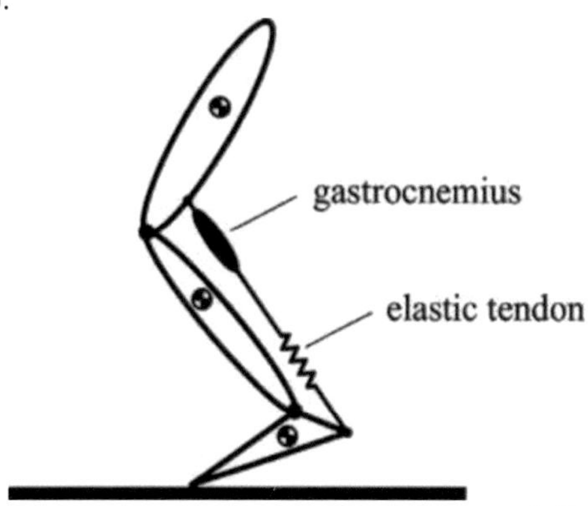

Fig. 1. Biarticular muscle gastrocnemius passing the knee and the ankle joints. It acts as a knee flexor and ankle extensor. Gastrocnemius muscle is connected to the foot by an elastic tendon.

In jumping, the activation of the biarticular gastrocnemius muscle prior to the end of the push-off enables the transportation of the power generated by the knee extensors from the knee to the ankle joint. This transfer of mechanical energy by gastrocnemius can be explained using the following example. During the push-off phase of the jump, the knee joint is rapidly extended as a result of the positive work done by the knee extensor muscles. If the biarticular gastrocnemius muscle contracts isometrically (its length does not change), the additional mechanical work is done at the ankle joint because of the gastrocnemius muscle, which contributes no mechanical work by itself. A part of the energy generated by the knee extensors appears as mechanical work at the ankle joint and the height of the jump is significantly increased. This is because, as the jump proceeds and the knee straightens, the angular position changes of the knee have progressively less effect on vertical velocity of the jumper's centre of gravity. By gastrocnemius muscle activation, a rapid extension of the foot is produced. This extension has a greater effect on the vertical velocity than the extension of the almost straightened knee. The energy is more effectively translated into vertical velocity and a greater height of the jump is achieved. However, the timing of the gastrocnemius muscle activation is critical to obtain a maximum effect. This was demonstrated by an articulated physical model of the vertical jump by Bobbert et al (1986).

Besides biarticularity, the gastrocnemius muscle has one more interesting feature. It is connected to the foot by an elastic tendon (see Fig. 1). The elasticity in the muscle fibers and tendons plays an important role in enhancing the effectiveness and the efficiency of human performance. An enhanced performance of human motion has been most effectively demonstrated for jumping and running (Cavagna, 1970; Bobbert et al., 1996; Shorten, 1985). An important feature of elastic tissues is the ability to store elastic energy when stretched and to recoil this energy afterwards as a mechanical work (Asmussen and Bonde-Petersen, 1974). Beside this feature, oscillatory movements performed at the natural frequency of muscle-tendon complex could maximize the performance. A countermovement vertical jump can be treated as one period of oscillatory movement and from this point of view the natural frequency, as well as parameters that define the natural frequency, can be determined.

The purpose of our research was twofold. First to analyse the human vertical jump and to show that for each and every subject there exists an optimal triceps surae muscle-tendon complex stiffness that ensures the maximal possible height of the vertical jump. We defined the influence of the m. gastrocnemius activation timing and the m. gastrocnemius and Achilles tendon stiffness on the height of the vertical jump and established the methodology for analysis and evaluation of the vertical jump. We monitored kinematics, dynamics and m. gastrocnemius electrical activity during the maximum height countermovement jump of human subjects and measured viscoelastic properties of the m. gastrocnemius and Achilles tendon using the free-vibration technique. Based on the findings of the biomechanical study of the human vertical jump we performed a simulation study of the humanoid robot vertical jump. As a result of our research we propose a new human inspired structure of the lower extremity mechanism by which a humanoid robot would be able to efficiently perform fast movements such as running and jumping.

2. Biorobotic Model of Vertical Jump

Biorobotic model of the vertical jump consists of the dynamic model of the musculoskeletal system and of the mathematical model used for the motion control of the model. The results of the modelling are differential equations and a diagram for simulation and optimization of the vertical jump.

2.1 Dynamic Model of Musculoskeletal System

Vertical jump is an example of a movement that can be satisfactorily observed and analyzed in just a sagital plane. Therefore we built a model of the musculoskeletal system in a two dimensional space of the sagital plane. Because both lower extremities perform the same movement during the vertical jump, we joined both extremities in one extremity with three rigid body segments. Trunk, head and upper extremities were modeled as one rigid body with a common COG, mass and moment of inertia. The model of the musculoskeletal system is therefore a planar model composed of four segments that represent the foot, shank, thigh and trunk together with the head and both upper extremities. Segments of the model are joined together by frictionless rotational hinges whose axes are perpendicular to the sagital plane. The contact between the foot and the ground is modeled as a rotational joint between the tip of the foot and the ground. A model, whose foot is connected to the ground by a rotational joint, is applicable only for the push-off and landing phases of the vertical jump and is not applicable for the flight. As the motion of the COG during the flight is simply described and depends only on the speed vector of the COG just prior to the takeoff, this simplification does not present any limitations.

Fig. 2 shows the planar model of the musculoskeletal system, composed of four rigid bodies that represent the foot, shank, thigh and trunk together with the head and both upper extremities. The origin of the base coordinate system is in the center of the virtual joint that connects the foot with the ground.

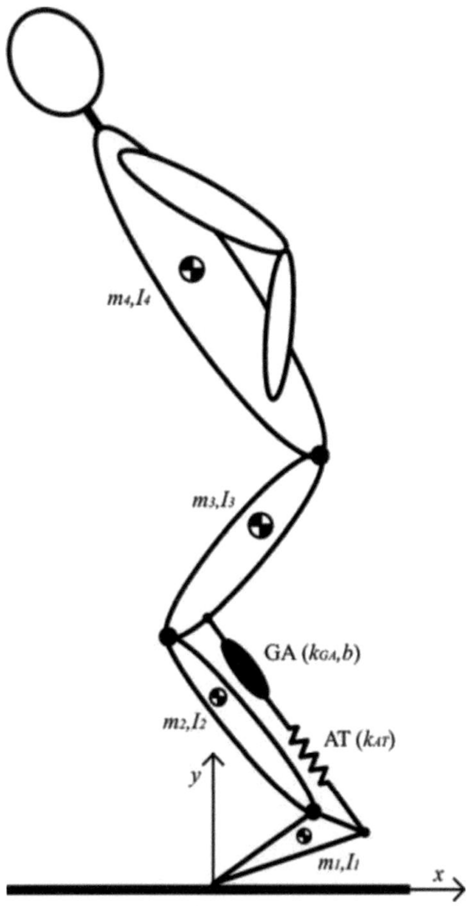

Fig. 2. Planar model of the musculoskeletal system

Passive constraints in the hip, knee and ankle that define the range of motion of these joints were modeled as simple nonlinear springs (Audu and Davy, 1985; Davy and Audu, 1987).
Fig. 2 also shows the model of the biarticular link that consists of the gastrocnemius muscle (GA) with stiffness k_{GA} and damping b and the Achilles tendon (AT) with stiffness k_{AT}. Contrary to the real gastrocnemius muscle, biarticular link can not contract. It can only be enabled or disabled at different moments during the push-off phase of the jump. High pennation angle of the gastrocnemius muscle fibers suggest that the predominant role of the m. gastrocnemius is not in its contraction but in its ability to bear high forces and to enable the energy transportation from the knee to the ankle (Bogert et al., 1989; Legreneur et al., 1997). Therefore our simulated biarticular link that can not contract and can only be enabled or disabled is considered as an appropriate model for this study.
Insertions of the biarticular link on the foot (B) and thigh (C) have been determined from the muscle data provided by Brand et al. (1982) and Delp (1990).

Vector of the force in the biarticular link f is

$$f = k \cdot (BC - BC_0) - b \cdot BC \,, \tag{1}$$

where k represents the stiffness of the m. gastrocnemius and Achilles tendon connected in series, $\boldsymbol{BC}$ is the vector between the insertions of the biarticular link on the foot and thigh. $\boldsymbol{BC}_0$ is the vector $\boldsymbol{BC}$ in the moment of the gastrocnemius muscle activation. Force in the biarticular link f causes a torque in the ankle joint

$$Q_{bl2} = -\|r_B \times f\| \,, \tag{2}$$

where $\boldsymbol{r}_B$ is the moment arm vector from the center of the ankle joint to the insertion of the biarticular link on the foot and a torque in the knee joint

$$Q_{bl3} = \|r_C \times f\| \,, \tag{3}$$

where $\boldsymbol{r}_C$ is the moment arm vector from the center of the knee joint to the insertion of the biarticular link on the thigh.

Motion of the musculoskeletal system is written with a system of dynamic equations of motion

$$\mathbf{H}(q)\ddot{q} + h(q,\dot{q}) + G(q) = Q_{mov} + Q_{pas} + Q_{bl} \,, \tag{4}$$

where $\boldsymbol{Q}_{pas}$ is the vector of joint torques caused by the passive constraints and $\boldsymbol{Q}_{bl}$ is the vector of joint torques caused by the biarticular link. $\boldsymbol{Q}_{mov}$ is the vector of joint torques caused by muscles and represents the input to the direct dynamic model of the musculoskeletal system. The output from the direct dynamic model is the vector of joint displacements $\boldsymbol{q}$. We determined parameters of (4) $\mathbf{H}(q), h(q,\dot{q}), G(q)$ using the equations published by Asada and Slotine (1986). Simulation diagram of the direct dynamic model of the musculoskeletal system is shown in the shaded rectangle of the Fig. 3.

2.2 Motion Control

Motion controller of the musculoskeletal system was designed to meet the following four requirements:

1. Perpendicular projection of the body's COG on the ground coincides with the virtual joint that connects the foot with the ground during the entire push-off phase of the vertical jump. Therefore balance of the jumper and verticality of the jump is assured. Equation that describes this requirement is

$$x_T^d(t) = 0 \,, \tag{5}$$

where $x_T^d(t)$ is the distance between the desired perpendicular projection of the body's COG on the ground from the origin of the base coordinate system in time t.

2. Motion controller assures the desired vertical movement of the body's COG relative to the ankle $y_{TA}^d(t)$. By controlling the movement of the body's COG relative to the ankle, we excluded the influence of the biarticular link on the motion $y_{TA}^d(t)$. Therefore parameters of the biarticular link can be varied and optimized for a certain desired motion of the body's COG relative to the ankle.

3. Motion controller assures a constant angle of the foot relative to the ground q_1 before the biarticular link activation occurs. Thus the number of degrees of freedom of the model remains constant during the push-off phase of the vertical jump

4. In the moment, when the biarticular link activates, motion controller sets the torque in the ankle joint Q_2 to zero and thus enable a free motion of the foot relative to the ground. By setting the torque in the ankle joint to zero, the motion in the ankle joint is only a function of the motion in the knee joint that is transmitted to the ankle by the biarticular link.

Motion controller that considers the requirement (5) and enables the desired vertical motion of the COG relative to the ankle $y_{TA}^d(t)$ in the base coordinate system is

$$\ddot{x}_T^c = \begin{bmatrix} 0 \\ \ddot{y}_{TA}^d + \ddot{y}_A \end{bmatrix} + k_p \left(\begin{bmatrix} 0 \\ y_{TA}^d + y_A \end{bmatrix} - x_T \right) + k_d \left(\begin{bmatrix} 0 \\ \dot{y}_{TA}^d + \dot{y}_A \end{bmatrix} - \dot{x}_T \right), \tag{6}$$

where k_p and k_d are coefficients of the PD controller, $\ddot{x}_T^c$ is the vector of the control acceleration of the COG in the base coordinate system, y_A is the current height of the ankle joint relative to the ground, $\dot{x}_T, x_T$ are the vectors of the current speed and position of the COG in the base coordinate system.

The relation between the vector of the control speed of the COG in the base coordinate system $\dot{x}_T^c$ and the vector of the control angular velocities in the joints $\dot{q}_c$ is

$$\dot{x}_T^c = \mathbf{J}_T \dot{q}_c , \tag{7}$$

where $\mathbf{J}_T$ is the Jacobian matrix of the COG speed in the base coordinate system. Equation (7) represents an under determined system of equations. From the requirement that the motion controller assures a constant angle of the foot relative to the ground q_1 before the biarticular link activation occurs, follows the condition

$$\dot{q}_{c1} = 0 . \tag{8}$$

An additional condition that abolishes the under determination of (7) is the relationship of the knee and hip joint angles

$$\dot{q}_{c4} = n \cdot \dot{q}_{c3} , \tag{9}$$

where n is the coefficient that describes the relationship.

By substitution of (8) and (9) into (7) we get a relation between the vector of the control speed of the COG in the base coordinate system $\dot{x}_T^c$ and the vector of the control angular velocities in the ankle and knee joints $\dot{q}'_c$

$$\dot{x}_T^c = \mathbf{J}'_T \dot{q}'_c , \tag{10}$$

where $\mathbf{J}'_T$ is a new Jacobian matrix of the center of gravity speed in the base coordinate system. Differentiation of (10) with respect to time yields

$$\ddot{q}'_c = \begin{bmatrix} \ddot{q}_{c2} \\ \ddot{q}_{c3} \end{bmatrix} = \mathbf{J}'^{-1}_T (\ddot{x}_T^c - \dot{\mathbf{J}}'_T \dot{q}') , \tag{11}$$

where

$$\dot{q}' = \begin{bmatrix} \dot{q}_2 \\ \dot{q}_3 \end{bmatrix} . \tag{12}$$

On the basis of conditions (8) and (9) and relation (11) we define control angular accelerations in all four joints

$$\ddot{q}_c = \begin{bmatrix} 0 \\ \ddot{q}_{c2} \\ \ddot{q}_{c3} \\ n\ddot{q}_{c3} \end{bmatrix}. \quad (13)$$

By substitution of (13) into a system of dynamic equations of motion (4) we get control torques in joints Q_{mov} that we need to control the model of the musculoskeletal system

$$Q_{mov} = \mathbf{H}(q)\ddot{q}_c + h(q,\dot{q}) + G(q). \quad (14)$$

Direct dynamic model of the musculoskeletal system together with the motion controller compose the biorobotic model of the vertical jump. Simulation diagram for the simulation of the vertical jump is shown in Fig. 3. Inputs into the simulation diagram are the desired trajectory of COG relative to the ankle y_{TA}^d and a signal for biarticular link activation a. Output from the simulation diagram is the vector of body's COG position x_T.

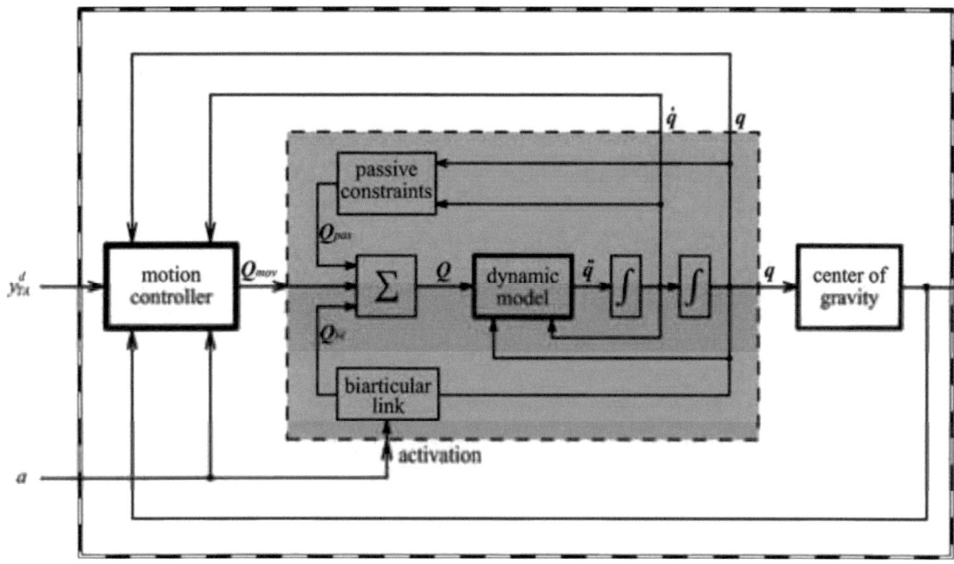

Fig. 3. Simulation diagram for the simulation of the vertical jump. Inputs into the simulation diagram are the desired trajectory of the COG relative to the ankle and a signal for biarticular link activation. Output from the simulation diagram is the vector of jumper's COG position.

3. Biomechanical Analysis of Human Vertical Jump

3.1 Subjects and Experimental Protocol

Ten trained male subjects (age 26 ± 4 years, height 180.3 ± 6.56 cm, body mass 77.1 ± 7.24 kg) participated in the study. Informed consent was obtained from all of them. The protocol of the study was approved by the National Ethics Committee of the Republic of Slovenia. The experiments comply with the current laws of the Republic of Slovenia.

After a warm-up procedure and three practice jumps, subjects subsequently performed four countermovement vertical jumps. They were instructed to keep their hands on the hips and to jump as high as possible. At the beginning of each vertical jump, subjects stood on their toes and in the erected position. During each jump, position of anatomical landmarks over epicondylus lateralis and fifth metatarsophalangeal joint were monitored, ground reaction forces were measured and electromyogram of m. gastrocnemius was recorded. After the jumping, we determined viscoelastic properties of the triceps surae muscle tendon complex of all subjects. Details on methods and procedures are provided in the following sections.

3.2 Anthropometric Measurements

Segmental anthropometric parameters, such as masses, moments of inertia about the transverse axes, lengths and locations of the centers of gravity were estimated using regression equations (Zatsiorsky and Seluyanov 1983; Leva 1996). Position of the first metatarsophalangeal joint and the Achilles tendon insertion on the calcaneus were determined by palpation for each subject. Insertion of the m. gastrocnemius on femur was determined using the muscle data collected by Brand et al. (1982) and Delp (1990).

3.3 Kinematics and Dynamics

To determine the motion of the body's COG during the vertical jump we measured the vertical component of the ground reaction force caused by the subject during the jump. Subjects performed vertical jumps on a force platform (Kistler 9281CA) that is capable to measure ground reaction forces with the frequency of 1000 Hz. We zeroed the force platform just prior to the vertical jump when the subject was standing still in the erected position. Thus we enabled the precise determination of the subject's body mass. Body mass of the subject standing on the force platform is therefore equal to the quotient between the negative ground reaction force during the flight phase of the jump and the ground acceleration. The vertical position of the body's COG relative to the ground in time t $y_T^m(t)$ was obtained by double integrating with respect to time the vertical ground reaction force in time t $F(t)$

$$y_T^m(t) = \frac{1}{m}\int_0^t\int_0^t F(t)dtdt + y_T^m(0) \tag{15}$$

where m is the body mass of the subject and $y_T^m(0)$ is the initial height of the body's COG relative to the ground. To determine the vertical position of the body's COG relative to the ankle in time t $y_{TA}^m(t)$ we measured the motion of the ankle during the vertical jump by means of the contactless motion capture system (eMotion Smart). The vertical position of the body's COG relative to the ankle is therefore

$$y_{TA}^m(t) = y_T^m(t) - y_A^m(t) \tag{16}$$

where $y_A^m(t)$ is the vertical position of the ankle in time t.

3.4 Electromyography

The activity of the m. gastrocnemius was recorded using a pair of surface electrodes put over the medial head of the m. gastrocnemius. Analog EMG signals were amplified and filtered with a band-pass filter with cut off frequencies at 1 Hz and 200 Hz. The signals were then digitalized with 1000 Hz sampling frequency and full-wave rectified. To reduce the variability of sampled EMG signal, the signal was then smoothed with a digital low-pass Butterworth filter. Finally the EMG signal was normalized with respect to the maximum value attained during the vertical jump.

3.5 Measurements of Muscle-Tendon Viscoelastic Properties

Triceps surae muscle-tendon complex viscoelastic properties of both legs were measured for each subject using the free-vibration method described by Babič and Lenarcič (2004). The measurement device and the procedure have been designed in such a manner that as few human body segments move as possible during the measurement. Thus the measurement uncertainty due to the approximation of the properties of the human body segments was minimized. The results of the measurements are the elastic stiffness k_{GA} and viscosity b of the m. gastrocnemius and the elastic stiffness k_{AT} of the Achilles tendon.

3.6 Treatment of Data

For the purposes of analysis and optimization of the vertical jump we adjusted the biomechanical model of the musculoskeletal system with segmental anthropometric parameters, such as masses, moments of inertia about the transverse axes, lengths and locations of the centers of gravity of each subject. Parameters of the biarticular link, such as insertion of the m. gastrocnemius on femur, insertion of the Achilles tendon on calcaneus, elastic stiffness and viscosity were adjusted to match the measured parameters of each subject.

To simulate the vertical jump of the individual subject we used the measured trajectory of the body's COG as the input into the biomechanical model of the vertical jump. Biarticular link activation that is also an input into the biomechanical model of the vertical jump was determined from the EMG signal of the m. gastrocnemius. The moment of biarticular link activation was determined as the moment when the rectified, normalized and filtered EMG signal of the m. gastrocnemius increased to 95% of its maximum value. After the activation, the biarticular link remains active during the entire push-off phase of the jump.

3.7 Results

To determine the optimal timing of the biarticular link activation that results in the highest vertical jump, a series of the countermovement vertical jump simulations have been performed for each subject. Each simulation was performed with a different timing of the biarticular link activation.

All subjects activated their m. gastrocnemius slightly before the optimal moment, determined by means of simulations. In average, the difference between the optimal and measured knee angle when the m. gastrocnemius was activated was 6.4 ± 2.22°. Because the dynamic model of the musculoskeletal system does not include the monoarticular muscle soleus, the measured heights of the jumps were higher than the jump heights determined with the simulations for 4.3 ±1.12 % in average. The primary motive for omitting the m. soleus from the modelling is that we wanted to control the motion of the body's COG relative to the ankle so that the parameters of the biarticular link could be varied and optimized for a certain measured motion of the subject's COG relative to the ankle. If the dynamic model of the musculoskeletal system would include the m. soleus, the motion of the ankle would be fully determined by the force of the m. soleus and we would not be able to control it with regard to the desired body's COG relative to the ankle. Moreover if the dynamic model of the musculoskeletal system would include the m. soleus, force produced by the m. soleus would be another input into the biomechanical model of the vertical jump and we would have to accurately measure the force produced by the m. soleus of subjects performing the vertical jump. An additional cause for the differences between the measured heights of the jumps and the jump heights determined by means of simulations can be the simplified model of the foot that we modeled as one rigid body. The arch of the foot is linked up by elastic ligaments that can store elastic energy when deformed and later reutilize it as the mechanical work (Alexander, 1988). Ker et al. (1987) measured the deformation of the foot during running and determined the amount of the energy stored during the deformation. They showed that the elasticity of the foot significantly contribute to the efficiency of the human movement. To be able to compare the measurements and the simulation results, we corrected the simulation results by adding the contribution of the m. soleus to the height of the vertical jump. Such corrected heights of the vertical jumps at the optimal moments of m. gastrocnemius activation are insignificantly larger than the measured heights of the vertical jumps for all subjects. In average the height difference is only 1.6 ± 0.74 cm.

To determine the optimal stiffness of the Achilles tendon regarding to the height of the vertical jump, a series of the countermovement vertical jump simulations have been performed, each with a different stiffness of the Achilles tendon. Optimal timing of the biarticular link has been determined for each stiffness of the Achilles tendon as described in the previous paragraph. The measured values of the Achilles tendon stiffness for all subjects were always higher than the optimal values determined by means of simulations. By considering the elastic properties of the arch of the foot, we can assume that the optimal values of the Achilles tendon stiffness would increase and therefore the differences between the optimal and measured values would decrease.

Results of the measurements, simulations and optimizations of the human vertical jumps are presented in Fig. 4. Subjects are arranged with regard to the ratio between the optimal stiffness of the Achilles tendon determined by means of simulations and the measured stiffness of the Achilles tendon. If we want to evaluate the contribution of the viscoelastic properties to the height of the jump, ranking of the subjects with regard to the height of the jump is not appropriate because the main parameter that influences the height of the vertical jump is the power generated by muscles during the push-off phase of the jump. The elasticity of the Achilles tendon has the secondary influence on the height of the jump. Results show that for the same power generated by an individual subject during the push-off phase of the jump, the height of the vertical jump varies with the Achilles tendon stiffness. Therefore the appropriate criterion for ranking the subjects have to consider the elasticity of the Achilles tendon.

4. Simulation and Optimization of Vertical Jump

4.1 Simulation Results

Simulations of the counter movement vertical jumps were performed. At the beginning of the simulation the legs were fully straightened. A counter movement was performed until the desired minimal knee angle was reached. The flexion of the joints was such that the centre of gravity of the robot dropped with the acceleration of the gravity. Lifting of the foot from the floor would occur if faster flexions would be applied. Flexion was immediately followed by a rapid extension which provided the energy for the jump. An example trajectory of the robot center of gravity during the countermovement vertical jump is shown in Fig. 5.

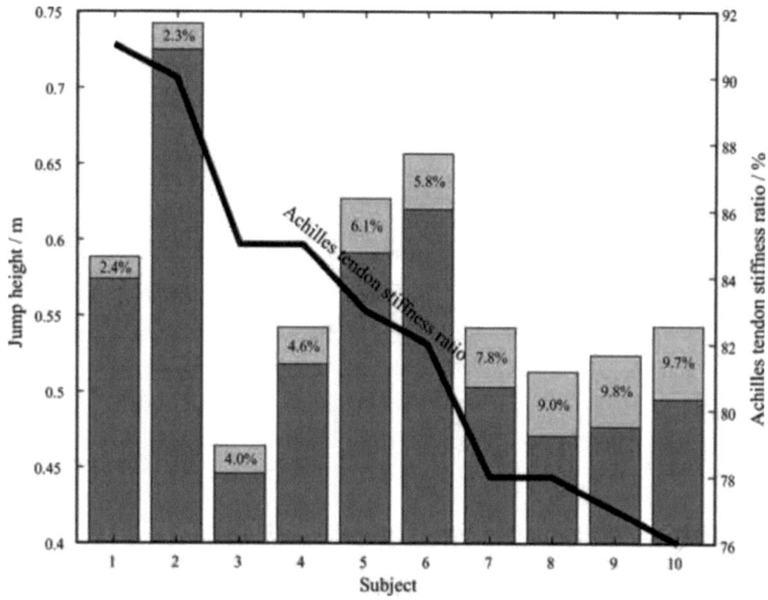

Fig. 4. Results of the measurements, simulations and optimizations of the vertical jumps. Whole bars represent the maximal heights of the vertical jumps achieved with the optimal values of the Achilles tendon stiffness and determined by means of simulations. The dark shaded parts of the bars represent the measured heights of the vertical jumps while the light shaded tops represent the differences between the maximal and measured heights of the vertical jumps. The differences are also shown as percentages relative to the measured heights. Bold line represents the ratio between the optimal stiffness of the Achilles tendon determined by means of simulations and the measured stiffness of the Achilles tendon for each subject.

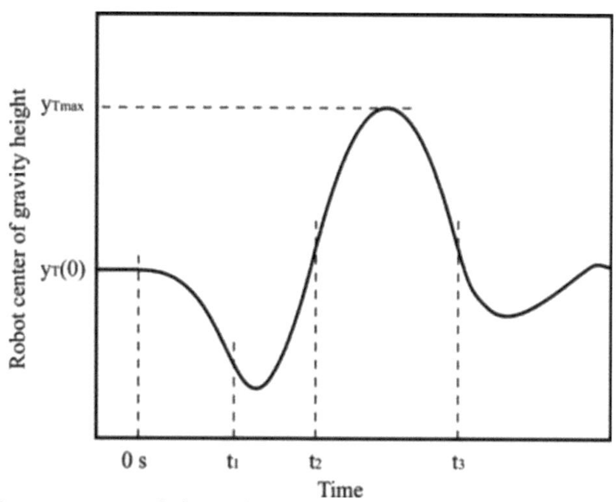

Fig. 5. An example trajectory of the robot center of gravity during the countermovement vertical jump. Time interval between 0s and t_2 represents the push-off phase of the jump while the time interval between t_2 and t_3 represents the flight phase of the jump. t_1 is the time when robot center of gravity starts to decelerate in the downward direction.

The height of the vertical jump can be calculated with the following equation

$$y_{T\,max} = y_{TA}(t_2) + y_A(t_2) - \frac{1}{2g}\left(\dot{y}_{TA}(t_2) + \dot{y}_A(t_2)\right)^2 . \quad (17)$$

As evident from (17), there are four parameters whose values in time t_2 influence the height of the vertical jump:
the height of the robot center of gravity relative to the ankle joint $y_{TA}(t_2)$,
the height of the ankle joint relative to the ground $y_A(t_2)$,
the velocity of the robot center of gravity relative to the ankle joint $\dot{y}_{TA}(t_2)$ and
the velocity of the ankle joint relative to the ground in the vertical direction $\dot{y}_A(t_2)$.
The height and the velocity of the robot center of gravity relative to the ankle joint in time t_2 depend only on the power of the actuators in the hip and knee joints while the height and the velocity of the ankle joint relative to the ground in time t_2 is the consequence of the feet motion dictated by the biarticular link. Therefore the height and the velocity of the ankle joint relative to the ground in time t_2 depend on the biomechanical parameters of the biarticular link and its activity during the push-off.

4.2 Optimization of Vertical Jump

To determine the optimum timing of the biarticular link activation that results in the highest vertical jump, a series of the countermovement vertical jump simulations have been performed, each with a different timing of the biarticular link activation. The influence of the biarticular link elasticity was eliminated by increasing the biarticular link stiffness towards the infinity. Thus we achieved a stiff biarticular link with no elastic properties. The relationship of the jump height and the biarticular link activation moment is presented in Fig. 6.

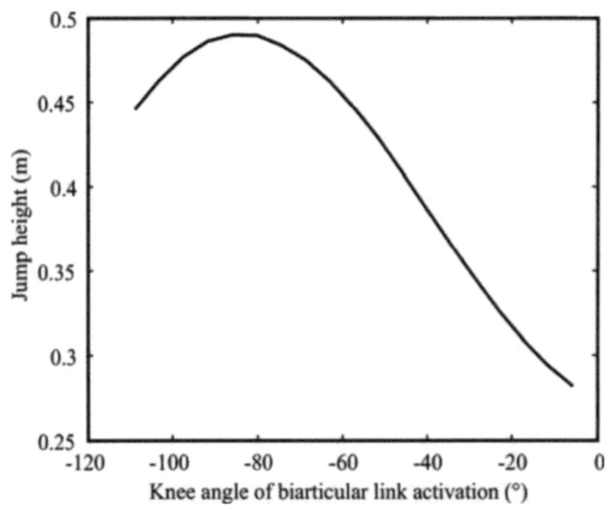

Fig. 6. Jump height as a function of the knee angle when the biarticular link is activated.

Biarticular link activation moment is presented with the knee angle when the biarticular link is activated. The highest jump has been achieved when the biarticular link was activated in the moment when the knee angle was approximately -84°. With this optimal timing, the robot jumped almost twice a high as when the biarticular link was not actuated or was actuated when the knee was almost fully extended. If the biarticular link activates too early in the push-off phase, the robot looses its ground contact too early when its center of gravity has not reached the maximal velocity. If the biarticular link activates too late in the push-off phase, the rotational energy of the thigh and shank increases uselessly. The optimal timing represents a compromise between these two cases.

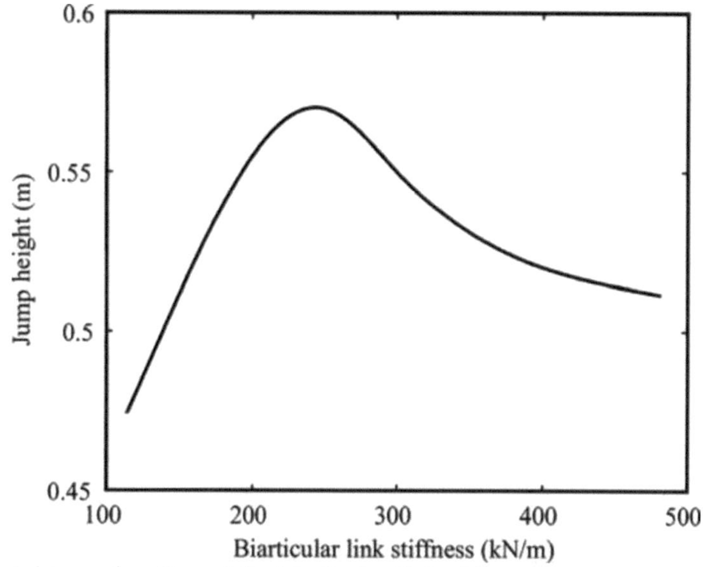

Fig. 7. Jump height as a function of the biarticular link stiffness.

To determine the optimum stiffness of the biarticular link regarding to the height of the vertical jump, a series of the countermovement vertical jump simulations have been performed, each with a different stiffness of the biarticular link. For each and every stiffness of the biarticular link the optimum timing has been determined as described in the previous paragraph. The relationship of the jump height and the biarticular link stiffness is presented in Fig. 7.

The highest jump has been achieved when the stiffness of the elastic biarticular link was approximately 245 kN/m. The highest jump was approximately 16% higher than the highest jump with the stiff biarticular link. We can thus conclude that the reutilization of the elastic energy stored during the countermovement contributes 16% to the height of the vertical jump.

4.3 Sensitivity Analysis

The purpose of the sensitivity analysis described in this section is to determine the sensitivity of the vertical jump height to the variations of the biarticular link parameters. We performed local sensitivity analysis that refers to variations of parameter around the values that ensure the maximal possible height of the vertical jump. These optimal values were determined with the optimization methods as described in the previous section. Changes of the vertical jump height that result from variations of biarticular link stiffness k, biarticular link damping b and ratio between the moment arms of the biarticular link r_C/r_B are presented in Fig. 8. Diagram in Fig. 8 also shows whether a certain parameter change results in an increase or in a decrease of the vertical jump height.

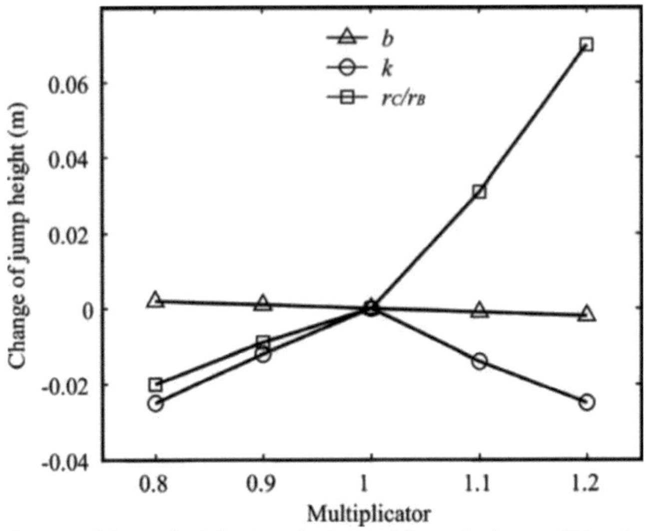

Fig. 8. Changes of vertical jump height in relation with variations of biarticular link stiffness *k*, biarticular link damping *b* and ratio between the moment arms of the biarticular link r_C/r_B. Multiplicator represents relative changes of parameter values.

As the sensitivity analysis refers to the model with optimal biarticular link stiffness, both positive and negative change of the biarticular link stiffness *k* cause a decrease of the jump height. An increase of the ratio between the moment arms of the biarticular link r_C/r_B causes an increase of the jump height. Higher ratio between the moment arms means higher angular speed in the ankle joint as a result or the biarticular link activation. Beside the increase of the angular speed in the ankle joint, also the force in the biarticular link increases. Variations of the biarticular link damping *b* only have negligible effect on the height of the vertical jump.

5. Discussion

By building an efficient biorobotic model which includes an elastic model of the biarticular muscle gastrocnemius and by simulation of the vertical jump we have demonstrated that biarticular links contribute a great deal to the performance of the vertical jump. Besides, we have shown that timing of the biarticular link activation and stiffness of the biarticular link considerably influence the height of the jump.

Methodology and results of our study offer an effective tool for the design of the humanoid robot capable of performing vertical jumps. We propose a new human inspired structure of the lower extremity mechanism that includes an elastic biarticular link and by which a humanoid robot would be able to efficiently perform fast movements such as running and jumping. However, it has to be considered that this study deals only with one biarticular link. Although the biarticular link we included in our study has the most distinctive elastic properties among all biarticular muscles of the human leg, other biarticular muscles such as the recuts femoris and the hamstrings should be included in humanoid robot design. A special challenge would be to design a humanoid lower extremity that includes all biarticular muscles of the human lower extremity and to demonstrate their joint effect on the vertical jump performance.

6. Acknowledgement

This investigation was supported by the Slovenian Ministry of Education, Science and Sport.

7. References

Alexander, R.McN. (1988). *Elastic mechanisms in animal movement,* Cambridge University Press, Cambridge

Asada, H. & Slotine, J.J.E. (1986). *Robot analysis and control,* John Wiley and sons, New York

Asmussen, E. & Bonde-Petersen, F. (1974). Storage of elastic energy in skeletal muscles in man. *Acta Physiologica Scandinavica,* 91, 385-392

Audu, M.L. & Davy, D.T. (1985). The influence of muscle model complexity in musculoskeletal motion modeling. *Journal of Biomechanical Engineering,* 107, 147-157

Babič, J. & Lenarčič, J. (2004). In vivo determination of triceps surae muscle-tendon complex viscoelastic properties. *European Journal of Applied Physiology,* 92, 477-484

Babič, J.; Karčnik, T. & Bajd, T. (2001). Stability analysis of four-point walking, *Gait & Posture,* 14, 56-60

Bobbert, M.F.; Gerritsen, K.G.M.; Litjens, M.C.A. & Soest, A.J. van (1996). Why is countermovement jump height greater than squat jump height? *Medicine & Science in Sports & Exercise,* 28, 1402-1412

Bobbert, M.F.; Hoed, E.; Schenau, G.J. van.; Sargeant, A.J. & Schreurs, A.W. (1986). A model to demonstrate the power transportating role of bi-articular muscles, *Journal of Physiology,* 387 24

Bogert, A.J. van den; Hartman, W.; Schamhardt, H.C. & Sauren, A.A. (1989). In vivo relationship between force, EMG and length change in the deep digital flexor muscle in the horse, In *Biomechanics XI-A,* G. de Groot; A.P. Hollander; P.A. Huijing; G.J. van Ingen Schenau (Eds.), pp. 68-74, Free University Press, Amsterdam

Brand, R.A.; Crowninshield, R.D.; Wittstock, C.E.; Pederson, D.R.; Clark, C.R. & Krieken, F.M. van (1982). A model of lower extremity muscular anatomy. *Journal of Biomechanics,* 104, 304-310

Cavagna, G.A. (1970). Elastic bounce of the body, *Journal of Applied Physiology,* 29, 279-282

Davy, D.T. & Audu, M.L. (1987). A dynamic optimization technique for predicting muscle forces in the swing phase of gait. *Journal of Biomechanics,* 20, 187-201

Delp, S.L. (1990). Surgery simulation: A computer graphics system to analyze and design musculoskeletal reconstructions of the lower limb, Doctoral Dissertation, Stanford University, Palo Alto

Ker, R.F.; Bennett, M.B.; Bibby, S.R.; Kester, R.C. & Alexander, R.McN. (1987). The spring in the arch of the human foot. *Nature,* 325, 147-149

Legreneur, P.; Morlon, B. & Hoecke, J. van (1997). Joined effects of pennation angle and tendon compliance on fibre length in isometric contractions: A simulation study. *Archives of Physiology and Biochemistry,* 105, 450-455

Leva, P. de (1996). Adjustments to Zatsiorsky-Seluyanov's segment inertia parameters. *Journal of Biomechanics,* 29, 1223-1230

Schenau, G.J. van. (1989). From rotation to translation: constraints of multi-joint movements and the unique action of bi-articular muscles, *Human Movement Science,* 8, 301-337

Shorten, M.R. (1985). Mechanical energy changes and elastic energy storage during treadmill running, In: *Biomechanics IXB*, D.A. Winter; R.W. Norman; R. Wells; K.C. Hayes; A. Patla (Eds.), pp. 65-70, Human Kinetics Publ, Champaign

Zatsiorsky, V. & Seluyanov, V. (1983). The mass and inertia characteristics of the main segments of the human body, In *Biomechanics VIII-B,* H. Matsui; K. Kobayashi (Eds.), pp. 1152-1159, Human Kinetics, Illinois

32

Planning Versatile Motions for Humanoid in a Complex Environment

Tsai-Yen Li, Pei-Zhi Huang
Computer Science Department, National Chengchi University
Taipei, Taiwan

1. Introduction

The potential market of service and entertainment humanoid robots has attracted a great amount of research interests in the recent years. Several models of humanoid robots have been designed in research projects. However, it remains a great challenge to make humanoid robots move autonomously. Motion planning is one of the key capabilities that an autonomous robot should have. An autonomous robot should be able to accept high-level commands and move in a real-life environment without colliding with environmental obstacles. A high-level command is something like "Move to location A on the second floor" while the robot is currently at some location B on the first floor, for example. It is an interesting problem for humanoid robots since locomotion capability possessed by a humanoid robot is usually much better than a mobile robot. Like a human, a humanoid robot should be able to step upstairs or downstairs and stride over small obstacles or a narrow deep gap in a complex environment such as the one shown in Fig. 1. Similar needs for the planning capability also arise in the domain of computer animation in generating motions for autonomous characters.

The motion for a humanoid robot to achieve a given goal is typically very complex because of the degrees of freedom involved and the contact constraint that needs to be maintained. Therefore, it is common to take a two-level planning approach to solve this problem. In our previous work (Li et al, 2003), we have been able to plan efficient humanoid walking motions in a layered environment. The approach used in the planner decomposed the planning problem into subproblems of global and local planning, each of which is easier to solve. The global planner assumes some basic properties of a humanoid and uses an approximated geometric shape to define the path planning problem. The path generated by the global planner is then passed to the local planner to realize the path with appropriate walk motions. However, in previous work, the locomotion of a humanoid is usually limited to forward walking only, and a humanoid cannot pass a deep gap even if it can stride over it. In this chapter, we will describe a motion planning system adopting the two-level approach and extending the work to overcome the above two limitations. We will present an efficient implementation of the planner in terms of space and time such that it can be used in an on-line manner.

The rest of the chapter will be organized as follows. In the next section, we will review the researches pertaining to our work. In the third section, we will give a formal description of the problem we consider in this chapter. We will then propose the planning algorithm in the fourth section and present some implementation details and experimental results in the fifth section. In the last section, we will conclude the chapter with some future directions.

Fig.1. Example of a layered virtual environment with deep narrow gaps (colored columns between two large platforms)

2. Related Work

Motion planning problems have been studied for more than three decades. An overview of motion planning algorithms can be found in Latombe's book (Latombe, 1991). The researches pertaining to humanoid motion planning or simulation can be found in the fields of robotics and computer animation (Kuffner et al., 2001; Polland et al., 2002; Sun & Dimitris, 2001). These researches differ mainly on the way they perform global path planning and how locomotion is taken into account. For example, early work in computer animation focused on generating human walking motion to achieve a high-level goal (Bruderlin & Calvert, 1989). However, no planning was done to generate the global path automatically. Since the application was mainly for computer graphics, how to simulate human walking with realistic looking was the main concern (Sun & Dimitris, 2001). A dynamic filtering algorithm was proposed in (Yamane & Nakamura, 2000) to ensure that the generated motions could be transformed into a dynamically feasible one.

In an early paper by Kuffner (1998), a gross motion planner utilizing graphics hardware was proposed to generate humanoid body motions on a flat ground in real time. Captured locomotion was used in this case to move the humanoid along the generated global path. In (Kalisiak & Panne, 2001), a stochastic search approach with versatile locomotion was proposed to generate humanoid motions in an unstructured environment where a set of predefined grasp points served as contact constraints. In (Choi et al, 2003), sequences of valid footprints were searched through augmented probabilistic roadmap with a posture transition graph, and then each pairs of footprints were substituted by corresponding motion clips. Since the motion clips were captured in advance, the locomotion may not be flexibly adapted to uneven terrain to avoid collisions. In (Shiller et al, 2002), a multi-layer grid was used to represent the configuration space for a humanoid with different types of locomotion such as walking and crawling, and the humanoid may change its posture along the global path.

In (Pettre et al., 2003), a digital actor was modelled with active (all degrees of freedom attached to the legs) and reactive (attached to the upper parts of the body) degrees of freedom. A collision-free trajectory was computed for the active part of the digital actor. Then the motion warping technique (Witkin & Poppvic, 1995) was applied to the motion capture data along the path when the reactive part of the digital actor collided with obstacles. The planner proposed in (Chestnutt et al., 2003) evaluated footstep locations for viability using a collection of heuristic metrics of relative safety, effort required, and overall motion complexity. At each iteration of the search loop, a feasible footstep was selected from the footstep transition sets. The global path of this approach was a sequence of footstep locations toward a given goal state.

In (Kuffner et al., 2001a; Kuffner et al., 2001b), a humanoid robot with real-time vision and collision detection abilities was presented. The robot could plan its footsteps amongst obstacles but could not step onto them. Considering locomotion directly in global path planning may generate more complete result but, on the other hand, it limits the flexibility of locomotion. In our previous work (Li et al., 2003), we were able to plan humanoid motions in real time with a given general description of the objects in the workspace. A more detailed description of this approach will be given in the next two sections.

3. Problem Description

3.1 Problem Overview

According to motion granularity, the motion-planning problem usually can be classified into global (gross) motion planning and local (fine) motion planning. In the global motion-planning problem we concern with working out the body logistics, such as planning a collision-free path amongst trees in a forest to reach some destination. In contrast, for the local motion-planning problem we focus more on limb logistics, such as planning hand motion to grab an awkwardly placed object. For the problem of walking on a layered complex environment for a humanoid robot, both types of planning needs to be considered in order to ensure that the desired task can be accomplished with feasible motion plans.

The inputs to a typical motion planner for a humanoid robot include the initial and goal configurations of the robot, the kinematics description and locomotion abilities of the robot, and a geometric description of objects in the environment. In our approach, the planning problem is decomposed into two subproblems: the global motion planning for moving the body trunk of a humanoid and the local motion planning for realizing the global motion plan with the chosen locomotion. The planners at the two levels can be linked together to solve the problem in sequence as well as to feed back failure situation for further replanning as shown in Fig. 2. The global motion planning will be the main concern of this chapter.

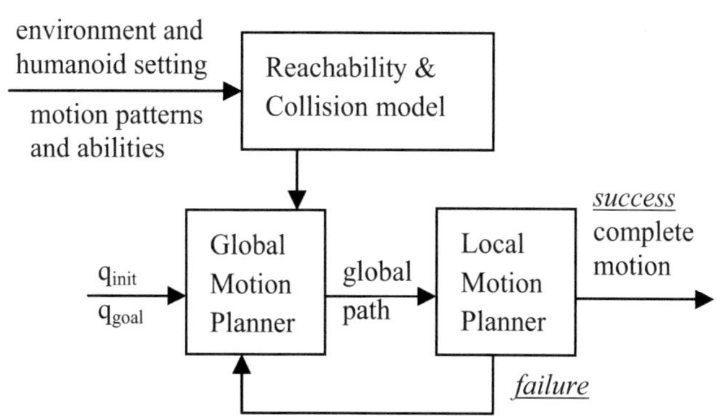

Fig.2. Planning loop for a typical query of humanoid motion.

3.2 Modeling the Humanoid

The kinematics description of a humanoid robot includes the maximal gait size, the maximal step height (h), and the bounding cylinder for each type of locomotion that the robot can adopt. Unlike traditional motion planning problems where the obstacles are given explicitly, the definition of obstacles in our case depends on the kinematics properties of the humanoid as well as the geometry of the objects. For example, an object is an obstacle to a humanoid if there is no way for the humanoid to step onto or pass under the object for a given locomotion due to the height and leg length of the humanoid.

In the global motion planning for a humanoid, we assume that the humanoid can be modeled as a bounding cylinder to simplify collision detection. In our system, we uses a large and a small cylinders to model a humanoid for frontal and lateral walking, respectively, as shown in Fig. 3. The radius of the inner cylinder is the size of the minimal region for a stable stance. The radius of the outer cylinder is determined according to the type of locomotion. We use the largest lateral width orthogonal to the moving direction of the humanoid to determine the radius. For example, the cylinder used in planning the side-walk motion is smaller than the one for regular walking motion. The height of the cylinders, denoted by H, is also related to the height and locomotion type of the humanoid. If we allow the humanoid to bend its upper body during the walking cycle, the actual height may be lower than the height of the humanoid. When using cylinders to model the geometry of a humanoid, we actually ignore its orientation at the planning time. We assume that we can recover the orientation of a humanoid in a postprocessig step according to the generated path and the adopted locomotion.

We assume that a humanoid robot can perform several types of locomotion and choose the most appropriate one according to the environment. Although many types of locomotion can be considered and implemented, only frontal walking, side walking, and jumping are demonstrated in this work. In addition, we assume that the local planner can generate necessary motion transition from one type of locomotion to another at a given configuration along a path.

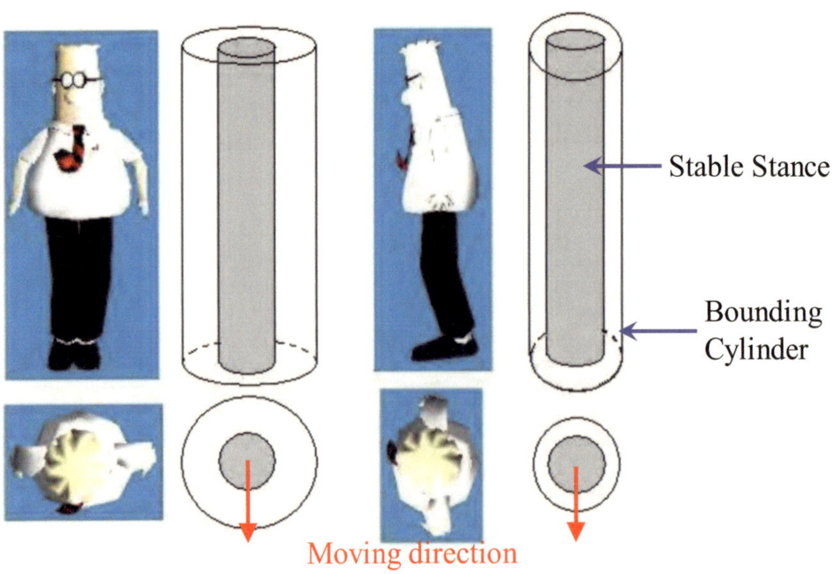

Fig.3. Frontal and lateral walking model for the humanoid.

4. Motion Planning for Humanoid

As described in the previous section, we use a decoupled approach to reduce the complexity of the motion-planning problem. In this section, we will focus on addressing the global path-planning problem for a humanoid with multiple types of locomotion including striding. We will first describe how we model the environment and prepare the search space according to the given environmental description and the data of a humanoid. Then we will present the planning algorithm that is used to search for a feasible path with these abilities.

4.1 Modeling the Environment

We assume that we are given the geometric description of the objects in the workspace such as the one shown in Fig. 4(a) and 4(b). Each object is described by a height and an offset from a reference level (such as the ground) in addition to its polygonal description. According to this geometric description, one can discretize the workspace into a grid of cells with an appropriate resolution. The resolution is chosen such that the area of a cell can allow a foot of the humanoid to step onto. Each cell in the workspace is given an offset value representing the distance from the bottom of an object to the referenced ground. Each cell is also assigned a height value above the offset. According to the offset values of the objects, we separate the 3D workspace into multiple connected 2D layers. For each layer, we compute a height map containing the elevation value of each cell above the layer in the workspace grid as shown in Fig. 4(c) and 4(d). Each cell in the layer l is referred to the same offset value d^l representing the distance from the base of the layer l to the referenced ground. Each cell i is also assigned a height value α_i^l above the layer l. The cell i of the height map for the layer l is represented as $c_i^l = \alpha_i^l$. Therefore, in a layered environment, a point may have different offset and height values on these different layers.

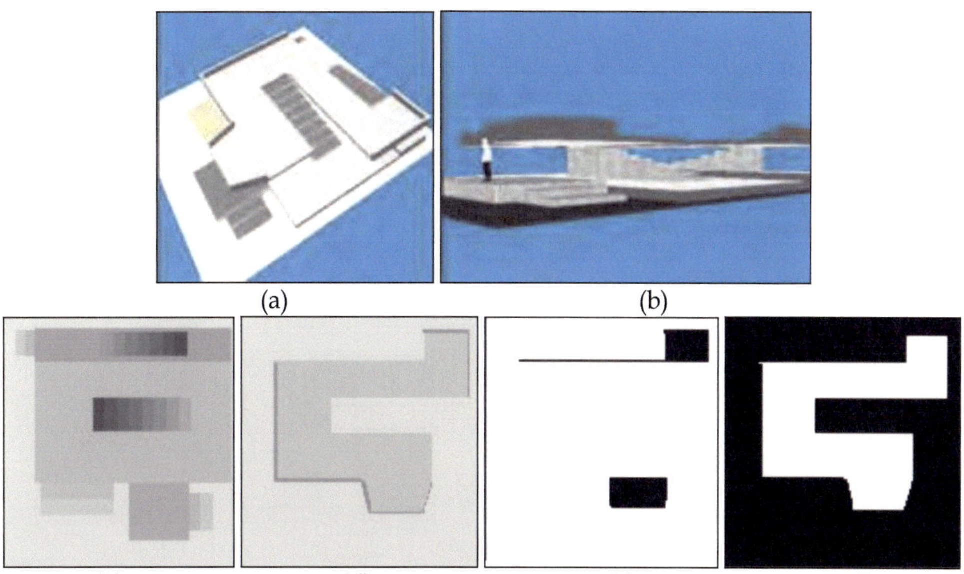

Fig. 4. (a) Top view of the workspace, (b) side view of the workspace, (c) and (d) are the height maps (the darker, the higher), (e) and (f) are the reachability maps of first and second layers (reachable regions in white).

The layered representation approach is appropriate for the gross motion-planning problem of a humanoid robot since the humanoid needs to stay on the ground all the time, even when climbing up or stepping down stairs. Therefore, a configuration q in workspace can be represented as (x, y, l), which means that the humanoid is standing at position (x, y) and on the "ground" of layer l.

4.2 Basic Reachable Region

Given an initial configuration of the humanoid, we compute a map, called *reachability map*, to represent the reachable regions from the initial configuration. The unreachable cells are marked as obstacle regions in this map. A reachability map may consist of several slices with one slice for each layer. For example, Fig. 4(e) and 4(f) show the reachability map for the two layers of the scene in Fig. 4(a). The black regions are the unreachable regions marked as obstacles. These slices could be "connected" if there exists an object whose height is large enough to bring the humanoid to step onto some neighboring cells of the above or below layer. We compute this map by a wave propagation algorithm similar to the one used to construct NF1 potential fields (Latombe, 1991). Suppose that the current cell under propagation is i. The algorithm advances to a neighboring cell i' at the layer l or a neighboring layer l' only if the height difference between them is less than h ($|c_i^l - c_{i'}^l| < h$ or $|(d^l + c_i^l) - (d^{l'} + c_{i'}^{l'})| < h$) and the humanoid height H can fit into the clearance above the cell i' at the layer l or l' ($|(d^{l+} - d^l - c_{i'}^l)| > H$ or $|(d^{l'+} - d^{l'} - c_{i'}^{l'})| > H$, where l+ denotes the layer above l).

Fig.5. (a) height map for the environment in Fig. 1., (b) map in (a) after applying dilation, (c) map in (b) after applying erosion. After the closing operator, the columns in the discontinuous region are connected. (d) instability map

4.3 Reachable Region for Striding Ability

For a humanoid with versatile motion abilities, the reachable region in the workspace is affected by several factors. For the walking motion, we need to take the maximal gait size and leg length of the humanoid as well as the height of the bounding cylinder into account. In the previous subsection, we have described a method to compute the reachability map for an environment with several layers. However, for an environment with deep narrow gaps such as the one shown in Fig. 1, large height differences between neighbouring cells prevent a humanoid to move across the region occupied by the columns. The main difficulty comes from the fact that we only check the neighbours that are one cell away although the maximal gait size of the humanoid could occupy several cells.

A straightforward approach to overcome this problem is extending the search algorithm to visit all the neighbours that are a few cells away from the current configuration and take the legal ones for further processing. An obvious drawback of this approach is that the number of neighbours increases rapidly as the gait size increases. The planning time will suffer as the number of visited cells soars. In addition, if we do not have a good idea about the reachable region, it would be difficult to compute an effective potential field to guild the search.

In this work, we propose a method for reachability computation that can account for the striding ability of a humanoid to handle the discontinuous regions mentioned above. The idea is that we can connect together the separated regions within the gait range by using the "Closing" operator, an important morphological operator in image processing (Gonzale & Woods, 2002). Closing tends to smooth sections of the image. It generally fuses narrow breaks and long thin gulfs, eliminates small holes, and fills gaps in the image. Closing of set A by a structuring element B, denoted $A \bullet B$, is defined as:

$$A \bullet B = (A \oplus B)\Theta B \tag{1}$$

The equation, in words, says that the closing of A by B is simply the dilation of A by B, followed by the erosion of the result by B. The height map is A in our case, and we select the circle with a radius of half of gait size as the structuring element B. The dilation operation ($A \oplus B$) will extend the edge of objects in bitmap by 1/2 gait size, which fills small gaps or holes between two disconnect regions if these regions are near. After dilation, the resulting bitmap shows that two regions are connected if their distance is within one foot step (See Fig. 5(b)). But for the extended edges which do not connect to other regions, we apply erosion ($A\Theta B$) to eliminate these useless edges. The erosion operation will corrode the edges in bitmap by 1/2 gait size, which dose not break the connected regions. Fig. 5(c) shows the result after erosion. Note that the unnecessary edges are removed after the operation and the map gets back to its original shape as in Fig. 5(a) except for the small discontinuous regions that are connected now. This modified height map after the closing operation is then used to determine the reachability of each cell for the walking motion with the method described in the previous subsection.

Since we have more than one type of locomotion available for the humanoid to use, we will need to compute a reachability map for each type of locomotion according to the parameters of the humanoid such as the step size and the size of the bounding cylinder for that type of locomotion. We can then take the union of these reachability maps to form the final reachability map. In other words, a cell is reachable if there exists at least one type of locomotion that can bring the humanoid to the cell. We record the available locomotion types for each cell such that we know how to choose an appropriate motion later in the search process.

4.4 Instability Map

In order to know which part of the region is collision-free, we can convert the height map, built in the workspace, into the corresponding C-space by growing the obstacle regions with the radius of the bounding cylinder of the humanoid for a given locomotion. However, not all collision-free regions allow the humanoid to stay for long. For example, we may allow the humanoid to walk on a stair for trespassing purposes but do not want it to stay at the edge of a stair for too long. We need to identify these regions where unstable situation might occur. A map describing the regions is called *instability map* (see Fig. 5(d) for an example). A cell in the instability map is defined as unstable if and only if the region covered by the enclosing circle of a humanoid contains cells with different heights and this height difference is smaller than the maximal step height of the humanoid. Otherwise, if this region contains an object whose height difference is larger than the maximal step height of the humanoid, we will consider the cell a *forbidden* cell. Similarly, a cell is defined as *gap* if and only if the region in the height map is reachable after the closing operation. A humanoid can enter the unstable or gap regions for trespassing purpose but the transition time usually needs to be short. Therefore, we have set an upper bound for each type of these regions in the search process to prevent the humanoid from staying in these regions for too long.

```
STABLE_BFP()
install q_i in T;
INSERT(q_i, OPEN); mark q_i visited;
SUCCESS ← false;
while ¬ EMPTY(OPEN) and ¬ SUCCESS do
   q←FIRST(OPEN);
   for every neighbor q' of q in the grid do
       if q' is stable then
           mark q' visited;
       if LEGAL(q', q) then
         install q' in T with a pointer toward q;
         INSERT(q', OPEN);
         if q' = q_g then SUCCESS ← true;
if SUCCESS then
   return the backtracked feasible path
else return failure;
```

Fig.6. The STABLE_BFP algorithm

4.5 Path Planning Algorithm

The planning algorithm that we use to compute the humanoid motion is shown Fig. 6. The STABLE_BFP algorithm is similar to the classical Best-First Planning algorithm (Barraquand & Latombe, 1991) that was used to solve path-planning problems with low DOF's. In each iteration of the search loop, we use the FIRST operation to select the most promising configuration q from the list of candidates (*OPEN*) for further exploration. We visit each neighbor q' of q, check their validity (via the LEGAL operation) and maintain their stability (through STABLE operation) for further consideration. The number of neighbors is related to the number of major locomotion, n. That is, if we visited m neighbors for each locomotion, we have to visited $m*n$ neighbors in each iteration. The procedures for checking stability and legality are shown in Fig. 7. A configuration is legal if it is collision-free, marked unvisited, and temporarily stable. It is temporarily stable if and only if the humanoid has not entered the unstable regions or the gap regions for longer than some threshold. This duration is kept as an instability counter (*q'.cnt*) in each cell in the unstable region and step counter (*q'.step*) in the gap region when we propagate nodes into it. Note that the validity of a configuration in the unstable or gap regions depends on the counter of its parent configuration. If there are more than one possible parent configurations, we cannot exclude any of them. Therefore, in the STABLE_BFP algorithm, we do not mark a configuration visited if it is in the unstable or gap region. A configuration in these regions can be visited multiple times as long as the counter does not exceed the maximal bound. In a gap region, we need to check the gap width with the gait size. We keep the start point of a gap (*gap_begin*) when we first enter the gap region and the end point of a gap (*gap_end*) when a stable region is reached. These two points are used to compute the gap width, and the procedure is to ensure that the gap does not exceed the gait size and the humanoid can stride from *gap_begin* to *gap_end*.

```
STABLE (q' ,q)
  if q' is unstable then
     q'.cnt= q.cnt+1;
  else if q' is gap then
     if q.ingap then
        q'.step= q.step+1;
     else
        q'.step=1;
        q'.ingap=true;
        gap_begin= q;
  else if q' is stable and q.ingap then
     gap_end= q';
```

```
LEGAL (q' ,q)
  STABLE(q' ,q)
  if q' is visited or forbidden then
     return false;
  if q' is gap and q'.step>N then
     return false;
  if q' is unstable and q'.cnt>M then
     return false;
  if q' is stable and q.ingap then
    check distance between gap_begin and gap_end;
    if the distance is large than gait size, then
     return false;
  return true;
```

Fig. 7. The STABLE and LEGAL procedures.

In the STABLE_BFP algorithm, we use the FIRST operation to select the most promising configuration for further exploration. In general BFP planners, the artificial potential field is usually the only index for the goodness criterium. Planners with this approach can usually yield short paths. In our case, the height difference could be an important index as well since one may prefer climbing up or stepping down stairs to taking a longer path. Preference on each available locomotion is also an important factor. For example, generally speaking, we prefer walking to crawling. Therefore, in the FIRST operation, we use a linear combination of these criteria, whose weights are specified by the user. In general, unstable regions and gap regions have lower preference, and staying in these regions is not preferred. Therefore, we use instability (*q.cnt*) or step counter (*q.step*) as a penalty measurement to avoid motions over difficult areas whenever possible.

5. Implementation and Experiments

We have implemented the global and local motion planners in Java and connected the planners to a VRML browser to display the final simulation results. All the planning times reported below are taken from experiments run on a regular 1.6GHz PC. The size of the workspace is 25.6m x 25.6m, and the height and gait size of the humanoid is 180cm and 60cm, respectively. The width of the shoulder (for enclosing cylinder) and the foot length is 60cm and 37cm, respectively. The resolutions for the grid workspace and configuration space are all 256x256 in the global planner. In the following subsection, we will describe how to optimize storage space by merging several layers with sparse objects in them to fewer layers. We will also use two examples to demonstrate the ability and efficiency of the planner in later subsections.

5.1 Merging Layers with Sparse Objects

In the proposed system, we use the offset value of an object to separate the workspace into several 2D grid layers. However, for workspace like gyratory stairs, such as the one shown in Fig. 8, each stair may have a unique offset value, and we need a layer for each offset. The result is that each of the layered maps only contains sparse objects. As the number of layers increases, not only the storage space will increase, but the search performance will degrade as well. In our implementation, we first sort the objects in workspace by their offsets in ascending order. Then we add objects into the first layer one by one until an object overlapped with other objects in this layer. A new layer is then created on demand. After this process, each object is assigned to a specified layer, and the resulting number of merged layers is usually much smaller. Fig. 8 is the example of gyratory stairs, which need 25 layers if we use offsets to separate the workspace directly. If we use the merge approach mentioned above, only two layers are needed to represent the workspace.

5.2 Example

In Fig. 9, we use an example to illustrate the features of this planning system. The environment, which is the same as the one in Fig. 1, consists of two layers of objects with various sizes and heights scattered on the two main platforms connected by several columns of various heights. The C-obstacles (configuration space obstacles) with different bounding cylinders for forward walking and side walking are shown in Fig. 9 (a) and (b), respectively. Note that the narrow passage exists only when side-walk locomotion (Fig. 9. (b)) is used. The global path found by the planner is shown in Fig. 9 (c) and (d). In this example, the initial configuration is on the ground, and the goal is at some location on the second layer that is reachable only through the following passages: climbing upstairs to the left platform, changing locomotion to side walk to pass the narrow passage, crossing the columns to the right platform, and climbing upstairs again to the second layer on the right platform. The planning time for a typical run consists of two parts: preprocessing and search. In this example, the total preprocessing time is 251ms (constructing layer map takes 10ms, computing the reachable region takes 35ms, computing the potential field takes 74ms, and computing the instability map takes 132ms) and the search time is 15ms.

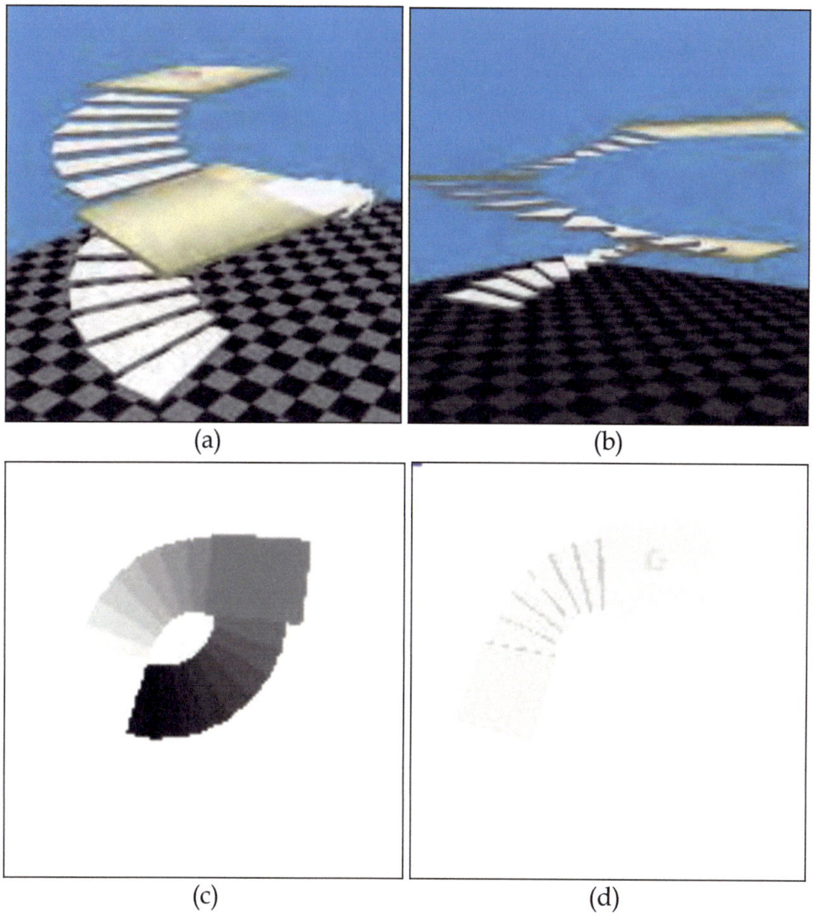

Fig. 8. Merging sparse layers. (a) and (b) are the 3D workspace with spiral stairs from different views (c) and (d) are the merged results for layer 1 and layer 2, respectively.

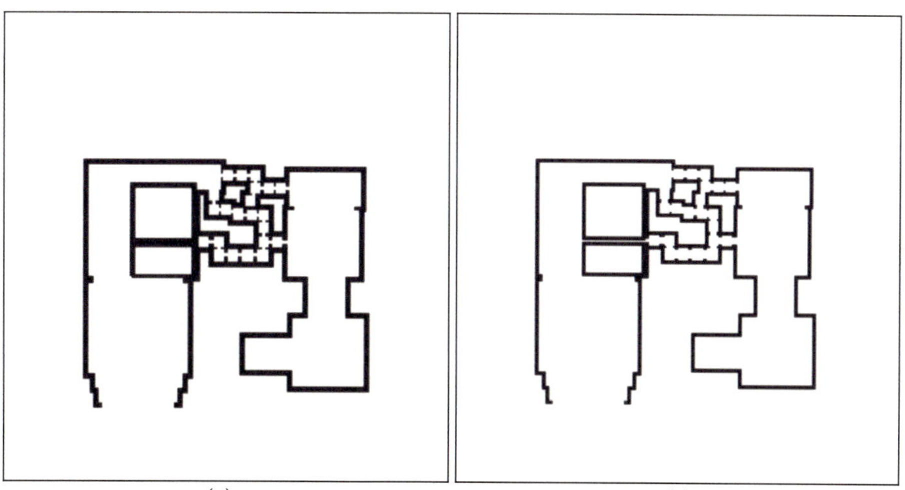

(a) (b)

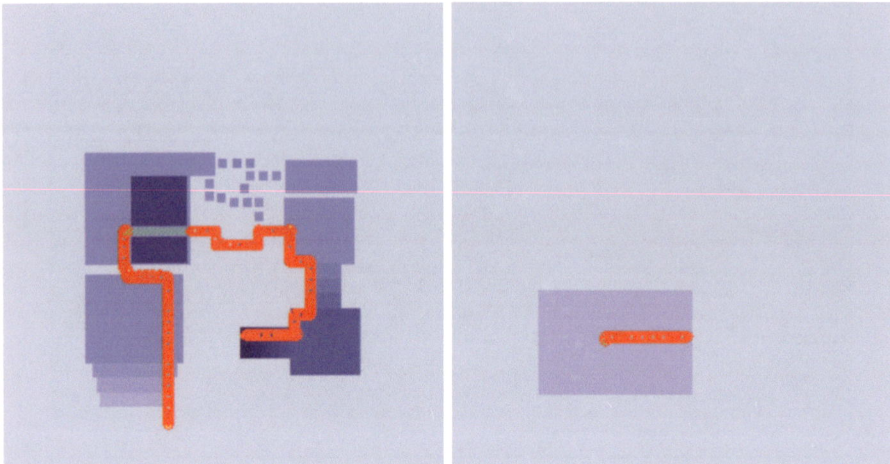

(c) (d)

(e) (f)

Fig. 9. (a) the C-obstacles for forward walking and (b) side walking. (c) the search result of the global path planner on layer 1 and (d) layer 2, (e) passing the narrow passage with side-walk motion, (c) feasible local motions generated to walk across the columns.

In Fig. 10, we use an outdoor environment to illustrate the use of several types of locomotion. The environment, as shown in Fig. 10(a), is partitioned into three blocks by the river. Upper and lower blocks are connected through the lily pads floating on the river. The broken bridge between the left and the right block is the shortest path between them. However, the distance is too large for the humanoid to stride over. In this case, the humanoid can only use jumping to move over the broken bridge. In addition, the passage to the house is too narrow for the humanoid to walk through without switching to the side-walk locomotion because the narrow passage exists only when side-walk locomotion is used. The global path found by the planner is shown in Fig. 10 (b), the red portion of the path is frontal walking, yellow portion is jumping, and green portion is lateral walking. In this example, the initial configuration is on the left bottom of the lower block, and the goal is in the yard of the house at the center of the right block that is reachable only through the following passages: climbing upstairs to the platform, detouring round the woods in the forest (Fig. 10(c)), striding over the lily pads (Fig. 10(d)), jumping over the broken bridge between the left and the right block (Fig. 10(e)), changing locomotion to side walk to pass the narrow fence (Fig. 10(f)), and finally reach the goal. The total planning time is 312ms (281ms for preprocessing and 31ms for path searching).

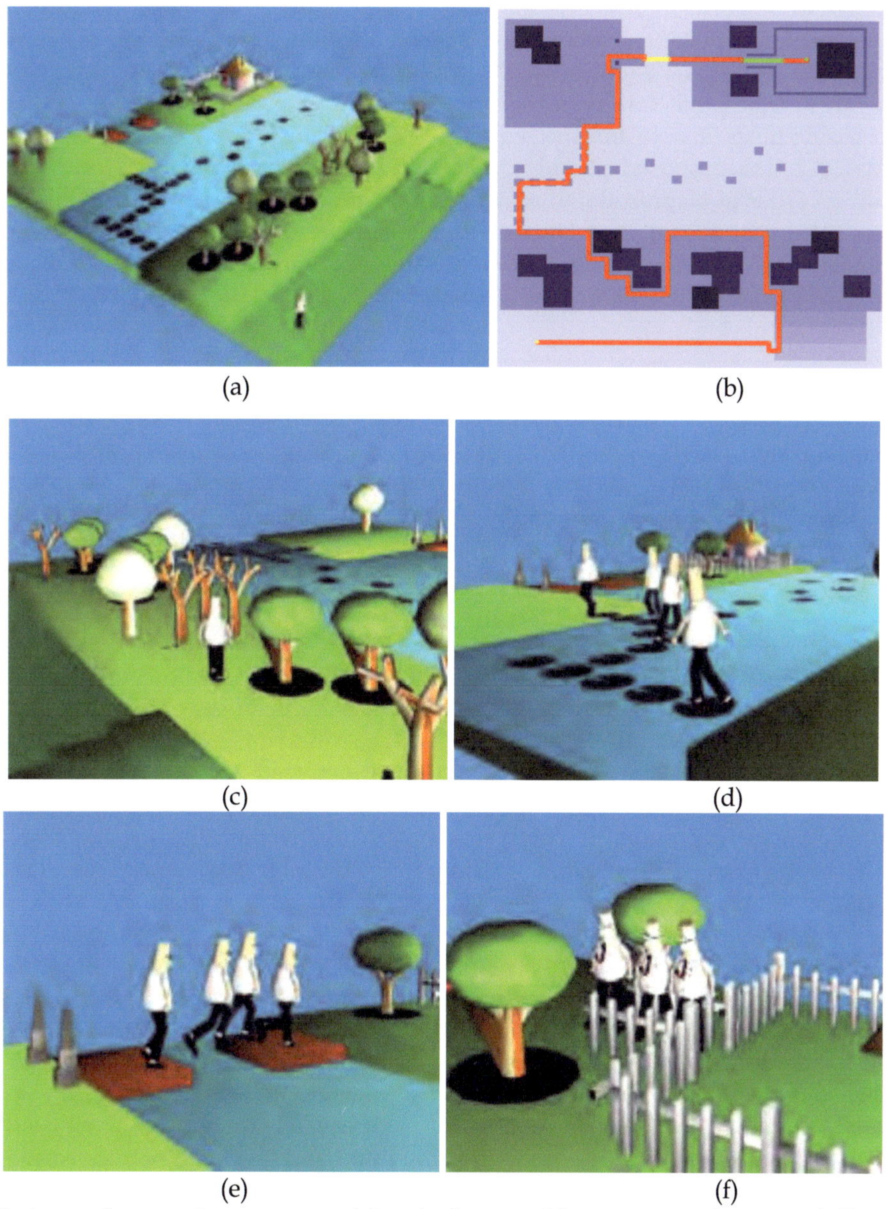

Fig. 10. An outdoor environment requiring the humanoid to use several motion skills to reach the goal.

5.3 Planning Performance Compared to the Traditional Approach

As mentioned in the previous sections, traditional straightforward search algorithms may also be able to find paths with the striding ability by extending the range of the neighbor search. Compared to this traditional approach, our approach is superior in performance and extensibility. Assume that a normal walking gait size is 60cm, which is equal to 6 cells in search space with the current resolution. In the traditional approach, the number of neighbors that needs to be covered is 64 for each configuration. This number could be even larger if we allow the humanoid to have a larger gait size or if we use finer resolution for the search space. The search time for an example similar to the one shown in Fig. 9 is 1246ms with the traditional approach where 64 neighbors are visited at a time and totally 47952 nodes are visited. With our approach, the preprocessing takes 263ms, and the search takes only 34ms. The total number of visited nodes is 2612. The proposed planning approach is faster than the traditional approach by about an order of magnitude. In addition to the reduced number of visited neighbors, the new approach is fast also because the potential field computed with a correct reachability map is more effective. The traditional approach can only use the distance to the goal as a heuristic since the reachable regions in the reachability map are discontinuous.

6. Conclusions and Future Work

Building autonomous humanoid robots has been the goal of many applications in robotics and computer animation. Spatial reasoning is a key capability to enable a robot to accept high-level commands and move autonomously. In this chapter, we have described a motion-planning system that can generate feasible humanoid animations with versatile locomotion in a complex virtual environment. The focus has been put on designing a global path planner that can generate a gross motion plan respecting the environmental and the humanoid constraints. The reachability can be defined by the properties of humanoid and can be extended through locomotion abilities. With the new problem definition, we modify the Best-First planning algorithm to account for user preference on horizontal or vertical distances that a humanoid travels. We have extended the planning system to consider such abilities and demonstrated the efficiency and effectiveness of the planner by simulation examples. We believe that this work will inspire further studies on the interesting problem of computing humanoid motion automatically for movie generation or autonomous humanoid robots.

In the global planner, collision detections for transition between different types of locomotion are done by checking the potential collisions with both types of motions. In fact, this assumption has over-simplified the transition problem since the bounding cylinder of a transition motion could be larger than either motion at both ends. Besides, not every transition between different types of locomotion is possible. In (Choi et al., 2003), a predefined transition graph is used to check whether the transition between different locomotion is valid or not. In the future, we would like to construct a transition graph; every node is a type of locomotion and the edge between two nodes exists only if the transition is possible. Corresponding models for collision checks should be attached to the edges to ensure that the global planner can properly check the collisions.

A real human usually can avoid obstacles with various kinds of locomotion and body motions. In the future, we would also like to enable the humanoid robot with more locomotion abilities when moving to the goal. For example, a human can crawl or stoop to avoid upper-layer obstacles. In addition, in a complex environment, there could exist movable objects that can be moved away by pulling or pushing by the humanoid robot to make ways for itself to reach the goal. We will also consider this kind of manipulation and reasoning capability in the future.

7. References

Bruderlin, A. & Calvert, T.W. (1989). Goal-Directed, Dynamic Animation of Human Walking, *Proceedings of ACM SIGGRAPH 1989*

Barraquand, J. & Latombe, J.C. (1991). Robot Motion Planning: A Distributed Representation Approach, *Intl J. of Robotics Research*, 10:628-649

Chestnutt, J.; Kuffner, J.; Nishiwaki, K. & Kagami, S. (2003) Planning Biped Navigation Strategies in Complex Environments, *Proceedings of IEEE International Conference on Humanoid Robotics*

Choi, M.G.; Lee, J. & Shin, S.Y. (2003). Planning Biped Locomotion Using Motion Capture Data and Probabilistic Roadmaps, *ACM Transactions on Graphics*, Vol. 22, No. 2, pp. 182–203

Gonzale, R.C. & Woods, R.E. (2002). *Digital Image Processing*, Second Edition, Prentice Hall

Kalisiak, M. & Panne, M. (2001). A Grasp-Based Motion Planning Algorithm for Character Animation, *Journal of Visual Computer and Animation*, pp. 117-129

Kuffner, J. (1998). Goal-Directed Navigation for Animated Characters Using Real-time Path Planning and Control, *Proceedings of CAPTECH'98 Workshop on Modeling and Motion capture Techniques for Virtual Environments*, Springer-Verlag

Kuffner, J.J.; Nishiwaki, K.; Kagami, S. Inaba, M. & Inoue, H. (2001). Footstep Planning Among Obstacles for Biped Robots, *Proceedings of 2001 IEEE International Conference on Intelligent Robots and Systems (IROS 2001)*

Kuffner, J.J.; Nishiwaki, K.; Kagami, S.; Inaba, M. & Inoue, H. (2001). Motion Planning for Humanoid Robots under Obstacle and Dynamic Balance Constraints, *Proceedings of 2001 IEEE International Conference on Robotics and Automation*

Latombe, J.C. (1991). *Robot Motion Planning*, Kluwer, Boston, MA

Li, T.Y.; Chen, P.F. & Huang, P.Z. (2003). Motion Planning for Humanoid Walking in a Layered Environment, *Proceedings of the 2003 International Conference on Robotics and Automation*

Pettre, J.; Laumond, J.P. & Simeon, T. (2003). A 2-Stage Locomotion Planner for Digital Actors, *Proceedings of Eurographics/SIGGRAPH Symposium on Computer Animation*

Pollard, N.S.; Hodgins, J.K.; Riley, M.J. & Atkeson, C.G. (2002). Adapting Human Motion for the Control of a Humanoid Robot, *Proceedings of 2002 IEEE International Conference on Robotics and Automation*, pp. 2265-2270

Shiller, Z.; Yamane, K. & Nakamura, Y. Planning Motion Patterns of Human Figures Using a Multi-Layered Grid and the Dynamics Filter, *Proceedings of IEEE International Conference on Robotics and Automation*, pp.1-8

Sun, H.C. & Dimitris, N.M. (2001). Automating Gait Generation, *Proceedings of ACM SIGGRAPH 2001*

Witkin, A. & Popovic, Z. (1995). Motion Warping, *Computer Graphics Proceedings, SIGGRAPH95*, pp. 105-108

Yamane, K. & Nakamura, Y. (2000). Dynamics Filter - Concept and Implementation of On-Line Motion Generator for Human Figures, *Proceedings of IEEE International Conference on Robotics and Automation*, pp. 688-695